K. Erickson

D1209968

Elementary Particle Physics

ELEMENTARY PARTICLE PHYSICS

STEPHEN GASIOROWICZ
Professor of Physics
University of Minnesota

John Wiley & Sons, Inc., New York | London | Sydney

Copyright © 1966 by John Wiley & Sons, Inc.

All Rights Reserved
This book or any part thereof
must not be reproduced in any form
without the written permission of the publisher.

Library of Congress Catalog Card Number: 66-17637
Printed in the United States of America

Preface

Our fund of knowledge of the properties of elementary particles has grown a great deal in the last two decades, and the rate of accumulation of new data promises to increase. There is, as yet, no theory that can quantitatively describe the enormous variety of phenomena discovered in the laboratory, so that the theoretical physicist concerned with these matters cannot afford to ignore any of the approaches that have been *partly* successful in clarifying some of these phenomena. This book was written to present a fairly comprehensive survey of the theoretical ideas that have proved useful in this sense. These ideas have been developed in a framework of certain fundamental assumptions, which are (*a*) that quantum theory is applicable to elementary particle physics, (*b*) that the description of all phenomena must be consistent with the special theory of relativity, and (*c*) that certain "internal symmetries" restrict the phenomena.

The fusion of the first two of these principles in the quantum theory of fields has been spectacularly successful. Quantum field theory has predicted a large class of phenomena, such as the connection between spin and statistics, the existence of antiparticles, and the necessity for multiple-particle production in high-energy collisions, which could not be expected from experience with classical theory or nonrelativistic quantum mechanics. Furthermore, in the one class of phenomena in which the technical complications of making quantitative predictions has been surmounted (quantum electrodynamics) the theory is in excellent agreement with experiment. The theory also provides a framework for the formulation of a variety of "internal symmetries," although it does not require them. I have therefore prefaced the main body of the book with an introduction to this subject, and have vindicated my point of view with several applications in later chapters, notably the forward-scattering dispersion relations for pion-nucleon scattering and the sum rule for the axial coupling constant in nuclear beta decay.

In spite of these indications of the richness of field theory, it is not possible to claim for it more than the role of a plausible framework at this time. Difficulties with carrying out calculations when the coupling is strong have largely limited the usefulness of field theory for most of elementary particle physics. Whenever the interactions are strong, a more or less sophisticated phenomenology has been substituted for theory. The form of transition matrix elements in such cases has been restricted by Lorentz invariance, the consequences of the unitarity of the S-matrix, and certain functional forms of the energy (or angle) dependence coming from analyticity properties derived from, or suggested by, field theory, or otherwise postulated (but never inconsistent with it). This approach has been very successful in that, once these restrictions are imposed, a variety of phenomena can be described with the introduction of but a few parameters which must be adjusted to experiment. Notable among the successes of this approach is the qualitative prediction of the quantum numbers of a number of resonances, a fairly detailed understanding of low-energy pion-nucleon scattering, and an understanding of the main features of electron-nucleon scattering as well as of the form factors in the decay of the strange particles.

The role of symmetries in restricting the phenomena in elementary particle physics has gained complete acceptance, and this topic is discussed in a variety of connections. In our discussion of Lorentz invariance the helicity representation of one-particle states is introduced. This tool is used repeatedly in the discussion of the "angular" physics of particle reactions such as the decay of unstable particles of unknown spin. The restriction on the observables imposed by invariance under space reflections and time reversal, as well as charge conjugation, are discussed in connection with the weak interactions, including the decay of the neutral K-mesons. The internal symmetries that we discuss are the well-established charge independence of the strong interactions, and $SU(3)$, which is approached from a point of view that does not presuppose any knowledge of group theory. The table of contents indicates the areas of interaction between symmetries and dynamics.

The principal omissions (within the framework of the goals stated at the beginning of the Preface) are a more detailed exposition of the application of Regge Pole Theory to high energy physics and all reference to attempts to combine internal symmetries with Lorentz invariance ($SU(6)$, $\tilde{U}(12), \ldots$). The former omission is mitigated by the existence of several excellent and detailed monographs; the second topic was omitted because it could not be fitted into the ordinary framework of relativistic quantum theory. Whether the higher symmetries point to the "theory of the future" is a debatable question, and my decision does, in a sense, reflect my opinion on this matter.

The material in this book has served as subject matter for a one-year course in elementary particle physics. Not all of the topics can be covered in a year, of course, but the range of material allows for a variety of plans. A good background in quantum mechanics is obviously required by the student who wishes to profit from the book. In particular, some familiarity with the Heisenberg picture and the idea of the representations of states as vectors in a Hilbert space is necessary for a quick assimilation of the material in Part I.

To conclude, I should like to express both regret and pleasure. The reason for the first is that the range of topics treated, and the pace of the development of the subject, have made the writing of an adequate bibliography impossible. It would, of course, be possible to list all the papers on a variety of subjects, for example, up to August 1965. Such a companion volume would, however, be useless. A critical annotated bibliography would take another year to write. I have therefore chosen to refer primarily to those articles from which I happened to have learned certain things. Inevitably such articles are concentrated in the American journals to which I subscribe. Consequently, inadequate recognition is frequently given to the originators of certain ideas, fundamental or technical. The brief bibliography is designed to remedy this, but apologies are undoubtedly due to a number of my colleagues, especially those in the Soviet Union and Japan, whose work may not be given adequate recognition.

The pleasure comes from the opportunity to express appreciation to those who contributed to the existence and final form of the book. My thanks go to Professors D. Yennie and M. Hamermesh who provided the necessary encouragement for the writing of the book. Particular thanks go to Professor W. Theis who carefully read much of the manuscript and made countless suggestions for improvements. Professor J. D. Jackson read the first part of the manuscript and I am grateful to him for a number of very helpful comments. Specific topics are treated better than they might have been because of discussions with Professors D. Geffen, D. Yennie, H. Suura, and R. Haag, and Drs. J. Meyer, L. Simmons, J. Uretsky, and C. Weil. I have received much stimulation and instruction during periodic visits to the Argonne National Laboratory, and I am grateful to Professor R. G. Sachs for providing me with this opportunity. Ultimately, my greatest thanks go to my wife whose constant encouragement was essential to the completion of the work.

Minneapolis, Minnesota STEPHEN GASIOROWICZ
March 1966

Acknowledgments

I express my thanks to the editors of *The Physical Review*, *Physical Review Letters*, *The Reviews of Modern Physics*, *Physics Letters*, *Il Nuovo Cimento*, Annual Reviews Inc., The Stanford University Press, The CERN Scientific Information Service, and Messrs. Oliver & Boyd, Publishers for their kind permission to reproduce illustrations.

Contents

PART II. INTRODUCTION

PART III. INTRODUCTION

Notation

The notation is defined in the text. A few symbols have been overlooked or insufficiently stressed. These brief remarks are designed to remedy this deficiency.

Whenever covariant notation is used, the metric defined by

$$g_{00} = 1, g_{11} = g_{22} = g_{33} = -1$$

$$g_{\mu\nu} = 0 \qquad \mu \neq \nu$$

will be used. We define

$$x^\mu \equiv (x^0, \mathbf{x}) \equiv (t, \mathbf{x})$$

and

$$x_\mu = g_{\mu\nu} x^\nu = (t, -\mathbf{x})$$

We shall also use the notation

$$\partial_\mu \equiv \frac{\partial}{\partial x^\mu} = g_{\mu\nu}\, \partial^\nu$$

The symbol $\Box$ will be used for $\partial_\mu\, \partial^\mu$. The symbol K_x will be used for $(\Box + \mu^2)$. In general, the scalar product

$$a_\mu b^\mu = a_0 b^0 - \mathbf{a} \cdot \mathbf{b}$$

will be denoted by $a \cdot b$. In particular, the scalar product px, which appears in the four-dimensional Fourier transforms, stands for

$$px = p_0 x^0 - \mathbf{p} \cdot \mathbf{x}$$

since the energy-momentum four-vector is

$$p^\mu = (p^0, \mathbf{p})$$

The symbol $\overleftrightarrow{\partial}_\mu$ appears in the combination

$$a \overleftrightarrow{\partial}_\mu b \equiv a(\partial_\mu b) - (\partial_\mu a)b$$

We use $^+$ for hermitian conjugate, when operators are considered, and * for complex conjugation of ordinary functions.

Integrals over four-dimensional volumes in space-time or momentum space are denoted by

$$\int dx$$

and

$$\int dp$$

unless danger of confusion requires $\int d^4x$ or $\int d^4p$. Three-dimensional integrals will always be denoted by

$$\int d^3x$$

and

$$\int d^3p$$

The notation

$$\dot{a} \equiv \frac{da}{dt} \equiv \frac{da}{dx^0}$$

will sometimes be used to save space.

State vectors will be denoted by Ψ or Φ.

Finally, we use natural units, in which

$$\hbar = c = 1$$

The symbol $e_{ijk}(i, j, k) = 1, 2, 3$ stands for the totally antisymmetric tensor

$$\begin{aligned} e_{ijk} &= 1 \text{ when } i, j, k \text{ is an even permutation of } 1, 2, 3 \\ &= -1 \text{ when } i, j, k \text{ is an odd permutation of } 1, 2, 3 \\ &= 0 \text{ otherwise} \end{aligned}$$

$\epsilon_{\alpha\beta\gamma\delta}$, with $(\alpha, \beta, \gamma\ \delta) = 0, 1, 2, 3$ is 4-dimensional generalization of this, with $\epsilon_{0123} = 1$.

Note. The word *lepton* has not been defined in the text. It refers to the *electron* (mass 0.511 MeV, spin $\frac{1}{2}$), *muon* (mass ~105 MeV, spin $\frac{1}{2}$), and the *neutrinos*.

Elementary Particle Physics

PART I

Introduction

The most challenging area of research in physics has always been the one concerned with the structure of matter and the deduction of the laws of interaction between the atomic constituents of matter. The quantum theory enabled one to see that most of the properties of matter are a consequence of the electrical forces between charged particles. Since these forces are known, a good quantitative understanding of most properties of matter could be arrived at.

In the nuclear domain, characterized by distances of 10^{-13} cm (in contrast to atomic distances of 10^{-8} cm), no such quantitative understanding exists. Over the last three decades, very active experimentation has gone on in the search of clues to what is hoped will be a fundamentally simple law of interaction between the nuclear particles. As the form of the "nuclear potential" was probed by scattering experiments carried on at ever increasing energies, it turned out that a simple picture of an interaction potential was untenable because some of the available energy was often materialized in the form of quanta, photons or mesons of one sort or another. Furthermore, the target particle was often left in a different state, which was loosely called a "new" elementary particle. Surprisingly enough, these curious new effects found a ready-made framework for their description in the *quantum theory of fields*. This theoretical development grew out of an application of the Heisenberg quantization rules to the classical electromagnetic field. This field, in the absence of matter, is easily decomposed into contributions from independent simple harmonic oscillators, and the canonical quantization procedure replaces these oscillators by their quantum counterparts. The state of the electromagnetic field could be described by stating which oscillator states were occupied. It turned out that interaction with matter could change the occupation number, which, in effect, led to the appearance and disappearance of photons. The extension of these ideas to other fields provided the framework into which all of the new experimental phenomena could be fitted.

The quantum theory of fields, restricted by the requirement of invariance under Lorentz transformations imposed on the observables, turned out to be a very tight framework indeed, so that if quantized fields exist, they must have some very special properties: e.g., to each particle there corresponds an antiparticle, and particles of half-odd integral spin must obey the Pauli exclusion principle. These consequences are in agreement with what is observed. Once field equations are assumed, a formal perturbation solution of these equations can be written down, and in the case of the interaction of the electromagnetic field with the electron the perturbation theory results agree with experiment. It is therefore quite possible that quantum field theory is the correct framework for all of the physics of elementary particles, and for that reason we consider it an indispensable subject in learning about elementary particle physics. The first part of this book is devoted to an introduction to quantum field theory.

Our aims here are very limited. The discussion will be formal and culminate in the derivation of the Feynman rules for the perturbation solution of the equations of motion resulting from a given Lagrangian function. Although it is very unlikely that the perturbation series converges, even for quantum electrodynamics where the expansion parameter is very small, the Feynman series does provide a formal solution of the equations of motion, and as such has been a rich source of conjectures about the properties of the true solution, be it in the symmetry under interchange of variables (crossing symmetry) or in the analyticity properties of the matrix elements. Not all of this material will appear in the first part of this book: questions of analyticity will not be brought up until Part 3. Only those aspects of field theory which are directly relevant to our choice of topics in particle physics will be discussed, and derivations, unless they provide some physical insight, will be abbreviated or omitted (except for Chapter 8). We omit a number of topics which are perhaps too technical to be discussed in an introductory treatment. Notable among them is the very physical discussion of collision theory due to Haag and the potentially important Bethe-Salpeter equation. We have also made no mention of the development of perturbation theory due to Schwinger and Tomonaga.

We begin our treatment with a discussion of the quantization of a classical field without interaction. We treat fields with different transformation properties, whose quanta are seen to have spin 0, $\frac{1}{2}$, and 1. Although this is a topic that is adequately treated in every book on field theory, it is necessary to cover this material to set up the framework for further discussion of fields in interaction. Free (i.e., noninteracting) fields play this important role because most forces occurring in nature have a short range of interaction. Thus particles that interact do so only when they are close together, and when they move out of each others range they are free. Scattering amplitudes are thus transition amplitudes between an incoming configuration of free particles and an outgoing configuration of free particles, and it is these configurations we learn to describe in the first few chapters. The general characterization of one-particle states, as restricted by Lorentz invariance, is discussed in Chapter 4.

Fields in interaction are generally described by nonlinear equations of motion. Criteria that might be useful in choosing the form of such equations of motion make up the content of Chapter 5. Even when the solutions of the equations of motion are given, it is not obvious what one should do with them. The expression of transition amplitudes in terms of the field operators are discussed in Chapters 6 and 7. Chapter 8 deals with a technique for obtaining the solutions of the equations of motion when the interaction is weak. This technique yields the Feynman rules which are described in Chapter 9. The problem of what to do when the Feynman graph rules yield infinite integrals is postponed to Chapter 13 in Part 2.

Further useful material on quantum field theory may be found in the following books:

A. I. Akhiezer and V. B. Berestetski, *Quantum Electrodynamics*, Wiley, New York, 1965.

J. D. Bjorken and S. D. Drell, *Relativistic Quantum Mechanics*, McGraw-Hill, New York (1964) and *Relativistic Quantum Fields* (1965).

N. N. Bogoliubov and D. V. Shirkov, *Introduction to the Theory of Quantized Fields*, Interscience, New York (1959).

C. De Witt and R. Omnes, *Dispersion Relations and Elementary Particles*, Wiley, New York (1961).

M. L. Goldberger and K. M. Watson, *Collision Theory*, Wiley, New York (1964).

W. Heitler, *The Quantum Theory of Radiation*, Clarendon, Oxford (1954).

J. M. Jauch and F. Rohrlich, *The Theory of Photons and Electrons*, Addison-Wesley Cambridge, Mass. (1955).

G. Källén, "Quantenelektrodynamik," *Handbuch der Physik*, **V**/1 Springer-Verlag, Berlin (1958).

F. Mandl, *An Introduction to Quantum Field Theory*, Interscience, New York (1960).

S. S. Schweber, *An Introduction to Relativistic Quantum Field Theory*, Harper and Row, New York (1961).

H. Umezawa, *Quantum Field Theory*, North-Holland, Amsterdam (1956).

G. Wentzel, *Quantum Theory of Fields*, Interscience, New York (1949).

1

The Scalar Field

The quantum theory of fields was created by the application of a form of the Heisenberg quantization rules to the classical electromagnetic field.[1] There are special complications which arise in that case, caused in part by the fact that the electromagnetic field has several components, and in part by the fact that electromagnetic waves move with the velocity of light. The basic concepts of field quantization are more easily studied in connection with the scalar field, which has only one component and which will be denoted by $\phi(x)$. The field $\phi(x)$ is assumed to have a Fourier transform

$$\phi(x) = \frac{1}{(2\pi)^{3/2}} \int dq e^{-iqx} \tilde{\phi}(q) \tag{1.1}$$

If this field is to describe the motion of a packet whose frequency and wave number are related by

$$q^0 = \sqrt{(q^k)^2 + \mu^2} = \sqrt{\mathbf{q}^2 + \mu^2} \equiv \omega_q \tag{1.2}$$

which, using the De Broglie relations between wave and particle attributes, describes the relation between the energy and momentum for a particle of mass μ, then $\tilde{\phi}(q)$ must have the form

$$\begin{aligned} \tilde{\phi}(q) &= \delta(q_\mu q^\mu - \mu^2)\chi(q) \\ &= \delta(q^2 - \mu^2)\chi(q) \end{aligned} \tag{1.3}$$

Thus the field $\phi(x)$ must satisfy the Klein-Gordon equation

$$(\partial_\mu \partial^\mu + \mu^2)\, \phi(x) \equiv (\Box + \mu^2)\, \phi(x) = 0 \tag{1.4}$$

[1] The historical development of field theory may be followed through the collection of papers published under the title *Quantum Electrodynamics*, J. Schwinger (ed.), Dover, New York (1958).

For our purposes it will be enough to consider plane wave solutions of this equation. We shall use the notation

$$f_q(x) = \frac{1}{(2\pi)^{3/2}} e^{-iqx} \tag{1.5}$$

where

$$\begin{aligned} qx &= q^\mu x_\mu = q^0 x_0 - \mathbf{q} \cdot \mathbf{x} \\ q^\mu &= (\omega_q, \mathbf{q}) \end{aligned} \tag{1.6}$$

The plane wave solutions will always be understood as the limiting case of wave packet solutions, so that we will allow ourselves integrations by parts of the spatial coordinates. The solutions (1.5) satisfy the following normalization condition

$$\begin{aligned} i\int d^3x \left[f_{q'}^*(x) \frac{\partial}{\partial x_0} f_q(x) - \frac{\partial}{\partial x_0} f_{q'}^*(x) f_q(x) \right] \\ \equiv i\int d^3x f_{q'}^*(x) \overleftrightarrow{\partial^0} f_q(x) = 2\omega_q\, \delta(\mathbf{q} - \mathbf{q}') \end{aligned} \tag{1.7}$$

Also

$$\int d^3x f_{q'}(x) \overleftrightarrow{\partial^0} f_q(x) = \int d^3x f_{q'}^*(x) \overleftrightarrow{\partial^0} f_q^*(x) = 0 \tag{1.8}$$

We shall call the solutions $f_q(x)$ positive frequency solutions and the solutions $f_q^*(x)$ negative frequency solutions.

In contrast to the functions that satisfy the Schrödinger equation, $|\phi(x)|^2$ may not be used as a density, for in general the integral of $|\phi(x)|^2$ over all space is not independent of time. A suitable density may be found with the help of the observation that the four-current

$$j^\mu(x) = i\,\phi^*(x) \overleftrightarrow{\frac{\partial}{\partial x_\mu}} \phi(x) \equiv i\,\phi^*(x) \overleftrightarrow{\partial^\mu} \phi(x) \tag{1.9}$$

satisfies the condition

$$\begin{aligned} \partial_\mu j^\mu(x) &= i\,\partial_\mu(\phi^*(x)\,\partial^\mu \phi(x) - \partial^\mu \phi^*(x)\,\phi(x)) \\ &= i(\phi^*(x) \,\Box\, \phi(x) - \Box\, \phi^*(x)\,\phi(x)) = 0 \end{aligned} \tag{1.10}$$

Hence, with $j^\mu(x) = (\rho(x), \mathbf{j}(x))$, we get, on integrating by parts,

$$\int d^3x\, \partial^0 \rho(x) = -\int d^3x\, \nabla \cdot \mathbf{j}(x) = 0$$

so that

$$\int d^3x\, \rho(x) = \text{constant.}$$

Let us now turn to the quantization of the scalar field. The term "quantization" means exactly what it does in the transition from a classical particle system to the corresponding quantum system. To the classical canonical variables, corresponding *operators* are associated, with well-defined "canonical" commutation rules

$$[p_i(t), q_j(t)] = -\delta_{ij}$$

In ordinary quantum mechanics these operators act in a vector space (Hilbert space of square integrable functions), and they are assumed to obey equations of motion which formally look exactly like the classical equations of motion. In order to establish the quantum theory which corresponds (in the above sense) to the classical scalar field we must determine (1) what is the momentum canonically conjugate to the "variable" $\phi(\mathbf{x}, t)$ and (2) what is the generalization of the Heisenberg commutation rules applicable to this system. We follow the procedure in the mechanical case, which is to find a Lagrangian that leads to the equations of motion (1.4) and from it to obtain the canonical momentum and the Hamiltonian. The equations below show the parallelism between a system of N free particles and the scalar field. The particle Lagrangian

$$L(t) = \frac{1}{2} m \sum_{i=1}^{3N} (\dot{q}_i(t))^2$$

contains a sum over the labels for each degree of freedom. If we recognize that the variable $\mathbf{x}$ in $\phi(\mathbf{x}, t)$ is also a labeling over a continuously varying index [a more transparent notation might be $\phi_{\mathbf{x}}(t)$], we see that the field Lagrangian must be an integral over a Lagrangian density

$$L(t) = \int d^3x \, \mathfrak{L}(\mathbf{x}, t) \tag{1.11}$$

The equations of motion for the mechanical system are obtained from

$$\frac{d}{dt}\left(\frac{\partial L}{\partial \dot{q}_i(t)}\right) = \frac{\partial L}{\partial q_i(t)} \qquad (i = 1 \cdots 3N)$$

which follow from the principle of least action. The corresponding field equations are

$$\partial^\mu \frac{\partial \, \mathfrak{L}(\phi(x), \partial^\alpha \phi(x))}{\partial \, (\partial^\mu \phi(x))} = \frac{\partial \mathfrak{L}}{\partial \, \phi(x)} \tag{1.12}$$

The canonical momentum for the particle system is

$$p_i(t) = \frac{\partial L}{\partial \, \dot{q}_i(t)}$$

and the corresponding quantity for the field is

$$\pi(x) = \frac{\partial \mathcal{L}}{\partial\,(\partial^0\,\phi(x))} = \frac{\partial \mathcal{L}}{\partial\,\dot{\phi}(x)} \tag{1.13}$$

Finally, the Hamiltonian is defined by

$$H = \sum_{i=1}^{3N} p_i(t)\,\dot{q}_i(t) - L(t)$$

and the field Hamiltonian is

$$H = \int d^3x\;\mathcal{H}(x) \tag{1.14}$$

with

$$\mathcal{H}(x) = \pi(x)\,\partial^0\,\phi(x) - \mathcal{L}(x) \tag{1.15}$$

The Heisenberg commutation relations

$$[p_i(t), q_j(t)] = -i\,\delta_{ij}$$

$$[p_i(t), p_j(t)] = [q_i(t), q_j(t)] = 0$$

have as their field counterparts

$$[\pi(\mathbf{x}, t), \phi(\mathbf{y}, t)] = -i\,\delta(\mathbf{x} - \mathbf{y}) \tag{1.16}$$

with all other pairs of operators commuting. If there are several classical fields to be quantized, e.g., $\phi^*(x)$ and $\phi(x)$, the Lagrangian density will be a function of $\phi^*(x)$ and $\partial^\mu\phi^*(x)$ as well as $\phi(x)$ and $\partial^\mu\,\phi(x)$; the equation

$$\partial^\mu \frac{\partial \mathcal{L}}{\partial(\partial^\mu\,\phi^*(x))} = \frac{\partial \mathcal{L}}{\partial\,\phi^*(x)}$$

will also be satisfied. The field $\phi^*(x)$ will have its canonically conjugate momentum

$$\pi^*(x) = \frac{\partial \mathcal{L}}{\partial(\partial^0\,\phi^*(x))}$$

The Hamiltonian density will be

$$\mathcal{H}(x) = \pi(x)\,\partial^0\,\phi(x) + \pi^*(x)\,\partial^0\,\phi^*(x) - \mathcal{L}$$

and the additional commutation relation

$$[\pi^*(\mathbf{x}, t), \phi^*(\mathbf{y}, t)] = -i\,\delta(\mathbf{x} - \mathbf{y})$$

will be assumed to hold. All commutators involving starred with unstarred fields vanish at equal times, since these are independent fields.

It is important to note that the commutation relations are only defined for equal times. Once these are given, the commutation relations at different times are determined by the development of the system in time, i.e., by the solutions of the equations of motion. Furthermore, we note that in (1.16), say, the times were set equal but not otherwise specified. This implies an assumption, which will be discussed later, that a change in the origin in time has no physical consequences.

One requirement to be imposed on a theory that will be used in describing the behavior of particles at high energies is that it be consistent with the special theory of relativity. In classical electrodynamics this requirement is satisfied by the covariance of the equations under Lorentz transformations: if the space-time coordinates are changed according to the law

$$x'^{\mu} = \Lambda^{\mu}{}_{\nu} x^{\nu} \tag{1.17}$$

with

$$g_{\mu\nu} \Lambda^{\mu}{}_{\rho} \Lambda^{\nu}{}_{\sigma} = g_{\rho\sigma} \tag{1.18}$$

(so that $x'_{\mu} x'^{\mu} = x_{\mu} x^{\mu}$), there is a transformation law for the field quantities, so that the transformed fields $F'_{\mu\nu}(x')$ satisfy the same equations in the new coordinate system. The quantized theory will then also be Lorentz invariant if (as is indeed the case) the commutation relations transform covariantly. Actually, in quantum theory, it is possible to discuss Lorentz invariance in a way divorced from the special form of the equations of motion.

Let us consider the coordinate system to be fixed and imagine some apparatus that serves to prepare a physical state Ψ_A. Consider now another, similar, apparatus related to the first one by a Lorentz transformation, which prepares the physical state $\Psi_{A'}$. Apparatus A may, for example, be a black box that emits electrons through an aperture; apparatus A' will be the same source, rotated through 90° about some axis, say, and moving with some fixed velocity relative to the same apparatus. Consider, similarly, a measuring apparatus M, which is being used to make measurements on the state Ψ_A and another measuring apparatus M', which differs from M only in that it is shifted relative to M by the same Lorentz transformation that connects A' with A. The statement of relativistic invariance is that the measurements made by M on the state Ψ_A yield the same results as those made by M' on the state $\Psi_{A'}$.

To obtain the formal consequences of this statement, we recall that in a quantum mechanical measurement we generally determine the probability that the physical system is in some state Φ; e.g., we may ask for the probability that the electrons emitted have a momentum $\mathbf{p}$. The probability of that happening will be $|(\Phi_{\mathbf{p}}, \Psi_A)|^2$, where $\Phi_{\mathbf{p}}$ describes the state in which just this particular momentum is found for the electron. For the transformed source and measuring apparatus, the corresponding probability is $|(\Phi_{\mathbf{p}'}, \Psi_{A'})|^2$, where $\Phi_{\mathbf{p}'}$ is the state for which the electron has the momentum $\mathbf{p}'$ connected to $\mathbf{p}$ by the same Lorentz transformation as connects A and A'. Because the vector space of states contains all possible physical states, Ψ_A and $\Psi_{A'}$ must be related by some transformation $U(\Lambda)$ that depends on the Lorentz transformation Λ. Because the measuring apparatus M and M' are connected by the same Lorentz transformation, we must have both

$$\Psi_{A'} = U(\Lambda)\Psi_A$$

and

$$\Phi_{\mathbf{p}'} = U(\Lambda)\Phi_{\mathbf{p}}$$

The invariance requirement implies that

$$|(\Phi_{\mathbf{p}'}, \Psi_{A'})|^2 = |(\Phi_{\mathbf{p}}, \Psi_A)|^2 \tag{1.19}$$

From this we can deduce that $U(\Lambda)$ must be a unitary transformation.[1]

Consider now the scalar field. Let us consider the measurement of the expectation value of the field $\phi(x)$. For the state Ψ_A, this will be $(\Psi_A, \phi(x)\Psi_A)$, and for the state $\Psi_{A'}$ it will be the measurement of the expectation value of the field at the transformed point, i.e. $(\Psi_{A'}, \phi(x')\Psi_{A'})$. We thus have

$$(\Psi_A, \phi(x)\Psi_A) = (U(\Lambda)\Psi_A, \phi(x')\, U(\Lambda)\Psi_A) \tag{1.20}$$

Therefore the scalar field in a Lorentz invariant theory should transform according to

$$\phi(x') = U(\Lambda)\, \phi(x)\, U^{-1}(\Lambda) \tag{1.21}$$

with $x' = \Lambda x$. In the case discussed in this chapter the equations satisfied by the field operator and the transform are evidently the same[2]; we shall prove the same for the commutation relations a little later.

We may add to the homogeneous Lorentz transformations displacements in space and time. It is reasonable to assume that locally at least (in the laboratory or in the galaxy) space is homogeneous, so that there is no "origin" in space or time. The homogeneous Lorentz transformations and the displacements together form the inhomogeneous Lorentz transformations, also called Poincaré transformations

$$x'^{\mu} = \Lambda^{\mu}{}_{\nu} x^{\nu} + a^{\mu} \tag{1.22}$$

For a displacement the analog of (1.21) is

$$\phi(x + a) = U(a)\, \phi(x)\, U^{-1}(a) \tag{1.23}$$

[1] Actually $U(\Lambda)$ can also be an antiunitary transformation. For a definition of an antiunitary transformation and an application, see page 26.

[2] This follows from the invariance of $\Box = \partial_{\mu}\partial^{\mu}$ under Lorentz transformations.

We shall again deal with the unitary operators $U(\Lambda)$ and $U(a)$ in this and later chapters.

Let us now return to the explicit quantization of the pair of scalar fields $\phi(x)$ and $\phi^*(x)$. It is easily checked that (1.4) and its counterpart for $\phi^*(x)$ may be derived from the Lagrangian density

$$\mathcal{L}(x) = \partial_\mu \phi(x)\, \partial^\mu \phi^*(x) - \mu^2 \phi(x)\, \phi^*(x) \tag{1.24}$$

The canonical momenta conjugate to $\phi(x)$ and $\phi^*(x)$ are

$$\begin{aligned} \pi(x) &= \partial_0 \phi^*(x) = \partial^0 \phi^*(x) \\ \pi^*(x) &= \partial^0 \phi(x) = \partial_0 \phi(x) \end{aligned} \tag{1.25}$$

respectively. The Hamiltonian density is

$$\begin{aligned} \mathcal{H}(x) &= \pi(x)\, \partial^0 \phi(x) + \pi^*(x)\, \partial^0 \phi^*(x) - \mathcal{L}(x) \\ &= \pi(x)\, \pi^*(x) + \pi^*(x)\, \pi(x) - \partial_\mu \phi(x)\, \partial^\mu \phi^*(x) + \mu^2 \phi(x)\, \phi^*(x) \\ &= \pi(x)\, \pi^*(x) + \nabla \phi(x) \cdot \nabla \phi^*(x) + \mu^2 \phi(x)\, \phi^*(x) \end{aligned} \tag{1.26}$$

As in ordinary quantum mechanics, there is an ambiguity in the ordering of operators in the quantum expression for a classical quantity. We shall later want to rearrange some terms in $\mathcal{H}(x)$.

We may use the commutation relations to obtain

$$\begin{aligned} [H, \phi(x)] &= \int_{x'_0 = x_0} d^3x' [\mathcal{H}(x'), \phi(x)] \\ &= -i \int_{x_0 = x'_0} d^3x'\, \pi^*(x')\, \delta(\mathbf{x} - \mathbf{x}') \\ &= -i\, \pi^*(x) \\ &= -i\, \partial^0 \phi(x) \end{aligned} \tag{1.27}$$

Similarly

$$\begin{aligned} [H, \pi(x)] &= -i\, \partial^0 \pi(x) = -i(\nabla^2 - \mu^2)\, \phi^*(x) \\ [H, \phi^*(x)] &= -i\, \partial^0 \phi^*(x) \\ [H, \pi^*(x)] &= -i\, \partial^0 \pi^*(x) = -i(\nabla^2 - \mu^2)\, \phi(x) \end{aligned}$$

These are just the Heisenberg equations of motion for the field operators. We may use them to derive

$$[H, F(x)] = -i\, \partial^0 F(x) \tag{1.28}$$

where $F(\phi(x), \ldots)$ is any polynomial of field operators. We may also construct the operator

$$P^k = \int d^3x (\pi(x)\, \partial^k \phi(x) + \pi^*(x)\, \partial^k \phi^*(x)) \tag{1.29}$$

which has the following commutation relation with H:

$$[H, P^k] = 0 \tag{1.30}$$

This result may be established with the help of the canonical commutation relations. We omit the details, for a simpler proof will emerge when both the energy H and the momentum $\mathbf{P}$ are expressed in terms of an operator representing a number density for quanta. With $\mathbf{P}$ shown to be constant, we may evaluate

$$\begin{aligned}[\mathbf{P}, \phi(x)] &= -\int_{x'_0=x_0} d^3x'[\pi(x')\,\nabla'\,\phi(x'), \phi(x)] \\ &= i\,\nabla\,\phi(x)\end{aligned} \tag{1.31}$$

etc. If we write

$$H \equiv P^0 \tag{1.32}$$

we may summarize (1.28), (1.30), and (1.31) by the formulas

$$[P^\mu, P^\nu] = 0 \tag{1.33}$$

and

$$[P^\mu, F(x)] = -i\,\partial^\mu\,F(x) \tag{1.34}$$

The commutation relations (1.33) and (1.34) may be used to obtain an explicit representation for the operator $U(a)$ which generates displacements in the operators (1.23). If we write

$$U(a) = e^{i\eta} \tag{1.35}$$

then for infinitesimal a_μ (1.23) becomes

$$\phi(x) + a_\mu\,\partial^\mu\,\phi(x) \simeq \phi(x) + i[\eta, \phi(x)]$$

or

$$ia_\mu[P^\mu, \phi(x)] = i[\eta, \phi(x)]$$

so that we can make the identification $\eta = a_\mu P^\mu$ and write the unitary operator $U(a)$ for arbitrary displacements in the form

$$U(a) = e^{ia_\mu P^\mu} \tag{1.36}$$

The Hamiltonian generates displacements in time, and the operator $\mathbf{P}$, which will be seen as the operator representing the momentum of the field, generates spatial displacements.

Let us now turn to the solution of the quantum mechanical problem of satisfying both the equations of motion and the commutation relations. This is done by diagonalizing the Hamiltonian, as in the familiar harmonic oscillator problem, and, as in the simple case, the important step is to find a convenient basis for the still unmentioned state vectors. Let us first expand the field operators $\phi(x)$ and $\phi^*(x)$ in a complete set of solutions of the Klein-Gordon equation. There will be positive and negative frequency terms, associated with $f_q(x)$ and $f_q^*(x)$, respectively. We write

$$\phi(x) = \int \frac{d^3q}{2\omega_q} [f_q(x) A(q) + f_q^*(x) B^+(q)]$$
$$\phi^*(x) = \int \frac{d^3q}{2\omega_q} [f_q(x) B(q) + f_q^*(x) A^+(q)] \quad (1.37)$$

with $f_q(x)$ as defined in (1.5). The orthonormality conditions (1.7) and (1.8) may be used to invert these expansions. We get

$$A(q) = i \int d^3x\, f_q^*(x)\, \overleftrightarrow{\partial^0}\, \phi(x)$$
$$B(q) = i \int d^3x\, f_q^*(x)\, \overleftrightarrow{\partial^0}\, \phi^*(x)$$
$$A^+(q) = i \int d^3x\, \phi^*(x)\, \overleftrightarrow{\partial^0}\, f_q(x) \quad (1.38)$$
$$B^+(q) = i \int d^3x\, \phi(x)\, \overleftrightarrow{\partial^0}\, f_q(x)$$

The commutation relations satisfied by the operators $A(q), \ldots, B^*(q)$ are easily obtained. For example,[1]

$$[A(q), A^+(q')] = \int d^3x \int d^3y [f_q^*(x)\, \overleftrightarrow{\partial^0}\, \phi(x), f_{q'}(y)\, \overleftrightarrow{\partial^0}\, \phi^*(y)]$$
$$= \int d^3x \int d^3y \{-f_q^*(x) \dot{f}_{q'}(y)\, [\dot{\phi}(x), \phi^*(y)]$$
$$- \dot{f}_q^*(x) f_{q'}(y)\, [\phi(x), \dot{\phi}^*(y)]\} + \cdots$$

We choose $x_0' = y_0$ and use the equal time commutation to get

$$[A(q), A^+(q')] = i \int d^3x (f_q^*(x) \dot{f}_{q'}(x) - \dot{f}_q^*(x) f_{q'}(x))$$
$$= 2\omega_q\, \delta(\mathbf{q} - \mathbf{q}') \quad (1.39)$$

Similarly, we find that

$$[B(q), B^+(q')] = 2\omega_q\, \delta(\mathbf{q} - \mathbf{q}') \quad (1.40)$$

with all other pairs of operators commuting.

Let us digress briefly to calculate the commutator between fields at different times. As was pointed out before, these can only be calculated

[1] For conciseness we use the notation $\dot{F}(x) \equiv \partial^0 F(x)$.

if we have solutions to the equations of motion. For the free field we are in this fortunate situation. We have

$$[\phi(x), \phi^*(y)] = \int \frac{d^3q}{2\omega_q} \int \frac{d^3q'}{2\omega_{q'}} \{f_q(x) f_{q'}^*(y)[A(q), A^+(q')] + f_q^*(x) f_{q'}(y)[B^+(q), B(q')]\}$$
$$= \int \frac{d^3q}{2\omega_q} [f_q(x) f_q^*(y) - f_q(y) f_q^*(x)]$$
$$= \frac{1}{(2\pi)^3} \int \frac{d^3q}{2\omega_q} [e^{-iq(x-y)} - e^{iq(x-y)}]$$

This may be rewritten in the form[1]

$$[\phi(x), \phi^*(y)] = \frac{1}{(2\pi)^3} \int dq e^{-iq(x-y)} \delta(q^2 - \mu^2)(\theta(q) - \theta(-q))$$

With the notation

$$\Delta(x; \mu) \equiv -\frac{i}{(2\pi)^3} \int dq e^{-iqx} \delta(q^2 - \mu^2)\, \epsilon(q) \tag{1.41}$$

we get

$$[\phi(x), \phi^*(y)] = i\, \Delta(x - y;\ \mu) \tag{1.42}$$

The singular function $\Delta(x; \mu)$ has the following properties, easily read off from the representation (1.41).

(1) It satisfies the Klein-Gordon equation

$$(\Box + \mu^2)\, \Delta(x; \mu) = 0 \tag{1.43}$$

with the singular initial conditions

$$\begin{aligned} \Delta(\mathbf{x}, 0; \mu) &= 0 \\ \partial^0 \Delta(\mathbf{x}, x^0; \mu)\big|_{x_0=0} &= -\delta(\mathbf{x}) \end{aligned} \tag{1.44}$$

2. It is an odd function of x,

$$\Delta(-x) = -\Delta(x) \tag{1.45}$$

3. It is manifestly a Lorentz invariant function. The factor $\epsilon(q)$ which distinguishes between the positive and negative frequency parts would normally not be an invariant factor. It is however multiplied by $\delta(q^2 - \mu^2)$ which vanishes outside the region where q^0 is timelike and where the distinction between positive and negative frequencies is meaningful. A consequence of the invariance is that for *spacelike* x

$$\Delta(-x) = \Delta(x)$$

[1] We use the standard notation $\theta(x) = 1, x_0 > 0$; $\theta(x) = 0$, $x_0 < 0$ and $\epsilon(x) = \theta(x) - \theta(-x) = x_0/|x_0|$.

This follows from Lorentz invariance and the fact that a spacelike x can be rotated into $-x$. Hence to satisfy (1.45) we must have

$$\Delta(x) = 0 \qquad x^2 < 0 \tag{1.46}$$

Thus the commutator $[\phi(x), \phi^*(y)]$ vanishes when the separation between the points is spacelike. We shall have more to say about this important result.

Let us return to the diagonalization of H. Some algebraic manipulations give us H in terms of the operators $A(q), \ldots, B^+(q)$,

$$H = \int \frac{d^3q}{2\omega_q} [\omega_q A^+(q)\, A(q) + \omega_q B(q)\, B^+(q)] \tag{1.47}$$

From this we find, with the help of the commutation relations (1.39) and (1.40), that

$$[A(q), H] = \omega_q\, A(q) \tag{1.48}$$

and

$$[A^+(q), H] = -\omega_q\, A^+(q) \tag{1.49}$$

Similarly

$$[B(q), H] = \omega_q\, B(q) \tag{1.50}$$

and

$$[B^+(q), H] = -\omega_q\, B^+(q) \tag{1.51}$$

If the state Ψ_E is an eigenstate of H with energy E, then

$$H\, A(q)\Psi_E = A(q)H\Psi_E + [H, A(q)]\Psi_E = (E - \omega_q)A(q)\Psi_E \tag{1.52}$$

Similarly

$$H\, A^+(q)\Psi_E = (E + \omega_q)A^+(q)\Psi_E \tag{1.53}$$

and analogously for the states $B(q)\Psi_E$ and $B^+(q)\Psi_E$. Thus the state $A(q)\Psi_E$ is again an eigenstate of H, but with energy lowered by an amount ω_q. The state $A^+(q)\Psi_E$ is also an eigenstate of H, but with energy raised by ω_q; similar statements hold for $B(q)\Psi_E$ and $B^+(q)\Psi_E$. We may speak of $A(q)$ and $B(q)$ as *annihilation operators*, for they take away energy from the system, and $A^+(q)$ and $B^+(q)$ as *creation operators*, for they add energy to the system. The Hamiltonian is a positive definite operator, and there must be a state of lowest energy. By adding a constant to H, if necessary, we may shift the energy scale so that the lowest energy is zero. We assume now that there is a *unique* state of zero energy, denoted by Ψ_0 and called the *vacuum state*. It must satisfy

$$H\Psi_0 = 0 \tag{1.54}$$

and

$$A(q)\Psi_0 = B(q)\Psi_0 = 0 \qquad \text{for all } q. \tag{1.55}$$

We shall normalize it so that

$$(\Psi_0, \Psi_0) = 1 \tag{1.56}$$

The vacuum expectation value of H, with the help of (1.55), is infinite. Thus to shift the energy scale we must add an infinite constant or, alternately, make use of our freedom to reorder the operators in H. If we define H as

$$H = \int \frac{d^3q}{2\omega_q} [\omega_q A^+(q) A(q) + \omega_q B^+(q) B(q)] \tag{1.57}$$

there are no difficulties. The fact that the additive constant is infinite is the first manifestation of trouble we encounter in a much less benign form later: frequently matrix elements of products of operators at the same point lead to infinities, and only sometimes will we be able to get rid of them. To make sure that the most trivial infinities do not occur, we shall always define quantities given in terms of free field operators (such as H) as *normal ordered*, i.e., as decomposed into free creation and annihilation operators and then reordered so that the annihilation operators stand to the right of the creation operators. This instruction is denoted by colons. Thus H, as it appears in (1.57), is written

$$H = :\int d^3x[\pi(x)\pi^*(x) + \nabla\, \phi(x) \cdot \nabla\, \phi^*(x) + \mu\phi(x)\, \phi^*(x)]: \tag{1.58}$$

in x-space.

Consider now the states

$$\Psi_{q_A} \equiv A^+(q)\Psi_0 \tag{1.59}$$

and

$$\Psi_{q_B} \equiv B^+(q)\Psi_0 \tag{1.60}$$

In both cases we find that the states have energy ω_q:

$$\begin{aligned} H\Psi_{q_A} = HA^+(q)\Psi_0 &= [H, A^+(q)]\Psi_0 \\ &= \omega_q\, A^+(q)\Psi_0 \\ &= \omega_q \Psi_{q_A} \end{aligned} \tag{1.61}$$

The orthonormality properties of the states are

$$\begin{aligned} (\Psi_{q_A'}, \Psi_{q_A}) = (\Psi_0, A(q')A^+(q)\Psi_0) &= (\Psi_0, [A(q'), A^+(q)]\Psi_0) \\ &= 2\omega_q\, \delta(\mathbf{q} - \mathbf{q}') \end{aligned} \tag{1.62}$$

Similarly,

$$(\Psi_{q_B'}, \Psi_{q_B}) = 2\omega_q\, \delta(\mathbf{q} - \mathbf{q}') \tag{1.63}$$

and

$$(\Psi_{q_A'}, \Psi_{q_B}) = 0 \tag{1.64}$$

A generalization of the states (1.59) and (1.60) is

$$\Psi_{q_1 \cdots q_m;\, k_1 \cdots k_n} = \frac{1}{\sqrt{m!\, n!}} A^+(q_1) \cdots A^+(q_m) B^+(k_1) \cdots B^+(k_n) \Psi_0 \tag{1.65}$$

The energy of the state is calculated as follows

$$\begin{aligned}
& H\, A^+(q_1)\, A^+(q_2) \cdots B^+(k_n)\Psi_0 \\
&\quad = ([H, A^+(q_1)] + A^+(q_1)H) A^+(q_2) \cdots B^+(k_n)\Psi_0 \\
&\quad = \omega_{q_1} A^+(q_1) \cdots B^+(k_n)\Psi_0 + A^+(q_1) H\, A^+(q_2) \cdots B^+(k_n)\Psi_0 \\
&\quad = \omega_{q_1} A^+(q_1) \cdots B^+(k_n)\Psi_0 + A^+(q_1)\, A^+(q_2) H \cdots B^+(k_n)\Psi_0 \\
&\qquad + A^+(q_1)[H, A^+(q_2)] \cdots B^+(k_n)\Psi_0 \\
&\quad = (\omega_{q_1} + \omega_{q_2})\, A^+(q_1) \cdots B^+(k_n)\Psi_0 + A^+(q_1)\, A^+(q_2) H \cdots B^+(k_n)\Psi_0 \\
&\quad = \cdots \\
&\quad = (\omega_{q_1} + \omega_{q_2} + \cdots + \omega_{k_n})\, A^+(q_1) \cdots B^+(k_n)\Psi_0
\end{aligned} \tag{1.66}$$

The energy is therefore the sum of the energies created by the various operators. This suggests the interpretation that A^+ and B^+ create quanta carrying energy. To test this interpretation we consider the operator

$$\mathcal{N}_A(q) = \frac{1}{2\omega_q} A^+(q)\, A(q) \tag{1.67}$$

which appears in H and in $\mathbf{P}$. With the help of the commutation relations we can show that

$$\mathcal{N}_A(q)\Psi_{q_1 \ldots, k_1 \ldots} = [\delta(\mathbf{q} - \mathbf{q}_1) + \cdots + \delta(\mathbf{q} - \mathbf{q}_m)]\Psi_{q_1 \ldots, k_1 \ldots} \tag{1.68}$$

Thus the states $\Psi_{q_1, \ldots, k_1, \ldots}$ are eigenstates of $\mathcal{N}_A(q)$. Moreover, the operator

$$N_A(\Omega) = \int_\Omega d^3q\, \mathcal{N}_A(q) \tag{1.69}$$

has the eigenvalue ν, where ν is the number of vectors among $\mathbf{q}_1, \mathbf{q}_2, \cdots \mathbf{q}_m$ which lie in the volume Ω in momentum space. We may interpret $\mathcal{N}_A$ as a *number density* operator, the number referring to the number of quanta created by the operator $A^+(q)$. Similarly

$$\mathcal{N}_B(q) = \frac{1}{2\omega_q} B^+(q)\, B(q) \tag{1.70}$$

is the number density operator for the quanta created by the operator $B^+(q)$. The Hamiltonian may now be rewritten in the form

$$H = \int d^3q [\omega_q \mathcal{N}_A(q) + \omega_q \mathcal{N}_B(q)] \tag{1.71}$$

Thus the energy in a volume element d^3q in momentum space is ω_q times the number of quanta of types A and B in that volume of momentum space. Simple manipulations show that the operator **P**, defined in (1.29) may be written as

$$\mathbf{P} = \int d^3q[\mathbf{q}\,\mathcal{N}_A(q) + \mathbf{q}\,\mathcal{N}_B(q)] \tag{1.72}$$

If we accept the quantum interpretation, this expression shows that we may identify **P** with the momentum operator. It has the correct additivity properties. If we apply H and **P** to a one particle state of type A or B, we obtain the eigenvalues ω_q and **q**, respectively,

$$H\Psi_q = \omega_q\Psi_q$$
$$\mathbf{P}\Psi_q = \mathbf{q}\Psi_q$$

from which we deduce that both quanta of type A and B have mass μ.

To find the intrinsic spin of the particles that are the quanta of the scalar field we shall explore (1.21) which describes the transformation properties of the scalar field under Lorentz transformations; among these transformations are rotations, and how the field transforms under rotations will yield the answer. It is sufficient to consider an infinitesimal Lorentz transformation

$$x^\mu \to x'^\mu = \Lambda^\mu{}_\nu x^\nu = (\delta^\mu{}_\nu + \alpha^\mu{}_\nu)x^\nu \tag{1.73}$$

In what follows we shall only keep terms linear in $\alpha^\mu{}_\nu$. The condition that $\Lambda^\mu{}_\nu$ be a Lorentz transformation is that

$$g_{\mu\nu}x^\mu x^\nu = g_{\mu\nu}x'^\mu x'^\nu = g_{\mu\nu}\Lambda^\mu{}_\alpha\Lambda^\nu{}_\beta x^\alpha x^\beta$$

i.e.,

$$g_{\mu\nu}\Lambda^\mu{}_\alpha\Lambda^\nu{}_\beta = g_{\alpha\beta} \tag{1.74}$$

For the infinitesimal transformations this implies that

$$g_{\mu\beta}\alpha^\mu{}_\sigma + g_{\sigma\nu}\alpha^\nu{}_\beta = 0$$

i.e.,

$$\alpha_{\beta\sigma} + \alpha_{\sigma\beta} = 0 \tag{1.75}$$

If we write $U(\Lambda) = e^{i\eta}$, where η is hermitian and reduces to zero for the identity transformation, we get, for the infinitesimal transformation

$$\phi(x) + i[\eta, \phi(x)] + \cdots = \phi(x^\mu + \alpha^\mu{}_\nu x^\nu)\cdots$$

Expanding the right-hand side in terms of α we obtain

$$\begin{aligned} i[\eta, \phi(x)] &\cong \phi(x) + \alpha^\mu{}_\nu x^\nu\,\partial_\mu\,\phi(x) - \phi(x) \\ &\cong \alpha^\mu{}_\nu x^\nu\,\partial_\mu\,\phi(x) \\ &\cong \tfrac{1}{2}\alpha^{\mu\nu}(x_\nu\,\partial_\mu - x_\mu\,\partial_\nu)\,\phi(x) \end{aligned} \tag{1.76}$$

where in the last line we have used the antisymmetry of $\alpha^{\mu\nu}$. Now if we write

$$\eta = \tfrac{1}{2}\alpha^{\mu\nu}M_{\mu\nu} \tag{1.77}$$

we obtain

$$[M_{\mu\nu}, \phi(x)] = -\frac{1}{i}(x_\mu\,\partial_\nu - x_\nu\,\partial_\mu)\,\phi(x) \equiv L_{\mu\nu}\,\phi(x) \tag{1.78}$$

Note that for μ, $\nu = 1, 2, 3$ the quantities L_{23}, L_{31}, L_{12} are just the differential operators representing the orbital angular momentum. Consider now a special transformation, a rotation about the z-axis, of magnitude $\delta\theta$. The transformation is

$$\begin{aligned} x'_1 &= x_1 - \delta\theta x_2 \\ x'_2 &= x_2 + \delta\theta x_1 \\ x'_3 &= x_3 \\ x'_0 &= x_0 \end{aligned} \tag{1.79}$$

so that $\alpha^1{}_2 = -\alpha^2{}_1 = -\delta\theta$ with all other $\alpha^\mu{}_\nu$ vanishing. Thus in this special case

$$[\eta, \phi(x)] = \delta\theta[M_{12}, \phi(x)] = \delta\theta L_3\,\phi(x)$$

Consider now the wave function of a single meson in a state Ψ. It is given by the matrix element

$$(\Psi_0, \phi(x)\Psi)$$

If the state Ψ is now rotated by $\delta\theta$ about the z-axis, the new wave function will be

$$\begin{aligned} (\Psi_0, \phi(x)U\Psi) &= (\Psi_0, UU^{-1}\,\phi(x)U\Psi) \\ &= (\Psi_0, e^{-i\eta}\phi(x)e^{i\eta}\Psi) \\ &= (\Psi_0, (\phi(x) - i[\eta, \phi(x)])\Psi) \\ &= (1 - i\delta\theta L_3)(\Psi_0, \phi(x)\Psi) \end{aligned}$$

If the meson is at rest, then

$$\mathbf{P}\Psi = 0$$

Consequently, with the help of (1.23) and (1.36), we have

$$\begin{aligned} (\Psi_0, \phi(x)\Psi) &= (\Psi_0, e^{iP_\mu x^\mu}\,\phi(0)e^{-iP_\mu x^\mu}\Psi) \\ &= (\Psi_0, \phi(0)e^{-iP_0 x^0}\Psi) \end{aligned}$$

so that $(\Psi_0, \phi(x)\Psi)$ is independent of $\mathbf{x}$. Therefore

$$L_3(\Psi_0, \phi(x)\Psi) = 0 \tag{1.80}$$

Thus the wave function of a meson at rest is unaltered by a rotation, from which we deduce that the angular momentum of a one-meson state in the rest system, i.e., its spin, is zero.

Particles A and B have the same mass and are both spinless. Moreover, they are related to each other in that the operators $A(q)$ and $B^{+}(q)$ are linked by appearing in the same field operator. We may therefore expect that there is a relationship between the two types of quanta.

To clarify the relationship, let us consider the current

$$j^{\mu}(x) = i\!:\!\phi^{*}(x)\,\overleftrightarrow{\partial^{\mu}}\,\phi(x)\!: \tag{1.81}$$

which, as we know, satisfies the conservation law

$$\partial_{\mu}\, j^{\mu}(x) = 0 \tag{1.82}$$

The zero component of this current $j^{0}(x)$ is a density that may be used to construct the time independent operator Q

$$Q = \int d^{3}x\, j^{0}(x) = i\int d^{3}x\!:\!\phi^{*}(x)\,\overleftrightarrow{\partial^{0}}\,\phi(x)\!: \tag{1.83}$$

Some algebra leads to the result that

$$Q = \int d^{3}q(\mathcal{N}_{A}(q) - \mathcal{N}_{B}(q)) \tag{1.84}$$

It is clear from this that

$$[Q, H] = 0 \tag{1.85}$$

which comes as no surprise, because the charge is conserved [(1.10) ff.], and also

$$Q\Psi_{0} = 0 \tag{1.86}$$

In general

$$Q\Psi_{q_{1}\cdots q_{n},k_{1}\cdots k_{m}} = (n - m)\Psi_{q_{1}\cdots q_{n},k_{1}\cdots k_{m}} \tag{1.87}$$

We call Q the *charge operator*, and we see that the distinction between the A-quanta and the B-quanta is that they have opposite charge. We capsule their connection through their appearance in the same field operator by calling them *antiparticles* of one another. Antiparticles have the same mass and spin as their corresponding particles, but they differ in charge; it should be noted that Q does not necessarily represent electrical charge; it could equally well represent another additive quantum number such as hypercharge, which takes on the values ± 1 for the electrically neutral K^{0} and $\bar{K}^{0}$. For a hermitian field

$$\phi(x) = \phi^{*}(x) \tag{1.88}$$

so that $A(q) \equiv B(q)$ and therefore $Q = 0$.

If there is an operator C with the property that

$$CA(q)C^{-1} = B(q)$$
$$CB(q)C^{-1} = A(q)$$
$$CC^{+} = C^{+}C = 1 \tag{1.89}$$

i.e.,

$$C\phi(x)C^{-1} = \phi^{*}(x) \tag{1.90}$$

it follows that

$$CQC^{-1} = -Q \tag{1.91}$$

and we call it a *charge conjugation* operator. For the free field such a C can be constructed. We leave it to the reader to check that

$$C = \exp\left[\pi i\int \frac{d^3q}{2\omega_q}(B^{+}(q) - A^{+}(q))(B(q) - A(q))\right] \tag{1.92}$$

has the required properties. It is clear that in the free field theory both the equations of motion and the commutation relations are invariant under the transformation $\phi(x) \leftrightarrow \phi^{*}(x)$, so that the theory is invariant under charge conjugation.

By now several questions may have arisen in the reader's mind. The first one is, why do we bother to deal with the field $\phi(x)$ and $\phi^{*}(x)$ at all, when all the relevant quantities are so simply expressed in terms of the operators $A(q) \cdots B^{+}(q)$? The significance of the rather special combination of Fourier transforms of these operators, the fields, lies in the property of the commutator. We have

$$[\phi(x), \phi^{*}(y)] = 0 \qquad (x - y)^2 < 0 \tag{1.93}$$

a property not shared by operators like

$$\phi^{(+)}(x) = \frac{1}{(2\pi)^{3/2}}\int \frac{d^3q}{2\omega_1} A(q)e^{-iqx}$$

This property, called "local commutativity" or "microscopic causality" follows from the equations of motion in the free particle case, but its significance transcends the special theory we are studying. If we express *causality* by the requirement that no effect may propagate with velocity larger than that of light (Haag calls this condition "Einstein causality"), we must require that any two observables, described by the operators $M(R)$ and $N(R')$ (R and R' refer to the regions of space-time in which the local observations are made), be independent of one another when the regions R and R' are spacelike with respect to each other. This implies that the operators $M(R)$ and $N(R')$ must commute when R and R' are spacelike with respect to each other. Local fields are mathematical idealizations, but for any weighted average of a field over a small space-time region

$$\phi(f) = \int_R dx\, \phi(x) f(x)$$

we should require that[1]

$$[\phi(f_1), \phi(f_2)] = 0 \qquad R_1 \sim R_2$$

if the fields are measurable. We shall see that in most cases the fields are not measurable, but bilinear combinations, such as the current, should be measurable. It is very difficult to satisfy the causality condition for the bilinear combinations without requiring local commutativity (or, as we shall see in Chapter 2, local anticommutativity) for the fields. It is causality, therefore, that endows the fields with special significance. Moreover, we shall find, when we consider fields in interaction that it is difficult to construct Lorentz invariant interactions other than by the use of local fields.

A second question that comes to mind is (a) under what circumstances may we expect to find a conserved charge Q in a theory, and (b) how dependent is the form of the operator Q on the equations of motion? To answer this question, let us first express the properties of the charge in a way that is as independent as possible of the special dynamics. We should require that

$$Q\Psi_0 = 0 \tag{1.94}$$

The charge conservation condition is that

$$[Q, H] = 0 \tag{1.95}$$

In the free field case we found that $A(q)$ and $B^+(q)$ lowered the charge by one unit [$B^+(q)$ creates a unit of the opposite charge], so that $\phi(x)$ is a charge-lowering operator. Similarly, $\phi^*(x)$ is a charge-raising operator. If we anticipate that the fields in interaction will play the same role, we may write quite generally

$$\begin{aligned} [Q, \phi(x)] &= -\phi(x) \\ [Q, \phi^*(x)] &= \phi^*(x) \end{aligned} \tag{1.96}$$

[1] The notation $R_1 \sim R_2$ is meant to indicate that the two regions are spacelike with respect to each other, i.e., the interval connecting any point in R_1 to any other point in R_2 is spacelike.

If this is so, the unitary transformation $e^{i\alpha Q}$ has the following effect on $\phi(x)$:

$$
\begin{aligned}
e^{i\alpha Q}\,\phi(x)e^{-i\alpha Q} &= \phi(x) + i\alpha[Q, \phi(x)] + \tfrac{1}{2}(i\alpha)^2[Q, [Q, \phi(x)]] + \cdots \\
&= \phi(x)\left[1 - i\alpha + \frac{(i\alpha)^2}{2!} - \frac{(i\alpha)^3}{3!} + \cdots\right] \\
&= e^{-i\alpha}\,\phi(x)
\end{aligned} \tag{1.97}
$$

Similarly,

$$
e^{i\alpha Q}\,\phi^*(x)e^{-i\alpha Q} = e^{i\alpha}\,\phi^*(x) \tag{1.98}
$$

Charge conservation requires that

$$
e^{i\alpha Q}He^{-i\alpha Q} = H \tag{1.99}
$$

Hence H, and therefore the Lagrangian, must be invariant under the substitutions

$$
\begin{aligned}
\phi(x) &\rightarrow e^{-i\alpha}\,\phi(x) \\
\phi^*(x) &\rightarrow e^{i\alpha}\,\phi^*(x)
\end{aligned} \tag{1.100}
$$

This transformation is called a *gauge transformation of the first kind.* Our conclusion is that only a gauge invariant theory, in the above sense, can have conserved charge.

We shall now see that gauge invariance of the Lagrangian or, more generally, of the equations of motion implies the existence of a conserved charge and gives us the form of the charge operator. The way we proceed is to consider the implications of gauge invariance for the action

$$
I = \int_{t_1}^{t_2} dx\, \mathcal{L}(\phi(x), \partial^\mu\, \phi(x)) \tag{1.101}
$$

The equations of motion are deduced from the assumption that the action is stationary with respect to variations in the field which vanish at t_1 and t_2. We shall not be concerned with such variations but rather with the infinitesimal version of (1.100), namely,

$$
\begin{aligned}
\phi(x) &\rightarrow \phi(x) + \delta\,\phi(x) = \phi(x) - i\alpha\,\phi(x) \\
\phi^*(x) &\rightarrow \phi^*(x) + \delta\,\phi^*(x) = \phi^*(x) + i\alpha\,\phi^*(x)
\end{aligned} \tag{1.102}
$$

We get

$$
\delta I = \int_{t_1}^{t_2} dx \left[\frac{\partial \mathcal{L}}{\partial \phi}\,\delta\phi + \frac{\partial \mathcal{L}}{\partial(\partial^\mu \phi)}\,\partial^\mu(\delta\phi) + \text{hermitian conjugate}\right]
$$

[We have used $\partial^\mu(\delta\phi) = \delta(\partial^\mu\phi)$.] However, from the equation of motion,

$$
\partial^\mu \frac{\partial \mathcal{L}}{\partial(\partial^\mu\phi)} - \frac{\partial \mathcal{L}}{\partial \phi} = 0
$$

it follows that

$$\frac{\partial \mathfrak{L}}{\partial \phi}\,\delta\phi + \frac{\partial \mathfrak{L}}{\partial(\partial^\mu \phi)}\,\partial^\mu(\delta\phi) = \partial^\mu\left(\frac{\partial \mathfrak{L}}{\partial(\partial^\mu \phi)}\,\delta\phi\right)$$

Hence

$$\delta I = -i\alpha \int_{t_1}^{t_2} dx\, \partial^\mu \left[\frac{\partial \mathfrak{L}}{\partial(\partial^\mu \phi)}\,\phi - \frac{\partial \mathfrak{L}}{\partial(\partial^\mu \phi^*)}\,\phi^*\right]$$

Invariance of the action implies that $\delta I = 0$, so that[1]

$$\int_{t_1}^{t_2} dx\, \partial^\mu \left[\frac{\partial \mathfrak{L}}{\partial(\partial^\mu \phi)}\,\phi - \text{h.c.}\right] = 0 \tag{1.103}$$

The spatial part of the divergence vanishes if, as is usually assumed, the field vanishes as $|\mathbf{x}| \to \infty$. We are then left with

$$\int_{t_1}^{t_2} dx_0 \frac{\partial}{\partial x_0} \int d^3x \left[\frac{\partial \mathfrak{L}}{\partial(\partial^0 \phi)}\,\phi - \text{h.c.}\right] = 0$$

Using the usual connection

$$\frac{\partial \mathfrak{L}}{\partial(\partial^0 \phi)} = \pi(x)$$

we see that gauge invariance implies that the operator

$$Q = -i \int d^3x(\pi(x)\,\phi(x) - \pi^*(x)\,\phi^*(x)) \tag{1.104}$$

satisfies the conservation condition

$$\frac{\partial}{\partial x_0}\,Q = 0 \tag{1.105}$$

Hence

$$[Q, H] = 0$$

Given the canonical commutation relations, Q will also satisfy the relations in (1.96). To make sure that the vacuum state has zero charge, it is enough to define Q suitably normal-ordered. Quite generally, for hermitian fields, Q is identically zero.

We must prolong this already long chapter with a discussion of the properties of the fields under space and time inversions. We first consider *space inversions*, and, as in our discussion of Lorentz transformations, we deal with two states: one is the physical state represented by Ψ and the other is the inverted state, i.e., the state prepared by the mirror image of the apparatus which prepared Ψ. We denote the inverted state by Ψ'.

[1] h.c. stands for hermitian conjugate.

If the theory is invariant under space inversions, or, in common usage, parity is conserved, Ψ and Ψ' are related by a unitary transformation U_s, which is independent of the states Ψ and Ψ':

$$\Psi' = U_s\Psi \tag{1.106}$$

U_s must be such that the expectation values of the observables in the state Ψ, say, must be equal to the expectation values of the space-inverted observables in the state Ψ'. Thus U_s must be such that

$$(\Psi, \mathbf{P}\Psi) = -(\Psi', \mathbf{P}\Psi') = -(\Psi, U_s^{-1}\mathbf{P}U_s\Psi) \tag{1.107}$$

$$(\Psi, \mathbf{J}\Psi) = (\Psi', \mathbf{J}\Psi') = (\Psi, U_s^{-1}\mathbf{J}U_s\Psi) \tag{1.108}$$

$$(\Psi, Q\Psi) = (\Psi', Q\Psi') = (\Psi, U_s^{-1}QU_s\Psi) \tag{1.109}$$

In our theory this may be achieved by defining U_s such that[1]

$$\begin{aligned} U_s^{-1} A(q)U_s &= \eta\, A(-q) \\ U_s^{-1} B(q)U_s &= \eta^*\, B(-q) \\ U_s^{-1} A^+(q)U_s &= \eta^*\, A^+(-q) \\ U_s^{-1} B^+(q)U_s &= \eta\, B^+(-q) \end{aligned} \tag{1.110}$$

with $\eta\eta^* = 1$, so that η, η^* are phase factors. As long as all observables are defined as bilinear combinations of $\phi(x)$ and $\phi^*(x)$, the phase cannot be fixed. If the field should be observable—and for this it has to be hermitian—the phase must have the property that

$$\eta^2 = 1 \tag{1.111}$$

because two successive inversions should not change an observable quantity. Note that we have chosen the phases of A and B related in a particular way; this was done to give the fields the well-defined transformation properties

$$\begin{aligned} U_s^{-1}\, \phi(\mathbf{x}, t)U_s &= \eta\, \phi(-\mathbf{x}, t) \\ U_s^{-1}\, \phi^*(\mathbf{x}, t)U_s &= \eta^*\, \phi^*(-\mathbf{x}, t) \end{aligned} \tag{1.112}$$

For a free field it is obvious that η may be chosen anything we like. For a field in interaction, the defining equations (1.110) will lead to the proper transformation properties of $\mathbf{P}$, $\mathbf{J}$ and Q, because these quantities depend only on the $(\partial_\mu\phi)(\partial^\mu\phi^*)$ part of the Lagrangian, and this will be taken over unaltered when we discuss fields in interaction (Chapter 5). It may or may not be possible to find a choice of phases η such that

$$U_s^{-1}HU_s = H \tag{1.113}$$

[1] It should be understood that in $A(-q), \ldots$ only a sign change in the spatial momentum $\mathbf{q}$ is implied.

If it is not, we say that parity is not conserved. The states Ψ and Ψ'' are still related by a unitary transformation (for they are vectors in the same space of all physical states), but this transformation depends on Ψ. If parity is conserved, an arbitrary choice of phase factors for various fields will in general not be consistent, and we say that the parity of a field, relative to the particular choice of parities of other fields, is fixed. It is always possible[1] to choose $\eta = \pm 1$. If $\eta = +1$ we call the field (and the quanta) *scalar*, and if $\eta = -1$ we call it *pseudoscalar*. We return to parity in Chapter 2, when the question of superselection rules is raised.

In conclusion, we shall consider *time reversal*. Time reversal is a symmetry operation just like space inversion, and, as pointed out before, invariance is obtained if, for any pair of states Ψ and Φ and Ψ' and Φ', in which the latter are the time-reversed counterparts of Ψ and Φ (e.g., velocities reversed, spin directions reversed),

$$|(\Psi', \Phi')|^2 = |(\Psi, \Phi)|^2$$

Wigner pointed out that this requires that

$$\Psi' = U\Psi$$

$$\Phi' = U\Phi$$

where U is unitary (as for space inversions and Lorentz transformations) or *antiunitary*. An antiunitary transformation is one of the form

$$U = VK \tag{1.114}$$

where V is a unitary transformation and K, the complex conjugation operator, has the properties that

$$K^+ = K \tag{1.115}$$

$$K^2 = 1 \tag{1.116}$$

$$K(\alpha\Psi_1 + \beta\Psi_2) = \alpha^* K\Psi_1 + \beta^* K\Psi_2 \tag{1.117}$$

$$(K\Psi_1, K\Psi_2) = (\Psi_2, \Psi_1) \tag{1.118}$$

The equation of motion

$$\frac{d\,\phi(\mathbf{x}, t)}{dt} = i[H, \phi(\mathbf{x}, t)] \tag{1.119}$$

forces an antiunitary time reversal transformation on us. To see this, suppose that T were unitary. With the choice

$$T\,\phi(\mathbf{x}, t)T^{-1} = \phi(\mathbf{x}, -t)$$

[1] See G. Feinberg and S. Weinberg, *Nuovo Cimento* **14,** 571 (1959) for a general discussion of phases. Under certain circumstances, which do not appear to hold in the real world, it might be necessary to have imaginary parities for spinless mesons.

or

$$T\,\phi(\mathbf{x}, t)T^{-1} = \phi^*(\mathbf{x}, -t)$$

we get

$$T\,\frac{d\,\phi(\mathbf{x}, t)}{dt}\,T^{-1} = i[THT^{-1}, \phi(\mathbf{x}, -t)]$$

i.e.,

$$\frac{d}{d(-t)}\,\phi(\mathbf{x}, -t) = i[-THT^{-1}, \phi(\mathbf{x}, -t)]$$

in one case and

$$\frac{d}{d(-t)}\,\phi^*(\mathbf{x}, -t) = i[-THT^{-1}, \phi^*(\mathbf{x}, -t)]$$

in the other. In either case we find that (1.119) can be invariant only if

$$THT^{-1} = -H$$

This, however, is an unacceptable condition, because time reversal cannot change the spectrum of H, which consists of positive energies only. If T is taken to be antiunitary, the K operator changes the i to $-i$ in (1.119) and the trouble does not occur. With the form $T = VK$ and

$$V\,\phi(\mathbf{x}, t)V^{-1} = \phi^*(\mathbf{x}, -t)$$

We are led to

$$T\,\phi(\mathbf{x}, t)T^{-1} = \phi(\mathbf{x}, -t) \tag{1.120}$$

Consider the time inverse of the matrix element

$$(\Psi, \phi(\mathbf{x}_1, t_1) \cdots \phi(\mathbf{x}_n, t_n)\Phi)$$

This is given by

$$\begin{aligned}
(T\Psi, T\,\phi(\mathbf{x}_1, t_1) &\cdots \phi(\mathbf{x}_n, t_n)\Phi)\\
&= (KV\Psi, KV\,\phi(\mathbf{x}_1, t_1) \cdots \phi(\mathbf{x}_n, t_n)\Phi)\\
&= (KV\Psi, K\,\phi^*(\mathbf{x}_1, -t_1) \cdots \phi^*(\mathbf{x}_n, -t_n)V\Phi)\\
&= (\phi^*(\mathbf{x}_1, -t_1) \cdots \phi^*(\mathbf{x}_n, -t_n)V\Phi, V\Psi)\\
&= (V\Phi, \phi(\mathbf{x}_n, -t_n) \cdots \phi(\mathbf{x}_1, -t_1)V\Psi)
\end{aligned} \tag{1.121}$$

This derivation illustrates the manipulations with the K operator. Note that $V\Phi$ and $V\Psi$ are the "motion-reversed" counterparts of the states Φ and Ψ, because the unitary operator V changes t into $-t$ and can easily be shown to change the sign of the momentum operator.

The use of the K operator may be avoided, provided the antiunitary operation is defined to reverse the order of the field operators in a product, as well as the states. This will again save the Heisenberg equations of motion, not by changing i to $-i$ but by reversing the order of factors in a commutator. If we now take

$$T\,\phi(\mathbf{x}, t)T^{-1} = \eta\,\phi^*(\mathbf{x}, -t) \tag{1.122}$$

and apply our rule to $(\Psi, \phi(\mathbf{x}_1, t_1) \cdots \phi(\mathbf{x}_n, t_n)\Phi)$, we obtain

$$(T\Psi, T\,\phi(\mathbf{x}_1, t_1) \cdots \phi(\mathbf{x}_n, t_n)\Phi) = \eta^n(T\Phi, \phi^*(\mathbf{x}_n, -t_n) \cdots \phi^*(\mathbf{x}_1, -t_1)T\Psi) \tag{1.123}$$

where $T\Phi$ and $T\Psi$ represent the "motion-reversed" states. This result is apparently different from that obtained in (1.121). It turns out, however, that for matrix elements of *hermitian operators* the results are the same. This is evident if $\phi(\mathbf{x}, t)$ is a hermitian field, and it is also true for a matrix element such as $(\Phi, j_\mu(x_1)j_\nu(x_2) \cdots \Psi)$, which can easily be checked by the reader. We use the second definition of the antiunitary operation in the rest of the book. Incidentally, the transformation law (1.122), in which η is a phase factor, implies that for the creation and annihilation operators.

$$\begin{aligned} T\,A(\mathbf{q})T^{-1} &= \eta A^+(-\mathbf{q}) \\ T\,B(\mathbf{q})T^{-1} &= \eta^* B^+(-\mathbf{q}) \end{aligned} \tag{1.124}$$

The phase factors for the two quanta are related because we require the fields to transform with a definite phase factor. We see that the free Hamiltonian is manifestly invariant under time reversal.

PROBLEMS[1]

1. Consider the Hamiltonian for a neutral scalar field interacting with a static source,

$$H = \tfrac{1}{2}\int d^3x(\pi^2(x) + (\nabla\,\phi(x))^2 + \mu^2\,\phi^2(x)) + g\int d^3x\,\phi(x)\,\rho(|\mathbf{x}|)$$

   Find the eigenstates of H.

2. Suppose that there are two static sources, so that $\rho(|\mathbf{x}|)$ is replaced by $\rho(|\mathbf{x} - \mathbf{x}_1|) + \rho(|\mathbf{x} - \mathbf{x}_2|)$ in the above Hamiltonian. Calculate the lowest energy eigenvalue. Prove that the energy eigenvalues can depend only on $|\mathbf{x}_1 - \mathbf{x}_2|$.

[1] The collection of problems at the ends of chapters have been chosen, in most instances to supplement material that could not be developed in the text because of space limitations. At the end of the book are listed some references to the literature where answers to some of the problems can readily be found. This collection does not include the many derivations which have been left for the reader to complete in the text itself.

3. The operator $M_{\mu\nu}$ satisfies the commutation relations

$$[M_{\mu\nu}, \phi(x)] = i(x_\mu \partial_\nu - x_\nu \partial_\mu)\, \phi(x)$$

Find a representation of $M_{\mu\nu}$ in terms of the field operators $\phi(x)$ and $\pi(x)$. (*Hint.* see Problem 4.)

4. The canonical energy momentum tensor corresponding to a Lagrangian density $\mathcal{L}$ is defined as

$$T^{\mu\nu} = \partial^\mu\left(\frac{\partial\mathcal{L}}{\partial(\partial_\nu\, \phi(x))}\right) - g^{\mu\nu}\mathcal{L}.$$

Show that P^μ and $M^{\mu\nu}$ may be calculated in terms of the $T^{\mu\nu}$:

$$P^\mu = \int d^3x T^{\mu 0}$$

$$M^{\mu\nu} = \int d^3x(x^\mu T^{\nu 0} - x^\nu T^{\mu 0})$$

5. Consider the Hamiltonian

$$H = \tfrac{1}{2}\int d^3x(\pi^2(x) + (\nabla\, \phi(x))^2 + \mu^2\, \phi^2(x) + U(|\mathbf{x}|)\, \phi^2(x))$$

with $U(|\mathbf{x}|) \to 0$ rapidly as $|\mathbf{x}| \to \infty$. How would you modify the quantization procedure outlined in this chapter to deal with this Hamiltonian? What special problems arise when $U(|\mathbf{x}|) \leq 0$?

6. Show that the matrix elements of the current operator

$$j_k(\mathbf{x}, t) = -i\phi^*(\mathbf{x}, t)\overleftrightarrow{\partial_k}\phi(\mathbf{x}, t)$$

have the proper transformation properties under time reversal, whether one uses the K operation or the order reversal for the antiunitary operation.

2

The Dirac Field

In the discussion of the scalar field we saw that the density function $j^0(x)$, whose space-integral we interpreted as a charge, is not positive definite. In the early days of the development of relativistic quantum mechanics this appeared to be a difficulty, for one was still thinking in terms of one-particle theories, i.e., relativistic generalizations of the Schrödinger equation with an accompanying probability interpretation.[1] Dirac noted that the indefinite density arose from the presence of a higher-than-first time derivative in the equation of motion and began a search for an equation that contained only a first time derivative, yet preserved its form under Lorentz transformations. He was led to the remarkable equation that bears his name and that soon turned out to be well suited to the description of particles of spin $\frac{1}{2}$. In this chapter we start with a brief discussion of the Dirac equation, as a classical field equation, and then proceed to the quantization of this field.

The Dirac equation is an equation for a four-component quantity $\psi_\alpha(x)$ called a spinor. Its transformation properties are different from that of a four-vector and we shall study them later in this chapter. The equation has the form

$$-i(\alpha^k)_{\rho\sigma}\frac{\partial}{\partial x^k}\psi_\sigma(x) + M(\beta)_{\rho\sigma}\psi_\sigma(x) = i\frac{\partial}{\partial x^0}\psi_\rho(x) \tag{2.1}$$

Here α^k and β are 4×4 matrices, which are required to satisfy the conditions

$$\begin{aligned} \alpha^k\alpha^l + \alpha^l\alpha^k \equiv \{\alpha^k, \alpha^l\} &= 2\delta^{kl} \\ \{\alpha^k, \beta\} &= 0 \\ \beta^2 &= \mathbf{1} \end{aligned} \tag{2.2}$$

[1] The one-particle theory is discussed in some detail in J. D. Bjorken and S. D. Drell, *Relativistic Quantum Mechanics*, McGraw-Hill Book Co., New York (1964).

in order that the Klein-Gordon equation,

$$-\frac{\partial^2}{\partial {x^0}^2}\,\psi_\rho(x) = \left(M^2 - \frac{\partial^2}{\partial {x^k}^2}\right)\psi_\rho(x) \tag{2.3}$$

which will guarantee the relativistic energy-momentum relation, will also hold true. We shall omit the spinor subscripts whenever there is no danger of confusion: $\psi(x)$ will always stand for a column to the right of the 4×4 matrices, and $\psi^+(x)$ for a row to the left of the matrices.

It is actually never necessary to have a specific representation of the matrices α^k and β; nevertheless some calculations are made more transparent by the choice of a canonical form. We take

$$\alpha^k = \begin{pmatrix} 0 & \sigma^k \\ \sigma^k & 0 \end{pmatrix} \qquad \beta = \begin{pmatrix} \mathbf{1} & 0 \\ 0 & -\mathbf{1} \end{pmatrix} \tag{2.4}$$

where the submatrices are the usual Pauli spin matrices and the 2×2 unit matrix:

$$\sigma^1 = \begin{pmatrix} 0 & 1 \\ 1 & 0 \end{pmatrix} \qquad \sigma^2 = \begin{pmatrix} 0 & -i \\ i & 0 \end{pmatrix} \qquad \sigma^3 = \begin{pmatrix} 1 & 0 \\ 0 & -1 \end{pmatrix} \qquad \mathbf{1} = \begin{pmatrix} 1 & 0 \\ 0 & 1 \end{pmatrix} \tag{2.5}$$

With the help of the matrices

$$\rho_1 = \begin{pmatrix} 0 & 1 \\ 1 & 0 \end{pmatrix} \qquad \rho_2 = \begin{pmatrix} 0 & -i \\ i & 0 \end{pmatrix} \qquad \rho_3 = \begin{pmatrix} 1 & 0 \\ 0 & -1 \end{pmatrix} \tag{2.6}$$

which look just like the Pauli matrices but act on a different set of indices (with two components), we may write α^k and β as outer products of the σ and the ρ matrices. Thus (2.4) is equivalent to writing

$$\boldsymbol{\alpha} = \rho_1 \otimes \boldsymbol{\sigma} \equiv \rho_1\boldsymbol{\sigma} \tag{2.7}$$

and

$$\beta = \rho_3 \otimes \mathbf{1} \equiv \rho_3 \tag{2.8}$$

When we use this form, the four-component spinor ψ will have to be viewed as a pair of two-component spinors

$$\psi = \begin{pmatrix} a_1 \\ a_2 \\ \hline b_1 \\ b_2 \end{pmatrix}$$

and the σ matrices act to rearrange a_1 and a_2 (and b_1, b_2) among themselves, whereas the ρ matrices act to rearrange the upper and lower pairs as units among themselves.

If we introduce the matrices

$$\gamma^k \equiv \beta\alpha^k \qquad \gamma^0 \equiv \beta \tag{2.9}$$

we may rewrite (2.1) in the more symmetric form

$$(i\gamma^\mu\, \partial_\mu - M)\, \psi(x) = 0 \tag{2.10}$$

It is easily checked that (2.2) is replaced by the anticommutation rule

$$\{\gamma^\mu, \gamma^\nu\} = 2g^{\mu\nu} \tag{2.11}$$

Both the preceding anticommutation condition and the hermiticity properties

$$\gamma^{0+} = \gamma^0; \;\; \gamma^{k+} = -\gamma^k \tag{2.12}$$

are abstracted from the special representation exhibited. The equation for the conjugate field $\psi^+(x)$ (written as a row) is somewhat complicated, but because

$$\gamma^0\gamma^{\mu+}\gamma^0 = \gamma^\mu \tag{2.13}$$

the spinor

$$\bar{\psi}(x) \equiv \psi^+(x)\gamma^0 \tag{2.14}$$

obeys the simple equation

$$i\partial_\mu\, \bar{\psi}(x)\gamma^\mu + M\, \bar{\psi}(x) = 0 \tag{2.15}$$

Equations 2.10 and 2.15 may be used to show that the current defined by

$$j^\mu(x) = \bar{\psi}(x)\gamma^\mu\, \psi(x) \tag{2.16}$$

is conserved:

$$\begin{aligned} \partial_\mu j^\mu(x) &= \partial_\mu\, \bar{\psi}(x)\gamma^\mu\, \psi(x) + \bar{\psi}(x)\gamma^\mu\partial_\mu\, \psi(x) \\ &= iM\, \bar{\psi}(x)\, \psi(x) - iM\, \bar{\psi}(x)\, \psi(x) = 0 \end{aligned} \tag{2.17}$$

The density $j^0(x)$ is given by

$$j^0(x) = \bar{\psi}(x)\gamma^0\, \psi(x) = \psi^+(x)\, \psi(x) = \sum_\rho |\psi_\rho(x)|^2 \tag{2.18}$$

This is a positive expression and, because of the conservation law, the space integral of $j^0(x)$ is time-independent. In this respect the quantity $\psi(x)$ resembles the Schrödinger wave function, and the Dirac equation may serve as a one-particle equation. In that role, however, the coefficient of $-ix^0$ in the decomposition

$$\psi(x) = \int dp\; \phi(p) e^{-ipx}$$

plays the role of the energy, and there is no reason why negative energies should be excluded. The difficulties of the Dirac equation as a one-particle equation were only resolved by a quantization of the field.

Consider the plane wave solutions of the Dirac equation. We shall treat positive and negative frequency terms separately and therefore write

$$\psi(x) = \frac{1}{(2\pi)^{3/2}}(u(p)e^{-ipx} + v(p)e^{ipx}) \tag{2.19}$$

Since $\psi(x)$ also satisfies the Klein-Gordon equation

$$(\Box + M^2)\,\psi(x) = 0$$

it is necessary that

$$p^2 - M^2 = 0 \tag{2.20}$$

so that

$$p^0 = +\sqrt{\mathbf{p}^2 + M^2} \equiv E_p \tag{2.21}$$

Conventionally, we shall call the term with $e^{-iE_px_0}$ the positive frequency solution. From the Dirac equation it follows that

$$(i\gamma^\mu(-ip_\mu) - M)\,u(p)e^{-ipx} + (i\gamma^\mu(ip_\mu) - M)v(p)e^{ipx} = 0$$

or

$$\begin{aligned}(\gamma^\mu p_\mu - M)u(p) &= 0\\ (\gamma^\mu p_\mu + M)v(p) &= 0\end{aligned} \tag{2.22}$$

because the positive and negative frequency solutions are independent.

It is useful to introduce the notation

$$\gamma^\mu p_\mu = \gamma_\mu p^\mu = \gamma^0 p^0 - \boldsymbol{\gamma}\cdot\mathbf{p} \equiv \not{p} \tag{2.23}$$

The "slash" quantities satisfy

$$\{\not{a}, \not{b}\} = a_\mu b_\nu\{\gamma^\mu, \gamma^\nu\} = 2a_\mu b^\mu \equiv 2a\cdot b \tag{2.24}$$

The Dirac equation for the plane wave solutions may thus be written as

$$\begin{aligned}(\not{p} - M)\,u(p) &= 0\\ (\not{p} + M)\,v(p) &= 0\end{aligned} \tag{2.25}$$

It is easily checked that

$$\begin{aligned}\bar{u}(p)(\not{p} - M) &= 0\\ \bar{v}(p)(\not{p} + M) &= 0\end{aligned} \tag{2.26}$$

When $\mathbf{p} = 0$, $p_0 = M$, the equations take the form

$$\begin{aligned}(\gamma^0 - 1)M\,u(0) &= 0\\ (\gamma^0 + 1)M\,v(0) &= 0\end{aligned} \tag{2.27}$$

There are, therefore, two positive and two negative frequency solutions, which we take to be

$$u^{(1)}(0) = \begin{pmatrix} 1 \\ 0 \\ 0 \\ 0 \end{pmatrix} \qquad u^{(2)}(0) = \begin{pmatrix} 0 \\ 1 \\ 0 \\ 0 \end{pmatrix} \qquad v^{(1)}(0) = \begin{pmatrix} 0 \\ 0 \\ 1 \\ 0 \end{pmatrix} \qquad v^{(2)}(0) = \begin{pmatrix} 0 \\ 0 \\ 0 \\ 1 \end{pmatrix} \tag{2.28}$$

Since

$$(\not{p} + M)(\not{p} - M) = p^2 - M^2 = 0 \tag{2.29}$$

we may write the solution for arbitrary p [still satisfying (2.20)] in the form

$$\begin{aligned} u^{(r)}(p) &= C(M + \not{p})\, u^{(r)}(0) \\ v^{(r)}(p) &= C'(M - \not{p})\, v^{(r)}(0) \end{aligned} \tag{2.30}$$

Here $r = 1, 2$; C and C' are normalization constants. We determine them by requiring that

$$\bar{u}^{(r)}(p)\, u^{(s)}(p) = \bar{u}^{(r)}(0)\, u^{(s)}(0) = \delta_{rs} \tag{2.31}$$

and

$$\bar{v}^{(r)}(p)\, v^{(s)}(p) = \bar{v}^{(r)}(0)\, v^{(s)}(0) = -\delta_{rs} \tag{2.32}$$

Thus

$$\begin{aligned} \bar{u}^{(r)}(p)\, u^{(s)}(p) &= |C|^2\, \bar{u}^{(r)}(0)(M + \not{p})(M + \not{p})\, u^{(s)}(0) \\ &= 2M\, |C|^2\, \bar{u}^{(r)}(0)(M + \not{p}) u^{(s)}(0) \\ &= 2M\, |C|^2\, \bar{u}^{(r)}(0)(M + \gamma^0 p_0 + \boldsymbol{\alpha} \cdot \mathbf{p}\beta)\, u^{(s)}(0) \\ &= 2M\, |C|^2\, (M + E_p)\, \bar{u}^{(r)}(0)\, u^{(s)}(0) \\ &= 2M\, |C|^2\, (M + E_p)\, \delta_{rs} \end{aligned}$$

Hence

$$C = \frac{1}{\sqrt{2M(M + E_p)}} \tag{2.33}$$

Similarly, we find that

$$C' = \frac{1}{\sqrt{2M(M + E_p)}} \tag{2.34}$$

Note that because $\bar{u}(p)$ and $u(p)$ satisfy the Dirac equation it follows that

$$\bar{u}^{(r)}(p)\{(-\not{p} + M), \gamma^\mu\} u^{(s)}(p) = 0$$

so that

$$2M\, \bar{u}^{(r)}(p)\, \gamma^\mu\, u^{(s)}(p) - 2p^\mu\, \bar{u}^{(r)}(p)\, u^{(s)}(p) = 0$$

Choosing $\mu = 0$, we get

$$u^{(r)*}(p)\, u^{(s)}(p) = \frac{E_p}{M}\, \delta_{rs} \tag{2.35}$$

To obtain the completeness properties of the solutions we consider the positive and negative frequency solutions separately. We use the explicit solutions already obtained.

$$(\Lambda_+)_{\alpha\beta} \equiv \sum_r u_\alpha^{(r)}(p)\, \bar{u}_\beta^{(r)}(p)$$

$$= \frac{1}{2M(M+E_p)} \left\{ \sum_r (\not{p} + M)\, u^{(r)}(0)\, \bar{u}^{(s)}(0)(\not{p} + M) \right\}_{\alpha\beta}$$

$$= \frac{1}{2M(M+E_p)} \left\{ (M + \not{p}) \frac{1+\gamma^0}{2} (M + \not{p}) \right\}_{\alpha\beta}$$

$$= \frac{1}{2M(M+E_p)} \left\{ M(\not{p} + M) + \frac{1}{2}(\not{p} + M)((M - \not{p})\gamma^0 + 2E_p) \right\}_{\alpha\beta}$$

$$= \frac{1}{2M} (\not{p} + M)_{\alpha\beta} \tag{2.36}$$

Similarly, if we define Λ_- by

$$(\Lambda_-)_{\alpha\beta} = -\sum_r v_\alpha^{(r)}(p)\, \bar{v}_\beta^{(r)}(p) \tag{2.37}$$

we get

$$(\Lambda_-)_{\alpha\beta} = \frac{1}{2M}(M - \not{p})_{\alpha\beta} \tag{2.38}$$

The completeness relation is that

$$\Lambda_+ + \Lambda_- = \sum_r (u^{(r)}(p)\, \bar{u}^{(r)}(p) - v^{(r)}(p)\, \bar{v}^{(r)}(p)) = \mathbf{1} \tag{2.39}$$

The separate matrices, Λ_+ and Λ_-, have the properties of *projection operators*, since

$$\Lambda_\pm{}^2 = \Lambda_\pm$$
$$\Lambda_+\Lambda_- = \Lambda_-\Lambda_+ = 0 \tag{2.40}$$

The operators $\Lambda_\pm$ project positive and negative frequency solutions, respectively, but because there are four solutions, there must be still another projection operator, which separates the $r = 1, 2$ solutions. This projection operator must be such that

$$\Pi^{(r)}\, \Pi^{(s)} = \delta_{rs}\, \Pi^{(r)} \tag{2.41}$$

and

$$[\Pi^{(r)}, \Lambda_\pm] = 0 \tag{2.42}$$

Since the two solutions have something to do with the two possible polarization directions of a spin $\frac{1}{2}$ particle, we may expect the operator to be some sort of generalization of the nonrelativistic operator

$$P_{\hat{n}} = \frac{1 + \hat{\mathbf{n}} \cdot \boldsymbol{\sigma}}{2}$$

which projects out the state polarized in the direction $\hat{\mathbf{n}}$ for a two-component spinor. We shall leave it to the reader to verify that the following generalization satisfies the conditions (2.41) and (2.42): define a four-vector n^μ which has the properties

$$n_\mu n^\mu = -1 \tag{2.43}$$

and

$$n_\mu p^\mu = 0 \tag{2.44}$$

It is thus of unit length and spacelike; in the rest frame of the particle whose four-momentum is p it has no zero component. The operators

$$\Pi_n^{(+)} = \tfrac{1}{2}(1 \pm \gamma_5 \not{n}) \tag{2.45}$$

where[1]

$$\gamma_5 = \gamma^5 = i\gamma^0\gamma^1\gamma^2\gamma^3 = \rho_1 \tag{2.46}$$

are the ones we want; with the canonical choice of representation of the matrices, they reduce to the nonrelativistic form $P_{\hat{n}}$.

We may examine the explicit form of the solution of the Dirac equation in the canonical representation. We have

$$\begin{aligned} u^{(r)}(p) &= \frac{M + \not{p}}{\sqrt{2M(M+E_p)}}\, u^{(r)}(0) = \frac{M + \rho_3 E_p - i\rho_2 \boldsymbol{\sigma}\cdot\mathbf{p}}{\sqrt{2M(M+E_p)}}\, u^{(r)}(0) \\ &= \frac{1}{\sqrt{2M(M+E_p)}} \begin{pmatrix} M + E_p & -\boldsymbol{\sigma}\cdot\mathbf{p} \\ \boldsymbol{\sigma}\cdot\mathbf{p} & M - E_p \end{pmatrix} \begin{pmatrix} u^{(r)}(0) \\ 0 \end{pmatrix} \\ &= \sqrt{\frac{E_p + M}{2M}} \begin{pmatrix} u^{(r)}(0) \\ \dfrac{\boldsymbol{\sigma}\cdot\mathbf{p}}{E_p + M}\, u^{(r)}(0) \end{pmatrix} \end{aligned} \tag{2.47}$$

For low momenta the upper two components are a great deal larger than the lower ones. When evaluating matrix elements of $\boldsymbol{\rho}$ and $\boldsymbol{\sigma}$ between low energy solutions, the diagonal ones, i.e., ρ_3 and 1, will evidently be much larger than the off-diagonal ones, ρ_1 and ρ_2.

It is sometimes more convenient to choose a different representation of the Dirac matrices. If, for example,

$$\gamma^0 = \rho_1 \qquad \boldsymbol{\alpha} = \rho_3 \boldsymbol{\sigma} \tag{2.48}$$

is chosen, then

$$\not{p} - M = \gamma^0 E_p + \boldsymbol{\alpha}\gamma^0 \cdot \mathbf{p} - M = -M + i\rho_2 \boldsymbol{\sigma}\cdot\mathbf{p} + \rho_1 E_p$$

[1] If we define the antisymmetric tensor of rank 4, $e_{\mu\nu\rho\sigma}$, with the properties: $e_{\mu\nu\rho\sigma} = \pm 1$ if μ,ν,ρ,σ form an even (odd) permutation of 0,1,2,3, and $e_{\mu\nu\rho\sigma} = 0$ otherwise, then we may write $\gamma_5 = \frac{i}{4!}\, e_{\mu\nu\rho\sigma}\gamma^\mu\gamma^\nu\gamma^\rho\gamma^\sigma$.

and the Dirac equation becomes

$$\begin{pmatrix} -M & \boldsymbol{\sigma}\cdot\mathbf{p} + E_p \\ E_p - \boldsymbol{\sigma}\cdot\mathbf{p} & -M \end{pmatrix} \psi = 0 \tag{2.49}$$

In this representation

$$\gamma_5 = \rho_3 \tag{2.50}$$

i.e., γ_5 is diagonal. Also, for very large energies, so that the mass M can be neglected, the equation splits into a pair of two-component equations; with $\psi(p) = \begin{pmatrix} a \\ b \end{pmatrix}$ they are

$$(E_p - \boldsymbol{\sigma}\cdot\mathbf{p})a \cong 0; \quad (E_p + \boldsymbol{\sigma}\cdot\mathbf{p})b \cong 0 \tag{2.51}$$

The first one represents a particle, massless, with spin pointing in the direction of the momentum, and the second represents a particle with spin pointing in the direction opposite that of the momentum.

We shall discuss the relativistic transformation properties of the Dirac field after quantization. We can, however, learn much about these properties by considering the unquantized equation, which will still be obeyed by matrix elements of the quantized $\psi(x)$. In nonquantized theory, the usual way of ensuring covariance under Lorentz transformations is to require that the equation be form invariant. With the transformation

$$x^\mu \rightarrow x'^\mu = \Lambda^\mu{}_\nu x^\nu$$

we do not expect form invariance with

$$\psi'(x') = \psi(x)$$

(as is the case for the Klein-Gordon equation for the scalar field), since a general Lorentz transformation contains rotations, and $\psi(x)$ is supposed to describe a field with spin. We therefore allow for a rearrangement of components and write

$$\psi'(x') = S(\Lambda)\, \psi(x) \tag{2.52}$$

The equation will be form invariant provided there exists a matrix $S(\Lambda)$ such that

$$(i\gamma^\mu \partial_\mu' - M)\, \psi'(x') = 0$$

This, however, is equivalent to

$$(i\gamma^\mu \Lambda_\mu{}^\nu\, \partial_\nu - M)\, S(\Lambda)\, \psi(x) = 0$$

If we multiply by $S^{-1}(\Lambda)$ from the left, we get

$$(iS^{-1}\gamma^\mu S \Lambda_\mu{}^\nu\, \partial_\nu - M)\, \psi(x) = 0$$

The equation therefore is form-invariant, provided we can find $S(\Lambda)$ such that

$$S^{-1}(\Lambda)\gamma^{\mu} S(\Lambda)\Lambda_{\mu}{}^{\nu} = \gamma^{\nu} \tag{2.53}$$

To find $S(\Lambda)$ we resort to the trick of considering an infinitesimal Lorentz transformation [(1.73), (1.75)]. Let

$$S(\Lambda) = 1 - \frac{i}{2}\alpha_{\mu\nu}\Sigma^{\mu\nu} \tag{2.54}$$

Then (2.53) reduces, after a little algebra, to the condition

$$(\Sigma^{\mu\nu}, \gamma^{\beta}] = -i(g^{\mu\beta}\gamma^{\nu} - g^{\nu\beta}\gamma^{\mu}) \tag{2.55}$$

A solution is seen to be

$$\Sigma^{\mu\nu} = -\frac{1}{4i}[\gamma^{\mu}, \gamma^{\nu}] \tag{2.56}$$

Incidentally

$$S^{+}(\Lambda) = \gamma^{0} S^{-1}(\Lambda)\gamma^{0} \tag{2.57}$$

The form for $S(\Lambda)$ when Λ is not infinitesimal is

$$S(\Lambda) = e^{-(i/2)\alpha_{\mu\nu}\Sigma^{\mu\nu}} \tag{2.58}$$

For a rotation

$$\alpha_{i0} = 0 \qquad \alpha_{ij} = a_k$$

and because

$$\Sigma^{ij} = \tfrac{1}{2}e_{ijk}\sigma_k$$

we get

$$S(\Lambda) = e^{-i\boldsymbol{\sigma}\cdot\mathbf{a}/2} \tag{2.59}$$

which shows the connection between the α_{ij} and the parameters characterizing the rotation.

For a pure Lorentz transformation

$$\alpha_{ij} = 0\,(i, j = 1, 2, 3) \qquad \alpha_{i0} = b_i$$

We have

$$\begin{aligned} S(\Lambda) &= e^{\frac{1}{2}\mathbf{b}\cdot\boldsymbol{\alpha}} \\ &= 1 + \frac{1}{2}\mathbf{b}\cdot\boldsymbol{\alpha} + \frac{1}{2!}\left(\frac{b^2}{4}\right) + \frac{1}{3!}\left(\frac{b^2}{4}\right)\frac{\mathbf{b}\cdot\boldsymbol{\alpha}}{2} + \cdots \\ &= \cosh\frac{b}{2} + \hat{\mathbf{b}}\cdot\boldsymbol{\alpha}\sinh\frac{b}{2} \end{aligned} \tag{2.60}$$

We may find the connection between $\mathbf{b}$ and the velocity $\mathbf{v}$ characterizing the pure Lorentz transformation by looking at (2.53) for a special case. We shall leave it to the reader to convince himself that

$$\cosh b = \frac{1}{\sqrt{1 - v^2}} \qquad \hat{\mathbf{b}} = \frac{\mathbf{v}}{v} \tag{2.61}$$

An important byproduct of this is that the transformation properties of bilinear combinations can easily be found. We have

$$\begin{aligned}\bar{\psi}'(x')\gamma^\mu\,\psi'(x') &= \bar{\psi}(x)\,S^{-1}(\Lambda)\gamma^\mu\,S(\Lambda)\,\psi(x)\\ &= \Lambda^\mu{}_\alpha\,\bar{\psi}(x)\,\gamma^\alpha\psi(x)\end{aligned}\tag{2.62}$$

Thus under a Lorentz transformation, the quantity $\bar{\psi}(x)\gamma^\mu\,\psi(x)$ transforms according to

$$V'^\mu = \Lambda^\mu{}_\alpha V^\alpha$$

i.e., like a contravariant four-vector. The knowledge of the transformation properties of bilinear combinations will turn out to be important when interacting fields are considered.

We may also show that the Dirac equation is invariant under space reflections. We write

$$\psi'(x') = \eta A\,\psi(x)\tag{2.63}$$

where η is a phase factor and A is some matrix to be determined so that the equation is form-invariant. We have $x' = (x^0, -\mathbf{x})$, so that

$$\begin{aligned}(i\gamma^\mu\,\partial'_\mu - M)\,\psi'(x') &= (i\gamma^0\,\partial_0 - i\boldsymbol{\gamma}\cdot\boldsymbol{\nabla}' - M)\eta A\,\psi(x)\\ &= (i\gamma^0\,\partial_0 + i\boldsymbol{\gamma}\cdot\boldsymbol{\nabla} - M)\eta A\,\psi(x) = 0\end{aligned}$$

If A is chosen such that

$$\begin{aligned}A^{-1}\gamma^0 A &= \gamma^0\\ A^{-1}\boldsymbol{\gamma}A &= -\boldsymbol{\gamma}\end{aligned}$$

we recover the Dirac equation. A solution is $A = \gamma^0$, so that

$$\psi'(x') = \eta\gamma^0\,\psi(x)\tag{2.64}$$

and

$$\bar{\psi}'(x') = \eta^*\,\bar{\psi}(x)\gamma^0\tag{2.65}$$

To carry out the *quantization* we must proceed in a way slightly different from that discussed in Chapter 1. As was done there, the quantities $\psi(x)$ and $\bar{\psi}(x)$ are expanded in terms of solutions of the Dirac equation. We write

$$\begin{aligned}\psi(x) &= \frac{1}{(2\pi)^{3/2}}\sum_r\int\frac{d^3p}{E_p/M}\,[\mathbf{a}^{(r)}(p)\,u^{(r)}(p)e^{-ipx} + \mathbf{b}^{(r)+}(p)v^{(r)}(p)e^{ipx}]\\ \bar{\psi}(x) &= \frac{1}{(2\pi)^{3/2}}\sum_r\int\frac{d^3p}{E_p/M}\,[\mathbf{b}^{(r)}(p)\,\bar{v}^{(r)}(p)e^{-ipx} + \mathbf{a}^{(r)+}(p)\,\bar{u}^{(r)}(p)e^{ipx}]\end{aligned}\tag{2.66}$$

and immediately turn to the Hamiltonian. The Dirac equation may be derived from the Lagrangian density

$$\mathfrak{L}(x) = \bar{\psi}(x)(i\gamma^\mu\,\partial_\mu - M)\,\psi(x)\tag{2.67}$$

from which it follows that

$$\pi(x) = \frac{\partial \mathfrak{L}}{\partial(\partial^0 \psi(x))} = i\,\bar{\psi}(x)\gamma_0$$

and[1]

$$\bar{\pi}(x) = 0$$

so that

$$\begin{aligned}\mathcal{H}(x) &= i\,\bar{\psi}(x)\gamma_0\,\partial^0\,\psi(x) - \bar{\psi}(x)(i\gamma^0\,\partial_0 + i\boldsymbol{\gamma}\cdot\boldsymbol{\nabla} - M)\,\psi(x)\\ &= -i\,\bar{\psi}(x)\boldsymbol{\gamma}\cdot\boldsymbol{\nabla}\,\psi(x) + M\,\bar{\psi}(x)\,\psi(x)\end{aligned} \tag{2.68}$$

and

$$H = \int d^3x\,\psi^+(x)(-i\boldsymbol{\alpha}\cdot\boldsymbol{\nabla} + \beta M)\,\psi(x) \tag{2.69}$$

In terms of the operators $\mathbf{a}^{(r)}(p)\cdots\mathbf{b}^{(r)+}(p)$ this becomes

$$\begin{aligned}H = \frac{1}{(2\pi)^3}\int\frac{d^3p}{E_p/M}\int\frac{d^3p'}{E_{p'}/M}\sum_{r,s}\int d^3x\{[\mathbf{a}^{(s)+}(p')\mathbf{a}^{(r)}(p)e^{i(p'-p)x}\\ \times u^{+(s)}(p')(\boldsymbol{\alpha}\cdot\mathbf{p}+\beta M)u^{(r)}(p)] + [\mathbf{b}^{(s)}(p')\mathbf{a}^{(r)}(p)e^{-i(p+p')x}\\ \times v^{+(s)}(p')(\boldsymbol{\alpha}\cdot\mathbf{p}+\beta M)u^{(r)}(p)] + [\mathbf{a}^{+(s)}(p')\mathbf{b}^{(r)+}(p)e^{i(p+p')x}u^{(s)+}(p')\\ \times(-\boldsymbol{\alpha}\cdot\mathbf{p}+\beta M)v^{(r)}(p)] + [\mathbf{b}^{(s)+}(p')\mathbf{b}^{(r)+}(p)e^{-i(p'-p)x}v^{(s)+}(p')\\ \times(-\boldsymbol{\alpha}\cdot\mathbf{p}+\beta M)v^{(r)}(p)]\}\end{aligned}$$

With the help of the Dirac equation written in the form

$$\begin{aligned}(\boldsymbol{\alpha}\cdot\mathbf{p}+\beta M)\,u^{(r)}(p) &= E_p\,u^{(r)}(p)\\ (-\boldsymbol{\alpha}\cdot\mathbf{p}+\beta M)\,v^{(r)}(p) &= -E_p\,v^{(r)}(p)\end{aligned} \tag{2.70}$$

and using

$$\begin{aligned}u^{(s)+}(p)u^{(r)}(p) = v^{(s)+}(p)v^{(r)}(p) = \delta_{rs}\frac{E_p}{M}\\ u^{(s)+}(p)v^{(r)}(p) = 0\end{aligned} \tag{2.71}$$

the following expression is obtained:

$$H = \sum_{r=1}^{2}\int\frac{d^3p}{E_p/M}\,[E_p\mathbf{a}^{(r)+}(p)\,\mathbf{a}^{(r)}(p) - E_p\mathbf{b}^{(r)}(p)\mathbf{b}^{(r)+}(p)] \tag{2.72}$$

This Hamiltonian bears a strong resemblance to the one obtained for the scalar field (1.47). There is one profound difference, however. The energy here, more like the *density* in the scalar case, is not positive definite. Thus negative energies are not prohibited, and there is no lower bound to the energy. If we impose commutation relations such as (1.39), say, on the

[1] Since the quantization will not follow canonical rules, the vanishing of $\bar{\pi}(x)$ does not present any problems.

operators, no improvement of the situation is possible; we may at best add an infinite constant to the energy. A solution to the difficulty is to require that an interchange of $\mathbf{b}^{(r)}(p)$ and $\mathbf{b}^{(r)+}(p)$ also change the sign of the product, i.e., that the operators satisfy *anticommutation relations.* If this is done, the Hamiltonian becomes (aside from a possible additive infinite constant which we subtract away)

$$H = \sum_{r=1}^{2} \int \frac{d^3p}{E_p/M} \left[E_p \mathbf{a}^{(r)+}(p)\, \mathbf{a}^{(r)}(p) + E_p\, \mathbf{b}^{(r)+}(p)\, \mathbf{b}^{(r)}(p)\right] \tag{2.73}$$

This now looks exactly like the scalar field Hamiltonian except that the invariant volume integral is replaced by E_p/M, simply a change in normalization. Thus a *number density* operator would be

$$\mathcal{N}_a^{(r)}(\mathbf{p}) = \frac{\mathbf{a}^{(r)+}(p)\, \mathbf{a}^{(r)}(p)}{E_p/M} \tag{2.74}$$

We write the following anticommutation relations

$$\begin{aligned} \{\psi_\alpha(x), \psi_\beta(y)\}_{x_0=y_0} &= \{\psi_\alpha^{\,+}(x), \psi_\beta^{\,+}(y)\}_{x_0=y_0} = 0 \\ \{\psi_\alpha(x), \psi_\beta^{\,+}(y)\}_{x_0=y_0} &= \delta_{\alpha\beta}\, \delta(\mathbf{x} - \mathbf{y}) \end{aligned} \tag{2.75}$$

and explore their consequences. The expansions (2.66) may be inverted with the help of the orthonormality properties of the solutions of Dirac's equation. We obtain

$$\begin{aligned} \mathbf{a}^{(r)}(p) &= \frac{1}{(2\pi)^{3/2}} \int d^3x e^{ipx}\, \bar{u}^{(r)}(p)\gamma^0\, \psi(x) \\ \mathbf{b}^{(r)}(p) &= \frac{1}{(2\pi)^{3/2}} \int d^3x e^{ipx}\, \bar{\psi}(x)\gamma^0\, v^{(r)}(p) \end{aligned} \tag{2.76}$$

etc. From this we can show that the commutation relations for the momentum space operators are

$$\begin{aligned} \{\mathbf{a}^{(r)}(p), \mathbf{a}^{(s)}(p')\} &= \{\mathbf{b}^{(r)}(p), \mathbf{b}^{(s)}(p')\} = \{\mathbf{a}^{(r)}(p), \mathbf{b}^{(s)}(p')\} \\ &= \{\mathbf{a}^{(r)}(p), \mathbf{b}^{(s)+}(p')\} = \cdots = 0 \\ \{\mathbf{a}^{(r)}(p), \mathbf{a}^{(s)+}(p')\} &= \{\mathbf{b}^{(r)}(p), \mathbf{b}^{(s)+}(p')\} = \delta_{rs} \frac{E_p}{M}\, \delta(\mathbf{p} - \mathbf{p}') \end{aligned} \tag{2.77}$$

With the help of these relations we see that

$$\begin{aligned} [\mathbf{a}^{(r)}(p), H] &= \sum_s \int \frac{d^3p'}{E_{p'}/M} [\mathbf{a}^{(r)}(p) E_{p'}\, \mathbf{a}^{(s)+}(p')\, \mathbf{a}^{(s)}(p') \\ &\qquad - E_{p'}\, \mathbf{a}^{(s)+}(p')\, \mathbf{a}^{(s)}(p')\, \mathbf{a}^{(r)}(p)] \\ &= \sum_s M \int d^3p' [\mathbf{a}^{(r)}(p)\, \mathbf{a}^{(s)+}(p')\, \mathbf{a}^{(s)}(p') + \mathbf{a}^{(s)+}(p')\, \mathbf{a}^{(r)}(p)\, \mathbf{a}^{(s)}(p')] \\ &= E_p\, \mathbf{a}^{(r)}(p) \end{aligned} \tag{2.78}$$

Similarly,

$$[\mathbf{a}^{(r)+}(p), H] = -E_p\, \mathbf{a}^{(r)+}(p) \tag{2.79}$$

and

$$[\mathbf{b}^{(r)}(p), H] = E_p\, \mathbf{b}^{(r)}(p)$$
$$[\mathbf{b}^{(r)+}(p), H] = -E_p\, \mathbf{b}^{(r)+}(p) \tag{2.80}$$

We follow the arguments of Chapter 1 to interpret $\mathbf{a}^{(r)+}(p)$ and $\mathbf{b}^{(r)+}(p)$ as *creation operators* and $\mathbf{a}^{(r)}(p)$ and $\mathbf{b}^{(r)}(p)$ as *annihilation* operators. If we define the vacuum state Ψ_0 for this system as before, then

$$\mathbf{a}^{(r)}(p)\Psi_0 = \mathbf{b}^{(r)}(p)\Psi_0 = 0 \qquad \text{all } \mathbf{p},\, r = 1, 2 \tag{2.81}$$

We may construct one-particle states

$$\Psi_{p(r)} = \mathbf{a}^{(r)+}(p)\Psi_0 \tag{2.82}$$

and, as for the scalar case, many-particle states. An arbitrary state of a system will consist of a superposition of zero, one, two, . . . -particle states. If we restrict ourselves to particles of type **a** for simplicity, we can write

$$\Psi = c_0\Psi_0 + \sum_{r_1 \cdots r_n} \int \frac{d^3p_1}{E_{p_1}/M} \cdots \frac{d^3p_n}{E_{p_n}/M}\, C_n(p_1 \cdots p_n)\Psi_{p_1,r_1; \cdots p_n,r_n;} \tag{2.83}$$

where

$$\Psi_{p_1,r_1;p_2,r_2; \ldots ; p_n,r_n} = \mathbf{a}^{(r_1)+}(p_1) \cdots \mathbf{a}^{(r_n)+}(p_n)\Psi_0 \tag{2.84}$$

Since

$$\{\mathbf{a}^{(r_i)+}(p_i), \mathbf{a}^{(r_j)+}(p_j)\} = 0$$

it follows that

$$\Psi_{p_1 \cdots p_i \cdots p_j \cdots p_n} = -\Psi_{p_1 \cdots p_j \cdots p_i \cdots p_n} \tag{2.85}$$

Since the state vector changes sign under the interchange of any two particles, the functions $C_n(p_1(r_1), \ldots, p_n(r_n))$ in (2.83), which are the probability amplitudes for finding n quanta with the labeled momenta and spin projections in the state Ψ, are antisymmetric functions of their arguments. This implies that the **a**-quanta and the **b**-quanta satisfy the *Pauli exclusion principle.* We did not find this to be the case for the scalar mesons; the exclusion principle is intimately related to the need to use anticommutation relations for the quantization of the Dirac field. For the scalar field we did not need to use anticommutation rules. In fact, had we imposed anticommutation relations, we would have found that the commutator does not vanish outside the light cone, so that with the "wrong" quantization rules we would have had a violation of the natural physical requirement of causality. These observations about the free field theory are special cases of the *general theorem on the connection between spin and statistics:* it can be shown quite generally that for a local field theory, if we impose a choice of quantizing with commutation or anticommutation rules, consistency demands that fields whose quanta have half odd integral spin must be quantized with anticommutation relations (and thus satisfy Fermi statistics). These predictions of local field theory are in perfect accord with experience.[1]

With the help of the creation and annihilation operator commutation relations we can calculate the anticommutator for unequal times. We have

$$\begin{aligned}\{\psi(x), \bar{\psi}(y)\} &= \frac{1}{(2\pi)^3} \sum_{rs} \int \frac{d^3p}{E_p/M} \int \frac{d^3p'}{E_{p'}/M} \\ &\quad \times [\{\mathbf{a}^{(r)}(p), \mathbf{a}^{(s)+}(p')\}\, u^{(r)}(p)\bar{u}^{(s)}(p')e^{-ipx+ip'y} \\ &\quad + \{\mathbf{b}^{(r)+}(p), \mathbf{b}^{(s)}(p')\}\, v^{(r)}(p)\, \bar{v}^{(s)}(p')e^{ipx-ip'y}] \\ &= (M + i\gamma^\mu\, \partial_\mu) \frac{1}{(2\pi)^3} \int \frac{d^3p}{2E_p} (e^{-ip(x-y)} - e^{ip(x-y)}) \\ &= i(M + i\gamma^\mu\, \partial_\mu)\, \Delta(x - y, M) \\ &\equiv -iS(x - y, M) \qquad (2.86)\end{aligned}$$

The anticommutator thus vanishes outside the light cone. We shall soon see the significance of this. At this point, we examine the charge operator to obtain the connection between the **a**-quanta and the **b**-quanta. With the charge density obtained in (2.18), we define

$$Q = \int d^3x\, j^0(x) = \int d^3x\, \bar{\psi}(x)\gamma^0\psi(x) \qquad (2.87)$$

In terms of the operators $\mathbf{a}^{(r)}(p), \ldots, \mathbf{b}^{(r)+}(p)$ we readily see that

$$Q = \sum_r \int \frac{d^3p}{E_p/M} [\mathbf{a}^{(r)+}(p)\, \mathbf{a}^{(r)}(p) - \mathbf{b}^{(r)+}(p)\, \mathbf{b}^{(r)}(p)] \qquad (2.88)$$

so that we are again led to the interpretation of the **b**-quanta as the antiparticles of the **a**-quanta. Since

$$[\mathbf{a}^{(r)}, Q] = \sum_s \int \frac{d^3p'}{E_{p'}/M} [\mathbf{a}^{(r)}(p), \mathbf{a}^{(s)+}(p')\, \mathbf{a}^{(s)}(p')] = \mathbf{a}^{(r)}(p)$$

we see that if

$$Q\Psi = q\Psi$$

[1] A complete survey of the history of the spin-statistics theorem may be found in "Das Pauli-Prinzip und Die Lorentz-Gruppe," by R. Jost in the Pauli memorial volume, *Theoretical Physics in the Twentieth Century*, M. Fierz and V. F. Weisskopf, editors, Interscience Publishers, New York (1960). See also R. Streater and A. Wightman, *PCT, Spin and Statistics and All That*, W. A. Benjamin, New York (1964).

then

$$Q\,\mathbf{a}^{(r)}(p)\Psi = [\mathbf{a}^{(r)}(p)Q - \mathbf{a}^{(r)}(p)]\Psi = (q-1)\,\mathbf{a}^{(r)}(p)\Psi$$

so that **a** is a charge lowering operator.

We may again construct a charge conjugation operator C which has the property that

$$C\mathbf{a}^{(r)}(p)C^{-1} = \mathbf{b}^{(r)}(p)$$
$$C\mathbf{a}^{(r)+}(p)C^{-1} = \mathbf{b}^{(r)+}(p) \tag{2.89}$$

From this we see that $C\,\psi(x)C^{-1}$ must be related to $\bar{\psi}(x)$. In fact, if we write

$$C\psi_\alpha(x)C^{-1} = \bar{\psi}_\beta(x)\mathcal{C}_{\beta\alpha}$$
$$C\,\bar{\psi}_\alpha(x)C^{-1} = -\mathcal{C}_{\alpha\beta}{}^{-1}\,\psi_\beta(x) \tag{2.90}$$

where $\mathcal{C}$ is a 4×4 matrix, then $\mathcal{C}$ must have the property that

$$u^{(r)}(p) = \bar{v}^{(r)}(p)\mathcal{C}$$
$$v^{(r)}(p) = \bar{u}^{(r)}(p)\mathcal{C} \tag{2.91}$$

We shall leave it to the reader to prove that as a consequence of this $\mathcal{C}$ has the properties that

$$\mathcal{C}^T = -\mathcal{C} \tag{2.92}$$

and

$$\mathcal{C}^{-1}\gamma^\mu\mathcal{C} = -(\gamma^\mu)^T \tag{2.93}$$

We have, up to this point, implied that the Dirac field describes particles of spin $\frac{1}{2}$, with the index (r) related to the polarization state. The confirmation of this comes from the study of the transformation laws of the Dirac field under homogeneous Lorentz transformations. The transformation properties under a space-time displacement are no problem. As in the scalar case (1.23) we have the definition of $U(a)$,

$$\psi(x+a) = U(a)\,\psi(x)\,U^{-1}(a) \tag{2.94}$$

and the commutation relation

$$[P^\mu, \psi(x)] = -i\,\partial^\mu\,\psi(x) \tag{2.95}$$

where $P^\mu = (H, \mathbf{P})$, and

$$\mathbf{P} = -\int d^3x\,\pi(x)\,\nabla\,\psi(x) = \frac{1}{i}\int d^3x\,\bar{\psi}(x)\gamma^0\,\nabla\,\psi(x) \tag{2.96}$$

so that

$$U(a) = e^{iP^\mu a_\mu} \tag{2.97}$$

For the homogeneous Lorentz transformations the law must be more complicated than the scalar transformation law (1.21), since the particle has spin. One way of arriving at the transformation law is to consider the expectation value of $\psi(x)$ in some state Ψ. This expectation value is now a c-number, and we may expect it to transform like the unquantized Dirac field, i.e., according to (2.52). Thus we require that

$$(\Psi', \psi(\Lambda x)\Psi') = S(\Lambda)(\Psi, \psi(x)\Psi) \tag{2.98}$$

where Ψ' is the Lorentz transformed state

$$\Psi' = U(\Lambda)\Psi$$

From this we get the generalization of (1.21), namely

$$U^{-1}(\Lambda)\,\psi(\Lambda x)\,U(\Lambda) = S(\Lambda)\,\psi(x)$$

i.e.

$$U(\Lambda)\,\psi(x)\,U^{-1}(\Lambda) = S^{-1}(\Lambda)\,\psi(\Lambda x) \tag{2.99}$$

Let us now consider an infinitesimal transformation characterized by the six parameters ($\alpha^{\mu\nu}$, $\alpha^{\mu\nu} = -\alpha^{\nu\mu}$). As in the scalar case we have

$$U(\Lambda) = 1 + \frac{i}{2}\,\alpha^{\mu\nu}M_{\mu\nu}$$

Thus

$$\begin{aligned}
\psi(x) + \frac{i}{2}\,\alpha^{\mu\nu}[M_{\mu\nu}, \psi(x)] &= \left(1 + \frac{i}{2}\,\alpha^{\mu\nu}\Sigma_{\mu\nu}\right)\psi(x^\mu + \alpha^\mu{}_\nu x^\nu) \\
&= \left(1 + \frac{i}{2}\,\alpha^{\mu\nu}\Sigma_{\mu\nu}\right)(\psi(x) + \alpha^\mu{}_\nu x^\nu\,\partial_\mu\,\psi(x)) \\
&= \left(1 + \frac{i}{2}\,\alpha^{\mu\nu}\Sigma_{\mu\nu}\right) \\
&\quad\times\left(\psi(x) + \frac{i}{2}\,\alpha^{\mu\nu}\left(\frac{1}{i}\,x_\nu\,\partial_\mu - \frac{1}{i}\,x_\mu\,\partial_\nu\right)\psi(x)\right)
\end{aligned}$$

This leads to the relation

$$[M_{\mu\nu}, \psi(x)] = \left[\Sigma_{\mu\nu} + \frac{1}{i}\,(-x_\mu\,\partial_\nu + x_\nu\,\partial_\mu)\right]\psi(x) \tag{2.100}$$

If we now consider a rotation about the z-axis, for example, we obtain

$$[M_{12}, \psi(x)] = (\Sigma_{12} + L_3)\,\psi(x)$$

Using (2.56) it is easy to check that the quantities $\Sigma_{12} \equiv \frac{1}{2}\sigma_3$, $\Sigma_{31} \equiv \frac{1}{2}\sigma_2$ and $\Sigma_{23} \equiv \frac{1}{2}\sigma_1$ satisfy the commutation relations

$$[\tfrac{1}{2}\sigma_i, \tfrac{1}{2}\sigma_j] = ie_{ijk}(\tfrac{1}{2}\sigma_k)$$

and that

$$\sum_i \left(\frac{1}{2}\sigma_i\right)^2 = \frac{3}{4} = \frac{1}{2}\left(\frac{1}{2}+1\right)$$

If we now ask what happens to the wave function of a spinor particle at rest $(\Psi_0, \psi(x)\Psi_{\mathbf{p}=0})$ when the system is rotated, we find, as in the scalar case, that

$$\begin{aligned}(\Psi_0, \psi(x)\, U_z(\delta\theta)\Psi_{\mathbf{p}=0}) &\cong (1 - \tfrac{1}{2}i\,\delta\theta\sigma_3 - i\,\delta\theta L_3)(\Psi_0, \psi(x)\Psi_{\mathbf{p}=0}) \\ &\cong (1 - \tfrac{1}{2}i\,\delta\theta\sigma_3)(\Psi_0, \psi(x)\Psi_{\mathbf{p}=0}) \\ &\cong e^{-i/2\sigma_3\delta\theta}(\Psi_0, \psi(x)\Psi_{\mathbf{p}=0})\end{aligned}$$

i.e., the L_3 term does not contribute and the remaining transformation matrix is characteristic of a particle of spin $\frac{1}{2}$.

The parity transformation will be of the form

$$U_s\psi(\mathbf{x}, t)U_s^{-1} = \eta A\psi(-\mathbf{x}, t) \tag{2.101}$$

The form invariance of the Dirac equation requires that

$$A = \gamma^0 \tag{2.102}$$

as seen in (2.64).

It is perhaps unexpected that the parity of a particle-antiparticle system is opposite to that of a two-particle system in the same state. To see this consider the operator $\psi(x)\psi^c(y)$, where

$$\psi^c(y) = C\,\psi(y)C^{-1} = \bar{\psi}(y)\mathcal{C} \tag{2.103}$$

This operator has the property that when it acts on the vacuum state it creates a particle-antiparticle pair. Under a spatial inversion this transforms as follows (we use the notation $x' = (-\mathbf{x}, t)$:

$$\begin{aligned}U_s(\psi_\alpha(x)\,\psi_\rho{}^c(y))U_s^{-1} &= A_{\alpha\rho}\,\psi_\rho(x')\,\bar{\psi}_\sigma(y')A_{\sigma\lambda}{}^+\mathcal{C}_{\lambda\beta} \\ &= \eta(\gamma^0\,\psi(x'))(-\eta^*\,\bar{\psi}(y')\mathcal{C}\gamma^{0T}) \\ &= -(\eta\gamma^0\,\psi(x')(\eta^*\gamma^0\,\psi^c(y'))\end{aligned} \tag{2.104}$$

As an example, the electron positron system in an orbital S-state has *odd* parity, and this fact will have measurable consequences as far as the properties of the photon state, resulting from its annihilation, are concerned.[1]

[1] The polarization correlation of the two photons in the decay of positronium from the ${}^1S_0$ ground state has been measured by C. S. Wu and I. Shaknow, *Phys. Rev.* **77**, 136 (1950), who found that the correlation was characteristic of the decay of an initial $J = 0$ *odd* parity state. This last point is discussed in connection with the π^0 decay in Chapter 14.

The phase factor η is the intrinsic parity of the particle created by $\psi(x)$. If the field were measurable, we would require that $\eta^2 = 1$. Clearly $\psi(x)$ is not hermitian, but the combinations

$$\begin{aligned} \psi_1(x) &= \frac{1}{\sqrt{2}}(\psi(x) + \psi^+(x)) \\ \psi_2(x) &= \frac{i}{\sqrt{2}}(\psi^+(x) - \psi(x)) \end{aligned} \tag{2.105}$$

are. Are these ever measurable? Wick, Wightman, and Wigner[1] pointed out that in quantum field theory, in contrast to quantum mechanics, we are not actually dealing with a single vector space $\mathcal{H}$ (Hilbert space) but with many, which are independent of each other in the sense that a state vector $\sum_i \Psi_{(i)}$, where $\Psi_{(i)}$ lies in $\mathcal{H}^{(i)}$, is indistinguishable from a state vector $e^{i\alpha_1}\Psi_{(1)} + \cdots + e^{i\alpha_n}\Psi_{(n)}$ with α_i an arbitrary phase.

Consider, for example, a state that is a superposition of the vacuum state and a one-fermion state

$$\Psi = c_0\Psi_0 + \int dx\, f(x)\, \bar{\psi}(x)\Psi_0 \tag{2.106}$$

Under a rotation of 2π about some axis, the application of (2.99) shows that

$$U_{2\pi}\, \bar{\psi}(x) U_{2\pi}^{-1} = -\bar{\psi}(x) \tag{2.107}$$

so that the state Ψ and the state

$$\Psi' = c_0\Psi_0 - \int dx\, f(x)\, \bar{\psi}(x)\Psi_0 \tag{2.108}$$

are physically indistinguishable, for no observable can depend on whether the system has been rotated by 2π. This, however, implies that the relative phase of the term with no fermions (or, more generally, an even number of fermions) and the term with one fermion (or an odd number of fermions) is not measurable. Consequently, the matrix element of any operator that changes the "fermion number" by an odd integer, i.e., connects state vectors of even and odd fermion number, *will have an undetermined phase* ± 1.

This example shows that the superposition principle must be restricted to hold in each of a set of subspaces only, but that states built up out of states from different *superselected subspaces* cannot possibly be physically realizable. If we introduce an operator N whose eigenvalues are ± 1 when acting on a state with an even (odd) number of fermions, then

$$[\Theta, N] = 0$$

[1] G. C. Wick, A. S. Wightman, and E. P. Wigner, *Phys. Rev.* **88,** 101 (1952).

for all observables Θ. In quantum mechanics any operator that commutes with a complete set of observables must be (up to a factor) unity. In quantum field theory this is no longer true. There may be such operators that are not "unity."

It is generally believed that all observables commute with the charge operator, which therefore plays a role analogous to N. All of our experience indicates that physically realizable states can be classified by their charge and that there are no states that consist of a superposition of states of charge q_1 and charge q_2. Thus it is believed to be a fact of nature that states of fixed charge lie in separate subspaces, and consequently operators that change the charge, such as $\phi(x)$, are not measurable. An equivalent way of saying this is that all observables must be invariant under gauge transformations. Similarly, it is our experience that *baryon number*, and very likely two types of lepton numbers can be used to classify independent subspaces of the space of all physical states.

A consequence is that the parity of a particle that carries charge, or baryon number, or has half-odd integral spin, is arbitrary. It is still meaningful to assign a conventional parity to the proton and the neutron, say, and to determine the parity of the π^+ relative to this choice by a study of the reaction

$$p \leftrightarrow \pi^+ + n$$

which takes place within the same subspace of charge 1.

The fact that $\psi(x)$ is not measurable could also have been deduced from the commutation rules. A calculation, left to the reader, shows that $[\psi(x), \bar{\psi}(y)]$ is an operator whose matrix elements (e.g., vacuum expectation value) do not vanish outside the light cone, in apparent conflict with causality. This conflict is only apparent. For a quantity which we might expect to be measurable, e.g., the current, we have

$$\begin{aligned}[j_\mu(x), j_\nu(y)] &= (\gamma_\mu)_{\alpha\beta}(\gamma_\nu)_{\rho\sigma}[\bar{\psi}_\alpha(x)\,\psi_\beta(x), \bar{\psi}_\rho(y)\,\psi_\sigma(y)] \\ &= (\gamma_\mu)_{\alpha\beta}(\gamma_\nu)_{\rho\sigma}[\bar{\psi}_\alpha(x)\,\psi_\sigma(y)\{\bar{\psi}_\rho(y), \psi_\beta(x)\} \\ &\qquad - \bar{\psi}_\rho(y)\,\psi_\beta(x)\{\psi_\sigma(y), \bar{\psi}_\alpha(x)\}] \end{aligned} \quad (2.109)$$

We have already shown that $\{\psi(x), \bar{\psi}(y)\}$ vanishes outside the light cone, so that the causality condition is satisfied.

We conclude this chapter with a discussion of the transformation properties of the Dirac field under time reversal. We must again take the transformation to be antiunitary to preserve the positivity of the energy, and for the single field we require

$$T\,\psi_\alpha(\mathbf{x}, t)T^{-1} = \eta\,\bar{\psi}_\beta(\mathbf{x}, -t)B_{\beta\alpha} \quad (2.110)$$

The matrix B is included to allow for the reversal of the spin, which is part of the time inversion. We choose B such that

$$BB^+ = 1 \tag{2.111}$$

and determine it from the form invariance of the Dirac equation. We get

$$\begin{aligned}
(i\gamma^\mu \partial_\mu - M)T\, \psi(\mathbf{x}, t)T^{-1} &= 0 \\
&= \left(i\gamma^0 \frac{\partial}{\partial t} - i\boldsymbol{\gamma}\cdot\boldsymbol{\nabla} - M\right)\bar{\psi}(\mathbf{x}, -t)B \\
&= \left(-i\gamma^0 \frac{\partial}{\partial(-t)} - i\boldsymbol{\gamma}\cdot\boldsymbol{\nabla} - M\right)\bar{\psi}(\mathbf{x}, -t)B \\
&= -\Bigg[i\frac{\partial}{\partial(-t)}\bar{\psi}(\mathbf{x}, -t)B\gamma^{0T}B^{-1} \\
&\qquad + i\,\boldsymbol{\nabla}\bar{\psi}(\mathbf{x}, -t)B\boldsymbol{\gamma}^T B^{-1} + M\bar{\psi}(\mathbf{x}, -t)\Bigg]B \\
&= -[i\,\partial_\mu{}'\,\bar{\psi}(x')\gamma^\mu + M\,\bar{\psi}(x')]B \\
&= 0
\end{aligned} \tag{2.112}$$

provided

$$\begin{aligned} B\gamma^{0T}B^{-1} &= \gamma^0 \\ B\boldsymbol{\gamma}^T B^{-1} &= -\boldsymbol{\gamma} \end{aligned} \tag{2.113}$$

This implies that

$$B\gamma^{\mu *}B^{-1} = \gamma^\mu \tag{2.114}$$

We leave it to the reader to verify that

$$BC^{-1} = \gamma_0\gamma_5 \tag{2.115}$$

PROBLEMS

1. Work out the transformation properties under P, C, and T (time reversal) of the bilinear operator combinations

(*a*) $[\bar{\psi}(x), \gamma_5\psi(x)] \equiv \bar{\psi}(x)\gamma_5\psi(x) - \gamma_5\psi(x)\bar{\psi}(x)$,
(*b*) $[\bar{\psi}(x), \gamma^\alpha\psi(x)]$
(*c*) $[\bar{\psi}(x), \sigma^{\alpha\beta}\psi(x)]$
(*d*) $[\bar{\psi}(x), i\gamma^\alpha\gamma_5\psi(x)]$

2. The equation for a spin $\frac{1}{2}$ particle in an external electromagnetic field is

$$(i\gamma^\mu(\partial_\mu - ieA_\mu^{\text{ext}}(x)) - M)\psi(x) = 0$$

Solve this for the case of a constant, uniform magnetic field, when $A_\mu^{\text{ext}}(x) = (0, -\frac{1}{2}\mathbf{r}\times\mathbf{H})$. How would you quantize the Dirac field in the presence of such an external magnetic field? Work out, as far as you are able, the anticommutator between $\psi(x)$ and $\bar{\psi}(y)$ for unequal times in the presence of the constant field.

3. Consider the explicit solution (2.47). Since a spinor for arbitrary momentum p may be generated from the spinor at rest by a Lorentz transformation, it should be possible to write the solution in the form

$$e^{-i\alpha_{io}\Sigma^{io}}u(0)$$

with Σ^{io} given by (2.56). Find α_{io}.

4. Express the solution of the Dirac equation in eigenfunctions of the total angular momentum. Use the eigenfunctions of J^2 and J_z to separate the equation of a Dirac particle in a Coulomb field. The equation is that of problem 2.
5. Show that under time reversal $u(\mathbf{p}) \rightarrow \bar{u}(-\mathbf{p})B$ and $v(\mathbf{p}) \rightarrow \bar{v}(-\mathbf{p})B$. Use this to show that under PT together

$$\bar{u}(p')\gamma_{\alpha_1}\gamma_{\alpha_2}\cdots\gamma_{\alpha_n}u(p) \rightarrow \bar{u}(p)\gamma_{\alpha_n}\cdots\gamma_{\alpha_2}\gamma_{\alpha_1}u(p')$$

6. Show that tr $\gamma^\mu\gamma^\nu = 4g^{\mu\nu}$ and use $\{\gamma^\mu, \gamma^5\} = 0$ to prove that tr $(\gamma^{\alpha_1}\gamma^{\alpha_2}\cdots\gamma^{\alpha_{2n+1}}) = 0$.

3

Vector Mesons and Photons

In addition to particles of spin 0 and $\frac{1}{2}$, particles of spin 1 appear to play a fundamental role in particle physics. In this chapter we discuss briefly the vector field and the electromagnetic field. The electromagnetic field, whose quanta are massless, is described by a vector potential. Although this description leads to some formal complications, it is one that leads to the simplest equations of motion for the interaction of photons with matter and is therefore to be preferred in the context of the pedestrian use we make of field theory.

Let us begin by considering a field that transforms as a four-vector under Lorentz transformations. The quantized field is required to transform according to

$$U(\Lambda)\,\phi^\mu(x)\,U^{-1}(\Lambda) = (\Lambda^{-1})^\mu{}_\nu\,\phi^\nu(\Lambda x) \tag{3.1}$$

The reader may convince himself that it is $(\Lambda^{-1})^\mu{}_\nu$, which appears in (3.1), by considering the special case of a rotation or by investigating the requirement that the expectation value of the field transform according to

$$(\Psi', \phi^\mu(\Lambda x)\Psi') = \Lambda^\mu{}_\nu(\Psi, \phi^\nu(x)\Psi) \tag{3.2}$$

We shall also require that the vector field satisfy the Klein-Gordon equation

$$(\Box + m^2)\,\phi^\mu(x) = 0 \tag{3.3}$$

This, we may say from experience, will imply that the quanta of the field, $\phi^\mu(x)$ will have mass m. In addition, a subsidiary condition is required: we are trying to describe a field whose quanta have spin 1, i.e., three independent polarization states (two, when $m = 0$), and the field $\phi^\mu(x)$ has four components. It is natural to impose the simplest invariant condition, namely

$$\partial_\mu\,\phi^\mu(x) = 0 \tag{3.4}$$

We shall, of course, have to check that this condition does what it is supposed to do.

To proceed with the quantization we must write down a Lagrangian that leads to (3.3) and, if possible, also contains the subsidiary condition. It turns out that the inclusion of the subsidiary condition is most easily accomplished by the introduction of auxiliary fields (in this case $f^{\mu\nu}$ and $f^{\mu\nu+}$) and then writing the Lagrangian in such a way that it involves only first-order derivatives. It is easily checked that with the Lagrangian density

$$\mathcal{L} = -\tfrac{1}{2}f^{\mu\nu+}(\partial_\mu\phi_\nu - \partial_\nu\phi_\mu) - \tfrac{1}{2}f^{\mu\nu}(\partial_\mu\phi_\nu^{+} - \partial_\nu\phi_\mu^{+}) + \tfrac{1}{2}f^{\mu\nu+}f_{\mu\nu} + m^2\phi_\mu\phi^{\mu+} \tag{3.5}$$

the following equations emerge:

$$\partial_\alpha \frac{\partial\mathcal{L}}{\partial(\partial_\alpha f^{\mu\nu+})} = \frac{\partial\mathcal{L}}{\partial(f^{\mu\nu+})} \tag{3.6}$$

leads to

$$f_{\mu\nu} = \partial_\mu\phi_\nu - \partial_\nu\phi_\mu \tag{3.7}$$

Also

$$\partial_\alpha \frac{\partial\mathcal{L}}{\partial(\partial_\alpha\phi_\mu^{+})} = \frac{\partial\mathcal{L}}{\partial\phi_\mu^{+}} \tag{3.8}$$

leads to

$$\partial_\mu f^{\mu\nu} + m^2\phi^\nu = 0 \tag{3.9}$$

The other equations are

$$f_{\mu\nu}^{+} = \partial_\mu\phi_\nu^{+} - \partial_\nu\phi_\mu^{+} \tag{3.10}$$

and

$$\partial_\mu f^{\mu\nu+} + m^2\phi^{\nu+} = 0 \tag{3.11}$$

Equation (3.9) when differentiated with respect to x^ν gives

$$\partial_\mu\partial_\nu f^{\mu\nu} = -m^2\partial_\nu\phi^\nu = 0 \tag{3.12}$$

The last step follows from the antisymmetry of $f^{\mu\nu}$ in its two indices.

With the help of the subsidiary condition (3.12), the equation of motion (3.9) may be rewritten as follows

$$\begin{aligned}\partial_\mu f^{\mu\nu} + m^2\phi^\nu &= \partial_\mu(\partial^\mu\phi^\nu - \partial^\nu\phi^\mu) + m^2\phi^\nu \\ &= -\partial^\nu(\partial_\mu\phi^\mu) + (\Box + m^2)\phi^\nu \\ &= (\Box + m^2)\phi^\nu = 0\end{aligned}$$

Thus the Klein-Gordon equation is satisfied. We may now use (3.9) to eliminate one of the four components of the vector field. We shall choose $\phi^0(x)$ to be eliminated. The required expressions are

$$\phi^0(x) = -\frac{1}{m^2}\partial_\mu f^{\mu 0}(x); \qquad \phi^{0+}(x) = -\frac{1}{m^2}\partial_\mu f^{\mu 0+}(x) \tag{3.13}$$

Consider now the momentum canonically conjugate to $\phi^\mu(x)$. We have

$$\pi^\mu(x) = \frac{\partial \mathcal{L}}{\partial(\partial_0 \phi_\mu(x))} = -f^{0\mu+}(x) = f^{\mu 0+}(x) \tag{3.14}$$

Equivalently

$$\pi_\mu(x) = -f_{0\mu}{}^+(x) = f_{\mu 0}{}^+(x)$$

In the same way we show that

$$\pi^{\mu+}(x) = f^{\mu 0}(x) \tag{3.15}$$

or equivalently

$$\pi_\mu{}^+(x) = f_{\mu 0}(x)$$

It is clear from this that π^0 and π^{0+} vanish identically; since, however, ϕ^0 and ϕ^{0+} are also to be eliminated, this does not present the problem of a lack of a complementary variable for the quantization.

The Hamiltonian density is

$$\mathcal{H}(x) = \sum_k (\pi^k(x)\, \partial_0\, \phi^k(x) + \pi^{k+}(x)\, \partial_0\, \phi^{k+}(x)) - \mathcal{L} \tag{3.16}$$

In the above expression we only include the "dynamic variables" $\phi^k(x)$ and $\pi^k(x)$. It is understood that $\mathcal{L}$ is also expressed only in terms of these variables. When this is done, we get

$$\begin{aligned}\mathcal{H}(x) = \pi^k(x)\pi^{k+}(x) &+ (\nabla \times \boldsymbol{\phi}(x)) \cdot (\nabla \times \boldsymbol{\phi}^+(x)) \\ &+ m^2 \phi^k(x)\phi^{k+}(x) + \frac{1}{m^2}(\nabla \cdot \boldsymbol{\pi}(x))(\nabla \cdot \boldsymbol{\pi}^+(x)) \\ &+ \frac{1}{m^2}\nabla \cdot [\boldsymbol{\pi}(x)(\nabla \cdot \boldsymbol{\pi}^+(x)) + \boldsymbol{\pi}^+(x)(\nabla \cdot \boldsymbol{\pi}(x))]\end{aligned} \tag{3.17}$$

We used

$$\begin{aligned}\phi^0(x) &= -\frac{1}{m^2}\partial_\mu\, \pi^{\mu+}(x) \\ \phi^{0+}(x) &= -\frac{1}{m^2}\partial_\mu\, \pi^\mu(x)\end{aligned} \tag{3.18}$$

in getting the simple form (3.17).

When this is inserted in

$$H = \int d^3x\, \mathcal{H}(x) \tag{3.19}$$

the last term in $\mathcal{H}(x)$ disappears as it is a divergence of a vector which presumably vanishes at infinity.

The canonical commutation relations are straightforward extensions of those written down for the scalar field. They are

$$[\pi^i(x), \phi^j(y)]_{x_0=y_0} = -i\,\delta_{ij}\,\delta(\mathbf{x}-\mathbf{y})$$
$$[\pi^{i+}(x), \phi^{j+}(y)]_{x_0=y_0} = -i\,\delta_{ij}\,\delta(\mathbf{x}-\mathbf{y})$$

with all other equal time commutations involving $\phi^i(x)$, $\phi^{i+}(x)$, $\pi^i(x)$, and $\pi^{i+}(x)$ vanishing. One can proceed without ever bringing in the components $\phi^0(x)$ and $\phi^{0+}(x)$, for they play no role in the dynamics. When treating the vector field in a "manifestly covariant form," it is convenient to be able to write down expressions involving all four components, which are treated formally on the same footing. It is useful, therefore, to write down the commutation relations involving the zeroth components. We have

$$\begin{aligned}[\phi^{0+}(x), \phi^i(y)]_{x_0=y_0} &= \frac{1}{m^2}[\partial_k\pi^k(x), \phi^i(y)]_{x_0=y_0} \\ &= -\frac{i}{m^2}\partial_i\,\delta(\mathbf{x}-\mathbf{y}) \qquad (3.21)\end{aligned}$$

Also

$$[\phi^0(x), \phi^i(y)]_{x_0=y_0} = 0 \qquad (3.22)$$

With the help of these equations we may compute commutation relations involving time derivatives of the fields. Thus

$$\begin{aligned}[\partial^0\phi^{i+}(x), \phi^k(y)]_{x_0=y_0} &= [\pi^i(x) + \partial^i\,\phi^{0+}(x), \phi^k(y)]_{x_0=y_0} \\ &= -i\,\delta_{ik}\,\delta(\mathbf{x}-\mathbf{y}) - i\,\frac{\partial^i\,\partial_k}{m^2}\,\delta(\mathbf{x}-\mathbf{y}) \\ &= i\left(g^{ik} + \frac{\partial^i\,\partial^k}{m^2}\right)\delta(\mathbf{x}-\mathbf{y}) \\ [\partial^0\phi^{0+}(x), \phi^i(y)]_{x_0=y_0} &= -[\partial_k\phi^{k+}(x), \phi^i(y)]_{x_0=y_0} = 0 \\ [\partial_0\phi^{0+}(x), \phi^0(y)]_{x_0=y_0} &= -\left[-\partial_k\phi^{k+}(x), \frac{1}{m^2}\partial_i\pi^{i+}(y)\right]_{x_0=y_0} \\ &= \frac{-1}{m^2}\frac{\partial}{\partial x^k}\frac{\partial}{\partial y^i}[\pi^{i+}(y), \phi^{k+}(x)]_{x_0=y_0} \\ &= -\frac{i}{m^2}\nabla_x^{\,2}\,\delta\,(\mathbf{x}-\mathbf{y})\end{aligned}$$

We shall leave it to the reader to check that all of these commutation relations are consistent with the formula

$$[\phi^{\mu+}(x), \phi^\nu(y)] = -i\left(g^{\mu\nu} + \frac{1}{m^2}\partial^\mu\,\partial^\nu\right)\Delta(x-y;m) \qquad (3.23)$$

Let us now expand the field operators in terms of solutions of the Klein-Gordon equation. If we write

$$\begin{aligned}\phi^k(x) &= \frac{1}{(2\pi)^{3/2}} \sum_{\lambda=1}^{3} \int \frac{d^3q}{2\omega_q}\, e^k(q,\lambda)[A(q,\lambda)e^{-iqx} + B^+(q,\lambda)e^{iqx}] \\ \phi^{k+}(x) &= \frac{1}{(2\pi)^{3/2}} \sum_{\lambda=1}^{3} \int \frac{d^3q}{2\omega_q}\, e^k(q,\lambda)[B(q,\lambda)e^{-iqx} + A^+(q,\lambda)e^{iqx}]\end{aligned} \tag{3.24}$$

we may, by a proper choice of the c-number functions $e^k(q, \lambda)$ arrange matters so that the operators $A(q, \lambda) \cdots B^+(q, \lambda)$ satisfy the simple commutation relations

$$\begin{aligned}[A(q,\lambda), A^+(q',\lambda')] &= [B(q,\lambda), B^+(q',\lambda')] = 2\omega_q\, \delta_{\lambda\lambda'}\, \delta(\mathbf{q} - \mathbf{q}') \\ [A(q,\lambda), A(q',\lambda')] &= [A(q,\lambda), B(q',\lambda)'] \\ &= [A(q,\lambda), B^+(q',\lambda')] = 0\end{aligned} \tag{3.25}$$

The functions $e^k(q, \lambda)$ must be chosen such that

$$\sum_{\lambda=1}^{3} e^k(q,\lambda)e^l(q,\lambda) = \delta_{kl} + \frac{q^k q^l}{m^2} \tag{3.26}$$

so that (3.23) may be satisfied for $\mu, \nu = 1, 2, 3$. A solution is obtained by writing

$$e^k(q,\lambda) = \delta_{k\lambda} + \alpha q^k q^\lambda$$

Then the condition (3.26) requires that

$$2\alpha + \alpha^2\mathbf{q}^2 = \frac{1}{m^2}$$

so that

$$e^k(q,\lambda) = \delta_{k\lambda} + \frac{q^k q^\lambda}{m(m + \omega_q)} \tag{3.27}$$

If the momentum is taken along the z-axis, so that $\mathbf{q} = (0, 0, q)$, then

$$\begin{aligned}e^k(q,1) &= \delta_{k1} \\ e^k(q,2) &= \delta_{k2} \\ e^k(q,3) &= 0 \qquad k = 1, 2 \\ e^3(q,3) &= 1 + \frac{q^2}{m(\omega_q + m)} = \frac{\omega_q}{m}\end{aligned} \tag{3.28}$$

If we tentatively identify $A(q, \lambda) \cdots$ as annihilation operators and creation operators for spin-1 quanta with momentum q and with polarization in the direction λ, we see that the longitudinal polarization vector $e^k(q, 3)$ is normalized differently from the transverse ones. If we now define

$$\phi^\mu(x) = \frac{1}{(2\pi)^{3/2}} \sum_{\lambda=1}^{3} \int \frac{d^3q}{2\omega_q} e^\mu(q, \lambda)[A(q, \lambda)e^{-iqx} + B^+(q, \lambda)e^{iqx}] \quad (3.29)$$

then to satisfy the condition (3.4) we must define

$$e^0(q, \lambda) = \frac{q^\lambda}{m} \quad (3.30)$$

It is easy to check that the following is true

$$\sum_{\lambda=1}^{3} e^\mu(q, \lambda)e^\nu(q, \lambda) = -\left(g^{\mu\nu} - \frac{q^\mu q^\nu}{m^2}\right) \quad (3.31)$$

There is no need to go through the construction of the vector space or through the diagonalization of the Hamiltonian. The results are completely analogous to those obtained for the scalar field, the only difference being that the quanta are characterized by an additional index, the polarization.

In conclusion, we shall show that the vector mesons indeed have spin 1. Our procedure is analogous to that used to show that the quanta of the Dirac field have spin $\frac{1}{2}$. We consider an infinitesimal Lorentz transformation $\Lambda^\mu{}_\nu \simeq \delta^\mu{}_\nu + \alpha^\mu{}_\nu$ and study the consequences of the condition (3.1). We have

$$\phi^\rho(\Lambda x) = U(\Lambda)\phi^\mu(x)\, U^{-1}(\Lambda)\Lambda_\mu{}^\rho$$

Thus with

$$U(\Lambda) \cong 1 + \frac{i}{2}\alpha^{\mu\nu}M_{\mu\nu}$$

we obtain, after some simple algebraic manipulations, the result

$$[M_{\mu\nu}, \phi_\sigma(x)] = i(x_\mu \partial_\nu - x_\nu \partial_\mu)\, \phi_\sigma(x) + i(g_{\mu\sigma}\, \phi_\nu(x) - g_{\nu\sigma}\, \phi_\mu(x)) \quad (3.32)$$

If we restrict ourselves to rotations, i.e., choose $\mu, \nu = 1, 2, 3$, we get

$$[M_{ij}, \phi_\sigma(x)] = -i(x^i \partial_j - x^j \partial_i)\, \phi_\sigma(x) + i(g_{i\sigma}\, \phi_j(x) - g_{j\sigma}\, \phi_i(x)) \quad (3.33)$$

If we write this in the form

$$[M_{ij}, \phi_\sigma(x)] = e_{ijk}[L_k\, \phi_\sigma(x) + (S_k)_{\sigma\rho}\, \phi_\rho(x)] \quad (3.34)$$

then we find that

$$\begin{aligned} L_1 &= -i(x^2\, \partial_3 - x^3\, \partial_2) \\ L_2 &= -i(x^3\, \partial_1 - x^1\, \partial_3) \\ L_3 &= -i(x^1\, \partial_2 - x^2\, \partial_1) \end{aligned} \quad (3.35)$$

and

$$(S_1)_{\rho\sigma} = -i(\delta_{\rho 2}\,\delta_{\sigma 3} - \delta_{\rho 3}\,\delta_{\sigma 2})$$
$$(S_2)_{\rho\sigma} = -i(\delta_{\rho 3}\,\delta_{\sigma 1} - \delta_{\rho 1}\,\delta_{\sigma 3})$$
$$(S_3)_{\rho\sigma} = -i(\delta_{\rho 1}\,\delta_{\sigma 2} - \delta_{\rho 2}\,\delta_{\sigma 1}) \tag{3.36}$$

This may be summarized by

$$(S_k)_{\rho\sigma} = -ie_{k\rho\sigma} \qquad (k,\, \rho,\, \sigma = 1, 2, 3) \tag{3.37}$$

It is easily checked that S_k satisfy the commutation relations characteristic of an angular momentum

$$[S_k, S_l] = ie_{klm}S_m \tag{3.38}$$

and further that

$$(S^2)_{ab} = (S_k)_{ac}(S_k)_{cb} = (-ie_{kac})(ie_{kbc}) = 2\delta_{ab} \tag{3.39}$$

which shows that the intrinsic spin of the vector meson is 1.

It is clear from the way in which the mass of the vector meson appears in the various formulas [e.g. (3.12)] that it is not possible in this case—in contrast to the spin-$\frac{1}{2}$ case—to discuss the massless vector meson field, the electromagnetic field, as a straightforward limiting case of a neutral vector meson theory. The fact is (and we shall prove this generally in Chapter 4) that massless fields differ essentially from massive ones; in our case the difference is that photons have one state of polarization less, two instead of the three we would expect on the basis of the value of the spin. This fact leads to some complications of a formal nature when we try to describe the photon by a four-vector field $A^\mu(x)$. One can, of course, treat the photon field by quantizing the field quantities $\mathbf{E}(x)$ and $\mathbf{B}(x)$ directly, but, as will be seen in Chapter 5, the equations of motion for the electromagnetic field coupled to charges are much simpler when written in terms of the vector potentials. We shall therefore face the formal difficulties right here and stay with the vector potentials.

In the case of the vector field, condition (3.4) eliminated in a natural way one of the four components. In the massless case only two components are dynamically independent and one more has to be eliminated. Thus there must exist an additional constraint over and above the divergence condition. The manner in which this constraint works is already apparent in the classical theory, so that it is useful, for clarification of this point, to review the classical equations of motion. There, the Maxwell equations, may be derived from the Lagrangian density

$$\mathcal{L} = \tfrac{1}{4}F_{\mu\nu}F^{\mu\nu} - \tfrac{1}{2}F^{\mu\nu}(\partial_\mu A_\nu - \partial_\nu A_\mu) \tag{3.40}$$

and they are

$$F_{\mu\nu}(x) = \partial_\mu\, A_\nu(x) - \partial_\nu\, A_\mu(x)$$
$$\partial^\mu\, F_{\mu\nu}(x) = 0 \tag{3.41}$$

Only the quantities $F_{\mu\nu}(x)$ are physically significant, so that there is no change in the content of the theory if the "gauge transformation of the second kind,"

$$A_\mu(x) \to A_\mu'(x) = A_\mu(x) - \partial_\mu \chi(x) \tag{3.42}$$

is applied. In the classical theory it is possible to constrain the vector potentials by the Lorentz condition

$$\partial^\mu A_\mu(x) = 0 \tag{3.43}$$

Because $A_\mu(x)$ and $A_\mu(x) - \partial_\mu \chi(x)$ are physically indistinguishable, the vector potentials are still undetermined, provided that the gauge function $\chi(x)$ is restricted to be a solution of the wave equation

$$\Box\, \chi(x) = 0 \tag{3.44}$$

Let us now examine how the combination of the divergence condition and the remaining freedom in the choice of gauge act to eliminate two of the four components of the vector potential. The expression

$$A_\mu(x) = \frac{1}{(2\pi)^{3/2}} \sum_{\lambda=0}^{3} \int \frac{d^3k}{2k}\, e_\mu(k, \lambda)[a(k, \lambda)e^{-ikx} + \text{h.c.}] \tag{3.45}$$

automatically satisfies the equation of motion ($\omega_k = |\mathbf{k}| \equiv k$), which is just the zero mass Klein-Gordon equation [combine the two equations in (3.41) with (3.43)]. The $e_\mu(k, \lambda)$ are polarization vectors and the $a(k, \lambda)$ are amplitudes which will ultimately be quantized. The Lorentz condition implies that

$$k^\mu e_\mu(k, \lambda) = 0 \tag{3.46}$$

In a theory with mass it was possible to choose the $e_\mu(k, \lambda)$ such that

$$k^\mu e_\mu(k, \lambda) = 0 \qquad \lambda = 1, 2, 3 \tag{3.47}$$

so that $a(k, \lambda)$, ($\lambda = 1, 2, 3$) were independent. Here, because $m = 0$ this is no longer possible. The most general form for $e_\mu(k, \lambda)$ is

$$e_\mu(k, \lambda) = a\, \delta_\mu^\lambda + bk_\mu k^\lambda$$

and the condition (3.47) would imply that

$$k^\mu e_\mu(k, \lambda) = ak^\lambda = 0$$

since $k^2 = 0$. This implies that $a = 0$, leading to a nonsensical result that the vector potential (3.45) is always a gradient of a scalar function. We may, however, choose

$$k^\mu e_\mu(k, \lambda) = 0 \qquad \lambda = 1, 2 \tag{3.48}$$

so that $a(k, 1)$ and $a(k, 2)$ are unrestricted. We do this with the following choice of polarization vectors:

For $\lambda = 1, 2$, we choose the polarization vectors transverse, with no zero components

$$k^i e_i(k, \lambda) = 0 \qquad e_0(k, \lambda) = 0 \qquad \lambda = 1, 2 \tag{3.49}$$

We also choose

$$\begin{aligned} e^i(k, 3) &= k^i/k & e_0(k, 3) &= 0 \\ e^i(k, 0) &= 0 & e_0(k, 0) &= 1 \end{aligned} \tag{3.50}$$

Then

$$k^\mu e_\mu(k, 3) = -k$$

and

$$k^\mu e_\mu(k, 0) = k$$

so that the Lorentz condition (3.46) implies that

$$a(k, 3) - a(k, 0) = 0 \tag{3.51}$$

Thus, given the longitudinal component $a(k, 3)$, the "scalar" component $a(k, 0)$ is determined. Next we consider a gauge function which is also a solution of the wave equation (3.44). It may be written in the form

$$\chi(x) = \frac{1}{(2\pi)^{3/2}} \int \frac{d^3k}{2k} [\tilde{\chi}(k)e^{-ikx} + \tilde{\chi}^*(k)e^{ikx}] \tag{3.52}$$

Then the transformed vector potential is

$$\begin{aligned} A_\mu'(x) &= A_\mu(x) - \partial_\mu \chi(x) \\ &= \frac{1}{(2\pi)^{3/2}} \int \frac{d^3k}{2k} \left[\left(\sum_{\lambda=0}^{3} e_\mu(k, \lambda)\, a(k, \lambda) + ik_\mu\, \tilde{\chi}(k)\right) e^{-ikx} + \text{c.c.}\right] \end{aligned} \tag{3.53}$$

The relation between the new amplitude and the old is

$$\sum_{\lambda=0}^{3} e_\mu(k, \lambda)(a'(k, \lambda) - a(k, \lambda)) = ik_\mu\, \tilde{\chi}(k) \tag{3.54}$$

This equation is satisfied with the choice

$$a'(k, \lambda) = a(k, \lambda) \qquad \lambda = 1, 2 \tag{3.55}$$

$$a'(k, 3) = a(k, 3) + ik\, \tilde{\chi}(k) \tag{3.56}$$

and

$$a'(k, 0) = a(k, 0) + ik\, \tilde{\chi}(k) \tag{3.57}$$

It is clear, therefore, that the freedom of choice of $\tilde{\chi}(k)$ is just such that it allows us to eliminate both $a'(k, 3)$ and $a'(k, 0)$ (which are equal). In summary, we see that gauge invariance, in addition to the Lorentz condition, is necessary to make the theory consistent, in the sense of allowing only two independent amplitudes. To translate these constraints into the quantum theory is not a trivial task.

The first obstacle in this translation is that with the Lagrangian density of (3.40), the canonical momentum is

$$\pi^\mu(x) = \frac{\partial \mathcal{L}}{\partial(\partial_0 A_\mu(x))} = -F^{0\mu}(x) \tag{3.58}$$

so that $\pi^0(x)$ vanishes identically. Thus canonical quantization cannot be applied to this component. This is in contrast to the vector field with mass, in which the Lorentz condition was an operator equation contained in the Lagrangian and in which $\phi^0(x)$ could be eliminated directly. Fermi suggested that the Lagrangian

$$\mathcal{L} = -\tfrac{1}{2}(\partial_\mu A_\nu(x))(\partial^\mu A^\nu(x)) \tag{3.59}$$

be used. This differs from that of (3.40) by terms involving $(\partial_\mu A^\mu(x))^2$. The equation of motion that follows from (3.59) is

$$\Box\, A_\mu(x) = 0 \tag{3.60}$$

and this is correct only if supplemented in some way by the condition

$$\partial^\mu A_\mu(x) = 0 \tag{3.61}$$

Now the canonical momentum is

$$\pi^\mu(x) = -\partial^0 A^\mu(x) \tag{3.62}$$

The commutation relations can now be written as

$$[\pi^\mu(x), A^\nu(y)]_{x_0=y_0} = ig^{\mu\nu}\, \delta(\mathbf{x} - \mathbf{y}) \tag{3.63}$$

with all other pairs of operators commuting. From this we can find, in the standard way, the unequal time commutation relations. These are

$$[A^\mu(x), A^\nu(y)] = ig^{\mu\nu}\, D(x - y) \tag{3.64}$$

where

$$D(x - y) = \Delta(x - y, 0) = -\frac{i}{(2\pi)^3}\int dk\, \delta(k^2)\, \epsilon(k_0)\, e^{-ik(x-y)} \tag{3.65}$$

It is clear now that we cannot impose the subsidiary condition in the form (3.61): it follows from the commutation relations that

$$[\partial_\mu A^\mu(x), A^\nu(y)] = i\partial^\nu D(x - y) \tag{3.66}$$

and the right side does not vanish. We may think of imposing the constraint in another way by restricting the class of state vectors to those for which

$$\partial^\mu A_\mu(x)\Psi = 0 \tag{3.67}$$

is satisfied. This, too, is inconsistent with the commutation relations, for

$$(\Psi, [\partial^\mu A_\mu(x), A^\nu(y)]\Psi) = i\,\partial^\nu\, D(x-y)(\Psi, \Psi) \neq 0$$

The condition

$$(\partial_\mu\, A^\mu(x))^{(+)}\Psi = 0 \tag{3.68}$$

does not suffer from this defect. Here$^{(+)}$ denotes the positive frequency part, i.e., the annihilation part, of the operator. The implications of this condition are the following: by working with the four-component vector potential and quantizing according to (3.64) we are in effect allowing for the existence of four different kinds of particles, and we must somehow neutralize, as far as any physical consequences are concerned, the effect of two of them. The Lorentz condition from the classical theory suggests (3.68) which says that the "unphysical" longitudinal and scalar particles must enter into any acceptable state in a special combination. Two remarks should be made in connection with this subsidiary condition. The first is that the expectation value of $\partial_\mu\, A^\mu(x)$ vanishes in all acceptable states. Equation 3.68 implies that

$$(\Psi, (\partial_\mu\, A^\mu(x))^{(-)}\Psi) = ((\partial_\mu\, A^\mu(x))^{(+)}\Psi, \Psi) = 0$$

so that

$$\begin{aligned}(\Psi, \partial_\mu\, A^\mu(x)\Psi) &= (\Psi, (\partial_\mu\, A^\mu(x))^{(-)} + (\partial_\mu\, A^\mu(x))^{(+)}\Psi)\\ &= 0\end{aligned} \tag{3.69}$$

The second remark anticipates a question. We shall see, when fields in interaction are discussed, that the separation into positive and negative frequency parts, i.e., into "creation" and "annihilation" parts, is ambiguous and does not make sense. What happens then to (3.68)? It will be seen that the photon field in interaction satisfies

$$\Box\, A^\mu(x) = e\, j^\mu(x) \tag{3.70}$$

with a conserved current

$$\partial_\mu\, j^\mu(x) = 0 \tag{3.71}$$

Hence

$$\Box\, (\partial_\mu\, A^\mu(x)) = 0 \tag{3.72}$$

so that $\partial_\mu\, A^\mu(x)$ still satisfies a free-field equation, and the separation for $\partial_\mu\, A^\mu(x)$ does make sense.

The Hamiltonian density is

$$\begin{aligned}\mathcal{H}(x) &= \pi^\mu(x)\,\partial_0\, A_\mu(x) - \mathcal{L}\\ &= \tfrac{1}{2}(\partial_i\, A_\mu)(\partial^i A^\mu) - \tfrac{1}{2}(\partial_0 A_\mu)(\partial^0 A^\mu)\end{aligned} \tag{3.73}$$

If we use (3.45) and note that $e_\mu(k, \lambda)$ defined by (3.48) to (3.50) satisfy

$$\sum_{\mu=0}^{3} e^\mu(k, \lambda)e_\mu(k, \lambda') = g_{\lambda\lambda'} \tag{3.74}$$

we get after some calculations

$$H = -\sum_{\lambda\lambda'} g_{\lambda\lambda'} \int \frac{d^3k}{2k}\, k\, a^+(k, \lambda)\, a(k, \lambda') \tag{3.75}$$

Here we have used the normal ordering for the operators. We also get, from (3.64), the commutation rules

$$[a(k, \lambda), a^+(k', \lambda')] = -2kg_{\lambda\lambda'}\, \delta(\mathbf{k} - \mathbf{k}') \tag{3.76}$$

It appears that the energy is not positive definite. If, however, we consider the expectation value of the energy

$$(\Psi, H\Psi) = \int \frac{d^3k}{2k}\, k(\Psi, (a^+(k, 1)a(k, 1) + a^+(k, 2)a(k, 2)$$
$$+ a^+(k3)a(k, 3) - a^+(k, 0)a(k, 0))\Psi) \tag{3.77}$$

Since the states Ψ are restricted to those for which

$$a(k, 3)\Psi = a(k, 0)\Psi \tag{3.78}$$

we obtain

$$H = \sum_{\lambda=1}^{2} \int \frac{d^3k}{2k}\, k\, a^+(k, \lambda)\, a(k, \lambda) \tag{3.79}$$

Thus the energy depends on the contribution from the transverse photons alone, and this is true for all observables.

It should be pointed out for the sake of completeness that the theory as it stands has a flaw. Consider a state containing one scalar photon, of the form

$$\Psi_1 = \int \frac{d^3k}{2k}\, C_0(k)a^+(k, 0)\Psi_0 \tag{3.80}$$

The norm of the state is obtained from

$$(\Psi_1, \Psi_1) = \int \frac{d^3k}{2k} \int \frac{d^3k'}{2k'}\, C_0{}^*(k')\, C_0(k)(\Psi_0, a(k', 0)\, a^+(k, 0)\Psi_0)$$
$$= \int \frac{d^3k}{2k} \int \frac{d^3k'}{2k'}\, C_0{}^*(k')C_0(k)(\Psi_0, [a_0(k', 0), a^+(k, 0)]\Psi_0$$
$$= -\int \frac{d^3k}{2k}\, |C_0(k)|^2 \tag{3.81}$$

and the result is negative! Were it not for the fact that the timelike and longitudinal photons play no role in determining observable quantities, this would be a very serious difficulty because the norm in the vector space is connected with the probability interpretation of quantum theory and must be positive or zero. In this case the flaw is a purely formal one. To overcome it, Gupta and Bleuler[1] modified the vector space by introducing a modification in the definition of the scalar product, which instead of

$$(\Phi, \Psi)$$

is defined as

$$(\eta\Phi, \Psi) \tag{3.82}$$

where η is called a *metric operator*. Since the norm is to be real, we require that

$$\begin{aligned}(\eta\Phi, \Phi) &= (\eta\Phi, \Phi)^* = (\Phi, \eta^+\Phi)\\ &= (\eta^+\Phi, \Phi)\end{aligned} \tag{3.83}$$

Hence

$$\eta = \eta^+ \tag{3.84}$$

It is also convenient to take

$$\eta\eta^+ = \eta^2 = 1 \tag{3.85}$$

The expectation value of an operator is now

$$\bar{F} = (\eta\Phi, F\Phi) = (\Phi, \eta F\Phi) \tag{3.86}$$

and it is clear that $\bar{F}$ need no longer be real for a hermitian F, for

$$\bar{F}^* = (\Phi, F^+\eta\Phi) = (\Phi, F\eta\Phi) \neq (\Phi, \eta F\Phi) \tag{3.87}$$

Because we want states of the type (3.80) for $\lambda = 1, 2, 3$ to be unaffected, we must have

$$[\eta, a(k, \lambda)] = 0; \qquad \lambda = 1, 2, 3 \tag{3.88}$$

The difficulty with timelike photons will be resolved if

$$\{\eta, a(k, 0)\} = 0 \tag{3.89}$$

which can be satisfied with the choice

$$\eta = (-1)^{n_0} \tag{3.90}$$

where n_0 is the number of timelike scalar photons in the state.

For a completely consistent use of the indefinite metric we should really go back to the start of the discussion of the electromagnetic field

[1] S. Gupta, *Proc. Phys. Soc.* (*London*) **A63,** 681 (1951); K. Bleuler, *Helv. Phys. Acta* **23,** 567 (1950).

quantization and change a number of formulas. For example, because $A_0(x)$ should have a real expectation value, we must have

$$(\Psi, \eta\, A_0 \Psi)^* = (\Psi, A_0^+ \eta \Psi) = -(\Psi, \eta\, A_0^+ \Psi)$$

i.e.,

$$A_0^+(x) = -A_0(x)$$

which affects the definition of the zeroth component of the polarization vector. Because our concern here is with practical calculations, we need not worry about these matters; we have already seen that the photon field may be treated in the same manner as a vector meson field. In the study of the consistency of field theory this matter becomes important. This field of study is one in which great delicacy is required, and many of those who concern themselves with such matters prefer not to enlarge the physical Hilbert space but would rather work in the so-called Coulomb gauge. These matters, however, are beyond the scope of this introductory treatment.

PROBLEMS

1. Discuss the properties of vector mesons under *P* and *T*.
2. The equation of motion of a vector meson in an external electromagnetic field is obtained by replacing $\partial_\mu \phi^\nu$ by $(\partial_\mu - ieA_\mu)\phi^\nu$ and $\partial_\mu \phi^{\nu+}$ by $(\partial_\mu + ieA_\mu)\phi^{\nu+}$ in the Lagrangian. Discuss the motion of a vector meson in a weak constant magnetic field.

4

Lorentz Invariance and Spin

In the last three chapters we discussed in some detail the quantum theory of free fields, whose quanta had spin 0, $\frac{1}{2}$, and 1. Our emphasis on fields was motivated by the belief that quantum field theory is very likely the correct way of fusing quantum theory with special relativity and that interactions among particles of low spin may play a special role in the structure of elementary particles. In this discussion we found that state vectors describing single particles could be constructed by operating with the creation parts of the appropriate field operators on the vacuum state. It is important to know whether it is necessary to have a corresponding field to describe a state of given mass and spin. In view of the recent proliferation of particles, including some of higher spin, it would be very disturbing if we could not avoid introducing a new field for each particle. The answer to this question is the following: (a) recent research in field theory has shown that the intuitive idea that all particles may be "composites" of a few "fundamental" entities can be satisfactorily formulated and it is not necessary to have an interacting field for each particle[1]; (b) the state vectors describing one-particle states can be discussed independently, without the introduction of the appropriate free fields; rather, the free fields can be constructed once the transformation properties of the one-particle state vectors are understood.[2]

[1] R. Haag, *Phys. Rev.* **112,** 669 (1958); K. Nishijima, *Phys. Rev.* **111,** 995 (1958); W. Zimmermann, *Nuovo. Cimento,* **10,** 59 (1958).

[2] Almost all of the material contained in this chapter appears explicitly or implicitly in the pioneering paper of E. P. Wigner, *Ann. Math.* **40,** 149 (1939). See also V. Bargmann, *Ann. Math.* **48,** 568 (1947), and R. F. Streater and A. S. Wightman, *PCT, Spin and Statistics and All That,* W. A. Benjamin, New York (1964). In all of these papers a number of subtle points, which we gloss over, are discussed. Among them are the distinction between the group of Lorentz transformations and the "covering group" of 2 × 2 complex matrices, which is really the one being studied. A detailed coverage of these matters would take up a prohibitive amount of space and would not add anything to our understanding of the basic physical consequences of Lorentz invariance discussed in this chapter.

In Chapters 1 through 3 we found that the characteristics of the particles described by a field were determined by the transformation properties of the field (and therefore of the state vectors) under homogeneous Lorentz transformations and displacements in space and time, i.e., inhomogeneous Lorentz transformations. Our more general discussion therefore leans heavily on the properties of the operators P_μ and $M_{\mu\nu}$, which generate these transformations in the vector space spanned by all physical states.

The inhomogeneous Lorentz transformations are of the form

$$x'^\mu = \Lambda^\mu{}_\nu x^\nu + a^\mu \tag{4.1}$$

corresponding to a homogeneous transformation, followed by a displacement. The matrix $\Lambda^\mu{}_\nu$ must satisfy the condition

$$g_{\mu\nu}\Lambda^\mu{}_\alpha\Lambda^\nu{}_\beta = g_{\alpha\beta} \tag{4.2}$$

Two successive transformations lead to

$$\begin{aligned} x^\mu \rightarrow x''^\mu &= \tilde{\Lambda}^\mu{}_\nu x'^\nu + \tilde{a}^\mu \\ &= \tilde{\Lambda}^\mu{}_\nu(\Lambda^\nu{}_\alpha x^\alpha + a^\nu) + \tilde{a}^\mu \\ &= (\tilde{\Lambda}^\mu{}_\nu\Lambda^\nu{}_\alpha)x^\alpha + (\tilde{\Lambda}^\mu{}_\nu a^\nu + \tilde{a}^\mu) \end{aligned} \tag{4.3}$$

This transformation is easily seen to be a Lorentz transformation, and one may use (4.3) to construct the inverse to the transformation (4.1). Lorentz transformations thus form a group, called the Poincaré group or group of inhomogeneous Lorentz transformations.

It follows from (4.2) that the matrix $\Lambda^\mu{}_\nu$ has determinant ± 1. Since the identity transformation has determinant $+1$, and we shall want to use infinitesimal transformations, we shall restrict ourselves to the transformations for which $\det \Lambda = +1$; this is no serious restriction, because the others can be reached by a space inversion. It also follows from (4.2) that

$$(\Lambda^0{}_0)^2 = 1 + \sum_{i=1}^{3} (\Lambda^0{}_i)^2$$

so that $\Lambda^0{}_0 \geqslant 1$ or $\Lambda^0{}_0 \leqslant -1$. Only the first set of transformations can be used as infinitesimal transformations; the second set can be reached by a time inversion. We shall therefore be discussing the set of proper ($\det \Lambda = +1$), orthochronous ($\Lambda^0{}_0 \geqslant 1$) Lorentz transformations.

It was already pointed out in Chapter 1 that Lorentz invariance carried with it the implication that in the space of state vectors there exists a unitary transformation associated with each transformation of the type

(4.1); this transformation, denoted by $U(a, \Lambda)$, when acting on a given state, yields the state vector for the transformed state. It follows from definition (4.1) that

$$U(a, \Lambda) = U(a, 1)\, U(0, \Lambda) \equiv U(a)\, U(\Lambda) \tag{4.4}$$

Equation 4.3 implies the following multiplication law

$$U(a', \Lambda')\, U(a, \Lambda) = U(a' + \Lambda' a, \Lambda'\Lambda) \tag{4.5}$$

If we write (as in Chapter 1)

$$U(a) = e^{iP^\mu a_\mu} \tag{4.6}$$

and

$$U(\Lambda) = e^{\frac{i}{2}\alpha_{\mu\nu}M^{\mu\nu}} \tag{4.7}$$

with $\alpha_{\mu\nu} = -\alpha_{\nu\mu}$, we may use the multiplication law to obtain the commutation relations among the ten generators P^μ and $M^{\mu\nu}$. For example, when $\Lambda = \Lambda' = 1$, (4.5) reads

$$U(a')\, U(a) = U(a + a') \tag{4.8}$$

from which it immediately follows that

$$[P^\mu, P^\nu] = 0 \tag{4.9}$$

If we substitute (4.4) in (4.5), we get

$$U(a')\, U(\Lambda')\, U(a)\, U(\Lambda) = U(a')\, U(\Lambda' a)\, U(\Lambda')\, U(\Lambda)$$

from which we deduce that

$$U(\Lambda')\, U(a)\, U^{-1}(\Lambda') = U(\Lambda' a) \tag{4.10}$$

This implies that

$$U(\Lambda) e^{iP^\mu a_\mu}\, U^{-1}(\Lambda) = e^{iP^\mu \Lambda_\mu{}^\nu a_\nu}$$

i.e.

$$U(\Lambda) P^\mu\, U^{-1}(\Lambda) = (\Lambda^{-1})^\mu{}_\nu P^\nu \tag{4.11}$$

The implication of this is that the operator P^μ transforms as a four-vector. If we use (4.7) with infinitesimal $\alpha_{\mu\nu}$, we obtain the commutation relations between $M^{\mu\nu}$ and P^σ. A little computation leads to the result

$$[M^{\mu\nu}, P^\sigma] = -i(P^\mu g^{\nu\sigma} - P^\nu g^{\mu\sigma}) \tag{4.12}$$

If we introduce the quantities J_i and K_i defined by

$$J_i = -\tfrac{1}{2}e_{imn}M^{mn} \qquad K_i = M_{i0} \tag{4.13}$$

these commutation relations read as follows:

$$\begin{aligned}[][J_i, P_k] &= ie_{ikl}P^l \\ [J_i, P_0] &= 0 \\ [K_i, P_k] &= i P_0 g_{ik} \\ [K_i, P_0] &= -i P_i\end{aligned} \tag{4.14}$$

To find the commutation relations among the $M^{\mu\nu}$ we make use of the fact that

$$U(\Lambda)\, U^{-1}(\Lambda) = 1 = U(\Lambda\Lambda^{-1}) = U(\Lambda)\, U(\Lambda^{-1}) \tag{4.15}$$

so that

$$U(\Lambda^{-1}) = U^{-1}(\Lambda) \tag{4.16}$$

If we set $a = a' = 0$ in (4.5), we get with the help of (4.16)

$$U(\Lambda)\, U(\Lambda')\, U^{-1}(\Lambda) = U(\Lambda\Lambda'\Lambda^{-1}) \tag{4.17}$$

When this is worked out for infinitesimal transformations, the following commutation relations result after a little algebra

$$[M^{\mu\nu}, M^{\rho\sigma}] = i(M^{\mu\rho}\, g^{\nu\sigma} + M^{\nu\sigma}\, g^{\mu\rho} - M^{\nu\rho}\, g^{\mu\sigma} - M^{\mu\sigma}\, g^{\nu\rho}) \tag{4.18}$$

These imply the following relations among the J_i and K_i:

$$\begin{aligned}[][J_m, J_n] &= ie_{mnk}J_k \\ [J_m, K_n] &= ie_{mnk}K_k \\ [K_m, K_n] &= -ie_{mnk}J_k\end{aligned} \tag{4.19}$$

It is important to notice that J_m as well as $\mathfrak{J}_m^{\pm} \equiv \frac{1}{2}(J_m \pm i\, K_m)$ have commutation relations characteristic of angular momentum operators.

Since our interest is in the inhomogeneous Lorentz group, we do not digress into a discussion of the finite-dimensional representations of the homogeneous transformations; this material appears in a brief appendix to this chapter.

One may check that the operator

$$\mathcal{M}^2 = P_\mu P^\mu \tag{4.20}$$

commutes with all the generators. This follows from (4.9) and from the relations in (4.12):

$$\begin{aligned}[][P_\mu P^\mu, M^{\rho\sigma}] &= P_\mu[P^\mu, M^{\rho\sigma}] + [P^\mu, M^{\rho\sigma}]P_\mu \\ &= iP_\mu(P^\rho g^{\mu\sigma} - P^\sigma g^{\mu\rho}) + i(P^\rho g^{\mu\sigma} - P^\sigma g^{\mu\rho})P_\mu = 0\end{aligned}$$

In analogy with the usual procedure in the theory of angular momentum, we may classify irreducible representations of the group by the value of this invariant operator. With the identification of P^μ with the energy-momentum four-vector, it is clear that this invariant is the square of the mass of the state being described.

We know from experience that one-particle states are also characterized by the *spin*, the value of the angular momentum of the particle about an arbitrary axis when the particle is at rest. To find the invariant operator that describes the spin, we should go to the rest frame. Actually, for greater generality, we shall deal with the subspace of states of a given mass m, which is characterized by a fixed value of $\mathring{p}^\mu$, the eigenvalue of P^μ. The state vector thus satisfies

$$P^\mu \Phi_{\mathring{p}} = \mathring{p}^\mu \Phi_{\mathring{p}} \tag{4.21}$$

The transformations belonging to the subset that leaves the eigenvalues of P^μ invariant are said to belong to the "little group," and they are restricted by the condition

$$\mathring{\Lambda}^\mu{}_\nu \mathring{p}^\nu = \mathring{p}^\mu \tag{4.22}$$

For infinitesimal transformations

$$\mathring{\Lambda}^\mu{}_\nu = \delta^\mu{}_\nu + \mathring{\alpha}^\mu{}_\nu$$

this implies that

$$\mathring{\alpha}^\mu{}_\nu \mathring{p}^\nu = 0 \tag{4.23}$$

A general expression for an antisymmetric $\mathring{\alpha}_{\mu\nu}$ satisfying (4.23) is

$$\mathring{\alpha}_{\mu\nu} = \epsilon_{\mu\nu\rho\sigma} \mathring{p}^\rho n^\sigma \tag{4.24}$$

where n^σ is an arbitrary four-vector and $\epsilon_{\mu\nu\rho\sigma}$ is the totally antisymmetric tensor whose value is

+1 for μ, ν, ρ, σ, an even permutation of 0, 1, 2, 3;
−1 for μ, ν, ρ, σ, an odd permutation of 0, 1, 2, 3;
0 otherwise.

The unitary transformation $U(\mathring{\Lambda})$ appropriate to the Lorentz transformations belonging to the little group is, in infinitesimal form,

$$U(\mathring{\Lambda}) \cong 1 + \frac{i}{2} \mathring{\alpha}_{\mu\nu} M^{\mu\nu} = 1 + \frac{i}{2} \epsilon_{\mu\nu\rho\sigma} \mathring{p}^\rho n^\sigma M^{\mu\nu}$$

We may write this as

$$U(\mathring{\Lambda}) \cong 1 - i n^\sigma W_\sigma$$

where the operator W_σ is defined by

$$W_\sigma = -\tfrac{1}{2} \epsilon_{\mu\nu\rho\sigma} M^{\mu\nu} P^\rho \tag{4.25}$$

There should be no confusion arising from the replacement of $\overset{\circ}{p}^\rho$ by the operator P^ρ in (4.25), since the transformation $U(\overset{\circ}{\Lambda})$ of the little group only acts on states satisfying (4.21). The following properties of W_σ are easily established

$$W_\sigma P^\sigma = 0 \tag{4.26}$$

Thus for a system at rest the eigenvalue of $\mathbf{P}$ is zero, and W_0, acting on a state at rest, must yield zero.

$$[W_\sigma, P^\mu] = 0 \tag{4.27}$$

It follows from the definition (4.25) that W_σ is a four-vector; it must therefore have the same commutation relations with $M^{\mu\nu}$ as P_σ does, so that

$$[M_{\mu\nu}, W_\sigma] = -i(W_\mu g_{\nu\sigma} - W_\nu g_{\mu\sigma}) \tag{4.28}$$

These relations may be used to prove that

$$[W_\lambda, W_\sigma] = i\epsilon_{\lambda\sigma\alpha\beta} W^\alpha P^\beta \tag{4.29}$$

Finally, $W_\sigma W^\sigma$ is a scalar; it therefore commutes with all the $M^{\mu\nu}$ and, by virtue of (4.27), with all the P^μ. It is therefore an invariant operator, and, like $\mathscr{M}^2$, it may be used to label irreducible representations. The eigenvalues of $W_\sigma W^\sigma$ may be found by going to the rest frame. If we start with the state having momentum p^μ, represented by the state vector Φ_p, the state vector for the same system at rest may be obtained by operating on Φ_p with the unitary transformation $U(L^{-1}(p))$. Here $L(p)$ is the Lorentz transformation which takes the four-vector $(m, 0, 0, 0)$ into p^μ, and $L^{-1}(p)$ is its inverse. Because $W_\sigma W^\sigma$ is an invariant operator, we choose to calculate its eigenvalues in the rest system, i.e., calculate $W_\sigma W^\sigma \Phi_{\text{rest}}$. Now

$$\begin{aligned} W_\sigma W^\sigma \Phi_{\text{rest}} &= W_\sigma W^\sigma\, U(L^{-1}(p))\Phi_p \\ &= U(L^{-1}(p))\, W_\sigma' W'^\sigma\, \Phi_p \end{aligned} \tag{4.30}$$

where

$$W_\sigma' = U(L(p))\, W_\sigma\, U^{-1}(L(p)) \tag{4.31}$$

Since W_σ is a four-vector, its transformation properties are the same as those of P_σ. Hence we may use (4.11) to obtain

$$W_\sigma' = (L^{-1}(p))_\sigma{}^\rho W_\rho \tag{4.32}$$

The matrix $L(p)$ is given by

$$(L(p))_\sigma{}^\rho = \left(\begin{array}{c|c} \dfrac{E_p}{m} & -\dfrac{p^j}{m} \\ \hline \dfrac{p_i}{m} & \delta_i{}^j - \dfrac{p_i p^j}{m(E_p + m)} \end{array}\right) \tag{4.33}$$

and its inverse is obtained by changing the sign of the spatial components. Thus

$$W_0' = \frac{1}{m}(E_p W_0 - \mathbf{p} \cdot \mathbf{W}) = 0$$

$$\mathbf{W}' = -\frac{1}{m} W_0 \mathbf{p} + \mathbf{W} + \frac{\mathbf{p}}{m} \frac{\mathbf{p} \cdot \mathbf{W}}{E_p + m} \tag{4.34}$$

$$= \mathbf{W} - \frac{\mathbf{p}}{E_p} \frac{\mathbf{W} \cdot \mathbf{p}}{E_p + m} = \mathbf{W} - \frac{W_0 \mathbf{p}}{E_p + m} \tag{4.35}$$

Now the commutation relations of the operators

$$S_i = \frac{1}{m} W_i' \tag{4.36}$$

may be calculated from those of W_σ, given in (4.29). After some computation we find that since we want to calculate $W_\sigma' W^{\sigma\prime} \Phi_p$ we use

$$[W_\mu, W_\nu] = i\epsilon_{\mu\nu\rho\sigma} W^\rho p^\sigma \tag{4.37}$$

and obtain

$$[S_i, S_j] = i e_{ijk} S_k \tag{4.38}$$

These are the commutation relations for angular momentum operators, and it is known that the eigenvalues of S^2 must be of the form $s(s + 1)$, where $s = 0, \frac{1}{2}, 1, \frac{3}{2}, \ldots$. Thus equation (4.30) reads

$$W_\sigma W^\sigma \Phi_{\text{rest}} = m^2 s(s + 1)\, U(L^{-1}(p))\Phi_p = m^2 s(s + 1)\Phi_{\text{rest}} \tag{4.39}$$

The irreducible representations of the Poincaré group are thus characterized by two invariants, the mass and the spin.

We state without proof that $P_\mu P^\mu$ and $W_\sigma W^\sigma$ are the only independent invariants that can be constructed. The irreducible representations labeled by $[m; s]$ alone will be associated with one-particle states; different states within an irreducible representation will be distinguished by the value of the momentum and by the eigenvalue of one component of the spin, usually the z-component. Compound systems have additional labels because of their higher degeneracy, e.g., a two-particle system has a total momentum and mass and a total angular momentum in its rest system. There are, however, many possible directions for the relative momentum vector, or alternately, there are many possible values of the orbital angular momentum, which, together with the spins of the individual particles, make up the total angular momentum in the rest system.

The foregoing discussion is not applicable when the mass of the system is zero, since there is no rest system. If we again consider an arbitrary system Φ_p, we can by a simple rotation obtain the standard state Φ_R, which will be characterized by the four-momentum $(p, 0, 0, p)$.[1]

Equation (4.26) implies that

$$p(W_3 - W_0)\Phi_p = 0 \tag{4.40}$$

and the commutation relations, as a little algebra shows, are

$$\begin{aligned} [W_1, W_2]\Phi_R &= 0 \\ [W_3, W_1]\Phi_R &= ip\, W_2\, \Phi_R \\ [W_3, W_2]\Phi_R &= -ip\, W_1\, \Phi_R \end{aligned} \tag{4.41}$$

We shall assert without proof that these are the commutation relations for the generators of translations (W_1, W_2) and rotations (W_3) in a plane. There does not appear to be any fundamental significance in this, but it shows that $W_1^2 + W_2^2$ can take on any value. This would imply that $W_\sigma W^\sigma$ is not quantized; because there are no physical states corresponding to "continuous spin," we restrict the physical massless states Φ_R to be such that

$$W_1\, \Phi_R = W_2\, \Phi_R = 0 \tag{4.42}$$

We may thus write

$$W_\mu\, \Phi_R \propto P_\mu\, \Phi_R \tag{4.43}$$

and, because P_μ and W_μ transform in the same way under Lorentz transformations, we have the operator equation

$$W_\mu = -\lambda P_\mu \tag{4.44}$$

for all massless states. The quantity λ is an invariant. We may write

$$\lambda = -\frac{W_\mu n^\mu}{P_\mu n^\mu} = \frac{1}{2}\frac{\epsilon_{\mu\nu\rho\sigma}\, M^{\nu\rho}P^\sigma n^\mu}{P_\mu n^\mu}$$

where n^μ is an arbitrary four-vector. If we choose it to be $(1, 0, 0, 0)$, we obtain

$$\lambda = \frac{M_{12}P_3 + \cdots}{P_0} = \frac{\mathbf{J}\cdot\mathbf{P}}{|\mathbf{P}|} \tag{4.45}$$

λ is thus the component of the total angular momentum $\mathbf{J}$ along the direction of motion; it is called the *helicity*.[2]

A state of zero mass is thus described by a single component, in contrast to the $(2s + 1)$ components which are required to completely describe a

[1] Because, for a massless state, $\mathbf{p}^2 = p_0^2$, the form $(p, 0, 0, p)$ can be obtained by a transformation among the components of $\mathbf{p}$ alone.

[2] The effect of a rotation of 2π about the direction of motion on a zero mass state of helicity λ is to multiply the state vector by $e^{2\pi i\lambda}$. This must be $+1$ (-1) for single-(double)-valued representations, so that λ is an integer (odd half-integer). The absolute value of the helicity is often called the spin of the massless particle.

massive state of spin s. If we include spatial inversions among the allowed transformations, so that det Λ can also be negative, we have a doubling of states: since $\mathbf{J}$ is even under inversions and $\mathbf{p}$ odd, the helicity is a pseudoscalar, i.e., changes sign under an inversion. The photon, for example, interacts with charged matter in such a way that photons of both negative and positive helicity can be produced.

Massive states may also be labeled by the helicity rather than by the component of the angular momentum in some arbitrarily fixed "z-direction." For such states, Lorentz transformations do mix up different helicity states, e.g., a velocity-changing Lorentz transformation could reverse the motion of a particle, thereby changing the sign of its helicity. Under rotations, however, the helicity is unchanged, and this sometimes makes the helicity description of states very convenient. We shall have several occasions to use the helicity description, which is discussed later in this chapter.

For the sake of completeness we will briefly discuss the transformation properties of state vectors describing particles with mass. In order to accommodate the many indices, we shall use the Dirac bra and ket notation. A state of mass m, spin s, momentum $\mathbf{p}$, and z-component of angular momentum σ will be described by $|[m, s], \mathbf{p}, \sigma\rangle$. We have, with the S_i defined in (4.36),

$$S_3 \,|[m, s], \mathbf{p}, \sigma\rangle_{\mathbf{p}=0} = \sigma \,|[m, s], \mathbf{p}, \sigma\rangle_{\mathbf{p}=0} \tag{4.46}$$

and we use the usual phase convention, according to which

$$(S_1 \pm iS_2)\,|[m, s], 0, \sigma\rangle = ((s \mp \sigma)(s \pm \sigma + 1))^{1/2}\,|[m, s], 0, \sigma\rangle \tag{4.47}$$

The state with momentum $\mathbf{p}$ is given by $U(L(p))\,|[m, s], 0, \sigma\rangle$ as discussed earlier.

The states $|[m, s], \mathbf{p}, \sigma\rangle$ are normalized according to

$$\langle[m, s], \mathbf{p}', \sigma'\,|[m, s], \mathbf{p}, \sigma\rangle = 2p_0\, \delta(\mathbf{p} - \mathbf{p}')\, \delta_{\sigma\sigma'} \tag{4.48}$$

The transformation properties of the state vector under displacements are simple to find. We have

$$U(a)\,|[m, s], \mathbf{p}, \sigma\rangle = e^{ip^\mu a_\mu}|[m, s], \mathbf{p}, \sigma\rangle \tag{4.49}$$

where

$$p_0 = (m^2 + \mathbf{p}^2)^{1/2}$$

The transformation properties under the homogeneous transformation $U(\Lambda)$ are a little more complex. We have

$$\begin{aligned} U(\Lambda)\,|[m, s], \mathbf{p}, \sigma\rangle &= U(\Lambda)\, U(L(p))\,|[m, s], 0, \sigma\rangle \\ &= U(\Lambda L(p))|[m, s], 0, \sigma\rangle \\ &= U(L(\Lambda p))\, U(L^{-1}(\Lambda p)\Lambda\, L(p))\,|[m, s], 0, \sigma\rangle \end{aligned}$$

The nature of the transformation $L^{-1}(\Lambda p)\Lambda\, L(p)$ is easy to determine: when applied to the four vector $(m, 0, 0, 0)$ it yields in successive stages, p, Λp, and then $(m, 0, 0, 0)$ again. It is therefore a rotation. It is generally called the *Wigner rotation* and is denoted by $R(\Lambda p, p)$. Now a state of angular momentum s (in the rest system) has well-defined transformation properties under a rotation. We have

$$U(R(\Lambda p, p))\,|[m, s], 0, \sigma\rangle = \sum_{\sigma'} D^{(s)}_{\sigma'\sigma}(R(\Lambda p, p))\,|[m, s], 0, \sigma'\rangle \quad (4.50)$$

Thus we get

$$U(\Lambda)\,|[m, s], \mathbf{p}, \sigma\rangle = \sum_{\sigma'} D^{(s)}_{\sigma'\sigma}(R(\Lambda p, p))\,|[m, s], \mathbf{\Lambda p}, \sigma'\rangle \quad (4.51)$$

The form of the matrices $D^{(s)}_{\sigma'\sigma}(R)$ is discussed in many books on the theory of angular momentum. If $\mathbf{J}^{(s)}$ denotes the $(2s + 1) \times (2s + 1)$ dimensional angular momentum matrices, and if the rotation R is of magnitude θ about a direction $\hat{\mathbf{n}}$,

$$D^{(s)}_{\sigma'\sigma}(R) = (e^{-i\theta\hat{\mathbf{n}}\cdot\mathbf{J}^{(s)}})_{\sigma'\sigma} \quad (4.52)$$

If the rotation is described by the Euler angles (α, β, γ), then

$$R = e^{-i\alpha J_3{}^{(s)}}e^{-i\beta J_2{}^{(s)}}e^{-i\gamma J_3{}^{(s)}}$$

and

$$D^{(s)}_{\sigma'\sigma}(R) = e^{-i\alpha\sigma' - i\gamma\sigma}\, d^{(s)}_{\sigma'\sigma}(\beta)$$

where

$$d^{(s)}_{\sigma'\sigma}(\beta) = (e^{-i\beta J_2{}^{(s)}})_{\sigma'\sigma} \quad (4.53)$$

In general, the rotation $R(\Lambda p, p)$ is quite complex. If, however, Λ is a pure Lorentz transformation, which transforms p to p' according to

$$\mathbf{p}' = \mathbf{p} + \mathbf{u}\left(\frac{\mathbf{p}\cdot\mathbf{u}}{u}\frac{\gamma_u - 1}{u} + \gamma_u p_0\right)$$

$$p_0{}' = \gamma_u(p_0 + \mathbf{p}\cdot\mathbf{u}) \quad (4.54)$$

then Λ is of the form

$$\left(\begin{array}{c|c} \gamma_u & -u^i\gamma_u \\ \hline u_j\gamma_u & \delta_j{}^i - \dfrac{u_j u^i}{u^2}(\gamma_u - 1) \end{array}\right) \quad (4.55)$$

The Wigner rotation may be computed without any difficulty. We find, as expected, that $R_i{}^0 = R_0{}^i = 0$, $R_0{}^0 = 1$, and after some computation

$$R_i{}^j = \delta_i{}^j + \frac{u_i u^j(\gamma_u - 1)}{u^2} + \frac{\gamma_u}{E_p + m}u_i p^j - \frac{\gamma_u}{E_{p'} + m}p_i{}'u^j - (\gamma_u - 1)\frac{p_i{}'p^j}{(E_p + m)(E_{p'} + m)} \quad (4.56)$$

From this we find that

$$R_i^{\ j} p_j = \frac{E_p + \gamma_u m}{E_{p'} + m}\, p_i' - \gamma_u\, m u_i \tag{4.57}$$

Thus if the energy is ultrarelativistic both before and after the transformation, i.e., if $E_p, E_{p'} \gg m$, then

$$R_i^{\ j}\left(\frac{p_j}{E_p}\right) \simeq \frac{p_i'}{E_{p'}} \tag{4.58}$$

Thus here R is just the rotation undergone by the velocity vector of the particle. If we take $\mathbf{p}$ to lie along the z-axis, and $\mathbf{p}'$ in the x, z-plane, making an angle θ with the z-direction, the rotation is represented by $e^{-i\theta J_2^{(s)}}$ and

$$D^{(s)}_{\sigma'\sigma}(R) \simeq d^{(s)}_{\sigma'\sigma}(\theta) \tag{4.59}$$

In the nonrelativistic limit all velocities are small, and $R_{ij} \simeq \delta_{ij}$. Thus there is no rotation, and

$$D^{(s)}_{\sigma'\sigma}(R) \simeq \delta_{\sigma'\sigma}$$

With these remarks we complete our discussion of the *canonical basis* of state vectors.

For the construction of *field operators* for noninteracting particles of arbitrary spin, other bases for the state vectors are necessary. Since if we write the canonical state vector in the form of a creation operator acting on the vacuum state,

$$|[m, s], \mathbf{p}, \sigma\rangle \equiv A_\sigma^+(m, s; \mathbf{p})\, |0\rangle \tag{4.60}$$

then

$$U(\Lambda)\, A_\sigma^+(m, s; \mathbf{p})\, U^{-1}(\Lambda) = \sum_{\sigma'} A_{\sigma'}^+(m, s; \mathbf{\Lambda p})\, D^{(s)}_{\sigma'\sigma}\,(R(\Lambda p, p)) \tag{4.61}$$

i.e., the creation operators transform in a way which depends not only on Λ, but also on the momentum of the particle being created. This is quite different from the transformation properties implied by (2.99), for example, for the local fields. If, however, we consider the basis $\Phi_{p,A}$ defined by[1]

$$\Phi_{pA} = \sum_\sigma |[m, s], \mathbf{p}, \sigma\rangle\, D^{(s)}_{\sigma A}(L^{-1}(p)) \tag{4.62}$$

where $D^{(s)}_{\sigma A}(L^{-1}(p))$ is the $(2s + 1) \times (2s + 1)$ dimensional matrix representation[2] of the homogeneous Lorentz transformation $L^{-1}(p)$ (see

[1] H. Joos, *Fortschr. Physik*, **10,** 65 (1962); S. Weinberg, *Phys. Rev.*, **133,** B1318 (1964).
[2] $D^{(s)}_{\sigma A}(\Lambda)$, as written above, stands for the representation $D^{(s,0)}$ as defined in the appendix. From the form defined in (A4.6) is follows that $D^{(0,s)}_{\sigma A}(\Lambda) = D^{(s,0)*}_{A\sigma}(\Lambda^{-1})$.

appendix), the states transform as follows

$$U(\Lambda)\,\Phi_{pA} = \sum_{\sigma'\alpha} |[m, s], \mathbf{\Lambda p}, \sigma'\rangle\, D^{(s)}_{\sigma'\alpha}(L^{-1}(\Lambda p))\, D^{(s)}_{\alpha A}(\Lambda)$$
$$= \sum_{\alpha} \Phi_{\Lambda p,\alpha}\, D^{(s)}_{\alpha A}(\Lambda) \qquad (4.63)$$

since

$$D^{(s)}\,(L^{-1}(\Lambda p)\,\Lambda L(p))\,D^{(s)}\,(L^{-1}(p)) = D^{(s)}\,(L^{-1}(\Lambda p))\,D^{(s)}\,(\Lambda) \qquad (4.64)$$

For states in this representation (called *spinor* representation) the creation operators transform more simply:

$$U(\Lambda)\,\mathbf{a}_A{}^{+}(p)\,U^{-1}(\Lambda) = \sum_{\alpha} \mathbf{a}_\alpha{}^{+}(\Lambda p)\,D^{(s)}_{\alpha A}(\Lambda) \qquad (4.65)$$

Another spinor basis can be constructed. We take advantage of the fact that for a *rotation*

$$D^{(s)}_{\sigma'\sigma}(R) = (D^{*}_{\sigma\sigma'}(R))^{-1} = D^{*}_{\sigma\sigma'}(R^{-1}) \qquad (4.66)$$

so that

$$D^{(s)}_{\sigma'\sigma}(R(\Lambda p,\, p)) = D^{(s)*}_{\sigma\sigma'}(L^{-1}(p)\Lambda^{-1}\,L(\Lambda p))$$

Hence if we define

$$\Phi_{p\dot{A}} \equiv \sum_{\sigma} D^{(s)*}_{\dot{A}\sigma}(L(p))\,|[m, s], \mathbf{p}, \sigma\rangle \qquad (4.67)$$

We find that this transforms according to

$$U(\Lambda)\,\Phi_{p,\dot{A}} = \sum_{\dot{B}} D^{(s)*}_{\dot{A}\dot{B}}(\Lambda^{-1})\Phi_{\Lambda p,\dot{B}} \qquad (4.68)$$

The creation operators for such a state are then seen to transform according to

$$U(\Lambda)\,\mathbf{b}_{\dot{A}}{}^{+}(p)\,U^{-1}(\Lambda) = \sum_{\dot{B}} D^{(s)*}_{\dot{A}\dot{B}}(\Lambda^{-1})\,\mathbf{b}_{\dot{B}}{}^{+}(\Lambda p) \qquad (4.69)$$

We may, with the help of the operators $\mathbf{a}_A{}^{+}(p)$ and $\mathbf{b}_{\dot{A}}{}^{+}(p)$, construct local field operators that have simple transformation properties. We do not go into details but refer the reader to the papers of Joos and Weinberg, in which these matters are further discussed.

Let us now turn to the so-called *helicity basis*.[1] Here the states of a particle will be classified by the momentum and λ, the component of the angular momentum along the direction of motion. This description of the states is frequently convenient, and, for massless particles, it is essential. Since we are only peripherally interested in massless particles, we shall construct the states for particles with mass.

[1] The usefulness of the helicity representation in collision theory involving massive as well as massless particles was first stressed in a very influential paper by M. Jacob and G. C. Wick, *Ann. Phys.* (N.Y.) **7,** 404 (1959).

We begin with a particle at rest. The z-component of its spin is λ, and it is described by $|[m, s], 0, \lambda\rangle$. If we now set the new particle in motion in the z-direction, with a momentum of magnitude $|\mathbf{p}|$, we have a helicity state. We describe this mathematically by

$$\Psi_{\mathring{p},\lambda} = U(L(\mathring{p}))\,|[m, s], 0, \lambda\rangle \tag{4.70}$$

We shall reserve the notation $\Psi_{p\lambda}$ for helicity states in this chapter. If the state $\Psi_{\mathring{p}\lambda}$ is now rotated so that the momentum is $\mathbf{p}$, the helicity does not change—it is a pseudoscalar. Thus if $R_{p,\mathring{p}}$ denotes the rotation which takes the z-axis ($\mathring{\mathbf{p}}$) into $\mathbf{p}$, then[1]

$$\Psi_{p,\lambda} = U(R_{p\mathring{p}})\,\Psi_{\mathring{p},\lambda} \tag{4.71}$$

We may use this expression to obtain the relation between the helicity basis and the canonical basis. We have

$$\Psi_{p,\lambda} = U(R_{p\mathring{p}})\,|[m, s], \mathring{\mathbf{p}}, \lambda\rangle \tag{4.72}$$

and the right-hand side may be determined from (4.51) to be given by

$$U(R_{p\mathring{p}})\,|[m, s]\,\mathring{\mathbf{p}}, \lambda\rangle = \sum_{\sigma} |[m, s], \mathbf{p}, \sigma\rangle\; D^{(s)}_{\sigma\lambda}(L^{-1}(p)R_{p\mathring{p}}\,L(\mathring{p}))$$

so that

$$\Psi_{p,\lambda} = \sum_{\sigma} |[m, s], p, \sigma\rangle\; D^{(s)}_{\sigma\lambda}(\tilde{R}) \tag{4.73}$$

with

$$\tilde{R} \equiv L^{-1}(p)\,R_{p\mathring{p}}\,L(\mathring{p}) \tag{4.74}$$

We may use this expression to obtain the transformation properties of the helicity states under Lorentz transformations. We get

$$\begin{aligned} U(\Lambda)\,\Psi_{p,\lambda} &= \sum_{\sigma} U(\Lambda)\,|[m, s], \mathbf{p}, \sigma\rangle\; D^{(s)}_{\sigma\lambda}\,(L^{-1}(p)\,R_{p\mathring{p}}\,L(\mathring{p}))\\ &= \sum_{\sigma\tau} |[m, s], \mathbf{\Lambda p}, \tau\rangle\; D^{(s)}_{\tau\lambda}(L^{-1}(\Lambda p)\,\Lambda R_{p\mathring{p}}\,L(\mathring{p}))\\ &= \sum_{\tau\mu} |[m, s], \mathbf{\Lambda p}, \tau\rangle\; D^{(s)}_{\tau\mu}(L^{-1}(\Lambda p)R_{\Lambda p,\mathring{p}}\,L(\mathring{p}))\\ &\quad\times\; D^{(s)}_{\mu\lambda}(L^{-1}(\mathring{p})R^{-1}_{\Lambda p,\mathring{p}}\Lambda R_{p\mathring{p}}\,L(\mathring{p}))\\ &= \sum_{\mu} \Psi_{\Lambda p,\mu}\; D^{(s)}_{\mu\lambda}(L^{-1}(\mathring{p})R^{-1}_{\Lambda p,\mathring{p}}\Lambda R_{p\mathring{p}}\,L(\mathring{p})) \end{aligned} \tag{4.75}$$

If we now use the relation

$$R_{p\mathring{p}}\,L(\mathring{p}) = L(p)R_{p\mathring{p}} \tag{4.76}$$

[1] The rotation $R_{p\mathring{p}}$ is given by $R(\phi, \theta, -\phi) \equiv e^{-i\phi J_3}\,e^{-i\theta J_2}\,e^{i\phi J_3}$ where $\mathbf{p}$ points in the direction (θ, ϕ) with respect to the z-axis $\mathring{\mathbf{p}}$.

which states that an "acceleration" in a given direction, followed by a rotation, can be replaced by a rotation and then an acceleration to the final momentum, and equivalently

$$L^{-1}(\mathring{p})R^{-1}_{\Lambda p,\mathring{p}} = R^{-1}_{\Lambda p,\mathring{p}}\, L^{-1}(\Lambda p)$$

We obtain

$$U(\Lambda)\Psi_{p,\lambda} = \sum_{\mu} \Psi_{\Lambda p,\mu}\, D^{(s)}_{\mu\lambda}(\mathcal{R}) \tag{4.77}$$

where[1]

$$\mathcal{R} = R^{-1}_{\Lambda p,\mathring{p}}\, R(\Lambda p, p) R_{p\mathring{p}}$$

and $R(\Lambda p, p)$ is the Wigner rotation.

These relations are of interest if we wish to develop a relativistic generalization of the familiar process of adding angular momenta. The product of two states, say $\Phi_{p_1A_1} \otimes \Phi_{p_2A_2}$, transforms as a reducible representation of the Poincaré group. The product may be reduced to give a sum of irreducible representations, characterized by a mass M, with

$$M^2 = (p_1 + p_2)^2$$

and a total angular momentum J, made up by the addition of the spins and an orbital angular momentum that serves as a degeneracy parameter in the decomposition of the product. The procedure may be found in the paper by Joos cited earlier in this chapter and in Macfarlane's more detailed exposition of this procedure.[2] Since we shall work in the center of mass system, this decomposition need not be treated here.

Two particle states, describing particles with momentum and helicity p_1, λ_1 and p_2, λ_2, respectively, are denoted as follows:

$$|\mathbf{p}_1\lambda_1; \mathbf{p}_2\lambda_2\rangle = \Psi_{p_1,\lambda_1} \otimes \Psi_{p_2,\lambda_2} \tag{4.78}$$

If we define

$$\begin{aligned} P &= p_1 + p_2 \\ k &= \tfrac{1}{2}(p_1 - p_2) \end{aligned} \tag{4.79}$$

and work in the center of mass system in which $P = (W, 0, 0, 0)$, the two-particle state will be denoted by $|W, \mathbf{p}_1; \lambda_1\lambda_2\rangle$, with $\mathbf{p}_1$ denoting the momentum of one of the particles in the center of mass system. The construction of the two-particle state involves a small problem of phases. We begin with a standard state for one of the particles, $\Psi^{(1)}_{\mathring{p},\lambda_1}$ in which the particle has momentum $\mathring{\mathbf{p}} = (0, 0, p_1)$. The state of the particle moving in the direction of the negative z axis is defined by

$$\Psi^{(2)}_{-\mathring{p},\lambda_2} \equiv (-1)^{s_2-\lambda_2} e^{-i\pi J_2} \Psi^{(2)}_{\mathring{p},\lambda_2} \tag{4.80}$$

[1] Note that $R_{\Lambda p,\mathring{p}}$ is represented by $R(\phi', \theta', -\phi')$ where $\mathbf{\Lambda p}$ points in the direction (θ', ϕ').

[2] A. Macfarlane, *Rev. Mod. Phys.*, **34**, 41 (1962).

This choice is made so that in the limit as $|\mathbf{p}_1| \to 0$ we obtain[1]

$$\begin{aligned}(\Psi'^{(2)}_{\mathring{p},-\lambda_1}, \Psi'^{(2)}_{-\mathring{p},\lambda_2}) &\to (-1)^{s_2-\lambda_2}\langle s_2, -\lambda_1 | e^{-i\pi J_2} | s_2, \lambda_2\rangle \\ &= (-1)^{s_2-\lambda_2} d^{(s_2)}_{-\lambda_1\lambda_2}(\pi) \\ &= \delta_{\lambda_1\lambda_2} \qquad (4.81)\end{aligned}$$

The state $|W, \mathbf{p}_1, \lambda_1\lambda_2\rangle$ is now obtained by rotating $|W, \mathring{\mathbf{p}}, \lambda_1\lambda_2\rangle$ with the rotation $R(\phi, \theta, -\phi)$, where (θ, ϕ) are the spherical angles of $\mathbf{p}_1$.

It is very useful to construct states of definite angular momentum out of states containing particles of definite helicity. If such an angular momentum eigenstate is denoted by $|W; J, M; \lambda_1\lambda_2\rangle$ then it must transform under rotations according to

$$U(R(\alpha\beta\gamma))\,|W, J, M; \lambda_1\lambda_2\rangle = \sum_{M'} |W, J, M'; \lambda_1\lambda_2\rangle\, D^{(J)}_{M'M}(R(\alpha\beta\gamma)) \qquad (4.82)$$

If we now write

$$|W; J, M; \lambda_1\lambda_2\rangle = \int d\alpha \int \sin\beta\, d\beta \int d\gamma\; D^{(J)*}_{MM'}(R(\alpha\beta\gamma)) \times U(R(\alpha\beta\gamma))\,|W, \mathring{\mathbf{p}}, \lambda_1\lambda_2\rangle \qquad (4.83)$$

or schematically

$$|JM\rangle = \int dR\; D^{(J)*}_{MM'}(R)\, U(R)\, |\mathring{\mathbf{p}}, \lambda\rangle \qquad (4.84)$$

(in which dR denotes an integration over the range of the parameters of the rotation), then

$$\begin{aligned}U(R')\,|JM\rangle &= \int dR\; D^{(J)*}_{MM'}(R)\, U(R'R)\, |\mathring{\mathbf{p}}, \lambda\rangle \\ &= \sum_{M''} \int dR\; D^{(J)*}_{MM''}(R'^{-1})\, D^{(J)*}_{M''M'}(R'R)\, U(R'R)\, |\mathring{\mathbf{p}}, \lambda\rangle \\ &= \sum_{M''} \int d(R'R)\; D^{(J)*}_{M''M'}(R'R)\, U(R'R)\, |\mathring{\mathbf{p}}, \lambda\rangle\, D^{(J)}_{M''M}(R') \\ &= \sum_{M''} |JM''\rangle\, D^{(J)}_{M''M}(R')\end{aligned}$$

which shows that (4.84) is indeed the correct construction.

Because

$$R(\alpha\beta\gamma) = R(\alpha\beta 0)e^{-i\gamma J_3}$$

we have

$$\begin{aligned}D^{(J)*}_{MM'}(R(\alpha\beta\gamma)) &= e^{iM'\gamma}\, D^{(J)*}_{MM'}(R(\alpha\beta 0)) \\ U(R(\alpha\beta\gamma))\,|\mathring{\mathbf{p}}, \lambda\rangle &= e^{-i\gamma(\lambda_1-\lambda_2)}\, U(R(\alpha\beta 0))\,|\mathring{\mathbf{p}}, \lambda\rangle\end{aligned}$$

[1] Strictly speaking, with our continuum normalization, this scalar product is not defined in the limit. We do, however, wish to keep the proper phases in case we should want to normalize the one-particle states in a box with periodic boundary conditions.

Furthermore,

$$\int d\gamma e^{i(M'-\lambda_1+\lambda_2)\gamma} \propto \delta_{M',\lambda_1-\lambda_2}$$

so that we may write

$$|W; J,M, \lambda_1\lambda_2\rangle = N_J\int d\alpha\int \sin\beta\, d\beta\, D^{(J)}_{M,\lambda_1-\lambda_2}(R(\alpha\beta 0)) \times U(R(\alpha\beta 0))\,|W,\mathring{\mathbf{p}}, \lambda_1\lambda_2\rangle \quad (4.85)$$

Noting that

$$\begin{aligned}|W,\mathbf{p}_1; \lambda_1\lambda_2\rangle &= U(R(\phi, \theta, -\phi))\,|W, \mathring{\mathbf{p}}, \lambda_1\lambda_2\rangle \\ &= U(R(\phi, \theta, 0)\,|W, \mathring{\mathbf{p}}, \lambda_1\lambda_2\rangle \end{aligned} \quad (4.86)$$

we finally obtain

$$|W; J, M; \lambda_1\lambda_2\rangle = N_J\int_0^{2\pi} d\phi\int_0^{\pi} d\theta \sin\theta\, D^{(J)*}_{M,\lambda_1-\lambda_2}(R(\phi, \theta, 0))\,|W, \mathbf{p}_1; \lambda_1\lambda_2\rangle \quad (4.87)$$

where the variable of integration was changed from (α, β) to (θ, ϕ).[1]

To determine the constant N_J, we must discuss the normalization of the states appearing in (4.87). If the one-particle states are normalized according to

$$(\Psi_{p_2,\lambda_2}, \Psi_{p_1,\lambda_1}) = 2p_{10}\,\delta(\mathbf{p}_1 - \mathbf{p}_2)\,\delta_{\lambda_1\lambda_2} \quad (4.88)$$

the normalization of the two-particle states may be written in the form

$$\begin{aligned}\langle \mathbf{p}_1'\lambda_1'; \mathbf{p}_2'\lambda_2' \,|\mathbf{p}_1\lambda_1; \mathbf{p}_2\lambda_2\rangle &= (2p_{10})(2p_{20})\,\delta(\mathbf{p}_1 - \mathbf{p}_1')\,\delta(\mathbf{p}_2 - \mathbf{p}_2')\,\delta_{\lambda_1\lambda_1'}\,\delta_{\lambda_2\lambda_2'} \\ &= 4p_{10}p_{20}\,\delta(\mathbf{P} - \mathbf{P}')\,\delta(\mathbf{k} - \mathbf{k}')\,\delta_{\lambda_1\lambda_1'}\,\delta_{\lambda_2\lambda_2'} \\ &= \frac{4p_{10}p_{20}}{k^2}\,\delta(\mathbf{P} - \mathbf{P}')\,\delta(k - k') \\ &\qquad \times \delta(\hat{k} - \hat{k}')\,\delta_{\lambda_1\lambda_1'}\,\delta_{\lambda_2\lambda_2'}\end{aligned}$$

We use

$$\delta(k - k') = \delta(P_0 - P_0')\,\frac{dP_0}{dk}$$

to write this in the form

$$\langle P', \mathbf{k}', \lambda_1'\lambda_2' \,|P, \mathbf{k}, \lambda_1\lambda_2\rangle = \frac{4p_{10}p_{20}}{k^2}\frac{dP_0}{dk}\,\delta^{(4)}(P_\mu - P_\mu')\,\delta(\hat{k} - \hat{k}')\delta_{\lambda_1\lambda_1'}\,\delta_{\lambda_2\lambda_2'} \quad (4.89)$$

[1] This range of integration is correct for integral as well as half-odd integral spin. We do not want to get involved in the intricacies of the theory of rotations, but for the reader who might worry about this matter, we point out that for the double-valued representations (half-odd integral J) one must integrate over the domain of parameters of the "covering group," which is twice as large. This means that we take $0 \leq \gamma \leq 4\pi$. Now the integration is $\int_0^{4\pi} d\gamma\, e^{-i(\lambda-M')\gamma}$ which is just what is required to obtain $\delta_{\lambda M'}$.

where $|P, \mathbf{k}, \lambda_1\lambda_2\rangle$ denotes the same two particle state, with a different labeling.

In the center of mass system, $k = p_1$, so that

$$P_0 = p_{10} + p_{20} \equiv W$$

$$\frac{dP_0}{dp_1} = \left(\frac{p_1}{p_{10}} + \frac{p_1}{p_{20}}\right) = \frac{p_1 P_0}{p_{10}p_{20}} = \frac{p_1 W}{p_{10}p_{20}}$$

If we introduce the state vectors $|W, \hat{\mathbf{p}}_1, \lambda_1\lambda_2\rangle$ defined by

$$|W, \hat{\mathbf{p}}_1, \lambda_1\lambda_2\rangle = \left(\frac{p_1}{4W}\right)^{1/2} |P, \mathbf{p}, \lambda_1\lambda_2\rangle_{\mathbf{P}=0} \tag{4.90}$$

these are seen to be normalized according to

$$\langle W, \hat{\mathbf{p}}_1', \lambda_1'\lambda_2' \mid W, \hat{\mathbf{p}}_1, \lambda_1\lambda_2\rangle = \delta(\hat{\mathbf{p}}_1' - \hat{\mathbf{p}}_1)\, \delta_{\lambda_1\lambda'}\, \delta_{\lambda_2\lambda_2'} \tag{4.91}$$

after the $\delta^{(4)}(P_\mu - P_\mu')$ has been split off. From this it follows that

$$|W, \hat{\mathbf{p}}, \lambda_1\lambda_2\rangle = \sum_{JM} N_J\, D^{(J)}_{M,\lambda_1-\lambda_2}(R(\phi, \theta, 0))\, |W; J, M, \lambda_1\lambda_2\rangle \tag{4.92}$$

If we require that the angular momentum states $|W, J, M; \lambda_1\lambda_2\rangle$ satisfy

$$\langle W, J', M', \lambda_1'\lambda_2' \mid W, J, M, \lambda_1\lambda_2\rangle = \delta_{JJ'}\, \delta_{MM'}\, \delta_{\lambda_1\lambda_1'}\, \delta_{\lambda_2\lambda_2'} \tag{4.93}$$

then

$$\int_0^{2\pi} d\phi \int_{-1}^{1} d(\cos\theta)\, D^{(J)*}_{M\lambda}(R(\phi, \theta, 0))\, D^{(J')}_{M'\lambda}(R(\phi, \theta, 0)) = \frac{4\pi}{2J+1}\, \delta_{JJ'}\, \delta_{MM'} \tag{4.94}$$

may be used to show that

$$N_J = \left(\frac{2J+1}{4\pi}\right)^{1/2} \tag{4.95}$$

The formalism developed above is very useful in the decomposition of the scattering operator S, which will be discussed in detail in succeeding chapters. The operator S has the property that it commutes with the operators P_μ. If we consider

$$R = S - 1$$

then the "submatrix" $R(P_\mu)$ is defined by

$$\begin{aligned}&\langle \mathbf{p}_1'\lambda_1', \mathbf{p}_2'\lambda_2' \,|R|\, \mathbf{p}_1\lambda_1, \mathbf{p}_2\lambda_2\rangle \\ &\qquad = \delta(P_\mu' - P_\mu)\, \langle \mathbf{p}_1'\lambda_1', \mathbf{p}_2'\lambda_2' \,|R(P_\mu)|\, \mathbf{p}_1\lambda_1, \mathbf{p}_2\lambda_2\rangle\end{aligned} \tag{4.96}$$

In the center of mass system we have

$$\begin{aligned}&\langle \mathbf{p}_1'\lambda_1', \mathbf{p}_2'\lambda_2' \,|R(P_\mu)|\, \mathbf{p}_1\lambda_1, \mathbf{p}_2\lambda_2\rangle \\ &\qquad = \left(\frac{(4W)^2}{p_1 p_1'}\right)^{1/2} \langle W, \hat{\mathbf{p}}_1', \lambda_1'\lambda_2' \,|R(W)|\, W, \hat{\mathbf{p}}_1, \lambda_1\lambda_2\rangle\end{aligned} \tag{4.97}$$

with the states defined in (4.90). We may now write

$$\langle W, \hat{\mathbf{p}}_1', \lambda_1'\lambda_2' \,|R(W)|\, W, \hat{\mathbf{p}}_1, \lambda_1\lambda_2\rangle$$
$$= \sum_{JM} \sum_{J'M'} N_J N_{J'}\, D^{(J')}_{M',\lambda_1'-\lambda_2'}(R(\phi', \theta', 0))\, D^{(J)*}_{M,\lambda_1-\lambda_2}(R(\phi, \theta, 0))$$
$$\times \langle W, J', M', \lambda_1'\lambda_2' \,|R(W)|\, W, J, M, \lambda_1\lambda_2\rangle$$

Since R is invariant under rotations, it follows that

$$\langle W, J', M', \lambda_1'\lambda_2' \,|R(W)|\, W, J, M, \lambda_1\lambda_2\rangle = \delta_{JJ'}\,\delta_{MM'}\,R_J(\lambda_1'\lambda_2'; \lambda_1\lambda_2) \tag{4.98}$$

We thus obtain

$$\langle W, \hat{\mathbf{p}}_1', \lambda_1'\lambda_2' \,|R(W)|\, W, \hat{\mathbf{p}}_1, \lambda_1\lambda_2\rangle$$
$$= \sum_{JM} N_J^{\,2}\, D^{(J)}_{M,\lambda_1'-\lambda_2'}(R(\phi', \theta', 0))\, D^{(J)}_{M,\lambda_1-\lambda_2}(R(\phi, \theta, 0))$$
$$\times R_J(\lambda_1'\lambda_2'; \lambda_1\lambda_2)$$

If we choose $\hat{\mathbf{p}}_1$ as z-axis, then

$$D^{(J)}_{M,\lambda_1-\lambda_2}(R(0, 0, 0)) = \delta_{M,\lambda_1-\lambda_2}$$

and

$$\langle W, \hat{\mathbf{p}}_1', \lambda_1'\lambda_2' \,|R(W)|\, W, \hat{\mathbf{p}}_1, \lambda_1\lambda_2\rangle$$
$$= \sum_J \frac{2J+1}{4\pi}\, D^{(J)}_{\lambda_1-\lambda_2,\lambda_1'-\lambda_2'}(R_{\hat{\mathbf{p}}'})R_J(\lambda_1'\lambda_2'; \lambda_1\lambda_2) \tag{4.99}$$

We anticipate a result of Chapter 9 which shows that the differential cross section for the process (1) + (2) → (1′) + (2′) is given by

$$\frac{d\sigma}{d\Omega} = \frac{4\pi^2}{p_1^{\,2}}\, |\langle W, \hat{\mathbf{p}}_1', \lambda_1'\lambda_2' \,|R(W)|\, W, \hat{\mathbf{p}}_1, \lambda_1\lambda_2\rangle|^2$$
$$= \left| \sum_J \left(J + \frac{1}{2}\right) D^{(J)}_{\lambda_1-\lambda_2,\lambda_1'-\lambda_2'}(R(\phi', \theta', 0)) \times \frac{1}{p_1} R_J(\lambda_1'\lambda_2', \lambda_1\lambda_2) \right|^2 \tag{4.100}$$

Not all of the $R_J(\lambda_1'\lambda_2'; \lambda_1\lambda_2)$ are independent. With a little attention to phases, it is not hard to show that parity conservation implies that

$$R_J(-\lambda_1', -\lambda_2'; -\lambda_1, -\lambda_2) = \eta(-1)^{s_1'+s_2'-s_1-s_2} R_J(\lambda_1'\lambda_2'; \lambda_1\lambda_2) \tag{4.101}$$

where $s_1, \ldots, s_2'$ are the spins of the particles involved and η is the product of their intrinsic parities. Similarly, time-reversal invariance implies that

$$R_J(\lambda_1'\lambda_2'; \lambda_1\lambda_2) = R_J(\lambda_1\lambda_2; \lambda_1'\lambda_2') \tag{4.102}$$

Thus for pion-nucleon scattering, for example, where the pion has spin 0 and the nucleon spin $\frac{1}{2}$, there are only two independent R_J functions, $R_J(\frac{1}{2}, \frac{1}{2}) = R_J(-\frac{1}{2}, -\frac{1}{2})$ and $R_J(\frac{1}{2}, -\frac{1}{2}) = R_J(-\frac{1}{2}, \frac{1}{2})$.

A useful parametrization follows from the property of S that

$$S^+S = 1 \tag{4.103}$$

i.e.

$$R^+ + R = -R^+R$$

This implies that

$$\begin{aligned}\langle \mathbf{p}_1\lambda_1, \mathbf{p}_2\lambda_2 \,|R(P_\mu)|\, \mathbf{p}_1'\lambda_1', \mathbf{p}_2'\lambda_2' \rangle^* &+ \langle \mathbf{p}_1'\lambda_1', \mathbf{p}_2'\lambda_2' \,|R(P_\mu)|\, \mathbf{p}_1\lambda_1, \mathbf{p}_2\lambda_2 \rangle \\ - \sum_{\lambda_1''\lambda_2''} \int \frac{d^3p_1''}{2p_{10}''} \int \frac{d^3p_2''}{2p_{20}''} &\delta(p_1'' + p_2'' - p_1 - p_2) \\ \times \langle \mathbf{p}_1''\lambda_1'', \mathbf{p}_2''\lambda_2'' \,|R(P_\mu)|\, \mathbf{p}_1'\lambda_1', \mathbf{p}_2'\lambda_2' \rangle^* & \\ \times \langle \mathbf{p}_1''\lambda_1'', \mathbf{p}_2''\lambda_2'' \,|R(P_\mu)|\, \mathbf{p}_1\lambda_1, \mathbf{p}_2\lambda_2 \rangle &\end{aligned}$$

The integration of $d^3p_1''/2p_{10}''$ and $d^3p_2''/2p_{20}''$ in the center of mass frame can be done without great difficulty. With the help of the expansion (4.99), we can show that under the conditions following (4.103) are correct (only two-particle unitarity[1])

$$\begin{aligned}R_J(\lambda_1'\lambda_2'; \lambda_1\lambda_2) &+ R_J^*(\lambda_1\lambda_2; \lambda_1'\lambda_2') \\ &= -\sum_{\lambda_1''\lambda_2''} R_J^*(\lambda_1''\lambda_2''; \lambda_1'\lambda_2')R_J(\lambda_1''\lambda_2''; \lambda_1\lambda_2)\end{aligned} \tag{4.104}$$

This takes a simple form if R_J is diagonalized. For each eigenvalue r_J^α

$$r_J^\alpha + r_J^{\alpha^+} = -|r_J^\alpha|^2$$

so that

$$r_J^\alpha = e^{2i\delta_J^\alpha} - 1 \tag{4.105}$$

with δ_J^α real. For pion-nucleon scattering, for example, the eigenvalues of the R_J matrix are

$$\begin{aligned}R(\tfrac{1}{2}; \tfrac{1}{2}) + R(\tfrac{1}{2}; -\tfrac{1}{2}) &= e^{2i\delta_{J+}} - 1 \\ R(\tfrac{1}{2}; \tfrac{1}{2}) - R(\tfrac{1}{2}; -\tfrac{1}{2}) &= e^{2i\delta_{J-}} - 1\end{aligned} \tag{4.106}$$

We shall leave it to the reader to convince himself that this leads to the same form of the scattering amplitude for pion-nucleon scattering as does the usual procedure [Chapter 23 (23.28)].

Actually our main use of the helicity formalism is in the discussion of the decay of particles. If a state of angular momentum J and z-component

[1] Here again we anticipate material from Chapter 21; in general $R + R^+ = -R^+R$ implies $R_{ab} + R_{ba}^* = -\sum_n R_{nb}^*R_{na}$, and for "$n$" we only take the states containing two particles.

M decays into two particles in the center of mass system, and if these particles are described in a helicity basis, the angular distribution is particularly simple. It is given by the absolute square of

$$(\Phi_{\mathbf{p},\lambda_1,-\mathbf{p},\lambda_2}, S\Psi_{JM}) = \mathcal{S}_J(\lambda_1, \lambda_2) N_J \, D^{(J)*}_{M,\lambda_1-\lambda_2}(R(\phi, \theta, 0)) \quad (4.107)$$

APPENDIX

A long footnote on finite dimensional representations of the homogeneous Lorentz group.

The operator representing a homogeneous Lorentz transformation in the Hilbert space of physical states is given by

$$U(\Lambda) = e^{(i/2)\alpha_{\mu\nu}M^{\mu\nu}} \quad (A4.1)$$

In terms of the **J** and **K** operators defined in (4.13), this reads

$$U(\Lambda) = e^{-i\mathbf{a}\cdot\mathbf{J}+i\mathbf{b}\cdot\mathbf{K}} \quad (A4.2)$$

If we introduce the operators $\mathfrak{J}_+$ and $\mathfrak{J}_-$ defined by

$$\begin{aligned} \mathbf{J} &= \mathfrak{J}_+ + \mathfrak{J}_- \\ \mathbf{K} &= -i\mathfrak{J}_+ + i\mathfrak{J}_- \end{aligned} \quad (A4.3)$$

we can easily check that $\mathfrak{J}_\pm$ obeys angular momentum commutation relations

$$[\mathfrak{J}_{\pm a}, \mathfrak{J}_{\pm b}] = i\epsilon_{abc}\mathfrak{J}_{\pm c} \quad (A4.4)$$

and

$$[\mathfrak{J}_{+a}, \mathfrak{J}_{-b}] = 0 \quad (A4.5)$$

Thus formally $U(\Lambda)$ looks like the product of two rotations. We may, therefore write down matrix representations of $U(\Lambda)$ in the form

$$e^{\mathfrak{J}_+\cdot(\mathbf{b}-i\mathbf{a})}e^{-\mathfrak{J}_-\cdot(\mathbf{b}+i\mathbf{a})} \quad (A4.6)$$

where $\mathfrak{J}_\pm$ are represented by $(2s_\pm + 1) \times (2s_\pm + 1)$ dimensional angular momentum matrices. These are only unitary representations if $\mathbf{b} = 0$, i.e., if the transformation is a rotation. An irreducible representation thus has two labels (s_+, s_-), and we denote it by $D^{(s_+,s_-)}(\Lambda)$. *Under rotations alone* this representation is not irreducible; it is of the form of a product of a $(2s_+ + 1)$ dimensional representation and a $(2s_- + 1)$ dimensional representation. It may be decomposed into a sum of representations characterized by angular momenta $(s_+ + s_-)$, $(s_+ + s_- - 1) \cdots |s_+ - s_-|$. The representations $D^{(s,0)}(\Lambda)$ and $D^{(0,s)}(\Lambda)$ are irreducible under rotations as well, and it is these which are used in the spinor representation [(4.62)].

Under space inversions, **J** does not change sign, whereas **K** does, so that

$$U_s\mathfrak{J}_\pm U_s^{-1} = \mathfrak{J}_\mp \quad (A4.7)$$

Thus under an inversion a matrix representation changes from $D^{(s_+,s_-)}$ to $D^{(s_-,s_+)}$. For the construction of wave functions, which transform under the matrices $D^{(s_+,s_-)}(\Lambda)$, the requirement of spatial inversion symmetry implies that we must construct them to transform under $D^{(s,s)}(\Lambda)$ or we must double the number of components and have them transform under $D^{(s,0)} \oplus D^{(0,s)}$ in the simplest case.

PROBLEMS

1. Use

$$d^J_{m'm}(\theta) = \sum_s \frac{(-1)^s (J+m)!\,(J-m)!\,(J+m')!\,(J-m')!}{s!\,(J-s-m')!\,(J+m-s)!\,(m'-m+s)!} \times \left(\cos\frac{\theta}{2}\right)^{2J+m-m'-2s} \left(-\sin\frac{\theta}{2}\right)^{m'-m+2s}$$

to show that

$$\begin{aligned} d^J_{m'm}(\theta) &= d^J_{-m,-m'}(\theta) \\ d^J_{mm'}(\theta) &= (-1)^{m'-m}\, d^J_{m'm}(\theta) \\ d^J_{m'm}(\theta) &= (-1)^{J+m'}\, d^J_{m',-m}(\pi-\theta) \end{aligned}$$

2. Calculate $d^{1/2}_{m'm}(\theta)$ and $d^1_{m'm}(\theta)$ with the help of the formula

$$d^J_{m'm}(\theta) = (\Psi_{Jm'}, e^{-i\theta J_2}\Psi_{Jm})$$

and the representations

$$J_2 = \frac{1}{2}\begin{pmatrix} 0 & -i \\ i & 0 \end{pmatrix} \qquad J_2 = \frac{1}{\sqrt{2}}\begin{pmatrix} 0 & -i & 0 \\ i & 0 & -i \\ 0 & i & 0 \end{pmatrix}$$

for $J = \frac{1}{2}$ and $J = 1$, respectively.

3. Calculate $D^{1/2}(L^{-1}(p))$ and $D^{1/2}(L(p))^*$ with the help of the representations of the homogeneous Lorentz group given in the appendix to Chapter 4. How are these related to the spinors $u(p), \ldots$?

4. Work out the R matrix parametrization [analogous to (4.106)] for Compton scattering

$$e + \gamma \to e + \gamma$$

5

Fields in Interaction

In relativistic quantum mechanics, as in classical field theory, the concept of an "action at a distance" has been replaced by the simpler one of a *local interaction.* We do not, therefore, attempt to construct relativistic potentials to be inserted into the free-particle equations, but modify the Lagrangian instead by adding to it nonlinear terms involving the field operators. A very useful illustration of how this is done comes from quantum electrodynamics. The validity of classical electromagnetic theory suggests that the quantum equations have the same form as the classical ones: this is consistent with the correspondence principle. We are therefore tempted to look for equations of motion for the field operators, which take the form

$$\Box\, A_\mu(x) = -ej_\mu(x) \tag{5.1}$$

where the inhomogeneous term is the current that acts as a source of the field. The current is conserved, and if we are looking at the interaction of the electromagnetic field with electrons alone, it is natural to try

$$j_\mu(x) = :\bar{\psi}(x)\,\gamma_\mu\,\psi(x): \tag{5.2}$$

There is unfortunately no correspondence principle to guide us in the construction of the equation for $\psi(x)$. If, however, we assume that the field equations come from a Lagrangian, the equation for the photon field already suggests the form of the Lagrangian density: it should be

$$\mathfrak{L} = \mathfrak{L}(A^\mu) + \mathfrak{L}(\psi, \bar{\psi}) + eA^\mu(x)\,j_\mu(x) \tag{5.3}$$

with

$$\mathfrak{L}(A^\mu) = -\tfrac{1}{2}(\partial_\mu A_\nu(x))(\partial^\mu A^\nu(x)) \tag{5.4}$$

and

$$\mathfrak{L}(\psi, \bar{\psi}) = \bar{\psi}(x)(i\gamma^\mu\,\partial_\mu - M)\,\psi(x) \tag{5.5}$$

The equation for $\psi(x)$ follows

$$(i\gamma^\mu\,\partial_\mu - M)\,\psi(x) + e\gamma^\mu A_\mu(x)\,\psi(x) = 0 \tag{5.6}$$

It should be noted that the Lagrangian now contains products of more than two operators at the same space-time point. Because the operators are very singular—we still assume something like canonical commutation relations—we might expect difficulties of a mathematical nature. We do, indeed, find such difficulties, which appear in the form of infinite integrals, and only an appeal to physical interpretation of ill-defined quantities will allow us to make sense of the theory.[1] Another point is that for fields in interaction, the connection between the fields and the particles is loosened. It could happen that in a given form of the interaction between fields of spin $\frac{1}{2}$ and spin 0, only stable particles of spin $\frac{3}{2}$ are actually produced, with the masses of all entities of spin $\frac{1}{2}$ lying well above the mass of the particle of spin $\frac{3}{2}$. This possibility is an asset of field theory; it would be undesirable to require that each particle be represented by a new field. Since we do not, at this time, know how to solve field equations except in perturbation theory (when the interaction necessarily has to be assumed to leave the spectrum the same as for the noninteracting system), this possibility has not been tested.

If we believe that the properties of particles are determined by the equation of motion for some field (or fields), it is obviously essential to find both the basic fields and the Lagrangian density for these basic fields. We have, at this time, no good idea of the nature of basic fields. The only statement we can make is that at least one of the basic fields must have half-odd integral spin, since among the real particles there are some with spin $\frac{1}{2}$. Furthermore, at least one of the fields must be electrically charged, at least one must have baryon number, but it is not settled at this point that the charge and baryon number carried by the basic fields could not be fractional. If, for example, the basic field carried charge $\frac{1}{3}$, all charged particles would have to be "bound states" of three of the basic entities. This is no place to speculate on this point. We do, however, have some guidelines for the construction of Lagrangians. It is useful to discuss these, since one is not merely interested in finding THE Lagrangian, but often sets up equations of motion for fields representing known real particles in order to study the theoretical implications of different physical assumptions. Furthermore, there are some relatively weak interactions in nature, whose effect on the real particles may be represented phenomenologically by a Lagrangian composed of fields which stand in a one-to-one correspondence with the particles. An example is the Lagrangian (5.3) describing the interaction of photons with electrons or muons.

In all models to be described by Lagrangians we shall insist on the form invariance of the equations of motion under Lorentz transformations.

[1] See Chapter 13.

Thus the Lagrangian density $\mathcal{L}$ must be a hermitian[1] *scalar* under Lorentz transformations. We shall require that

$$U(\Lambda)\,\mathcal{L}(\phi(x), A_\mu(x), \ldots)\,U^{-1}(\Lambda) = \mathcal{L}(\phi(\Lambda x), (\Lambda^{-1})^\nu{}_\mu\,A_\nu(\Lambda x), \ldots) \quad (5.7)$$

If the physical system to be described is to be invariant under reflections, i.e., if parity is conserved, the Lagrangian must have this property.[2] If the system is to be invariant under charge conjugation, i.e., under the replacement of each particle by its antiparticle, then this too must be built into the Lagrangian. It is a very deep consequence of Lorentz invariance alone that if a system is invariant under space inversions and charge conjunction it is automatically invariant under time reversal. More generally, any system described by a Lorentz invariant Lagrangian is invariant under the product of the three operations CPT. This theorem is discussed in detail in Chapter 30.

In order to impose conservation of some additive quantum number such as charge, baryon number, and the like, use can be made of the connection between the existence of a conserved "charge" and a form of gauge invariance. As discussed in Chapter 1, a Lagrangian composed of non-hermitian fields, invariant under the transformation

$$\chi(x) \to e^{-i\alpha}\,\chi(x) \quad (5.8)$$

$$\chi^+(x) \to e^{i\alpha}\,\chi^+(x) \quad (5.9)$$

[where $\chi(x)$ stands for a general field], will incorporate the conservation of a charge Q carried by the field $\chi(x)$ with the commutation relations

$$[Q, \chi(x)] = \chi(x) \quad (5.10)$$

Thus, if we have a theory involving nucleons and mesons, baryon conservation implies invariance under

$$\begin{aligned} \psi(x) &\to \psi(x)e^{-i\beta} \\ \bar{\psi}(x) &\to \bar{\psi}(x)e^{i\beta} \end{aligned} \quad (5.11)$$

and charge conservation implies

$$\begin{aligned} \psi_{\text{ch}}(x) &\to \psi_{\text{ch}}(x)e^{-i\alpha} \\ \bar{\psi}_{\text{ch}}(x) &\to \bar{\psi}_{\text{ch}}(x)e^{i\alpha} \\ \phi_{\text{ch}}(x) &\to \phi_{\text{ch}}(x)e^{-i\alpha} \\ \phi_{\text{ch}}{}^*(x) &\to \phi_{\text{ch}}{}^*(x)e^{i\alpha} \end{aligned} \quad (5.12)$$

[1] So that the Hamiltonian is hermitian.

[2] Sometimes it is harder to construct a Lagrangian that violates a symmetry. For example, the coupling for $\pi^0 \to 3\gamma$, which violates C (and P), is of the form

$$G\pi^0\,\partial^\alpha F_{\mu\nu}\,\partial_\rho F_{\alpha\beta}\,\partial^\nu\,\partial^\beta F^{\rho\mu}$$

This implies that $\bar{\psi}(x)$ and $\psi(x)$ must appear together in bilinear combinations, and $\phi_{\text{ch}}(x)$ must be multiplied by either $\phi_{\text{ch}}{}^*(x)$ or $\bar{\psi}_{\text{ch}}(x)\,\psi_{\text{neutral}}(x)$.

The interaction of charged fields with the electromagnetic field has a special element of simplicity. If we look back at (5.3) we notice that the Lagrangian may be written in the form

$$\mathcal{L} = \mathcal{L}(A^\mu) + \bar{\psi}(x)(i\gamma^\mu(\partial_\mu - ie\,A_\mu(x)) - M)\,\psi(x) \tag{5.13}$$

Similarly, the Lagrangian describing the interaction of a charged boson field (spin 0 for simplicity) will again be, using

$$j_\mu(x) = i\,\phi^*(x)\,\overleftrightarrow{\partial}_\mu\,\phi(x) \tag{5.14}$$

of the form

$$\mathcal{L} = \mathcal{L}(A^\mu) + (\partial_\mu - ie\,A_\mu(x))\,\phi(x)(\partial^\mu + ie\,A^\mu(x))\,\phi^*(x) - m^2\,\phi^*(x)\,\phi(x) \tag{5.15}$$

which differs from

$$\mathcal{L} = \mathcal{L}(A^\mu) + \mathcal{L}(\phi) + e\,A^\mu(x)\,j_\mu(x)$$

only by terms quadratic in A_μ. In both cases the interacting Lagrangian is obtained from the free one by changing the terms involving the gradient of the field in the particular way just shown. We shall now show that this form is intimately connected with the gauge invariance of the second kind, discussed in Chapter 3, in which it was pointed out that the vector potentials were undetermined to within a gradient of a scalar function.

Suppose that we require the Lagrangian to be invariant under x-dependent phase transformations, e.g., under

$$\begin{aligned} \phi(x) &\to \phi(x)\,e^{-i\alpha(x)} \\ \phi^*(x) &\to \phi^*(x)\,e^{i\alpha(x)} \end{aligned} \tag{5.16}$$

Clearly, the free Lagrangian does not possess this property, since, for example,

$$\partial_\mu\phi\,\partial^\mu\phi^*(x)$$

becomes

$$(\partial_\mu - i\,\partial_\mu\,\alpha(x))\,\phi(x)(\partial^\mu + i\,\partial^\mu\alpha(x))\,\phi^*(x)$$

If, however, the gradient terms were

$$(\partial_\mu - ie\,A_\mu(x))\,\phi(x)(\partial^\mu + ie\,A^\mu(x))\,\phi^*(x)$$

the requirement (5.16) could be met, provided at the same time A_μ changed according to

$$e\,A_\mu(x) \to e\,A_\mu(x) - \partial_\mu\,\alpha(x) \tag{5.17}$$

a change that leaves the electromagnetic field strengths unaltered.

The coupling to the electromagnetic field, obtained by replacing

$$\partial_\mu \chi \rightarrow (\partial_\mu - ie\, A_\mu)\chi \quad \text{and} \quad \partial_\mu \chi^+ \rightarrow (\partial_\mu + ie\, A_\mu)\chi^+$$

in the original Lagrangian, is called a *minimal coupling*. This is unfortunately not a unique prescription, for the free Lagrangian is undetermined to within a divergence of a four-vector; the equations of motion are unaltered when $\mathcal{L}$ is replaced by $\mathcal{L} + \partial_\mu M^\mu$. To illustrate this nonuniqueness, consider the free Lagrangian for a Dirac field, with the term $\mathcal{L}'$ added to it:

$$\begin{aligned} \mathcal{L}' &= \gamma\, \partial_\mu(\bar{\psi}\, \sigma^{\mu\nu}\, \partial_\nu \psi - \partial_\nu \bar{\psi} \sigma^{\mu\nu} \psi) \\ &= \gamma(\partial_\mu \bar{\psi} \sigma^{\mu\nu}\, \partial_\nu \psi - \partial_\nu \bar{\psi} \sigma^{\mu\nu}\, \partial_\mu \psi) \end{aligned} \tag{5.18}$$

(use has been made of the antisymmetry of $\sigma^{\mu\nu}$). The additional minimal coupling is

$$\begin{aligned} 2\gamma(\partial_\mu &+ ie\, A_\mu)\bar{\psi}\sigma^{\mu\nu}(\partial_\nu - ie\, A_\nu)\psi - 2\gamma\, \partial_\mu \bar{\psi}\sigma^{\mu\nu}\, \partial_\nu \psi \\ &= 2ie\gamma(A_\mu \bar{\psi}\sigma^{\mu\nu}\, \partial_\nu \psi - A_\nu\, \partial_\mu \bar{\psi}\sigma^{\mu\nu}\psi) \\ &= \partial_\nu(2ie\gamma A_\mu \bar{\psi}\sigma^{\mu\nu}\psi) + 2ie\gamma F_{\mu\nu}\bar{\psi}\sigma^{\mu\nu}\psi \end{aligned} \tag{5.19}$$

The first term will not affect the equations of motion, but the second term adds a coupling which would appear if the spin $\frac{1}{2}$ particle had a rigid magnetic dipole moment in excess of that normal for a Dirac particle. Such additional terms do not appear to be necessary in the treatment of the electrodynamics of spin $\frac{1}{2}$ particles, but they cannot be logically excluded, and should make one cautious in ascribing a deep significance to the one known case of a successful minimal coupling prescription.

We conclude this chapter by giving several illustrations of how interaction Lagrangians are written down with the help of the above guidelines. Consider first the coupling of a neutral pseudoscalar meson to the nucleon. By the use of analogy with electromagnetism or experimental observation (e.g., that mesons can be created singly), the interaction will be taken linear in the meson field $\phi(x)$. Since the meson is assumed to be pseudoscalar and parity is conserved (again by experimental check), the meson field must be multiplied by a bilinear combination of $\bar{\psi}(x)$ and $\psi(x)$ (to conserve baryon number), which transforms like a pseudoscalar. We are thus led to two possibilities

$$\begin{aligned} \mathcal{L}_1 &= ig\, \bar{\psi}(x)\gamma_5\, \psi(x)\, \phi(x) \\ \mathcal{L}_1' &= \frac{if}{m}\, \bar{\psi}(x)\gamma^\mu\gamma_5\, \psi(x)\, \partial_\mu\, \phi(x) \end{aligned}$$

The i is present to make $\mathcal{L}$ hermitian, a condition necessary to lead to hermitian Hamiltonian. The two interactions will lead to different predictions, and the formalism will also be different, for the definition of canonical momentum in the first case will be the same as that for the free meson field, whereas in the second case the canonical momentum will be the time derivative of the meson field together with a term that depends on the coupling.

Another example is the coupling of a vector meson to a nucleon. The coupling may be either to the current

$$f\,\bar{\psi}(x)\gamma^{\mu}\,\psi(x)\,\phi_{\mu}(x)$$

or a magnetic moment type of coupling

$$\gamma\,\bar{\psi}(x)\sigma^{\mu\nu}\,\psi(x)(\partial_{\mu}\,\phi_{\nu}(x) - \partial_{\nu}\,\phi_{\mu}(x))$$

Still another example involves the existence of internal symmetries. We shall learn that to a good approximation there is an internal symmetry, similar to rotational invariance, according to which the three-pseudoscalar pions transform like a three-vector (*i*-spin 1) and the nucleons like a two-component internal spinor (*i*-spin $\frac{1}{2}$). The pion-nucleon coupling, which is invariant under this group and linear in the pion field, must be of the form

$$ig\sum_{i=1}^{3}\sum_{A,B=1}^{2}\bar{\psi}_{A}(x)(\tau_{i})_{AB}\gamma_{5}\,\psi_{B}(x)\,\phi_{i}(x)$$

where the τ_i are 2×2 matrices with the same form as the Pauli spin matrices.

We shall again use these guidelines when we construct phenomenological Lagrangians for the strong interactions, invariant under the internal symmetry called $SU(3)$ (Chapter 18), and when we construct the weak interaction Lagrangians.

As a consequence of our inability to solve field equations when the coupling is strong, Lagrangians have not played an important role in recent approaches to elementary particle problems. They have been used more as a symmetry bookkeeping device, as a way of describing decay processes in a compact way, e.g.,

$$\mathcal{L}_1 = \frac{f}{m}\,\bar{\psi}_{e}(x)\gamma^{\alpha}\gamma_{5}\,\psi_{\nu}(x)\,\partial_{\alpha}\,\phi(x)$$

for $\pi \rightarrow e + \nu$ and as a way of counting the number of derivatives necessary in a description of a certain process. For example, a decay

$$\eta^{0} \rightarrow \pi^{+} + \pi^{-} + \pi^{0}$$

asymmetric in $\pi^{+}\pi^{-}$ must be described by

$$\mathcal{L} = \eta^{0}\,\partial_{\lambda}\pi^{0}(g\pi^{+}\,\partial^{\lambda}\pi^{-} + g^{*}\,\partial^{\lambda}\pi^{+}\pi^{-})$$

with g complex and two derivative factors, for the hermitian form without derivatives

$$f\eta^0\pi^0\pi^+\pi^-$$

is necessarily symmetric in $\pi^+\pi^-$.[1]

PROBLEMS

1. Write down two interaction Lagrangians for the coupling of a vector meson to a spin $\frac{1}{2}$ field.
2. Consider the most general nonderivative coupling of charged and neutral mesons to neutrons and protons

$$\begin{aligned}\mathcal{L}_I = {} & g_1\bar{P}P\,\phi_0 + ig_2\bar{P}\gamma_5 P\,\phi_0 + g_3\,\bar{N}N\,\phi_0 + ig_4\bar{N}\gamma_5 N\,\phi_0 \\ & + g_5\bar{P}N\,\phi + g_5{}^*\bar{N}P\,\phi^* + ig_6\bar{P}\gamma_5 N\,\phi + ig_6{}^*\bar{N}\gamma_5 P\,\phi^*\end{aligned}$$

Show that the requirement that the coupling is invariant under CP (but not under C and P separately), together with the requirement that it be invariant under i-spin rotations, i.e., under

$$\begin{pmatrix} P \\ N \end{pmatrix} \to (1 + i\boldsymbol{\tau}\cdot\mathbf{u})\begin{pmatrix} P \\ N \end{pmatrix} \qquad \boldsymbol{\phi} \to \boldsymbol{\phi} - 2\mathbf{u}\times\boldsymbol{\phi}$$

$$\boldsymbol{\phi} = \left(\frac{\phi + \phi^*}{\sqrt{2}}, \frac{\phi - \phi^*}{\sqrt{2}i}, \phi_0\right)$$

$$\boldsymbol{\tau} = \left(\begin{pmatrix} 0 & 1 \\ 1 & 0 \end{pmatrix}, \begin{pmatrix} 0 & -i \\ 1 & 0 \end{pmatrix}, \begin{pmatrix} 1 & 0 \\ 0 & -1 \end{pmatrix}\right)$$

implies that the coupling is, in fact, invariant under C and P separately.
3. Consider the Lagrangian $\mathcal{L}(\psi_i(x))$. Show that corresponding to any local gauge transformation

$$\psi_i(x) \to \psi_i(x) + \Lambda(x)\,F_i(\psi_1(x), \psi_2(x), \ldots)$$

a current can be defined, whose divergence is given by $\delta\mathcal{L}/\delta\Lambda$.
4. Consider the Lagrangian

$$\begin{aligned}\mathcal{L}(x) = {} & +\bar{\psi}(x)[i\gamma^\mu\,\partial_\mu + g_0(\sigma(x) + i\gamma_5\boldsymbol{\tau}\cdot\boldsymbol{\phi}(x))]\,\psi(x) \\ & + \tfrac{1}{2}(\partial_\mu\boldsymbol{\phi}(x))(\partial^\mu\boldsymbol{\phi}(x)) + \tfrac{1}{2}(\partial_\mu\,\sigma(x))(\partial^\mu\,\sigma(x)) - \tfrac{1}{2}\mu_0{}^2(\boldsymbol{\phi}^2(x) + \sigma^2(x)) \\ & - \lambda\left(\boldsymbol{\phi}^2(x) + \sigma^2(x) - \frac{1}{4f_0{}^2}\right)^2 - \frac{\mu_0{}^2}{2f_0}\,\sigma(x)\end{aligned}$$

Construct the current corresponding to the gauge transformation

$$\begin{aligned}\psi(x) &\to (1 + i\boldsymbol{\tau}\cdot\mathbf{w}(1 + \gamma_5))\,\psi(x) \\ \boldsymbol{\phi}(x) &\to \boldsymbol{\phi}(x) - 2\mathbf{w}\,\sigma(x) - 2\mathbf{w}\times\boldsymbol{\phi}(x) \\ \sigma(x) &\to \sigma(x) + 2\mathbf{w}\cdot\boldsymbol{\phi}(x)\end{aligned}$$

What is the divergence of this current?

[1] This question arises in recent investigations of a possibility of C and T violations in certain processes.

6

The Scattering Matrix

In the last chapter guidelines for the construction of interaction Lagrangians were laid down. Whatever the Lagrangian is, one ultimately arrives at a set of equations of motion which for a meson interacting with a nucleon say, might have the form

$$
\begin{aligned}
(\Box + \mu^2)\, \phi(x) &= j(x) \\
(-i\gamma^\mu\, \partial_\mu + M)\, \psi(x) &= f(x) \\
i\, \partial_\mu\, \bar{\psi}(x)\gamma^\mu + M\, \bar{\psi}(x) &= \bar{f}(x)
\end{aligned}
\tag{6.1}
$$

where $j(x)$, $f(x)$, and $\bar{f}(x)$ are polynomials in the field operators. These are supplemented by commutation relations which might be the canonical

$$
\begin{aligned}
[\pi(x), \phi(y)]_{x_0=y_0} &= -i\, \delta(\mathbf{x} - \mathbf{y}) \\
\{\psi(x), \psi^+(y)\}_{x_0=y_0} &= \delta(\mathbf{x} - \mathbf{y})
\end{aligned}
\tag{6.2}
$$

but which is any case must satisfy the microscopic causality conditions that

$$
\begin{aligned}
&[\phi(x), \phi(y)] = 0 \\
&\{\psi(x), \bar{\psi}(y)\} = 0 \qquad \text{for} \quad (x - y)^2 < 0
\end{aligned}
\tag{6.3}
$$

The solution of these equations of motion can, at best, be obtained in the form of a perturbation series, and even this approach has its mathematical difficulties which arise from the lack of a proper definition of the meaning of products of operators which appear in the source terms $j(x)$ and $f(x)$. Aside from these problems, there is still the question of how we would use the solutions of the equations to make physical predictions.

Although, in principle, all hermitian operators are measurable—subject to the restriction arising from superselection rules—in practice all measurements in elementary particle physics are made by scattering experiments. What we are interested in, therefore, is the calculation of the outcome of a collision process in which a finite number of particles converge to a finite space-time region and then emerge from it with different momenta, energies, and spin states. The process is often accompanied by the disappearance of some of the original particles and the creation of some new ones. Intensity limitations restrict present-day experiments to those with two-particle initial states, but we are really interested in the more general scattering problem.

The quantity of interest, the scattering matrix element $S_{\beta\alpha}$, is the probability amplitude that a certain initial configuration of particles in a state α, say, ends up as a different configuration of particles in a state β. The collection of all the matrix elements $S_{\beta\alpha}$ yields the *S-matrix*. The description of initial and final configurations is made easy by the fact that interactions are generally short-range. Thus both before the collision and after it the configuration consists of free particles. Even in processes like

$$p + p \to \pi^+ + d$$

say, this assertion can be made, provided that we treat the deuteron as a particle in this context, and not as a bound neutron-proton system. The scattering matrix element is thus given by

$$S_{\beta\alpha} = (\Psi_\beta(t \to +\infty), \Psi_\alpha(t \to -\infty)) \tag{6.4}$$

according to the general rules of quantum mechanics. Because we are working in the Heisenberg representation in which the state vectors are time-independent, the meaning of the labeling in (6.4) needs some explanation. The point is that a state vector Ψ for a given physical system contains all the available information about the system—it describes, so to speak, the whole history of the system. The state of the system may be unambiguously determined by measuring a complete set of commuting observables for the state. This determination can be made at any time, and we may, therefore, label the state by the results of this "experiment" at a given time t, and then the state, i.e., both its past and its future, are completely determined. Since we know, physically, that as $t \to \pm\infty$ the state consists of noninteracting particles, the construction of the state vectors $\Psi_\alpha(t \to -\infty)$ and $\Psi_\beta(t \to +\infty)$ is particularly easy. Thus $\Psi_\alpha(t \to -\infty)$, which we shall call Ψ_α^{in} from now on, is constructed by applying free-particle creation operators to the vacuum state. It is assumed that the set of "in" states

$$\begin{gathered}
\Psi_0 \\
a_{\text{in}}^+(p)\,\Psi_0 \\
a_{\text{in}}^+(p_1)\,a_{\text{in}}^+(p_2)\,\Psi_0 \\
\cdot \\
\cdot \\
\cdot
\end{gathered} \tag{6.5}$$

is a complete set even for interacting particles. The operators a_{in}^+ are free-field creation operators which satisfy the free-field commutation relations

$$[a_{\text{in}}(p_1), a_{\text{in}}^+(p_2)] = 2\omega_{p_1}\,\delta(\mathbf{p}_1 - \mathbf{p}_2)$$

etc., for bosons, and

$$\{\mathbf{a}_{\text{in}}^{(r)}(p_1), \mathbf{a}_{\text{in}}^{(s)+}(p_2)\} = \frac{E_{p_1}}{M}\,\delta_{rs}\,\delta(\mathbf{p}_1 - \mathbf{p}_2)$$

etc., for fermions, and the labeling "in" on the operators is to remind us that we are constructing the complete set of states described by what their configuration was in the remote past. It should be noticed that there is no "in" label on the vacuum state Ψ_0. This is because the vacuum is a stationary state in that a measurement on it will yield the same result at all times—a vacuum always remains a vacuum. The same is true of the one-particle states Ψ_{p_1}: a state consisting of a single particle of a given momentum and spin state cannot change since it has nothing to interact with to change its state of motion. The same is *not* true of a two-particle state: the two particles may collide and their direction of motion may be changed or some of their kinetic energy may be used up to create new particles.

The complete set of states Ψ_β^{out} (the label "out" replaces $(t \to +\infty)$) may similarly be constructed with the help of free-particle operators. It is

$$\begin{array}{c} \Psi_0 \\ a_{\text{out}}^+(p_1)\Psi_0 \\ a_{\text{out}}^+(p_1)\,a_{\text{out}}^+(p_2)\Psi_0 \\ \cdot \\ \cdot \\ \cdot \end{array} \tag{6.6}$$

and again the free-particle operators satisfy the commutation relations

$$[a_{\text{out}}(p_1), a_{\text{out}}^+(p_2)] = 2\omega_{p_1}\,\delta(\mathbf{p}_1 - \mathbf{p}_2)$$

etc., or

$$\{\mathbf{a}_{\text{out}}^{(r)}(p_1), \mathbf{a}_{\text{out}}^{(s)+}(p_2)\} = \frac{E_{p_1}}{M}\,\delta_{rs}\,\delta(\mathbf{p}_1 - \mathbf{p}_2)$$

Since Ψ^{out} and Ψ^{in} form a complete set each, they must be related by a *unitary transformation*:

$$\Psi_\alpha^{\text{in}} = \sum_\beta S_{\beta\alpha}\Psi_\beta^{\text{out}} \tag{6.7}$$

Hence

$$(\Psi_\beta^{\text{out}}, \Psi_\alpha^{\text{in}}) = S_{\beta\alpha} \tag{6.8}$$

The operator S whose matrix elements are $S_{\beta\alpha}$ may equivalently be defined by

$$\begin{aligned}\mathbf{a}_{\text{out}}(p) &= S^+ \, \mathbf{a}_{\text{in}}(p) S \\ \mathbf{a}^+_{\text{out}}(p) &= S^+ \, \mathbf{a}^+_{\text{in}}(p) S\end{aligned} \tag{6.9}$$

It follows from (6.9) that

$$\begin{aligned}(\Psi^{\text{out}}_{p_1 \cdots p_n}, \Psi^{\text{in}}_{q_1 \cdots q_m}) &= (\Psi_0, a_{\text{out}}(p_n) \cdots a_{\text{out}}(p_1) \Psi^{\text{in}}_{q_1 \cdots q_m}) \\ &= (\Psi_0, S^+ a_{\text{in}}(p_n) S S^+ \cdots a_{\text{in}}(p_n) S \Psi^{\text{in}}_{q_1 \cdots q_m}) \\ &= (\Psi^{\text{in}}_{p_1 \cdots p_n}, S \, \Psi^{\text{in}}_{q_1 \cdots q_m})\end{aligned} \tag{6.10}$$

Here use has been made of the unitarity of S

$$S^+S = SS^+ = 1 \tag{6.11}$$

and of the fact that

$$S\Psi_0 = S^+\Psi_0 = \Psi_0 \tag{6.12}$$

On cursory reading the development of the last few pages always leads to the question: how can one ever learn anything about the dynamics of interacting particles from free fields alone? The answer to this question is that the "in" and "out" states serve only as a descriptive framework, particularly constructed to fit the experimental initial and final conditions in a collision experiment. The "in" and "out" fields are like the asymptotes to a trajectory in a classical scattering problem: to find which initial "asymptotic" state will lead to which final "asymptotic" state, we must know the potential and solve the equation of motion. Here, too, the descriptive framework must be fitted around a dynamical theory.

In order to express the scattering matrix elements in terms of the field operators (for which we have equations of motion), we must obtain a relation between the "in" and "out" fields $\phi_{\text{in}}(x)$, $\psi_{\text{in}}(x)$, . . .—constructed as in Chapters 1 through 3 with the help of the creation and annihilation operators—and the fields $\phi(x)$, $\psi(x)$, ... Our discussion of the definition of the scattering matrix leaned heavily on the notion that a physical system behaves, for large times as a system of free particles. These asymptotic properties must somehow be reflected in the properties of the field operators themselves. Let us first examine this problem in a very formal way. If we consider the equation

$$(\Box + \mu^2)\, \phi(x) = j(x)$$

then, as is easily checked, a formal solution may be written in the form

$$\phi(x) = -\int_{x_0'=T} d^3x' \left[\Delta(x - x', \mu) \frac{\partial}{\partial x_0'} \phi(x') + \frac{\partial}{\partial x_0} \Delta(x - x', \mu)\, \phi(x') \right] - \int_T^{x_0} dx' \, \Delta(x - x', \mu)\, j(x') \tag{6.14}$$

The first term satisfies the homogeneous equation; it represents a solution, in the absence of the source term $j(x)$, satisfying certain initial conditions at time T. The second term contains the effect of the interaction with $j(x)$. If we let $T \to -\infty$ and define

$$\begin{aligned} \Delta_{\text{ret}}(x, \mu) &= -\Delta(x, \mu) \qquad x_0 > 0 \\ &= 0 \qquad\qquad\quad\;\; x_0 < 0 \\ &\equiv -\theta(x_0)\, \Delta(x, \mu) \end{aligned} \tag{6.15}$$

we may write (6.14) in the form[1]

$$\phi(x) = \phi_{\text{in}}(x) + \int dx'\, \Delta_{\text{ret}}(x - x', \mu)\, j(x') \tag{6.16}$$

where $\phi_{\text{in}}(x)$ satisfies the free-field equation. We shall not prove that $\phi_{\text{in}}(x)$ satisfies the free-field commutation relations[2] since proof requires a sophistication far beyond what is taken for granted at this level. We shall however accept that

$$[\phi_{\text{in}}(x), \phi_{\text{in}}(y)] = i\, \Delta(x - y, \mu) \tag{6.17}$$

If we let $T \to +\infty$ in (6.14) and define

$$\begin{aligned} \Delta_{\text{adv}}(x, \mu) &= 0 \qquad\qquad\; x_0 > 0 \\ &= \Delta(x, \mu) \qquad x_0 < 0 \\ &\equiv \theta(-x_0)\, \Delta(x, \mu) \end{aligned} \tag{6.18}$$

we may write similarly

$$\phi(x) = \phi_{\text{out}}(x) + \int dx'\, \Delta_{\text{adv}}(x - x', \mu)\, j(x') \tag{6.19}$$

We see that in some sense the field operators approach the free field operators in the distant past and in the remote future, provided that we concentrate on the behavior of $\Delta_{\text{ret}}(x - x')$, $\Delta_{\text{adv}}(x - x')$ and not worry about anything else. Actually this procedure is quite unjustified. First of, all, as was stressed in Chapter 1, the fields themselves are idealizations and only their averages, with some weight function, over some space-time region can have any significance. Thus a statement of asymptotic behavior must be about $\int_R \phi(x)\, g(x)\, dx$ and $\int_R \phi_{\text{in}}(x)\, g(x)\, dx$ or, more conveniently,

[1] The "in" and "out" operators were first introduced by G. Källén, *Arkiv. Fysik* **2,** 187, 371 (1950) and by C. N. Yang and D. Feldman, *Phys. Rev.* **79,** 972 (1950).

[2] Details can be found in the paper by W. Zimmermann, *Nuovo. Cimento,* **10,** 597 (1958).

about the quantities

$$
\begin{aligned}
A(q, t) &= i\int d^3x\, f_q{}^*(x)\frac{\overleftrightarrow{\partial}}{\partial x_0}\,\phi(x) \\
A_{\text{in}}(q, t) &= i\int d^3x\, f_q{}^*(x)\frac{\overleftrightarrow{\partial}}{\partial x_0}\,\phi_{\text{in}}(x) \\
&\cdots
\end{aligned}
\tag{6.20}
$$

Here the $f_q(x)$ are wave-packet solutions of the Klein-Gordon equation with positive frequencies. The operators A_{in} are free-field annihilation operators, and the operators $A(q, t)$ are analogously defined quantities for the interacting fields. The corresponding equations for spinor fields are

$$
\begin{aligned}
\psi(x) &= \psi_{\text{in}}(x) - \int dx' S_{\text{ret}}(x - x', M)\, f(x') \\
\psi(x) &= \psi_{\text{out}}(x) - \int dx' S_{\text{adv}}(x - x', M)\, f(x')
\end{aligned}
\tag{6.21}
$$

and the asymptotic statements will be made about

$$
\begin{aligned}
\mathbf{a}(p, t) &= \int d^3x\, \bar{u}_p(x)\gamma_0\, \psi(x) \\
\mathbf{a}_{\text{in}}(p, t) &= \int d^3x\, \bar{u}_p(x)\gamma_0\, \psi_{\text{in}}(x) \\
&\cdots
\end{aligned}
\tag{6.22}
$$

where $u_p(x)$ are positive frequency wave-packet solutions of the Dirac equation.

Secondly, the meaning of "approach in the limit as $t \to \pm\infty$ has to be clarified. When we are talking about operators in an infinite-dimensional vector space, a statement like "O_1 approaches O_2 in some limit" can be made in two different ways, which have quite different consequences.[1] We may require *strong convergence*: this implies that in the limit, the difference in the "length" of the state vectors $O_1\Psi$ and $O_2\Psi$ (for all Ψ) approaches zero. In other words, with

$$\Phi = O_1\Psi - O_2\Psi$$

we have

$$(\Phi, \Phi) \to 0$$

in the limit. We may, on the other hand, require *weak convergence*: this means that we merely require that in the limit

$$(\Psi', (0_1 - 0_2)\Psi) \to 0 \tag{6.23}$$

[1] The relevance of the two kinds of "convergence" to field theory was first pointed out by R. Haag (unpublished Copenhagen Lectures, 1953).

for any pair of vectors Ψ and Ψ'. Weak convergence does not necessarily imply that $(\Phi, \Phi) \to 0$. There is no difference in the two types of convergence for finite-dimensional vector spaces. That is why these concepts are perhaps unfamiliar.

It was first pointed out by Lehmann, Symanzik, and Zimmermann[1] that the correct statement of the asymptotic properties of the fields is the weak convergence statement

$$\begin{aligned} \lim_{t\to\pm\infty} (\Phi, A(p, t)\Psi) &= (\Phi, A_{\substack{\text{out}\\\text{in}}}(p)\Psi) \\ \lim_{t\to\pm\infty} (\Phi, A^{+}(p, t)\Psi) &= (\Phi, A^{+}_{\substack{\text{out}\\\text{in}}}(p)\Psi) \\ \lim_{t\to\pm\infty} (\Phi, \mathbf{a}(p, t)\Psi) &= (\Phi, \mathbf{a}_{\substack{\text{out}\\\text{in}}}(p)\Psi) \\ &\cdots \end{aligned} \tag{6.24}$$

where Φ and Ψ stand for any pair of normalizable states in the vector space of physical states. It goes without saying that we shall allow ourselves to use plane wave states with the mental reservation that these represent very large wavepackets.

One of the limitations of weak convergence is that we cannot infer that the commutation relations of the interacting fields are those of the free fields, since

$$\lim_{t\to-\infty} (\Phi, [A(p, t), A^{+}(p', t)]\Psi) \neq (\Phi, [A_{\text{in}}(p), A_{\text{in}}^{+}(p')]\Psi)$$

If we had strong convergence, the commutation relations would be the same. The fact that they are not implies that there is no transformation of the type

$$A(p, t) = U^{+}(t)\, A_{\text{in}}(p)\, U(t)$$

that is unitary.

To illustrate the use of the asymptotic condition, as here postulated,[2] in relating matrix elements of S to field operators, let us consider the scattering amplitude from an initial two-particle state to an arbitrary final state α. We write

$$\begin{aligned} S_{\alpha;pq} &= (\Psi_{\alpha}^{\text{out}}, \Psi_{pq}^{\text{in}}) \\ &= (\Psi_{\alpha}^{\text{out}}, A_{\text{in}}^{+}(p)\Psi_{q}) \\ &= \lim_{t\to-\infty} (\Psi_{\alpha}^{\text{out}}, A^{+}(p, t)\Psi_{q}) \\ &= \lim_{t\to-\infty} \left(\Psi_{\alpha}^{\text{out}}, i\int_{x_0=t} d^3x\, \phi^{*}(x)\, \overleftrightarrow{\partial^0} f_p(x)\Psi_{q}\right) \end{aligned} \tag{6.25}$$

[1] H. Lehmann, K. Symanzik, and W. Zimmermann, *Nuovo Cimento*, **1**, 425 (1955). This paper will be referred to as LSZ.

[2] These asymptotic conditions can actually be proved rigorously from general properties of field theory. K. Hepp, Comm. Math. Phys. **1** (1965).

Now we write

$$\int_{x_0=-\infty} d^3x[\ ] = \int_{x_0=+\infty} d^3x[\ ] - \int dx \frac{\partial}{\partial x_0}[\ ] \tag{6.26}$$

The first term yields

$$\begin{aligned}\lim_{t\to+\infty}\left(\Psi_\alpha^{\text{out}}, i\int_{x_0=t} d^3x\, \phi^*(x)\,\overleftrightarrow{\partial^0} f_p(x)\Psi_q\right) &\\ = \lim_{t\to+\infty}(\Psi_\alpha^{\text{out}}, A^+(p,t)\Psi_q) &\qquad (6.27)\\ = (\Psi_\alpha^{\text{out}}, A_{\text{out}}^+(p)\Psi_q) = (\Psi_\alpha^{\text{out}}, \Psi_{pq}^{\text{out}}) &\end{aligned}$$

The last step follows since $\Psi_q^{\text{in}} = \Psi_q = \Psi_q^{\text{out}}$. We may thus write

$$\begin{aligned}(S-1)_{\alpha;pq} &= -i\int dx \frac{\partial}{\partial x_0}(\Psi_\alpha^{\text{out}}, \phi^*(x)\Psi_q)\overleftrightarrow{\frac{\partial}{\partial x_0}} f_p(x)\\ &= -i\int dx\left[(\Psi_\alpha^{\text{out}}, \phi^*(x)\Psi_q)\frac{\partial^2}{\partial x_0^2} f_p(x)\right.\\ &\qquad\left. - \left(\Psi_\alpha^{\text{out}}, \frac{\partial^2}{\partial x_0^2}\phi^*(x)\Psi_q\right) f_p(x)\right]\end{aligned}$$

At this point we make use of the fact that $f_p(x)$ is a solution of the Klein-Gordon equation, so that

$$\frac{\partial^2}{\partial x_0^2} f_p(x) = (\nabla^2 - \mu^2) f_p(x)$$

Also we are allowed to integrate by parts, since $f_p(x)$ is a wavepacket solution, which will only be replaced by $e^{-ipx}/(2\pi)^{3/2}$ at the end. Thus we obtain

$$(S-1)_{\alpha;pq} = i\int dx (\Psi_\alpha^{\text{out}}, j^*(x)\Psi_q) f_p(x) \tag{6.28}$$

where

$$(\Box + \mu^2)\,\phi^*(x) = j^*(x) \tag{6.29}$$

Thus, if we knew the solution of the equation of motion, we could compute the scattering matrix elements. We list the corresponding formula for the spinor case: if we write

$$(\Psi_\alpha^{\text{out}}, \Psi_{pq}^{\text{in}}) = (\Psi_\alpha^{\text{out}}, a_{\text{in}}^+(p)\Psi_q)$$

where

$$a_{\text{in}}^+(p) = \int d^3x\, \bar{\psi}(x)\gamma^0\, u_p(x)$$

and $u_p(x)$ is a wave-packet solution of the Dirac equation, which will be replaced at the end by $u(p)e^{-ipx}/(2\pi)^{3/2}$, then

$$(S-1)_{\alpha;pq} = i\int dx(\Psi_\alpha^{\text{out}}, \bar{f}(x)\Psi_q)\, u_p(x) \tag{6.30}$$

where

$$\bar{f}(x) = i\,\partial_\mu\,\bar{\psi}(x)\gamma^\mu + M\,\bar{\psi}(x) \equiv \bar{\psi}(x)\,\overleftarrow{\mathfrak{D}}_x \tag{6.31}$$

Other more useful expressions relating the S-matrix elements to matrix elements of products of field operators are derived in Chapter 7.

7

Reduction Formulas

In this chapter we explore in more detail the consequences of the asymptotic conditions as formulated by Lehmann, Symanzik, and Zimmermann (LSZ). We shall continue with the procedure used in deriving (6.28) and (6.30) and obtain reduction formulas for elements of the S operator in terms of certain field quantities. The case of meson-nucleon scattering will be worked out in great detail, and several equivalent forms will be obtained. These different forms will be used later in the book in the discussion of the unitarity of the S matrix.

We consider the amplitude for the scattering of a pion (momentum q, charge state α) by a nucleon (momentum p, spin state s, charge state A) to give a pion (momentum q', charge state β) and a nucleon (momentum p', spin state s', charge state B). The extension to bosons with more complicated spin structure will be obvious. We begin with the expression obtained in Chapter 6 (6.28) for the matrix element of

$$R = S - 1 \tag{7.1}$$

It is[1]

$$\langle p'q' |R| pq\rangle = i\int dx(\Psi^{\text{out}}_{p'q'}, \phi_\alpha(x)\Psi_p)\overleftarrow{K}_x f_q(x) \tag{7.2}$$

or alternately[2]

$$\langle p'q' |R| pq\rangle = i\int dx(\Psi^{\text{out}}_{p'q'}, \bar{\psi}_A(x)\Psi_q)\,\overleftarrow{\mathfrak{D}}_x\, u_p^{(s)}(x) \tag{7.3}$$

The next step is to "contract" another one of the field operators, and there are a number of possibilities available. We may use any one of the

[1] We treat the ϕ_α as hermitian fields ($\alpha = 1, 2, 3$).

[2] Again, we recall $\overrightarrow{\mathfrak{D}}\psi = (-i\gamma^\mu\,\partial_\mu + M)\psi$ and $\bar{\psi}\overleftarrow{\mathfrak{D}} = M\bar{\psi} + i\,\partial_\mu\bar{\psi}\gamma^\mu$.

following expressions:

$$\langle p'q'\,|R|\,pq\rangle = i\int dx(\Psi^{\text{out}}_{p'q'}, \phi_\alpha(x)a_{\text{in}}^{+}(p)\Psi_0)\overleftarrow{K}_x f_q(x) \tag{7.4a}$$

$$= i\int dx(\Psi_{p'}, A_\beta^{\text{out}}(q')\,\phi_\alpha(x)\Psi_p)\overleftarrow{K}_x f_q(x) \tag{7.4b}$$

$$= i\int dx(\Psi_{q'}, a_{\text{out}}(p')\,\phi_\alpha(x)\Psi_p)\overleftarrow{K}_x f_q(x) \tag{7.4c}$$

$$= i\int dx(\Psi^{\text{out}}_{p'q'}, \bar{\psi}_A(x)\,A_\alpha^{\text{in}+}(q)\Psi_0)\overleftarrow{\mathfrak{D}}_x\, u_p^{(s)}(x) \tag{7.4d}$$

$$= i\int dx(\Psi_{p'}, A_\beta^{\text{out}}(q')\,\bar{\psi}_A(x)\Psi_q)\overleftarrow{\mathfrak{D}}_x\, u_p^{(s)}(x) \tag{7.4e}$$

$$= i\int dx(\Psi_{q'}, a_{\text{out}}(p')\,\bar{\psi}_A(x)\Psi_q)\overleftarrow{\mathfrak{D}}_x\, u_p^{(s)}(x) \tag{7.4f}$$

In each case we may replace the "in" or "out" operators by the limits

$$0^{\text{in}} = \lim_{t\to-\infty} 0(t) \qquad 0^{\text{out}} = \lim_{t\to+\infty} 0(t) \tag{7.5}$$

We are allowed to do this only if we consider matrix elements of these operators connecting normalizable states. The kind of states that occur are states like

$$\Psi' = \int dx\, f_q{}^*(x)\overrightarrow{K}_x\, \phi_\alpha(x)\Psi^{\text{out}}_{p'q'}$$

and

$$\Phi' = \int dx\, f_q(x)\overrightarrow{K}_x\, \phi_\alpha(x)\Psi_p$$

Since it is assumed that the field operators—here properly smoothed with a wavepacket solution—do give physical (normalizable) states when acting on any state in the vector space, the LSZ asymptotic condition may be used.

Let us first work with the (7.4*b*). We have

$$(\Psi_{p'}, A_\beta^{\text{out}}(q')\,\phi_\alpha(x)\Psi_p) = \lim_{t\to+\infty} (\Psi_{p'}, A_\beta(q't)\,\phi_\alpha(x)\Psi_p)$$

$$= \lim_{t\to\infty} i\int d^3y\, f_{q'}{}^*(y)\,\overleftrightarrow{\partial}_{y_0}(\Psi_{p'}, \phi_\beta(y)\,\phi_\alpha(x)\Psi_p) \tag{7.6}$$

We canot usefully apply the identity

$$\int_{t=+\infty} d^3y[\;] = \int_{t=-\infty} d^3y[\;] + \int dy\, \frac{\partial}{\partial y_0}[\;]$$

at this point, because the first term on the right-hand side would not lead to anything easily evaluated. Two harmless modifications of (7.6) are possible at this stage, one leading to expressions in terms of the so-called R-products and the other in terms of the so-called T-products. Let us consider the T-form first. We rewrite

$$\begin{aligned}\lim_{y_0\to+\infty} \phi_\beta(y)\,\phi_\alpha(x) &= \lim_{y_0\to+\infty} (\theta(y_0 - x_0)\,\phi_\beta(y)\,\phi_\alpha(x) \\ &\qquad\qquad + \theta(x_0 - y_0)\,\phi_\alpha(x)\,\phi_\beta(y)) \\ &\equiv \lim_{y_0\to+\infty} T(\phi_\beta(y)\,\phi_\alpha(x))\end{aligned} \tag{7.7}$$

Here the step functions ensure that the expression is unaltered. In general, a *T-product* of operators is defined as an instruction: write the operators in temporal order, with the operator parametrized by the latest time index on the left. For fermion operators, in order to arrive at the proper temporal ordering, a number of permutations of operators are necessary. The instruction is to keep count of the number of permutations and insert a minus sign if the number is odd. It should be noted that no account of the noncommutativity of operators is to be taken into consideration, except for the $\pm$ sign for the fermions. We have the formulas

$$T(A_1(x_1)\cdots A_n(x_n)) = \pm A_{i_1}(x_{i_1})\cdots A_{i_n}(x_{i_n}) \qquad t_{i_1} > t_{i_2} > \cdots > t_{i_n} \tag{7.8}$$

and, as a special case, for fermion operators

$$T(\psi(x)\,\bar\psi(y)) = \theta(x_0 - y_0)\,\psi(x)\,\bar\psi(y) - \theta(y_0 - x_0)\,\bar\psi(y)\,\psi(x) \tag{7.9}$$

which differs in sign from (7.7). We now have

$$\begin{aligned}(\Psi_{p'}, A_\beta^{\text{out}}(q')\,\phi_\alpha(x)\Psi_p) &= \lim_{y_0\to+\infty} i\int d^3y\, f_{q'}{}^*(y)\,\overleftrightarrow{\partial}_{y_0}(\Psi_{p'}, T(\phi_\beta(y)\,\phi_\alpha(x))\Psi_p) \\ &\quad + i\int dy\,\partial_{y_0}\Big[f_{q'}{}^*(y)\,\overleftrightarrow{\partial}_{y_0}(\Psi_{p'}, T(\phi_\beta(y)\,\phi_\alpha(x))\Psi_p)\Big]\end{aligned}$$

The first term on the right-hand side can be worked backwards to yield

$$(\Psi_{p'}, \phi_\alpha(x)\,A_\beta{}^{\text{in}}(q')\Psi_p)$$

Since the state Ψ_p does not contain anything except a nucleon,

$$A_\beta{}^{\text{in}}(q')\Psi_p = 0,$$

the term vanishes, and we are left with

$$i\int dy\, \partial_{y_0}\left[f_{q'}^*(y)\overleftrightarrow{\partial}_{y_0}(\Psi_{p'}, T(\phi_\beta(y)\,\phi_\alpha(x))\Psi_p)\right]$$

$$= i\int dy\left\{ f_{q'}^*(y)\,\partial_{y_0}{}^2(\Psi_{p'}, T(\cdots)\Psi_p) - \partial_{y_0}{}^2 f_{q'}^*(y)(\Psi_{p'}, T(\cdots)\Psi_p)\right\}$$

$$= i\int dy\, f_{q'}^*(y)\overrightarrow{K}_y(\Psi_{p'}, T(\phi_\beta(y)\,\phi_\alpha(x))\Psi_p) \tag{7.10}$$

In obtaining the last line we used the fact that $f_{q'}^*(y)$ is a solution of the Klein-Gordon equation with appropriate behavior for large $|\mathbf{y}|$ so that integration by parts is justified. We obtain the result

$$\langle p'q'\, |R|\, pq\rangle = i^2\int dx\int dy\, f_{q'}^*(y)\overrightarrow{K}_y(\Psi_{p'}, T(\phi_\beta(y)\,\phi_\alpha(x))\Psi_p)\overleftarrow{K}_x f_q(x) \tag{7.11}$$

Another harmless modification which leads to an alternate but equivalent expression for the scattering amplitude is to first replace $(\Psi_{p'}, A_\beta^{\text{out}}(q')\,\phi_\alpha(x)\Psi_p)$ by $(\Psi_{p'}, [A_\beta^{\text{out}}(q'), \phi_\alpha(x)]\,\Psi_p)$. These are equivalent, since

$$A_\beta^{\text{out}}(q')\,\Psi_p = 0 \tag{7.12}$$

Therefore we can write

$$(\Psi_{p'}, A_\beta^{\text{out}}(q')\,\phi_\alpha(x)\Psi_p)$$

$$= \lim_{y_0\to+\infty} i\int d^3y\, f_{q'}^*(y)\overleftrightarrow{\partial}_{y_0}(\Psi_{p'}, [\phi_\beta(y), \phi_\alpha(x)]\,\Psi_p) \tag{7.13}$$

$$= \lim_{y_0\to+\infty} i\int d^3y\, f_{q'}^*(y)\,\overleftrightarrow{\partial}_{y_0}(\Psi_{p'}, \theta(y_0 - x_0)[\phi_\beta(y), \phi_\alpha(x)]\Psi_p)$$

Here again the step-function could be inserted with impunity, for in the region of interest, when $y_0 \to \infty$, it is always equal to 1. If we write this in terms of a contribution at $-\infty$ and the integral, we find that the contribution at $y_0 = -\infty$ vanishes because of the step function. We are then left with

$$i\int dy\, \partial_{y_0}\left[f_{q'}^*(y)\,\overleftrightarrow{\partial}_{y_0}(\Psi_{p'}, \theta(y_0 - x_0)[\phi_\beta(y), \phi_\alpha(x)]\Psi_p)\right.$$

$$= i\int dy\, f_{q'}^*(y)\overrightarrow{K}_y(\Psi_{p'}, \theta(y_0 - x_0)[\phi_\beta(y), \phi_\alpha(x)]\Psi_p)$$

which gives the alternate form

$$\langle p'q'\, |R|\, pq\rangle = i^2\int dx\int dy\, f_{q'}^*(y)\overrightarrow{K}_y$$

$$\times\, (\Psi_{p'}, \theta(y_0 - x_0)[\phi_\beta(y), \phi_\alpha(x)]\Psi_p)\overleftarrow{K}_x f_q(x) \tag{7.14}$$

The expression involves an R-product, i.e., a retarded commutator. A general definition for boson operators is

$$R(A_1(x_1)\cdots A_n(x_n)) = i^n \sum_{P(2\cdots n)} \theta(x_1 - x_2)\,\theta(x_2 - x_3)\cdots\theta(x_{n-1} - x_n)$$

$$\times\ [[[\cdot\cdot\cdot[A_1(x_1), A_2(x_2)], A_3(x_3)]\cdots], A_n(x_n)] \quad (7.15)$$

the sum being over all permutations of the coordinates $x_2, x_3, \ldots, x_n$. Both in the R-product and in the T-product step-functions appear, and the question of whether this is compatible with relativistic invariance might arise. In general a function

$$\theta(x_0) = \begin{cases} 1 & x_0 > 0 \\ 0 & x_0 < 0 \end{cases}$$

depends for its definition on a choice of a $t = 0$ "surface," and this choice can be altered by a Lorentz transformation. Thus a step-function is clearly not a covariant quantity. In the R-product, however, the step-function is multiplied by a commutator, as in $\theta(y_0 - x_0)\,[\phi_\beta(y), \phi_\alpha(x)]$, and, provided the commutator vanishes outside of the light cone,

$$[\phi_\beta(y), \phi_\alpha(x)] = 0 \qquad (x - y)^2 < 0$$

the only role of the step-function is to distinguish between the inside of the future light cone and the inside of the past light cone. This is a Lorentz invariant distinction. In the T-products, we have

$$\begin{aligned} T(\phi_\beta(y)\,\phi_\alpha(x)) &= \theta(y_0 - x_0)\,\phi_\beta(y)\,\phi_\alpha(x) + \theta(x_0 - y_0)\,\phi_\alpha(x)\,\phi_\beta(y) \\ &= \theta(y_0 - x_0)\,\phi_\beta(y)\,\phi_\alpha(x) + [1 - \theta(y_0 - x_0)]\,\phi_\alpha(x)\,\phi_\beta(y) \\ &= \theta(y_0 - x_0)[\phi_\beta(y), \phi_\alpha(x)] + \phi_\alpha(x)\,\phi_\beta(y) \end{aligned} \quad (7.16)$$

and

$$\begin{aligned} T(\psi(x)\,\bar{\psi}(y)) &= \theta(x_0 - y_0)\,\psi(x)\,\bar{\psi}(y) - \theta(y_0 - x_0)\,\bar{\psi}(y)\,\psi(x) \\ &= (1 - \theta(y_0 - x_0))\,\psi(x)\,\bar{\psi}(y) - \theta(y_0 - x_0)\,\bar{\psi}(y)\,\psi(x) \\ &= \psi(x)\,\bar{\psi}(y) - \theta(y_0 - x_0)\{\psi(x), \bar{\psi}(y)\} \end{aligned} \quad (7.17)$$

so that here, too, the step function is multiplied by a quantity that vanishes outside of the light cone.

We shall now list the relations corresponding to (7.11) and (7.14) for the forms (a), (e), and (f) of (7.4), because they will prove useful in later work.

The relations are

$$\langle p'q'\,|R|\,pq\rangle = i^2\iint dx\,dy(\Psi^{\text{out}}_{p'q'}, T(\phi_\alpha(x)\,\bar{\psi}_A(y)\Psi_0)$$
$$\times \overleftarrow{\mathfrak{D}}_y\overleftarrow{K}_x\,u_p^{(s)}(y)\,f_q(x)$$
$$= i^2\iint dx\,dy(\Psi^{\text{out}}_{p'q'}, \theta(x-y)[\phi_\alpha(x), \bar{\psi}_A(y)]\Psi_0)$$
$$\times \overleftarrow{\mathfrak{D}}_y\overleftarrow{K}_x\,u_p^{(s)}(y)\,f_q(x) \quad (7.18)$$

$$\langle p'q'\,|R|\,pq\rangle = i^2\iint dx\,dy\,f_{q'}{}^*(y)\overrightarrow{K}_y(\Psi_{p'}, T(\phi_\beta(y)\,\bar{\psi}_A(x))\Psi_q)\overleftarrow{\mathfrak{D}}_x\,u_p^{(s)}(x)$$
$$= i^2\iint dx\,dy\,f_{q'}{}^*(y)\overrightarrow{K}_y(\Psi_{p'}, \theta(y-x)$$
$$\times [\phi_\beta(y), \bar{\psi}_A(x)]\Psi_q)\overleftarrow{\mathfrak{D}}_x\,u_p^{(s)}(x) \quad (7.19)$$

$$\langle p'q'\,|R|\,pq\rangle = i^2\iint dx\,dy\,\bar{u}_{p'}^{(s')}(y)\overrightarrow{\mathfrak{D}}_y$$
$$\times (\Psi_{q'}, T(\psi_B(y)\,\bar{\psi}_A(x))\Psi_q)\overleftarrow{\mathfrak{D}}_x\,u_p^{(s)}(x)$$
$$= i^2\iint dx\,dy\,\bar{u}_{p'}^{(s')}(y)\overrightarrow{\mathfrak{D}}_y$$
$$\times (\Psi_{q'}, \theta(y-x)\{\psi_B(y), \bar{\psi}_A(x)\}\Psi_q)\overleftarrow{\mathfrak{D}}_x\,u_p^{(s)}(x) \quad (7.20)$$

In the actual evaluation of the matrix elements the wave-packet solutions of the classical equations for the free particles are usually replaced by the plane wave-solutions.

We can obtain, in a similar way, the forms for more complicated matrix elements of the S-operator, and we can also continue with the "contractions" of particles until all the particle coordinates appear in terms of fields and the states are only vacuum states. To illustrate this, we shall obtain the expression for a general S-matrix element in terms of the vacuum expectation value of a T-product of operators. To simplify the writing, we shall deal with a theory of a single scalar field, but the extension to other cases will be obvious.

We consider the matrix element

$$(q_1'q_2'\cdots q_l'^{\text{out}}\,|\,q_1q_2\cdots q_m{}^{\text{in}}) = \lim_{t\to-\infty}(q_1'\cdots q_l'^{\text{out}}\,|a_{q_1}{}^+(t)|\,q_2\cdots q_m{}^{\text{in}})$$
$$= (q_1'\cdots q_l'^{\text{out}}\,|a_{q_1}^{+\text{out}}|\,q_1\cdots q_m{}^{\text{in}})$$
$$-\int_{-\infty}^{\infty} dx_0\frac{\partial}{\partial x_0}(q_1'\cdots q_m'^{\text{out}}\,|a_{q_1}{}^+(x_0)|\,q_2\cdots q_m{}^{\text{in}})$$

(We are using the Dirac bra and ket notation because there are many indices). The first term gives zero if the momenta in the final state are all different from those in the initial state. If this is not the case, this term becomes (if, say $q_1 = q_1'$)

$$2\omega_{q_1}\,\delta(\mathbf{q}_1 - \mathbf{q}_1')(q_2' \cdots q_l'^{\text{out}} \mid q_2 \cdots q_m^{\text{in}})$$

This term corresponds to the situation in which the actual interaction involves $m - 1$ particles going in and $l - 1$ particles going out, with one particle not interacting with the rest of the system. We shall assume that the final momenta are all different from the initial ones, so that this special situation does not arise. We thus get, from the second term alone,

$$i\int dx_1 (q_1' \cdots q_l'^{\text{out}} \,|\phi(x_1)|\, q_2 \cdots q_m^{\text{in}})\overleftarrow{K}_x f_{q_1}(x_1) \tag{7.21}$$

As in the derivation of (7.11), we go on to the next stage by writing

$$\begin{aligned}
&(q_1' \cdots q_l'^{\text{out}} \,|\phi(x_1)|\, q_2 \cdots q_m^{\text{in}}) \\
&\quad = \lim_{x_{20}\to-\infty} (q_1' \cdots q_l'^{\text{out}} \,|\phi(x_1)a_{q_2}^{+}(x_{20})|\, q_3 \cdots q_m^{\text{in}}) \\
&\quad = \lim_{x_{20}\to-\infty} (q_1' \cdots q_l'^{\text{out}} \,|T(\phi(x_1)a_{q_2}^{+}(x_{20})|\, q_3 \cdots q_m^{\text{in}}) \\
&\quad = i\int_{x_{20}=-\infty} d^3x_2 (q_1' \cdots q_l'^{\text{out}} \,|T(\phi(x_1)\phi(x_2))|\, q_3 \cdots q_m^{\text{in}}) \frac{\overleftrightarrow{\partial}}{\partial x_{20}} f_{q_2}(x_2) \\
&\quad = i\int_{x_{20}=+\infty} d^3x_2 (q_1' \cdots q_l'^{\text{out}} \,|T(\phi(x_1)\phi(x_2))|\, q_3 \cdots q_m^{\text{in}}) \frac{\overleftrightarrow{\partial}}{\partial x_{20}} f_{q_2}(x_2) \\
&\qquad - i\int dx_2 \frac{\partial}{\partial x_{20}}\left[(q_1' \cdots q_l'^{\text{out}} \,|T(\phi(x_1)\,\phi(x_2))|\, q_3 \cdots q_m^{\text{in}}) \frac{\overleftrightarrow{\partial}}{\partial x_{20}} f_{q_2}(x_2)\right]
\end{aligned} \tag{7.22}$$

Again we use the inequality between the initial and final momenta to show that the $x_{20} = +\infty$ contribution vanishes, so that only the integral over all x_2 space-time is left. After the usual integration by parts, we get

$$i^2\int dx_1\, dx_2 (q_1' \cdots q_l'^{\text{out}} \,|T(\phi(x_1)\,\phi(x_2))|\, q_3 \cdots q_m^{\text{in}})\overleftarrow{K}_{x_1}\overleftarrow{K}_{x_2} f_{q_1}(x_1) f_{q_2}(x_2) \tag{7.23}$$

This procedure may be continued until all the particles appear as fields. The final formula is

$$\begin{aligned}
&(q_1' \cdots q_l'^{\text{out}} \mid q_1 \cdots q_m^{\text{in}}) \\
&\quad = i^{l+m}\int dy_1 \cdots dy_l\, dx_1 \cdots dx_m f_{q_1'}^{*}(y_1) \cdots f_{q_l'}^{*}(y_l)\overrightarrow{K}_{y_1} \cdots \overrightarrow{K}_{y_l} \\
&\qquad\quad \times \tau(y_1 \cdots y_l, x_1 \cdots x_m)\overleftarrow{K}_{x_1} \cdots \overleftarrow{K}_{x_m} f_{q_1}(x_1) \cdots f_{q_m}(x_m)
\end{aligned} \tag{7.24}$$

Here we have used the standard notation

$$\tau(y_1 \cdots x_m) = (\Psi_0', T(\phi(y_1) \cdots \phi(x_m))\Psi_0) \tag{7.25}$$

Incidentally, under the time-reversal transformation

$$\begin{aligned}\tau(y_1 \cdots x_m)' &= (\Psi_0, T(\phi(y_1) \cdots \phi(x_m))\Psi_0)' \\ &= (\Psi_0, T(\phi(x_m') \cdots \phi(y_1'))\Psi_0)\end{aligned} \tag{7.26}$$

where $y' = (-y_0, \mathbf{y})$. If we use

$$f_q(x) = \frac{e^{-iq_0x^0 + i\mathbf{q}\cdot\mathbf{x}}}{(2\pi)^{3/2}}$$

we can by a change of variables prove that[1]

$$(q_1' \cdots q_l'^{\text{out}} \mid q_1 \cdots q_m^{\text{in}}) = (-q_1 \cdots -q_m^{\text{out}} \mid -q_1' \cdots -q_l'^{\text{in}}) \tag{7.27}$$

Invariance under time reversal implies that, quite generally

$$S_{\alpha\beta} = S_{-\beta,-\alpha} \tag{7.28}$$

where $(-\beta)$ is the "motion-reversed" version of (β); when there is spin, not only the momenta but the directions of the spin are reversed. We leave the proof of this to the reader. Equation 7.28 leads to the principle of detailed balance, which always involves squares of transition matrices, *summed* over spins.

Whenever there is an incident fermion, the combination

$$i\,\phi(x_k)\overleftarrow{K}_{x_k} f_{q_k}(x_k) \tag{7.29a}$$

with the field appearing inside the T-product is to be replaced by

$$i\,\bar{\psi}(x_k)\overleftarrow{\mathfrak{D}}_{x_k}\, u_{q_k}(x_k) \tag{7.29b}$$

and for an emerging fermion,

$$if^*_{q_n'}(y_n)\overrightarrow{K}_{y_n}\,\phi(y_n) \tag{7.29c}$$

is to be replaced by

$$i\,u_{q_n'}(y_n)\overrightarrow{\mathfrak{D}}_{y_n}\,\psi(y_n) \tag{7.29d}$$

with the operators appearing inside the T-products.

An expression like (7.24) may be represented pictorially by a "black box" into which m appropriately labeled lines feed and from which l lines

[1] Here $-q_i$ stands for the four-vector $(q_{i0}, -\mathbf{q}_i)$

emerge [Fig. (7.1)]. The possible processes represented by Fig. 7.2 have been omitted.

In general, the expression in (7.24) is reducible, in that the expression we wrote down still contains products of "submatrix elements" corresponding to a subset of the incident particles scattering into a subset of the final particles, quite independently of the rest. Thus among the terms described by (7.24) there will be some that describe two or more independent processes as illustrated in Fig. 7.3. In a terminology borrowed from statistical mechanics the form we have contains "unlinked" as well as "linked" terms. In a calculation one is interested only in the linked terms. We are thus more interested in the functions $\varphi(x_1 \cdots x_n)$ which are related to the $\tau(x_1 \cdots x_n)$ as follows:

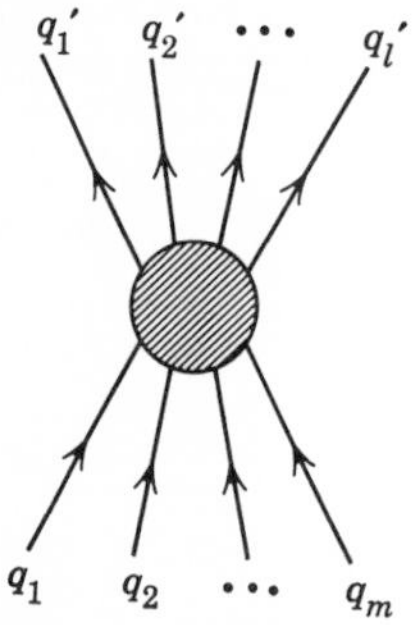

$$\varphi(x_1x_2) = \tau(x_1x_2)$$

$$\varphi(x_1x_2x_3) = \tau(x_1x_2x_3)$$

$$\varphi(x_1 \cdots x_4) = \tau(x_1 \cdots x_4) - \tau(x_1x_3)\,\tau(x_2x_4) - \tau(x_1x_2)\,\tau(x_3x_4) - \tau(x_1x_4)\tau(x_2x_3) \quad (7.30)$$

or, generally,

$$\tau(x_1 \cdots x_n) = \varphi(x_1 \cdots x_n) + \sum_A \varphi(x_{i_1} \cdots x_{i_k}) \cdots \varphi(x_{i_r} \cdots x_{i_n}) \quad (7.31)$$

Fig. 7.1. Graph representing amplitude $(q_1' \cdots q_l'^{\text{out}} | q_1 \cdots q_m^{\text{in}})$.

Here A stands for all possible partitions of the indices $1 \cdots n$ into distinct classes $(i_1 \cdots i_k) \cdots (i_r \cdots i_n)$.

As a matter of fact, even the "irreducible" functions are not exactly what we are interested in when we calculate the multiple-particle scattering matrix elements, because these contain terms that consist of two (or more) scattering matrix elements connected by a single *real* particle, which correspond to two independent scatterings, with one of the outgoing particles in Process 1 then taking part as an incident particle in Process 2.

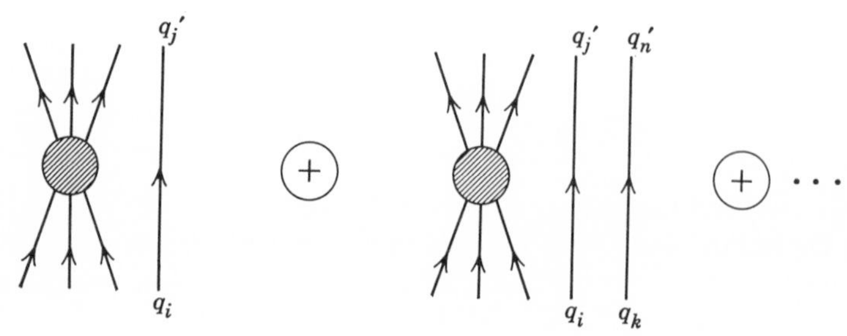

Fig. 7.2. Graphs omitted in the expression (7.24).

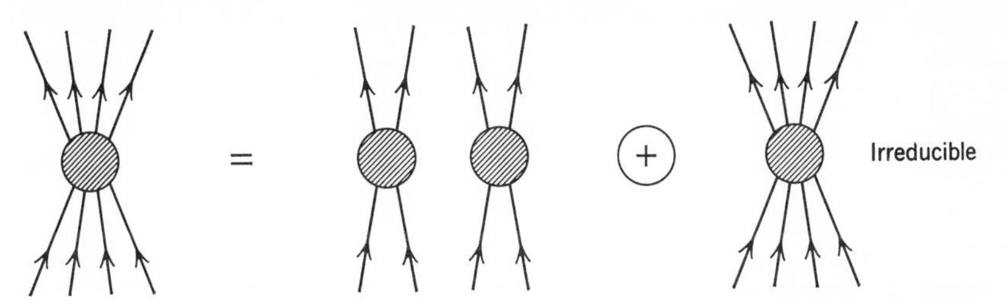

Fig. 7.3. Illustration of decomposition of a matrix element into "irreducible" parts.

This is illustrated in Fig. 7.4. It is the "one-particle irreducible" part that describes the specifically three-particle part of the S-matrix element. This separation is actually necessary when writing a Schrödinger equation for three-particle scattering and is not just characteristic of field theory. We shall be working with two-particle initial states so that these matters will not concern us.

The reduction formulas derived express the S-matrix elements in terms of operators $\phi(x)$ and $\psi(x)$, whose *scale* is set by the LSZ asymptotic conditions (6.24). If we specialize to states which are stationary, i.e., the vacuum and the one-particle states, the asymptotic condition of LSZ implies that since

$$\phi(x) = e^{iP_\mu x^\mu}\,\phi(0)e^{-iP_\mu x^\mu} \tag{7.32}$$

it follows that

$$\begin{aligned}(\Psi_0, \phi(x)\Psi_q) = e^{-iqx}(\Psi_0, \phi(0)\Psi_q) &= e^{-iqx}(\Psi, \phi^{\text{in}}(0)\Psi_q)\\ &= \frac{1}{(2\pi)^{3/2}}\,e^{iqx}\end{aligned} \tag{7.33}$$

On the other hand, the asymptotic conditions are weak convergence conditions and the field operators do not, in general, satisfy the canonical

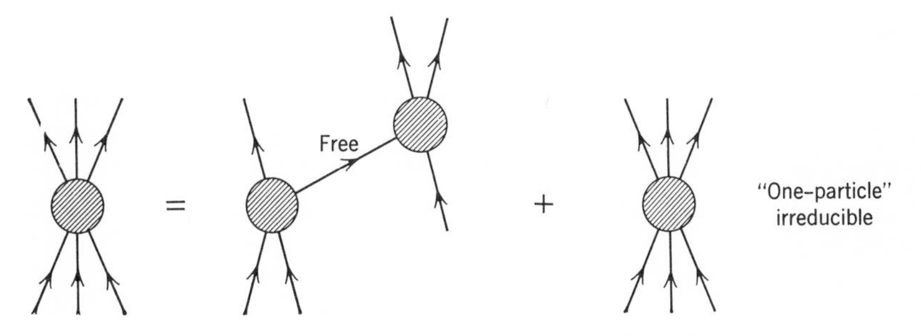

Fig. 7.4. Example of a "one-particle" reducible part in S_{33}.

commutation relations satisfied by the "in" or "out" fields. We can still write[1]

$$\{\psi(x), \psi^+(y)\}_{x0=y0} = \frac{1}{Z_2}\,\delta(\mathbf{x}-\mathbf{y})$$
$$[\dot{\phi}_\alpha(x), \phi_\beta(y)]_{x0=y0} = -\frac{i}{Z_3}\,\delta_{\alpha\beta}\,\delta(\mathbf{x}-\mathbf{y}) \tag{7.34}$$

where Z_2 and Z_3 are certain constants to be calculated later.

In order to keep the Hamiltonian simple and to keep the standard form of the Heisenberg equations of motion, it is sometimes convenient to work with operators that satisfy the canonical commutation relations and are therefore the above operators multiplied by certain constants. If we choose the operators to be

$$\hat{\psi}(x) = Z_2^{1/2}\,\psi(x)$$
$$\hat{\phi}(x) = Z_3^{1/2}\,\phi(x) \tag{7.35}$$

then these operators satisfy the canonical commutation relations, but the asymptotic conditions now read

$$\lim_{t\to-\infty}\left(\Phi, i\int_{x0=t} d^3x\,\hat{\phi}(x)\,\overleftrightarrow{\partial^0} f_\alpha(x)\Psi\right) = Z_3^{1/2}(\Phi, A_\alpha^{\text{in}}\Psi) \tag{7.36}$$

and so on. Thus the reduction formulas for the S-matrix, when expressed in terms of the operators $\hat{\psi}(x), \hat{\phi}(x), \ldots$, the *unrenormalized operators*, differ from the ones obtained by constant coefficients. Because of the correspondence

$$\phi_{\text{in}}(x) \leftrightarrow \phi(x) \leftrightarrow Z_3^{-1/2}\,\hat{\phi}(x) \tag{7.37}$$

etc., the expressions for the S-matrix elements are unaltered, except that the expression is to be multiplied by a $Z_2^{-1/2}$ for each fermion in the initial or final state and by $Z_3^{-1/2}$ for each boson in the initial or final state. For example, (7.24) reads

$$(q_1' \cdots q_l'^{\text{out}} \,|\, q_1 \cdots q_m^{\text{in}})$$
$$= i^{l+m}(Z_3^{-1/2})^{l+m}\int\cdots\int dy_1 \cdots dx_m f_{q_1'}^*(y_1)\cdots\overrightarrow{K}_{y_1}\cdots$$
$$\times\,(\Psi_0, T(\hat{\phi}(y_1)\cdots\hat{\phi}(x_m))\Psi_0)\overleftarrow{K}_{x_1}\cdots f_{q_m}(x_m) \tag{7.38}$$

The factors $Z_3^{-1/2}, \ldots$ can in principle be calculated from the commutation relations (7.34). It will turn out, however, that in perturbation theory, at least, these factors drop out and never have to be calculated.

[1] See, for example, H. Lehmann, *Nuovo Cimento* **11**, 342 (1954).

8

Perturbation Theory

In the last chapter we saw how to calculate the scattering matrix elements in terms of vacuum expectation values of T-products of the field operators. The difficult problem is, of course, the solution of the equations of motion; there is no known method, other than perturbation theory, that may be used to solve these equations. In perturbation theory we divide the Hamiltonian into two parts

$$H = H_0 + H_1 \tag{8.1}$$

where H_0 is a Hamiltonian for which we know how to solve the equations of motion and H_1 is a perturbing interaction. Since the only soluble field theory is the free-field theory, we take for H_0 the sum of all the free-particle Hamiltonians, *with the physical masses appearing in them.* This is an important point: the quantities μ, M, etc., which in the free Lagrangians played the role of the masses of the free particles, are no longer equal to the masses when interactions are present because of the possibility of self-interaction. Thus, if H_0 is to be written in terms of the physical masses, compensating terms have to be included in H_1 so that the total Hamiltonian H is not altered. For example, if

$$H = \frac{1}{2}\int d^3x(\pi^2(x) + (\nabla\,\phi(x))^2 + \mu_0{}^2\,\phi^2(x)) + \tilde{H}_1$$

we can choose for H_0

$$H_0 = \frac{1}{2}\int d^3x(\pi^2(x) + (\nabla\,\phi(x))^2 + \mu_{\text{exp}}^2\,\phi^2(x))$$

and then

$$H_1 = \tilde{H}_1 - \frac{1}{2}\int d^3x(\mu_{\text{exp}}^2 - \mu_0{}^2)\,\phi^2(x)$$

We say that H_1 contains a *mass-renormalization counterterm.*[1] The problem

[1] In this chapter we shall ignore this part of H_1 and only consider its implications in Chapter 13.

now is, given H_1, to find an expression for the τ-functions. In the formal development we shall consider, for the sake of simplicity, a theory involving only one scalar field, $A(x)$. We denote the solution of the free-field equations with the canonical commutation relations, involving the physical masses μ_{exp}, by $A^{(0)}(x)$.

The formal solutions of the Heisenberg equations of motion are ($H = H\{A\}$)

$$A(x) = e^{iHt} A(\mathbf{x}, 0)e^{-iHt} \tag{8.2}$$

and ($H_0 = H_0\{A^{(0)}\}$)

$$A^{(0)}(x) = e^{iH_0t} A^{(0)}(\mathbf{x}, 0)e^{-iH_0t} \tag{8.3}$$

We would like to choose the "initial" conditions on the free field such that it will coincide with the interacting field at $t = 0$, but this can be done only if we work with the "unrenormalized" field $\hat{A}(x)$,[1] in which case we may choose ($\hat{A} = Z^{1/2}A$)

$$\begin{aligned} \hat{A}(\mathbf{x}, 0) &= A^{(0)}(\mathbf{x}, 0) \\ \frac{\partial}{\partial t}\hat{A}(\mathbf{x}, 0) &= \frac{\partial}{\partial t} A^{(0)}(\mathbf{x}, 0) \end{aligned} \tag{8.4}$$

We may then express the unrenormalized field at any time in terms of the free-field operator at the same time:

$$\begin{aligned} \hat{A}(x) &= e^{iHt} \hat{A}(\mathbf{x}, 0)e^{-iHt} \\ &= e^{iHt} A^{(0)}(\mathbf{x}, 0)e^{-iHt} \\ &= e^{iHt}e^{-iH_0t} A^{(0)}(x)e^{iH_0t}e^{-iHt} \end{aligned}$$

i.e.,

$$\hat{A}(x) = e^{iH\{Z^{-1/2}\hat{A}\}t}e^{-iH_0\{A^{(0)}\}t} A^{(0)}(x)e^{iH_0\{A^{(0)}\}t}e^{-iH\{Z^{-1/2}\hat{A}\}t} \tag{8.5}$$

We have been careful to indicate the dependence of H and H_0 on the appropriate operators. For the free problem, H_0 is independent of time, and we may choose

$$H_0\{A^{(0)}(x)\} = H_0\{A^{(0)}(\mathbf{x}, 0)\} \tag{8.6}$$

Also H is independent of time, and we may choose

$$\begin{aligned} H\{Z^{-1/2}\hat{A}(x)\} &= H\{Z^{-1/2}\hat{A}(\mathbf{x}, 0)\} \\ &= H_0\{A^{(0)}(\mathbf{x}, 0)\} + H_1\{Z^{-1/2}A^{(0)}(\mathbf{x}, 0)\} \end{aligned} \tag{8.7}$$

The last step may seem a little puzzling because the conventional form for $H_0(A^{(0)})$ is

$$H_0 = \frac{1}{2}\int d^3x[\pi^{(0)}(x)^2 + (\nabla A^{(0)}(x))^2 + \mu_{\text{exp}}^2 A^{(0)}(x)^2]$$

[1] See the remarks in Chapter 7, p. 112.

and with H of the form

$$H = \frac{1}{2}\int d^3x[\pi^2(x) + (\nabla A(x))^2 + \mu_{\text{exp}}^2 A^2(x)] + H_1\{A\}$$

the last step in (8.7) appears to be wrong. The point is that to preserve the equations of motion, such as

$$\frac{d}{dt} A(x) = i[H, A(x)]$$

in face of the renormalized commutation relations (7.34), it is necessary to have H in the form

$$H\{A\} = \frac{1}{2} Z \int d^3x[\pi^2(x) + (\nabla A(x))^2 + \mu_{\text{exp}}^2 A^2(x)] + H_1\{A\}$$

so that

$$H\{Z^{-1/2}\hat{A}\} - H_1\{Z^{-1/2}\hat{A}\} = \frac{1}{2}\int d^3x[\hat{\pi}^2(x) + (\nabla \hat{A}(x))^2 + \mu_{\text{exp}}^2 \hat{A}^2(x)]$$

If we write this in the form

$$H = \frac{1}{2}\int d^3x[\hat{\pi}^2(x) + (\nabla \hat{A}(x))^2 + \mu_{\text{exp}}^2 A^2(x)] + H_1(Z^{-1/2}\hat{A})$$

we see that, by requiring that the Heisenberg equations have the same form as for the free fields, we are forced to write the free Hamiltonian in terms of the unrenormalized fields. This suggests that it is best to write the total Hamiltonian (hence the Lagrangian) in terms of the *unrenormalized field operators*, $\hat{A}(x)$, $\hat{\psi}(x)$, ... We shall show in Chapter 13 that the coupling parameters which appear in the Lagrangian are not directly measurable, so that replacing $H_1(Z^{-1/2}\,\hat{A})$ by $H_1(\hat{A})$ to give

$$H = H_0(\hat{A}) + H_1(\hat{A})$$

merely amounts to a redefinition of some parameters which do not in any case appear in expressions for scattering matrix elements.

Let us now define

$$U(t, 0) = e^{iH_0\{A^{(0)}(\mathbf{x},0)\}t}e^{-iH\{Z^{-1/2}\hat{A}(\mathbf{x},0)\}t} \tag{8.8}$$

Differentiation with respect to t if we pay attention to the order of the operators, yields,

$$\begin{aligned}\frac{dU(t, 0)}{dt} &= ie^{iH_0t}H_0e^{-iHt} - ie^{iH_0t}He^{-iHt}\\ &= -ie^{iH_0t}(H - H_0)e^{-iH_0t}e^{iH_0t}e^{-iHt}\\ &= -ie^{iH_0t}(H_0\{A^{(0)}\} + H_1\{Z^{-1/2}A^{(0)}\} - H_0\{A^{(0)}\})e^{-iH_0t}U(t)\end{aligned} \tag{8.9}$$

We shall denote

$$e^{iH_0\{A^{(0)}\}t} H_1\{Z^{-1/2} A^{(0)}(\mathbf{x},0)\} e^{-iH_0\{A^{(0)}\}t} = H_1\{Z^{-1/2} A^{(0)}(\mathbf{x},t)\} \equiv H_1(t) \quad (8.10)$$

As pointed out in the remarks preceding (8.8), we may replace $H_1(Z^{-1/2}A^{(0)}(\mathbf{x}, t))$ by $H_1(A^{(0)}(\mathbf{x}, t))$, as this merely changes the value of an unobservable parameter.
We solve (8.9),

$$\frac{dU(t, 0)}{dt} = -i\, H_1(t)\, U(t, 0)$$

for $U(t, 0)$ subject to the condition

$$U(0, 0) = 1 \quad (8.11)$$

We do this by rewriting the equation with the initial condition in the form of an integral equation

$$U(t, 0) = 1 - i\int_0^t dt'\, H_1(t')\, U(t', 0) \quad (8.12)$$

An iterative solution of this equation is given by

$$\begin{aligned} U(t, 0) &= 1 - i\int_0^t dt'\, H_1(t')\left(1 - i\int_0^{t'} dt''\, H_1(t'')\, U(t'', 0)\right) \\ &= 1 - i\int_0^t dt'\, H_1(t') + (-i)^2\int_0^t dt' \int_0^{t'} dt''\, H_1(t')H_1(t'')\, U(t'', 0) \\ &= 1 - i\int_0^t dt'\, H_1(t') \\ &\quad + (-i)^2\int_0^t dt' \int_0^{t'} dt''\, H_1(t')\, H_1(t'') \\ &\quad + (-i)^3\int_0^t dt' \int_0^{t'} dt'' \int_0^{t''} dt'''\, H_1(t')\, H_1(t'')\, H_1(t''') + \cdots \end{aligned} \quad (8.13)$$

It is important to note that because the $H_1(t)$ at different times do not commute [since the free fields $A^{(0)}(\mathbf{x}, t)$ do not] the order of the factors in the integrand is important. We may, however, simplify the expression by introducing the T-product instruction. With its help we may rewrite (8.13) in the form

$$U(t, 0) = \sum_{n=0}^{\infty}(-i)^n \int_0^t dt_1 \int_0^{t_1} dt_2 \ldots, \int_0^{t_{n-1}} dt_n\, T(H_1(t_1) \cdots H_1(t_n)) \quad (8.14)$$

Since, however, $T(H_1(t_1)\, H_1(t_2) \cdots H_1(t_n))$ is symmetric in the variables $t_1, t_2, \ldots t_n$, we may extend the intervals of integration to t in all cases and take care of the repetition by dividing by $n!$. For example,

$$\frac{1}{2!}\int_0^t dt_1 \int_0^t dt_2\, T(H_1(t_1)\, H_1(t_2))$$
$$= \frac{1}{2!}\int_0^t dt_1 \int_0^{t_1} dt_2\, H_1(t_1)\, H_1(t_2) + \frac{1}{2!}\int_0^t dt_1 \int_{t_1}^t dt_2\, H_1(t_2)\, H_1(t_1)$$
$$= \frac{1}{2!}\int_0^t dt_1 \int_0^{t_1} dt_2\, H_1(t_1)\, H_1(t_2) + \frac{1}{2!}\int_0^t dt_2 \int_0^{t_2} dt_1\, H_1(t_2)\, H_1(t_1)$$
$$= \int_0^t dt_1 \int_0^{t_1} dt_2\, H_1(t_1)\, H_1(t_2),$$

and so on. Thus we get

$$U(t, 0) = \sum_{n=0}^{\infty} \frac{(-i)^n}{n!} \int_0^t dt_1 \cdots \int_0^t dt_n\, T(H_1(t_1) \cdots H_1(t_n))$$
$$= T\left(e^{-i\int_0^t H_1(t')\,dt'}\right) \tag{8.15}$$

The operator $U(t, 0)$ is unitary since

$$U(t, 0)\, U^+(t, 0) = e^{iH_0t}e^{-iHt}e^{iHt}e^{-iH_0t} = 1 \tag{8.16}$$

If we write $T\left(e^{-i\int_0^t H_1(t')\,dt'}\right)$ in the form

$$\lim_{\Delta\to 0} e^{-i\int_{t-\Delta}^t H_1\,dt'} e^{-i\int_{t-2\Delta}^{t-\Delta} H_1\,dt'} \cdots e^{-i\int_{n\Delta}^{(n+1)\Delta} H_1\,dt'} \cdots e^{-i\int_0^\Delta H_1\,dt'}$$

we see that the inverse of this expression must be

$$\lim_{\Delta\to 0} e^{i\int_0^\Delta H_1\,dt'} e^{i\int_\Delta^{2\Delta} H_1\,dt'} \cdots e^{i\int_{n\Delta}^{(n+1)\Delta} H_1\,dt'} \cdots e^{i\int_{t-\Delta}^t H_1\,dt'}$$

If we introduce the instruction T^*—an antichronological ordering prescription—we may write

$$U^+(t, 0) = T^*(e^{i\int_0^t H_1(t')\,dt'}) \tag{8.17}$$

The quantity $U(t, t')$ defined by

$$U(t, t') \equiv U(t, 0)\, U^+(t', 0) \tag{8.18}$$

is given by

$$U(t, t') = T\left(e^{-i\int_0^t H_1(t'')\,dt''}\right) T^*\left(e^{i\int_0^{t'} dt''\,H_1(t'')}\right) = T\left(e^{-i\int_{t'}^t dt''\,H_1(t'')}\right) \tag{8.19}$$

Let us now consider the vacuum expectation value of a product of operators, $(\Psi_0, \hat{A}(x_1) \cdots \hat{A}(x_n)\Psi_0)$. We shall leave off the spatial coordinates in the following and write

$$
\begin{aligned}
(\Psi_0, \hat{A}(t_1) \cdots \hat{A}(t_n)\Psi_0) \\
&= (\Psi_0, U^+(t_1, 0)\, A^{(0)}(t_1)\, U(t_1, 0)\, U^+(t_2, 0)\, A^{(0)}(t_2)\, U(t_2, 0)\, U^+(t_3, 0) \\
&\qquad \cdots A^{(0)}(t_n)\, U(t_n, 0)\Psi_0) \qquad (8.20) \\
&= (\Psi_0, U^+(t_1, 0)\, A^{(0)}(t_1)\, U(t_1, t_2)\, A^{(0)}(t_2)\, U(t_2, t_3) \cdots A^{(0)}(t_n)\, U(t_n, 0)\Psi_0)
\end{aligned}
$$

Because we are dealing with free-field operators, it would be useful to express the physical vacuum state Ψ_0, which is the lowest energy state for the Hamiltonian H in terms of the state vectors characteristic of the free-field theory. In particular, it will be useful to obtain Ψ_0 in the form of an operator acting on Φ_0, the free-particle vacuum state, for which

$$
A_+^{(0)}(x)\Phi_0 = 0 \tag{8.21}
$$

where $A_+^{(0)}(x)$ is the annihilation part of the operator $A^{(0)}(x)$.

To obtain this relation, consider the quantity

$$
\lim_{t\to\pm\infty} (\Phi_1, e^{iHt}\Phi_2)
$$

in which Φ_1 and Φ_2 are any two normalizable state vectors constructed with the help of the free-field operators. We may rewrite this relation, using the completeness of the state vectors Ψ_n which are eigenstates of H:

$$
\begin{aligned}
\lim_{t\to\pm\infty} (\Phi_1, e^{iHt}\Phi_2) &= \lim_{t\to\pm\infty} \sum_n (\Phi_1, \Psi_n)e^{iE_nt}(\Psi_n, \Phi_2) \\
&= \lim_{t\to\pm\infty} \left[(\Phi_1, \Psi_0)(\Psi_0, \Phi_2) + \sum_{n\neq 0} e^{iE_nt}(\Phi_1, \Psi_n)(\Psi_n, \Phi_2)\right]
\end{aligned} \tag{8.22}
$$

If we now assume that there are no zero mass particles in the theory, so that the lowest energy in the second term is μ_{exp}, then, as $|t| \to \infty$, the oscillations in the exponential factor become so rapid that they make the second term vanish relative to the first one. We do not know under what conditions this generalization of the Riemann-Lesbegue lemma holds, and we have to assume that the theory is sufficiently free of pathological behavior that we may drop the second term in (8.22). We thus have

$$
\lim_{t\to\pm\infty} (\Phi_1, e^{iHt}\Phi_2) = (\Phi_1, \Psi_0)(\Psi_0, \Phi_2) \tag{8.23}
$$

for all free-field state vectors Φ_1, so that we may write the *weak* limit

$$
\lim_{t\to\pm\infty} e^{iHt}\Phi_2 = (\Psi_0, \Phi_2)\Psi_0
$$

or, alternately,

$$
\Psi_0 = \frac{1}{(\Psi_0, \Phi)} \lim_{t\to\pm\infty} e^{iHt}\Phi \tag{8.24}
$$

In particular,

$$\Psi_0 = \lim_{t\to-\infty} \frac{e^{iHt}\Phi_0}{(\Psi_0, \Phi_0)}$$

$$= \lim_{t\to-\infty} \frac{e^{iHt}e^{-iH_0t}\Phi_0}{(\Psi_0, e^{-iH_0t}\Phi_0)}$$

$$= \lim_{t\to-\infty} \frac{U^+(t, 0)\Phi_0}{(\Psi_0, U^+(t, 0)\Phi_0)} \tag{8.25}$$ [1]

Similarly, we obtain

$$\Psi_0 = \lim_{t\to+\infty} \frac{U^+(t, 0)\Phi_0}{(\Psi_0, U^+(t, 0)\Phi_0)} \tag{8.26}$$

We finally get

$$(\Psi_0, \hat{A}(t_1) \cdots \hat{A}(t_n)\Psi_0)$$

$$= \lim_{\substack{T\to+\infty \\ T'\to-\infty}} \frac{(\Phi_0, U(T, t_1)\, A^{(0)}(t_1)\, U(t_1, t_2)\, A^{(0)}(t_2) \cdots U(t_n, T')\Phi_0)}{(\Phi_0, U(T, 0)\Psi_0)(\Psi_0, U^+(T', 0)\Phi_0)} \tag{8.27}$$

The denominator may be treated as follows: we note that

$$\lim_{T'\to-\infty} (\Psi_0, U^+(T', 0)\Phi_0) = \lim_{T'\to-\infty} (\Psi_0, e^{iHT'}e^{-iH_0T'}\Phi_0)$$

and

$$\lim_{T\to+\infty} (\Phi_0, U(T, 0)\Psi_0) = \lim_{T\to+\infty} (\Phi_0, e^{iH_0T}e^{-iHT}\Psi_0)$$

With the help of (8.24) we may rewrite these equations as

$$\lim_{\substack{T\to+\infty \\ T'\to-\infty}} (\Phi_0, U(T, 0)\Psi_0)(\Psi_0, U^+(T', 0)\Phi_0)$$

$$= \lim_{\substack{T\to\infty \\ T'\to-\infty}} (\Phi_0, e^{iH_0T}e^{-iHT}\Psi_0)(\Psi_0, e^{iHT'}e^{-iH_0T'}\Phi_0)$$

$$= \lim_{\substack{T\to\infty \\ T'\to-\infty}} \sum_n (\Phi_0, e^{iH_0T}e^{-iHT}\Psi_n)(\Psi_n, e^{iHT'}e^{-iH_0T'}\Phi_0)$$

$$= \lim_{\substack{T\to\infty \\ T'\to-\infty}} (\Phi_0, U(T, 0)\, U^+(T', 0)\Phi_0)$$

$$= \lim_{\substack{T\to+\infty \\ T'\to-\infty}} (\Phi_0, U(T, T')\Phi_0) \tag{8.28}$$

[1] This formula was first derived by M. Gell-Mann and F. E. Low, *Phys. Rev.* **84,** 340 (1951). This particular derivation was suggested to me by R. Haag.

We can now write

$$
\begin{aligned}
(\Psi_0, T(\hat{A}(x_1)\,\hat{A}(x_2)\cdots\hat{A}(x_n))\Psi_0)
&= \lim_{\substack{T'\to\infty\\ T''\to-\infty}} \frac{(\Phi_0, T(U(T'T'')\,A^{(0)}(x_1)\cdots A^{(0)}(x_n))\Phi_0)}{(\Phi_0, U(T', T'')\Phi_0)} \\
&= \lim_{\substack{T'\to\infty\\ T''\to-\infty}} \frac{\left(\Phi_0, T\left(e^{-i\int_{T''}^{T'} H_1(t)\,dt} A^{(0)}(x_1)\cdots A^{(0)}(x_n)\right)\Phi_0\right)}{\left(\Phi_0, T\left(e^{-i\int_{T''}^{T'} H_1(t)\,dt}\right)\Phi_0\right)}
\end{aligned}
\tag{8.29}
$$

If there are spin $\frac{1}{2}$ particles involved, a quantity of the type

$$
\lim_{\substack{T'\to\infty\\ T''\to-\infty}} \frac{\left(\Phi_0, T\left(e^{-i\int_{T''}^{T'} H_1(t)\,dt} A^{(0)}(x_1)\cdots\psi^{(0)}(y_1)\cdots \psi^{(0)}(y_k)\,\bar{\psi}^{(0)}(z_1)\cdots\bar{\psi}^{(0)}(z_k)\right)\Phi_0\right)}{(\Phi_0, U(T', T'')\Phi_0)}
\tag{8.30}
$$

has to be evaluated. It should be noted that the number of $\psi^{(0)}$'s and $\bar{\psi}^{(0)}$'s is usually the same because all known spin $\frac{1}{2}$ particles carry some attribute represented by a conserved quantum number (charge, baryon number, hypercharge, etc.). For all known spin $\frac{1}{2}$ particles the antiparticles are distinct from the particles.[1] Thus the scattering matrix expressions are invariant under

$$
\psi \to e^{-i\alpha}\psi \qquad \bar{\psi} \to e^{i\alpha}\bar{\psi}
$$

Our next task is to develop techniques for evaluating expressions of the kind shown in (8.30). We assume a simple Yukawa model in which the interaction Hamiltonian is of the form

$$
H_1(t) = -\int d^3x\, j^{(0)}(x)\, A^{(0)}(x) \tag{8.31}
$$

where the current $j^{(0)}(x)$ does not contain any $A^{(0)}(x)$ operators. Actually $H_1(t)$ will also contain mass-renormalization counterterms, but they will not be considered at this point. The quantity of interest is

$$
\begin{aligned}
&\left(\Phi_0, T\left(e^{-i\int_{-\infty}^{\infty} dt H_1(t)}\, A^{(0)}(x_1)\cdots A^{(0)}(x_m)\,\psi^{(0)}(y_1)\cdots\bar{\psi}^{(0)}(z_k)\right)\Phi_0\right) \\
&\quad = \sum_n \frac{(-i)^n}{n!}\int\cdots\int du_1\cdots du_n (\Phi_0, T(\mathcal{H}_1(u_1)\cdots\mathcal{H}_1(u_n)) \\
&\qquad\qquad \times A^{(0)}(x_1)\cdots A^{(0)}(x_m)\,\psi^{(0)}(y_1)\cdots\bar{\psi}^{(0)}(z_k))\Phi_0)
\end{aligned}
\tag{8.32}
$$

where

$$
\mathcal{H}_1(u) = -j^{(0)}(u)\,A^{(0)}(u) \tag{8.33}
$$

[1] In principle, it is possible to have neutral spin $\frac{1}{2}$ particles which are their own antiparticles, but such "Majorana neutrinos" do not seem to exist.

Since $j^{(0)}(x)$ does not contain $A^{(0)}(x)$, and these are all free-field operators, it follows that

$$[j^{(0)}(u), A^{(0)}(x)] = 0 \tag{8.34}$$

Thus the expression (8.32) may be written as

$$\sum_{n=0}^{\infty} \frac{i^n}{n!} \int \cdots \int du_1 \cdots du_n (\Phi_0, T(A^{(0)}(u_1) \cdots A^{(0)}(u_n)\, A^{(0)}(x_1) \cdots A^{(0)}(x_m))\, \Phi_0)$$
$$\times\, (\Phi_0, T(j^{(0)}(u_1) \cdots j^{(0)}(u_n)\, \psi^{(0)}(y_1) \cdots \psi^{(0)}(y_k)\, \bar{\psi}^{(0)}(z_1) \cdots \bar{\psi}^{(0)}(z_k))\Phi_0) \tag{8.35}$$

To evaluate the first factor, let us introduce the functional

$$\mathcal{F}(J) \equiv \left(\Phi_0, T\left(e^{-i\int dx J(x) A^{(0)}(x)}\right)\Phi_0\right) \tag{8.36}$$

The vacuum expectation values of the T-products may be obtained by functional differentiation with respect to $J(x)$. Thus

$$(\Phi_0, T(A^{(0)}(x_1) \cdots A^{(0)}(x_k))\Phi_0) = \left(i^k \frac{\delta}{\delta J(x_1)} \cdots \frac{\delta}{\delta J(x_k)} \mathcal{F}(J)\right)_{J=0} \tag{8.37}$$

If we can calculate the value of the functional $\mathcal{F}(J)$, all T-products are easily obtained. The calculation of the functional is actually possible, since the field $A^{(0)}(x)$ is a free field about which much is known. The procedure is to decompose the field into positive and negative frequency components[1] $A_+^{(0)}(x)$ and $A_-^{(0)}(x)$ and to use the fact that for the vacuum state

$$A_+^{(0)}(x)\Phi_0 = 0 \tag{8.38}$$

Let us consider the functional

$$\mathcal{F}(J;t) = \left(\Phi_0, T\left(e^{-i\int_{-\infty}^{t} dx_0 \int d^3x J(x) A^{(0)}(x)}\right)\Phi_0\right) \tag{8.39}$$

We next define a quantity $W(t)$ by the following relation

$$T\left(e^{-i\int_{-\infty}^{t} dx J(x) A^{(0)}(x)}\right) = T\left(e^{-i\int_{-\infty}^{t} dx J(x) A_-^{(0)}(x)}\right) W(t) \tag{8.40}$$

Differentiation with respect to t leads to the equation

$$-i\int_{x_0=t} d^3x\, J(x)\, A^{(0)}(x)\, T\left(e^{-i\int_{-\infty}^{t} dx J(x) A_-^{(0)}(x)}\right) W(t)$$
$$= -i\int_{x_0=t} d^3x\, J(x)\, A_-^{(0)}(x)\, T\left(e^{-i\int_{-\infty}^{t} dx J(x) A_-^{(0)}(x)}\right) W(t)$$
$$+\, T\left(e^{-i\int_{-\infty}^{t} dx J(x) A_-^{(0)}(x)}\right) \frac{dW}{dt}$$

[1] Conventionally, the term containing the annihilation operator and the plane wave factor e^{-ikx} is called the *positive* frequency term.

Now, recalling that two free field creation operators commute, we get for all times

$$[A_-^{(0)}(x), A_-^{(0)}(y)] = 0$$

As a consequence of this we are permitted to remove the T "instruction" and get, after some rearrangement,

$$i\frac{dW(t)}{dt} = \left(e^{i\int_{-\infty}^{t} dx J(x) A_-^{(0)}(x)}\right)\int_{y_0=t} d^3y\, J(y)\, A_+^{(0)}(y) \times \left(e^{-i\int_{-\infty}^{t} dx J(x) A_-^{(0)}(x)}\right) \times W(t) \quad (8.41)$$

$$= \int_{y_0=t} d^3y\, J(y)\left[A_+^{(0)}(y) + i\int_{-\infty}^{t} dx\, J(x)[A_-^{(0)}(x), A_+^{(0)}(y)]\right] W(t)$$

The initial condition on $W(t)$ is

$$W(-\infty) = 1 \quad (8.42)$$

Thus the solution of the differential equation (8.41) is

$$W(t) = e^{-i\int_{-\infty}^{t} dy J(y) A_+^{(0)}(y)}\, e^{\int_{-\infty}^{t} dy \int_{-\infty}^{y_0} dx J(y) J(x) [A_-^{(0)}(x), A_+^{(0)}(y)]} \quad (8.43)$$

and

$$W(\infty) = e^{-i\int dy J(y) A_+^{(0)}(y)}\, e^{\int\int dy dx \theta(y_0 - x_0) J(y) J(x) [A_-^{(0)}(x), A_+^{(0)}(y)]} \quad (8.44)$$

Hence using (8.38) we obtain

$$\mathcal{F}(J) = \left(\Phi_0, e^{-i\int dy J(y) A_-^{(0)}(y)}\, e^{-i\int dy J(y) A_+^{(0)}(y)}\Phi_0\right) \times e^{\int\int dx dy J(x) J(y) \theta(y_0 - x_0)[A_-^{(0)}(x), A_+^{(0)}(y)]}$$

$$= e^{-\int\int dy dx J(y) J(x) \theta(y_0 - x_0)[A_+^{(0)}(y), A_-^{(0)}(x)]} \quad (8.45)$$

We can proceed further by calculating the commutator. We get

$$[A_+^{(0)}(y), A_-^{(0)}(x)] = \frac{1}{(2\pi)^3}\int \frac{d^3k_1}{2\omega_{k_1}}\, e^{-ik_1 y}\int \frac{d^3k_2}{2\omega_{k_2}}\, e^{ik_2 x}[A(k_1), A^+(k_2)]$$

$$= \frac{1}{(2\pi)^3}\int \frac{d^3k}{2\omega_k}\, e^{-ik(y-x)}$$

Next

$$\theta(y_0 - x_0)[A_+^{(0)}(y), A_-^{(0)}(x)]$$

$$= \frac{1}{2\pi i}\int \frac{d\lambda}{\lambda - i\epsilon}\, e^{i\lambda(y_0 - x_0)} \frac{1}{(2\pi)^3}\int \frac{d^3k}{2\omega_k}\, e^{-i\omega_k(y_0 - x_0)} e^{i\mathbf{k}\cdot(\mathbf{y}-\mathbf{x})}$$

$$= \frac{1}{(2\pi)^4 i}\int_{-\infty}^{\infty} \frac{d\lambda}{\lambda - i\epsilon}\int \frac{d^3k}{2\omega_k}\, e^{-i(\omega_k - \lambda)(y_0 - x_0)}\, e^{i\mathbf{k}\cdot(\mathbf{y}-\mathbf{x})}$$

$$= \frac{-i}{(2\pi)^4}\int dp\, e^{-ip(y-x)} \frac{1}{2\omega_p} \frac{1}{\omega_p - p_0 - i\epsilon}$$

This quantity is multiplied by a function symmetric in x and y, namely, $J(y)\,J(x)$. We may therefore write the exponent of (8.45) in the symmetrized form

$$
\begin{aligned}
-\int dx \int dy\, J(x)\, J(y) \frac{(-i)}{(2\pi)^4} \int dp e^{-ip(y-x)} \\
\times \frac{1}{4\omega_p}\left(\frac{1}{\omega_p - p_0 - i\epsilon} + \frac{1}{\omega_p + p_0 - i\epsilon}\right) \\
= \frac{i}{4}\iint dx\, dy\, J(x)\, J(y) \frac{1}{(2\pi)^4}\int dp e^{-ip(y-x)} \frac{2}{\mu_{\text{exp}}^2 - p^2 - i\epsilon} \\
= -\frac{i}{2}\iint dx\, dy\, J(x)\, J(y) \frac{1}{(2\pi)^4}\int dp \frac{e^{-ip(y-x)}}{p^2 - \mu_{\text{exp}}^2 + i\epsilon} \\
\equiv -\frac{1}{4}\iint dx\, dy\, J(x)\, J(y)\, \Delta_F(y - x; \mu_{\text{exp}}^2)
\end{aligned}
\tag{8.46}
$$

Here we define the so-called "causal" Feynman propagator[1]

$$
\Delta_F(y - x; \mu_{\text{exp}}^2) = \frac{2i}{(2\pi)^4}\int dp \frac{e^{-ip(y-x)}}{p^2 - \mu_{\text{exp}}^2 + i\epsilon} \tag{8.47}
$$

We will in general suppress the dependence on the variable μ_{exp}^2. Thus

$$
\mathcal{F}(J) = e^{-\frac{1}{4}\iint dy dx J(y)\, \Delta_F(y-x)\, J(x)} \tag{8.48}
$$

It is clear that the vacuum expectation value of an odd number of $A^{(0)}(x)$ vanishes. The first two nontrivial expectation values are

$$
\begin{aligned}
(\Phi_0, T(A^{(0)}(x_1)\, A^{(0)}(x_2))\Phi_0) &= \tfrac{1}{2}\, \Delta_F(x_1 - x_2) \\
&= \tfrac{1}{2}\, \Delta_F(x_2 - x_1)
\end{aligned}
\tag{8.49}
$$

and

$$
\begin{aligned}
(\Phi_0, T(A^{(0)}(x_1)\, A^{(0)}(x_2)\, A^{(0)}(x_3)\, A^{(0)}(x_4))\Phi_0) \\
= \tfrac{1}{4}\, \Delta_F(x_1 - x_2)\, \Delta_F(x_3 - x_4) + \tfrac{1}{4}\, \Delta_F(x_1 - x_3)\, \Delta_F(x_2 - x_4) \\
+ \tfrac{1}{4}\, \Delta_F(x_1 - x_4)\, \Delta_F(x_2 - x_3)
\end{aligned}
\tag{8.50}
$$

We may give a graphical prescription for the expression of

$$
(\Phi_0, T(A^{(0)}(x_1)\, A^{(0)}(x_2) \cdots A^{(0)}(x_n))\Phi_0)
$$

in terms of $\frac{1}{2}\, \Delta_F(x_i - x_j)$: for each variable $x_1, x_2, \ldots, x_n$, mark a point on a sheet of paper. Connect these points in pairs (x_1, x_{i_1}), (x_2, x_{i_2}), etc.

[1] The definition of the causal functions follows that of S. Schweber. Note that a different convention is used by Bjorken and Drell.

(recall that n must be even). For such a configuration write down the product

$$(\tfrac{1}{2})\,\Delta_F(x_1 - x_{i_1})(\tfrac{1}{2})\,\Delta_F(x_2 - x_{i_2})(\tfrac{1}{2})\,\Delta_F(x_3 - x_{i_3})\cdots$$

Now $(\Phi_0, T(A^{(0)}(x_1)\cdots A^{(0)}(x_n))\,\Phi_0)$ is given by the sum of these terms corresponding to all possible configurations, i.e., all possible pairings. It should be noted that a pairing (x_i, x_j) is *not* to be treated as distinct from (x_j, x_i). We leave it to the reader to convince himself that this procedure can be extended to photons and vector mesons, provided that $\frac{1}{2}\,\Delta(x_i - x_j)$ is replaced by $(\Phi_0, T(A_\mu^{(0)}(x_i)\,A_\nu^{(0)}(x_j))\Phi_0)$.

We must next obtain an expression for

$$(\Phi_0, T(j^{(0)}(u_1)\cdots j^{(0)}(u_n)\,\psi^{(0)}(y_1)\cdots\psi^{(0)}(y_k)\,\bar{\psi}^{(0)}(z_1)\cdots\bar{\psi}^{(0)}(z_k))\Phi_0) \quad (8.51)$$

We shall take for $j^{(0)}(u)$ the form

$$j^{(0)}(u) = g\,\bar{\psi}_\alpha^{(0)}(u)\Gamma_{\alpha\beta}\,\psi_\beta^{(0)}(u) \quad (8.52)$$

where $\Gamma_{\alpha\beta}$ is some 4×4 matrix that need not be further specified at this point. We define the current with a normal ordering so that

$$(\Phi_0, j^{(0)}(u)\Phi_0) = 0$$

We again construct a generating functional which couples the field to classical sources. In this case we write

$$\mathcal{F}(\eta, \bar{\eta}) = (\Phi_0, Te^{-i\int dx[\bar{\eta}(x)\psi^{(0)}(x)+\bar{\psi}^{(0)}(x)\eta(x)]}\Phi_0) \quad (8.53)$$

The fields $\psi^{(0)}$ and $\bar{\psi}^{(0)}$ intrinsically *anticommute*, so that their order in a T-product determines the sign of the T-product. For example,

$$(\Phi_0, T(\psi^{(0)}(x)\,\bar{\psi}^{(0)}(y))\Phi_0) = -(\Phi_0, T(\bar{\psi}^{(0)}(y)\,\psi^{(0)}(x))\Phi_0) \quad (8.54)$$

This can be reproduced with the functional differentiation method only if the sources $\bar{\eta}$ and η also have anticommutation properties.[1] We require that

$$\{\eta(x), \eta(y)\} = \{\bar{\eta}(x), \eta(y)\} = \{\eta(x), \bar{\eta}(y)\} = \{\bar{\eta}(x), \bar{\eta}(y)\} = 0$$

so that

$$\frac{\delta^2}{\delta\eta(x)\,\delta\eta(y)} = -\frac{\delta^2}{\delta\eta(y)\,\delta\eta(x)}$$
$$\frac{\delta^2}{\delta\bar{\eta}(x)\,\delta\eta(y)} = -\frac{\delta^2}{\delta\eta(y)\,\delta\bar{\eta}(x)}\cdots \quad (8.55)$$

[1] As a result of their anticommutative properties, the η, $\bar{\eta}$ are not really classical sources; they could not be, for they transform like spinors to keep $\bar{\eta}\psi$ and $\bar{\psi}\eta$ scalar. This does not really matter because we need them only for bookkeeping purposes.

The T-product of interest (8.51) can now be written in the form

$$(\Phi_0, T(\bar{\psi}^{(0)}(u_1)\,\psi^{(0)}(u_1)\cdots\bar{\psi}^{(0)}(u_n)\,\psi^{(0)}(u_n)\,\psi^{(0)}(y_1)\cdots\bar{\psi}^{(0)}(z_k))\Phi_0)$$
$$= i\frac{\delta}{\delta\eta(z_k)}\cdots i\frac{\delta}{\delta\bar{\eta}(y_1)}\cdots i\frac{\delta}{\delta\bar{\eta}(u_1)}\,i\frac{\delta}{\delta\eta(u_1)}\,\mathcal{F}(\eta,\bar{\eta})\Big|_{\eta=\bar{\eta}=0} \tag{8.56}$$

Actually this expression is not complete, since the normal ordering of the current has not been taken into account. This will be done at a later stage. We evaluate the functional by again decomposing the free field into the annihilation and creation parts

$$\psi^{(0)}(x) = \psi_+^{(0)}(x) + \psi_-^{(0)}(x) \qquad \bar{\psi}^{(0)}(x) = \bar{\psi}_+^{(0)}(x) + \bar{\psi}_-^{(0)}(x)$$

and writing the functional in such a form that the vacuum expectation value takes on a particularly simple form. We define $G(t)$ by

$$T\left(e^{-i\int_{-\infty}^{t}dx(\bar{\eta}(x)\psi^{(0)}(x)+\bar{\psi}^{(0)}(x)\eta(x))}\right) = T\left(e^{-i\int_{-\infty}^{t}dx(\bar{\eta}(x)\psi_-^{(0)}(x)+\bar{\psi}_-^{(0)}(x)\eta(x))}\right)G(t) \tag{8.57}$$

Manipulations completely parallel to those leading to (8.41) give us

$$i\frac{dG(t)}{dt} = \bigg[\int_{y_0=t} d^3y(\bar{\eta}(y)\,\psi_+^{(0)}(y) + \bar{\psi}_+^{(0)}(y)\,\eta(y))$$
$$+ i\int_{-\infty}^{t} dx\int_{y_0=t} d^3y\,\bar{\eta}(x)\{\psi_-^{(0)}(x),\,\bar{\psi}_+^{(0)}(y)\}\,\eta(y)$$
$$- i\int_{-\infty}^{t} dx\int_{y_0=t} d^3y\,\bar{\eta}(y)\{\psi_+^{(0)}(y),\,\bar{\psi}_-^{(0)}(x)\}\,\eta(x)\bigg]\,G(t) \tag{8.58}$$

This equation is easily integrated. We have

$$G(t) = e^{-i\int_{-\infty}^{t}dy(\bar{\eta}(y)\psi_+^{(0)}(y)+\bar{\psi}_+^{(0)}(y)\eta(y))}$$
$$\times\, e^{-\int_{-\infty}^{t}dx\int_{-\infty}^{x_0}dy\bar{\eta}(y)\{\psi_+^{(0)}(y),\bar{\psi}_-^{(0)}(x)\}\eta(x)}$$
$$\times\, e^{\int_{-\infty}^{t}dx\int_{-\infty}^{x_0}dy\bar{\eta}(y)\{\psi_-^{(0)}(y),\bar{\psi}_+^{(0)}(x)\}\eta(x)}$$

where in the last term we interchanged the dummy variables x and y. Consequently

$$\mathcal{F}(\eta,\bar{\eta}) = \left(\Phi_0,\,T\left(e^{-i\int dx(\bar{\eta}(x)\psi_+^{(0)}(x)+\bar{\psi}_+^{(0)}(x)\eta(x))}\right)G(\infty)\Phi_0\right)$$
$$= e^{\int dx\int dy\bar{\eta}(y)[\theta(x_0-y_0)\{\psi_-^{(0)}(y),\bar{\psi}_+^{(0)}(x)\}-\theta(y_0-x_0)\{\psi_+^{(0)}(y),\bar{\psi}_-^{(0)}(x)\}]\eta(x)} \tag{8.59}$$

We can now evaluate the terms in the exponent. First of all[1]

$$\{\psi_+^{(0)}(y), \bar{\psi}_-^{(0)}(x)\} = \frac{1}{(2\pi)^3}\int \frac{d^2p_1}{E_{p_1}/M}\int \frac{d^3p_2}{E_{p_2}/M} \times e^{-ip_1y}\, u^{(s)}(p_1)\, e^{ip_2x}\, \bar{u}^{(r)}(p_2)\{\mathbf{a}^{(s)}(p_1), \mathbf{a}^{(r)+}(p_2)\}$$

$$= \frac{1}{(2\pi)^3}\int d^3p e^{-ip(y-x)}\frac{M + \not{p}}{2E_p} \tag{8.60}$$

Similarly,

$$\{\psi_-^{(0)}(y), \bar{\psi}_+^{(0)}(x)\} = -\frac{1}{(2\pi)^3}\int d^3p e^{ip(y-x)}\frac{M - \not{p}}{2E_p} \tag{8.61}$$

Hence the term in the square bracket in (8.59) is

$$-\theta(y_0 - x_0)\frac{1}{(2\pi)^3}\int \frac{d^3p}{2E_p}\, e^{-ip(y-x)}(M + \not{p}) - \theta(x_0 - y_0)\int \frac{d^3p}{2E_p}\, e^{ip(y-x)}(M - \not{p})$$

$$= \frac{i}{(2\pi)^4}\int dp \frac{e^{-ip(y-x)}}{2E_p(E_p - p_0 - i\epsilon)}(M + i\gamma_0 E_p - i\boldsymbol{\gamma}\cdot\mathbf{p}) + \frac{i}{(2\pi)^4}\int dp \frac{e^{-ip(y-x)}}{2E_p(E_p + p_0 - i\epsilon)}(M - i\gamma_0 E_p - i\boldsymbol{\gamma}\cdot\mathbf{p})$$

$$= -\frac{i}{(2\pi)^4}\int dp e^{-ip(y-x)}\frac{M + \not{p}}{p^2 - M^2 + i\epsilon} \tag{8.62}$$

as can be shown after half a page of algebra, with the help of the representation for the step function

$$\theta(y_0 - x_0) = \frac{1}{2\pi i}\int_{-\infty}^{\infty} d\lambda \frac{e^{i\lambda(y_0 - x_0)}}{\lambda - i\epsilon}$$

If we now introduce the notation

$$S_F(x) \equiv -(i\gamma^\mu \partial_\mu + M)\Delta_F(x) \tag{8.63}$$

We find that

$$\mathcal{F}(\eta, \bar{\eta}) = e^{\frac{1}{2}\int\int dy\, dx \bar{\eta}(y) S_F(y-x)\eta(x)} \tag{8.64}$$

When the functional differentiation is carried out as in (8.56), we must keep track of the order of η and $\bar{\eta}$. Thus to obtain $(\Phi_0, T(\psi^{(0)}(y)\, \bar{\psi}^{(0)}(x))\Phi_0)$

[1] Here M stands for the experimental mass.

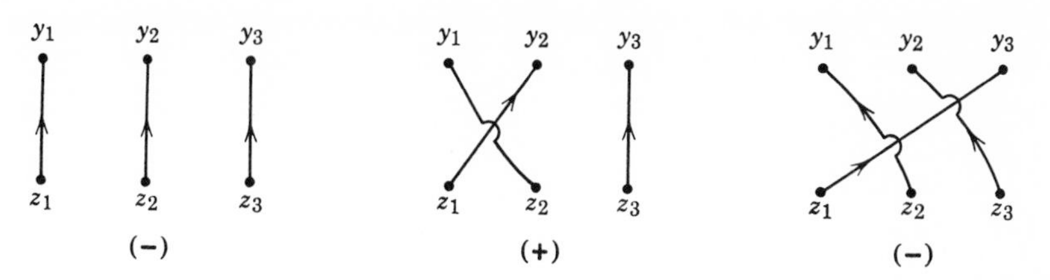

Fig. 8.1. Some of the graphs and associated signs for the six-point fermion function.

we require

$$i\frac{\delta}{\delta\eta(x)}\,i\frac{\delta}{\delta\bar{\eta}(y)}\,e^{\frac{1}{2}\iint du\,dv\,\bar{\eta}(u)S_F(u-v)\eta(v)}\Big|_{\eta=\bar{\eta}=0}$$

$$= i\frac{\delta}{\delta\eta(x)}\,i\frac{\delta}{\delta\bar{\eta}(y)}\left(\frac{1}{2}\iint du\,dv\,\bar{\eta}(u)\,S_F(u-v)\,\eta(v)\right)$$

$$= -\frac{\delta}{\delta\bar{\eta}(y)}\left(\frac{1}{2}\int du\,\bar{\eta}(u)\,S_F(u-x)\right) = -\frac{1}{2}\,S_F(y-x) \quad (8.65)$$

We can work out the four-point vacuum expectation value in a similar way:

$$(\Phi_0, T(\psi^{(0)}(x_1)\,\psi^{(0)}(x_2)\,\bar{\psi}^{(0)}(x_3)\,\bar{\psi}^{(0)}(x_4))\,\Phi_0)$$
$$= \tfrac{1}{4}\,S_F(x_1-x_4)\,S_F(x_2-x_3) - \tfrac{1}{4}\,S_F(x_1-x_3)\,S_F(x_2-x_4) \quad (8.66)$$

The general procedure can again be represented graphically. We mark points associated with $\psi^{(0)}$'s by $y_1, y_2, y_3, \ldots$ and points associated with $\bar{\psi}^{(0)}$'s by $z_1, z_2, z_3, \ldots$ on a sheet of paper. There will be as many $\psi^{(0)}$ points as there are $\bar{\psi}^{(0)}$ points; otherwise, the vacuum expectation value would vanish. We now connect each $\psi^{(0)}$ point with a $\bar{\psi}^{(0)}$ point. The line joining the points is directed, with the arrow pointing from the $\bar{\psi}^{(0)}$ point to the $\psi^{(0)}$ point. With each line, connecting $(z_i \to y_j)$ we associate a factor

$$-\tfrac{1}{2}\,S_F(y_j - z_i)$$

The particular configuration has a $\pm$ sign associated with it, depending on whether the number of interchanges of operators in

$$(\Phi_0, T(\psi^{(0)}(y_1)\,\psi^{(0)}(y_2)\cdots\bar{\psi}^{(0)}(z_1)\cdots)\,\Phi_0)$$

needed to bring this into the form

$$(\Phi_0, T(\psi^{(0)}(y_1)\,\bar{\psi}^{(0)}(z_{i_1})\,\psi^{(0)}(y_2)\,\bar{\psi}^{(0)}(z_{i_2})\cdots)\Phi_0)$$

is even or odd. The products associated with all such graphs are then added. Figure 8.1 illustrates this for

$$(\Phi_0, T(\psi^{(0)}(y_1)\,\psi^{(0)}(y_2)\,\psi^{(0)}(y_3)\,\bar{\psi}^{(0)}(z_1)\,\bar{\psi}^{(0)}(z_2)\,\bar{\psi}^{(0)}(z_3))\Phi_0)$$

It will be observed that the expression can also be written in the form of a determinant, which takes the signs into account:[1]

$$(\Phi_0, T(\psi^{(0)}(y_1) \cdots \psi^{(0)}(y_n)\, \bar{\psi}^{(0)}(z_1) \cdots \bar{\psi}^{(0)}(z_n))\Phi_0)$$

$$= \pm \begin{vmatrix} -\frac{1}{2} S_F(y_1 - z_1) & -\frac{1}{2} S_F(y_1 - z_2) & -\frac{1}{2} S_F(y_1 - z_3) \cdots \\ -\frac{1}{2} S_F(y_2 - z_1) & -\frac{1}{2} S_F(y_2 - z_2) & -\frac{1}{2} S_F(y_2 - z_3) \cdots \\ -\frac{1}{2} S_F(y_3 - z_1) \cdots & & \\ \cdot & & \\ \cdot & & \\ \cdot & & \end{vmatrix} \tag{8.67}$$

The effect of taking the current in normal order is to eliminate lines which leave a point u_i and return to the same point.

In the expression

$$(\Phi_0, T(j^{(0)}(u_1) \cdots j^{(0)}(u_0)\, \psi^{(0)}(y_1) \cdots \bar{\psi}^{(0)}(z_1) \cdots)\Phi_0)$$

the points $u_1, u_2, \ldots u_n$ serve both as end points of lines [because of the $\psi^{(0)}(u_i)$] and as starting points of lines [because of the $\bar{\psi}^{(0)}(u_i)$], so that these points serve as junction points, with the directed fermion lines going in and out. Figure 8.2 shows the graph for $(\Phi_0, T(j^{(0)}(u)\, \psi^{(0)}(y)\, \bar{\psi}^{(0)}(z))\Phi_0)$.

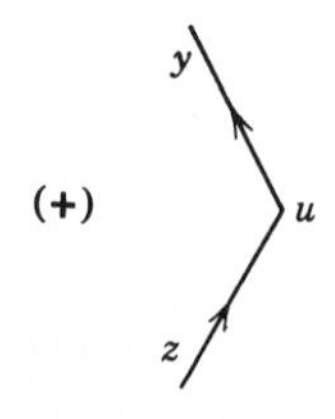

Fig. 8.2. Graph showing role of $j^{(0)}(u)$ as junction.

Now that the rules for evaluating the two components of the expression (8.35) have been obtained, it is easy to see how these are to be put together. A graphical prescription involves labeling points on a graph with the coordinates $u_1, u_2, \ldots, u_n, y_1, \ldots, z_1, \ldots, z_k$ and drawing all possible connecting lines in all possible ways. To each such graph there corresponds a mathematical expression consisting of the appropriate product of $\frac{1}{2}S_F$ and $\frac{1}{2}\Delta_F$ functions associated with it. The coordinates $u_1, \ldots, u_n$ are to be integrated over. They are therefore dummy variables, so that all graphs, which differ only in that the coordinates $u_1, \ldots, u_n$ are permuted, are equivalent. There are $n!$ such permutations, and we may *consider only distinct graphs* and drop the $1/n!$ in (8.35). The factors $\Gamma_{\alpha\beta}$ which enter into the definition of the currents, appear at the junction points u_i, between the S_F function describing the emerging line and the S_F function describing the line coming into u_i.

To evaluate the scattering matrix element, we must still act on (8.35) with the required Klein-Gordon and/or Dirac differential operators,

[1] The over-all sign is $(-1)^{n/2}$ when n is even and $(-1)^{n/2-1/2}$ when n is odd, as consideration of the diagonal term shows.

which act on the coordinates $\{x_i\}$, $\{y_i\}$, and $\{z_i\}$. Their effect is to remove some of the S_F and D_F functions, replacing them with delta functions, for, as is clear from (8.47) and (8.63),

$$K_x \Delta_F(x-\xi) = (\Box_x + \mu^2)\frac{2i}{(2\pi)^4}\int dp\, \frac{e^{-ip(x-\xi)}}{p^2-\mu^2+i\epsilon}$$
$$= -2i\,\delta(x-\xi) \tag{8.68}$$

and

$$(-i\gamma^\mu\,\partial_\mu + M)\,S_F(y-\eta) = -(\Box + M^2)\,\Delta_F(y-\eta) = 2i\,\delta(y-\eta)$$
$$S_F(\xi - z)(\overleftarrow{M + i\gamma\,\partial}) = 2i\,\delta(\zeta - z) \tag{8.69}$$

After this "amputation of the external lines," as this process is sometimes called, we must still multiply the expression by the wave functions for the incoming and outgoing particles and carry out the final integration, e.g., as indicated in (7.11). As an illustration, we shall consider

$$\int\cdots\int dx_2\, dy_2\, dx_1\, dy_1\, \frac{e^{iq'x_1}e^{ip'y_2}}{(2\pi)^3}\,\overrightarrow{K}_{x_2}\,\overrightarrow{\mathfrak{D}}_{y_2}$$
$$\times\, (\Psi_0, T(\psi(y_2)\,\phi_\beta(x_2)\,\bar\psi(y_1)\,\phi_\alpha(x_1))\Psi_0)\overleftarrow{K}_{x_1}\,\overleftarrow{\mathfrak{D}}_{y_1}\,\frac{e^{-ipy_1}e^{-iqx_1}}{(2\pi)^3} \tag{8.70}$$

the matrix element for meson nucleon scattering, to fourth order in the coupling constant g which appears in the interaction Hamiltonian

$$H_1 = ig\!:\!\int \bar\psi(x)\gamma_5\tau_\alpha\,\psi(x)\,\phi_\alpha(x)\,d^3x\!: \tag{8.71}$$

The graphs for the appropriate T-product,

$$(\Phi_0, T(j^{(0)}_{\alpha_1}(u_1)\cdots j^{(0)}_{\alpha_4}(u_4)\,\phi^{(0)}_{\alpha_1}(u_1)\cdots$$
$$\times\, \phi^{(0)}_{\alpha_4}(u_4)\,\psi^{(0)}(y_2)\,\phi^{(0)}_\beta(x_2)\,\phi^{(0)}_\alpha(x_1)\,\bar\psi^{(0)}(y_1))\Phi_0)$$

are shown in Fig. 8.3. The details of the calculation (up to a point) are carried out below for the special case of the graph labeled ($c-1$). This contribution is

$$(ig)^4\int\cdots\int du_1\cdots du_4\left(-\frac{1}{2}\right)S_F(y_2-u_4)\gamma_5\tau_{\alpha_4}\left(-\frac{1}{2}\right)S_F(u_4-u_3)\gamma_5\tau_{\alpha_3}$$
$$\times\left(-\frac{1}{2}\right)S_F(u_3-u_2)\gamma_5\tau_{\alpha_2}\left(-\frac{1}{2}\right)S_F(u_2-u_1)\gamma_5\tau_{\alpha_1}\left(-\frac{1}{2}\right)S_F(u_1-y_1)$$
$$\times\,\delta_{\alpha_1\alpha_4}\left(\frac{1}{2}\right)\Delta_F(u_4-u_1)\,\delta_{\beta\alpha_3}\left(\frac{1}{2}\right)\Delta_F(x_2-u_3)\,\delta_{\alpha\alpha_2}\left(\frac{1}{2}\right)\Delta_F(u_2-x_1)$$
$$\tag{8.72}$$

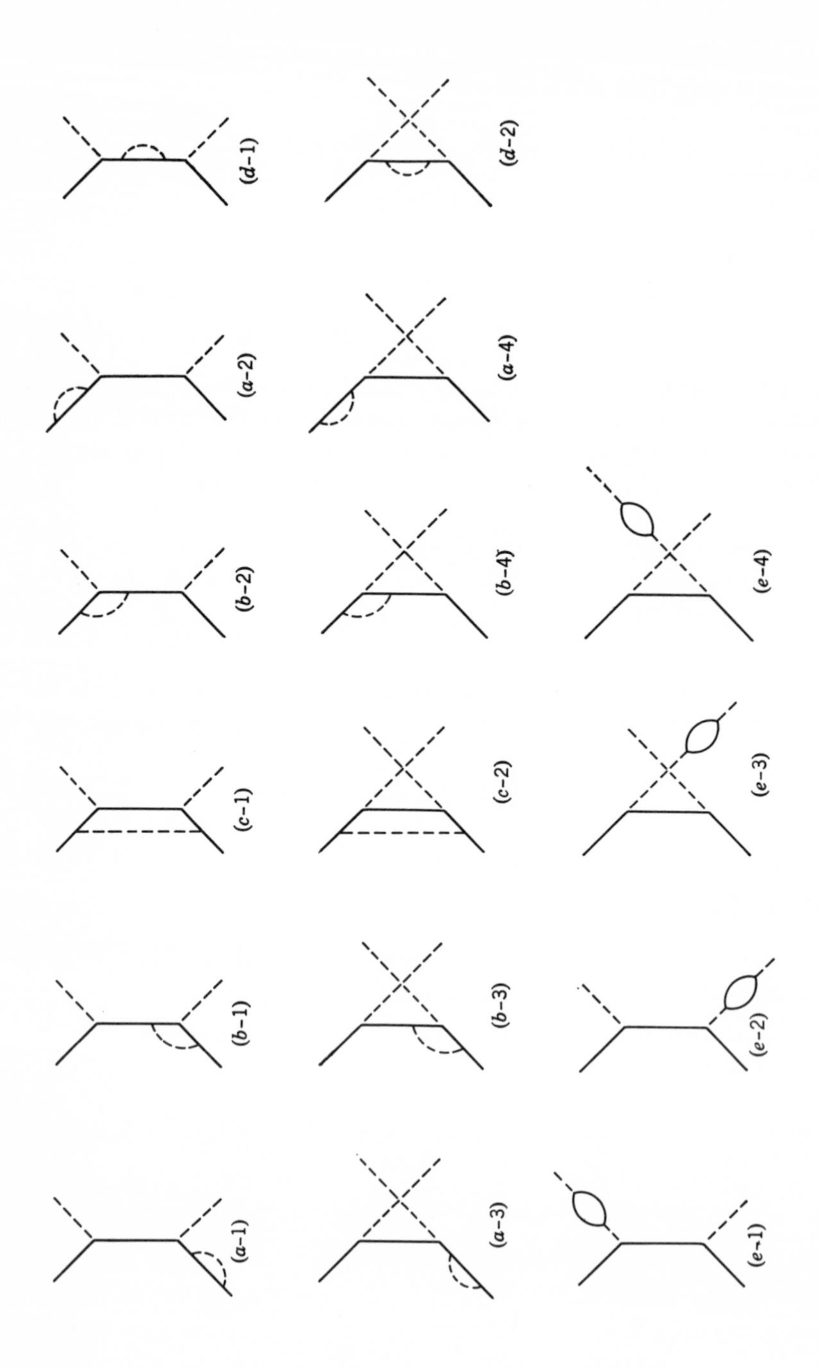
(d-1)
(d-2)
(a-2)
(a-4)
(b-2)
(b-4)
(e-4)
(c-1)
(c-2)
(e-3)
(b-1)
(b-3)
(e-2)
(a-1)
(a-3)
(e-1)

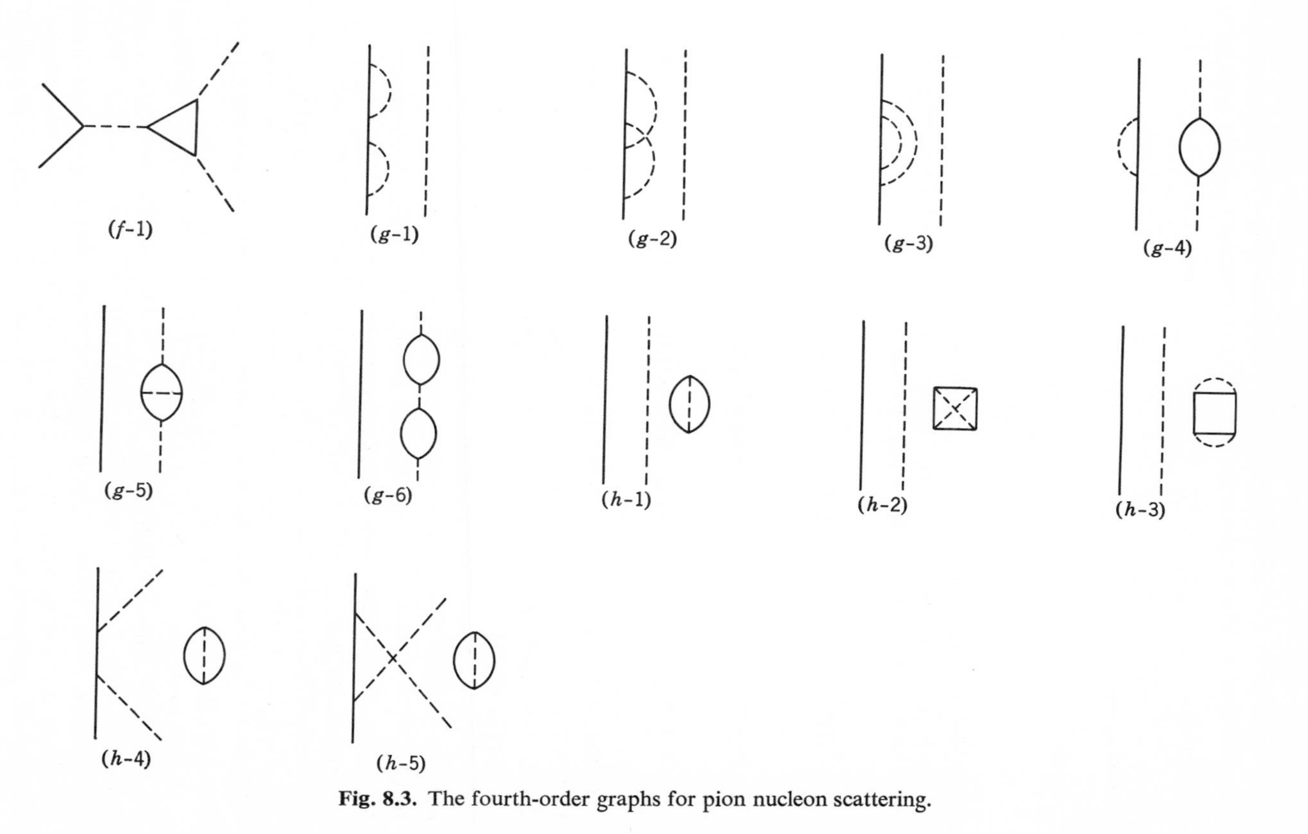

Fig. 8.3. The fourth-order graphs for pion nucleon scattering.

Operation with the Dirac and Klein-Gordon operators eliminates the lines leading to the points not being integrated over (hence the name "amputation"). What the amputation does is to replace $\frac{1}{2}\Delta_F(x-\xi)$ with $i\,K_x(\frac{1}{2}\Delta_F(x-\xi)) = \delta(x-\xi)$, $-\frac{1}{2}S_F(y-\eta)$ with $-\frac{1}{2}i\overrightarrow{\mathfrak{D}}_y\,S_F(y-\eta) = \delta(y-\eta)$, and $-\frac{1}{2}S_F(\zeta-z)$ with $i(-\frac{1}{2}S_F(\zeta-z))\overleftarrow{\mathfrak{D}}_z = \delta(\zeta-z)$. We therefore get

$$(ig)^4\gamma_5\tau_{\alpha_1}(-\tfrac{1}{2}S_F(y_2-x_2))\gamma_5\tau_\beta(-\tfrac{1}{2}S_F(x_2-x_1))\gamma_5\tau_\alpha(-\tfrac{1}{2}S_F(x_1-y_1))$$
$$\times\ \gamma_5\tau_{\alpha_1}(\tfrac{1}{2}\Delta_F(y_2-y_1))$$

Note that the repeated indices are summed over (in this case the indices are on τ_{α_1}). This expression must now be multiplied by the wave functions and the final integration performed, to lead to

$$\frac{1}{(2\pi)^6}(ig)^4\int\cdots\int dx_2\,dy_2\,dx_1\,dy_1\,e^{iq'x_2}e^{ip'y_2}e^{-iqx_1}e^{-ipy_1}$$
$$\times\ \bar{u}(p')\gamma_5\tau_{\alpha 1}\left(-\frac{1}{2}S_F(y_2-x_2)\right)\gamma_5\tau_\beta\left(-\frac{1}{2}S_F(x_2-x_1)\right)\gamma_5\tau_\alpha$$
$$\times\ \left(-\frac{1}{2}S_F(x_1-y_1)\right)\gamma_5\tau_{\alpha 1}\,u(p)\,\frac{1}{2}\Delta_F(y_2-y_1) \tag{8.73}$$

In the next chapter the Feynman rules for calculating scattering matrix elements will be given. Not all the steps leading to them are exhibited in detail; we might just mention some of the prescriptions with reference to the graphs of Fig. 8.3.

1. Graphs of the type (*h*) actually do not appear, if one takes into account the fact that the expression for the S-matrix includes a division by $(\Phi_0, U(\infty, -\infty)\Phi_0)$ in perturbation theory. When the denominator is expanded, these graphs are canceled by products of lower-order terms in the numerator [e.g. (*h* — 4, 5)] with denominator terms. Effectively, this means that vacuum fluctuations do not contribute to the scattering.

2. Graphs of type (*g*) describe the meson and the nucleon moving independently, and they do not actually appear in the expression for $S - 1 = R$. In general, only connected graphs are of interest.

3. For some theories not all graphs contribute. In the example outlined we can show that the contribution of (f — 1) vanishes on grounds of symmetry. The reason here is parity conservation. The closed loop with three meson lines feeding into it would, if it did not vanish, lead to a transition from a one-meson to a two-meson state. Now, one meson has angular momentum and parity 0^-. A two-meson state, with orbital angular momentum l has parity $(-1)^l$, and conservation of angular

momentum implies that the parities do not match. Graphs may vanish also if internal symmetries are involved.

In conclusion, the following statement will be made without proof. In cases in which the interaction Lagrangian contains derivatives of field operators (as in the case of interaction between vector mesons and spin 0 mesons), the canonical momentum contains, in addition to the time derivative of the field, terms which depend on the interaction. This fact leads to certain noncovariant terms that appear in the interaction Hamiltonian and certain technical complications in the perturbation expansion. The complications can be avoided if for $U(t, t')$ the following expression is written

$$U(t, t') = T\left(e^{i\int_{t'}^{t} dx\,\mathfrak{L}_I(x)}\right) \tag{8.74}$$

It should also be kept in mind that the expression for the matrix element of R already obtained has to be multiplied by a $Z_3^{-1/2}$ for each external meson line and $Z_2^{-1/2}$ for each external fermion line. These quantities will play their proper role in canceling parts of the graphs (*a*), (*b*), (*e*) when taken together with the mass-renormalization counterterms which have been ignored here. We defer consideration of this to Chapter 13.

PROBLEMS

1. Show that

$$T\left(e^{-i\int_{-\infty}^{\infty} ds(X(s)+Y(s))}\right)$$

may be written in the form

$$T\left(e^{-i\int_{-\infty}^{\infty} dsX(s)}\right)\,T\left(e^{-i\int_{-\infty}^{\infty} W(s)ds}\right)$$

with

$$W(s) = T^*\left(e^{i\int_{-\infty}^{s} ds'X(s')}\right)\;Y(s)\;T\left(e^{-i\int_{-\infty}^{s} ds'X(s')}\right)$$

2. Use the method of functionals to calculate the matrix element for the emission of n photons by a classical current $J_\mu(x)$. Show that the total probability for radiating n photons follows the Poisson distribution

$$P_n = \frac{e^{-N}N^n}{n!}$$

where N is the average number of photons, $N = \sum_{n=0}^{\infty} nP_n$. Obtain an expression for N in terms of $J_\mu(x)$. *Hint.* Evaluate

$$(\Psi_{k_1k_2\cdots k_n}, e^{-i\int J_\mu(x)A^\mu(x)dx}\,\Psi_0)$$

with a classical current, for which $[J_\mu(x), J_\nu(y)] = 0$.

9

The Feynman Graph Rules

The material presented in Chapter 8 indicates that it is possible to associate a graph with each term of the perturbation expansion of the scattering matrix elements. Historically, the reverse development took place. Feynman developed a graphical representation of the perturbation series before the connection with field theory became clear. The graphical approach has played an important role not only in field theory but in many-body physics and in nuclear structure theory, because the graphs not only serve as a simple device that keeps track of all the terms to be calculated, but also have associated with them a definite physical picture of the process. Such pictorial representations are sometimes useful in stimulating guesses of possible important contributions in a certain energy range.

In what follows the Feynman graph rules for a few representative theories are written down. Their derivation is straightforward, given the material of Chapter 8; all that is missing are some minor details, which will be left to the reader to check. The Feynman rules for the interaction of a charged particle with an external field will also be written down, since these are of interest in problems of the interaction of electrons or radiation with matter. In addition, a number of results, peculiar to quantum electrodynamics and having to do with charge conservation and charge conjugation, will be obtained, so that the groundwork for the chapters on quantum electrodynamics will be laid. The rules listed in this chapter are the ones that lead to the scattering matrix elements and not to the τ-functions. We first consider

Interaction of Spin One Half Particles with Photons

The interaction Lagrangian is taken to be

$$\mathcal{L}_I = e:\bar{\psi}(x)\gamma^\mu \psi(x) A_\mu(x): = -\mathcal{H}_1 \tag{9.1}$$

for a spin $\frac{1}{2}$ particle whose charge is $-e$. To obtain the nth order term of the scattering matrix, draw all possible connected graphs joining the n points $x_1, x_2, x_3, \ldots x_n$ together. Each point has an incoming fermion line, an outgoing fermion line, and a photon line coming to it. With each graph we associate a mathematical expression determined as follows:

1. For an internal photon line joining the point x_i (and labeled with the polarization index μ_i) to the point x_j (labeled with μ_j) write

$$(\Phi_0, T(A_{\mu_i}(x_i)\, A_{\mu_j}(x_j)\Phi_0) = -\tfrac{1}{2}\, g_{\mu_i\mu_j}\, \Delta_F(x_j - x_i, 0)$$
$$= -\tfrac{1}{2}\, g_{\mu_i\mu_j}\, D_F(x_j - x_i)$$

2. For an internal fermion line directed from x_a to x_b write

$$-\tfrac{1}{2}\, S_F(x_b - x_a)$$

3. For a vertex at x_i, labeled with μ_i, at which two fermion lines and one photon line meet, write

$$e\gamma^{\mu_i}$$

4. For an external photon line joining to the point $x_i(\mu_i)$ write

$$e_{\mu_i}^{\lambda}[e^{-ikx_i}/(2\pi)^{3/2}]$$

if the line describes a photon of momentum k *coming in* with polarization λ; we write $e_{\mu_i}^{\lambda}[e^{ikx_i}/(2\pi)^{3/2}]$ if the external line describes a photon of polarization λ carrying off momentum k.

5. For an incoming electron of spin state s, bringing in momentum p to the point x_i, write $u^{(s)}(p)e^{-ipx_i}/(2\pi)^{3/2}$. If the electron of spin state s leaves from that point with momentum p, write $\bar{u}^{(s)}(p)e^{ipx_i}/(2\pi)^{3/2}$.

6. If the positron comes in at the point x_i, with spin state s, and brings in momentum p, write $\bar{v}^{(s)}(p)e^{-ipx_i}/(2\pi)^{3/2}$; if the positron carries off momentum p, write $v^{(s)}(p)e^{ipx_i}/(2\pi)^{3/2}$. In the graphs, an incoming positron line would have its arrow pointed in a direction opposite to that of an incoming electron, for which the momentum flow and charge flow are conventionally taken in the same direction. This should not cause any confusion when momenta are finally balanced.

The expression so obtained is to be integrated over all the coordinates $x_1, x_2, \ldots x_n$, which label the points. If only distinct graphs (which differ by more than a change of point labelings) are drawn, we need to multiply the result by i^n.

If there is interaction with an external field, as well as with the quantized radiation field, the interaction Lagrangian will contain the additional term

$$\mathcal{L}_I' = e\, \bar{\psi}(x)\gamma^{\mu}\, \psi(x)\, A_{\mu}^{\text{ext}}(x) \tag{9.2}$$

This implies that there may be vertices for which the photon is neither "free" nor "virtual." At such vertices we shall write $e\gamma^{\mu}\, A_{\mu}^{\text{ext}}(x)$.

To illustrate the use of the graph rules, consider the matrix element associated with the graph shown in Fig. 9.1. The Feynman rules lead to the expression

$$i^3 \iiint dx_1\, dx_2\, dx_3\, \bar{u}(p') \frac{e^{ip'x_3}}{(2\pi)^{3/2}} e\gamma^{\mu_3} \left(-\frac{1}{2} S_F(x_3 - x_2) \right)$$
$$(e\gamma^{\mu_2} A^{\text{ext}}_{\mu_2}(x_2)) \left(-\frac{1}{2} S_F(x_2 - x_1) \right)$$
$$\times e\gamma^{\mu_1} u(p) \frac{e^{-ipx_1}}{(2\pi)^3} \left(-\frac{1}{2} g_{\mu_1\mu_3} D_F(x_1 - x_3) \right) \quad (9.3)$$

Actually, it is only for problems involving interactions with an external field that the configuration-space rules are ever used. When there is no external field, it is much simpler to work in momentum space. To restate these rules we use the representations[1]

$$\frac{1}{2} \Delta_F(x - y) = \frac{i}{(2\pi)^4} \int dq \frac{e^{-iq(x-y)}}{q^2 - \mu^2 + i\epsilon} \quad (9.4)$$

$$\frac{1}{2} D_F(x - y) = \frac{i}{(2\pi)^4} \int dk \frac{e^{-ik(x-y)}}{k^2 + i\epsilon} \quad (9.5)$$

$$-\frac{1}{2} S_F(x - y) = \frac{i}{(2\pi)^4} \int dp e^{-ip(x-y)} \frac{\not{p} + M}{p^2 - M^2 + i\epsilon} \quad (9.6)$$

and carry out the spatial integrations. To illustrate the procedure consider again the graph shown in Fig. 9.1, with the difference that the external line now represents a photon of momentum k and polarization λ. This means that in configuration space (9.3) is altered in that $e\gamma^{\mu_2} A^{\text{ext}}_{\mu_2}(x)$ is replaced by $e\gamma^{\mu_2} e^{\lambda}_{\mu_2} e^{-ikx_2}/(2\pi)^{3/2}$, according to the rules. If we now insert (9.5) and (9.6) into the expression, we get

$$i^3 \iiint dx_1\, dx_2\, dx_3 \frac{\bar{u}(p')}{(2\pi)^{3/2}} e^{ip'x_3} e\gamma^{\mu_3} \left(\frac{i}{(2\pi)^4} \int dp_1 \frac{e^{-ip_1(x_3 - x_2)}}{\not{p}_1 - M + i\epsilon} \right)$$
$$\times e\gamma^{\mu_2} e^{\lambda}_{\mu_2} \frac{e^{-ikx_2}}{(2\pi)^{3/2}} \left(\frac{i}{(2\pi)^4} \int dp_2 \frac{e^{-ip_2(x_2 - x_1)}}{\not{p}_2 - M + i\epsilon} \right) e\gamma^{\mu_1} \frac{u(p)}{(2\pi)^{3/2}}$$
$$\times e^{-ipx_1}(-g_{\mu_1\mu_3}) \frac{i}{(2\pi)^4} \int dk' \frac{e^{-ik'(x_3 - x_1)}}{k'^2 + i\epsilon}$$
$$= \frac{e^3}{(2\pi)^{9/2}} \int dp_1 \int dp_2 \int dk'\, \bar{u}(p')\gamma^{\mu_3} \frac{1}{\not{p}_1 - M} \gamma^{\mu_2} e^{\lambda}_{\mu_2} \frac{1}{\not{p}_2 - M}$$
$$\gamma_{\mu_3} u(p) \frac{-1}{k'^2 + i\epsilon} \delta(p' - p_1 - k')\, \delta(p_1 - k - p_2)\, \delta(p_2 - p + k') \quad (9.7)$$

[1]We write μ^2 for the (physical mass)^2.

The transition to momentum space is thus made by labeling each of the lines in Fig. 9.1 with a momentum. At each vertex there will appear a delta function representing four-momentum conservation. It should be noted that although the photon line does not intrinsically have a direction associated with it, once a momentum is attached, it does acquire a direction in that momentum is carried *out* of one vertex or *into* another. Figure 9.2 shows the newly labeled version of Fig. 9.1.

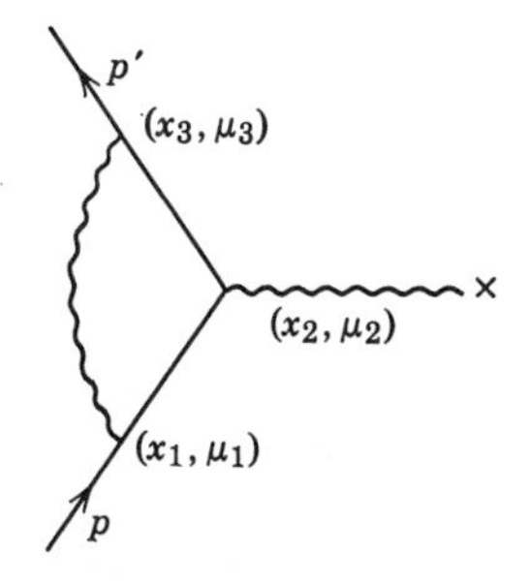

Fig. 9.1. Graph representing radiative correction to Coulomb scattering.

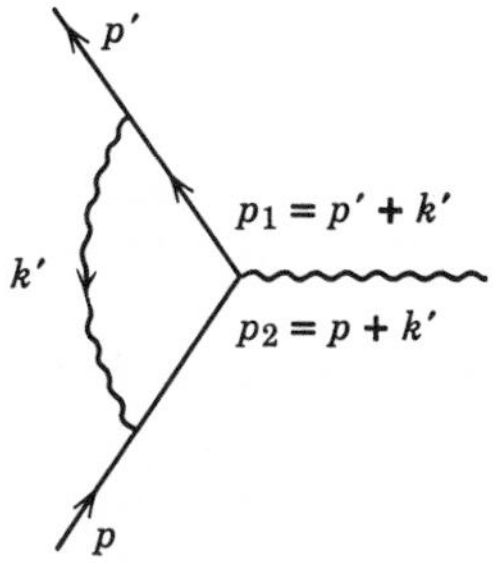

Fig. 9.2. Vertex correction graph in momentum space.

There are as many factors of $(2\pi)^{-3/2}$ as there are external lines; there is a $(2\pi)^4$ for each vertex (associated with the momentum conserving delta function) and a $(2\pi)^{-4}$ for each internal line. There is a factor of i for each order in perturbation theory, an e for each vertex and an i for each internal line. We generalize from this special example and again write the rules:

To calculate the nth order contribution, draw all distinct graphs connecting n-points; each point has an incoming fermion line, an outgoing fermion line, and a photon line attached to it. Only connected graphs are to be drawn. The contribution from each graph is calculated as follows:

1. For each internal photon line labeled with momentum k write $-g_{\mu\nu}/(k^2 + i\epsilon)$. Here μ and ν are the indices on the matrices at the vertices connected by the internal photon line.

2. At each vertex a factor $e\gamma^\mu$ is written, together with a delta function representing four-momentum conservation at the vertex.

3. For each internal fermion line a factor

$$\frac{1}{\not{p} - M + i\epsilon} = \frac{\not{p} + M}{p^2 - M^2 + i\epsilon}$$

is written, the momentum being chosen to flow in the same direction as the charge, i.e., in the direction of the arrow.

4. All internal momenta are integrated over. This will lead to an overall four-momenta conservation delta function for the external momenta and usually some remaining integrations as in (9.7).

5. For each external line write

   $u(p)$ for an incoming electron line,

   $\bar{v}(p)$ for an ingoing positron line (here the ingoing line will have an arrow pointing away from the graph, and momentum flowing opposite to the arrow direction),

   $\bar{u}(p')$ for an outgoing electron line,

   $v(p)$ for an outgoing positron line,

   $\epsilon_\mu{}^\lambda$ for an external photon of polarization λ entering a vertex with a γ^μ in it.

6. Each term is to be multiplied by $(i)^{n+e_i+b_i}$, where e_i is the number of internal fermion lines and b_i is the number of internal photon lines. Furthermore, each term is multiplied by $(2\pi)^{4(n-e_i-b_i)}(2\pi)^{-3/2(e_e+b_e)}$, where e_e is the number of external fermion lines and b_e is the number of external photon lines.

7. Following our discussion on page 127 the contribution from the graph must be multiplied by ± 1, the signature of the permutation of the fermions in the final state. It should be stressed that the effective antisymmetrization has to be carried out for particle-antiparticle pairs, and not only for particle-particle pairs as might be expected from the requirements of the exclusion principle familiar from nonrelativistic quantum mechanics. This is a consequence of field theory which treats electron creation and positron annihilation on the same footing and which requires, for consistency, that electron and positron operators anticommute. The Dirac "hole theory" of antiparticles, in which positrons were visualized as holes in a sea of negative-energy electrons (necessary to prevent electrons from making transitions to negative-energy states), makes this generalized Pauli principle more like the ordinary exclusion principle, but, because we do not approach the subject from this point of view, we prefer to rely on the anticommutativity.

8. There is an additional factor of $(-1)^L$, where L is the number of closed fermion loops in the graph. This comes about because closed loops correspond to terms involving $(\Phi_0, T(j_\mu^{(0)}(x_1)\, j_\nu^{(0)}(x_2) \cdots j_\lambda^{(0)}(x_n))\Phi_0)$. The rearrangement of the $\bar{\psi}^{(0)}$ and $\psi^{(0)}$ operators to yield the $\psi^{(0)}(x_i)\,\bar{\psi}^{(0)}(x_j)$ ordering (which leads to the appearance of $-\frac{1}{2}S_F(x_i - x_j)$) always involves an odd number of permutations.

9. Closed loops with an odd number of photons leading into them give a vanishing contribution. This follows from the invariance of the

interaction under charge conjugation. Under charge conjugation, Φ_0 is unaltered, whereas $j_\mu(x)$ changes sign. Hence

$$(\Phi_0, T(j_\mu(x_1)j_\nu(x_2)\cdots)\Phi_0) = (-1)^n(\Phi_0, T(j_\mu(x_1), j_\nu(x_2)\cdots)\Phi_0) \quad (9.8)$$

This is known as *Furry's theorem*.

These rules do not yet include the effect of mass renormalization—the inclusion of $\delta m\bar{\psi}\psi$ in the interaction Lagrangian—or the effect of multiplying the matrix element by the still undetermined constants $Z_2^{-e_e/2}Z_3^{-b_e/2}$. The rules also need amplification when some of the integrals turn out to be infinite. These topics are reserved for Chapter 13.

As a last remark it should be pointed out that because of gauge invariance the photon "propagator" $-g_{\mu\nu}/(k^2 + i\epsilon)$ may be replaced by

$$-\left(g_{\mu\nu} - \lambda\frac{k_\mu k_\nu}{k^2}\right)\frac{1}{k^2 + i\epsilon}$$

with λ constant, without changing any observable quantity. This is sometimes useful in calculations.

Interaction of Spin Zero Bosons with Photons

The interaction Lagrangian for the coupling of a charged meson field with the electromagnetic field is obtained from

$$\mathcal{L}(x) = g^{\mu\nu}(\partial_\mu\,\phi(x))(\partial_\nu\,\phi^*(x)) - \mu^2\,\phi(x)\,\phi^*(x)$$

by the replacement

$$\partial_\mu\,\phi(x) \to (\partial_\mu - ie\,A_\mu(x))\,\phi(x)$$
$$\partial_\nu\,\phi^*(x) \to (\partial_\nu + ie\,A_\nu(x))\,\phi^*(x)$$

so that

$$\mathcal{L}_I = -ie\,A_\mu(x)(\phi(x)\,\overleftrightarrow{\partial}^\mu\,\phi^*(x)) + e^2A_\mu(x)\,A^\mu(x)\,\phi(x)\,\phi^*(x)$$

The Feynman rules are the following:

1. Each internal photon line is again represented by $-g_{\mu\nu}/(k^2 + i\epsilon)$.

2. At each vertex at which a single photon interacts we write $e(q_1{}^\mu + q_2{}^\mu)$, where q_1 is the momentum of the meson leading into the vertex and q_2 the momentum of the meson leaving it. It should be noted that charged mesons have an arrow associated with the lines, just like the fermions.

3. The presence of the term $e^2A_\mu A^\mu\phi\phi^*$ implies that there are vertices at which two photon lines enter. In terms of creation and annihilation parts this has the form

$$e^2\,g^{\mu\nu}(A_\mu^{(+)} + A_\mu^{(-)})(A_\nu^{(+)} + A_\nu^{(-)})\phi\phi^*$$

Thus the vertex contribution is $2e^2g^{\mu\nu}$ when one photon comes in and the other leaves at a point. The same contribution will appear if both photons come out (or go in), provided we do not draw two graphs for the process shown in Fig. 9.3*a*. If there are closed loops of the type seen in Fig. 9.3*b*, a factor of $\frac{1}{2}$ should be included to avoid the duplication implied in the factor $2e^2$ at the four-point vertex.

4. Each external photon line that joins a vertex with the label μ is to be multiplied by $e_\mu{}^\lambda$ if the polarization of the photon is λ.

5. The factors of 2π are the same as before. Again there is an i for each power of $\mathcal{L}_I$ and $(i)^{e_i+b_i}$, but this means that care has to be taken when we work to a given order in e rather than in $\mathcal{L}_I$. There is no factor of $(-1)^L$ associated with the closed loops, since we are dealing with bosons here, but Furry's theorem that closed loops with an odd number of photon vertices give no contribution still holds true.

6. The propagator for the meson, of momentum q, is given by

$$\frac{1}{q^2 - \mu^2 + i\epsilon}$$

We may write down the Feynman rules for other theories, but we shall not do so here, for the main use we shall make of perturbation theory is to calculate scattering matrix elements for quantum electrodynamics. This does not mean that our efforts to arrive at the Feynman rules and the perturbation series have been a waste of time. The perturbation series is a formal solution of the equations of motion, and a term-by-term examination of the series has yielded much insight into the structure of the general term. An example is the study of the analyticity properties of matrix elements, a subject we briefly discuss in Chapter 22.

It remains to establish a connection between the scattering matrix elements we calculate and the experimentally measured transition rates or cross sections. The quantity calculated is the matrix element of the operator R defined by

$$S - 1 = R \tag{9.9}$$

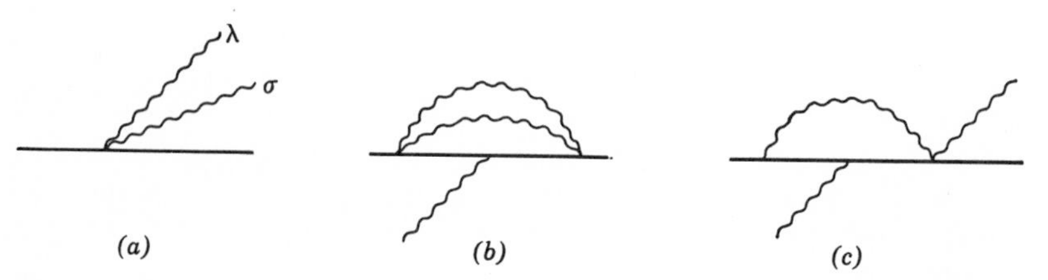

Fig. 9.3. Some graphs for photon-meson interactions. The solid lines represent mesons here.

If the initial state is denoted by i and the final state by f, we find quite generally that

$$R_{fi} = -(2\pi)^4 i\, \delta(P_f - P_i) T_{fi} \tag{9.10}$$

where P_f and P_i are the total energy-momentum four-vectors for the final and initial states. This follows from the Feynman rules but is really quite general. For example, if we insert the plane-wave solution in the right-hand side of the reduction formula (6.28) we find

$$\begin{aligned} R_{\alpha,pq} &= \frac{i}{(2\pi)^{3/2}} \int dx (\Psi_\alpha^{\text{out}}, j(x) \Psi_p)\, e^{-iqx} \\ &= i(2\pi)^4\, \delta(P_\alpha - p - q) \frac{(\Psi_\alpha^{\text{out}}, j(0)\Psi_p)}{(2\pi)^{3/2}} \end{aligned} \tag{9.11}$$

using

$$j(x) = e^{iPx} j(0) e^{-iPx}$$

Incidentally, if an interaction with an external field is being considered, there is no invariance under spatial displacements, hence no momentum conservation.[1] With a time-independent external field we find

$$R_{fi} = \delta(E_f - E_i) V_{fi} \tag{9.12}$$

To calculate the transition rate we must take the absolute square of R_{fi}. This presents an obvious but spurious problem arising from the fact that $(\delta(P_f - P_i))^2$ becomes $\delta(P_f - P_i)\, \delta(0)$ which has no meaning. The difficulty is in an idealization. In the plane-wave representation we deal with infinitely long wave trains, and with beams that are uniform over all space, with a density given by

$$(\Psi_q, i\, \phi^*(x) \overleftrightarrow{\partial}^0 \phi(x) \Psi_q) = \frac{2\omega_q}{(2\pi)^3} \tag{9.13}$$

for bosons. The normalization is such that we take $E_p/(2\pi)^3 M$ fermions per unit volume. If we were to put the system in a box of volume V and consequently keep the time of interaction finite (for a duration $2T$, with $T \ll V^{1/3}$ to avoid having to consider reflections at the boundary), the $(2\pi)^4\, \delta(P_f - P_i)$ in (9.11), for example, would be replaced by

$$\int_V d^3x \int_{-T}^{T} dx_0 e^{i(P_f - P_i)x} = V \delta_{\mathbf{P}_f \mathbf{P}_i} \frac{2}{|E_f - E_i|} \sin |T(E_f - E_i)|$$

[1] An external field is always used as an approximation for a very heavy source, for which recoil may be neglected. There is really momentum conservation, but it is assumed that the source can absorb momentum without a change of state.

The square of this is

$$4V^2\,\delta_{\mathbf{P}_f\mathbf{P}_i}\left|\frac{\sin T(E_f - E_i)}{E_f - E_i}\right|^2$$

If we now ask for the rate per unit volume, we must divide $R_{fi}{}^2$ by $2T$, to get a transition rate, and by V. Thus the rate is

$$\Gamma_{fi} = 2\,|T_{fi}|^2 V\,\delta_{\mathbf{P}_f\mathbf{P}_i}\,\frac{\sin^2 T(E_f - E_i)}{T(E_f - E_i)^2}$$

If we now let V and T increase to infinity and use the limit

$$V\,\delta_{\mathbf{P}_f\mathbf{P}_i} \rightarrow (2\pi)^3\,\delta(\mathbf{P}_f - \mathbf{P}_i)$$

and

$$\lim_{T\rightarrow\infty}\frac{2\sin^2 xT}{x^2T} = 2\pi\,\delta(x)$$

we obtain for the rate per unit volume

$$\Gamma_{fi} = (2\pi)^4\,\delta(P_f - P_i)\,|T_{fi}|^2 \tag{9.14}$$

The reader can check for himself that in the case of an interaction with an external field there is no difficulty with the volume (there is only one source, so to speak) and the interaction rate is

$$\Gamma_{fi} = \frac{\delta(E_f - E_i)}{2\pi}\,|V_{fi}|^2 \tag{9.15}$$

Whenever we have a two-particle initial state, or a single particle in the initial state for an external field problem, a useful quantity is the reaction cross section, which represents an effective area, over which the incident particle reacts. The area is defined in such a way that we need to renormalize the densities to one particle per unit volume.[1] When this is done, the cross section is just the rate divided by the flux of incident particles, which is now the velocity of the incident particle relative to the target particle or external source. Thus

$$\sigma_{fi} = \frac{(2\pi)^4\,\delta(P_f - P_i)\,|T_{fi}|^2}{\rho_i^{(1)}\rho_i^{(2)}\,|\mathbf{v}_{12}|}\,\frac{1}{\prod_j \rho_f^{(j)}} \tag{9.16}$$

[1] In calculating decay rates, we also deal with particle densities normalized to one per unit volume, so that

$$d\Gamma_{fi} = (2\pi)^4\,\delta(P_f - P_i)\,|T_{fi}|^2\,\frac{1}{\rho_i}\prod_n \frac{d^3q_f^{(n)}}{(2\pi)^3\rho_f^{(n)}}$$

with

$$\rho = \frac{2\omega}{(2\pi)^3} \qquad \text{for bosons}$$

$$\rho = \frac{E}{(2\pi)^3 M} \qquad \text{for fermions}$$

To find the cross section for the final particles j ending up in a momentum interval d^3q_j about $\mathbf{q}_j$ we must multiply σ_{fi} by

$$\prod_j \frac{d^3q_j}{(2\pi)^3} \tag{9.17}$$

The extra factor of $(2\pi)^{-3}$ comes from phase space when the volume is unity.

It might be expected that the cross section, representing an area normal to the incident beam, is relativistically invariant. We note that T_{fi} is invariant, and for the final particles

$$\int \frac{d^3q_j}{\rho_j} \propto \int \frac{d^3q}{\omega_q} = 2\int dq\, \delta(q^2 - \mu^2)\, \theta(q_0)$$

is also invariant. Also, the quantity

$$\rho^{(1)}\rho^{(2)}\, |\mathbf{v}_{12}| \propto \omega_1\omega_2 \left| \frac{\mathbf{q}_1}{\omega_1} - \frac{\mathbf{q}_2}{\omega_2} \right|$$

which in the center-of-mass system is equal to

$$\omega_1\omega_2 \left| \frac{\mathbf{q}_1}{\omega_1} + \frac{\mathbf{q}_1}{\omega_2} \right| = |\mathbf{q}_1|\, \omega_1\omega_2$$

can be written in terms of a quantity F introduced by Møller

$$F = ((q_1 \cdot q_2)^2 - q_1{}^2q_2{}^2)^{1/2} \tag{9.18}$$

Clearly in the center of mass system

$$F = [(-\mathbf{q}_1{}^2 - \omega_1\omega_2)^2 - \mu_1{}^2\mu_2{}^2]^{1/2} = |\mathbf{q}_1|\, (\omega_1 + \omega_2)$$

which established our result.

With this we conclude our introduction to field theory. In what follows we make use of much of these nine chapters. In particular, Feynman graphs appear frequently. There is, of course, much more to field theory, and only specialized treatises can hope to do justice to the subject. Some of the more recent practical developments appear in appropriate places, and a brief discussion of renormalization is given in Chapter 13, in which the radiative corrections to Coulomb scattering are explained; much, however, will have to remain unsaid.

PART II

Introduction

Quantum electrodynamics deals with the interaction between photons and electrons. Electrical forces are responsible for most of the manifestations of the behavior of matter, as we see it, so that an understanding of them is of great importance. Furthermore, the earliest "high energy physics" dealt with electrons and radiation, so that a number of techniques and a way of thinking about processes developed in this field. Both may prove useful in the less well understood domain of the strong interactions.

Inasmuch as a number of excellent and detailed books on quantum electrodynamics already exist, no attempt is being made to treat the subject exhaustively; this would, in any case, be inconsistent with our objectives. Rather, emphasis is given to topics usually not treated in detail. In Chapter 10 the Compton effect and related processes are discussed. Some of the commonly used calculational tricks are introduced in this chapter, but the Klein-Nishina formula is not derived. Instead, both in this chapter and in Chapter 11, which deals with electron scattering, much space is devoted to polarization phenomena. These phenomena are of interest in connection with the new experiments in the area of weak interactions. In Chapter 12 bremsstrahlung is discussed in the soft photon limit and in the Weizsäcker-Williams approximation.

Higher-order radiative corrections are discussed only for one process—Coulomb scattering. Thus no general treatment of the renormalization program is presented, although it is hoped that the concepts used in it will be clarified by the special example in Chapter 13. Our treatment is consciously old-fashioned in that no use is made of dispersion techniques. Our aim, after all, is to show that the Feynman rules, if interpreted properly, do give rise to finite answers with both the ultraviolet and the infrared divergences removed.

The reader may supplement our discussion by reference to the handbook article of Källén,[1] to the treatises of Heitler,[2] Jauch and Rohrlich,[3] or Akhieser and Berestetsky,[4] and, for the renormalization program, to the new book by Bjorken and Drell.[5]

[1] G. Källén, "Quantenelektrodynamik," *Handbuch der Physik*, Vol. V/I, Springer Verlag, Berlin.
[2] W. Heitler, *The Quantum Theory of Radiation*, Oxford University Press, Oxford, 1954.
[3] J. M. Jauch and F. Rohrlich, *The Theory of Photons and Electrons*, Addison-Wesley Publishing Co., Reading Mass., 1955.
[4] A. I. Akhiezer and V. B. Berestetsky, *Quantum Electrodynamics*, John Wiley and Sons, New York, 1963.
[5] J. D. Bjorken and S. D. Drell, *Relativistic Quantum Fields*, McGraw-Hill Book Co., New York, 1965.

10

The Compton Effect and Related Processes

The Compton effect is the name given to the process in which a photon is scattered by a free electron. We shall use the Feynman rules developed in Chapter 9 to calculate the cross section for this process. To lowest order there are two graphs whose contribution must be calculated (Fig. 10.1), and their contributions must be added coherently.

The matrix element for the process is given by

$$R = \delta(p' + q' - p - q)\mathfrak{R}$$

with

$$\begin{aligned}\mathfrak{R} &= -\frac{ie^2}{(2\pi)^2}\, e_\nu^{\lambda'} e_\mu^{\lambda}\, \bar{u}(p')\left(\gamma^\nu \frac{1}{\not p + \not q - m}\gamma^\mu + \gamma^\mu \frac{1}{\not p - \not q' - m}\gamma^\nu\right)u(p) \\ &= -\frac{ie^2}{(2\pi)^2}\,\bar{u}(p')\left[\not e' \frac{m + \not p + \not q}{(p+q)^2 - m^2}\not e + \not e\frac{m + \not p - \not q'}{(p - q')^2 - m^2}\not e'\right]u(p)\end{aligned} \tag{10.1}$$

This expression may be simplified somewhat with the help of the anti-commutation properties of the γ's, the fact that $u(p)$ and $\bar{u}(p')$ satisfy the Dirac equation, and the conditions

$$\begin{aligned} p^2 = p'^2 &= m^2 \\ q^2 = q'^2 &= 0 \\ q \cdot e^\lambda = q' \cdot e^{\lambda'} &= 0 \end{aligned} \tag{10.2}$$

A simplified form of the matrix element is

$$\begin{aligned}\mathfrak{R} &= -\frac{ie^2}{(2\pi)^2}\,\bar{u}(p')\left[\left(\not e'\frac{e\cdot p}{q\cdot p} + \frac{\not e'\not q\not e}{2q\cdot p}\right) + \left(-\not e\frac{e'\cdot p}{q'\cdot p} + \frac{\not e\not q'\not e'}{2p\cdot q'}\right)\right]u(p) \\ &\equiv -\frac{ie^2}{(2\pi)^2}\,\bar{u}(p')\mathcal{M}\,u(p)\end{aligned} \tag{10.3}$$

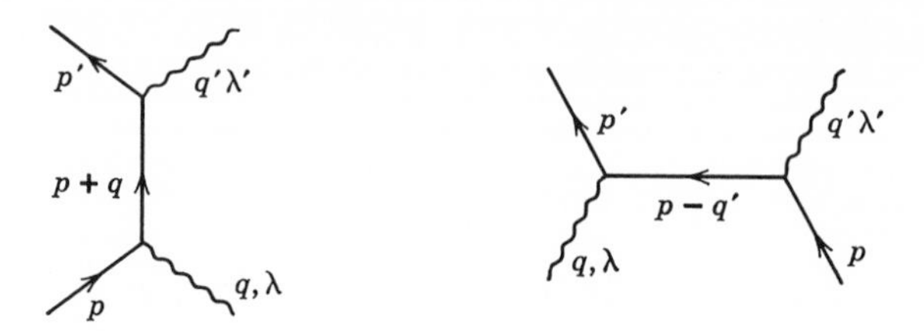

Fig. 10.1. The lowest order graphs for Compton scattering.

The terms corresponding to the two graphs have been kept separate because we now want to show that each term separately is not gauge-invariant but that the sum is. The gauge transformation

$$A_\mu(x) \to A_\mu(x) - \partial_\mu \Lambda(x)$$
$$\square \Lambda(x) = 0$$

leads to the transformation

$$\begin{aligned} \not{e} &\to \not{e} + \lambda \not{q} \\ \not{e}' &\to \not{e}' + \lambda \not{q}' \end{aligned} \tag{10.4}$$

with λ constant. The Lorentz conditions

$$e \cdot q = e' \cdot q' = 0$$

remain unaltered, since $q^2 = q'^2 = 0$. Under the transformation the separate terms in (10.3) become

$$(\not{e}' + \lambda \not{q}')\left(\frac{e \cdot p}{q \cdot p} + \lambda + \frac{\not{q}\not{e}}{2p \cdot q}\right)$$

and

$$-(\not{e} + \lambda \not{q})\left(\frac{e' \cdot p}{q' \cdot p} + \lambda - \frac{\not{q}'\not{e}'}{2p \cdot q'}\right)$$

The additional contributions therefore are

$$\bar{u}(p')\left[\lambda \not{q}'\left(\frac{e \cdot p}{q \cdot p} + \lambda + \frac{\not{q}\not{e}}{2q \cdot p}\right) + \lambda \not{e}' - \lambda \not{q} \times \left(\frac{e' \cdot p}{q' \cdot p} + \lambda - \frac{\not{q}'\not{e}'}{2q' \cdot p}\right) - \lambda \not{e}\right] u(p)$$

The coefficient of λ^2 immediately vanishes, since

$$\bar{u}(p')[\not{q}' - \not{q})\, u(p) = \bar{u}(p')(\not{p} - \not{p}')\, u(p) = 0$$

For the term linear in λ we use

$$\begin{aligned}\bar{u}(p')\not{q}'\not{q}\not{e}\,u(p) &= \bar{u}(p')(\not{p} + \not{q} - \not{p}')\not{q}\not{e}\,u(p)\\ &= \bar{u}(p')(\not{p} - m)\not{q}\not{e}\,u(p)\\ &= \bar{u}(p')(\not{q}\not{e}(\not{p} - m) - 2e\cdot p\not{q} + 2p\cdot q\not{e})\,u(p)\\ &= \bar{u}(p')(2p\cdot q\not{e} - 2e\cdot p\not{q})\,u(p)\end{aligned}$$

and

$$\bar{u}(p')\not{q}\not{q}'\not{e}'\,u(p) = \bar{u}(p')(2p\cdot e'\not{q}' - 2p\cdot q'\not{e}')\,u(p)$$

to show that they too cancel.

To compute the cross section we must first square the matrix element. This could be done directly, using explicit representations of the plane-wave spinors and the Dirac matrices. We can, however, save ourselves a great deal of labor with the help of a few computational tricks, and it is these that we now discuss. Consider the matrix element

$$a_{rs} = \bar{u}^{(r)}(p')A\,u^{(s)}(p) \tag{10.5}$$

where A is a 4×4 matrix, consisting of products of γ matrices. It will in general have the form

$$A = \sum_{i=0}^{n} C_i \not{a}_1 \not{a}_2 \cdots \not{a}_i \tag{10.6}$$

Since

$$a_{rs}^* = (u^{*(r)}(p')\gamma^0 A\,u^{(s)}(p))^* = u^{(s)}(p)^* A^+ \gamma^0\,u^{(r)}(p')$$

it follows that

$$|a_{rs}|^2 = \bar{u}^{(r)}(p')A\,u^{(s)}(p)\,\bar{u}^{(s)}(p)\gamma^0 A^+ \gamma^0\,u^{(r)}(p')$$

Now

$$\begin{aligned}\gamma^0 A^+ \gamma^0 &= \gamma^0 \sum_{i=0}^{n} C_i^* \not{a}_i^+ \not{a}_{i-1}^+ \cdots \not{a}_1^+ \gamma^0\\ &= \sum_i C_i^* \tilde{\not{a}}_i \tilde{\not{a}}_{i-1} \cdots \tilde{\not{a}}_1 \equiv \tilde{A}\end{aligned} \tag{10.7}$$

where

$$\tilde{\not{a}} = \gamma^\alpha a_\alpha^*$$

Thus

$$|a_{rs}|^2 = \bar{u}^{(r)}(p')A\,u^{(s)}(p)\,\bar{u}^{(s)}(p)\tilde{A}\,u^{(r)}(p') \tag{10.9}$$

If the initial electron spin states are averaged over and/or the final electron spin states are summed over, as is appropriate for an experiment not seeking spin information, we may use

$$\sum_s u^{(s)}(p)\,\bar{u}^{(s)}(p) = \frac{m + \not{p}}{2m}$$

and/or

$$\sum u^{(r)}(p')\,\bar{u}^{(r)}(p') = \frac{\not{p}' + m}{2m}$$

Thus, if no information regarding the electron polarization is sought, the absolute square of the matrix element is to be summed over the final spin states and averaged over the initial spin states. We therefore calculate

$$\frac{1}{2}\sum_{r,s}|a_{rs}|^2 = \frac{1}{2}\sum_{r}\bar{u}_\alpha^{(r)}(p')\left(A\frac{\not{p}+m}{2m}\tilde{A}\right)_{\alpha\beta}u_\beta^{(r)}(p')$$
$$= \frac{1}{2}\,\mathrm{tr}\left(A\frac{\not{p}+m}{2m}\tilde{A}\frac{\not{p}'+m}{2m}\right) \tag{10.10}$$

This is a very useful form, for traces of products of Dirac matrices are easily calculated with the help of the following rules:

$$\begin{aligned}
&\tfrac{1}{4}\,\mathrm{tr}\,\mathbf{1} = 1\\
&\tfrac{1}{4}\,\mathrm{tr}\,\gamma_\mu\gamma_\nu = g_{\mu\nu}\\
&\tfrac{1}{4}\,\mathrm{tr}\,(\gamma_{\mu_1}\cdots\gamma_{\mu_n}) = 0 \qquad (n\ \text{odd})\\
&\tfrac{1}{4}\,\mathrm{tr}\,(\not{a}\not{b}) = a\cdot b\\
&\tfrac{1}{4}\,\mathrm{tr}\,(\not{a}\not{b}\not{c}\not{d}) = a\cdot b\,c\cdot d - a\cdot c\,b\cdot d + a\cdot d\,b\cdot c\\
&\quad\cdot\\
&\quad\cdot\\
&\quad\cdot
\end{aligned} \tag{10.11}$$

Even if we do seek spin information we may still reduce the calculation to a trace calculation, provided we use the spin projection operators of the form

$$\tfrac{1}{2}(1 + \gamma_5\not{n})$$

as discussed on page 36.

If no information concerning either the initial or the final photon polarization is sought, the average or sum over polarization states must be calculated. This may be done explicitly: consider the coordinate system in which the photon momentum is $(q, 0, 0, q)$. The transverse polarization vector is of the form $\mathbf{e} = (0, \cos\varphi, \sin\varphi, 0)$, so that

$$\langle \mathbf{e}\,.\,\mathbf{M}\rangle = \langle M_1\cos\varphi\rangle + \langle M_2\sin\varphi\rangle = 0$$

However

$$\begin{aligned}
\langle \mathbf{e}\,.\,\mathbf{M}\mathbf{e}\,.\,\mathbf{N}\rangle &= \langle (M_1\cos\varphi + M_2\sin\varphi)(N_1\cos\varphi + N_2\sin\varphi)\rangle\\
&= M_1N_1\langle\cos^2\varphi\rangle + M_2N_2\langle\sin^2\varphi\rangle\\
&\quad + \tfrac{1}{2}(M_1N_2 + M_2N_1)\langle\sin 2\varphi\rangle\\
&= \tfrac{1}{2}(M_1N_1 + M_2N_2)\\
&= \tfrac{1}{2}(\mathbf{M}\cdot\mathbf{N} - \mathbf{M}\cdot\hat{\mathbf{q}}\mathbf{N}\cdot\hat{\mathbf{q}})
\end{aligned} \tag{10.12}$$

The same result can be obtained with the help of the covariant polarization relation

$$\sum_\lambda e_\mu^{(\lambda)}e_\nu^{(\lambda)} = -g_{\mu\nu}$$

With the help of this relation

$$\frac{1}{2}\sum_{\lambda} e_\mu{}^\lambda M^\mu e_\nu{}^\lambda N^\nu = -\frac{1}{2} M_\mu N^\mu \tag{10.13}$$

Gauge invariance however requires that

$$q^\mu M_\mu = q^\mu N_\mu = 0 \tag{10.14}$$

which implies that in the frame in which $q^\mu = (q, 0, 0, q)$, $M_3 = -M_0$ and $N_3 = -N_0$, so that the result is identical with that of (10.12).

To obtain the cross section for Compton scattering, we note from (10.3) that the transition matrix T_{fi}, related to R_{fi} by $-R_{fi} = (2\pi)^4 i T_{fi}$ is given by

$$T_{fi} = \frac{e^2}{(2\pi)^6} \bar{u}(p')\mathcal{M}\, u(p)$$

The cross section is given by [see (9.16)]

$$d\sigma = \frac{(2\pi)^4\, \delta(p' + q' - p - q) \sum |T_{fi}|^2}{\dfrac{2q}{(2\pi)^3}\dfrac{E_p/m}{(2\pi)^3}|\mathbf{v}_{\mathrm{rel}}|} \frac{d^3p'}{E_p'/m}\frac{d^3q'}{2q'} \tag{10.15}$$

In the center of mass system

$$|\mathbf{v}_{\mathrm{rel}}| = 1 + \frac{p}{E_p} = \frac{W}{E_p}$$

where W is the center of mass energy. Integration over the spatial momentum of the photon is trivially carried out, so that

$$d\sigma = \frac{(2\pi)^{10} m^2}{4q^2 W} \sum |T_{fi}|^2 \frac{d^3p'}{E_{p'}}\, \delta(E_{p'} + p' - W)$$

From this, noting that in the center of mass frame[1] $p = q = q' = p'$, we obtain

$$\begin{aligned}
\frac{d\sigma}{d\Omega} &= \frac{(2\pi)^{10} m^2}{4W} \int \frac{p'^2\, dp'}{q^2 E_{p'}} \sum |T_{fi}|^2\, \delta(E_{p'} + p' - W) \\
&= \frac{(2\pi)^{10} m^2}{4W^2} \sum |T_{fi}|^2 \\
&= \frac{e^4}{4\pi^2}\frac{m^2}{4W^2}\frac{1}{4m^2}\frac{1}{2} \operatorname{tr}(\mathcal{M}(p + m)\, \tilde{\mathcal{M}}(p' + m)) \\
&= \left(\frac{e^2}{4\pi}\right)^2 \frac{1}{2W^2}\frac{1}{4} \operatorname{tr}(\mathcal{M}(p + m)\, \tilde{\mathcal{M}}(p' + m))
\end{aligned} \tag{10.16}$$

[1] Here $p = |\mathbf{p}|$, etc.

The trace term is straightforwardly evaluated. We need not work it out in any particular reference frame. In general, it will turn out to be a function of various scalars, such as $p \cdot p'$, $p \cdot q$, $e \cdot p'$, $e' \cdot q$, ... We may also rewrite the kinematical terms in terms of invariants. For this purpose we introduce the variables,

$$\begin{aligned} s &= (p + q)^2 = (p' + q')^2 \\ t &= (q - q')^2 = (p' - p)^2 \\ u &= (p - q')^2 = (p' - q)^2 \end{aligned} \tag{10.17}$$

These are not independent, since

$$s + t + u = 2m^2 \tag{10.18}$$

In the center of mass system

$$s = (E_p + p)^2 = W^2$$

and

$$-t = (\mathbf{q} - \mathbf{q}')^2 = 2q^2(1 - \cos\theta)$$

where θ is the scattering angle in the center of mass frame. The cross section, as already pointed out, is an invariant. We now write

$$\begin{aligned} d\Omega &= 2\pi\, d(\cos\theta) \\ &= \frac{\pi}{q^2}\, dt = \frac{\pi\, dt}{(s - m^2)^2/4s} \end{aligned}$$

so that

$$\begin{aligned} \frac{d\sigma}{dt} &= \frac{4\pi s}{(s - m^2)^2} \frac{(e^2/4\pi)^2}{2s} \frac{1}{4} \operatorname{tr}\,(\mathcal{M}(\not{p} + m)\,\tilde{\mathcal{M}}(\not{p}' + m)) \\ &= 2\pi\left(\frac{e^2}{4\pi}\right)^2 \frac{1}{(s - m^2)^2} \frac{1}{4} \operatorname{tr}\,(\mathcal{M}(\not{p} + m)\,\tilde{\mathcal{M}}(\not{p}' + m)) \end{aligned} \tag{10.19}$$

The result of the trace calculation is (after polarization average)

$$\begin{aligned} \frac{1}{2}\sum_{\text{pol}}\frac{1}{4} \operatorname{tr}\,(\;) = \frac{m^2 - s}{u - m^2} + \frac{u - m^2}{m^2 - s} &+ 4m^2\left(\frac{1}{m^2 - s} - \frac{1}{u - m^2}\right) \\ &+ 4m^4\left(\frac{1}{m^2 - s} - \frac{1}{u - m^2}\right)^2 \end{aligned} \tag{10.20}$$

so that the differential cross section for unpolarized electrons and photons is,[1]

[1] J. M. Jauch and F. Rohrlich, *The Theory of Photons and Electrons.* Addison Wesley Publishing Co., Reading, Mass. (1955).

with $\alpha \equiv e^2/4\pi \cong \frac{1}{137}$, the fine structure constant

$$\frac{d\sigma}{dt} = \frac{2\pi\alpha^2}{(s - m^2)^2}\left[\left(\frac{s + m^2}{s - m^2} - \frac{2m^2}{s + t - m^2}\right)^2 + \frac{t}{s - m^2} + \frac{s - m^2}{s + t - m^2}\right] \tag{10.21}$$

The total cross section is obtained by integrating $d\sigma/dt$ over the range of t corresponding to $-1 \leq \cos\theta \leq 1$, i.e.,

$$\sigma_{\text{tot}}(s) = \int_{(s-m^2)^2/s}^{0} dt\, \frac{d\sigma}{dt} \tag{10.22}$$

At high energies

$$\sigma_{\text{tot}}(s) \approx \frac{2\pi\alpha^2}{s} \log\frac{s}{m^2} \tag{10.23}$$

The order of magnitude of the cross section is given by

$$\sigma_0 = \pi\left(\frac{\alpha}{m}\right)^2 = \pi\alpha^2\left(\frac{m_\pi}{m_e}\right)^2\left(\frac{\hbar}{m_\pi c}\right)^2 \approx 2.5 \times 10^{-25}\text{cm}^2$$

In the evaluation of this number we used the pion mass $m_\pi = 140$ MeV for which $m_\pi/m_e \approx 2/\alpha$, and $\hbar/m_\pi c \approx 1.4 \times 10^{-13}$ cm.

The angular distribution of the photons in the center of mass may be read off from (10.21). It is useful to realize that the dominant features of the angular distribution come from the denominators in the matrix elements. For the first graph, we have

$$\frac{1}{(p + q)^2 - m^2} = \frac{1}{s - m^2} \tag{10.24}$$

which only depends on the energy. For the second graph, we have

$$\frac{1}{(p - q')^2 - m^2} = -\frac{1}{2Eq + 2\mathbf{q}\cdot\mathbf{q}'} = -\frac{1}{2Eq}\frac{1}{1 + (q/E)\cos\theta} \tag{10.25}$$

This term depends on the scattering angle, and at high energies, when the electron velocity q/E is close to unity, there is considerable peaking in the backward direction. This is characteristic of exchange potentials, and the process described by the second graph is of an exchange type. The peak is quite narrow: for $\theta = \pi - \delta$,

$$\frac{1}{1 + v\cos\theta} = \frac{1}{1 - v\cos\delta} \approx \frac{1}{1 - v + \frac{1}{2}\delta^2} \approx \frac{2E^2}{m^2 + E^2\delta^2}$$

so that for $\delta \approx m/E$ the peak drops to half its size. It is easy to check that actually half of the cross section is contributed by scattering within an angle of $\approx(m/E)^2$ about the backward direction.

Although a separation of the amplitude into contributions from the two separate graphs is not unambiguous, depending as it does on the choice of gauge, some general deductions can still be made about the separate contributions of the two graphs. The first one is characterized by an intermediate state of total angular momentum $J = \frac{1}{2}$. Its contribution to the angular distribution must be very undramatic, whereas the second graph is of a potential type and represents scattering in all partial waves.[1]

The structure of the graphs can also give us some qualitative notions about the scattering of circularly polarized photons by polarized electrons. Consider, for example, right-circularly polarized photons, which have spin pointing in a direction opposite the direction of the momentum, i.e., they have negative helicity. If they collide with electrons also of negative helicity, taking the photon momentum direction in the center of mass frame as quantization axis, we have an initial state with $J_z = -\frac{1}{2}$, so that scattering with $J \geq \frac{1}{2}$ is allowed and both graphs contribute to the amplitude. If the electrons have positive helicity, $J_z = -\frac{3}{2}$, so that $J \geq \frac{3}{2}$ and only part of the second graph will contribute. We might thus be able to measure the circular polarization of photons by scattering them off polarized electrons. It is actually possible to polarize about 7% of the electrons in iron with strong magnetic fields. To obtain the dependence of the cross section on the polarization, we must look in detail at the Compton amplitude.[2] We shall do so in the laboratory frame.

In terms of the explicit representations of the Dirac spinors and Dirac matrices obtained in Chapter 2, we have

$$\mathfrak{R} = -\frac{ie^2}{4\pi^2}\frac{m - i\rho_2\boldsymbol{\sigma}\cdot\mathbf{p}' + \rho_3 E'}{\sqrt{2m(m+E')}}$$
$$\times\left\{i\rho_2\boldsymbol{\sigma}\cdot\mathbf{e}'\,\frac{m + \rho_3 m - i\rho_2\boldsymbol{\sigma}\cdot\mathbf{q} + \rho_3 q}{m^2 - s}\,i\rho_2\boldsymbol{\sigma}\cdot\mathbf{e}\right.$$
$$\left. + \, i\rho_2\boldsymbol{\sigma}\cdot\mathbf{e}\,\frac{m + \rho_3 m + i\rho_2\boldsymbol{\sigma}\cdot\mathbf{q}' - \rho_3 q'}{m^2 - u}\,i\rho_2\boldsymbol{\sigma}\cdot\mathbf{e}'\right\}\frac{1+\rho_3}{2}$$

[1] To this order in perturbation theory, changes in the gauge only mix up the $J = \frac{1}{2}$ parts of the two graphs.

[2] The most detailed calculations may be found in F. W. Lipps and H. A. Tolhoek, *Physica* **20,** 85,395 (1954).

which can be worked out to yield

$$\mathfrak{R} = -\frac{i\alpha}{\pi}\frac{1}{\sqrt{2m(m+E')}}$$
$$\times\left[\frac{1}{m^2-s}\{q(m+E')\boldsymbol{\sigma}\cdot\mathbf{e}'\boldsymbol{\sigma}\cdot\mathbf{e}+\boldsymbol{\sigma}\cdot\mathbf{p}'\boldsymbol{\sigma}\cdot\mathbf{e}'\boldsymbol{\sigma}\cdot\mathbf{q}\boldsymbol{\sigma}\cdot\mathbf{e}\}\right.$$
$$\left.-\frac{1}{m^2-u}\{q'(m+E')\boldsymbol{\sigma}\cdot\mathbf{e}\boldsymbol{\sigma}\cdot\mathbf{e}'+\boldsymbol{\sigma}\cdot\mathbf{p}'\boldsymbol{\sigma}\cdot\mathbf{e}\boldsymbol{\sigma}\cdot\mathbf{q}'\boldsymbol{\sigma}\cdot\mathbf{e}'\}\right] \quad (10.26)$$

with

$$\begin{aligned} s &= (m+q)^2 - \mathbf{q}^2 = m^2 + 2mq \\ u &= (m-q')^2 - \mathbf{q}'^2 = m^2 - 2mq' \end{aligned} \quad (10.27)$$

Note that at threshold $\mathfrak{R}$ reduces to the Thomson amplitude

$$\mathfrak{R} = \frac{i\alpha}{2m\pi}(\boldsymbol{\sigma}\cdot\mathbf{e}'\boldsymbol{\sigma}\cdot\mathbf{e}+\boldsymbol{\sigma}\cdot\mathbf{e}\boldsymbol{\sigma}\cdot\mathbf{e}') = \frac{i\alpha}{m\pi}\mathbf{e}'\cdot\mathbf{e} \quad (10.28)$$

The expression for $\mathfrak{R}$ is a 2×2 matrix, and the implication is that

$$|\chi_f^+\mathfrak{R}\chi_i|^2 = \chi_f^+\mathfrak{R}\chi_i\chi_i^+\mathfrak{R}^+\chi_f \quad (10.29)$$

is to be evaluated for the cross section. If the electron is polarized in the z-direction,

$$\chi_i\chi_i^+ = \begin{pmatrix}1\\0\end{pmatrix}(1\quad 0) = \begin{pmatrix}1&0\\0&0\end{pmatrix} = \frac{1+\sigma_3}{2}$$

More generally, if the electron has polarization P in the direction $\hat{\mathbf{n}}$,

$$\chi_i\chi_i^+ = \frac{1+P\boldsymbol{\sigma}\cdot\hat{\mathbf{n}}}{2} \quad (10.30)$$

If the polarization of the electron in the final state is not measured, what enters the cross section is

$$\sum_f \chi_f^+\mathfrak{R}\frac{1+P\boldsymbol{\sigma}\cdot\hat{\mathbf{n}}}{2}\mathfrak{R}^+\chi_f = \operatorname{tr}\left(\mathfrak{R}\frac{1+P\boldsymbol{\sigma}\cdot\hat{\mathbf{n}}}{2}\mathfrak{R}^+\right) \quad (10.31)$$

For an unpolarized target the cross section is proportional to

$$\tfrac{1}{2}\operatorname{tr}\mathfrak{R}\mathfrak{R}^+ \quad (10.32)$$

We see therefore that with a polarized target the cross section must be of the form[1]

$$\frac{d\sigma}{d\Omega} = \frac{d\sigma_0}{d\Omega} + P\frac{d\sigma_1}{d\Omega} \quad (10.33)$$

[1] Actually, to lowest order in perturbation theory there will be no dependence on P unless one other polarization is also measured. See the discussion on page 162.

If the correlation between the electron polarization and the photon polarization is to be studied, it is useful to rewrite the matrix element $\mathfrak{R}$ in terms of the *photon spin operator* **S**. **S** should play the same role for photons as $\boldsymbol{\sigma}$ does for electrons. It has the following properties:

1. It may be represented by a 3 × 3 matrix; this matrix stands between photon "spin vectors" **e** for the initial state and **e**′ for the final state, analogous to the way in which $\boldsymbol{\sigma}$ stands between χ_f^+ and χ_i.
2. **S** satisfies the conditions for an angular momentum, so that

$$\mathbf{S} \times \mathbf{S} = i\mathbf{S} \tag{10.34}$$

i.e.,

$$e_{ijk}(S_i)_{ab}(S_j)_{bc} = i(S_k)_{ac} \tag{10.35}$$

3. Since the photon has spin 1,

$$(\mathbf{S}^2)_{ac} = (S_i)_{ab}(S_i)_{bc} = 2\,\delta_{ac} \tag{10.36}$$

It is easily checked that a representation satisfying these conditions is

$$(S_i)_{ab} = -ie_{abi} \tag{10.37}$$

We may now compute some matrix elements of powers of **S**. For example,

$$\begin{aligned}(\mathbf{e}'\,|\mathbf{a}\cdot\mathbf{S}|\,\mathbf{e}) &= e_m' a_i(S_i)_{mn}e_n \\ &= -i\mathbf{a}\cdot\mathbf{e}'\times\mathbf{e}\end{aligned} \tag{10.38}$$

Similarly,

$$\begin{aligned}(\mathbf{e}'\,|\mathbf{a}\cdot\mathbf{S}\mathbf{b}\cdot\mathbf{S}|\,\mathbf{e}) &= e_m' a_i(S_i)_{mn}b_j(S_j)_{nk}e_k \\ &= -e_m' e_k a_i b_j e_{imn} e_{jnk} \\ &= e_m' e_k a_i b_j(\delta_{ij}\,\delta_{mk} - \delta_{ik}\,\delta_{mj}) \\ &= \mathbf{e}'\cdot\mathbf{e}\mathbf{a}\cdot\mathbf{b} - \mathbf{e}'\cdot\mathbf{b}\mathbf{e}\cdot\mathbf{a}\end{aligned} \tag{10.39}$$

In what follows we shall replace the dependence of the Compton scattering amplitude on **e** and **e**′ by a dependence on **S**, understood to be standing between the initial and final spin-state vectors. We first use

$$\begin{aligned}\boldsymbol{\sigma}\cdot\mathbf{q}\boldsymbol{\sigma}\cdot\mathbf{e} &= -\boldsymbol{\sigma}\cdot\mathbf{e}\boldsymbol{\sigma}\cdot\mathbf{q} \\ \boldsymbol{\sigma}\cdot\mathbf{q}'\boldsymbol{\sigma}\cdot\mathbf{e}' &= -\boldsymbol{\sigma}\cdot\mathbf{e}'\boldsymbol{\sigma}\cdot\mathbf{q}'\end{aligned}$$

which follow from $\boldsymbol{\sigma}\cdot\mathbf{q}\boldsymbol{\sigma}\cdot\mathbf{e} = \mathbf{q}\cdot\mathbf{e} + i\boldsymbol{\sigma}\cdot\mathbf{q}\times\mathbf{e}$ and $\mathbf{e}\cdot\mathbf{q} = \mathbf{e}'\cdot\mathbf{q}' = 0$ to juxtapose $\boldsymbol{\sigma}\cdot\mathbf{e}$ and $\boldsymbol{\sigma}\cdot\mathbf{e}'$. Next we note that

$$\begin{aligned}\boldsymbol{\sigma}\cdot\mathbf{e}\boldsymbol{\sigma}\cdot\mathbf{e}' &= \mathbf{e}\cdot\mathbf{e}' - i\boldsymbol{\sigma}\cdot\mathbf{e}'\times\mathbf{e} \\ &= (\mathbf{e}'\,|\mathbf{1}|\,\mathbf{e}) + (\mathbf{e}'\,|\boldsymbol{\sigma}\cdot\mathbf{S}|\,\mathbf{e}) \\ &= (\mathbf{e}'\,|\mathbf{1}+\boldsymbol{\sigma}\cdot\mathbf{S}|\,\mathbf{e})\end{aligned} \tag{10.40}$$

and

$$\boldsymbol{\sigma}\cdot\mathbf{e}'\boldsymbol{\sigma}\cdot\mathbf{e} = (\mathbf{e}'\,|\mathbf{1}-\boldsymbol{\sigma}\cdot\mathbf{S}|\,\mathbf{e}) \tag{10.41}$$

Thus the matrix element in the laboratory frame is

$$\mathcal{R} = \frac{-i\alpha}{\pi}\frac{1}{\sqrt{2m(m+E')}}\left\{\frac{1}{m^2-s}(q(m+E')(\mathbf{1}-\boldsymbol{\sigma}\cdot\mathbf{S})+\boldsymbol{\sigma}\cdot\mathbf{p}'(\boldsymbol{\sigma}\cdot\mathbf{S}-\mathbf{1})\boldsymbol{\sigma}\cdot\mathbf{q})\right.$$

$$\left. - \frac{1}{m^2-u}(q'(m+E')(\mathbf{1}+\boldsymbol{\sigma}\cdot\mathbf{S}) - \boldsymbol{\sigma}\cdot\mathbf{p}'(\boldsymbol{\sigma}\cdot\mathbf{S}+\mathbf{1})\boldsymbol{\sigma}\cdot\mathbf{q}')\right\} \quad (10.42)$$

It is understood that

$$(\mathbf{e}', \chi_f\,|\mathcal{R}|\,\mathbf{e}, \chi_i)$$

is to be calculated. If the photons in the initial state are circularly polarized, with for example, positive helicity, $|\mathbf{e})$ is to be replaced by

$$\frac{1+\hat{\mathbf{q}}\cdot\mathbf{S}}{2}|\mathbf{e}) = \frac{1}{2}(\mathbf{e} - i\mathbf{e}\times\hat{\mathbf{q}}) = \frac{1}{2}(\mathbf{e} + i\hat{\mathbf{q}}\times\mathbf{e}) \quad (10.43)$$

This, it should be noted, is just the direction of the electric vector for a left circularly polarized electromagnetic wave.[1] It can easily be shown that

$$\left(\frac{1+\mathbf{S}\cdot\hat{\mathbf{q}}}{2}\right)^2|\mathbf{e}) = \frac{1+\mathbf{S}\cdot\hat{\mathbf{q}}}{2}|\mathbf{e}) \quad (10.44)$$

If we have circularly polarized photons and a polarized electron target the cross section is proportional to

$$\frac{1}{4}\sum_{e'f}(\mathbf{e}', \chi_f|\,\mathcal{R}\,|\mathbf{e}, \chi_i)(\mathbf{e}\,\chi_i|\,\mathcal{R}^+\,|\mathbf{e}'\,\chi_f)$$

$$= \frac{1}{4}\sum_{e'}\sum_{f}(\mathbf{e}', \chi_f|\,\mathcal{R}\,\frac{1+P\boldsymbol{\sigma}\cdot\hat{\mathbf{n}}}{2}\frac{1\pm\mathbf{S}\cdot\hat{\mathbf{q}}}{2}\mathcal{R}^+\,|\mathbf{e}', \chi_f) \quad (10.45)$$

The sum over final electron states converts this into a trace over the electron spin variables, and the sum over the polarization states $\mathbf{e}'$ is carried out as in (10.12). The simple but lengthy algebraic expression necessary to obtain the correlation is not worked out here because a sample calculation of this kind is presented in our discussion of electron-positron annihilation. The results follow.

With circularly polarized photons and target electrons polarized in the direction of the incident photon momentum, the Compton cross section is

$$\frac{d\sigma}{d\Omega} = \frac{d\sigma_0}{d\Omega} + P'\frac{d\sigma_1}{d\Omega} \quad (10.46)$$

where

$$\frac{d\sigma_0}{d\Omega} = \frac{1}{2}\left(\frac{\alpha}{m}\right)^2\left(\frac{q'^2}{q^2}\right)\left[1+\cos^2\Theta + \frac{q-q'}{m}(1-\cos\Theta)\right] \quad (10.47)$$

[1] For a clear discussion of polarization, see J. D. Jackson, *Classical Electrodynamics*. John Wiley and Sons, New York (1962), Chapter 7.

is the polarization-independent cross section for Compton scattering (Klein-Nishina formula) which can be obtained from (10.21) by expressing the variables s, t, and u in terms of their laboratory frame values. In (10.47) q is the initial photon momentum and Θ is the laboratory scattering angle; q' is the final photon momentum. Furthermore,

$$\frac{d\sigma_1}{d\Omega} = -\frac{1}{2}\left(\frac{\alpha}{m}\right)^2\left(\frac{q'^2}{q^2}\right)\frac{q+q'}{m}\cos\Theta(1-\cos\Theta) \qquad (10.48)$$

and P' is the product of the photon polarization and the electron polarization. If we integrate over angles, the part of the total cross section that depends on the polarization is

$$\sigma_1 = (2\pi)\left(\frac{\alpha}{m}\right)^2\left[\frac{m}{q}\frac{m^2+4mq+5q^2}{(m+2q)^2} - \frac{m^2+mq}{2q^2}\log\left(1+\frac{2q}{m}\right)\right] \qquad (10.49)$$

Figure 10.2 shows the dependence of the ratio $(d\sigma_1/d\Omega)/(d\sigma_0/d\Omega)$ on angle

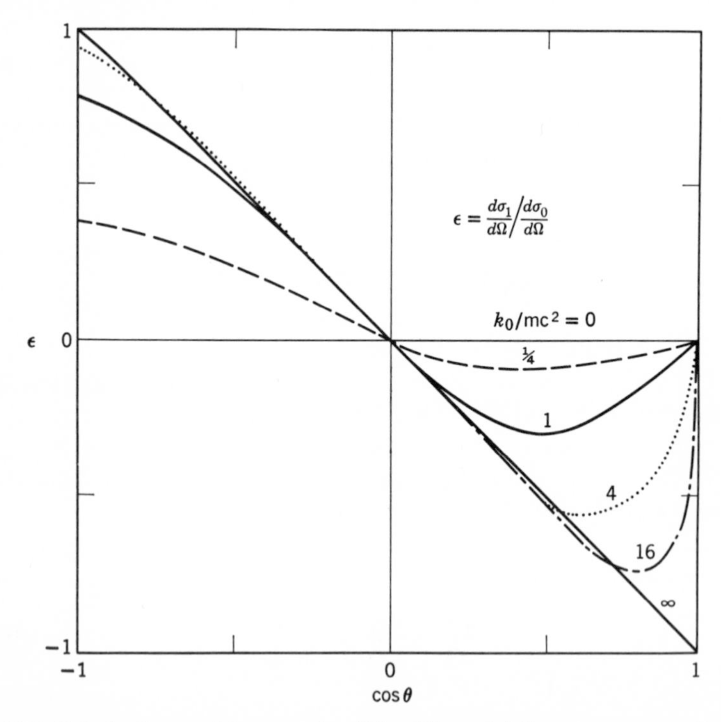

Fig. 10.2. Polarization-sensitive part $d\sigma_1$ of the differential Compton cross section for electrons polarized along the axis of incident circularly polarized photon. (From Gunst and Page, *Phys. Rev.* **90.** 970 (1953).) The curves are normalized to the polarization-insensitive cross section $d\sigma_0$ and plotted for several energies as a function of $\cos\theta$, where θ is the angle through which the photon is scattered.

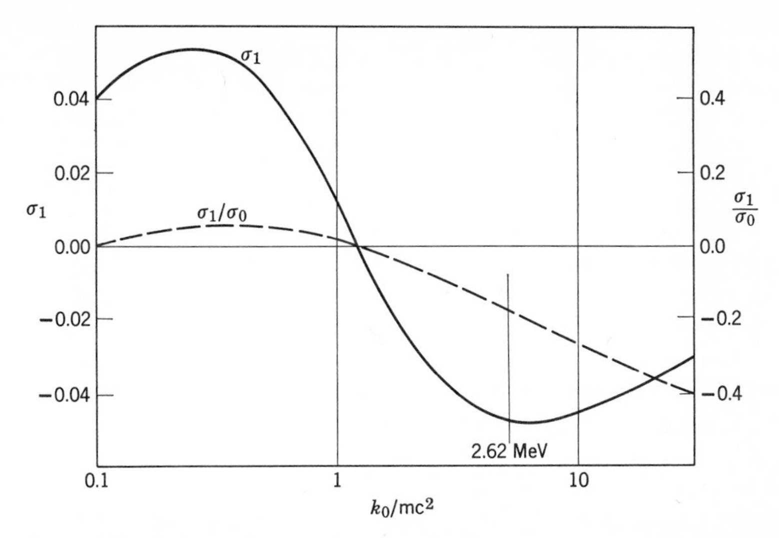

Fig. 10.3. Polarization-sensitive part σ_1 of total Compton cross section in units of $2\pi(\alpha/m)^2$ for the same electron and photon polarizations as in Fig. 10.2. (From Gunst and Page, *Phys. Rev.* **90,** 970 (1953).)

for a few energies, and Fig. 10.3 shows the dependence of the total cross section on polarization. The latter effect leads to a difference in the transmission of gamma rays through magnetized iron, dependent on the direction of magnetization. The prediction of the theory has been checked by Gunst and Page[1] and others.

It is instructive to consider the scattering of a photon by a spin 0 meson. There are now three Feynman graphs (Fig. 10.4) which have to be taken into account. The third one is the contribution of the term $e^2 A_\mu(x) A^\mu(x) \phi^* \phi$

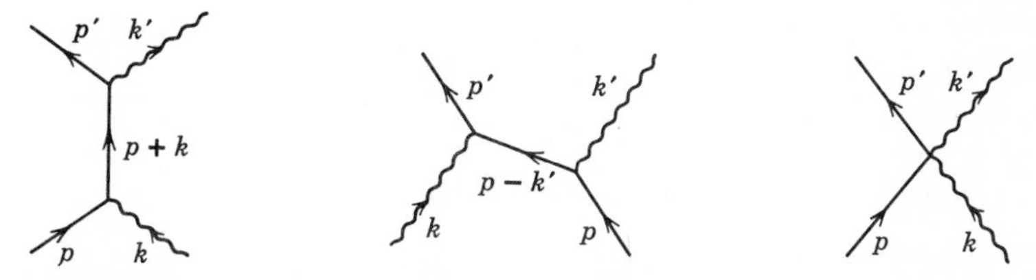

Fig. 10.4. Feynman graphs for Compton scattering by a spin 0 meson.

[1] S. B. Gunst and L. A. Page, *Phys. Rev.* **90,** 970 (1953). A brief discussion of this and subsequent experiments may be found in an excellent review article by H. A. Tolhoek, *Rev. Mod. Phys.* **28,** 277 (1956). A more up-to-date and experimentally oriented survey may be found in the review article of L. W. Fagg and S. S. Hanna, *Rev. Mod. Phys.* **31,** 711 (1959).

in the interaction Lagrangian. The matrix element to order e^2 is

$$\mathcal{R} = -\frac{ie^2}{(2\pi)^2}\left[e' \cdot (2p' + k')\frac{1}{(p+k)^2 - \mu^2}e \cdot (2p + k)\right.$$
$$\left. + e' \cdot (2p - k')\frac{1}{(p-k')^2 - \mu^2}e \cdot (2p' - k)\right] - \frac{2ie^2}{(2\pi)^2}e \cdot e' \tag{10.50}$$

First of all we notice that the energy denominators are the same as for electron-photon scattering, so that the backward peaking in the center of mass system at high energies will still be present. Next, we observe that at threshold, the contributions from the first two terms vanish. Only the last term remains, and it is just the Thomson amplitude. The term that contributes at threshold appears in the interaction Lagrangian as a consequence of gauge invariance, and its presence leads to the result that at threshold the scattering of photons has the same cross section for electrons as for mesons. That the threshold cross section should be independent of spin might have been expected: in the scattering of very long wavelength photons by a charged particle at rest, details of the structure, such as whether the particle has spin (and a magnetic moment), should not matter; all that determines the scattering is the charge and the mass, as in the classical calculation. We shall leave it to the reader to calculate the Compton cross section above threshold and to compare this with the electron Compton cross section.

We next discuss a process which, in a sense, is closely related to Compton scattering, namely, the *pair annihilation in flight* and its inverse process *pair production by two photons*. The Feynman graphs (Fig. 10.5) clearly show the relationship between these processes and Compton scattering. The graphs are just the Compton process graphs, except that one photon line and one electron line have been "crossed." The amplitudes for the two processes are said to be related by *crossing symmetry*. The concept of crossing will become very important when we discuss the strong interactions. The matrix element for the process is

$$\mathcal{R} = -\frac{ie^2}{(2\pi)^2}\bar{v}(p_2)\left(\not{e}'\frac{m + \not{p}_1 - \not{k}}{(p_1 - k)^2 - m^2}\not{e} + \not{e}\frac{m + \not{p}_1 - \not{k}'}{(p_1 - k')^2 - m^2}\not{e}'\right)u(p_1) \tag{10.51}$$

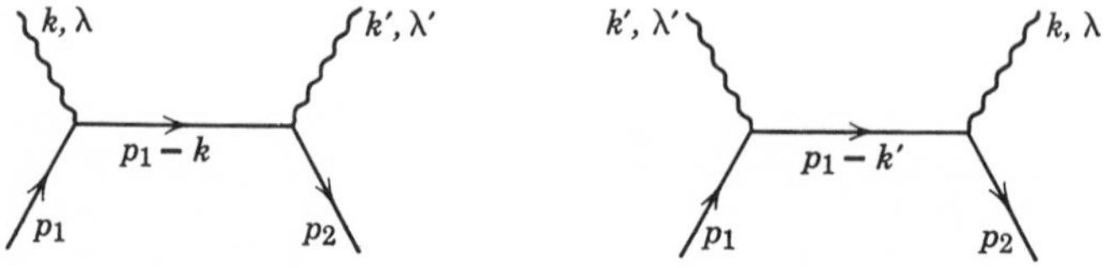

Fig. 10.5. Feynman graphs for pair annihilation.

The calculation of the cross section proceeds in the standard way. We merely quote the result

$$\sigma(s) = \pi\left(\frac{\alpha}{m}\right)^2 \frac{\gamma}{(\gamma+1)(\gamma^2-1)} \times \left[\left(\gamma + 4 + \frac{1}{\gamma}\right)\log\left(\gamma + \sqrt{\gamma^2-1}\right) - \frac{\sqrt{\gamma^2-1}}{\gamma}(\gamma+3)\right] \tag{10.52}$$

where γm is the positron energy in the laboratory frame. Note that

$$\gamma = \frac{s}{2m^2} - 1$$

with

$$s = (p_1 + p_2)^2$$

We observe the following:

1. The final state consists of two identical particles. Hence the center of mass angular distribution must be symmetric about 90°.
2. The denominators are $-2p_1 \cdot k$ and $-2p_1 \cdot k'$, respectively. In the center of mass system their values are

$$-2Ek(1 \pm \mathbf{v}\cdot\hat{\mathbf{k}}) = -2E^2(1 \pm v\cos\theta) \tag{10.53}$$

so that there will be both forward and backward peaking at high energies.
3. The annihilation process depends on the positron polarization and may be used to measure it. To see this in some detail, let us examine the matrix element $\mathfrak{R}$ in the laboratory frame, with the electron at rest. In this case the denominators are $-2mk$ and $-2mk'$, respectively, and we get

$$\mathfrak{R} = \frac{i\alpha}{2\pi m}\,\bar{v}(p_2)\left[\frac{1}{k}(-i\rho_2\boldsymbol{\sigma}\cdot\mathbf{e}')(m + m\rho_3 - \rho_3 k + i\rho_2\boldsymbol{\sigma}\cdot\mathbf{k})(-i\rho_2\boldsymbol{\sigma}\cdot\mathbf{e}) + \frac{1}{k'}(-i\rho_2\boldsymbol{\sigma}\cdot\mathbf{e})(m + m\rho_3 - \rho_3 k' + i\rho_2\boldsymbol{\sigma}\cdot\mathbf{k}')(-i\rho_2\boldsymbol{\sigma}\cdot\mathbf{e}')\right] \tag{10.54}$$

The representation for the positron spinor is

$$v(p) = \frac{m - \not{p}}{\sqrt{2m(m+E)}}\begin{pmatrix}0\\ \chi\end{pmatrix} = \frac{m - \not{p}}{\sqrt{2m(m+E)}}\,\rho_1\begin{pmatrix}\chi\\ 0\end{pmatrix} = \frac{m - \rho_3 p_0 + i\rho_2\boldsymbol{\sigma}\cdot\mathbf{p}}{\sqrt{2m(m+E)}}\,\rho_1 \tag{10.55}$$

so that

$$\bar{v}(p_2) = \rho_1\,\frac{m - \rho_3 E_2 - i\rho_2\boldsymbol{\sigma}\cdot\mathbf{p}_2}{\sqrt{2m(m+E_2)}}\,\rho_3 = \frac{\boldsymbol{\sigma}\cdot\mathbf{p}_2 - i\rho_2 m - \rho_1 E_2}{\sqrt{2m(m+E_2)}} \tag{10.56}$$

Thus after some straightforward algebraic manipulations we get

$$\mathcal{R} = -\frac{i\alpha}{2\pi m}\frac{1}{\sqrt{2m(m+E_2)}} \times (2\mathbf{e}\cdot\mathbf{e}'\boldsymbol{\sigma}\cdot\mathbf{p}_2 + i(E_2+m)(\boldsymbol{\sigma}\cdot\mathbf{e}'\boldsymbol{\sigma}\cdot\hat{\mathbf{k}}\times\mathbf{e} + \boldsymbol{\sigma}\cdot\mathbf{e}\boldsymbol{\sigma}\cdot\hat{\mathbf{k}}'\times\mathbf{e}')) \tag{10.57}$$

In the high energy limit ($|E_2| \gg m$) this simplifies to

$$\mathcal{R} = -\frac{i\alpha}{2\pi m}\sqrt{\frac{E_2}{2m}}(2\mathbf{e}\cdot\mathbf{e}'\boldsymbol{\sigma}\cdot\hat{\mathbf{p}}_2 + i\boldsymbol{\sigma}\cdot\mathbf{e}'\boldsymbol{\sigma}\cdot\hat{\mathbf{k}}\times\mathbf{e} + i\boldsymbol{\sigma}\cdot\mathbf{e}\boldsymbol{\sigma}\cdot\hat{\mathbf{k}}'\times\mathbf{e}') \tag{10.58}$$

If the positron is polarized longitudinally, the square of the matrix element is

$$\sum|\mathcal{R}|^2 = \frac{1}{2}\,\mathrm{tr}\left(\mathcal{R}^+\frac{1+P\boldsymbol{\sigma}\cdot\hat{\mathbf{p}}_2}{2}\mathcal{R}\right) \tag{10.59}$$

This turns out to be independent of P. It is easily understood if we recall that the polarization is a pseudoscalar quantity (it appears with $\boldsymbol{\sigma}\cdot\hat{\mathbf{p}}_2$), and in a cross section it must be multiplied by another pseudoscalar. Now the only pseudoscalars we can construct, in the absence of a photon polarization, are of the form $\hat{\mathbf{k}}\times\hat{\mathbf{k}}'\cdot\mathbf{e}$, etc., all of which change sign under time reversal and are therefore inadmissible.[1] If we ask for the correlation between the positron polarization and the polarization of one of the photons, we must replace one of the polarization vectors, say $\mathbf{e}$, by

$$\tfrac{1}{2}(1 + P_\gamma\mathbf{S}\cdot\hat{\mathbf{k}})\mathbf{e} = \tfrac{1}{2}(\mathbf{e} + iP_\gamma\mathbf{k}\times\mathbf{e}) \tag{10.60}$$

With this modification, $\mathcal{R}$ may be written as

$$\mathcal{R} = -\frac{i\alpha}{2\pi m}\sqrt{\frac{E_2}{2m}}(\mathbf{A}\cdot\boldsymbol{\sigma} + iB)$$

with $\mathbf{A}$ and B complex, the imaginary parts in both cases being proportional to P_γ. Now

$$|\mathcal{R}|^2 = \left(\frac{\alpha}{2\pi m}\right)^2\frac{E_2}{2m}\frac{1}{4}\,\mathrm{tr}\,(\mathbf{A}^*\cdot\boldsymbol{\sigma} - iB^*)(1 + P\boldsymbol{\sigma}\cdot\hat{\mathbf{p}}_2)(\mathbf{A}\cdot\boldsymbol{\sigma} + iB)$$

$$= \left(\frac{\alpha}{2\pi m}\right)^2\frac{E_2}{4m}(|\mathbf{A}\cdot\mathbf{A}^*| + |B|^2 + iP\mathbf{A}^*\cdot\hat{\mathbf{p}}_2\times\mathbf{A}$$

$$+ iBP\mathbf{A}^*\cdot\hat{\mathbf{p}}_2 - iB^*P\mathbf{A}\cdot\hat{\mathbf{p}}_2)$$

[1] This statement only holds in Born approximation. See Chapter 30 for a detailed discussion of this point.

Some pages of algebra lead to the result that the part of $|\mathfrak{R}|^2$ dependent on $P_\gamma P$ is of the form

$$2P_\gamma P[\tfrac{1}{2} - 2z + \tfrac{1}{2}z^2 - z^3 + \tfrac{1}{4}\hat{\mathbf{p}}_2 \cdot \hat{\mathbf{k}}(4 + 2z - 4z^2 + 4z^3) \\ + \tfrac{1}{4}\hat{\mathbf{p}}_2 \cdot \hat{\mathbf{k}}'(-1 + z - z^2 - z^3)]$$

where $z = \hat{\mathbf{k}} \cdot \hat{\mathbf{k}}'$. Now at high energies we have

$$p_2 + m = k + k'$$
$$\mathbf{p}_2 = \mathbf{k} + \mathbf{k}'$$

so that if $k \approx p_2$, i.e., the polarized photon takes almost all of the energy, it follows from

$$p_2^2 = k^2 + k'^2 + 2kk'z$$

that $z \approx -k'/2k$ is small. Thus the cross section is largest when the high energy photon has the same helicity as the positron and moves in the forward direction.

The effect is of practical significance, because it permits the transfer of the positron helicity to a photon. The photon helicity, as pointed out earlier can be measured via the Compton effect. This method may be used to study positron helicities, and it was so used in the measurement of the helicity of the positron from Ga^{66}, an experiment of interest in the study of beta decay.[1]

4. At low energies the interaction between electrons and positrons, leading to two γ rays, is predominantly an S-state interaction. This is always the case near zero energy unless there is a selection rule prohibiting it. The matrix element in the S-state is independent of energy, similar to the Compton amplitude at threshold. The process is exothermic, however, so that the phase-space factor becomes constant at threshold,

$$\left(k^2 \frac{dk}{dW}\right)_{k=m} = \frac{1}{2} m^2$$

Thus the cross section becomes infinite as $1/v$, a characteristic behavior for exothermic reactions. The rate, i.e., the number of annihilations per second, remains finite. Dimensional arguments would lead us to expect that

$$\Gamma = (\text{const})\left(\frac{\alpha}{m}\right)^2 \rho$$

where ρ is the density of electrons met by the slow positrons. The constant turns out to be π.

[1] M. Deutsch, B. Gittleman, R. W. Bauer, L. Grodzins and A. W. Sunyar, *Phys. Rev.* **107,** 1733 (1957).

Annihilation in flight is not the main process in the absorption of positrons in matter. The positrons are brought to rest about 80% of the time and then form *positronium*, an unstable bound state of an electron and positron. Positronium decays into a two-photon or a three-photon state. The properties of positronium form a very interesting chapter in the field of quantum electrodynamics, but lack of space prevents us from digressing into this field.[1]

To complete this chapter we shall quote the total cross section for pair production by photon-photon collisions. This will be useful to us when we discuss the pair production in the Coulomb field of a nucleus, a process of great importance in the absorption of high energy radiation in matter. The cross section in the center of mass system is readily obtained from (10.52) by detailed balance arguments, and it is

$$\sigma = \frac{\pi}{2}\left(\frac{\alpha}{m}\right)^2(1 - \beta^2)\left[(3 - \beta^4)\log\frac{1+\beta}{1-\beta} - 2\beta(2 - \beta^2)\right] \tag{10.61}$$

where $\beta = (1 - m^2/q^2)^{1/2}$ and q is the photon energy.

PROBLEMS

1. Suppose the electron had an anomalous magnetic moment so that the interaction with the electromagnetic field contains an additional term

$$\mathcal{L}_I' = \frac{2ie\gamma}{m}\bar{\psi}\sigma^{\mu\nu}\psi F_{\mu\nu}$$

Compute the matrix element for Compton scattering in the nonrelativistic limit. Can you interpret the terms linear in γ and quadratic in γ?

2. Compare the Compton matrix element for spin 0 particles with that for a spin $\frac{1}{2}$ particle.

3. What are the R-matrix parameters for Compton scattering (to order e^2)? These are defined by the decomposition in Problem 4, Chapter 4.

[1] S. De Benedetti and H. C. Corben, *Ann. Rev. Nucl. Sci.* **4** (1954).

11

The Scattering of Electrons and Positrons

Electrons and positrons scatter in matter; both electrons and nuclei (the latter may be treated as fixed sources of a Coulomb field) serve as targets, and this chapter is devoted to consideration of both processes.

Let us first consider *electron-electron scattering*, first studied by Møller.[1] The Feynman graphs for this process are shown in Fig. 11.1; the corresponding matrix element is

$$\mathcal{R} = \frac{i\alpha}{\pi}\left\{\frac{[\bar{u}(p_1')\gamma^\mu u(p_1)][\bar{u}(p_2')\gamma_\mu u(p_2)]}{(p_1' - p_1)^2 + i\epsilon} - \frac{[\bar{u}(p_2')\gamma^\mu u(p_1)][\bar{u}(p_1')\gamma_\mu u(p_2)]}{(p_1 - p_2')^2 + i\epsilon}\right\} \tag{11.1}$$

Note the difference in sign between the two contributions. This comes about because of the exclusion principle, which requires that the amplitude be antisymmetric under the interchange $p_1' \leftrightarrow p_2'$, with the corresponding change in the spin labels. If we were to consider electron-proton scattering, only the first graph would give a contribution.[2] The cross section is easily evaluated. We again limit ourselves to quoting the result in the center of mass frame:

$$\frac{d\sigma}{d\Omega} = \left(\frac{\alpha}{m}\right)^2 \frac{1}{E^2}\frac{1}{(E^2-1)^2}\left\{\frac{(2E^2-1)^2}{\sin^4\theta} - \frac{2E^4 - E^2 - \frac{1}{4}}{\sin^2\theta} + \frac{(E^2-1)^2}{4}\right\} \tag{11.2}$$

Here E is the energy of one of the electrons in the center of mass system and θ is the scattering angle. The energy is measured in units of the electron rest mass m.

[1] C. Møller, *Ann. Physik*, **14,** 568 (1932).
[2] However, the proton vertex would be more complicated than just $-e\gamma^\mu$ (see Chapter 26).

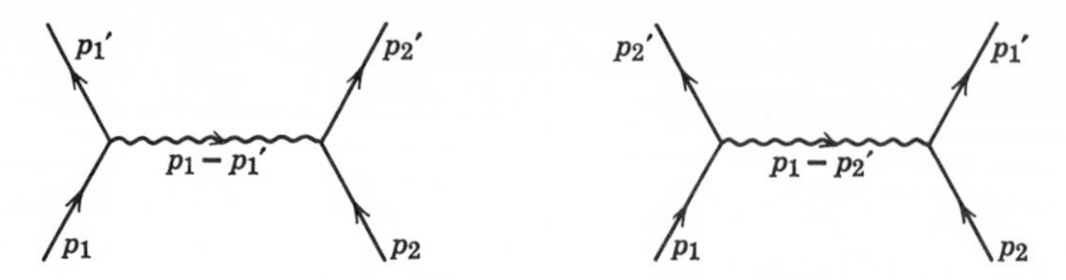

Fig. 11.1. Feynman graphs for electron-electron scattering.

The dominant features of the scattering amplitude are again contained in the denominators. In the center of mass system, with

$$p_1 = (E, \mathbf{p}) \qquad p_2 = (E, -\mathbf{p})$$
$$p_1' = (E, \mathbf{p}') \qquad p_2' = (E, -\mathbf{p}')$$

the denominators are

$$\frac{1}{(p_1 - p_1')^2} = -\frac{1}{(\mathbf{p} - \mathbf{p}')^2} = -\frac{1}{2p^2(1 - \cos\theta)} = -\frac{1}{4p^2 \sin^2 \theta/2} \tag{11.3}$$

for the first graph and

$$\frac{1}{(p_1 - p_2')^2} = -\frac{1}{(\mathbf{p} + \mathbf{p}')^2} = -\frac{1}{2p^2(1 + \cos\theta)} = -\frac{1}{4p^2 \cos^2 \theta/2} \tag{11.4}$$

for the second graph. Thus there is peaking both in the forward and backward directions; there is symmetry since we are dealing with identical particles.

In the nonrelativistic limit $\bar{u}(p')\gamma^\mu u(p) \to \delta_{\mu 0}$, etc., so that the amplitude becomes

$$\mathfrak{R} = -\frac{i\alpha}{4\pi p^2}\left(\frac{1}{\sin^2\theta/2} - \frac{1}{\cos^2\theta/2}\right) \tag{11.5}$$

and the cross section is

$$\frac{d\sigma}{d\Omega} = \left(\frac{\alpha m}{2p^2}\right)^2\left(\frac{1}{\sin^2\theta/2} - \frac{1}{\cos^2\theta/2}\right)^2 \tag{11.6}$$

If we try to evaluate the total cross section by integrating over all angles, we run into trouble, because the denominators make the integral diverge. The trouble stems from the masslessness of the photon; a consequence of this is that the electrical forces between two charged particles have an infinite range ($1/r^2$ in the nonrelativistic limit), and there is some scattering, no matter how far apart the particles are. This is not a difficulty that can be avoided. All of scattering theory as well as the asymptotic properties of the field operators can be formulated only when the forces have a finite range or, alternately, when all particles have a mass, no matter how small. In actuality, no experiment is ever carried out with two isolated charges and other charges act to screen the Coulomb field at large distances. There will be further occasion to discuss the complications arising from the masslessness of the photon.

Recent interest in electron-electron scattering has been in connection with the measurement of the longitudinal polarization of electrons emitted in beta decay. Detailed calculations by Bincer and others[1] showed that electron-electron and positron-electron cross sections depend on the longitudinal polarization of the incident beam, provided that the target electrons were also polarized. Briefly, for $e^- - e^-$ scattering, the cross section is smaller when the two electrons in the initial state are polarized in the same direction than when they have their spins antiparallel. The effect is most pronounced when the center of mass scattering angle is 90°. The reason for this lies in the exclusion principle: when the electron spins are parallel, the spatial part of the wave function must be antisymmetric; hence the amplitude must vanish at 90°. Bincer found that at 90°

$$\frac{\left(\dfrac{d\sigma}{d\Omega}\right)_{\text{parallel}}}{\left(\dfrac{d\sigma}{d\Omega}\right)_{\text{antiparallel}}} = \frac{v^4}{1 + 2v^2 + 5v^4} \tag{11.7}$$

The effect would not exist if there were a preponderance of events in which the spin flipped. Actually we can easily convince ourselves that at both low and high energies there is little spin flip. Consider the quantities $\bar{u}(p')\gamma^\mu u(p)$ which determine the behavior of the electron when it interacts with a photon (real or virtual). Explicitly

$$\begin{aligned} \bar{u}(p')\boldsymbol{\gamma}u(p) &= \frac{m - i\rho_2\boldsymbol{\sigma}\cdot\mathbf{p}' + \rho_3 E'}{\sqrt{2m(m+E')}}\, i\rho_2\boldsymbol{\sigma}\, \frac{m - i\rho_2\boldsymbol{\sigma}\cdot\mathbf{p} + \rho_3 E}{\sqrt{2m(m+E)}} \\ &= \frac{1}{2m}\sqrt{\frac{E'+m}{E+m}}\,(\mathbf{p} - i\boldsymbol{\sigma}\times\mathbf{p}) \\ &\quad + \frac{1}{2m}\sqrt{\frac{E+m}{E'+m}}\,(\mathbf{p}' + i\boldsymbol{\sigma}\times\mathbf{p}') \end{aligned} \tag{11.8}$$

[1] A. Bincer, *Phys. Rev.* **107**, 1434 (1957); K. Böckman, G. Kramer, and W. R. Theis, *Z. Physik* **150**, 201 (1958); for an experimentally oriented review, see L. Page, *Rev. Mod. Phys.* **31**, 759 (1959).

and

$$\bar{u}(p')\gamma^0 u(p) = \sqrt{\frac{(m+E)(m+E')}{4m^2}} + \frac{1}{2m}\frac{\mathbf{p}'\cdot\mathbf{p} + i\boldsymbol{\sigma}\cdot\mathbf{p}'\times\mathbf{p}}{\sqrt{(m+E)(m+E')}} \tag{11.9}$$

For low momenta we get, as mentioned before

$$\bar{u}(p')\gamma^\mu u(p) \approx \delta_{\mu 0}$$

which contains no spin dependence. At high energies it is convenient to use the representation (2.48) for the Dirac matrices. With this choice of representation, the spinor $u(p)$ breaks up into positive and negative helicity components in the limit in which we may neglect the mass. In this representation

$$\begin{aligned} \bar{u}(p')\boldsymbol{\gamma}\, u(p) &= u^*(p')\rho_3\boldsymbol{\sigma}\, u(p) \\ \bar{u}(p')\gamma^0\, u(p) &= u^*(p')\, u(p) \end{aligned} \tag{11.10}$$

so that there is no coupling between positive and negative helicity states. Thus electrons that are longitudinally polarized and parallel to one another before the collision will remain so afterward.

The Feynman graphs for *positron electron scattering*[1] (Fig. 11.2) show that there is a photon exchange term, as in $e^- - e^-$ scattering, and an annihilation term (with a $-$ sign). The matrix element is

$$\mathfrak{R} = \frac{i\alpha}{m}\left[\frac{[\bar{u}(p')\gamma^\mu\, u(p)][\bar{v}(q)\gamma_\mu\, v(q')]}{(p'-p)^2 + i\epsilon} - \frac{[\bar{u}(p')\gamma^\mu\, v(q')][\bar{v}(q')\gamma_\mu\, u(p)]}{(p+q)^2 + i\epsilon}\right] \tag{11.11}$$

The first term contributes to a strong forward peaking in the amplitude and the divergence of the total cross section. The intermediate state in the annihilation term is a one-photon state, so that this term only contributes to the $J = 1$ amplitude. The polarization dependence of the cross section for the scattering of longitudinally polarized positrons by a polarized target has also been computed, and the results may be found in the papers quoted on p. 167.

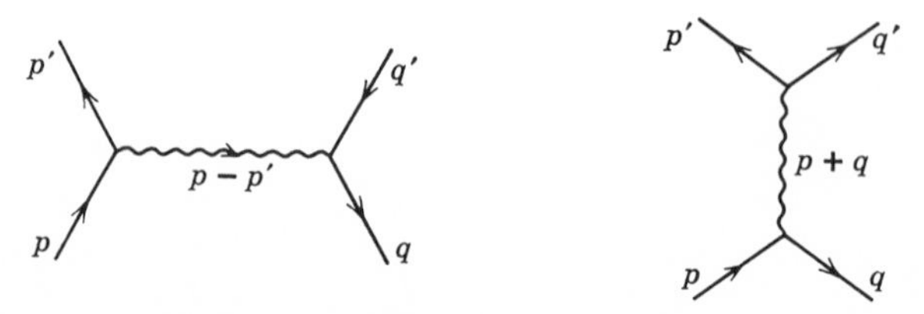

Fig. 11.2. Graphs for $e^- - e^+$ scattering.

[1] H. J. Bhabha, *Proc. Roy. Soc.* (*London*), **A154**, 195 (1935).

The next topic of interest is the scattering of electrons (or positrons) by an external field. In the presence of the external field represented by $A_\mu^{\text{ext}}(x)$, the interaction of (9.1) is changed by the addition of

$$H_1^{\text{ext}}(t) = -e\int_{x_0=t} d^3x\, j_\mu^{(0)}(x)\, A^{\mu\text{ext}}(x) \tag{11.12}$$

to H_1. The external field may be calculated in terms of the source density. We have

$$A_\mu^{\text{ext}}(x) = \delta_{\mu 0}\, \varphi(\mathbf{x}) \tag{11.13}$$

where

$$\nabla^2 \varphi(\mathbf{x}) = -Ze\, \rho(\mathbf{x})$$

i.e.,

$$\varphi(\mathbf{x}) = \frac{Ze}{4\pi}\int d^3x'\, \frac{\rho(\mathbf{x}')}{|\mathbf{x} - \mathbf{x}'|} \tag{11.14}$$

Here $Z\,\rho(\mathbf{x})$ is the charge density of the nucleus[1] so that we must have

$$\int \rho(\mathbf{x})\, d^3x = 1 \tag{11.15}$$

For a fixed source momentum is no longer conserved—rather, the source can absorb or contribute momentum without changing its state. Thus no momentum-conservation delta function appears.

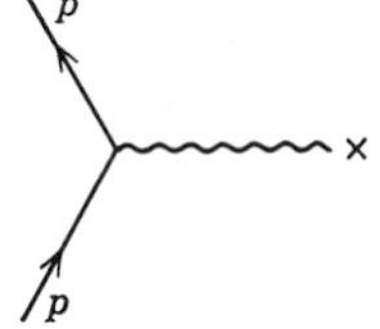

Fig. 11.3. Feynman graph for Coulomb scattering.

For low Z, the parameter $Z\alpha$, which plays the role of an expansion parameter, is still small, so that it is justifiable to treat $H_1^{\text{ext}}(t)$ as a perturbation, unless the fact that the electron may be bound by the Coulomb potential is of importance, as in the photoelectric effect. In high energy processes, however, we may use perturbation theory and replace the Coulomb field by an effective "virtual photon." Consider, for example, the single scattering of an electron by an external field (Fig. 11.3). To lowest order in Z, we have

$$\begin{aligned}(\Phi_{p'}, (S-1)\Phi_p) &= ie\int dx(\Phi_{p'},\, \bar\psi^{(0)}(x)\gamma^\mu\, \psi^{(0)}(x)\Phi_p)A_\mu^{\text{ext}}(\mathbf{x})\\ &= \frac{ie}{(2\pi)^3}\int dx\, \bar u(p')\gamma^\mu\, u(p)e^{-i(p-p')x}A_\mu^{\text{ext}}(\mathbf{x})\\ &= 2\pi i e\, \delta(E_p - E_{p'})\, \bar u(p')\gamma^\mu\, u(p)\\ &\qquad\times\int \frac{d^3x}{(2\pi)^3}\, e^{i(\mathbf{p}-\mathbf{p}')\cdot\mathbf{x}}A_\mu^{\text{ext}}(\mathbf{x})\end{aligned} \tag{11.16}$$

[1] The sign is such that the nucleus has charge opposite in sign to the electron charge which is $-e$.

For an external field given by (11.14) the relevant integral is

$$\frac{1}{(2\pi)^3}\int d^3x e^{i(\mathbf{p}-\mathbf{p}')\cdot\mathbf{x}}\frac{\delta_{\mu 0}Ze}{4\pi}\int d^3x'\,\frac{\rho(\mathbf{x}')}{|\mathbf{x}-\mathbf{x}'|}=\frac{Ze\delta_{\mu 0}}{(2\pi)^3}\frac{v(\mathbf{p}-\mathbf{p}')}{(\mathbf{p}-\mathbf{p}')^2} \tag{11.17}$$

where

$$v(\mathbf{k})\equiv\int d^3x e^{-i\mathbf{k}\cdot\mathbf{x}}\,\rho(\mathbf{x}) \tag{11.18}$$

is a *form factor* which characterizes the nuclear charge distribution. The normalization condition (11.15) implies that

$$v(0)=1 \tag{11.19}$$

For a point source $v(\mathbf{k})=1$ for all values of $\mathbf{k}$. The matrix element therefore is

$$(\Phi_{p'},(S-1)\Phi_p)=2\pi ie\,\delta(E_p-E_{p'})\,\bar{u}(p')\gamma^\mu\,u(p)$$

$$\times\,\delta_{\mu 0}\frac{Ze}{(2\pi)^3}\frac{v(\mathbf{p}-\mathbf{p}')}{(\mathbf{p}-\mathbf{p}')^2} \tag{11.20}$$

Now, if we were to follow the Feynman rules for the electron alone, we would get for its part of the matrix element

$$2\pi ie\,\delta(p'-p-k)\,\bar{u}(p')\gamma^\mu\,u(p)$$

Here k is the four-momentum transferred to the electron. To establish a rule for calculating $S-1$ in external fields we therefore (a) treat the electron as usual; (b) replace the four-dimensional delta function at the vertex by an energy-conservation delta function; (c) multiply the usual vertex factor $e\gamma^\mu$ by the quantity

$$\delta_{\mu 0}\frac{Ze}{(2\pi)^3}\frac{v(\mathbf{k})}{\mathbf{k}^2} \tag{11.21}$$

The transition rate per unit time is

$$\Gamma=\frac{\delta(E_{p'}-E_p)}{2\pi}\,|\,\mathfrak{R}_C|^2$$

where

$$S-1=\delta(E'-E)\mathfrak{R}_C$$

Here

$$\mathfrak{R}_C=i\,\bar{u}(p')\gamma^0\,u(p)\,\frac{Ze^2}{4\pi^2}\frac{v(\mathbf{p}'-\mathbf{p})}{(\mathbf{p}'-\mathbf{p})^2}$$

The cross section is

$$d\sigma = \frac{\delta(E_{p'} - E_p)}{2\pi} |\mathcal{R}_C|^2 \frac{1}{E_{p'}/(2\pi)^3 m} \frac{1}{(p/E_p)} \frac{d^3p'}{E_{p'}/m}$$

so that

$$\frac{d\sigma}{d\Omega} = (2\pi)^2 \frac{m^2}{p} \int p'^2 \, dp' \, |\mathcal{R}_C|^2 \frac{\delta(E_p - E_{p'})}{E_{p'}}$$

$$= (2\pi)^2 m^2 |\mathcal{R}_C|^2 \qquad (11.22)$$

If we average over initial electron spin states and sum over the final spin states, we require

$$\frac{1}{2} \sum_r \sum_s \left(\frac{Z\alpha}{\pi}\right)^2 \left|\frac{v(\mathbf{p}' - \mathbf{p})}{(\mathbf{p}' - \mathbf{p})^2}\right|^2 |u^{(r)*}(p')\, u^{(s)}(p)|^2$$

$$= \left(\frac{Z\alpha}{\pi}\right)^2 \left|\frac{v(\mathbf{p}' - \mathbf{p})}{(\mathbf{p}' - \mathbf{p})^2}\right|^2 \frac{1}{8m^2} \operatorname{tr}\left((\not{p}' + m)\gamma_0(\not{p} + m)\gamma_0\right)$$

This leads to

$$\frac{d\sigma}{d\Omega} = \left(\frac{Z\alpha m}{2p^2 \sin^2 \theta/2}\right)^2 |v(\mathbf{p}' - \mathbf{p})|^2 \left(1 + \frac{p^2}{m^2} \cos^2 \frac{\theta}{2}\right) \qquad (11.23)$$

For a point source $|v(\mathbf{p} - \mathbf{p}')|^2$ is replaced by unity. This differs from the classical Rutherford formula in the $(p^2/m^2) \cos^2 \theta/2$ term, which disappears in the limit $\mathbf{p} \to 0$. This term arises from the interaction of the magnetic moment of the electron with the magnetic field seen in the electron rest frame.

Since the electron has spin $\frac{1}{2}$, there is scattering in orbital angular momentum states $l = J \pm \frac{1}{2}$ for each value of J. In general such a situation would lead to a polarization of the outgoing electron. It turns out, however, that to lowest order in $Z\alpha$ there is no polarization. In higher order calculations, the electron does turn out to be polarized after the collision, with polarization of order $Z\alpha$. The reason for the absence of polarization is easy to see: the relevant part of the Coulomb scattering matrix element is

$$\bar{u}(p')\gamma^0 u(p) = \frac{E + m}{2m} + \frac{E - m}{2m} (\cos \theta - i\boldsymbol{\sigma} \cdot \hat{\mathbf{n}} \sin \theta) \qquad (11.24)$$

with

$$\hat{\mathbf{n}} = \frac{\mathbf{p} \times \mathbf{p}'}{|\mathbf{p} \times \mathbf{p}'|}$$

Thus the spin-independent part of the amplitude and the spin-dependent part are exactly 90° out of phase and cannot therefore interfere in the absolute square of the matrix element. The polarization is just the interference term proportional to $\boldsymbol{\sigma} \cdot \hat{\mathbf{n}}$.

The next term is a $Z\alpha$ expansion, corresponding to the graph in Fig. 11.4, is given by

$$\mathcal{R}_C^{(2)} = 2\pi i\left(\frac{Ze^2}{(2\pi)^3}\right)^2 \int dq\, \delta(q_0 - E) \frac{1}{(\mathbf{p} - \mathbf{q})^2 - i\epsilon} \frac{1}{(\mathbf{p}' - \mathbf{q})^2 - i\epsilon} \times \bar{u}(p')e\gamma^0 \frac{\not{q} + m}{q^2 - m^2 + i\epsilon} e\gamma^0 u(p) \quad (11.25)$$

If positron scattering by a Coulomb field is to be calculated, $\bar{u}(p')$ and $u(p)$ are to be replaced by $v(p')$ and $\bar{v}(p)$, respectively. An evaluation of the integral in (11.25) requires the knowledge of integrals of the type

$$\int d^3q \frac{(1, \mathbf{q})}{(\mathbf{p}' - \mathbf{q})^2(\mathbf{p} - \mathbf{q})^2(\mathbf{p}^2 - \mathbf{q}^2 + i\epsilon)} \quad (11.26)$$

The major contributions to the integral come from the regions $\mathbf{q} \approx \mathbf{p}$ and $\mathbf{q} \approx \mathbf{p}'$. There the integrand is highly peaked because in those regions two of the three denominators become small: one of the Coulomb denominators and the electron propagator denominator tend to vanish. The two regions represent the situation in which one of the scattering interactions is responsible for almost all the momentum transfer, with the electron both before and after being almost on the mass shell, and undergoing the other Coulomb scattering with almost no momentum transfer—a process which we know to have a large matrix element from the denominator in (11.23), say. The integral actually diverges, and we have to resort to an artifice to get around this difficulty. We choose to give the photon a small mass μ. When this is done, the external field denominators in (11.21) and therefore in (11.25) have to be changed to $1/(\mathbf{k}^2 + \mu^2)$, etc. With this change, the integrals become finite. Details of the calculation may be found in a paper by Dalitz[1] who found that as the photon mass is allowed to approach zero the matrix

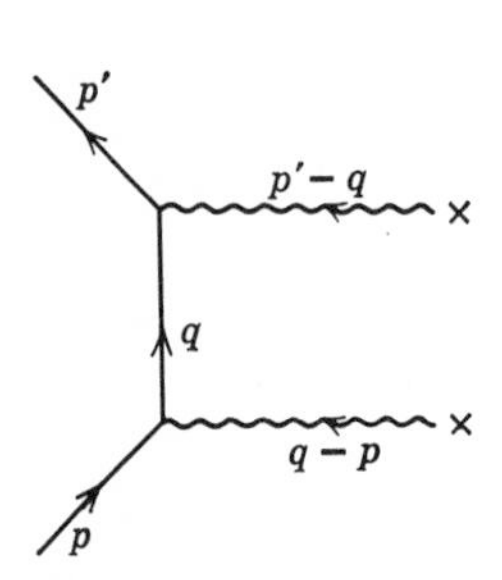

Fig. 11.4. Double scattering in a Coulomb field.

[1] R. H. Dalitz, *Proc. Roy. Soc. (London)*, **A206**, 509 (1951).

element can be written in the form of a finite matrix element, multiplied by the μ-dependent factor

$$1 - 2i\,\frac{Z\alpha E_p}{p}\log\frac{2p\sin\theta/2}{\mu} \tag{11.27}$$

which to this order acts like a phase factor and therefore does not affect the Coulomb cross section. The work of Yennie and collaborators[1] strongly supports Dalitz's stated conjecture that the term which diverges as the photon mass $\mu \to 0$ is, in fact, of the form

$$\exp\left[-2i\,\frac{Z\alpha}{v}\log\left(\frac{2p\sin\theta/2}{\mu}\right)\right] \tag{11.28}$$

If we leave out the phase factor, Dalitz's calculation for electrons/positrons yields

$$\begin{aligned}\mathcal{R}_C \propto \frac{1}{4p^2\sin^2\theta/2}\Bigg[\frac{E+m}{2m} &+ \frac{E-m}{2m}(\cos\theta - i\boldsymbol{\sigma}\cdot\hat{\mathbf{n}}\sin\theta)\Bigg] \\ &\pm \frac{Z\alpha}{4\pi^2 m}(1+\cos\theta - i\boldsymbol{\sigma}\cdot\hat{\mathbf{n}}\sin\theta)(X+iY)\end{aligned} \tag{11.29}$$

with

$$\begin{aligned} X &= \frac{\pi^3}{4p^2\cos^2\theta/2}\left(\operatorname{cosec}\frac{\theta}{2} - 1\right) \\ Y &= \frac{\pi^2}{4p^2\cos^2\theta/2}\log\left(\sin^2\frac{\theta}{2}\right)\end{aligned} \tag{11.30}$$

The scattered electron will be polarized in the direction $\hat{\mathbf{n}}$. To calculate this, let us choose the $\hat{\mathbf{n}}$ direction as quantization axis; the transitions caused by an amplitude of the form

$$\mathcal{R}_C = A + i\boldsymbol{\sigma}\cdot\hat{\mathbf{n}}B \tag{11.31}$$

[1] D. R. Yennie, S. C. Frautschi, and H. Suura, *Ann. Phys.* (*N.Y.*), **13**, 379 (1961) and references cited therein. The Coulomb scattering amplitude, which multiplies e^{ikr}/r in nonrelativistic scattering theory is

$$-\frac{Z\alpha m}{2p^2\sin^2\theta/2}\,e^{2i\sigma_0}\exp\left(-i\,\frac{Z\alpha m}{p}\log 2pr\sin^2\frac{\theta}{2}\right)$$

the last term coming from the infinite range of the potential. If the solution is matched to a plane-wave solution at some radius $r = p/\mu^2$ the effect of the small photon mass is reproduced.

are up-up $A + iB$, up-down 0, down-up 0, and down-down $A - iB$. Hence the polarization of the electron in the final state is

$$\frac{d\sigma^{\uparrow} - d\sigma^{\downarrow}}{d\sigma^{\uparrow} + d\sigma^{\downarrow}} = \frac{|A + iB|^2 - |A - iB|^2}{|A + iB|^2 + |A - iB|^2} = \frac{2 \operatorname{Im} AB^*}{|A|^2 + |B|^2} \tag{11.32}$$

In this expression the arrows correspond to "up" and "down" in relation to the direction $\hat{\mathbf{n}}$. The polarization was first calculated by Mott.[1]

With these remarks we conclude our discussion of the scattering of electrons and positrons. Electron scattering by nuclei has been used to study nuclear charge distributions, and is becoming an ever increasingly useful tool in the study of nuclear level structure. This subject, however, is outside the scope of this book.

PROBLEMS

1. Consider the interaction of a vector meson with an external electromagnetic field, if the interaction is given by

$$\begin{aligned}\mathcal{L} = {} & \tfrac{1}{2} f^{\mu\nu+}(\partial_\mu \phi_\nu - \partial_\nu \phi_\mu) + \tfrac{1}{2} f^{\mu\nu}(\partial_\mu \phi_\nu^+ - \partial_\nu \phi_\mu^+) - \tfrac{1}{2} f_{\mu\nu}^+ f^{\mu\nu} \\ & + m^2 \phi_\mu^+ \phi^\mu - \frac{ie}{2} f^{\mu\nu+}(A_\mu \phi_\nu - A_\nu \phi_\mu) + \frac{ie}{2} f^{\mu\nu}(A_\mu \phi_\nu^+ - A_\nu \phi_\mu^+) \\ & + \frac{ie\gamma}{2}(\phi_\mu^+ \phi_\nu - \phi_\nu^+ \phi_\mu) F^{\mu\nu} + \frac{ieq}{4m^2}(f_{\mu\nu}^+ \phi_\lambda - \phi_\lambda^+ f_{\mu\nu})\, \partial^\lambda F^{\mu\nu}\end{aligned}$$

Discuss the nonrelativistic limit and express it in terms of the vector spin operator $\mathbf{S}$, constructed in analogy with that for the photon in Chapter 10. Interpret the different terms. What is the magnetic moment of the vector meson described by the above interaction? What is its quadrupole moment?

2. Suppose the electron had an anomalous magnetic moment so that its interaction with the electromagnetic field had an additional term

$$\frac{2ie\gamma}{m} \bar{\psi} \sigma^{\mu\nu} \psi F_{\mu\nu}$$

Calculate the change in the electron-electron cross section to first order in γ.

[1] N. F. Mott, *Proc. Roy. Soc.* (*London*) **A124,** 425 (1929); a brief discussion may be found in H. A. Tolhoek, *Rev. Mod. Phys.* **28,** 277 (1956).

12

Bremsstrahlung and Related Processes

The most important mechanism by means of which high energy electrons are slowed down is *bremsstrahlung*, the radiation accompanying any change of state of motion of a charged particle. The process by which high energy radiation is preferentially absorbed in matter is *pair production* in the presence of an external field. This chapter will deal with both of these processes.

We shall first consider bremsstrahlung. The Feynman graphs for the process are shown in Fig. 12.1, and the corresponding matrix element is

$$-i\frac{Ze^3}{(2\pi)^{7/2}}\frac{v(\mathbf{\Delta})}{|\mathbf{\Delta}|^2}\bar{u}(p')\left\{\gamma^0\frac{1}{\not{p}-\not{k}-m}\not{e}+\not{e}\frac{1}{\not{p}'+\not{k}-m}\gamma^0\right\}u(p) \qquad (12.1)$$

where

$$\mathbf{\Delta} = \mathbf{p} - \mathbf{k} - \mathbf{p}' \qquad (12.2)$$

is the momentum transferred to the nucleus. With the help of the properties of the γ matrices we may transform this to

$$-i\frac{Ze^3}{(2\pi)^{7/2}}\frac{v(\mathbf{\Delta})}{|\mathbf{\Delta}|^2}\bar{u}(p')\left\{\gamma^0\left(\frac{e\cdot p'}{k\cdot p'}-\frac{e\cdot p}{k\cdot p}\right)+\frac{\not{k}\not{e}\gamma^0}{2p'\cdot k}-\frac{\gamma^0\not{e}\not{k}}{2p\cdot k}\right\}u(p) \qquad (12.3)$$

The matrix element is easily squared, and the various spin and polarization sums may be carried out. The resulting Bethe-Heitler cross section may be found in several textbooks.[1] We shall not reproduce the derivation, but rather discuss some limited aspects of the process. Consider, in particular, the *soft photon limit*, i.e., the matrix element for radiating a photon with energy small compared with all the other energies in the process, including $|\mathbf{\Delta}|$. In this limit, it is clear that the first two terms in (12.3) dominate.

[1] H. A. Bethe and W. Heitler, *Proc. Roy. Soc.* (*London*), **A146**, 85 (1934); for a detailed discussion, see W. Heitler, *The Quantum Theory of Radiation*, Clarendon Press, Oxford (1954).

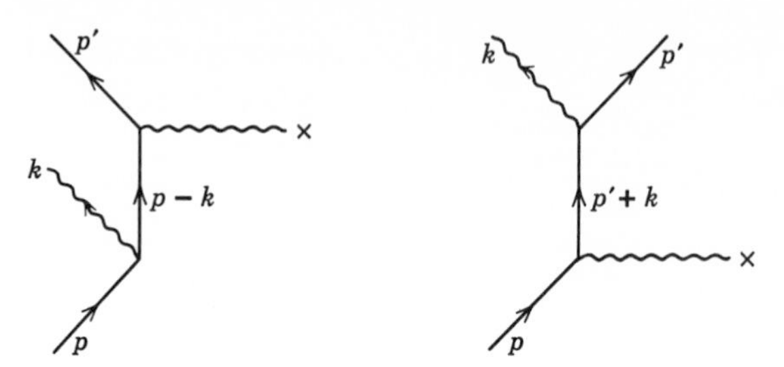

Fig. 12.1. The Feynman graphs for bremsstrahlung.

Comparison of this limit with the matrix element for Coulomb scattering shows that we may write the bremsstrahlung matrix element in the form

$$\mathcal{R}_B \cong -\frac{e}{(2\pi)^{3/2}}\left(\frac{e\cdot p'}{k\cdot p'}-\frac{e\cdot p}{k\cdot p}\right)\mathcal{R}_C \tag{12.4}$$

All delta functions have been left out in this expression. The cross section is easily calculated. It is given by

$$\begin{aligned} d\sigma_B &= \frac{\delta(E_{p'}+k-E_p)}{2\pi}|\mathcal{R}_B|^2\frac{(2\pi)^3 m}{E_p v}\frac{d^3p'}{E_{p'}/m}\frac{d^3k}{2k} \\ &\approx \frac{e^2}{(2\pi)^3}\left(\frac{e\cdot p'}{k\cdot p'}-\frac{e\cdot p}{k\cdot p}\right)^2\frac{d^3k}{2k}\left\{\frac{\delta(E_{p'}-E_p)}{2\pi}|\mathcal{R}_C|^2\frac{(2\pi)^3 m}{E_p v}\frac{d^3p'}{E_{p'}/m}\right\} \end{aligned} \tag{12.5}$$

We write this in the form

$$d\sigma_B \approx \frac{1}{2}\frac{e^2}{(2\pi)^3}\left(\frac{e\cdot p'}{k\cdot p'}-\frac{e\cdot p}{k\cdot p}\right)^2 k\,dk\,d\Omega_{\hat{k}}\,d\sigma_C(p',p) \tag{12.6}$$

The factor multiplying the Coulomb cross section is

$$\frac{\alpha}{\pi}\frac{d\Omega_{\hat{k}}}{4\pi}\frac{dk}{k}\left(\frac{\mathbf{v}'\cdot\mathbf{e}}{1-\mathbf{v}'\cdot\hat{\mathbf{k}}}-\frac{\mathbf{v}\cdot\mathbf{e}}{1-\mathbf{v}\cdot\hat{\mathbf{k}}}\right)^2 \tag{12.7}$$

and it contains all the information on soft photon radiation. There are several characteristic features to be observed.

1. The energy spectrum has a $1/k$ dependence; the cross section appears to diverge at the "infrared" end of the spectrum, an effect due to the zero mass of the photon.
2. The angular distribution of the radiation shows strong peaking in the direction of the initial and final electron velocity vectors. This is really a classical effect that dominates in the situation in which the electron motion is unaffected by the radiation emitted. We shall now show that the expression (12.7) is just what we expect when a classical current radiates.

The scattering matrix element for the radiation of photons by a classical current $J^\mu(x)$ is given by

$$\frac{\left(\Phi_{k_1\cdots k_n}, T\left(e^{-i\int dx\, J^\mu(x)A_\mu^{(0)}(x)}\right)\Phi_0\right)}{\left(\Phi_0, T\left(e^{-i\int dx\, J^\mu(x)A_\mu^{(0)}(x)}\right)\Phi_0\right)} = \left(\Phi_{k_1\cdots k_n}, T\left(e^{-i\int dx\, J^\mu(x)(A_\mu^{(0)}(x))_-}\right)\Phi_0\right) \quad (12.8)$$

where we have used (8.40) and (8.44). For a single photon we obtain

$$-i\int dx\, J^\mu(x)(\Phi_k, (A_\mu^{(0)}(x))_-\Phi_0) = \frac{i}{(2\pi)^{3/2}}\int dx_0\, d^3x\, \mathbf{e}\cdot\mathbf{J}(x)e^{i\omega x_0 - i\mathbf{k}\cdot\mathbf{x}} \quad (12.9)$$

With the classical current

$$\begin{aligned} \mathbf{J}(x) &= -e\mathbf{v}\,\delta(\mathbf{x}-\mathbf{v}x_0) \qquad x_0 < 0 \\ &= -e\mathbf{v}'\,\delta(\mathbf{x}-\mathbf{v}'x_0) \qquad x_0 > 0 \end{aligned} \quad (12.10)$$

the integral is easily evaluated and yields

$$-\frac{e}{(2\pi)^{3/2}}\frac{1}{k}\left(\frac{\mathbf{e}\cdot\mathbf{v}'}{1-\mathbf{v}'\cdot\hat{\mathbf{k}}} - \frac{\mathbf{e}\cdot\mathbf{v}}{1-\mathbf{v}\cdot\hat{\mathbf{k}}}\right) \quad (12.11)$$

If we assume that the photon has a small mass μ, (12.6) is altered because $k^2 = \mu^2$ rather than 0 in the electron propagator denominator, and the density is 2ω rather than $2k$. Thus we get

$$d\sigma_B \approx \frac{1}{2}\frac{e^2}{(2\pi)^3}\left(\frac{2e\cdot p'}{2k\cdot p' + \mu^2} - \frac{2e\cdot p}{2k\cdot p - \mu^2}\right)^2\frac{k^2\,dk}{\omega}\,d\Omega_{\hat{k}}\,d\sigma_C \quad (12.12)$$

The soft photon radiation factor may be written in the form

$$\frac{\alpha}{\pi}\frac{d\Omega_{\hat{k}}}{4\pi}\frac{k^2\,dk}{\omega}(e_\mu{}^\lambda I^\mu)(e_\nu{}^\lambda I^\nu)$$

with

$$I^\mu = \left(\frac{2p'^\mu}{2k\cdot p' + \mu^2} - \frac{2p^\mu}{2p\cdot k - \mu^2}\right)$$

If we sum over the polarization states of the photon, we get, using[1]

$$\sum_\lambda e_\mu{}^\lambda e_\nu{}^\lambda = -\left(g_{\mu\nu} - \frac{k_\mu k_\nu}{\mu^2}\right) \quad (12.13)$$

[1] This is the vector meson analog of the covariant photon polarization sum $\sum_\lambda e_\mu{}^\lambda e_\nu{}^\lambda = -g_{\mu\nu}$. As is clear from (12.14) the "longitudinal" contribution ${}^\mu k k_\nu/\mu^2$ vanishes as $\mu^2 \to 0$. It is only in that limit that the condition of gauge invariance $k_\mu I^\mu = 0$ is satisfied.

the factor

$$-\frac{\alpha}{\pi}\frac{d\Omega_{\hat{k}}}{4\pi}\frac{k^2\,dk}{\omega}\left[\frac{4m^2-\mu^2}{(2p'\cdot k+\mu^2)^2}+\frac{4m^2-\mu^2}{(2p\cdot k-\mu^2)^2}+\frac{-8p\cdot p'+2\mu^2}{(2p'\cdot k+\mu^2)(2p\cdot k-\mu^2)}\right] \tag{12.14}$$

Let us now calculate the cross section for Coulomb scattering accompanied by the emission of a photon of energy less than ΔE. The calculation involves an angular integration and an integration over photon energies from $\omega = \mu$ to ΔE. It is clear from (12.7) that this quantity diverges logarithmically when $\mu = 0$. We shall be interested in the leading term (as $\mu \to 0$) of the cross section.

For the first two terms we have an integral of the type

$$(4m^2-\mu^2)\frac{\alpha}{\pi}\int_{\mu}^{\Delta E}\sqrt{\omega^2-\mu^2}\,d\omega\int\frac{d\Omega_{\hat{k}}}{4\pi}\frac{1}{(2E'\omega-2\mathbf{p}'\cdot\mathbf{k}+\mu^2)^2}$$

$$\approx 4m^2\frac{\alpha}{\pi}\int_{\mu}^{\Delta E}\sqrt{\omega^2-\mu^2}\,d\omega\,\frac{1}{2}\int_{-1}^{1}d(\cos\theta)\frac{1}{(2E'\omega+\mu^2-2p'k\cos\theta)^2}$$

$$= 4m^2\frac{\alpha}{\pi}\int_{\mu}^{\Delta E}\sqrt{\omega^2-\mu^2}\,d\omega\,\frac{1}{(2E'\omega+\mu^2)^2-4k^2p'^2}$$

The leading contributions for the first two terms in (12.14) add up to

$$-\frac{\alpha}{\pi}\log\frac{2\,\Delta E}{\mu} \tag{12.15}$$

To carry out the angular integration in the last term we first combine the denominators using a trick invented by Feynman. This consists of writing

$$\frac{1}{ab}=\int_0^1 dx\,\frac{1}{[ax+b(1-x)]^2} \tag{12.16}$$

Thus we get

$$(8p\cdot p'-2\mu^2)\frac{\alpha}{\pi}\int_{\mu}^{\Delta E}\sqrt{\omega^2-\mu^2}\,d\omega\int\frac{d\Omega_{\hat{k}}}{4\pi}$$

$$\times\int_0^1 dx\,\frac{1}{[2k\cdot(p'x+p(1-x))-\mu^2(1-2x)]}$$

$$\approx 8p\cdot p'\frac{\alpha}{\pi}\int_{\mu}^{\Delta E}\sqrt{\omega^2-\mu^2}\,d\omega$$

$$\times\int_0^1 dx\,\frac{1}{(2E_x\omega-\mu^2(1-2x))^2-4k^2\mathbf{p}_x^{\,2}} \tag{12.17}$$

Here

$$E_x = xE' + (1 - x)E$$
$$\mathbf{p}_x^2 = (\mathbf{p}'x + \mathbf{p}(1 - x))^2 \tag{12.18}$$

If we carry out the ω integration first, we get for the most important contribution

$$2p \cdot p' \frac{\alpha}{\pi}\frac{1}{2}\int_0^1 \frac{dx}{E_x^2 - \mathbf{p}_x^2} \log \frac{(E_x^2 - \mathbf{p}_x^2)(\Delta E)^2}{E_x^2\mu^2} \tag{12.19}$$

When the momentum change is large, so that with

$$q = p' - p$$

we have $-q^2 \gg m^2$, (12.19) is much larger than the contributions (12.15). The expression in (12.19) diverges as $\mu \to 0$. We shall see in Chapter 13 that this is not a senseless result. This divergence in the Coulomb cross section with energy loss less than ΔE is compensated by a similar divergence in the radiative corrections to the purely elastic Coulomb scattering. The divergence only appears because of our perturbation approach and is not physical if sensible questions are asked.

Let us now turn to a calculation of bremsstrahlung by high energy electrons. The method we shall use is based on an observation by Fermi[1] that the field of a moving charge closely resembles a pulse of radiation which may be treated as a flux of photons distributed over the frequency spectrum with a certain density $n(\omega)$. If we view the bremsstrahlung process from the electron rest frame, the nucleus, which is the source of the Coulomb field, appears to move with a high velocity and its Coulomb field "looks" like a packet of photons. These photons undergo Compton scattering by the electron, and the scattered photons are actually the ones radiated by the electron.

The relationship of the process to Compton scattering is evident from Fig. 12.1. In fact, if $\mathcal{R}_B$ is the bremsstrahlung amplitude, $|\mathbf{\Delta}|^2 \mathcal{R}_B$ is just the Compton scattering amplitude, except for some factors and except for the fact that it is a *Compton scattering with one of the photons off the mass shell and no longer transversely polarized.* The Weizsäcker-Williams approximation, as this is called, consists of ignoring these last two deviations from free Compton scattering and in using the Klein-Nishina formula for describing the electron-photon interaction. It is interesting to note that the WW method has been revived in strong interaction

[1] E. Fermi, *Z. Physik* **29**, 315 (1924). The method was developed by K F. Weizsäcker, *Z. Physik* **88**, 612 (1934) and E. J. Williams, *Kgl. Danske Videnskab. Selskab, Mat.-Fys. Medd.*, **13**, No. 4 (1935). A good discussion may be found in J. D. Jackson, *loc. cit.*

physics under the name of "one-particle-exchange." There the conditions for the validity of the approximation are much harder to establish.

Limitations of space do not allow us to present a derivation of the result that the Coulomb field of a nucleus of charge Z moving with velocity $v \approx 1$ may be viewed as a distribution of photons with different frequencies, the number of photons per unit frequency interval being[1]

$$n(\omega)\, d\omega \cong \frac{2Z^2\alpha}{\pi\omega} \log \frac{183}{Z^{1/3}}\, d\omega \tag{12.20}$$

where the constant logarithmic factor is appropriate to a screened Coulomb field. We will now use this result to calculate the bremsstrahlung cross section by viewing the process in the frame in which the electron is at rest. The Klein-Nishina formula (10.47) for Compton scattering in the electron rest frame is

$$d\sigma(\omega^{*\prime}, \omega^*) = \frac{\alpha^2}{2m^2}\left(\frac{\omega^{*\prime}}{\omega^*}\right)^2\left(\frac{\omega^{*\prime}}{\omega^*} + \frac{\omega^*}{\omega^{*\prime}} - \sin^2\Theta^*\right) d\Omega^* \tag{12.21}$$

where ω^* is the energy of the incident photon, $\omega^{*\prime}$ is the energy of the outgoing photon, and the scattering angle is Θ^* related to them by energy momentum conservation in the form

$$1 - \cos\Theta^* = \frac{m}{\omega^{*\prime}} - \frac{m}{\omega^*} \tag{12.22}$$

The angular integration may be transformed into an integration over $\omega^{*\prime}$, since

$$d\Omega^* = 2\pi\, d(\cos\Theta^*) = 2\pi m \frac{d\omega^{*\prime}}{\omega^{*\prime 2}} \tag{12.23}$$

Thus the bremsstrahlung cross section is given by

$$\begin{aligned} d\sigma_B &= \int d\omega^*\, n(\omega^*) \int d\sigma(\omega^{*\prime}, \omega^*) 2\pi m \frac{d\omega^{*\prime}}{\omega^{*\prime 2}} \\ &= \frac{2Z^2\alpha^3}{m} \log\frac{183}{Z^{1/3}} \int \frac{d\omega^*}{\omega^*}\frac{m}{\omega^{*2}} \int d\omega^{*\prime} \\ &\quad \times \left(\frac{\omega^{*\prime}}{\omega^*} + \frac{\omega^*}{\omega^{*\prime}} - 1 + \left(1 + \frac{m}{\omega^*} - \frac{m}{\omega^{*\prime}}\right)^2\right) \end{aligned} \tag{12.24}$$

The integrals are simple; the limits of integration are restricted by energy and momentum conservation. We express the variables in terms of the laboratory variables. The usual Lorentz transformation formulas hold:

$$\begin{aligned} \omega' \sin\theta &= \omega^{*\prime} \sin\Theta^* \\ \omega' \cos\theta &= \gamma(\omega^{*\prime}\cos\Theta^* - v\omega^{*\prime}) \end{aligned} \tag{12.25}$$

[1] J. D. Jackson, *loc. cit.*; W. Heitler, *loc. cit.*

From this it follows that for $v \approx 1$,

$$\omega' = \gamma\omega^{*\prime}(1 - v \cos \Theta^*) \approx m\gamma\left(1 - \frac{\omega^{*\prime}}{\omega^*}\right)$$

It should be noted that in the laboratory frame the initial electron energy is $E = \gamma m$, and the final electron energy is given by

$$E' = E - \omega'$$

with ω' the energy of the radiated photon. It should also be noted that θ here is the angle the photon momentum vector makes with the direction from which the electron came.

The following relations apply immediately:

$$\frac{\omega^{*\prime}}{\omega^*} = 1 - \frac{\omega'}{m\gamma} = 1 - \frac{\omega'}{E} = \frac{E'}{E}$$

$$d\omega^{*\prime} = -\frac{\omega^*}{E}\, d\omega'$$

$$\omega' = \gamma\frac{\omega^{*\prime}}{\omega^*}\,\omega^*(1 - \cos \Theta^*) = \frac{E'}{m}\,\omega^*(1 - \cos \Theta^*) \qquad (12.26)$$

The last of these shows us that the range of the ω^* integration is ∞ to $\omega' m/2E'$. We finally write

$$d\sigma_B = 2Z^2\alpha^3 \log\frac{183}{Z^{1/3}}\left(-\frac{d\omega'}{E}\right)\int_\infty^{\omega' m/2E'} \frac{d\omega^*}{\omega^{*2}} \times \left[\frac{E'}{E} + \frac{E}{E'} - 1 + \left(1 + \frac{m}{\omega^*} - \frac{mE}{\omega^* E'}\right)^2\right]$$

which after a little algebra yields

$$d\sigma_B = \frac{4Z^2\alpha^3}{m^2}\log\frac{183}{Z^{1/3}}\left[1 - \frac{2}{3}\frac{E'}{E} + \left(\frac{E'}{E}\right)^2\right]\frac{d\omega'}{\omega'} \qquad (12.27)$$

This expression is in good agreement with exact calculations. We may use this cross section to calculate the energy loss in a distance dx. The expression is

$$\begin{aligned} dE &= -\rho\, dx\int_0^{\omega_{\max}} d\omega'\omega'\,\frac{d\sigma_B}{d\omega'} \\ &= -\rho\, dx\,\frac{4Z^2\alpha^3}{m^2}\log\frac{183}{Z^{1/3}}\int_0^{\omega_{\max}\approx E} d\omega' \\ &\quad\times\left(1 - \frac{2}{3}\frac{E - \omega'}{E} + \left(\frac{E - \omega'}{E}\right)^2\right) \end{aligned} \qquad (12.28)$$

Here ρ is the density of nuclei per unit volume. The result shows that

$$\frac{dE}{dx} = -\rho\left(\frac{4Z^2\alpha^3}{m^2}\log\frac{183}{Z^{1/3}}\right)E \tag{12.29}$$

i.e., energy is lost exponentially. We have

$$E(x) = E(0)e^{-x/l_R} \tag{12.30}$$

with

$$l_R = \left(\rho\frac{4Z^2\alpha^3}{m^2}\log\frac{183}{Z^{1/3}}\right)^{-1} \tag{12.31}$$

the *radiation length*, the distance in which the electron loses about $\frac{3}{4}$ of its energy. Some characteristic values are

Air (NTP)	330 m
Aluminum	9.7 cm
Lead	0.52 cm

Another important process which takes place in an external field is the materialization of a photon. In terms of our picture of the nuclear field as a collection of photons, the basic reaction is

$$\gamma + \gamma \rightarrow e^+ + e^-$$

We shall again use the WW method to discuss this process.

Let the photon energy in the laboratory system be ω_L. Then in the frame in which the nucleus is moving with velocity $v \approx 1$ toward the photon, this incident photon has frequency $\omega^* \approx \omega_L/2\gamma$. If the cross section for the production of a pair of photons of frequencies ω and ω^* moving toward each other collinearly is $\sigma(\omega, \omega^*)$, the quantity we want is

$$\sigma_{\text{pair}} = \int \frac{2Z^2\alpha}{\pi\omega}\log\frac{183}{Z^{1/3}}\,\sigma(\omega, \omega^*)\,d\omega \tag{12.32}$$

The total cross section for the process

$$\gamma + \gamma \rightarrow e^+ + e^-$$

can only be a function of the invariant variable $s = (k_1 + k_2)^2$, where k_1 and k_2 are the momenta of the two photons in the initial state. In our case

$$s = (\omega + \omega^*)^2 - (\omega - \omega^*)^2 = 4\omega\omega^*$$

so that

$$\sigma(\omega, \omega^*) = \sigma(\omega\omega^*)$$

We may therefore write

$$\sigma_{\text{pair}} = \frac{2Z^2\alpha}{\pi} \log \frac{183}{Z^{1/3}} \int \sigma(\omega\omega^*) \frac{d\omega}{\omega}$$

$$= \frac{2Z^2\alpha}{\pi} \log \frac{183}{Z^{1/3}} \int \sigma(s) \frac{ds}{s} \tag{12.33}$$

The integral may be carried out using the formula (10.61) for the cross section. Note that

$$\beta^2 = 1 - \frac{4m^2}{s}$$

so that $ds/s = 2\beta\, d\beta/(1 - \beta^2)$. The limits of integration are given by the condition $0 \leq \beta \leq 1$ (β is the velocity of the electron or positron). Thus we get

$$\sigma_{\text{pair}} = \frac{2Z^2\alpha}{\pi} \log \frac{183}{Z^{1/3}} \frac{\pi}{2} \frac{\alpha^2}{m^2} \int_0^1 2\beta\, d\beta \left[(3 - \beta^4) \log \frac{1+\beta}{1-\beta} - 2\beta(2 - \beta^2) \right]$$

$$= \frac{Z^2\alpha^3}{m^2} \cdot \frac{28}{9} \log \frac{183}{Z^{1/3}} \tag{12.34}$$

This is to be compared with the exact result of Bethe and Heitler

$$\sigma_{\text{pair}}^{\text{B.H.}} \cong \frac{Z^2\alpha^3}{m^2} \left(\frac{28}{9} \log \frac{183}{Z^{1/3}} - \frac{2}{27} \right) \tag{12.35}$$

As in bremsstrahlung, we can define a mean free path for pair production—the main mechanism for the absorption of high energy photons—and comparison of the cross section with that for bremsstrahlung shows that

$$l_{\text{pair}} = \left(\frac{28}{9} \frac{Z^2\alpha^3\rho}{m^2} \log \frac{183}{Z^{1/3}} \right)^{-1}$$

$$= \frac{9}{7} l_R \tag{12.36}$$

We conclude this chapter with some qualitative remarks. One of these has to do with the polarization of the photons radiated by high energy, longitudinally polarized electrons. This effect may be studied quite simply in the "hard photon" limit, when most of the energy is transferred to the photon. The factors appearing in (12.1) when written in terms of the momentum transfer

$$\Delta = (0, \mathbf{p} - \mathbf{k} - \mathbf{p}')$$

are proportional to

$$\bar{u}(p')\gamma_0 \frac{\not{p}' + \not{\Delta} + m}{\Delta^2 + 2p' \cdot \Delta} \boldsymbol{\gamma} \cdot \mathbf{e}\, u(p) \tag{12.37}$$

and

$$\bar{u}(p')\boldsymbol{\gamma} \cdot \mathbf{e} \frac{\not{p} - \not{\Delta} + m}{\Delta^2 - 2p \cdot \Delta} \gamma^0 u(p) \tag{12.38}$$

For small momentum transfers, when $E, E' \gg |\mathbf{\Delta}|$, we have

$$(m + \not{p} - \not{\Delta})\gamma_0 \, u(p) \approx 2p_0 \, u(p) \tag{12.39}$$

and

$$\bar{u}(p') \, \gamma^0(\not{p}' + \not{\Delta} + m) \approx 2p_0' \, \bar{u}(p') \tag{12.40}$$

Thus for both terms the spin dependence of the matrix element is determined by

$$\bar{u}(p')\boldsymbol{\gamma} \cdot \mathbf{e} \, u(p)$$

If the incident electron is longitudinally polarized, the spinor $u(p)$ may be replaced by $\frac{1}{2}(1 + \boldsymbol{\sigma} \cdot \hat{\mathbf{p}}) \, u(p)$. The matrix element, with $E \gg E' \gg m$ becomes

$$\frac{2E}{2\mathbf{p} \cdot \mathbf{\Delta} + \mathbf{\Delta}^2}\left(\bar{u}(p')\boldsymbol{\gamma} \cdot \mathbf{e} \, \frac{1 + \boldsymbol{\sigma} \cdot \hat{\mathbf{p}}}{2} \, u(p)\right) \tag{12.41}$$

With the help of the expression in (11.8) we write the second factor as

$$\bar{u}(p')\boldsymbol{\gamma} \cdot \mathbf{e} \, \frac{1 + \boldsymbol{\sigma} \cdot \hat{\mathbf{p}}}{2} \, u(p)$$

$$\cong \frac{1}{2m}\left[\sqrt{\frac{E'}{E}}\,(\mathbf{p} - i\boldsymbol{\sigma} \times \mathbf{p}) \cdot \mathbf{e} + \sqrt{\frac{E}{E'}}\,(\mathbf{p} + i\boldsymbol{\sigma} \times \mathbf{p}') \cdot \mathbf{e}\right]\frac{1 + \sigma \cdot \hat{\mathbf{p}}}{2} \tag{12.42}$$

To a good approximation, when the momentum transfer is small,

$$\mathbf{p} \cong \mathbf{k} + \mathbf{p}'$$

so that

$$\hat{\mathbf{p}} \cong \hat{\mathbf{k}} + \frac{E'}{E} \, \hat{\mathbf{p}}' \tag{12.43}$$

Hence

$$\hat{\mathbf{p}} \cdot \mathbf{e} \cong \frac{E'}{E} \, \hat{\mathbf{p}}' \cdot \mathbf{e}$$

Some fairly straightforward algebra leads to

$$\frac{\sqrt{EE'}}{4m} (\boldsymbol{\sigma} + \hat{\mathbf{p}}' + i\boldsymbol{\sigma} \times \hat{\mathbf{p}}') \cdot (\mathbf{e} - i\hat{\mathbf{k}} \times \mathbf{e}) \tag{12.44}$$

Now, if we note that

$$(\mathbf{e} \,|\, \frac{1 + \mathbf{S} \cdot \hat{\mathbf{k}}}{2} = \frac{1}{2} (\mathbf{e} - i\hat{\mathbf{k}} \times \mathbf{e})$$

[see (10.43)], we see that the photon has the same helicity as the incident electron.

Detailed calculations show that this phenomenon of "helicity transfer" also applies to pair production. In the creation of pairs by a longitudinally

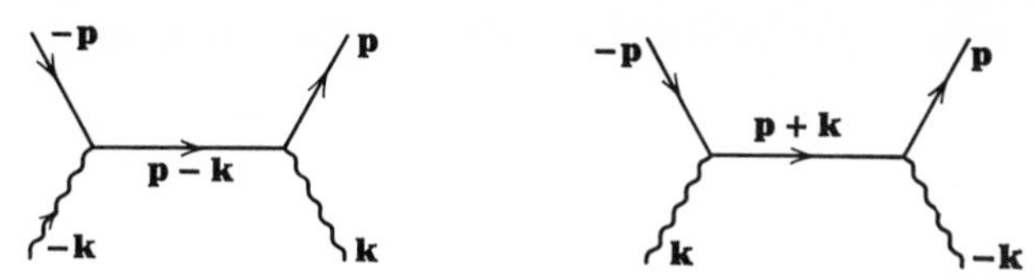

Fig. 12.2. Feynman graphs for pair production. The lines are labeled by the center of mass moments in this figure.

polarized photon the helicity is transferred to the faster of the electron-positron pair. To see this qualitatively we note that in the process

$$\gamma + \gamma \to e^+ + e^-$$

the energy denominators corresponding to the two graphs in Fig. 12.2 are $(\mathbf{p} \pm \mathbf{k})^2 + m^2$, which favor the electron or positron lining up with the direction of the incident photon in the center of mass system. In the frame in which one of the photons has a lot more energy than the other there will be a tendency for one of the members of the pair to carry off all of the energy and the other to remain almost at rest. The same argument which was used to show that the helicity is transferred from the electron to the photon in "hard" bremsstrahlung here shows that the helicity is transferred from the photon to the fast-moving charged particle. Qualitative estimates of this process were made by Dyson and McVoy.[1]

In conclusion we remark that the confirmation of theoretical predictions by experiments with high energy electrons does not in general represent a sensitive test of high energy quantum electrodynamics. In most cases, as our considerations of the Weizsäcker-Williams approximation show, only the validity of electrodynamics at low energies in the center of mass system is established. To check the validity of quantum electrodynamics at high energies we must perform experiments that test the behavior of electrons far off the mass shell or at high center of mass energies.[2]

PROBLEMS

1. Calculate the soft photon bremsstrahlung by a spin 0 particle. Compare your result with that for a spin $\frac{1}{2}$ particle.

2. A particle is known to decay into two π^0, whose decay mode is

$$\pi^0 \to 2\gamma$$

How large a liquid hydrogen bubble chamber would you need to study the process? (The density of liquid hydrogen is 0.07 g/cm^3.) How large a chamber containing Freon (CF_3Br), whose density is roughly 1.5 g/cm^3, would you need?

[1] K. W. McVoy and F. J. Dyson, *Phys. Rev.* **106,** 1360 (1957).
[2] S. D. Drell, *Ann. Phys.* (*N.Y.*) **4,** 75 (1958).

13

Higher Order Terms in Perturbation Theory

The methods developed in the last three chapters are quite adequate to predict results of quantum electrodynamic experiments with an accuracy of better than 10%. If we now turn to a discussion of radiative corrections, it is less with an aim of improving the accuracy of the calculations presented in this book than to clear up some real difficulties that arise in connection with the Feynman rules. When we calculate beyond lowest order, there are really two problems that arise and are a priori independent. The first one has to do with the meaning of the parameters m and e that appear in the Lagrangian. In lowest order perturbation theory we assumed that they were the mass and charge of the electron. This was supported by calculations and could well be established. For example, although we did not carry out the calculation, it is almost obvious that the expectation value of the Hamiltonian in a one-electron state is $E_p = (p^2 + m^2)^{1/2}$. Also, the lowest-order zero-frequency limit of the Compton cross section reduced to the classical Thomson cross section, provided e had the assigned meaning.

When higher-order corrections are considered, it is no longer true that the parameters m and e have this meaning. Rather, the physical mass of the electron has to be calculated to the higher order of precision required, using, for example

$$m_{\text{exp}} = \left.\frac{(\Psi_{p'}, H\Psi_p)}{(\Psi_{p'}, \Psi_p)}\right|_{\mathbf{p}=0}$$

Similarly, the physical charge of the electron should be calculated by defining the charge as the quantity which appears in an experimental result in a particular way; e.g., the charge may be defined as that parameter which appears in the Compton cross section at threshold. Thus to whatever order we calculate, the result should be[1]

$$\sigma = \frac{8\pi}{3}\left(\frac{e_{\text{exp}}^2}{4\pi m_{\text{exp}}}\right)^2$$

[1] The charge could also be defined by the requirement that the scattering cross section for an electron in an external Coulomb field be $d\sigma/d\Omega = (Z\alpha m/2p^2 \sin^2 \theta/2)^2$ as $p \to 0$. This definition will give the same result as the threshold requirement.

When all this is done, we should find relationships of the form

$$e = e(e_{\text{exp}})$$
$$m = m(e_{\text{exp}}, m_{\text{exp}})$$

and with their help eliminate the parameters e and m in terms of the experimental charge and mass. This elimination of the parameters in terms of the observable quantities is called *renormalization* and is part and parcel of any calculation that is to be confronted with experiment.

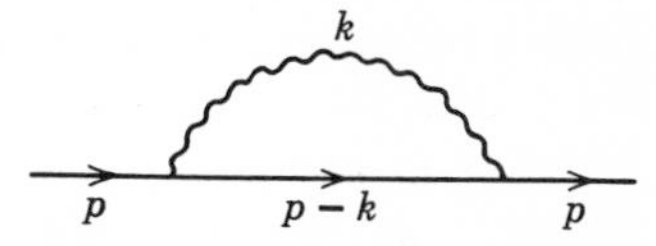

Fig. 13.1. Second order self-energy graph.

The second problem has to do with the appearance of infinite integrals when the Feynman rules are applied literally in higher order calculations. For example, the so-called self-energy correction to a fermion line shown in Fig. 13.1 leads to an expression that is proportional to the following integral

$$\int dk\gamma^{\mu} \frac{\not{p} - \not{k} + m}{(p - k)^2 - m^2} \gamma_{\mu} \frac{1}{k^2} \tag{13.1}$$

With help of the relation

$$\gamma^{\mu}\not{a}\gamma_{\mu} = -2\not{a}$$

this becomes

$$\int \frac{dk}{k^2 + i\epsilon} \frac{-2\not{p} + 2\not{k} + 4m}{(p - k)^2 - m^2} \tag{13.2}$$

and a counting of powers of k in the numerator and denominator shows that for large k the integrand behaves like

$$\int \frac{d^4k}{k^3} \sim \int dk$$

which means that if we had a cut-off on the integral,[1] it would have a leading term proportional to the cut-off. This is called a linearly divergent integral. The fact that it appears shows that the Feynman rules developed in Chapters 8 and 9 are incomplete, since they do not tell us what to do about the integrals. It turns out quite miraculously, however, that if we

[1] A covariant cut-off procedure, which does some violence to the theory at high energies, is to replace a propagator like $1/(q^2 - m^2)$ by $1/(q^2 - m^2)[-\Lambda^2/(q^2 - \Lambda^2)]$ with $\Lambda^2 \gg m^2$. Some caution must be used in applying this cut-off to graphs involving closed electron loops. See R. P. Feynman, *Phys. Rev.* **76**, 769 (1949), especially Sections 5 and 7 in this connection.

first renormalize the charge and mass of the electron, then, in all observable quantities no infinite integrals ever appear.[1] We shall show this in a simple, but nontrivial, calculation of the radiative corrections to the scattering of an electron in an external field. In doing so, we will still do violence to normal mathematical procedure: infinite integrals will be added, subtracted, or divided out, as if they were finite. The fact is that when these manipulations are carried out with care, the results are finite and agree with experiment to a remarkable degree, so that some truth lies hidden behind our present awkward method of extracting finite answers from seemingly infinite expressions. The difficulty seems to be that in a Lagrangian approach, one deals with an ill-defined product of several operators at the same point. In its simplest form this leads to the appearance of an infinite zero-point energy for a free-particle Hamiltonian, as noted in Chapter 1. Whether the infinities arise as a result of making an expansion in powers of e—the scattering matrix elements may not be analytic functions of e near $e = 0$—or whether they are a fundamental flaw of an otherwise plausible theory has not yet been answered to everybody's satisfaction. What has been shown is that *if* the theory is to be finite, rather irregular functions, for which interchanges of orders of integration are forbidden, for example, must appear in the matrix elements.[2]

We begin our treatment of the radiative corrections to the scattering of an electron in an external field by a discussion of *mass renormalization.* We could investigate the consequences of the requirement that for one-electron states the following condition holds:

$$(\Psi_{p'}, H\Psi_p) = E_p(\Psi_{p'}, \Psi_p) \tag{13.3}$$

but it is simpler to consider the asymptotic condition (6.24) that for the special states Ψ_0, the vacuum state, and $\Psi_{p'}$, a one-electron state, reads

$$(\Psi_0, \mathbf{a}_{\text{in}}(p)\Psi_{p'}) = \lim_{t\to-\infty} \int d^3x\, \bar{u}_p(x)\gamma^0(\Psi_0, \psi(x)\Psi_{p'}) \tag{13.4}$$

In the limit of plane wave solutions (rather than wave packets) the right-hand side is

$$\lim_{t\to-\infty} \frac{1}{(2\pi)^{3/2}} \int d^3x\, \bar{u}(p)e^{ipx}\, \gamma^0(\Psi_0, \psi(0)\Psi_{p'})e^{-ip'x}$$

$$= \lim_{t\to-\infty} \{(2\pi)^{3/2}\, \bar{u}(p)\, \gamma^0(\Psi_0, \psi(0)\Psi_{p'})\, \delta(\mathbf{p} - \mathbf{p}')\}$$

[1] Infinite integrals may arise from the zero mass of the photon, as we saw in the last chapter. Such divergences are called infrared divergences, in contrast to the high momentum ultraviolet ones. We shall see that the infrared divergence also disappears.

[2] G. Källén, *Kgl. Danske Videnskab. Selskab. Mat.-Fys. Medd.*, **29**, No. 17 (1954). K. Johnson, *Ann. Phys. (N.Y.)* **10**, 536 (1960).

The quantity inside the curly bracket is independent of time and we have

$$\delta(\mathbf{p} - \mathbf{p}')\, \bar{u}(p)\, \gamma^0(\Psi_0, \psi(0)\Psi_{p'}) = \frac{1}{(2\pi)^{3/2}}(\Psi_0, \mathbf{a}_{\text{in}}(p)\Psi_{p'})$$

The right-hand side, however, is just

$$\frac{1}{(2\pi)^{3/2}} \frac{E_p}{m}\, \delta(\mathbf{p} - \mathbf{p}')$$

Hence

$$\bar{u}(p)\, \gamma^0(\Psi_0, \psi(0)\Psi_p) = \frac{1}{(2\pi)^{3/2}} \frac{E_p}{m}$$

which implies that

$$(\Psi_0, \psi(0)\Psi_p) = \frac{1}{(2\pi)^{3/2}}\, u(p)$$

or, equivalently, that

$$(\Psi_0, \psi(x)\Psi_p) = \frac{1}{(2\pi)^{3/2}}\, u(p)e^{-ipx} = (\Psi_0, \psi_{\text{in}}(x)\Psi_p) \tag{13.5}$$

This means that

$$\begin{aligned}(\Psi_0, f(x)\Psi_p) &= -(i\gamma^\mu\, \partial_\mu - m_{\text{exp}})(\Psi_0, \psi(x)\Psi_p) \\ &= -(i\gamma^\mu\, \partial_\mu - m_{\text{exp}})(\Psi_0, \psi_{\text{in}}(x)\Psi_p) = 0 \end{aligned} \tag{13.6}$$

This is a condition on the interaction term which includes a mass renormalization counterterm (as pointed out in Chapter 8) and may be used to determine the difference between m_{exp} and the parameter m which appears in the Lagrangian.

We use the reduction formalism to write

$$\begin{aligned}(\Psi_0, f(x)\, \Psi_p) &= (\Psi_0, f(x)\, \mathbf{a}^{+}_{\text{in}}(p)\Psi_0) \\ &= \lim_{t\to-\infty} \int_{y_0=t} d^3y(\Psi_0, T(f(x)\, \bar{\psi}(y))\Psi_0)\gamma_0\, u_p(y) \\ &= -\int dy \frac{\partial}{\partial y_0}[(\Psi_0, T(f(x)\, \bar{\psi}(y))\Psi_0)\gamma_0\, u_p(y)] \\ &= i\int dy(\Psi_0, T(f(x)\, \bar{\psi}(y))\Psi_0)\, \overleftarrow{\mathfrak{D}}_y\, u_p(y)\end{aligned}$$

The right-hand side consists of two terms. In one of them we make use of

$$\bar{\psi}(y)\, \overleftarrow{\mathfrak{D}}_y = i\frac{\partial}{\partial y^\mu}\, \bar{\psi}(y)\gamma^\mu + m_{\text{exp}}\bar{\psi}(y) = \bar{f}(y) \tag{13.7}$$

and in the other we take the time derivative of the step functions appearing because of the T-product. We thus obtain

$$0 = i\int dy(\Psi_0, T(f(x)\bar{f}(y))\Psi_0)\, u_p(y)$$
$$-\int dy(\Psi_0, [-\delta(x_0 - y_0) f(x)\, \bar{\psi}(y) - \delta(y_0 - x_0)\, \bar{\psi}(y) f(x)]\Psi_0)\gamma_0 u_p(y)$$
$$= i\int dy(\Psi_0, T(f(x)\bar{f}(y))\Psi_0)\, u_p(y)$$
$$+\int d^3y(\Psi_0, \{f(x), \bar{\psi}(y)\}_{x_0=y_0}\Psi_0)\gamma_0\, u_p(y) \tag{13.8}$$

We now make use of the equations of motion in terms of the *unrenormalized* operators, since it is in terms of these that the Lagragian is written.[1]

$$(i\gamma^\mu \partial_\mu - m_{\text{exp}})\, \hat{\psi}(x) \equiv -\hat{f}(x)$$
$$= -e_1\gamma^\mu\, \hat{A}_\mu(x)\, \hat{\psi}(x) - (m_{\text{exp}} - m)\, \hat{\psi}(x) \tag{13.9}$$

The equation of motion, together with the equal time commutation relation

$$\{\hat{\psi}(x), \hat{\bar{\psi}}(y)\}_{x_0=y_0}\gamma^0 = \delta(\mathbf{x} - \mathbf{y}) \tag{13.10}$$

lead to a value of the equal time commutator term in (13.8) such that

$$Z_2^{-1}\,(m_{\text{exp}} - m)\, u_p(x) \equiv Z_2^{-1}\, \delta m\, u_p(x)$$
$$= -i\int dy\, (\Psi_0, T(f(x)\bar{f}(y))\Psi_0)\, u_p(y) \tag{13.11}$$

results.[2] The term involving $(\Psi_0, A_\mu(x)\Psi_0)$ vanishes.

Our next task is to evaluate $Z_2^{-1}\,\delta m$ in perturbation theory. We use the procedure of Chapter 8 to calculate

$$(\Psi_0,\ T(f(x)\,\bar{f}(y))\Psi_0) \simeq Z_2^{-1}e_1^2(\Phi_0, T(\gamma^\mu\, A_\mu^{(0)}(x)\, \psi^{(0)}(x) A_\nu^{(0)}(y)\, \bar{\psi}^{(0)}(y)\gamma^\nu)\Phi_0)$$
$$= Z_2^{-1}e_1^2(-\tfrac{1}{2}\gamma^\mu\, S_F(x - y)\gamma_\mu)(-\tfrac{1}{2}\, D_F(y - x)) \tag{13.12}$$

[1] The parameter e_1 is the coupling constant which appears in the Lagrangian. We reserve the letter e for the measured electric charge.

[2] This expression was first obtained by G. Källén, *Helv. Phys. Acta* **25,** 417 (1952).

Thus to this order in e_1

$$\frac{1}{(2\pi)^{3/2}}\,\delta m u(p) e^{-ipx} = -ie_1^{\,2}\int dy\gamma^\mu\left(-\frac{1}{2}S_F(x-y)\right)\gamma_\mu \times\left(-\frac{1}{2}D_F(y-x)\right)\frac{1}{(2\pi)^{3/2}}u(p)e^{-ipy}$$

$$= -\frac{ie_1^{\,2}}{(2\pi)^{3/2}}\frac{1}{(2\pi)^4}\int dk\gamma^\mu\frac{1}{\not p+\not k-m_{\text{exp}}} \times\gamma_\mu\frac{1}{k^2+i\epsilon}e^{-ipx}u(p)$$

i.e.

$$\delta m\, u(p) = -\frac{ie_1^{\,2}}{(2\pi)^4}\int dk\gamma^\mu\frac{\not p+\not k+m_{\text{exp}}}{(p+k)^2-m_{\text{exp}}^2+i\epsilon}\gamma_\mu\frac{1}{k^2+i\epsilon}u(p) \tag{13.13}$$

We can similarly obtain

$$\bar u(p')\delta m = -\frac{ie_1^{\,2}}{(2\pi)^4}\bar u(p')\int dk\gamma^\mu\frac{\not p'+\not k+m_{\text{exp}}}{(p'+k)^2-m_{\text{exp}}^2+i\epsilon}\gamma_\mu\frac{1}{k^2+i\epsilon} \tag{13.14}$$

A power count shows that in perturbation theory δm is represented by a divergent integral. It is not known whether this is a general feature of the theory.

The reason for our interest in δm is that it enters into the Feynman rules. In deriving and listing these rules in Chapters 8 and 9, we ignored the fact that the interaction term was

$$H_1 = -\int d^3x(\hat j^\mu(x)\,\hat A_\mu(x) + \delta m\,\hat{\bar\psi}(x)\,\hat\psi(x))$$

The extra term, we recall, was necessary to express the *free Hamiltonian* H_0 in terms of the physical mass m_{exp}, as a consequence of which the propagator for the electron and the free-field solutions $u(p)$, etc., were functions of m_{exp}. The additional term in H_1 implies that additional graphs have to be drawn. The rules can be derived in a way completely analogous to the ones in Chapter 8. The result is the following: whenever there is a fermion line, the propagator representing it should be replaced by

$$\frac{1}{\not p-m} = \frac{1}{\not p-m_{\text{exp}}+\delta m} = \frac{1}{\not p-m_{\text{exp}}} - \frac{1}{\not p-m_{\text{exp}}}\delta m\frac{1}{\not p-m_{\text{exp}}} + \cdots \tag{13.15}$$

to the necessary order. The leading term in δm is of the order e_1^2. As an example, we note that in addition to the graphs appearing in Fig. 8.3 the graphs in Fig. 13.2 should be added to order e_1^4. We shall see that with the identification (13.13) and (13.14) these graphs will cancel parts of the graphs of the kind seen in Fig. 13.1.

As was pointed out in Chapters 7 and 8 [e.g. (7.38)], the higher-order scattering matrix elements involve quantities like Z_2 and Z_3.[1] Because we are still dealing with the electron, we shall first study Z_2. We may do this through the commutation relations (7.34) or more directly through the relation

$$(\Psi_0, \hat{\psi}(0)\Psi_p) = Z_2^{1/2}(\Psi_0, \psi(0)\Psi_p) = \frac{Z_2^{1/2}}{(2\pi)^{3/2}}\, u(p) \tag{13.16}$$

We again use the reduction technique as in the determination of δm.

$$\begin{aligned}\frac{Z_2^{1/2}\, u(p)}{(2\pi)^{3/2}} &= \lim_{t\to-\infty} (\Psi_0, \hat{\psi}(0)\, \mathbf{a}^+(p, t)\Psi_0) \\ &= iZ_2^{-1/2}\int dx(\Psi_0, T(\hat{\psi}(0)\, \hat{\bar{\psi}}(x))\Psi_0)\, \overleftarrow{\mathfrak{D}}_x\, u_p(x) \end{aligned}\tag{13.17}$$

after some computation. Hence to order e_1^2 we get

$$\begin{aligned} Z_2\, u(p) &= i\int dx(\Psi_0, T(\hat{\psi}(0)\, \hat{\bar{\psi}}(x))\Psi_0)\, \overleftarrow{\mathfrak{D}}_x e^{-ipx}\, u(p) \\ &= i\int dx\Big[(\Phi_0, T(\psi^{(0)}(0)\, \bar{\psi}^{(0)}(x))\Phi_0) \\ &\quad + \frac{(ie_1)^2}{2!}\iint du_1\, du_2(\Phi_0, T(\bar{\psi}^{(0)}(u_1)\gamma^\lambda\psi^{(0)}(u_1)A_\lambda^{(0)}(u_1) \\ &\quad \times \bar{\psi}^{(0)}(u_2)\gamma^\rho\psi^{(0)}(u_2)A_\rho^{(0)}(u_2)\, \psi^{(0)}(0)\, \bar{\psi}^{(0)}(x))\Phi_0\Big]\, \overleftarrow{\mathfrak{D}}_x e^{-ipx}u(p) \\ &= i\int dx\Big[-\frac{1}{2}\, S_F(-x) - e_1^2\iint du_1\, du_2 \\ &\quad \times \Big(-\frac{1}{2}\, S_F(-u_1)\Big)\gamma^\lambda\Big(-\frac{1}{2}\, S_F(u_1 - u_2)\Big)\gamma^\rho \\ &\quad \times \Big(-\frac{1}{2}\, S_F(u_2 - x)\Big)\Big(-\frac{1}{2}\, g_{\lambda\rho}\, D_F(u_1 - u_2)\Big]\, \overleftarrow{\mathfrak{D}}_x e^{-ipx}u(p) \end{aligned}$$

In this expression the mass renormalization terms have not yet been inserted, because there are most easily put in at the end. With the help

[1] Z_3 is defined by $A_\mu(x) = Z_3^{-1/2}\, \hat{A}_\mu(x)$ in analogy with $\psi(x) = Z_2^{-1/2}\, \hat{\psi}(x)$.

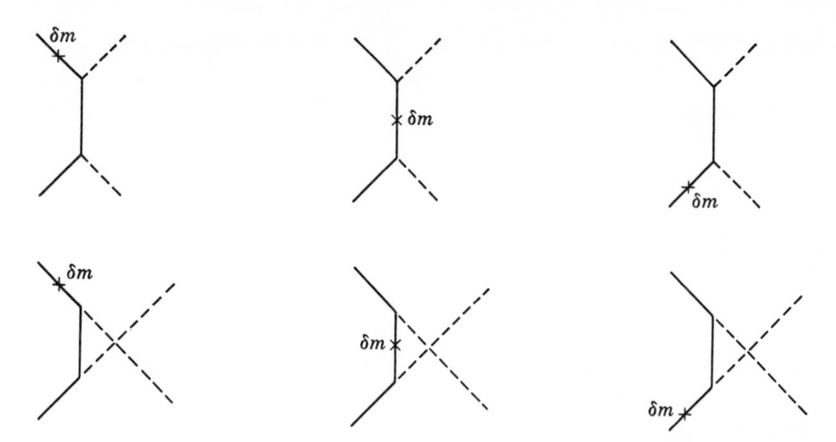

Fig. 13.2. Mass renormalization counterterm graphs in fourth order scattering.

of the relations (9.5) and (9.6), this finally gives

$$Z_2u(p) = \left[-\frac{1}{\not p - m_{\text{exp}}} + \frac{1}{\not p - m_{\text{exp}}}\,\delta m\,\frac{1}{\not p - m_{\text{exp}}}\right.$$
$$+\frac{ie_1{}^2}{(2\pi)^4}\frac{1}{\not p - m_{\text{exp}}}\int\frac{dk}{k^2+i\epsilon}\gamma^\lambda\frac{\not p+\not k+m_{\text{exp}}}{(p+k)^2-m_{\text{exp}}^2+i\epsilon}$$
$$\left.\times\,\gamma_\lambda\frac{1}{\not p - m_{\text{exp}}}\right]\overleftarrow{(-\not p+m_{\text{exp}})}u(p) \quad (13.18)$$

If we look back at (13.13), we see that apparently the δm term is canceled by the integral. This is not correct, however. A closer examination of the divergent integral

$$\sum(p) = -\frac{ie_1{}^2}{(2\pi)^4}\int\frac{dk}{k^2+i\epsilon}\gamma^\lambda\frac{\not p+\not k+m_{\text{exp}}}{(p+k)^2-m_{\text{exp}}^2+i\epsilon}\gamma_\lambda \quad (13.19)$$

shows that it is of the form[1]

$$\sum(p) = A + (\not p - m_{\text{exp}})B + (\not p - m_{\text{exp}})\,C(p)(\not p - m_{\text{exp}}) \quad (13.20)$$

Here A is a linearly divergent integral, B is a logarithmically divergent integral, and $C(p)$ is finite. Thus (13.13) leads to the identification

$$\delta m = A \quad (13.21)$$

and (13.18) reads

$$Z_2u(p) = \left[\frac{-1}{\not p - m_{\text{exp}}} + \frac{1}{\not p - m_{\text{exp}}}\,\delta m\,\frac{1}{\not p - m_{\text{exp}}} - \frac{1}{\not p - m_{\text{exp}}}\right.$$
$$\left.\times\,(A + (\not p - m_{\text{exp}})B + \cdots)\frac{1}{\not p - m_{\text{exp}}}\right]\overleftarrow{(-\not p+m_{\text{exp}})}u(p)$$

[1] J. M. Jauch and F. Rohrlich, *The Theory of Electrons and Photons* p. 178 ff.

The last term vanishes and $\delta m - A = 0$, so that we are left with

$$Z_2 u(p) = \left\{\left(1 - \left(B \frac{1}{\not{p} - m_{\text{exp}}}\right)\overleftarrow{(-\not{p} + m_{\text{exp}})}\right\} u(p)\right.$$

The factor $\overleftarrow{(m_{\text{exp}} - \not{p})}$ comes from the $\overleftarrow{\mathfrak{D}}_x$ factor in (13.17), and this unambiguously acts on what is to the left of it. Thus we get

$$Z_2 = 1 + B \tag{13.22}$$

It might be pointed out that by our treatment via the reduction formula we have avoided ambiguities in the identification of B which plagued earlier treatments.[1]

We are now ready to consider our illustrative problem—the calculation of the radiative corrections to the scattering of an electron by an external field. Figure 13.3 shows the graphs (including the lowest order graph) which have to be taken into account.

The Feynman rules give the following terms. *Graph* (*a*):

$$2\pi i e_1 \bar{u}(p')\gamma^\mu u(p)\left[\frac{Ze_1}{(2\pi)^3}\delta_{\mu 0}\frac{1}{(\mathbf{p} - \mathbf{p}')^2}\right] \tag{13.23}$$

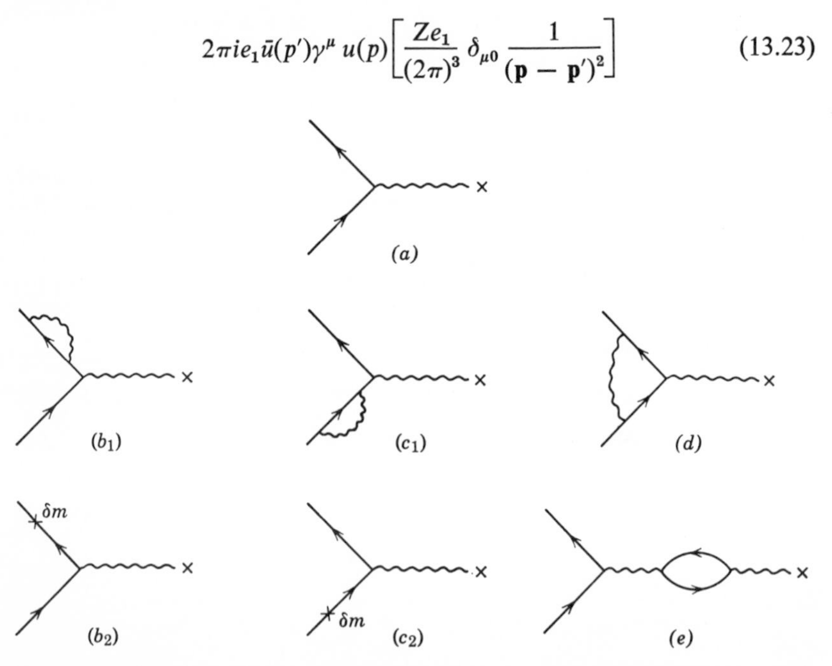

Fig. 13.3. Graphs for radiative corrections to electron scattering by an external field.

[1] See J. M. Jauch and F. Rohrlich, *op. cit.*, p. 185–186.

The factor in the brackets will be common to all graphs. We shall denote it by

$$\tilde{A}_\mu \equiv \frac{Ze_1}{(2\pi)^3}\,\delta_{\mu 0}\,\frac{1}{(\mathbf{p}-\mathbf{p}')^2} \tag{13.24}$$

so that the contribution from *graph a* will be written as

$$2\pi i e_1\,\bar{u}(p')\gamma^\mu\,\bar{u}(p)\,\tilde{A}_\mu \tag{13.25}$$

The contributions from *graph* (b_1) *to* (b_2) appear together:

$$\begin{aligned}
&-2\pi i\bar{u}(p')\delta m\,\frac{1}{\not{p}'-m_{\rm exp}}\,e_1\gamma^\mu\,u(p)\\
&\quad+\frac{1}{(2\pi)^3}\,\bar{u}(p')\int\frac{dk}{k^2}\,e_1\gamma^\lambda\,\frac{1}{\not{p}'-\not{k}-m_{\rm exp}}\,e_1\gamma_\lambda\,\frac{1}{\not{p}'-m_{\rm exp}}\,e_1\gamma^\mu\,u(p)\tilde{A}_\mu\\
&\quad=-2\pi i e_1\,\bar{u}(p')\left[\delta m+\frac{ie_1^{\,2}}{(2\pi)^4}\int\frac{dk}{k^2}\,\gamma^\lambda\,\frac{1}{\not{p}'-\not{k}-m_{\rm exp}}\,\gamma_\lambda\right]\\
&\qquad\qquad\times\frac{1}{\not{p}'-m_{\rm exp}}\,\gamma^\mu\,u(p)\tilde{A}_\mu
\end{aligned} \tag{13.26}$$

The contributions from *graphs* (c_1) *and* (c_2) are very similar:

$$-2\pi i e_1\,\bar{u}(p')\gamma^\mu\,\frac{1}{\not{p}-m_{\rm exp}}\left[\delta m+\frac{ie_1^{\,2}}{(2\pi)^4}\int\frac{dk}{k^2}\,\gamma^\lambda\,\frac{1}{\not{p}-\not{k}-m_{\rm exp}}\,\gamma_\lambda\right]u(p)\tilde{A}_\mu \tag{13.27}$$

The contribution from *graph* (d) is

$$-2\pi i e_1\bar{u}(p')\left(\frac{ie_1^{\,2}}{(2\pi)^4}\int\frac{dk}{k^2}\,\gamma^\lambda\,\frac{1}{\not{p}'-\not{k}-m_{\rm exp}}\,\gamma^\mu\,\frac{1}{\not{p}-\not{k}-m_{\rm exp}}\,\gamma_\lambda\right)u(p)\,\tilde{A}_\mu \tag{13.28}$$

and that from *graph* (e) is

$$\begin{aligned}
&2\pi i e_1\,\bar{u}(p')\gamma^\sigma\,u(p)\,\frac{1}{(p'-p)^2}\left(\frac{ie_1^{\,2}}{(2\pi)^4}\int dq\right.\\
&\qquad\left.\times\,\mathrm{tr}\left(\gamma_\sigma\,\frac{1}{\not{q}-m_{\rm exp}}\,\gamma^\mu\,\frac{1}{\not{q}+\not{p}'-\not{p}-m_{\rm exp}}\right)\right)\tilde{A}_\mu
\end{aligned} \tag{13.29}$$

(Note the extra minus sign coming from the closed loop.) Finally, the sum of all the contributions is to be multiplied by Z_2^{-1} since there are two external lines.

We begin by summing up the contributions from the first five graphs (and multiplying by Z_2^{-1}). We use (13.20) through (13.22) to write for the sum

$$2\pi i e_1 Z_2^{-1}\tilde{A}_\mu \bar{u}(p')\Big\{\gamma^\mu - (\delta m - A - (p\!\!\!/' - m_{\rm exp})B - (p\!\!\!/' - m_{\rm exp})$$

$$\times C(p')(p\!\!\!/' - m_{\rm exp}))\frac{1}{p\!\!\!/' - m_{\rm exp}}\gamma^\mu - \gamma^\mu \frac{1}{p\!\!\!/ - m_{\rm exp}}(\delta m - A$$

$$- B(p\!\!\!/ - m_{\rm exp}) - (p\!\!\!/ - m_{\rm exp})\, C(p)(p\!\!\!/ - m_{\rm exp}))\Big\}\, u(p)$$

$$= 2\pi i e_1 Z_2^{-1}(1 + 2B)\, \bar{u}(p')\gamma^\mu\, u(p)\tilde{A}_\mu$$

To this order, however,

$$1 + 2B \approx (1 + B)^2 = Z_2^{\,2} \tag{13.30}$$

so that the sum of graphs (*a*) through (c_2) is

$$2\pi i e_1 Z_2\, \bar{u}(p')\gamma^\mu\, u(p)\tilde{A}_\mu \tag{13.31}$$

The next step is to analyze the contribution from graph (*d*). We rationalize (13.28) and in anticipation of an infrared divergence give the photon a small mass. This gives

$$-2\pi i e_1 Z_2^{-1}\left(\frac{i e_1^{\,2}}{(2\pi)^4}\right)\int \frac{dk}{k^2 - \mu^2}\, \bar{u}(p')\gamma^\lambda \frac{p\!\!\!/' - k\!\!\!/ + m_{\rm exp}}{k^2 - 2p'\cdot k} \times \gamma^\mu \frac{p\!\!\!/ - k\!\!\!/ + m_{\rm exp}}{k^2 - 2p\cdot k}\gamma_\lambda u(p)\, \tilde{A}_\mu \tag{13.32}$$

We use the anticommutation properties of the Dirac matrices to write

$$(p\!\!\!/ - k\!\!\!/ + m_{\rm exp})\gamma_\lambda\, u(p) = (\gamma_\lambda k\!\!\!/ + 2p_\lambda - 2k_\lambda)\, u(p)$$

$$\bar{u}(p')\, \gamma^\lambda(p\!\!\!/' - k\!\!\!/ + m_{\rm exp}) = \bar{u}(p')(k\!\!\!/\gamma^\lambda + 2p'^\lambda - 2k^\lambda)$$

The contribution then becomes

$$-2\pi i e_1 Z_2^{-1}\left(\frac{i e_1^{\,2}}{(2\pi)^4}\right)\int \frac{dk}{k^2 - \mu^2}\, \bar{u}(p')$$

$$\times (-2k\!\!\!/\gamma^\mu k\!\!\!/ - 4mk^\mu + 4(p'^\mu + p^\mu)k\!\!\!/$$

$$+ 4\gamma^\mu(p\cdot p' - p'\cdot k - p\cdot k))\, u(p)\, \frac{1}{k^2 - 2p'\cdot k}\, \frac{1}{k^2 - 2p\cdot k}\, \tilde{A}_\mu$$

as a little computation shows. Next we combine the denominators with the help of a representation introduced by Feynman,

$$\frac{1}{ab} = \int_0^1 dx\, \frac{1}{[ax + b(1 - x)]^2}$$

$$\frac{1}{a^2 b} = \int_0^1 2x\, dx\, \frac{1}{[ax + b(1 - x)]^3} \tag{13.33}$$

Thus

$$\frac{1}{k^2 - \mu^2}\frac{1}{k^2 - 2p\cdot k}\frac{1}{k^2 - 2p'\cdot k}$$
$$= \int_0^1 dx \frac{1}{k^2 - \mu^2}\frac{1}{[k^2 - 2k\cdot(px + p'(1-x))]^2}$$
$$= \int_0^1 2y\, dy \int_0^1 dx \frac{1}{[k^2 - 2k\cdot(px + p'(1-x))y - \mu^2(1-y)]^3}$$
$$= \int_0^1 2y\, dy \int_0^1 dx \frac{1}{[(k - yp_x)^2 - y^2 p_x{}^2 - \mu^2(1-y)]^3}$$

with

$$p_x = px + p'(1-x)$$

Although the integral diverges logarithmically, it is permissible to shift the origin of integration. If we introduce the variable

$$q \equiv k - yp_x$$

we obtain

$$-2\pi i e_1 Z_2^{-1}\left(\frac{ie_1{}^2}{(2\pi)^4}\right)\int_0^1 2y\, dy\int_0^1 dx \int dq \frac{1}{[q^2 - y^2 p_x{}^2 - \mu^2(1-y)]^3}$$
$$\times\ \bar{u}(p')\{-2\not{q}\gamma^\mu\not{q} - 2y\not{p}_x\gamma^\mu\not{q} - 2\not{q}\gamma^\mu y\not{p}_x - 2y^2\not{p}_x\gamma^\mu\not{p}_x$$
$$- 4m_{\text{exp}}q^\mu - 4m_{\text{exp}}yp_x{}^\mu + 4(p + p')^\mu(\not{q} + y\not{p}_x)$$
$$+ 4\gamma^\mu(p\cdot p' - p'\cdot q - yp'\cdot p_x - p\cdot q - yp\cdot p_x)\}u(p)\tilde{A}_\mu \quad (13.34)$$

The denominator is now a function of q^2. Thus the terms linear in q give zero on grounds of symmetry. The numerator terms bilinear in q also simplify with the help of

$$\int dq \frac{q^\alpha q^\beta}{(q^2 - \Lambda^2)^3} = \frac{1}{4}g^{\alpha\beta}\int dq \frac{q^2}{(q^2 - \Lambda^2)^3} \quad (13.35)$$

If we now use the fact that the numerator stands between spinors which satisfy the Dirac equation, we can simplify matters further. The expression (13.34) becomes

$$-2\pi i e_1 Z_2^{-1}\left(\frac{ie_1{}^2}{(2\pi)^4}\right)\int_0^1 2y\, dy\int_0^1 dx \int dq\, \bar{u}(p')$$
$$\times\ (f_1\gamma^\mu + f_2 p^\mu + f_3 p'^\mu)\, u(p)\tilde{A}_\mu \quad (13.36)$$

where

$$f_1 = q^2 + 4p \cdot p'(1 - y + y^2x(1 - x)) + 2m_{\text{exp}}^2 y^2(1 - 2x + 2x^2) - 4m_{\text{exp}}^2 y$$

$$f_2 = 4m_{\text{exp}}y(1 - x - xy)$$

$$f_3 = 4m_{\text{exp}}y(x - y + xy)$$

Since the denominator involves

$$p_x^{\,2} = m_{\text{exp}}^2(x^2 + (1 - x)^2) + 2p \cdot p'x(1 - x)$$

which is symmetric under the interchange of x and $(1 - x)$ we can symmetrize f_2 and f_3 so that we finally get

$$-2\pi i e_1 Z_2^{-1}\left(\frac{ie_1^{\,2}}{(2\pi)^4}\right)\int_0^1 2y\,dy\int_0^1 dx\int dq\,\frac{1}{[q^2 - y^2p_x^{\,2} - \mu^2(1 - y)]^3}$$
$$\bar{u}(p')[\gamma^\mu f_1 + 2m_{\text{exp}}y(1 - y)(p^\mu + p'^\mu)]\,u(p)\tilde{A}_\mu \quad (13.37)$$

As a last step we make use of the fact that

$$(p' + p)^\mu\,\bar{u}(p')\,u(p) = \bar{u}(p')(2m_{\text{exp}}\gamma^\mu - i\sigma^{\mu\nu}Q_\nu)\,u(p) \quad (13.38)$$

the proof of which we leave to the reader. With this and the notation

$$Q = p' - p \quad (13.39)$$

we get

$$-2\pi i e_1 Z_2^{-1}\left(\frac{ie_1^{\,2}}{(2\pi)^4}\right)\int_0^1 2y\,dy\int_0^1 dx\int dq$$
$$\times\,\frac{1}{[q^2 - y^2m_{\text{exp}}^2 + y^2x(1 - x)Q^2 - \mu^2(1 - y)]^3} \quad (13.40)$$
$$\times\,\bar{u}(p')(\gamma^\mu(f_1 + 4m_{\text{exp}}^2y(1 - y)) - 2m_{\text{exp}}y(1 - y)i\sigma^{\mu\nu}Q_\nu)\,u(p)\tilde{A}_\mu$$

Let us write (13.40) in the form

$$2\pi i e_1 Z_2^{-1}\,\bar{u}(p')[\tilde{F}_1(Q^2)\gamma^\mu + i\sigma^{\mu\nu}Q_\nu\tilde{F}_2(Q^2)]\,u(p)\tilde{A}_\mu \quad (13.41)$$

We have not yet faced the fact that because of the presence of q^2 in f_1 the integral diverges logarithmically. If we write

$$\tilde{F}_1(Q^2) = \tilde{F}_1(0) + (\tilde{F}_1(Q^2) - \tilde{F}_1(0)) \quad (13.42)$$

then $\tilde{F}_1(0)$ contains the divergence, and $\tilde{F}_1(Q^2) - \tilde{F}_1(0)$ is finite. This follows from the fact that

$$\int dq\,\frac{1}{(q^2 - \Lambda^2)^2} - \int dq\,\frac{1}{(q^2 - \Lambda_0^{\,2})^2} = 2\int_{\Lambda_0^2}^{\Lambda^2} dZ\int\frac{dq}{(q^2 - Z)^3} \quad (13.43)$$

and[1]

$$\int dq\,\frac{1}{(q^2 - \Lambda^2 + i\epsilon)^3} = -\,\frac{\pi^2 i}{2\Lambda^2} \quad (13.44)$$

[1] See J. M. Jauch and F. Rohrlich, *op. cit.*, p. 454 ff.

There is in general no particular reason for isolating the value of $\tilde{F}_1$ at $Q^2 = 0$ rather than at any other point, since $\tilde{F}_1(Q^2) - \tilde{F}_1(\lambda)$ will also converge. For a photon vertex, however, the point $Q^2 = 0$ has a special meaning, since it corresponds to the vertex for the absorption of a real photon. It is conventional, in meson theories, to make the corresponding subtraction at $Q^2 = \mu^2$, where μ^2 is the meson (mass)^2. With the subtraction we may combine (13.41) with (13.31), the contribution from the first five graphs. The sum of the terms is

$$2\pi i e_1(Z_2 + Z_2^{-1}\,\tilde{F}_1(0))\,\bar{u}(p')\gamma^\mu\,u(p)\tilde{A}_\mu + 2\pi i e_1 Z_2^{-1}\,\bar{u}(p')((\tilde{F}_1(Q^2) - \tilde{F}_1(0))\gamma^\mu + i\sigma^{\mu\nu}Q_\nu\,\tilde{F}_2(Q^2))\,u(p)\tilde{A}_\mu$$

We can now prove that

$$Z_2 + \tilde{F}_1(0) = 1 \tag{13.45}$$

so that to the order to which we are working the contribution from all graphs but the last one is

$$2\pi i e_1\,\bar{u}(p')\{\gamma^\mu(1 + \tilde{F}_1(Q^2) - \tilde{F}_1(0)) + i\sigma^{\mu\nu}Q_\nu\,\tilde{F}_2(Q^2)\}\,u(p)\tilde{A}_\mu \tag{13.46}$$

To prove this result we note that with (13.19) and (13.20) we have

$$\begin{aligned}\bar{u}(p)\frac{\partial\sum}{\partial p_\mu}u(p) &= B\,\bar{u}(p)\gamma^\mu\,u(p)\\ &= -\frac{ie_1^2}{(2\pi)^4}\int\frac{dk}{k^2 - i\epsilon}\gamma^\lambda\frac{\partial}{\partial p_\mu}\frac{1}{p + k - m_{\text{exp}}}\gamma_\lambda\\ &= \frac{ie_1^2}{(2\pi)^4}\int\frac{dk}{k^2}\gamma^\lambda\frac{1}{p + k - m_{\text{exp}}}\gamma^\mu\frac{1}{p + k - m_{\text{exp}}}\gamma_\lambda\end{aligned}$$

Comparison with (13.32) evaluated at $p' = p$, i.e., at $Q = 0$ shows that[1]

$$B\,\bar{u}(p)\gamma^\mu\,u(p) = -\tilde{F}_1(0)\,\bar{u}(p)\gamma^\mu\,u(p)$$

or

$$\tilde{F}_1(0) + B = \tilde{F}_1(0) + Z_2 - 1 = 0 \tag{13.47}$$

We shall comment on this result later. Before we do this, we must still discuss the contribution of graph (e) as given by (13.29). This contribution, duly multiplied by Z_2^{-1}, is

$$2\pi i e_1 Z_2^{-1}\,\bar{u}(p')\gamma^\sigma\,n(p)\frac{1}{Q^2}\,\Pi_\sigma{}^\mu(Q)\tilde{A}_\mu$$

[1] This result, known as Ward's identity, is true to all orders in perturbation theory. What it establishes is that the structure of the charged particle does not enter into the description of the interaction with a zero frequency photon. J. C. Ward, *Phys. Rev.* **77,** 293 (1950); **78,** 182 (1950).

where we have introduced the notation

$$\Pi_\sigma{}^\mu(Q) = \frac{ie_1{}^2}{(2\pi)^4}\int dq \operatorname{tr}\left(\gamma_\sigma \frac{1}{\not{q} - m_{\text{exp}}}\gamma^\mu \frac{1}{\not{q} + \not{Q} - m_{\text{exp}}}\right) \tag{13.48}$$

When this is added to the contribution from all the other graphs, shown in (13.46), the total effect to the order we are considering is

$$2\pi i e_1\Big\{(1 + \tilde{F}_1(Q^2) - \tilde{F}_1(0))\,\bar{u}(p')\gamma^\sigma\, u(p) + i\tilde{F}_2(Q^2)\,\bar{u}(p')\sigma^{\sigma\lambda}Q_\lambda\, u(p)\Big\}\left(\delta_\sigma{}^\mu + \frac{1}{Q^2}\Pi_\sigma{}^\mu(Q)\right)\tilde{A}_\mu \tag{13.49}$$

Thus the effect of graph (*e*) is to change the external potential $\tilde{A}_\sigma$ to

$$\left(\delta_\sigma{}^\mu + \frac{1}{Q^2}\Pi_\sigma{}^\mu(Q)\right)\tilde{A}_\mu \tag{13.50}$$

The integral $\Pi_\sigma{}^\mu(Q)$ must have the form

$$\Pi_\sigma{}^\mu(Q) = \delta_\sigma{}^\mu\,\Pi_1(Q^2) + Q_\sigma Q^\mu\,\Pi_2(Q^2) \tag{13.51}$$

and since

$$Q^\mu\tilde{A}_\mu = 0$$

only the term Π_1 contributes. Incidentally, it is clear from (13.51) that $\Pi_1(Q^2)$ is given by

$$\Pi_1(Q^2) = \frac{1}{3}\left(\delta_\mu{}^\sigma + \frac{Q^\sigma Q_\mu}{Q^2}\right)\Pi_\sigma{}^\mu(Q) \tag{13.52}$$

The effect of the graph is therefore to change the external potential to

$$\left(1 + \frac{1}{Q^2}\Pi_1(Q^2)\right)\tilde{A}_\mu = \left(1 + \frac{1}{Q^2}\Pi_1(Q^2)\right)\frac{Ze_1}{(2\pi)^3}\,\delta_{\mu 0}\frac{1}{(\mathbf{p} - \mathbf{p}')^2} \tag{13.53}$$

The integral $\Pi_\sigma{}^\mu$ is so highly divergent that not much can be deduced about it from the form it takes. If we note that it appears in the correction to the photon propagator with momentum Q (Fig. 13.4), it is plausible, in analogy with the discussion of the radiative correction to the electron

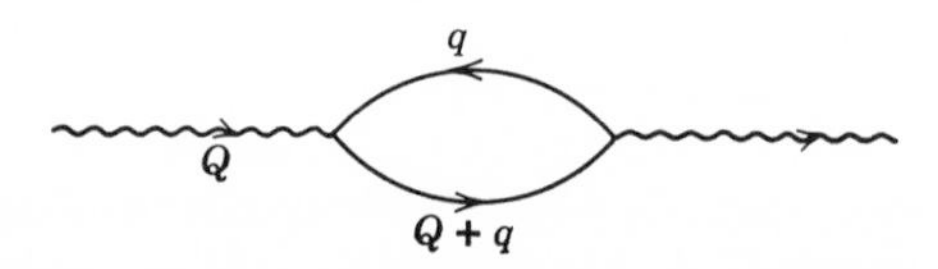

Fig. 13.4. Radiative correction to photon propagator.

propagator, that $\Pi_1(0)$ represents a mass renormalization term. This is borne out by a detailed examination of the expectation value of the Hamiltonian in a one-photon state. For lack of space we forgo this discussion and merely state that

$$\Pi_1(0) = 0 \tag{13.54}$$

since there is no parameter in the Lagrangian which corresponds to the photon mass in lowest order, and since we want the "renormalized" photon mass still to remain zero.[1]

We may use this result to rewrite (13.53) in the form

$$\begin{pmatrix}\text{External} \\ \text{potential}\end{pmatrix} = \left(1 + \frac{\Pi_1(Q^2) - \Pi_1(0)}{Q^2}\right)\frac{Ze_1}{(2\pi)^3(\mathbf{p} - \mathbf{p}')^2}$$

The integral $(\Pi_1(Q^2) - \Pi_1(0))$ is still logarithmically divergent. If however, we write

$$\Pi_1(Q^2) = \Pi_1(0) + Q^2\Pi_1'(0) + (Q^2)^2\, I(Q^2)$$

to order $e_1{}^2$ the effective potential is

$$(1 + \Pi_1'(0) + Q^2 I(Q^2))\frac{Ze_1}{(2\pi)^3(\mathbf{p} - \mathbf{p}')^2}$$
$$\cong (1 + \Pi_1'(0))(1 + Q^2 I(Q^2))\frac{Ze_1}{(2\pi)^3(\mathbf{p} - \mathbf{p}')^2} \tag{13.55}$$

The integral $I(Q^2)$ is finite, but because our aim here is to discuss renormalization rather than to calculate, we will not attempt to evaluate it.

A detailed calculation, which must be omitted for lack of space shows that when the Feynman rules are applied to the graph in Fig. 13.4 (without forgetting the $Z_3{}^{-1}$ for the two external photon lines), and if it is required that as $Q^2 \to 0$ the sum of that graph and the photon propagator yields $1/Q^2$, then

$$1 + \Pi_1'(0) = Z_3 \tag{13.56}$$

Furthermore, of we write

$$e_1^2(1 + \Pi_1'(0)) = Z_3 e_1^2 = e^2 \tag{13.57}$$

and call $-e$ the experimental charge of the electron, then to the order in which the calculation has been carried out, all divergent constants have

[1] A more careful definition of what is meant by a product of two field operators in a gauge invariant theory has led K. Johnson [*Nucl. Phys.* **25,** 431, 435 (1961)] to the conclusion that only $\Pi_1(Q^2) - \Pi_1(0)$ ever make their appearance in the theory.

disappeared in the course of the renormalization of the mass and the charge. The term $(1 + Q^2 I(Q^2))$ still modifies the Coulomb potential for large Q^2, i.e., at short distances, and this modification has been detected both in the Lamb shift[1] and in high precision proton-proton scattering experiments.[2] The experimental charge e is related to the Lagrangian parameter e_1 by a factor which only depends on the coupling of photons to matter. Z_3 is always the same, whether we are talking about the Coulomb scattering of electrons or protons. If (13.45) were not true, we could still get rid of infinities by defining

$$e_{\text{exp}} = e(Z_2 + \tilde{F}_1(0))$$

The renormalization would then depend on what kind of particle we are talking about, since Z_2 as well as $\tilde{F}_1(0)$ depend on the structure of the particle. It would then be impossible to explain the numerical equality of the electron and proton charges. With gauge invariance, i.e., current conservation, this equality is explained by requiring that the electromagnetic field is coupled *universally*, with the same e_1 to all charged particles, and then the physical charges will be the same.

Let us again consider the coupling, which is

$$\begin{aligned} 2\pi ie\{(1 + \tilde{F}_1(Q^2) - \tilde{F}_1(0))\, \bar{u}(p')\gamma^\mu\, u(p) \\ + i\tilde{F}_2(Q^2)\, \bar{u}(p')Q_\nu\sigma^{\mu\nu}\, u(p)\}(1 + Q^2\, I(Q^2))A_\mu^{\text{ext}} \end{aligned} \tag{13.58}$$

As $Q^2 \to 0$, we get, in the nonrelativistic approximation, the following form for the coupling:

$$-2\pi \frac{e}{2m_{\text{exp}}}(1 + 2m_{\text{exp}}\, \tilde{F}_2(0))\boldsymbol{\sigma} \times \mathbf{Q} \cdot \mathbf{A}_{\text{ext}} + (\text{charge terms}) \tag{13.59}$$

The effect of the radiative corrections is therefore to change the magnetic moment of the electron by a factor $(1 + 2m_{\text{exp}}\tilde{F}_2(0))$. It turns out that $\tilde{F}_2(0)$ is easy to calculate. From (13.40) using (13.44) we get

$$\begin{aligned} \tilde{F}_2(0) &= \frac{ie^2}{(2\pi)^4}\int_0^1 2y\,dy\int_0^1 dx\, 2m_{\text{exp}}y\, \frac{\pi^2}{2im_{\text{exp}}^2 y^2}(1-y) \\ &= \frac{e^2}{16\pi^2 m_{\text{exp}}} = \frac{1}{2m_{\text{exp}}}\frac{\alpha}{2\pi} \end{aligned} \tag{13.60}$$

Thus an anomalous moment of $\alpha/2\pi$ Bohr magnetons is predicted to this order. Higher corrections to the electromagnetic moment have been

[1] J. D. Bjorken and S. D. Drell, "Relativistic Quantum Mechanics."
[2] E. Eriksen and L. L. Foldy, *Phys. Rev.* **98**, 775 (1955).

calculated. The results of this calculation are[1]

$$\Delta\mu = \frac{\alpha}{2\pi} - 0.328\,\frac{\alpha^2}{\pi^2} = 0.0011596$$

to be compared with the experimental value[2]

$$\Delta\mu = 0.0011609 \pm 0.0000024$$

We conclude this long chapter with one more calculation. As was anticipated in the writing of (13.32), an infrared divergence develops because of the zero photon mass. We expect this to manifest itself in a singular dependence on μ, the photon mass. To isolate this dependence on μ we look back on the integral following (13.32). The infrared divergence comes from the region where $k \approx 0$; the dominant term will therefore be the one in which the numerator is independent of k. Thus the relevant term is

$$-\,2\pi ie\,\frac{ie_1^{\,2}}{(2\pi)^4}\,4p\cdot p'\,\bar{u}(p')\gamma^\mu\,u(p)\int\frac{dk}{k^2-\mu^2}\,\frac{1}{k^2-2p'\cdot k}\,\frac{1}{k^2-2p\cdot k}$$

$$= 2\pi ie\left(\frac{\alpha}{2\pi}\,p\cdot p'\,\bar{u}(p')\gamma^\mu\,u(p)\right)$$

$$\times\int_0^1 2y\,dy\int_0^1 dx\,\frac{1}{p_x^{\,2}y^2+\mu^2(1-y)} \tag{13.61}$$

After a little computation we can bring this into the form

$$-\,2\pi ie\,\bar{u}(p')\gamma^\mu\,u(p)\left(\frac{\alpha}{2\pi}\,p\cdot p'\int_0^1 dx\,\frac{1}{p_x^{\,2}}\log\frac{p_x^{\,2}}{\mu^2}\right) \tag{13.62}$$

If we compare this with the matrix element for Coulomb scattering, we see that as far as the dependence on μ is concerned, the cross section is corrected by a factor

$$\frac{d\sigma}{d\Omega} = \left(\frac{d\sigma_C}{d\Omega}\right)\left(1-\frac{\alpha}{\pi}\,p\cdot p'\int_0^1 dx\,\frac{1}{p_x^{\,2}}\log\frac{p_x^{\,2}}{\mu^2}\right) \tag{13.63}$$

This is still divergent. If we note, however, that in general the electron in the final state has an energy $E \pm \Delta E$, where ΔE depends on the experimental setup, there is no way of distinguishing between a pure Coulomb scattering and a scattering accompanied by a bremsstrahlung photon of

[1] The original calculation by R. Karplus and N. Kroll, *Phys. Rev.* **77,** 536 (1950) contained a numerical error. We quote the result of C. Sommerfield, *Ann. Phys.* (N.Y.) **5,** 26 (1958).

[2] A. A. Schupp, R. W. Pidd, and H. R. Crane, *Phys. Rev.* **121,** 1 (1961).

energy less than ΔE. Thus the observable cross section is

$$\left(\frac{d\sigma}{d\Omega}\right)_{\text{obs}} = \frac{d\sigma_C}{d\Omega}\left(1 - \frac{\alpha}{\pi}\, p \cdot p' \int_0^1 dx \frac{1}{p_x^{\,2}} \log \frac{p_x^{\,2}}{\mu^2}\right) + \frac{d\sigma_B}{d\Omega}(E_\gamma < \Delta E) \quad (13.64)$$

In Chapter 12 we derived the result that the bremsstrahlung cross section for soft photons may be written in the form of the Coulomb cross section multiplied by the factor (12.19)

$$\frac{\alpha}{\pi}\, p \cdot p' \int_0^1 dx \frac{1}{p_x^{\,2}} \log \frac{p_x^{\,2}(\Delta E)^2}{E_x^{\,2}\mu^2}$$

given in (12.19). Thus

$$\left(\frac{d\sigma}{d\Omega}\right)_{\text{obs}} = \frac{d\sigma_C}{d\Omega}\left(1 - \frac{\alpha}{\pi}\, p \cdot p' \int_0^1 \frac{dx}{p_x^{\,2}} \log \frac{E_x^{\,2}}{(\Delta E)^2}\right) \quad (13.65)$$

If we now use the fact that $E = E'$, so that $E_x = E$, and carry out the x-integration, for

$$-Q^2 = (\mathbf{p}' - \mathbf{p})^2 \gg m_{\text{exp}}^2$$

we get the result that

$$\left(\frac{d\sigma}{d\Omega}\right)_{\text{obs}} \cong \frac{d\sigma_C}{d\Omega}\left(1 - \frac{\alpha}{\pi} \log \frac{-Q^2}{m_{\text{exp}}^2} \log\left(\frac{E}{\Delta E}\right)^2\right) \quad (13.66)$$

Thus the infrared divergence no longer appears when sensible questions are asked of the theory. Although we do not show it, the correction given above turns out to be the most important radiative correction at high energies, and, if multiphoton effects are taken into account (an arbitrary number of photons can be radiated, and an arbitrary number of photons in all possible orders, are emitted by the incoming free electron and reabsorbed by the outgoing free electron), the formula becomes[1]

$$\left(\frac{d\sigma}{d\Omega}\right)_{\text{obs}} \cong \frac{d\sigma_C}{d\Omega}\, e^{-\frac{\alpha}{\pi} \log \frac{-Q^2}{m_{\text{exp}}^2} \log\left(\frac{E}{\Delta E}\right)^2}$$

The purpose of this chapter has been to show that with proper care in the interpretations of the parameters appearing in the theory, the Feynman rules give unambiguous answers to questions of physical interest. With this purpose served, it only remains to point out that, in practice, higher order electrodynamic calculations are sometimes carried out more easily by using dispersion techniques, which will be discussed in Chapter 26.

[1] D. R. Yennie, S. C. Frautschi, and H. Suura, *Ann. Phys.* (*N.Y.*), **13,** 379 (1961) contains the most detailed discussion of the infrared domain in the literature.

It should, of course, be mentioned that what was done to order e^2 in this chapter can be extended, without difficulty in principle, to any order, as was first proved by Dyson.[1]

PROBLEMS

1. Use induction to prove that

$$\frac{1}{a_1 a_2 \cdots a_n} = (n-1)! \int_0^1 d\alpha_1 \cdots \int_0^1 d\alpha_n \frac{\delta(\alpha_1 + \alpha_2 + \cdots + \alpha_n - 1)}{(\alpha_1 a_1 + \cdots + \alpha_n a_n)^n}$$

2. Evaluate the integral

$$\lim_{\epsilon \to 0} \int d^4k \frac{1}{(k^2 - \Lambda^2 - i\epsilon)^3}$$

Hint. First do the k_0 integral using Cauchy's formula.

3. Show that the matrix element for photon-photon scattering vanishes by gauge invariance when any one of the four-momenta of the incident of final photons goes to zero.

4. Feynman has introduced a relativistic cut-off in which the propagator

$$\frac{1}{p^2 - m^2 + i\epsilon}$$

is replaced by

$$\frac{1}{p^2 - m^2 + i\epsilon} - \frac{1}{p^2 - \Lambda^2 + i\epsilon} = -\frac{\Lambda^2}{(p^2 - m^2 + i\epsilon)(p^2 - \Lambda^2 + i\epsilon)}$$

For convergent integrals, the limit $\Lambda^2 \to \infty$ exists. Show that δm for the electron varies as $\log \Lambda^2$ for large Λ^2 and find the coefficient of that term.

5. Find the leading term, for large Λ, of the mass renormalization of a spin 0 particle. Use gauge invariance which allows the replacement of the photon propagator $-g_{\mu\nu}/k^2$ by

$$-\frac{1}{k^2}\left(g_{\mu\nu} - \lambda \frac{k_\mu k_\nu}{k^2}\right)$$

to simplify the calculation.

6. Calculate the leading term in Z_2 with the general photon propagator used in problem 5. Does Z_2 depend on the gauge? Discuss the lack of gauge invariance of Z_2.

[1] See J. D. Bjorken and S. D. Drell, *Relativistic Quantum Fields*, McGraw-Hill Book Co., New York (1965).

PART III

Introduction

The application of quantum theory to the interaction of the electromagnetic field with electrons has proved to be extremely successful from a practical standpoint. Until now, no disagreement between experiment and theory has been found, and the degree of accuracy of the agreement—thanks to remarkable advances in experimental techniques—is unprecedented in physics. Such excellent agreement could only be obtained because electrodynamic calculations could be carried out without a detailed understanding of the structure of the nucleons that serve as a source of the Coulomb field. Sooner or later, even in electrodynamic experiments, the structure of the nucleons begins to play a part, as in high energy, large angle, electron-proton scattering. Even before this structure came to be studied, it was inevitable that the forces which bind the nucleons together, and the mechanism responsible for these forces, should have attracted the attention of physicists as soon as some experimental information about nuclei came to light. This information, rudimentary as it was in the early 1930's, clearly indicated that the nuclear forces were of a short range, ($\leqslant 10^{-13}$ cm), and that they were strong: nuclear binding energies were typically measured in MeV whereas atomic binding energies are measured in electron volts.

It was Yukawa who in 1935 took the bold step to predict the existence of *heavy quanta*, which, assuming they played the same role in nuclear forces as photons play in electromagnetic ones, would give rise to short-range effects. The generalization of the equation describing the Coulomb field due to a point source

$$\nabla^2 \phi(\mathbf{r}) = -e\, \delta(\mathbf{r})$$

is

$$(\nabla^2 - \mu^2)\, \phi(\mathbf{r}) = -g\, \delta(\mathbf{r})$$

whose solution is the so-called Yukawa potential

$$\phi(\mathbf{r}) = \frac{g}{4\pi}\frac{e^{-\mu|\mathbf{r}|}}{|\mathbf{r}|}$$

with range $1/\mu$. From estimates of the range of nuclear forces, Yukawa predicted that the new particles should have a mass of the order of 200 to 300 electron masses. The order of magnitude ratio of nuclear binding energies to atomic binding energies indicated that the depth of the potential, or, equivalently, the coupling parameter $g^2/4\pi$, is significantly larger than the fine structure constant; in fact, we shall see that $g^2/4\pi \simeq 15$.

The Yukawa idea laid the groundwork for a field-theoretical exploration of the nuclear forces, but right from the start it became evident that there were two difficulties in the way: (a) there was no classical limit to the forces, so that there was no clue to the form of the interaction, and (b) the coupling, in contrast to the electromagnetic coupling, was so strong that perturbation calculations could not be trusted to give even a rough indication to the correctness of a particular trial form of the interaction. The second difficulty was, and still is, the more serious one, since it makes a trial and error type of theoretical exploration impossible. There has been one beneficial effect of this difficulty on the theoretical front—the extremely close collaboration between theorists and experimentalists in the pursuit of the most tenuous clues, both theoretical and experimental, to the benefit of both.

The theoretical difficulties, the variety of experiments, and the emerging richness of structure among the strongly interacting particles necessitate a treatment quite different from that used in quantum electrodynamics. Our discussion must proceed on several levels: we must first discover the quantum numbers, such as spin, parity, and "internal quantum numbers" of the many particles and the ever increasing number of resonances; we must look for some connection between all of these states, i.e., search for internal symmetries and/or dynamical mechanisms that might be responsible for the observed structure; finally, we must see how much of the known phenomena we can understand semiquantitatively, without a true theory of strong interactions.

We begin with an investigation of the most stable of the particles, the baryons and the pseudoscalar mesons. All evidence points to the conclusion that there is no fundamental difference between the stable particles and the highly unstable ones discussed later. Our separation is made only because more is known about the stable particles. The similarities among the baryons, and among the pseudoscalar mesons, lead us to a discussion of "internal symmetry." We shall examine the well-established charge independence of the strong interactions, the new concept of "strangeness," and the new "unitary symmetry" that generalizes the lower symmetries and has had just enough predictive power to warrant attention.

Thus prepared we can more profitably discuss the new resonances, the excited baryon states in Chapter 19 and the boson resonances in Chapter 20. Next we turn to some important properties of scattering matrix elements: in successive chapters (21 and 22) we discuss the consequences of the unitarity of the scattering matrix and the analyticity properties of some of them. The most successful application of these properties leads to the forward scattering dispersion relations for pion nucleon scattering, which are discussed in Chapter 23. The next five chapters deal with much more speculative matters; they are intended to give a picture of the current state of the art in particle physics, and they indicate what understanding has been achieved with the inadequate theoretical tools at hand. Among these tools are the partial wave dispersion relations that follow from the assumed analyticity properties. With their help some resonance phenomena and effective range formulas are discussed, and the technique is applied to the prediction of resonances. The work of Chew and Low is extended to SU(3), but only qualitative agreement results. The idea that one-particle states may dominate under certain circumstances is explored in the chapter on form factors and the one-particle-exchange model. Finally, the current situation in high energy elastic diffraction scattering is discussed in Chapter 28.

14

The Baryons

Fundamental to the understanding of the pattern of reactions among the strongly interacting particles (sometimes called *hadrons*) is the division of these particles into classes characterized by *baryon number*. The particles to be discussed in this chapter are a special subset of the particles of baryon number $+1$.

All particles of baryon number $+1$ share the following properties: (a) they interact strongly, (b) they have half-odd integral spin, and (c) they are in one way or another connected by a chain of interactions with the proton, the lightest particle of baryon number $+1$. We assume that baryon number is an additive quantum number:[1] the deuteron has $N_B = 2$. Since baryon number is assigned to the hadrons on the basis of the conservation of that quantum number, a field theoretical description of the strong interactions will require a theory invariant under gauge transformations of the first kind. A well-confirmed consequence of field theory is that for each baryon there should exist an antibaryon, with $N_B = -1$; the antiparticles corresponding to all the particles to be discussed in this chapter have been identified. Furthermore, in the annihilation of baryon and antibaryon, a state of baryon number 0 is always produced.

The quantum number N_B shares with electrical charge the attribute of being conserved in interactions other than the strong ones; baryon number seems to be universally conserved. The evidence for this is the following:

1. No case has been observed in which a heavier baryon, n, Λ^0, Σ^+, Σ^0, Σ^-, Ξ^0, Ξ^-, or an excited baryon state did not end up as a proton or a bound neutron.

2. Protons are stable. If protons decayed into lighter charged particles. e.g.

$$p \rightarrow e^+ + \pi^0$$

[1] If baryon number were only conserved modulo some integer N, heavy nuclei would be unstable. See M. Goldhaber and G. Feinberg, *Proc. Natl. Acad. Sci., U.S.* **45,** 1301 (1959).

the decay of a proton would give rise to a moving charged particle, a process accompanied by radiation. The occasional decay of a proton among the nuclei making up the matter in a large hydrogeneous scintillation counter, for example, would easily be detected. The absence of such decays yields a lower limit of 10^{23} years for the lifetime of the proton.[1]

The most familiar of the baryons are the *proton* and the *neutron*. These particles have charge 1 and 0 respectively, both have spin $\frac{1}{2}$ and their masses are

$$M_p = 938.256 \pm 0.005 \text{ MeV}$$

$$M_n = 939.550 \pm 0.005 \text{ MeV}$$

The neutron and proton attract one another; they form a bound state—the deuteron. They have magnetic moments which differ in value from what we would expect on the basis of the Dirac equation, which is 1 and 0 nuclear magnetons respectively. The actual values are[2]

$$\mu_p^{\text{total}} = 1 + \mu_p = 2.792816 \pm 0.000034$$

$$\mu_n^{\text{total}} = \mu_n = -1.913148 \pm 0.000066$$

respectively. It is believed, and calculations encourage if not establish this belief, that anomalous moments μ_p and μ_n are due to the electromagnetic interactions of the strongly interacting particles which, in the descriptive language associated with Feynman graphs, form the "cloud" surrounding the bare nucleon.[3]

The proton is stable, and the free neutron decays according to

$$n \rightarrow p + e^- + \bar{\nu}_e$$

with a mean life

$$\tau = (1.013 \pm 0.029) \times 10^3 \text{ sec}$$

This mean life is long partly because the phase space available for the three-particle decay is restricted by the small neutron-proton mass difference. The actual strength of the interaction responsible for the decay is the same as for most other decays which involve neutrinos. Since we shall analyze in some detail the interactions of the nucleons with other particles, we shall say no more about them at this point.

The group of particles that, together with the two nucleons, form a set of eight baryons are relatively new to physics. They are all unstable, with lifetimes on the order of, or less than, 10^{-10} sec so that great ingenuity had to be exercised to determine their properties. The remainder of this

[1] These matters are discussed more fully in M. Goldhaber and G. Feinberg, *Proc. Natl. Acad. Sci., loc. cit.*

[2] We reserve the notation μ_p and μ_n for the anomalous magnetic moments.

[3] See Chapter 26 for a more detailed discussion of this topic.

chapter is devoted to a description of these particles and some of their properties. Much of what is known about these six particles (sometimes called the hyperons) cannot be appreciated at this stage of our study; we will have to use a few results which will actually be discussed later, to be able to make the description more complete than would otherwise be possible.

The Λ^0 Hyperon

The Λ^0 particle was the first of the new "strange" particles to be studied in detail. It was first identified by the characteristic V-shaped tracks in cloud chambers; these tracks were found to come from the decay

$$\Lambda^0 \rightarrow p + \pi^-$$

The mass of the Λ^0 is

$$M_{\Lambda^0} = 1115.44 \pm 0.12 \text{ MeV}$$

Its lifetime is

$$\tau_{\Lambda^0} = (2.61 \pm 0.02) \times 10^{-10} \text{ sec}$$

Its main decay modes with branching percentages are

$$\begin{aligned} \Lambda^0 &\rightarrow p + \pi^- & 66.3 \pm 1.0\% \\ &\rightarrow n + \pi^0 & 33.6 \pm 1.0\% \\ &\rightarrow p + e^- + \nu & (0.88 \pm 0.08) \times 10^{-1}\% \\ &\rightarrow p + \mu^- + \nu & (1.5 \pm 1.2) \times 10^{-2}\% \end{aligned}$$

The beta decay is about fifteen times slower than we would expect if we assumed that the matrix elements for the decay are the same as for the beta decay of the neutron.

We shall discuss the decays of the Λ^0 later. At this point we shall present evidence that the *spin of the* Λ^0 *is* $\frac{1}{2}$. The method for the spin determination we shall apply makes use of its decay properties.[1] We consider the decay

$$\Lambda^0 \rightarrow p + \pi^-$$

in the Λ^0 rest frame. The Λ^0 is produced in some kind of reaction, say

$$\pi^- + p \rightarrow \Lambda^0 + K^0$$

[1] T. D. Lee and C. N. Yang, *Phys. Rev.* **109,** 1755 (1958); evidence that the Λ^0 spin was $\frac{1}{2}$ was obtained earlier, but the conclusion depended on the fact that the K spin is 0, a result not easy to establish if the Λ^0 spin is not known.

and a plane may be defined by the pion momentum and that of the Λ^0. We shall take the normal to this plane and use it as quantization axis for the angular momentum of the Λ^0. If the spin of the Λ^0 is J, its state vector will, in general, have the form

$$\Psi_\alpha = \sum_M C_M{}^\alpha \Psi_{J,M} \tag{14.1}$$

Here the index α labels the $(2J + 1)$ linearly independent, pure-state vectors. If the spin of the Λ^0 points in the direction of the quantization axis,[1] it is possible to restrict ourselves to $C_M{}^\alpha$ proportional to $\delta_{\alpha M}$. We shall keep the derivation more general. The final state will be described in the helicity representation[2] briefly discussed in Chapter 4.

Let $\mathbf{p}$ be the momentum of the proton in the Λ^0 rest system. We anticipate a result from the next chapter, in which it is established that the spin of the pion is 0, so that the final state, labeled by the proton helicity $\lambda = \pm\frac{1}{2}$, is $\Phi_{\mathbf{p},\lambda;-\mathbf{p},0}$. The decay amplitude is given by

$$(\Phi_{\mathbf{p},\lambda,-\mathbf{p},0}, S\Psi_\alpha) = \sum_M S_\lambda C_M{}^\alpha \sqrt{\frac{2J+1}{4\pi}}\, D^{(J)*}_{M\lambda}(R_p) \tag{14.2}$$

The absolute square of this is

$$\frac{2J+1}{4\pi} \sum_{MM'} C_M{}^\alpha C^{\alpha *}_{M'}\, |S_\lambda|^2\, D^{(J)*}_{M\lambda}(R_p)\, D^{(J)}_{M'\lambda}(R_p) \tag{14.3}$$

If (θ, ϕ) are the spherical angles which $\mathbf{p}$ makes relative to the quantization axis,

$$D^{(J)}_{M\lambda}(R_p) = e^{-iM\phi}\, d^J_{M\lambda}(\theta) \tag{14.4}$$

The angular distribution is given by

$$\begin{aligned} \frac{1}{2J+1} \sum_\alpha |(\Phi_{\mathbf{p}\lambda;\,-\mathbf{p},0}, S\Psi_\alpha|^2 &= \frac{1}{4\pi} \sum_\alpha \sum_M \sum_{M'} C_M{}^\alpha C^{\alpha *}_{M'}\, |S_\lambda|^2 e^{i(M'-M)\phi}\, d^J_{M\lambda}(\theta)\, d^J_{M'\lambda}(\theta) \\ &\equiv W_\lambda(\theta, \phi) \end{aligned} \tag{14.5}$$

The quantity

$$\rho_{MM'} = \sum_\alpha C_M{}^\alpha C^{\alpha *}_{M'} \tag{14.6}$$

[1] With our choice of quantization axis, this will in fact always be true, since a nonvanishing expectation value $\langle \mathbf{J}_\Lambda \cdot \mathbf{p}_\pi \rangle$, for example, is odd under the parity transformation and must vanish because of the conservation of parity in the strong production interaction.

[2] M. Jacob, *Nuovo Cimento*, **9**, 826 (1958).

is called the density matrix of the Λ^0 beam, provided the $C_M{}^\alpha$ are normalized so that $Tr\rho = 1$. Incidentally the S_λ are numerical coefficients which depend on the dynamics of the decay. We shall be interested in the θ dependence of the angular distribution and only consider[1]

$$W(\theta) = \frac{1}{2\pi}\int_0^{2\pi} d\phi \sum_\lambda W_\lambda(\theta, \phi)$$

$$= \frac{1}{4\pi}\sum_\lambda \sum_M \rho_{MM}\, |S_\lambda|^2 (d^J_{M\lambda}(\theta))^2 \tag{14.7}$$

The diagonal matrix elements of ρ, represent the probabilities of finding the Λ^0 in a state with $J_z = M$. We shall use the notation

$$\rho_{MM} = P(M)$$

If $P(M)$ is independent of M, we call the beam unpolarized. If $P(M)$ does depend on M, but $P(M) = P(-M)$, the beam is said to be aligned; otherwise it is polarized, with polarization $\sum_M M\, P(M)$.

We shall be interested in quantities like

$$\langle \cos\theta \rangle = \frac{\int_{-1}^{1} d(\cos\theta)\cos\theta\; W(\theta)}{\int_{-1}^{1} d(\cos\theta)\; W(\theta)} \tag{14.8}$$

or, more generally

$$\langle P_L(\cos\theta) \rangle = \frac{\int_{-1}^{1} d(\cos\theta) P_L(\cos\theta)\; W(\theta)}{\int_{-1}^{1} d(\cos\theta) W(\theta)} \tag{14.9}$$

We sh all use the formula[2]

$$(d^J_{M\lambda}(\theta))^2 = \sum_l (-1)^{M-\lambda} C(J, J, l; M, -M, 0)$$

$$\times\, C(J, J, l; \lambda, -\lambda, 0)\, P_l(\cos\theta) \tag{14.10}$$

and

$$C(J, J, 0; M, -M, 0) = (-1)^{M-J}\frac{1}{\sqrt{2J+1}}$$

[1] With our choice of quantization axis, there is actually no ϕ-dependence, for $\rho_{MM'} \propto \delta_{MM'}$ if $C_M{}^\alpha \propto \delta_{M\alpha}$.

[2] We use the notation in M. E. Rose, *Elementary Theory of Angular Momentum*, for the Wigner (Clebsch-Gordan) coefficients.

to obtain, after some computation

$$\langle P_L(\cos\theta)\rangle = \frac{2J+1}{2L+1}\sum_M\sum_\lambda P(M)\,|S_\lambda|^2(-1)^{M-\lambda}$$
$$\times\, C(J, J, L; M, -M, 0)C(J, J, L; \lambda, -\lambda, 0)$$
$$\times\left(\sum_M P(M)\sum_\lambda |S_\lambda|^2\right)^{-1} \quad (14.11)$$

We shall use

$$\sum_{M=-J}^{J} P(M) = 1 \quad (14.12)$$

We shall also work with normalized coefficients $\bar{S}$, defined by $\bar{S}_\lambda = S_\lambda \Big/ \left(\sum_\lambda |S_\lambda|^2\right)^{1/2}$, so that

$$\sum_\lambda |\bar{S}_\lambda|^2 = 1 \quad (14.13)$$

and introduce the parameter α by

$$\alpha = |\bar{S}_{1/2}|^2 - |\bar{S}_{-1/2}|^2 \quad (14.14)$$

The fact that

$$\sum_M P(M) \leq 1$$

and

$$|\alpha| \leq 1$$

has certain implications for $\langle P_L(\cos\theta)\rangle$. To determine these, we need to express quantities involving $P(M)$ and α in terms of the $\langle P_L(\cos\theta)\rangle$. To do this we first define the "testing functions"[1]

$$T^+_{JM} \equiv \frac{2}{2J+1}\sum_{L\,\text{even}}(-1)^{M-1/2}(2L+1)$$
$$\times\,\frac{C(J, J, L; M, -M, 0)}{C(J, J, L; \frac{1}{2}, -\frac{1}{2}, 0)}\langle P_L(\cos\theta)\rangle \quad (14.15)$$

and

$$T^-_{JM} = \frac{2}{2J+1}\sum_{L\,\text{odd}}(-1)^{M+1/2}(2L+1)\frac{C(J, J, L; M, -M, 0)}{C(J, J, L; \frac{1}{2}, -\frac{1}{2}, 0)}\langle P_L(\cos\theta)\rangle \quad (14.16)$$

With the help of (14.11) we calculate

$$T^+_{JM} = 2\sum_{L\,\text{even}}\sum_{M'}\sum_\lambda(-1)^{M-1/2}(-1)^{M'-\lambda}P(M')\,|\bar{S}_\lambda|^2$$
$$\times\, C(J, J, L; M', -M', 0)\, C(J, J, L; M, -M, 0)\,\frac{C(J, J, L; \lambda, -\lambda, 0)}{C(J, J, L; \frac{1}{2}, -\frac{1}{2}, 0)}$$

We can use

$$C(J, J, L; -M, M, 0) = (-1)^{2J-L}C(J, J, L; M, -M, 0) \quad (14.17)$$

[1] T. D. Lee and C. N. Yang, *loc. cit.* L. Durand, L. F. Landovitz and J. Leitner, *Phys. Rev.* **112,** 273 (1958).

to work out the λ-sum. We get (for even L and odd $2J$)

$$\sum_{\lambda=-\frac{1}{2}}^{\frac{1}{2}} (-1)^{-\frac{1}{2}-\lambda} |\bar{S}_\lambda|^2 \frac{C(J, J, L; \lambda, -\lambda, 0)}{C(J, J, L; \frac{1}{2}, -\frac{1}{2}, 0)} = -\sum_\lambda |\bar{S}_\lambda|^2 = -1$$

so that

$$\begin{aligned} T^+_{JM} &= 2\sum_{M'} \sum_{L\,\text{even}} (-1)^{M+M'+1} P(M')\, C(J, J, L; M, -M, 0) \\ &\qquad \times C(J, J, L; M', -M', 0) \\ &= \sum_{M'} \sum_{L} (1 + (-1)^L)(-1)^{M+M'+1} P(M')\, C(J, J, L; M, -M, 0) \\ &\qquad \times C(J, J, L; M', -M', 0) \end{aligned}$$

which, with the help of (14.17), may be written as

$$\begin{aligned} T^+_{JM} = \sum_{M'} \sum_{L} & P(M')(-1)^{M+M'+1} \\ & \times \{C(J, J, L; M, -M, 0)C(J, J, L; M', -M', 0) \\ & \qquad + (-1)^{2J} C(J, J, L; M, -M, 0)C(J, J, L; -M', M', 0)\} \end{aligned}$$

We now use

$$\sum_L C(J, J, L; M, -M, 0)C(J, J, L; \mu, -\mu, 0) = \delta_{\mu M} \tag{14.18}$$

to get

$$T^+_{JM} = P(M) + P(-M) \tag{14.19}$$

We can similarly show that

$$T^-_{JM} = -\alpha(P(M) - P(-M)) \tag{14.20}$$

From (14.12) through (14.14) it follows that

$$|T^-_{JM}| \leq |T^+_{JM}| \leq 1 \tag{14.21}$$

These relations may be used to obtain information about the value of J. In the special case of Λ^0 decay, it is found experimentally that

$$W(\theta) \propto 1 + A \cos\theta \tag{14.22}$$

This implies that

$$\langle P_L(\cos\theta)\rangle = 0 \qquad \text{for } L \geq 2 \tag{14.23}$$

The condition

$$|T^-_{JJ}| \leq T^+_{JJ} \tag{14.24}$$

now implies, as can be seen after some straightforward calculations, that

$$\langle \cos\theta \rangle \leq \frac{1}{6J} \tag{14.25}$$

The experimental result[1]

$$\langle \cos\theta \rangle \simeq 0.19 \tag{14.26}$$

shows that the spin of the Λ^0 is $\frac{1}{2}$.

[1] F. S. Crawford, M. Cresti, M. L. Good, M. L. Stevenson, H. K. Ticho, *Phys. Rev. Letters*, **2**, 114 (1959).

It should be pointed out that the mere existence of a nonvanishing expectation value for $\cos\theta$ is evidence for parity nonconservation in the decay. If parity were conserved, the formula quoted in (4.101) shows that (with η the pion parity)

$$\bar{S}_{-1/2} = \eta(-1)^{J-1/2}\bar{S}_{1/2}$$

i.e.

$$|\bar{S}_{1/2}|^2 = |\bar{S}_{-1/2}|^2$$

The quantity $\cos\theta$ is odd under reflections, and its expectation value must vanish when parity is conserved.

The fact that parity is not conserved in the Λ^0 decay implies that we cannot, by studying the final state in the reaction

$$\Lambda^0 \rightarrow p + \pi^-$$

deduce anything about the parity of the Λ^0. On the other hand, as we shall soon see, in parity conserving reactions involving the Λ^0, there is always another particle other than a nucleon or pion present. It will become clear in Chapter 16 that the parity of the Λ^0, or one other new particle (the K^0, say), must be defined conventionally. *We define the* Λ^0 *to have the same parity as the neutron and the proton.*

The only other remark which we will make about the Λ^0 is that, as expected, it has strong interactions with protons and neutrons. What is not known, a priori, is that the interaction is attractive. This was discovered through the identification of *hypernuclei.* These are nuclei with a Λ^0 attached to them. The lifetime of a hypernucleus is of the order of the Λ^0 lifetime, i.e., O (10^{-10} sec). A great variety of hypernuclei have been studied. We list a few of them, with the binding energy of the Λ^0, in the table below.

Table 14-1
Hypernuclear Binding Energies

Hypernucleus	$_\Lambda H^3$	$_\Lambda H^4$	$_\Lambda He^4$	$_\Lambda He^5$	$_\Lambda Li^8$	$_\Lambda Be^8$	$_\Lambda B^{12}$	$_\Lambda C^{13}$
Binding Energy B in MeV	0.21 ±0.20	2.11 ±0.10	2.40 ±0.11	3.10 ±0.07	6.37 ±0.23	6.38 ±0.50	9.75 ±0.42	10.5 ±0.5

($_\Lambda Z^A$ denotes a hypernucleus of charge Z, A-Z-1 neutrons and one Λ^0)

It is noticed that the binding energy increases almost monotonically with A. This is caused by the absence of an exclusion principle effect, so that the states available to the Λ^0 will not be restricted in any way. The Λ^0's will in general move in an S-orbit relative to the nuclear core and the attraction is, roughly speaking, proportional to the number of nucleons forming the core.

Lack of space prevents us from a detailed discussion of this interesting topic.[1] We shall, however, make occasional use of hypernuclear data and that is why this subject is being brought up at this time.

The Sigma Hyperons

There exist three particles, Σ^+, Σ^0, and Σ^-, collectively called the sigma hyperons. Their masses are

$$M_{\Sigma^+} = 1189.39 \pm 0.14 \text{ MeV}$$
$$M_{\Sigma^0} = 1192.3 \pm 0.2 \text{ MeV}$$
$$M_{\Sigma^-} = 1197.20 \pm 0.14 \text{ MeV}$$

The decay modes of the charged sigma particles with the branching ratios are given in the table below. The lifetimes are

$$\tau_{\Sigma^+} = (0.794 \pm 0.026) \times 10^{-10} \text{ sec}$$
$$\tau_{\Sigma^0} < 1.0 \times 10^{-14} \text{ sec}$$
$$\tau_{\Sigma^-} = (1.58 \pm 0.05) \times 10^{-10} \text{ sec}$$

Table 14-2

Decays of Charged Sigmas*

Particle	Decay mode	Branching ratio
Σ^+	$p + \pi^0$	$51.0 \pm 2.4\%$
Σ^+	$n + \pi^+$	$49.0 \pm 2.4\%$
Σ^+	$n + \pi^+ + \gamma$	$\approx 0.2 \times 10^{-2}\%$
Σ^+	$\Lambda^0 + e^+ + \nu$	$\approx 0.2 \times 10^{-2}\%$
Σ^+	$p + \gamma$	$(3.7 \pm 0.8) \times 10^{-2}\%$
Σ^+	$n + \mu^+ + \nu$	$<1.1 \times 10^{-2}\%$
Σ^+	$n + e^+ + \nu$	$<0.5 \times 10^{-2}\%$
Σ^-	$n + \pi^-$	100%
Σ^-	$n + \pi^- + \gamma$	$\approx 0.1 \times 10^{-2}\%$
Σ^-	$n + \mu^- + \nu$	$(0.66 \pm 0.15) \times 10^{-1}\%$
Σ^-	$n + e^- + \nu$	$(1.2 \pm 0.2) \times 10^{-1}\%$
Σ^-	$\Lambda^0 + e^- + \nu$	$(0.75 \pm 0.28) \times 10^{-2}\%$

* This table uses data from A. H. Rosenfeld, A. Barbaro-Galtieri, Walter H. Barkas, P. L. Bastien, J. Kirz, and M. Roos, *Rev. Mod. Phys.* **37,** 633 (1965), from which all of our data comes.

[1] See R. H. Dalitz, *Proceedings of the Rutherford Jubilee International Conference*, 1962, for a review in which references to the literature may be found.

The spin of the charged hyperons can be determined in the same way as that of the Λ^0. The spin of the Σ^+ was found by studying the decay following the production in the capture reaction

$$K^- + p \rightarrow \Sigma^+ + \pi^-$$

Measurements[1] strongly support the assignment of $\frac{1}{2}$ as the spin of the Σ^+. We shall argue in Chapter 16 that this implies that the other sigma particles also have spin $\frac{1}{2}$.

The lifetimes of the $\Sigma^\pm$ indicate that these particles decay by the same weak interactions as the Λ^0. The Σ^0, whose existence was predicted before it was ever found, decays according to

$$\Sigma^0 \rightarrow \Lambda^0 + \gamma$$

The lifetime for such a transition is very short, of the order of 10^{-19} to 10^{-21} sec. If the Σ^0 are polarized, a study of the correlation between the Σ^0 polarization, the Λ^0 polarization, and the photon polarization could yield information about the relative $(\Sigma^0 - \Lambda^0)$ parity. Since the photon, in the Σ^0 rest frame has 75 MeV energy, and it is very difficult to make polarization measurements on photons of such an energy, it has been suggested[2] that the internal conversion process

$$\Sigma^0 \rightarrow \Lambda^0 + e^+ + e^-$$

be studied instead. That the difference between even and odd Σ^0 parity may be obtainable from the study of this process is made plausible by the following:

1. If the "mass" of the electron-positron pair, denoted by $s^{1/2}$ with

$$s = (p_+ + p_-)^2 = (E_+ + E_-)^2 - (\mathbf{p}_+ + \mathbf{p}_-)^2$$

is small, i.e., in the range from the threshold, $2m_e$ to perhaps $5m_e$, the matrix element of the electromagnetic current $(\Psi_{\Lambda^0}, j_\mu \Psi_{\Sigma^0})$, which determines the decay, is evaluated for a momentum transfer not so very different from the value it takes for photon emission, namely

$$s = 0$$

Thus for small values of s this process gives no more information than the photon decay, except that the plane of the electron and positron, which is

[1] J. Leitner, P. Nordin, Jr., A. H. Rosenfeld, F. T. Solmitz, and R. D. Tripp, *Phys. Rev. Letters* **3**, 238 (1959).

[2] G. Feinberg, *Phys. Rev.* **109**, 1019 (1958); G. Feldman and T. Fulton, *Nucl. Phys.* **8**, 106 (1958); L. E. Evans, *Nuovo Cimento* **25**, 580 (1962); R. H. Dalitz, *Proceedings of the Aix-en-Provence Conference on Elementary Particles*, 1961 (C. E. N. Saclay, France, 1961).

correlated differently with the Σ^0 and Λ^0 polarizations for different Σ^0 parities, is easier to study than the photon polarization.

2. When s is large, however, the situation is different. Consider the maximum value of s, obtained when the Σ^0 decays into a Λ^0 at rest. In that limit all of the energy difference between the Σ^0 and the Λ^0 goes into the leptons, and the members of the pair have equal and opposite momenta. Also

$$q = p_\Sigma - p_\Lambda = (M_\Sigma - M_\Lambda, 0)$$

Since the current is conserved, i.e.

$$q^\mu j_\mu = 0$$

we have, in this case,

$$(\Psi_{\Lambda^0}, j_0\Psi_{\Sigma^0}) = 0$$

so that only the spatial components of the current lead to the transition. When $\mathbf{q} = 0$, however, the only quantity we can write down that transforms as a vector under rotations is $\chi_\Lambda{}^*\boldsymbol{\sigma}\chi_\Sigma$, and this is an axial vector,

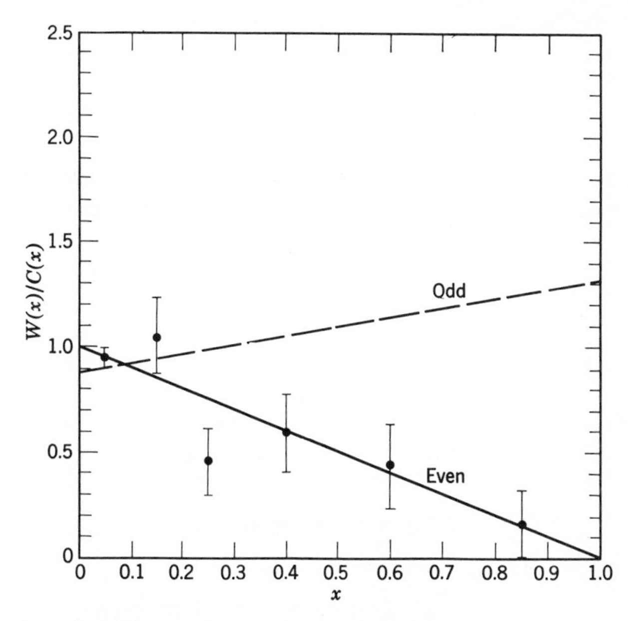

Fig. 14.1. The ratio of the number of events to the function

$$C(x) = \left(\frac{1}{x}\right)(1 - x)^{1/2}\left(1 - \frac{x_0}{x}\right)^{1/2}\left(1 + \frac{x_0}{2x}\right) \qquad x_0 = \frac{4m_e{}^2}{(M_\Sigma - M_\Lambda)^2}$$

plotted against x, the square of the invariant mass of the electron-positron pair. The theoretical predictions for odd and even parity are shown. (From Courant et al., *Phys. Rev. Letters*, **10**, 409 (1963).)

which only leads to parity-conserving transitions between states of opposite parity. We might thus expect, on the basis of this extreme situation, that there will be few electron-positron pairs produced with large "mass" for *even* relative $\Sigma^0 - \Lambda^0$ parity. This expectation is confirmed by detailed calculations, which we do not have space to cover. Experimental evidence for the suppression of electron-positron pairs with large "mass" has recently been obtained,[1] showing that the Σ^0 *has even parity* (Fig. 14.1). We shall later find indirect arguments for drawing the same conclusion about the charged Σ particles.

Σ hyperons do not appear to form hyperfragments. The reason for this is clearly that the reaction

$$\Sigma + \begin{pmatrix} p \\ n \end{pmatrix} \rightarrow \Lambda^0 + \begin{pmatrix} p \\ n \end{pmatrix}$$

and the like are *not* weak (10^{-10} sec) interactions, so that a Σ cannot exist in a nucleus. A potential exception is a $(\Sigma^- - n)$ bound state, which cannot decay into anything else; there is no evidence, to date, that such a bound state exists.

The Cascade Particles

There are two particles, designated by the symbols Ξ^- and Ξ^0, that are called the cascade particles, because of their principal decay modes

$$\Xi^- \rightarrow \Lambda^0 + \pi^-$$
$$\qquad \hookrightarrow p + \pi^-$$

and

$$\Xi^0 \rightarrow \Lambda^0 + \pi^0$$
$$\qquad \hookrightarrow p + \pi^-$$

Their masses are

$$M_{\Xi^-} = 1320.8 \pm 0.2 \text{ MeV}$$

and

$$M_{\Xi^0} = 1314.3 \pm 1.0 \text{ MeV}$$

respectively. Their lifetimes are

$$\tau_{\Xi^-} = (1.74 \pm 0.05) \times 10^{-10} \text{ sec}$$

and

$$\tau_{\Xi^0} = (3.05 \pm 0.38) \times 10^{-10} \text{ sec}$$

[1] H. Courant, H. Filthuth, P. Franzini, R. G. Glasser, A. MinguzziRanzi, A. Segar, W. Willis, R. A. Burnstein, T. B. Day, B. Kehoe, A. J. Herz, M. Sakitt, B. Sechi-Zorn, N. Seeman, G. Snow, *Phys. Rev. Letters*, **10**, 409 (1963).

So far the only decay mode observed for Ξ^0 has been

$$\Xi^0 \rightarrow \Lambda^0 + \pi^0$$

For the charged Ξ^-, the dominant mode is

$$\Xi^- \rightarrow \Lambda^0 + \pi^-$$

with the only other observed decay

$$\Xi^- \rightarrow \Lambda^0 + e^- + \nu$$

occurring with a branching ratio of $(3.0 \pm 1.7) \times 10^{-3}$. The decays have been studied thoroughly and will be discussed in Chapter 33.

Although the spin of the Ξ^- has not been unambiguously determined, there is evidence that *the spin is not* $\frac{3}{2}$. When this is combined with theoretical prejudices to be elaborated in later chapters, we are led to the tentative conclusion that the Ξ particles belong with the other baryons listed in this chapter, i.e., that they have spin $\frac{1}{2}$. The nature of the evidence against spin $\frac{3}{2}$ follows.

When the appropriate Wigner coefficients are evaluated, we find that [see (14.19) and (14.20)]

$$P(\tfrac{3}{2}) + P(-\tfrac{3}{2}) = T^{+}_{3/2,-3/2} = \tfrac{1}{2}(1 - 5\langle P_2(\cos\theta)\rangle)$$

and

$$\alpha(P(\tfrac{3}{2}) - P(-\tfrac{3}{2})) = T^{-}_{3/2,-3/2} = \tfrac{1}{2}(9\langle P_1(\cos\theta)\rangle - \tfrac{7}{3}\langle P_3(\cos\theta)\rangle)$$

Since α is known from an analysis of the Ξ^--decay (see Chapter 33)

$$\alpha_{\Xi^-} = -0.410 \pm 0.046$$

we may use it to construct

$$P(\pm\tfrac{3}{2}) = \tfrac{1}{4}\left\{1 - 5\langle P_2\rangle \pm \frac{1}{\alpha}(9\langle P_1\rangle - \tfrac{7}{3}\langle P_3\rangle)\right\}$$

The experimental data show that

$$P(-\tfrac{3}{2}) = -0.33 \pm 0.20$$

which violates the condition

$$P(M) \geq 0$$

by almost two standard deviations. Accumulations of more data will presumably lead to the sharpening of this conclusion.

There is at present no evidence that the parity of the Ξ is either the same or opposite to that of the Λ^0. Such evidence is obtainable from a study of the reaction

$$\Xi^- + p \rightarrow 2\Lambda^0$$

with slow Ξ^-. We shall argue in Chapter 15 that the capture reaction is most likely to take place from an atomic S state. It can either be a $^1S_0$ state or a $^3S_1$ state. The final two- Λ^0 state must be antisymmetric because of the exclusion principle, and this means, for values of $J = 0$ or 1, the states $^1S_0$ and $^3P_{1,0}$. Which it will be depends on the parity of the Ξ^-, as the table shows:

Capture state	Ξ-parity	Final two-Λ^0 state
$^1S_0$	+	$^1S_0$
	−	$^3P_0$
$^3S_1$	+	none
	−	$^3P_1$

Each of the final states leads to a different correlation between the polarizations of the two Λ^0's.[1] It is now generally believed that the parity of the Ξ^- is positive, and that the Ξ, together with the nucleons, the Λ^0-particle, and the Σ, form a supermultiplet of states which do not differ from each other in any essential way.

PROBLEMS

1. Write an effective Lagrangian to describe the decay $\Lambda^0 \to p + \pi^-$. Estimate the strength of the coupling from the measured decay rate.
2. If the matrix element for Σ beta decay is a constant of the same magnitude as that for neutron beta decay, so that the ratio of rates,

$$\Gamma(\Sigma^- \to n + e^- + \nu)/\Gamma(n \to p + e^- + \nu)$$

and

$$\Gamma(\Sigma^- \to n + \mu^- + \nu)/\Gamma(n \to p + e^- + \nu),$$

is determined by the ratio of the phase space factors alone, estimate the rates for the Σ decays and compare them with experiment. Use $m_\mu = 105.66$ MeV.
3. Use the helicity formalism to discuss the decay $\Sigma^0 — \Lambda^0 + \gamma$ assuming that parity is conserved in the process.

[1] S. B. Treiman, *Phys. Rev.*, **113**, 355 (1959).

15

The Pseudoscalar Mesons

The discovery of the pion in 1947 and the concurrent completion of several high (for that time) energy machines ushered in the present era of elementary particle physics. This discovery confirmed Yukawa's hypothesis that the short-range nuclear forces must be mediated by heavy quanta of baryon number zero; conversely, Yukawa's theory provided a rich source of qualitative predictions, which guided much of the experimental research undertaken with the new particles. The pions were followed by the K-mesons—these quite unexpected—and other bosons, more or less stable. In this chapter we discuss eight that share the property of being spinless and having odd parity (when the proton, neutron, and Λ^0 are defined to have even parity).

The Pions

The pions are the lightest of all the strongly interacting particles. They are known to have integral spin and baryon number zero as can be seen from the reaction

$$p + p \rightarrow d + \pi^+$$

which also shows that the pions, like the photons, may be created or annihilated singly in particle reactions. There are three particles which go by that name, $\pi^\pm$ and π^0. Their masses are

$$m_{\pi^\pm} = 139.580 \pm 0.015 \text{ MeV}$$
$$m_{\pi^0} = 134.974 \pm 0.015 \text{ MeV}$$

Both the charged and the neutral pions are unstable. The former decay primarily according to

$$\pi^\pm \rightarrow \mu^\pm + \nu$$

with a lifetime

$$\tau_{\pi^\pm} = (2.551 \pm 0.026) \times 10^{-8} \text{ sec}$$

The much rarer alternative decay modes and the branching ratios are listed below

Decay modes	Γ/Γ_{tot}
$\pi^\pm \to e^\pm + \nu$	$(1.24 \pm 0.03) \times 10^{-4}$
$\pi^\pm \to \mu^\pm + \gamma + \nu$	$(1.24 \pm 0.25) \times 10^{-4}$
$\pi^\pm \to \pi^0 + e^\pm + \nu$	$(1.13 \pm 0.09) \times 10^{-8}$

The neutral pion decays according to

$$\pi^0 \to 2\gamma$$

with a lifetime

$$\tau_{\pi^0} = (1.78 \pm 0.26) \times 10^{-16} \text{ sec}$$

Internal conversion also takes place, and the reactions

$$\pi^0 \to \gamma + e^+ + e^-$$

$$\pi^0 \to 2e^+ + 2e^-$$

are observed with a frequency of 1:83 and 1:29,000 respectively.

The spin of the charged pion has been determined by comparing the cross sections for the reaction

$$\pi^+ + d \to p + p$$

and the production

$$p + p \to \pi^+ + d$$

at the same center of mass energy. The absorption rate is given by

$$\Gamma_{\text{abs}} = 2\pi \overline{\sum_f{}' |M_{fi}|^2} \frac{4\pi p_f^{\,2}}{(2\pi)^3} \frac{dp_f}{dW} \tag{15.1}$$

Here p_f is the proton momentum, $W = 2(p_f^{\,2} + M_p^{\,2})^{1/2}$ is the center of mass energy, M_{fi} is the matrix element for the process, and the sum is over final states, the average over all initial states. The $'$ on the summation sign is there to remind us that since the two protons in the final state are identical particles, the angular integration is over a hemisphere. This may be converted to a full angular integration provided we divide by 2. If the pion has spin J, we get

$$\Gamma_{\text{abs}} = \frac{1}{2 \times 3 \times (2J + 1)} \sum_i \sum_f |M_{fi}|^2 \frac{W p_f}{4\pi} \tag{15.2}$$

since the deuteron has angular momentum 1. Dividing by the relative velocity

$$v_r = v_\pi + v_d = \frac{p_\pi}{\omega_\pi} + \frac{p_\pi}{E_d} = \frac{p_\pi W}{\omega_\pi E_d}$$

we get for the absorption cross section

$$\sigma_{\rm abs} = \frac{1}{6(2J+1)} \sum_{fi} |M_{fi}|^2 \frac{p_f \omega_\pi E_d}{4\pi p_\pi} \tag{15.3}$$

Similarly the production cross section is seen to be

$$\sigma_{\rm prod} = \left(\frac{1}{2}\right)^2 \sum_{fi} |M_{fi}|^2 \frac{\omega_\pi E_d p_\pi}{4\pi p_f} \tag{15.4}$$

the factor of $(\frac{1}{2})^2$ coming from the average over the initial state. The square of the matrix element, when summed over initial and final states, is the same in both cases, provided we consider the two processes for the same value of W. This is known as the principle of detailed balance and follows from the invariance of the interaction under time reversal.[1] Thus

$$\frac{\sigma_{\rm abs}}{\sigma_{\rm prod}} = \frac{2}{3(2J+1)} \left(\frac{p_f}{p_\pi}\right)^2 \tag{15.5}$$

a result completely independent of the details of the interaction.

The π^+ absorption in deuterium was studied at 24 MeV pion energy in the laboratory system.[2] The cross section which was obtained could be written in the form

$$\left(\frac{d\sigma}{d\Omega}\right)_{\rm abs} = 9(0.22 + \cos^2\theta) \times 10^{-28}\ {\rm cm}^2$$

and

$$\sigma_{\rm abs} = (3.1 \pm 0.3) \times 10^{-27}\ {\rm cm}^2$$

The production reaction was studied[3] with 341 MeV protons in the laboratory system which gives nearly the same center of mass energy as used in the absorption study. The data may be combined with (15.5) to predict the absorption cross sections

$$\left(\frac{d\sigma}{d\Omega}\right)_{\rm abs} = \frac{1}{2J+1}(11 \pm 4)(0.11 \pm 0.06 + \cos^2\theta) \times 10^{-28}\ {\rm cm}^2$$

and

$$\sigma_{\rm abs} = \frac{1}{2J+1}(3.0 \pm 1.0) \times 10^{-27}\ {\rm cm}^2$$

from which it is clear that *the spin of the π^+ must be zero.* The same must hold for the π^-, since it is clearly the antiparticle of the π^+.

[1] See (7.28), from which this result follows.

[2] R. Durbin, H. Loar, and J. Steinberger, *Phys. Rev.*, **83**, 646 (1951). D. L. Clark, A. Roberts, and R. Wilson, *Phys. Rev.*, **83**, 649 (1961).

[3] W. F. Cartwright, C. Richman, M. N. Whitehead, and H. A. Wilcox, *Phys. Rev.*, **91**, 677 (1953).

The spin of the π^0 cannot be as easily established; its lifetime is too short to make it a useful projectile in an absorption experiment. The decay mode

$$\pi^0 \to 2\gamma$$

shows that the π^0 cannot have spin 1. To see this in the simplest possible way, we note that the amplitude for the reaction must be linear in $\mathbf{e}_1$, the polarization vector of one of the photons, linear in $\mathbf{e}_2$, the polarization vector of the other photon, and otherwise it can only depend on the vector $\mathbf{k}$, the relative momentum of the two photons in the center of mass system. The amplitude must be symmetric under the interchange of the two identical photons in the final state, i.e., under the interchange

$$\begin{aligned} \mathbf{e}_1 &\leftrightarrow \mathbf{e}_2 \\ \mathbf{k} &\leftrightarrow -\mathbf{k} \end{aligned} \tag{15.6}$$

If we note that gauge invariance requires

$$\mathbf{e}_1 \cdot \mathbf{k} = \mathbf{e}_2 \cdot \mathbf{k} = 0 \tag{15.7}$$

it is clear that no vector can be constructed to satisfy these conditions. Since the amplitude for the decay of a vector particle at rest must be of the form $\boldsymbol{\eta} \cdot \mathbf{M}$, with $\boldsymbol{\eta}$ the polarization vector of the decaying particle, the absence of a possible $\mathbf{M}$ implies that a vector particle cannot decay into two photons. There is actually a great deal of evidence linking the two charged pions and the neutral pion into a multiplet having the same spin and parity, so that there is no feeling of urgency to study the rarer decay modes for correlations. There has been a study, however, of the reaction

$$\pi^0 \to 2e^+ + 2e^-$$

to obtain information about the parity of the π^0, assuming that the spin is 0. For a spinless π^0, two possible forms of the amplitude are

$$\begin{aligned} \mathbf{e}_1 \cdot \mathbf{e}_2 &\qquad \text{(even parity)} \\ \mathbf{k} \cdot \mathbf{e}_1 \times \mathbf{e}_2 &\qquad \text{(odd parity)} \end{aligned}$$

In the first case the polarization vectors of the two photons will tend to be parallel; in the second they will be perpendicular. These correlations carry over to the situation where the photons are both internally converted, in which case the planes of the electron-positron pairs are correlated. It was shown by Kroll and Wada[1] that the distribution of angles ϕ between the two planes formed by the pairs is

$$W(\phi) = 1 \pm 0.48 \cos 2\phi \tag{15.8}$$

[1] N. Kroll and W. Wada, *Phys. Rev.*, **98,** 1355 (1955).

depending on whether the parity is $\pm$. An experimental examination[1] of sixty-four events of a suitable character yielded

$$W(\phi) = 1 - (0.75 \pm 0.42) \cos 2\phi$$

from which one deduces that the π^0 *is pseudoscalar.*

The parity of the charged pions can be deduced from the occurrence of the reaction, at rest,

$$\pi^- + d \rightarrow 2n$$

If we assume that the π^- is captured from an S atomic orbit, the initial angular momentum and parity state is $1^\pm$ depending on whether the pion is a $0^\pm$ particle. The final state of the two neutrons is not arbitrary; it is restricted by the Pauli exclusion principle to be one of the antisymmetric states ${}^1S_0$, ${}^3P_{2,1,0}$, ${}^1D_2$, ... The only possible state with angular momentum 1 is the ${}^3P_1$ state, whose parity is odd. Thus the charged pion will be pseudoscalar, provided the reaction occurs and provided the capture does indeed take place from the S state.

The occurrence of the reaction was demonstrated by Panofsky et al.,[2] who found that the reactions

$$\pi^- + d \rightarrow 2n$$
$$\pi^- + d \rightarrow 2n + \gamma$$

occur, with a branching ratio of 2:1. Although the rather large electromagnetic mode is unexpected (1% to 5% would have been a normal guess), indicating a suppression of the capture reaction rate for $\pi^- + d \rightarrow 2n$, the process does occur. It remains to show that the π^- is captured from an S state.

First of all, we note that the capture does not take place in flight. The capture rate is

$$\Gamma = \sigma(\pi^- + d \rightarrow 2n)v_\pi N \tag{15.9}$$

For liquid deuterium the density is $N = 4 \times 10^{22}$ cm^{-3}; the cross section is

$$\sigma(\pi^- + d \rightarrow 2n) = \sigma(\pi^+ + d \rightarrow 2p) \approx 3 \times 10^{-27} \text{ cm}^2$$

at a typical energy of 24 MeV when $v \approx \frac{1}{2}c$. Thus

$$\Gamma \approx 1.8 \times 10^6 \text{ sec}^{-1}$$

On the other hand, the π^- has been shown by Wightman[3] to slow down to a velocity $v \simeq 0.05c$ in the short time of 4.8×10^{-10} sec losing its energy

[1] R. Plano, A. Prodell, N. Samios, M. Schwartz, and J. Steinberger, *Phys. Rev. Letters*, **3**, 525 (1959).
[2] W. K. H. Panofsky, R. L. Aamodt, and J. Hadley, *Phys. Rev.*, **81**, 565 (1951).
[3] A. S. Wightman, *Phys. Rev.*, **77**, 521 (1950).

by the ionization process. The capture of the pion into an atomic orbit is overwhelmingly more probable than capture in flight. This is generally true in elementary particle physics.

According to the calculations of Wightman, the pion takes about 3.7×10^{-12} sec to be captured into an atomic orbit with large n (principal quantum number).[1] De-excitation by the Auger process to an orbit with $n \approx 7$ takes another 0.8×10^{-12} sec. In this range of n values, a mechanism first proposed by Day, Sucher, and Snow[2] (DSS) becomes important. Instead of radiatively cascading to low n states, the pionic "atom," which is small (radius $\approx n/m_\pi\alpha$) and neutral, drifts about with a velocity of the order of 10^6 cm/sec, and in a very short time penetrates the electron cloud of a deuterium atom. The very strong local electric fields induce a mixing among the n^2 degenerate levels corresponding to a given n. Since capture from an S state is always favored over the higher l values, the DSS effect tends to increase the probability of S-state capture over what it would be if the pion were captured by undergoing radiative de-excitation to a $2P$ or $1S$ atomic state. Detailed calculations may be found in a paper by Leon and Bethe.[3] Experimental evidence for speeded up pionic capture was obtained by Fields et al.,[4] who studied the capture of π^- in hydrogen. They looked at some 80,000 pions, some of which were seen to decay into μ^-. From kinematical data the velocity of the pion at decay could be determined, and it was found that only two pions decayed with $v < 0.01c$. Since only events with the μ^- track going in the backward hemisphere relative to the pion direction were selected, the mean time taken by a π^- to get from $v/c = 0.01$ to nuclear capture is estimated to be

$$\tau = 2 \times \left(\frac{2}{80{,}000}\right) \times 2.5 \times 10^{-8}\ \text{sec} = 1.2^{+1.2}_{-0.5} \times 10^{-12}\ \text{sec}$$

If the Stark mixing mechanism were not available, the pion would have to cascade down from the $n \approx 7$ orbit, which would take about 16×10^{-12} sec. This observation provides indirect evidence for the DSS effect, which guarantees capture from the S orbit and confirms that the π^- *is pseudoscalar.* It is worth pointing out that the DSS mechanism does not operate only for pions, and that the expectation is that all long-lived ($\sim 10^{10}$ sec),

[1] This figure is for capture from $v/c = 0.05$; the time for capture from $v/c = 0.01$ is 1.2×10^{-12} sec according to the estimates of Wightman.

[2] T. B. Day, G. A. Snow, and J. Sucher, *Phys. Rev. Letters*, **3**, 61 (1960). See also G. A. Snow, *Proceedings of the* 1960 *Annual International Conference on High Energy Physics at Rochester*, Interscience Publishers, New York, p. 407.

[3] M. Leon and H. A. Bethe, *Phys. Rev.*, **127**, 636 (1962).

[4] T. H. Fields, G. G. Yodh, M. Derrick, and J. G. Fetkovitch, *Phys. Rev. Letters*, **5**, 69 (1960).

negatively charged, strongly interacting particles may be expected to be captured from S-orbital states in hydrogen.[1]

Much more has been learned about pions in the last decade; much of the material in later chapters will deal with the interactions of pions with pions and nucleons.

The K Mesons

K mesons of charge +1, 0, and −1 have been discovered. There are actually two neutral K mesons, denoted by K^0 and $\overline{K^0}$, so that there are actually four particles, K^+, K^0, and their antiparticles. We shall see in the next chapter that the K mesons are characterized by a new quantum number, *hypercharge*, and that is why the $\overline{K^0}$ is different from the K^0. Note that since the π^0 is self-conjugate, it must be neutral with respect to the hypercharge, as well as electrical charge.

The masses of the K mesons are

$$m_K{}^+ = 493.78 \pm 0.17 \text{ MeV}$$
$$m_K{}^0 = 497.7 \pm 0.3 \text{ MeV}$$

The charged K decays through the weak interactions, i.e., with a lifetime

$$\tau_{K^\pm} = (1.229 \pm 0.008) \times 10^{-8} \text{ sec}$$

The various partial modes with the percentage rates are listed below:

Decay mode	Branching ratio
$K^\pm \rightarrow \mu^\pm + \nu$	$63.2 \pm 0.4\%$
$K^\pm \rightarrow \pi^\pm + \pi^0$	$21.3 \pm 0.4\%$
$K^\pm \rightarrow \pi^\pm + \pi^+ + \pi^-$	$5.52 \pm 0.08\%$
$K^\pm \rightarrow \pi^\pm + \pi^0 + \pi^0$	$1.68 \pm 0.05\%$
$K^\pm \rightarrow \pi^0 + \mu^\pm + \nu$	$3.4 \pm 0.2\%$
$K^\pm \rightarrow \pi^0 + e^\pm + \nu$	$4.9 \pm 0.2\%$
$K^\pm \rightarrow \pi^\pm + \pi^\mp + e^\pm + \nu$	$(4.3 \pm 0.9) \times 10^{-3}\%$
$K^\pm \rightarrow \pi^\pm + \pi^\pm + e^\mp + \nu$	$<0.1 \times 10^{-3}\%$
$K^\pm \rightarrow \pi^\pm + \pi^0 + \gamma$	$(2.2 \pm 0.7) \times 10^{-2}\%$
$K^\pm \rightarrow e^\pm + \nu$	$<1.6 \times 10^{-1}\%$
$K^\pm \rightarrow \pi^\pm + e^+ + e^-$	$<1.1 \times 10^{-5}\%$
$K^\pm \rightarrow \pi^\pm + \pi^+ + \pi^- + \gamma$	$(9 \pm 4) + 10^{-4}\%$

[1] For experimental evidence of the existence of this effect for K^- capture in hydrogen, see R. Knop, R. A. Burnstein, and G. A. Snow, *Phys. Rev. Letters*, **14**, 767 (1965).

The neutral K meson also decays. It turns out, however, that its decay law is not the usual exponential law:[1] the K^0 decays like mixture of two particles. The short-lived one denoted by K_s has a lifetime

$$\tau_{K_s} = (0.909 \pm 0.015) \times 10^{-10} \text{ sec}$$

the main decay modes being

$$K_s \to \pi^+ + \pi^- \qquad (68.9 \pm 1.1\%)$$
$$K_s \to 2\pi^0 \qquad (31.1 \pm 1.1\%)$$

The long-lived component, denoted by K_L, has a lifetime

$$\tau_{K_L} = (5.70 \pm 0.65) \times 10^{-8} \text{ sec}$$

Its principal decay modes, with the rates, are listed below:

$K_L \to 3\pi^0$	$24.8 \pm 3.0\%$
$\pi^+\pi^-\pi^0$	$13.6 \pm 1.0\%$
$\pi^\pm\mu^\mp\nu$	$26.2 \pm 2.6\%$
$\pi^\pm e^\mp\nu$	$35.4 \pm 2.7\%$
$\pi^+\pi^-$	$0.21 \pm 0.03\%$

Because of their many decay modes, the K mesons are a source of much information about the interactions responsible for the decays, i.e., the weak interactions. K mesons, like the pions, are strongly coupled to the baryons, but because of their large mass they often play a less important role in mediating interactions than the pions do.

It is unfortunately impossible to determine the spin of the K meson by considering the reaction

$$K^- + p \to \pi^- + p$$

and its inverse, because this reaction does not occur. It is forbidden by a selection rule which will be discussed in Chapter 16. The reactions that do occur, e.g.

$$K^- + p \to \pi^- + \Sigma^+$$

and

$$K^- + n \to \pi^- + \Lambda^0$$

have inverse reactions which cannot be studied in the laboratory because of the short lifetime of the particles making up the state. The spin of the K meson can, however, be determined from its decay characteristics. This is not true of the parity of the K, as we shall see.

[1] This phenomenon will be discussed in detail in Chapter 32.

The decay mode

$$K^0 \to 2\pi^0$$

leads to a final state consisting of two identical bosons. The final state, symmetric under the interchange of the two particles, must have the spin-parity quantum numbers $0^+, 2^+, 4^+, \ldots$, since the orbital angular momentum must be even and the two pions are pseudoscalar. The corresponding decay for the charged K meson

$$K^+ \to \pi^+ + \pi^0$$

allows all values of the spin, but the zero spin of the pion requires that the parity be $(-1)^J$.

Much more detailed information can be obtained from a study of the decay into three pions, e.g.

$$K^+ \to \pi^+ + \pi^+ + \pi^-$$

The decay rate is given by

$$\Gamma = \iiint \frac{d^3q_1}{2\omega_{q1}} \frac{d^3q_2}{2\omega_{q2}} \frac{d^3q_3}{2\omega_{q3}} \delta(\mathbf{q}_1 + \mathbf{q}_2 + \mathbf{q}_3) \times \delta(\omega_{q1} + \omega_{q2} + \omega_{q3} - m_K)\, \overline{|M|^2} \quad (15.10)$$

Here $(\omega_{q_i}, \mathbf{q}_i)$ are the energies and momenta of the three pions in the rest system of the decaying K meson. $\overline{|M|^2}$ is the average over spin states of the absolute square of the decay matrix element. The latter is of the general form of a sum of angular factors like q_{1i}, q_{1j}, q_{2k} etc., multiplied by invariant functions. These invariant functions can only depend on scalars like ω_{q_i}. Scalars like $\mathbf{q}_1 \cdot \mathbf{q}_2$ are really not independent, since

$$\mathbf{q}_1 \cdot \mathbf{q}_2 = \frac{q_3{}^2 - q_1{}^2 - q_2{}^2}{2} = \frac{1}{2}(m_\pi{}^2 + \omega_{q3}^2 - \omega_{q1}^2 - \omega_{q2}^2)$$

Since the three energies add up to the K mass, we may write

$$\overline{|M|^2} = F(\omega_{q_1}, \omega_{q_2})$$

where ω_{q_1} and ω_{q_2} are the energies of any two of the three mesons. Then

$$\Gamma = \frac{1}{8} \int \cdots \int q_1\, d\omega_{q1} q_2\, d\omega_{q2}\, d\Omega_1\, d\Omega_2 \frac{d^3q_3}{\omega_{q3}} \delta(\mathbf{q}_1 + \mathbf{q}_2 + \mathbf{q}_3) \times \delta(\omega_{q1} + \omega_{q2} + \omega_{q3} - m_K) F(\omega_{q1}, \omega_{q2})$$

$$= \frac{1}{8} \int q_1\, d\omega_{q1} \int q_2\, d\omega_{q2} F(\omega_{q1}, \omega_{q2}) \int d\Omega_1 \int d\Omega_2 \times \frac{\delta(\omega_{q1} + \omega_{q2} + \sqrt{(\mathbf{q}_1 + \mathbf{q}_2)^2 + m_\pi{}^2} - m_K)}{\sqrt{(\mathbf{q}_1 + \mathbf{q}_2)^2 + m_\pi{}^2}}$$

so that

$$\frac{d^2\Gamma}{d\omega_{q1}\,d\omega_{q2}} = \frac{q_1q_2}{8}F(\omega_{q1},\omega_{q2})\int d\Omega_1\int d\Omega_2$$

$$\times\frac{\delta(\omega_{q1}+\omega_{q2}+\sqrt{\omega_{q1}^2+\omega_{q2}^2-m_\pi{}^2+2\mathbf{q}_1\cdot\mathbf{q}_2}-m_K)}{\sqrt{\omega_{q1}^2+\omega_{q2}^2-m_\pi{}^2+2\mathbf{q}_1\cdot\mathbf{q}_2}}$$

$$= \pi^2q_1q_2F(\omega_{q1},\omega_{q2})\int_{-1}^{1}d(\cos\theta)$$

$$\times\frac{\delta(\omega_{q1}+\omega_{q2}+\sqrt{\omega_{q1}^2+\omega_{q2}^2-m_\pi{}^2+2q_1q_2\cos\theta}-m_K)}{\sqrt{\omega_{q1}^2+\omega_{q2}^2-m_\pi{}^2+2q_1q_2\cos\theta}}$$

where θ is defined by

$$\mathbf{q}_1\cdot\mathbf{q}_2 = q_1q_2\cos\theta$$

A change of variable gives

$$\frac{d\Gamma}{d\omega_{q1}\,d\omega_{q2}} = \pi^2F(\omega_{q1}\omega_{q2})\int d(\sqrt{\omega_{q1}^2+\omega_{q2}^2-m_\pi{}^2+2q_1q_2\cos\theta})$$

$$\times\ \delta(\sqrt{\omega_{q1}^2+\omega_{q2}^2-m_\pi{}^2+2q_1q_2\cos\theta}+\omega_{q1}+\omega_{q2}-m_K)$$

$$= \pi^2F(\omega_{q1},\omega_{q2}) \tag{15.11}$$

where ω_{q_1} and ω_{q_2} lie in kinematically allowed regions.

Dalitz observed that a plot of decay events on a graph such as is shown on Fig. 15.1 yields information about the square of the matrix elements (from the density of points).[1]

The boundary of the Dalitz plot is given by the restriction that $|\cos\theta| \leq 1$ in the process, which implies that

$$-1 \leq \frac{(m_K-\omega_{q1}-\omega_{q2})^2-\omega_{q1}^2-\omega_{q2}^2+m_\pi{}^2}{2q_1q_2} \leq 1 \tag{15.12}$$

The usefulness of the Dalitz plot is enormous and has been demonstrated in high energy physics as well as in nuclear physics.

In the special case which we are considering, when all three particles in the final state have the same mass, it is more graphic to plot the points inside an equilateral triangle, with the kinetic energies of the pions ($T_i = \omega_{qi} - m_\pi$) measured inward from the sides, as shown in Fig. 15.2.

[1] R. H. Dalitz, *Phil. Mag.* **44,** 1068 (1953); *Phys. Rev.* **94,** 1046 (1954). See also E. Fabri, *Nuovo Cimento* **11,** 479 (1954).

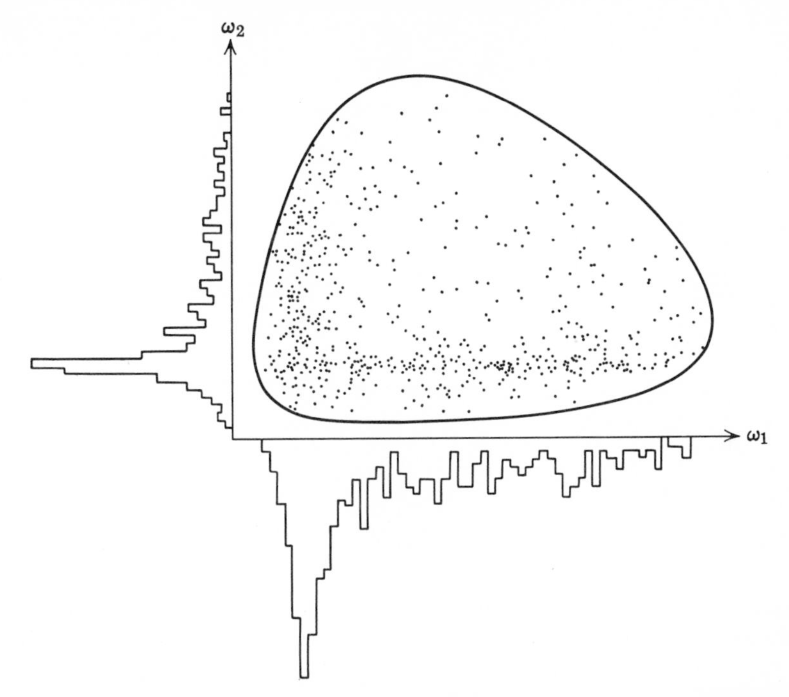

Fig. 15.1. A typical Dalitz plot. The density of points is proportional to the square of the matrix element. The projection of the density onto the two axes yields a mass distribution plot for the remaining two particles

It is easy to show that for any point inside the triangle,

$$T_1 + T_2 + T_3 = \text{const}$$

and the restriction

$$|\cos \theta| \leq 1$$

limits the points to the inside of the region bounded by the curve drawn inside the triangle. For low energy pions, which imply nonrelativistic kinematics, the curve is very nearly a circle. This is the situation in the K-decay, since the total Q-value of the reaction is 75 MeV to be shared by the three pions.

The curvilinear boundary corresponds to the limiting case $\cos \theta = \pm 1$, i.e., to decays in which pions "1" and "2" move parallel or antiparallel to one another. The regions around the lines AA', CC', and BB' correspond to events for which two of the pions have the same energy.

The regions in which we might expect depletion of events for different spin values of the K have been analyzed in all generality and detail by

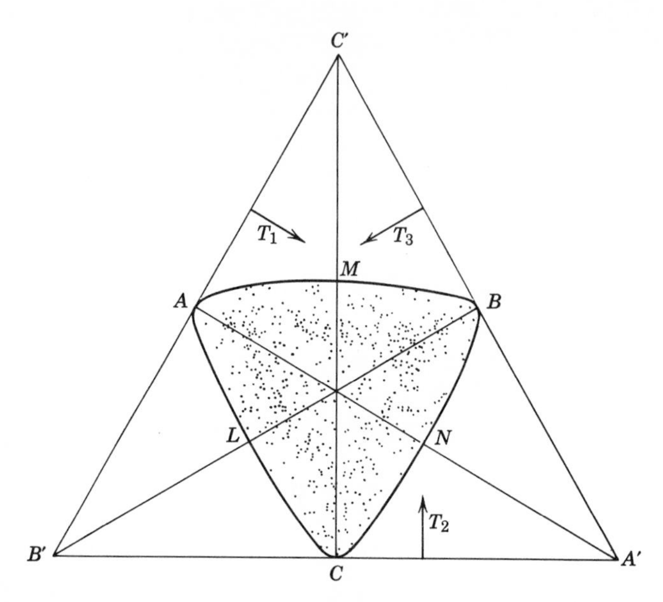

Fig. 15.2. An example of a symmetric Dalitz plot when the decay products are all pions. This particular plot is for the reaction $p + \bar{p} \rightarrow \pi^+ + \pi^- + \pi^0$ (from Alff et al. *Phys. Rev. Letters*, 9, 325 (1962)). T_1, T_2, and T_3 are the kinetic energies of the π^-, π^0 and π^+ respectively.

Zemach.[1] We shall content ourselves with a more qualitative examination of the distribution of points which may be expected for different values of the K-spin. We shall choose pions 1 and 3 to be positive and pion 2 to be negative in the decay of K^+. Thus the distribution must be identical in the triangles $B'C'C$ and $C'CA'$. It is convenient to plot all the points in one of the triangles, concentrating the points and making a visual estimate of their density distribution easier.

Let us consider the total angular momentum J of the three-pion state as the vector sum of two angular momenta: the orbital angular momentum of the two identical pions about their center of mass, l_+, and the angular momentum of the remaining pion about the center of mass of the two like pions, l_-. Thus

$$|l_+ - l_-| \leq J \leq l_+ + l_-$$

Because of the identity of the two positive pions, l_+ must be even. The possible values of J and the parity for different values of l_+ and l_- are shown in Table 15-1.

[1] C. Zemach, *Phys. Rev.* **133,** B1201 (1964).

Table 15-1

Values of J^P for Given (l_+, l_-)

l_-	$l_+ = 0$	$l_+ = 2$
0	0^-	2^-
1	1^+	$3^+, 2^+, 1^+$
2	2^-	$4^-, 3^-, 2^-, 1^-, 0^-$

If the angular momentum J is different from zero, at least one of l_+, l_- must be different from zero. If, for example, $l_+ \geq 2$, we would expect, because of centrifugal barrier effects, that the matrix element (hence the density of points on the plot) would vanish when the two positive pions are at rest relative to one another. This kinematical situation arises when the negative pion has its maximum possible energy. Thus, if $l_+ \geq 2$, we would expect that a depletion of events in the region about M on the plot should occur. How small such a depleted region might be depends on dynamical details that are not known; nevertheless, if there is no visible depletion, we would argue that $l_+ = 0$. We might qualitatively guess that the depletion effect decreases when the relative momentum between the two positive pions, q, is such that $qa \sim 1$, with a some effective radius of interaction usually taken to be some mean of the Compton wavelength of the particles involved in the reaction. Similarly, we argue that if $l_- \geq 1$, there should be a depletion of points in the region where the π^- is at rest, i.e., near C.
The data are shown in Fig. 15.3, and it is very hard to detect any departure from uniformity. From this we conclude that the *spin of the K meson is zero.*

Since it appears that $l_+ = l_- = 0$, the parity of the final state is odd. On the other hand, from the decay

$$K^+ \rightarrow \pi^+ + \pi^0$$

we conclude that if the spin is 0, the parity must be even! This result gave the first inkling that the decays of the K were inconsistent with parity conservation. The Dalitz analysis played the major role in sharpening what in 1955 to 1956 was the widely discussed $\tau - \theta$ paradox.[1] T. D. Lee

[1] The three-pion decay mode was called the τ-mode, and the two-pion was called the θ-mode. When the K-meson was first discovered it was not clear that the different decay products represented the debris from a single particle. The data became paradoxical only when mass and lifetime measurements clearly indicated that only one particle was involved.

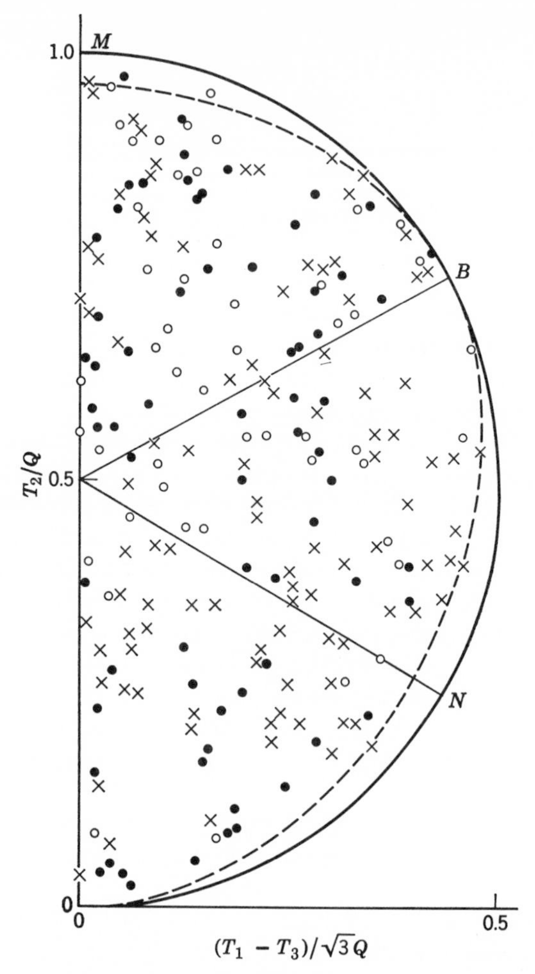

Fig. 15.3. Dalitz plot of 220 $K^+ \rightarrow \pi^+ + \pi^+ + \pi^-$ events. The dashed boundary is obtained if we use nonrelativistic kinematics. The events are plotted in half of the full plot, since there is symmetry between the two π^+. The axes are labeled as in Fig. 15.2, except that the kinetic energies are measured in units of the Q value, $m_K - 3m_\pi$. [From Orear et al., *Phys. Rev.*, 102, 1675 (1956).]

and C. N. Yang finally asked the crucial question of what the evidence for parity conservation in the weak interactions was. They found that there was none and suggested a number of experiments to test this point. (See Chapter 30.) The experiments showed that there was no paradox. This result does mean, however, that we cannot use the decay to determine the parity of the K-meson.

The parity of the K-meson was finally determined in an indirect way, through the observation of the reaction,

$$K^- + \mathrm{He}^4 \to {}_\Lambda\mathrm{H}^4 + \pi^0$$

If the spin of the ${}_\Lambda\mathrm{H}^4$ hyperfragment is 0, the orbital angular momentum in the capture must be the same as in the final state, and the parity of the K must be the same as that of the pion, i.e., odd, provided we make the usual conventional assignment of even parity to the Λ^0.

Evidence about the spin of ${}_\Lambda\mathrm{H}^4$ comes from the following considerations: In the two-body decay

$${}_\Lambda\mathrm{H}^4 \to \pi^- + \mathrm{He}^4$$

the π^- and He^4 are spinless. Thus the parity in the final state must be $-(-1)^J$ where J is the spin of ${}_\Lambda\mathrm{H}^4$. The spin J can be only zero or one.[1] The initial state has even parity, so that (a) if $J = 0$, the final state has odd parity and the decay must go through the $S_{1/2}$ channel, whereas (b) if $J = 1$, the decay must go through the $P_{1/2}$ channel. Now for the free decay of the Λ^0, it is known that the P-wave amplitude a_P is small. We discuss this analysis in Chapter 33, but now note that

$$\frac{|a_P|^2}{|a_S|^2 + |a_P|^2} = 0.11 \pm 0.03$$

Thus if $J = 1$, there would be a tendency for ${}_\Lambda\mathrm{H}^4$ to decay via three-body modes which are not constrained in any way by angular momentum conservation. We might expect that if $J = 1$, the quantity

$$R_4 \equiv \frac{\Gamma({}_\Lambda\mathrm{H}^4 \to \pi^- + \mathrm{He}^4)}{\Gamma_{\mathrm{tot}}}$$

would be small. The experiments of Ammar et al.[2] yield $R_4 = 0.67 \pm 0.06$ which strongly suggests that $J = 0$.

The only possibility which could vitiate our conclusion that the K *is pseudoscalar* is the possible existence of an excited state ${}_\Lambda\mathrm{H}^{4*}$ with $J = 1$, so that the process which actually takes place is

$$\begin{aligned} K^- + \mathrm{He}^4 \to {}&{}_\Lambda\mathrm{H}^{4*} + \pi^0 \\ &\hookrightarrow {}_\Lambda\mathrm{H}^4 + \gamma \end{aligned}$$

[1] This is because ${}_\Lambda\mathrm{H}^4$ consists of a Λ^0 in an S-orbit about a triton (spin $\frac{1}{2}$) core. If the singlet Λ^0-nucleon force is stronger than the triplet force, then the spin would be zero. With our conventional choice of Λ^0 parity, the parity is in any case even.

[2] R. G. Ammar, R. Levi-Setti, W. E. Slater, S. Limentani, P. E. Schlein, and P. H. Steinberg, *Nuovo Cimento* **19**, 20 (1961).

Calculations indicate that such a state does not exist, but since these calculations lean heavily on specific assumptions about the Λ^0-nucleon potential, it cannot be said that the K-parity is established with complete certainty. A search for γ-rays in the above experiment should remove some doubts. When polarized proton targets become available, different types of tests may become feasible.[1] We shall take the K to be pseudoscalar in what follows.

The η^0-Meson

The last particle to be considered in this chapter is the η^0-meson. This particle was discovered in the reaction[2]

$$\pi^+ + d \rightarrow p + p + \pi^+ + \pi^- + \pi^0$$

A plot of the distribution of the invariant mass of the three-pion system, i.e., of the number of events as a function of

$$M^2 = (\omega_{\pi^+} + \omega_{\pi^-} + \omega_{\pi^0})^2 - (\mathbf{q}_+ + \mathbf{q}_- + \mathbf{q}_0)^2$$

reveals a very sharp peak in the vicinity of 550 MeV. This is interpreted as the existence of a particle, called η^0, which in this experiment decays via

$$\eta^0 \rightarrow \pi^+ + \pi^- + \pi^0$$

Subsequent measurements showed that

$$m_{\eta^0} = 548.7 \pm 0.5 \text{ MeV}$$

Other decay modes have been seen:

$$\begin{aligned} \eta^0 &\rightarrow 2\gamma && 38.6 \pm 2.7\% \\ &\rightarrow 3\pi^0 \text{ or } 2\pi^0\gamma && 30.8 \pm 2.3\% \\ &\rightarrow \pi^+\pi^-\pi^0 && 25.0 \pm 1.6\% \\ &\rightarrow \pi^+\pi^-\gamma && 5.5 \pm 1.2\% \end{aligned}$$

The $\eta^0 \rightarrow 2\gamma$ decay, like that of the π^0 is clearly of electromagnetic origin. Since the other decay modes have comparable rates, they too must be electromagnetic (involving a virtual photon in some cases). Thus we might expect the lifetime to be of the same general order of magnitude as that of the π^0.

[1] A. Bohr, *Nucl. Phys.* **10,** 486 (1959); S. Bilenky, *Soviet Phys. JETP* **35,** 827 (1958).
[2] A. Pevsner, R. Kraemer, M. Nussbaum, C. Richardson, P. Schlein, R. Strand, T. Toohig, M. Block, A. Engler, R. Gessaroli, and C. Meltzer, *Phys. Rev. Letters* **7,** 421 (1961).

The decay mode

$$\eta^0 \to 2\gamma$$

allows us to conclude, as in the case of the π^0, that the spin cannot be 1. Furthermore, since by definition a one-photon state is odd under charge conjugation, the two photon decay of the η^0 shows that it, like the π^0, is even under charge conjugation, from which we can conclude that in the decay

$$\eta^0 \to \pi^0 + \pi^+ + \pi^-$$

the $(\pi^+\pi^-)$ state must be even under charge conjugation.[1] Because there

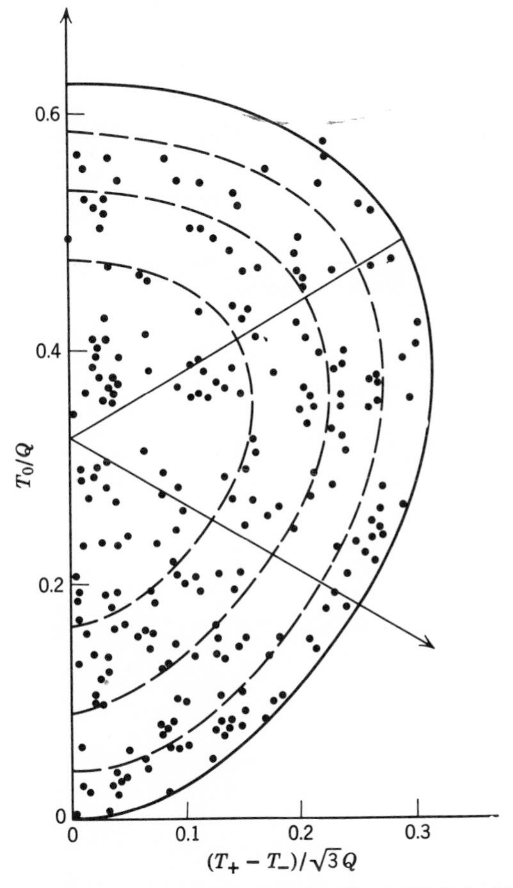

Fig. 15.4. Dalitz plot for $\eta^0 \to \pi^+ + \pi^0 + \pi^-$. The T's are labeled with the sign of the charge of the pion. [(From Alff et al., *Phys. Rev. Letters* **9**, 325 (1962)].

[1] An asymmetry in the $\pi^+\pi^-$ distribution would be evidence of a violation of charge conjugation invariance in the interaction responsible for the decay. See J. Bernstein, G. Feinberg and T. D. Lee, *Phys. Rev.* **139**, B1650 (1965).

is no distinction between charge conjugation and exchange for this pair of particles, this implies that the angular momentum of the $\pi^+\pi^-$ pair must be even. The same situation obtains in the K^+-decay, so that the charged mesons play the same role as the two positive mesons in the K-analysis. The Dalitz plot, shown in Fig. 15.4, again indicates no suppression of events in any of the regions, and we conclude that *the spin and parity of the η^0 are* 0^-.[1]

We see that there exist eight particles having the same spin and parity. We shall later study the question of whether there is some underlying symmetry scheme that treats these particles as members of a single family. Their unification is certainly not obvious, since their masses are so very different. This is not the case for the triplet of pions, π^+, π^0, π^-, and the two doublets of K mesons, K^+, K^0, and $\overline{K^0}$, K^-, which differ little in mass. The relationship between these particles, as well as the sets (p, n), $(\Sigma^+, \Sigma^0, \Sigma^-)$, and (Ξ^0, Ξ^-) which also split up into multiplets having almost the same mass, is the subject matter of Chapter 16.

[1] Since the interaction responsible for the η^0 decay is not weak, we may use parity conservation to determine the intrinsic parity of the η^0.

16

Charge Independence and Strangeness

The study of nuclear forces between protons and neutrons indicated, as early as the 1930's, that the proton-proton forces, the neutron-neutron forces, and the neutron-proton forces in the same state are more or less the same. For example, the binding energies of mirror nuclei, H^3 and He^3 (*nnp* and *ppn*) differ by an amount which can be accounted for by the difference in the charges of the constitutent particles, i.e., by the Coulomb energy. This indicates that there is an approximate equality in the potentials

$$V_{pp} = V_{nn} \tag{16.1}$$

Similarly, an analysis of the nucleon-nucleon scattering parameters at low energies suggested that

$$V_{np} = V_{pp} \tag{16.2}$$

in the same states.[1]

These observations led to the idea that the interactions between the nucleons must be such that, if we could somehow "switch off" the charge of the proton, there would be no way of distinguishing between the proton and the neutron. This carried with it the conjecture that if the proton charge were switched off, the difference in masses between the proton and the neutron, the difference in scattering lengths, etc., would disappear. The proton and the neutron could thus be considered as two states of a single entity, the nucleon.

The proposition that the nuclear forces do not distinguish between these two states, i.e., they are *charge independent*, is best explained by the existence of a symmetry in the theory. In analogy with our discussion of Lorentz invariance in Chapter 1, we will require that there exist a set of unitary transformations U in the Hilbert space of state vectors, such that the states Φ and $U\Phi$ describe the same phenomena when only the strong

[1] The proviso that the states be the same is required by the fact that the states symmetric under exchange ($^3S_1, {}^1P_0, \ldots$) cannot occur for the *pp* and the *nn* system.

interactions are taken into account. The nucleon field operators will transform according to

$$U\begin{pmatrix}\psi_p(x)\\ \psi_n(x)\end{pmatrix}U^{-1} = \begin{pmatrix}u_{11} & u_{12}\\ u_{21} & u_{22}\end{pmatrix}\begin{pmatrix}\psi_p(x)\\ \psi_n(x)\end{pmatrix} \equiv \mathcal{U}\begin{pmatrix}\psi_p(x)\\ \psi_n(x)\end{pmatrix} \tag{16.3}$$

The preservation of the commutation relations requires that the 2×2 matrix $\mathcal{U}$ be unitary. Such a 2×2 unitary matrix is characterized by four parameters; when the common phase factor is taken out, we have three parameters, and a conventional way of writing a general form for U is (omitting the phase factor)

$$\mathcal{U} = e^{\frac{1}{2}i\boldsymbol{\alpha}\cdot\boldsymbol{\tau}} \tag{16.4}$$

where the three traceless hermitian canonical 2×2 matrices

$$\tau_1 = \begin{pmatrix}0 & 1\\ 1 & 0\end{pmatrix} \qquad \tau_2 = \begin{pmatrix}0 & -i\\ i & 0\end{pmatrix} \qquad \tau_3 = \begin{pmatrix}1 & 0\\ 0 & -1\end{pmatrix} \tag{16.5}$$

are just the Pauli spin matrices. The close similarity between (16.3) and the way in which we would express rotational invariance[1] suggests a way of characterizing the invariance. We shall speak of an invariance under rotations in an "internal" space (internal, since this is not a spatial rotation). Equation (16.3) is consistent with the description of the nucleon as a spinor, or, to avoid confusion, an *isospinor*.

There will be an analog of the angular momentum; we shall call it *i-spin*, the rotational invariance implies that *i-spin is conserved.* If we denote the *i*-spin by the letter **T**,

$$U = e^{i\boldsymbol{\alpha}\cdot\mathbf{T}} \tag{16.6}$$

For an infinitesmal rotation, (16.3) reads

$$\psi(x) + i\alpha_i[T_i, \psi(x)] = \psi(x) + \tfrac{1}{2}i\alpha_i\tau_i\,\psi(x)$$

i.e.

$$[T_i, \psi(x)] = \tfrac{1}{2}\tau_i\psi(x) \tag{16.7}$$

[We represent $\begin{pmatrix}\psi_p(x)\\ \psi_n(x)\end{pmatrix}$ by $\psi(x)$.] It is easily checked that these relations are satisfied by

$$\mathbf{T} = \frac{1}{2}\int d^3x\, \psi^+(x)\boldsymbol{\tau}\,\psi(x) \tag{16.8}$$

Note that

$$T_3 = \frac{1}{2}\int d^3x(\psi_p^{\,+}(x)\,\psi_p(x) - \psi_n^{\,+}(x)\,\psi_n(x)) \tag{16.9}$$

[1] The major difference is that in (16.3) the spatial coordinates are not involved.

Thus the charge operator for nucleons Q may be written as

$$Q = \int d^3x \, \psi_p^+(x) \, \psi_p(x) = \int d^3x \, \psi(x) \frac{1 + \tau_3}{2} \psi(x) \tag{16.10}$$

We may introduce the baryon-number operator N_B by the definition

$$N_B = \int d^3x(\psi_p^+(x) \, \psi_p(x) + \psi_n^+(x) \, \psi_n(x) + \cdots) \tag{16.11}$$

The extra terms, not written down, are similar contributions from other fields carrying baryon number. Thus, if we consider only protons and neutrons,

$$Q = \tfrac{1}{2}N_B + T_3 \tag{16.12}$$

It follows from the easily derived commutation relations

$$[T_i, T_j] = ie_{ijk}T_k \tag{16.13}$$

that

$$[Q, T_i] \neq 0 \qquad i = 1, 2$$

so that charge violates *i*-spin conservation.

The formal identity of *i*-spin and ordinary spin makes available to us the whole machinery of the theory of angular momentum, and we shall use the same terminology to emphasize the parallelism. Thus we speak of the nucleon as yielding the simplest (spinor) representation of the transformation group. A product of two nucleon states is still a representation but not an irreducible one. The product of two spin $\frac{1}{2}$ objects decomposes into a spin 1 and a spin 0 combination, both of which transform among themselves.[1] The combinations characterized by $T = 1$ are

$$\begin{gathered} \psi_p^{(1)}\psi_p^{(2)} \\ \frac{1}{\sqrt{2}}(\psi_p^{(1)}\psi_n^{(2)} + \psi_n^{(1)}\psi_p^{(2)}) \\ \psi_n^{(1)}\psi_n^{(2)} \end{gathered} \tag{16.14}$$

and the $T = 0$ combination is

$$\frac{1}{\sqrt{2}}(\psi_p^{(1)}\psi_n^{(2)} - \psi_n^{(1)}\psi_p^{(2)}) \tag{16.15}$$

The same reduction of a product of a nucleon state and an antinucleon state can be expected. There is a difference, however, in the forms of the $T = 1$ and $T = 0$ combinations. This arises because the antinucleons

[1] This is what we mean by "irreducibility."

are created by the operators $\bar{\psi}_p\mathcal{C}$ and $\bar{\psi}_n\mathcal{C}$. Thus

$$\psi' = \mathcal{U}\psi$$

implies that

$$\bar{\psi}' = \mathcal{U}^*\bar{\psi} \tag{16.16}$$

Now, it is a special property of 2×2 unitary unimodular (i.e., such that $\det \mathcal{U} = 1$) matrices that $\mathcal{U}^*$ and $\mathcal{U}$ are related by a similarity transformation. In fact, with the standard form (16.4) it is easily seen that

$$\begin{aligned}\mathcal{U}^* = e^{-\frac{1}{2}i\boldsymbol{\alpha}\cdot\boldsymbol{\tau}^*} &= (-i\tau_2)e^{\frac{1}{2}i\boldsymbol{\alpha}\cdot\boldsymbol{\tau}}(i\tau_2)\\ &= (-i\tau_2)\mathcal{U}(i\tau_2)\end{aligned} \tag{16.17}$$

Thus if we write for the antinucleon

$$\chi = -i\tau_2(\bar{\psi}\mathcal{C}) = \begin{pmatrix} -\bar{\psi}_n\mathcal{C} \\ \bar{\psi}_p\mathcal{C} \end{pmatrix} = \begin{pmatrix} -\psi_{\bar{n}} \\ \psi_{\bar{p}} \end{pmatrix} \tag{16.18}$$

we get the transformation law

$$\chi' = \mathcal{U}\chi$$

Hence for a nucleon-antinucleon system, the $T = 1$ combinations are

$$\begin{gathered} -\psi_p\psi_{\bar{n}} \\ \frac{1}{\sqrt{2}}(\psi_p\psi_{\bar{p}} - \psi_n\psi_{\bar{n}}) \\ \psi_n\psi_{\bar{p}} \end{gathered} \tag{16.19}$$

and the $T = 0$ combination is

$$\frac{1}{\sqrt{2}}(\psi_p\psi_{\bar{p}} + \psi_n\psi_{\bar{n}}) \tag{16.20}$$

By working with the χ rather than the direct antinucleon doublet, we do not distinguish between particles and antiparticles as far as the *i*-spin transformation properties are concerned and thus need not modify the parallelism with the theory of angular momentum when antiparticles are involved.

The nucleon, as an entity, is represented by an eight-component operator, with four components describing the proton state and four the neutron state. Since we quantize all spin $\frac{1}{2}$ operators with anticommutation rules, we can introduce an extended Pauli principle: no two nucleons can be in the same state, where now the state has an additional label, the z component of the *i*-spin. Thus a two-nucleon wavefunction must be totally antisymmetric; for the deuteron which is a spatially symmetric two-nucleon state, the *i*-spin dependence must be such as to make the state antisymmetric, so that the deuteron must be a $T = 0$ state.

If we take account of our belief that the exchange of pions is (largely) responsible for the forces between nucleons, we must make the more fundamental assertion that the pion-nucleon interaction is charge independent. The large matrix element for the reaction

$$N \rightarrow N + \pi$$

which lies at the heart of the Yukawa hypothesis, implies that the pion must have $T = 1$ or 0. Since there are three pions, π^+, π^0 and π^-, and these have almost the same masses as well as the same spins and parities, it is natural to assign the value of $T = 1$ to the pion. The three pions thus have the same transformation properties as the three states (16.19). From the point of view of the internal symmetry, we may view the pions as "bound states" of a nucleon and an antinucleon, and this view is sometimes useful. It is not, however, necessary to take this view of the pion. It is merely required that the one-pion states form an irreducible representation, characterized by $T = 1$, of our symmetry.

The i-spin operator can be constructed from the requirement that the pion field operators transform as vectors under rotations in i-spin space

$$[T_i, \phi_j(x)] = ie_{ijk}\,\phi_k(x) \tag{16.21}$$

It is easily checked that the operator

$$\mathbf{T} = i\int d^3x\ \pi_i(x)(\mathbf{t})_{ij}\ \phi_j(x) \tag{16.22}$$

with

$$(t_a)_{ij} = -ie_{ija} \tag{16.23}$$

satisfies (16.21) (we use the canonical commutation rules), as well as

$$[T_i, T_j] = ie_{ijk}T_k$$

The $T = 1$ version of the τ-matrices is according to (16.23)

$$t_1 = \begin{pmatrix} 0 & 0 & 0 \\ 0 & 0 & -i \\ 0 & i & 0 \end{pmatrix} \qquad t_2 = \begin{pmatrix} 0 & 0 & i \\ 0 & 0 & 0 \\ -i & 0 & 0 \end{pmatrix} \qquad t_3 = \begin{pmatrix} 0 & -i & 0 \\ i & 0 & 0 \\ 0 & 0 & 0 \end{pmatrix}$$

It is often more convenient to work with a representation in which

$$t_3' = \begin{pmatrix} 1 & 0 & 0 \\ 0 & 0 & 0 \\ 0 & 0 & -1 \end{pmatrix} \tag{16.24}$$

Since the t_a must satisfy the same commutation relations as the i-spin operators T_a, which with

$$T_\pm = (T_1 \pm iT_2) \tag{16.25}$$

take the form

$$[T_3, T_\pm] = \pm T_\pm$$
$$[T_+, T_-] = 2T_3 \tag{16.26}$$

it is easy to find the 3×3 matrices t_1' and t_2' which go with t_3'. They are

$$t_1' = \frac{1}{\sqrt{2}}\begin{pmatrix} 0 & 1 & 0 \\ 1 & 0 & 1 \\ 0 & 1 & 0 \end{pmatrix} \qquad t_2' = \frac{1}{\sqrt{2}}\begin{pmatrix} 0 & -i & 0 \\ i & 0 & -i \\ 0 & i & 0 \end{pmatrix} \tag{16.27}$$

In (16.21) we introduced the pion field operators $\phi_i(x)$. It remains to establish which particular combinations of these, when acting on the vacuum state, create the π^+, π^0, and π^-. From (16.21) it follows that

$$\left[T_3, \frac{\phi_1 \pm i\phi_2}{\sqrt{2}}\right] = \pm\frac{\phi_1 \pm i\phi_2}{\sqrt{2}} \tag{16.28}$$

Thus, if we assign $T_3 = 1$ to the π^+, $T_3 = 0$ to the π^0, and $T_3 = -1$ to the π^-, we may identify

$$\frac{\phi_1 \pm i\phi_2}{\sqrt{2}}\Psi_0 = \alpha_\pm \Psi_{\pi^\pm} \tag{16.29}$$

The phase factor $\alpha_\pm$ will be chosen for convenience later. Incidentally, since $[T_3, \phi_3] = 0$ it is consistent to choose

$$\phi_3\Psi_0 = \Psi_{\pi^0} \tag{16.30}$$

If we now use the relation

$$[T_\pm, \phi_3] = \mp\sqrt{2}\,\frac{\phi_1 \pm i\phi_2}{\sqrt{2}} \tag{16.31}$$

which also follows from (16.21) and apply it to the vacuum state, we get

$$T_\pm\phi_3\Psi_0 = T_\pm\Psi_{\pi^0} = \mp\sqrt{2}\,\alpha_\pm\Psi_{\pi^\pm} \tag{16.32}$$

With the choice

$$\alpha_\pm = \mp 1 \tag{16.33}$$

we get

$$T_\pm\Psi_{\pi^0} = \sqrt{2}\,\Psi_{\pi^\pm} \tag{16.34}$$

with the coefficient positive for both cases. This is the standard convention used by physicists (the Condon-Shortley convention), according to which tables of Wigner coefficients are constructed. Equation (16.34) is a special case of the relation

$$T_{\pm}\Psi_{T',T_3'} = ((T' \mp T_3')(T' \pm T_3' + 1))^{1/2}\Psi_{T',T_3'\pm 1} \qquad (16.35)$$

which holds for any i-spin, and it shows that the operators $T_{\pm}$ are operators which raise (lower) the z component of i-spin of the state they act on, by one unit. This observation follows immediately from the first of the relations (16.26), which when applied to a state of given T' and T_3' yield

$$\begin{aligned} T_3 T_{\pm}\Psi_{T',T_3'} &= (T_{\pm}T_3 \pm T_{\pm})\Psi_{T',T_3'} \\ &= (T_3' \pm 1)T_{\pm}\Psi_{T',T_3'} \end{aligned} \qquad (16.36)$$

With the pion field $\phi_i(x)$ transforming as a vector under i-spin rotations, we can write down invariant interaction Lagrangians. The simplest form, which also takes into account the parity of the pion is

$$\mathcal{L}_1(x) = ig\bar{\psi}(x)\gamma_5\boldsymbol{\tau}\psi(x)\boldsymbol{\phi}(x) \qquad (16.37)$$

When this is written out explicitly, it takes the form

$$ig(\sqrt{2}\,\bar{\psi}_p\gamma_5\psi_n)\frac{\phi_1 - i\phi_2}{\sqrt{2}} + ig(\sqrt{2}\,\bar{\psi}_n\gamma_5\psi_p)\frac{\phi_1 + i\phi_2}{\sqrt{2}} + ig(\bar{\psi}_p\gamma_5\psi_p - \bar{\psi}_n\gamma_5\psi_n)\phi_3 \qquad (16.38)$$

The near equality of the masses of particles belonging to a multiplet may be considered to be a test of the validity of charge independence. For the pion,

$$\frac{m_{\pi^\pm} - m_{\pi^0}}{\bar{m}_\pi} \approx 3\%$$

which is considerably larger than the corresponding quantity for the nucleon. Presumably in the latter case, some cancelation takes place. The 3% figure seems to be the limit within which predictions made on the basis of exact symmetry may be expected to hold. Thus i-spin conservation predicts that for the reactions

$$p + d \begin{array}{l} \nearrow \pi^0 + \mathrm{He}^3 \\ \searrow \pi^+ + \mathrm{H}^3 \end{array}$$

the ratio

$$R = \frac{\sigma(p + d \to \pi^+ + \mathrm{H}^3)}{\sigma(p + d \to \pi^0 + \mathrm{He}^3)}$$

should be 2. The reason is that the deuteron has $T = 0$, so that the initial state has $T = \frac{1}{2}$. In the final state, the pion has $T = 1$ and (He^3, H^3)

form an isodoublet. We may use the rules of addition of angular momentum to decompose the final states into eigenstates of i-spin. We find that

$$\Psi_{\pi^0\text{He}^3} = \sqrt{\tfrac{2}{3}}\,\Psi_{\frac{3}{2}\frac{1}{2}} + \sqrt{\tfrac{1}{3}}\,\Psi_{\frac{1}{2}\frac{1}{2}}$$

$$\Psi_{\pi^+\text{H}_3} = \sqrt{\tfrac{1}{3}}\,\Psi_{\frac{3}{2}\frac{1}{2}} - \sqrt{\tfrac{2}{3}}\,\Psi_{\frac{1}{2}\frac{1}{2}} \qquad (16.39)$$

If the i-spin is conserved, only the $T = \frac{1}{2}$ final state is accessible. We find that the matrix elements of the scattering operator are

$$(\Psi_{\pi^0\text{He}^3}, S\Psi_{pd}) = \sqrt{\tfrac{1}{3}}\,M_{T=\frac{1}{2}}$$

$$(\Psi_{\pi^+\text{H}^3}, S\Psi_{pd}) = \sqrt{\tfrac{2}{3}}\,M_{T=\frac{1}{2}} \qquad (16.40)$$

whence the prediction. Coulomb corrections and mass difference effects have been estimated to change the prediction upward by about 4%. Experiments yield the ratio

$$R = 1.91 \pm 0.25 \text{ (Crewe et al.)}^{1}$$
$$= 2.26 \pm 0.11 \text{ (Harting et al.)}^{2}$$

confirming the conservation law to the expected degree.

Another confirmation of charge independence comes from the unsuccessful search for evidence for the reaction

$$d + d \rightarrow \text{He}^4 + \pi^0$$

which, since He^4, like the deuteron, is an isosinglet, is forbidden by charge independence. Akimov et al.[3] find that

$$\sigma(d + d \rightarrow \text{He}^4 + \pi^0) < 1.6 \times 10^{-32}\ \text{cm}^2$$

whereas the electromagnetic reaction

$$d + d \rightarrow \text{He}^4 + \gamma$$

does occur, with a cross section larger than 0.8×10^{-32} cm^2.

The extension of charge independence to the particles other than pions and nucleons, particles beginning to appear more and more frequently in experimental reports in the early 1950's, forms a fascinating chapter in the development of elementary particle physics. The first of the new particles

[1] A. V. Crewe, B. Ledley, E. Lillethus, S. M. Markowitz, and C. Rey, *Phys. Rev.* **118,** 1091 (1960).

[2] D. Harting, J. C. Kluyver, A. Kusumegi, R. Rigopoulos, A. M. Sachs, G. Tibell, G. Vanderhaege, and G. Weber, *Phys. Rev.* **119,** 1716 (1960).

[3] Yu. K. Akimov, O. V. Savchenko, and L. M. Soroko, *Proceedings of the 1960 Annual International Conference on High Energy Physics at Rochester*, Interscience Publishers, New York (1960).

to be discovered was the Λ^0. It was produced quite copiously (cross sections of the order of millibarns), and it decayed very slowly, with a lifetime of the order of 10^{-10} sec. This was a puzzle, since, if the production mechanism was

$$p + \pi^- \rightarrow \Lambda^0 + \text{pions}$$

the matrix element for

$$\Lambda^0 \rightarrow p + \pi^-$$

should be large enough to lead to a decay in less than 10^{-22} sec. A possible explanation that the Λ^0 was a high angular momentum ($J \geqslant \frac{13}{2}$) excited state of $p\pi^-$, whose decay is slowed down because of the enormous centrifugal barrier, foundered on the observation of the first hyperfragments, that showed a Λ^0 lived as long inside a nucleus as it does free, rather than undergoing instant internal conversion.

It was Pais who came up with the correct solution of this problem. Pais suggested that the Λ^0 had to be produced in association with another "new" particle, in order to be produced copiously, whereas in the decay, with no partner present, the Λ^0 had to decay slowly. Indeed, the reaction

$$\begin{array}{l} \pi^- + p \rightarrow \Lambda^0 + K^0 \\ \qquad\qquad\quad | \qquad \hookrightarrow \pi^+ + \pi^- \\ \qquad\qquad\quad \hookrightarrow p + \pi^- \end{array}$$

was soon observed. In due time the Σ^+ and the Σ^-, as well as the K^+ and the K^-, were discovered through the observation of certain of their decay modes

$$\Sigma^+ \rightarrow n + \pi^+$$
$$\Sigma^- = n + \pi^-$$
$$K^\pm \rightarrow \pi^\pm + \pi^0$$

For these particles, too, it was necessary to invoke Pais' associated production to avoid gross violations of the principle of detailed balance. Theoretical attempts to assign to the new particles a multiplicative quantum number ("internal parity") such that the pions and nucleons were assigned $+1$ and the new particles -1, with a consequent production of the new particles *modulo* 2 in ordinary reactions, had to be rejected with the discovery of yet another new particle Ξ^-, which decayed with a lifetime of the order of 10^{-10} sec according to

$$\Xi^- \rightarrow \Lambda^0 + \pi^-$$

but was not seen to decay into a neutron and a π^-. Thus no assignment of "internal parity" would work for the Ξ^-.

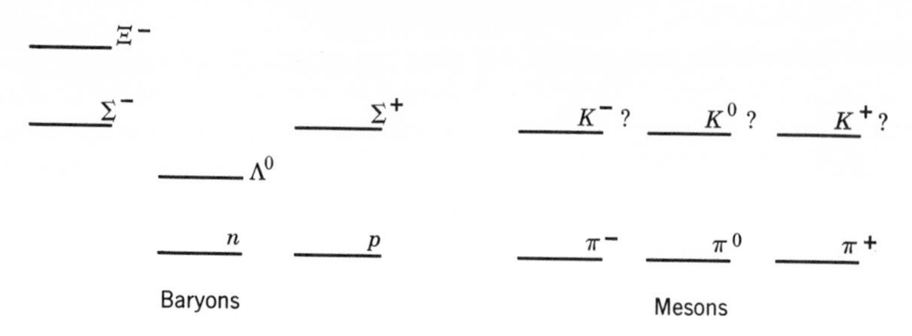

Fig. 16.1. The elementary particles, as known in 1953.

In 1953 Gell-Mann,[1] and independently Nishijima,[2] brought order into the rather jumbled situation by extending the notion of charge independence to the new particles. Recall that the pions form an *i*-spin triplet, with the charge of the individual members being given by

$$Q = T_3$$

For the nucleon we have already obtained

$$Q = \tfrac{1}{2}N_B + T_3$$

with N_B the baryon number. These two relations are consistent, since the pion carries no baryon number. In both cases the mass differences between members of a multiplet are of the order of a few MeV. If we were to look at the spectrum of the known particles, as it was known in 1953 (Fig. 16.1), we would find it difficult to treat Σ^+, Λ^0, and Σ^- as a triplet. Rather, in the absence of charged partners, the Λ^0 should be viewed as a $T = 0$ state. The relation between charge and T_3 no longer holds, and it becomes necessary to introduce a new quantity S, called *strangeness*, into the relation. With

$$Q = \tfrac{1}{2}N_B + \tfrac{1}{2}S + T_3 \tag{16.41}$$

we assign $S = 0$ to the pions and nucleons and $S = -1$ to the Λ^0. The decay

$$\Lambda^0 \rightarrow p + \pi^-$$

thus involves a change in strangeness, $|\Delta S| = 1$, and we shall associate strangeness nonconservation with the slowness of the decay. Since associated production is fast, the K mesons produced with the Λ^0 must have $S = +1$. Incidentally, this scheme forbids the reaction

$$n + p \rightarrow \Lambda^0 + \Lambda^0 + \pi^+$$

which was not forbidden by the general notion of associated production. The reaction has never been seen.

[1] M. Gell-Mann, *Phys. Rev.* **92,** 833 (1953).
[2] T. Nakano and K. Nishijima, *Progr. Theoret. Phys.* **10,** 581 (1953).

The Σ^+ and Σ^- have almost the same mass, and, if we exclude for the time being particles of charge larger than 1 in magnitude, Σ^+ and Σ^- must be members of a triplet. With this assignment we deduce from (16.41) that $S = -1$ for the triplet and that there must be a neutral partner Σ^0. The strangeness assignment explains why the Σ^0 was not seen: the Σ^0 may decay, with strangeness conservation, into a Λ^0, according to

$$\Sigma^0 \rightarrow \Lambda^0 + \gamma$$

The Σ^0 was subsequently looked for and found.

As far as the assignment of strangeness to the Ξ^- was concerned, a hint is obtained from the fact that the decay is

$$\Xi^- \rightarrow \Lambda^0 + \pi^-$$

If this is to be a $|\Delta S| = 1$ process, the Ξ^- has to have $S = -2$. This might provide an explanation for the absence of

$$\Xi^- \rightarrow n + \pi^-$$

since that would be a $|\Delta S| = 2$ process. With $S = -2$ the Ξ^- has $T_3 = -\frac{1}{2}$, and if we again exclude multiply charged particles, it will have only one partner, the Ξ^0, which must decay according to

$$\Xi^0 \rightarrow \Lambda^0 + \pi^0$$

The Gell-Mann—Nishijima scheme was soon confirmed by the observation of Ξ^- production in association with two K^+ particles and the discovery of the Ξ^0.

If we turn to the K-mesons, with $N_B = 0$, we find that for a K^+ with $S = 1$, the T_3-value must be $\frac{1}{2}$, so that we would expect the partner of the K^+ to be K^0. The antiparticle of the K^+, the K^-, will have strangeness $S = -1$, and it will have to have a partner we call $\overline{K^0}$. The scheme thus predicts two degenerate particles, K^0 and $\overline{K^0}$. A consequence of these strangeness assignments is that at low energies there should be a large excess of K^+ over K^- produced. The reason for this is that K^+ can be produced in processes like

$$p + p \rightarrow p + \Lambda^0 + K^+$$

with a threshold of about 1.5 BeV, whereas to produce K^--mesons and conserve S we must make a pair of K's, as in

$$p + p \rightarrow p + p + K^+ + K^-$$

with a threshold of about 2.5 BeV. Experiments at Brookhaven in 1956 showed this effect very clearly.

The idea of an additive quantum number, strangeness, conserved in the strong and electromagnetic interactions (as seen in associated photoproduction $\gamma + p \to \Lambda^0 + K^+$) has passed many tests and is undoubtedly correct. Since the weak interactions, responsible for the decays of the strange particles, do not conserve strangeness, we are faced here, as in the case of *i*-spin, with an imperfect symmetry. We shall see manifestations of this over and over again, when we discuss parity nonconservation in the weak interactions and when we discuss the higher symmetry that is now believed to unify the baryons and the pseudoscalar mesons among themselves.

We conclude this chapter with two remarks: one is a practical one and has to do with the fact that *i*-spin conservation allows us to introduce a useful quantum number called the *G-parity*,[1] defined by

$$G = Ce^{i\pi T_2} \tag{16.42}$$

where C is the charge conjugation operator. Under C, the pion field transforms as follows

$$C\phi C^{-1} = \phi^* \qquad C\phi_3 C^{-1} = \phi_3 \tag{16.43}$$

so that with $\phi = \dfrac{1}{\sqrt{2}}(\phi_1 + i\phi_2)$ we get

$$C\begin{pmatrix}\phi_1\\ \phi_2\\ \phi_3\end{pmatrix}C^{-1} = \begin{pmatrix}\phi_1\\ -\phi_2\\ \phi_3\end{pmatrix}$$

On the other hand

$$e^{i\pi T_2}\begin{pmatrix}\phi_1\\ \phi_2\\ \phi_3\end{pmatrix}e^{-i\pi T_2} = \begin{pmatrix}-\phi_1\\ \phi_2\\ -\phi_3\end{pmatrix}$$

so that

$$G\phi_i G^{-1} = -\phi_i \tag{16.44}$$

From this it follows that a state with an even number of pions cannot transform into a state with an odd number of pions, a result reminiscent of Furry's theorem in quantum electrodynamics. Since C is good both in the strong and the electromagnetic interactions, a violation of G invariance in a process that is not weak implies that the *i*-spin is not conserved, i.e., the process took place through the electromagnetic coupling. An example is the decay

$$\eta^0 \to \pi^+ + \pi^- + \pi^0 \tag{16.45}$$

[1] C. Goebel, *Phys. Rev.* **103,** 258 (1956); T. D. Lee and C. N. Yang, *Nuovo Cimento* **3,** 749 (1956).

Since the η^0 has no charged partners, we take it to be a $T = 0$ particle, and because it undergoes the decay

$$\eta^0 \to 2\gamma$$

it follows that

$$C\eta^0 C^{-1} = \eta^0$$

Hence

$$G\eta^0 G^{-1} = \eta^0$$

On the other hand, the final state in the reaction (16.45) has $G = -1$. Thus the decay must be electromagnetic. Since no photon is observed, the process must be made possible through the emission and reabsorption of a virtual photon, in the language of Feynman graphs. Thus the matrix element for the decay is reduced by a factor $\sim \dfrac{\alpha}{\pi}$ compared to a strong matrix element.

The second remark has to do with the observation that if we introduce the new quantum number *hypercharge* Y instead of strangeness, with

$$Y = \tfrac{1}{2}(N_B + S) \tag{16.46}$$

then there is a remarkable parallelism between the baryon states and the pseudoscalar meson states.

Y	T	Baryons	Mesons
$+1$	$\frac{1}{2}$	p, n	K^+, K^0
0	1	$\Sigma^+\Sigma^0\Sigma^-$	$\pi^+\pi^0\pi^-$
0	0	Λ^0	η^0
-1	$\frac{1}{2}$	Ξ^0, Ξ^-	$\overline{K^0}, K^-$

In Chapter 16 we pursue the idea of internal symmetry and discuss unitary symmetry, which counts among its many attractive features an explanation of the foregoing parallelism.

PROBLEMS

1. A new particle is observed to decay via the mode

$$Z_0 \to 3\pi^0$$

The mass of the Z_0 is approximately 1200 MeV, and the width for this decay mode is of the order of 20 MeV.
(a) What can you say about the *i*-spin of the Z_0?
(b) What are some of its possible values of spin and parity?

2. Some time later the decay mode

$$Z_0 \to 2\pi^0$$

is observed. It is *not* a weak decay. What is the simplest permitted spin and parity assignment for the Z_0? If there were no multiply charged Z particles (e.g., Z_{++}), what would you, *very roughly*, estimate the partial width of $Z_0 \to 2\pi^0$ to be? (Hint: The amplitude for a charge-independence violating process must involve electromagnetism, and if no real photon is emitted, the matrix element must involve the emission and reabsorption of a virtual photon, so that the amplitude is suppressed by a factor of the order of $\alpha = \frac{1}{137}$.)

3. Express all the possible pion-pion scattering amplitudes in terms of the amplitudes A_2, A_1, and A_0, where the subscripts denote the total (conserved) i-spin in the initial (and final) state.

4. There is good evidence that in the capture of antiprotons by protons the annihilation proceeds with 99% probability from the two S states, $^1S_0$ and $^3S_1$. Use the fact that i-spin is conserved, and invariance under C to construct a table showing which of the annihilation modes $p + \bar{p} \to n\pi$ $(n \leq 4)$ are allowed for the four possible initial states (two singlets with $T = 0, 1$ and the two corresponding triplets).

17

Unitary Symmetry

At the conclusion of the last chapter we drew attention to a certain pattern of similarity between the eight baryons and the eight pseudoscalar mesons. Had the η^0 been discovered five years earlier, this similarity might have served as a clue in the search for an underlying symmetry; as it was, the clue came from elsewhere. The search for such clues was one of the major occupations of theoretical physicists in the latter part of the 1950's. It was occasioned by the belief that there must be an underlying symmetry which would relate the many new particles and hopefully account for the existence of hypercharge. The search was hampered by the obvious fact that the symmetry was badly violated. In contrast to charge independence, for which the electromagnetic violations are so small as not to obscure the symmetry (e.g., $M_p \cong M_n$), the separation between the very strong interactions which presumably respected the symmetry, and the medium-strong interactions which violated it (leaving only *i*-spin conservation) was far from evident. The hope that the symmetry-violating terms, although large enough to drastically destroy any sign of mass degeneracy among particles of the same spin and parity, were nevertheless weak enough that some predictions could be tested led to the construction of a series of models. The "global symmetry" model of Gell-Mann[1] and Schwinger[2] was but the first of a series, all of which failed to give much insight into the problem and also failed the test of approximate agreement with experiment.

The clue came from Sakata's[3] model. He observed that, whereas in a world of nucleons and pions only nucleons (and antinucleons) were needed to make up all particles, the existence of strangeness implied that at least one strange particle had to be added to the proton and the neutron to

[1] M. Gell-Mann, *Phys. Rev.* **106,** 1296 (1957).
[2] J. Schwinger, *Ann. Phys.* (*N.Y.*) **2,** 407 (1957).
[3] S. Sakata, *Progr. Theoret. Phys.* **16,** 686 (1956).

make up all the known particles. Examples are

$$K^+ \equiv (\overline{\Lambda}^0 p)$$
$$\Sigma^+ \equiv (\Lambda^0 p \bar{n})$$
$$\Xi^- \equiv (\Lambda^0 \Lambda^0 \bar{p}) \tag{17.1}$$

The mathematical structure of the Sakata model was studied by a number of people,[1] in the idealized version in which the nucleons and the Λ^0 had the same mass. Whereas the study of the group of transformations among the proton and the neutron led to the appearance of *i*-spin, the group of unitary transformations among three particles (p, n, Λ^0) led to a higher symmetry, not very familiar to most physicists. It is the purpose of this chapter to familiarize the reader with the structure of the symmetry and to indicate the many points of contact with *i*-spin.

We begin by considering *i*-spin again. We recall that what was assumed was the invariance of the theory under the transformations of the fundamental doublet $\psi = \begin{pmatrix} \psi_p \\ \psi_n \end{pmatrix}$,

$$\psi \rightarrow \psi' = e^{\frac{1}{2} i \boldsymbol{\alpha} \cdot \boldsymbol{\tau}} \psi \tag{17.2}$$

These transformations form all unitary, unimodular, 2×2 matrix transformations. This group of transformations is conventionally called $SU(2)$. When $\boldsymbol{\alpha}$ is infinitesimal, we have

$$\psi' = (1 + \tfrac{1}{2} i \boldsymbol{\alpha} \cdot \boldsymbol{\tau}) \psi \tag{17.3}$$

The quantities $\frac{1}{2}\boldsymbol{\tau}$ are called the *generators of the infinitesimal transformations*, and we have seen that they satisfy the commutation relations

$$[\tfrac{1}{2}\tau_i, \tfrac{1}{2}\tau_j] = i e_{ijk} (\tfrac{1}{2}\tau_k) \tag{17.4}$$

These commutation relations can be abstracted from the special 2×2 representations of the generators. If we define the generators abstractly by giving their commutation relations

$$[T_i, T_j] = i e_{ijk} T_k \tag{17.5}$$

(the totally antisymmetric e_{ijk} are called *structure constants*), the complete exploration of *i*-spin really amounts to finding all higher dimensional matrices which satisfy (17.5). In the case of *i*-spin (or angular momentum) these matrices are just representations of the *i*-spin operators. The multiplets which are infinitesimally transformed by these matrices,

$$\chi' = (1 + i \boldsymbol{\alpha} \cdot \mathbf{T}) \chi \tag{17.6}$$

[1] References may be found in S. Gasiorowicz and S. L. Glashow, *Advances in Theoretical Physics*, Vol. II, Academic Press 1966. See also bibliography.

are then associated with the particles, such as π^+, π^0, π^- or Σ^+, Σ^0, Σ^-, for 3×3 matrix representations of the T_i. We can think of the T_i as operators acting on certain "states" χ.[1] It is the dimensionality and character of these *states* that are of primary interest to us.

A particularly simple, and well-known, technique for studying the states is by judicious use of the commutation relations[2] (17.5). With the introduction of

$$T_\pm = T_1 \pm iT_2$$

these may be written as

$$\begin{aligned}[T_3, T_\pm] &= \pm T_\pm \\ [T_+, T_-] &= 2T_3\end{aligned} \qquad (17.7)$$

It follows from these relations that the quantity C, defined by

$$C = \tfrac{1}{2}(T_+T_- + T_-T_+) + T_3^2 \equiv T^2 \qquad (17.8)$$

commutes with all three generators. It is therefore called an invariant operator, or a Casimir operator. We recognize it as the square of the *i*-spin. Another observation is that only one of the operators may be diagonalized; the conventional choice is T_3. By this statement we mean that the states may be labeled by the eigenvalue of T_3, which will be denoted by t_3.

It follows from (17.7) that

$$T_3T_\pm\chi_{t_3} = T_\pm(T_3 \pm 1)\chi_{t_3} = (t_3 \pm 1)T_\pm\chi_{t_3} \qquad (17.9)$$

so that given an eigenstate of T_3 we can find two others by applying the raising (T_+) and lowering (T_-) operators. Furthermore, since

$$T^2T_\pm\chi_{t_3} = T_\pm T^2\chi_{t_3} \qquad (17.10)$$

it follows that all the states generated from a particular one by repeated "raising" or "lowering" have the same value of T^2. Thus different representations may be labeled by the eigenvalue of T^2. We shall not go through the familiar procedure by which, starting with the state of highest t_3 in a given representation[3] it can be shown that (a) if the highest value of t_3 is taken to be t, there are $(2t + 1)$ states, so that t must be integral or half integral; (b) the possible values of t_3 are $t, t - 1, t - 2, \ldots, -t$; (c) the eigenvalue of T^2 is given by $t(t + 1)$; (d) there is only one state for

[1] We shall ignore for the time being that the χ represent field operators defined in the Hilbert space of all physical states and use the nomenclature "states."

[2] See, for example, M. E. Rose, *Elementary Theory of Angular Momentum*, John Wiley and Sons, New York (1957), p. 24.

[3] There must be a highest one if the representation is to have a finite number of components, and it must satisfy $T_+\chi_{\max} = 0$.

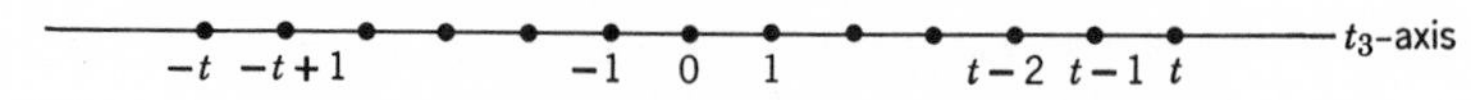

Fig. 17.1. Graphical representation of representation t.

each t_3 in a given irreducible representation (characterized by a given value of t); (e) that

$$T_{\pm}\chi_{t_3}\{t\} = [(t \mp t_3)(t \pm t_3 + 1)]^{1/2}\chi_{t_3\pm 1}\{t\} \tag{17.11}$$

and that we may choose

$$(\chi_{t_3}\{t\}, \chi_{t_3'}\{t\}) = \delta_{t_3t_3'} \tag{17.12}$$

A product of two irreducible representations $\chi\{t\}$ and $\chi\{t'\}$ will not, in general, give an irreducible representation. The decomposition of such a product is most easily seen graphically. We plot the states in a linear array, as in Fig. 17.1. For any member of the product, we have

$$T_3\chi_{t_3}\{t\}\chi_{t_3'}\{t'\} = (t_3 + t_3')\chi_{t_3}\{t\}\chi_{t_3'}\{t'\} \tag{17.13}$$

so that the sites occupied by the product states are generally multiply occupied (Fig. 17.2). The highest site is again singly occupied. The action of successive lowering operators on the highest site generates the representation $(t + t')$. The next site is doubly occupied; since one of the states has been used up by the $(t + t')$ representation, the state orthogonal to it can serve as the highest state of the next representation, which is labeled by $(t + t' - 1)$. Proceeding in this way we obtain the familiar reduction $(t + t'), (t + t' - 1), \ldots |t - t'|$.

The reason for going into so much detail on such a familiar problem is that these techniques yield the simplest approach to the much less familiar symmetry considered next, namely $SU(3)$. If we consider the fundamental fields to be a triplet (which in the Sakata model is p, n, Λ^0), we may assume invariance under the transformations

$$\chi_a \rightarrow \chi_a' = U_{ab}\chi_b \tag{17.14}$$

Here the U's are arbitrary, unitary, unimodular, 3×3 matrices. A canonical representation of such a matrix U is the form

$$U = \exp\left\{\tfrac{1}{2}i\sum_{k=1}^{8}\alpha_k\lambda_k\right\} \tag{17.15}$$

Fig. 17.2. The sites occupied by $(\chi(\frac{3}{2}))\cdot(\chi(1))$.

Here the λ_k play a role analogous to the τ_i matrices in $SU(2)$. A standard form, introduced by Gell-Mann,[1] is

$$\lambda_1 = \begin{pmatrix} 0 & 1 & 0 \\ 1 & 0 & 0 \\ 0 & 0 & 0 \end{pmatrix} \quad \lambda_2 = \begin{pmatrix} 0 & -i & 0 \\ i & 0 & 0 \\ 0 & 0 & 0 \end{pmatrix} \quad \lambda_3 = \begin{pmatrix} 1 & 0 & 0 \\ 0 & -1 & 0 \\ 0 & 0 & 0 \end{pmatrix}$$

$$\lambda_4 = \begin{pmatrix} 0 & 0 & 1 \\ 0 & 0 & 0 \\ 1 & 0 & 0 \end{pmatrix} \quad \lambda_5 = \begin{pmatrix} 0 & 0 & -i \\ 0 & 0 & 0 \\ i & 0 & 0 \end{pmatrix} \quad \lambda_6 = \begin{pmatrix} 0 & 0 & 0 \\ 0 & 0 & 1 \\ 0 & 1 & 0 \end{pmatrix}$$

$$\lambda_7 = \begin{pmatrix} 0 & 0 & 0 \\ 0 & 0 & -i \\ 0 & i & 0 \end{pmatrix} \quad \lambda_8 = \frac{1}{\sqrt{3}} \begin{pmatrix} 1 & 0 & 0 \\ 0 & 1 & 0 \\ 0 & 0 & -2 \end{pmatrix} \tag{17.16}$$

The matrices $\frac{1}{2}\lambda_i$ satisfy the commutation relations characteristic of the group, just as $\frac{1}{2}\tau_i$ satisfy the angular-momentum commutation relations. We have

$$[\tfrac{1}{2}\lambda_i, \tfrac{1}{2}\lambda_j] = if_{ijk}(\tfrac{1}{2}\lambda_k) \tag{17.17}$$

with the structure constants f_{ijk} taking on the values listed in the table given below. We can also derive the anticommutation relations, which, however, only hold for the 3×3 matrices

$$\{\lambda_i, \lambda_j\} = \tfrac{2}{3}\,\delta_{ij}\mathbf{1} + 2d_{ijk}\lambda_k \tag{17.18}$$

The f_{ijk} are antisymmetric and the d_{ijk} are symmetric under the interchange of any two indices. Their values are listed in Table 17-1.

As in the case of $SU(2)$, we define the generators by

$$F_i = \tfrac{1}{2}\lambda_i \tag{17.19}$$

which will satisfy the commutation relations

$$[F_i, F_j] = if_{ijk}F_k \tag{17.20}$$

We now proceed to generalize (17.7) by introducing the following quantities:

$$\begin{aligned} T_\pm &= F_1 \pm iF_2 \\ U_\pm &= F_6 \pm iF_7 \\ V_\pm &= F_4 \pm iF_5 \\ T_3 &= F_3 \\ Y &= \frac{2}{\sqrt{3}} F_8 \end{aligned} \tag{17.21}$$

[1] M. Gell-Mann, *Phys. Rev.* **125**, 1067 (1962).

Table 17-1

The Nonvanishing Values of f_{ijk} and d_{ijk}

(ijk)	f_{ijk}	(ijk)	d_{ijk}
123	1	118	$1/\sqrt{3}$
147	$\frac{1}{2}$	146	$\frac{1}{2}$
156	$-\frac{1}{2}$	157	$\frac{1}{2}$
246	$\frac{1}{2}$	228	$1/\sqrt{3}$
257	$\frac{1}{2}$	247	$-\frac{1}{2}$
345	$\frac{1}{2}$	256	$\frac{1}{2}$
367	$-\frac{1}{2}$	338	$1/\sqrt{3}$
458	$\sqrt{3}/2$	344	$\frac{1}{2}$
678	$\sqrt{3}/2$	355	$\frac{1}{2}$
		366	$-\frac{1}{2}$
		377	$-\frac{1}{2}$
		448	$-1/2\sqrt{3}$
		558	$-1/2\sqrt{3}$
		668	$-1/2\sqrt{3}$
		778	$-1/2\sqrt{3}$
		888	$-1/\sqrt{3}$

In terms of these, the commutation relations can easily be shown to be

$$\begin{aligned}
&[T_3, T_\pm] = \pm T_\pm \qquad [Y, T_\pm] = 0 \\
&[T_3, U_\pm] = \mp\tfrac{1}{2}U_\pm \qquad [Y, U_\pm] = \pm U_\pm \\
&[T_3, V_\pm] = \pm\tfrac{1}{2}V_\pm \qquad [Y, V_\pm] = \pm V_\pm
\end{aligned} \tag{17.22}$$

$$\begin{aligned}
&[T_+, T_-] = 2T_3 \\
&[U_+, U_-] = \tfrac{3}{2}Y - T_3 \equiv 2U_3 \\
&[V_+, V_-] = \tfrac{3}{2}Y + T_3 \equiv 2V_3
\end{aligned}$$

$$\begin{aligned}
&[T_+, V_+] = [T_+, U_-] = [U_+, V_+] = 0 \\
&[T_+, V_-] = -U_- \qquad [T_+, U_+] = V_+ \\
&[U_+, V_-] = T_- \qquad [T_3, Y] = 0
\end{aligned} \tag{17.24}$$

The unlisted commutation relations may be obtained with the use of

$$T_+ = (T_-)^+ \qquad U_+ = (U_-)^+ \qquad V_+ = (V_-)^+ \tag{17.25}$$

Our next step is to obtain the states that are transformed according to

$$\psi \rightarrow \psi' = (1 + i\boldsymbol{\alpha} \cdot \mathbf{F})\psi \tag{17.26}$$

where the F_i are represented by matrices which are not necessarily three-dimensional. It is a fact, illustrated in (17.16), that the commutation

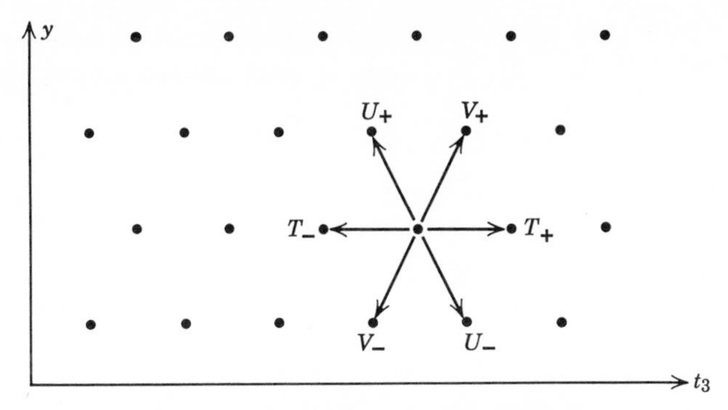

Fig. 17.3. The action of the shift operators.

relations allow only two of the eight generators to be represented by diagonal matrices.[1] We shall choose T_3 and Y to be the generators so singled out. Their eigenvalues will be denoted by t_3 and y, and the states in (17.26) will be labeled with these eigenvalues.

Similar to the situation in $SU(2)$, the commutation relations (17.22) show that the operators $T_\pm$, $U_\pm$, and $V_\pm$ are "raising" and "lowering" operators for both t_3 and y. In Fig. 17.3 the analog of Fig. 17.1 is drawn. Since the states are labeled by two quantum numbers, a two-dimensional grid needs to be plotted. The action of the operators $T_+, \ldots V_-$ is drawn on it. The intervals on the axes are scaled so that the lines forming the shift operators make angles which are multiples of 60° with one another.

All states of an irreducible representation may be generated by repeated application of the shift operators to any one of the states. For a finite-dimensional representation, only a finite number of sites can be occupied. The nature of the representation will be determined once the occupied sites, and the number of times each is occupied, are determined.

We start by considering the state with the largest value of t_3. We shall soon see that there is only one such state, i.e., we do not have two different values of y for $t_3 = (t_3)_{\text{max}}$. The particular state $\psi((t_3)_{\text{max}}y) \equiv \psi_{\text{max}}$ has the property that there is no neighboring state to the right of it, so that

$$T_+\psi_{\text{max}} = V_+\psi_{\text{max}} = U_-\psi_{\text{max}} = 0 \tag{17.27}$$

The boundary can be determined by repeated application of V_-, say, to this state, until a point is reached when V_- acting on the last of a sequence of states yields zero. We may find that

$$V_-^{p+1}\psi_{\text{max}} = 0 \tag{17.28}$$

[1] The mathematical expression of this is the statement that $SU(3)$ is a *group of rank* 2. In general $SU(n)$ is a group of rank $n - 1$.

When the state $(V_-)^p\psi_{\max}$ is reached, we can continue along the boundary by applying T_- repeatedly, etc., until we reach another corner, where

$$(T_-)^{q+1}(V_-)^p\psi_{\max} = 0 \tag{17.29}$$

In the last remarks we have assumed that the boundary is always convex. We can prove that this is so, with the help of the commutation relations. We shall present the proof as a protype of the kind of argument we use to establish the truth of a number of statements that will be made later on and not proved.

We shall consider the boundary of a distribution of sites shown in Fig. 17.4. The point M will be the one with the largest value of T_3; the state there will again be represented by $\psi_{\max}$. This is known as the state of highest weight. The application of V_- to $\psi_{\max}$ yields the state at the site N, which we label ψ_N. There is only one state at that site; the state defined by $U_-T_-\psi_{\max}$ is not independent of it, since

$$\begin{aligned} U_-T_-\psi_{\max} &= ([U_-, T_-] + T_-U_-)\psi_{\max} \\ &= V_-\psi_{\max} \end{aligned} \tag{17.30}$$

We leave it to the reader to convince himself that whatever path is taken along the lattice of sites, the same state results. We can similarly prove that the state ψ_B, defined by $V_-\psi_N$, is also unique, as is the state at the site A, $\psi_A \equiv V_-\psi_B$. If there is a state at site C, we will have

$$U_+\psi_C = \lambda\psi_A \tag{17.31}$$

Also

$$\lambda(\psi_A, \psi_A) = (V_-\psi_B, U_+\psi_C) = (\psi_B, V_+U_+\psi_C) \tag{17.32}$$

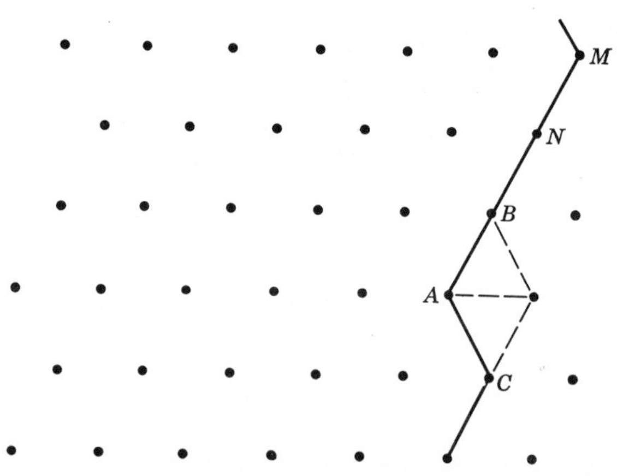

Fig. 17.4. A concave boundary section. The dotted lines represent forbidden steps.

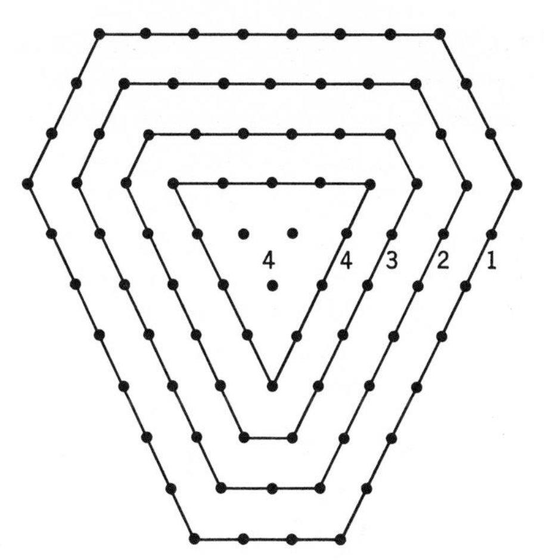

Fig. 17.5. A typical representation $[(p, q) = (7, 3)]$ with multiplicities of states at each site labeled on the graph.

Hence

$$\lambda = \frac{(\psi_B, ([V_+, U_+] + U_+V_+)\psi_C)}{(\psi_A, \psi_A)} = 0 \tag{17.33}$$

This must imply that there is no state at C.

Thus we see that a concavity is impossible; the outer boundary is convex, and there are no unfilled sites on the inside of any region surrounded by occupied sites.

With the help of the algebraic technique exhibited in establishing single occupancy of the boundary points and the absence of concavities, we can prove the following statements about irreducible representations:

1. The boundary will in general be a six-sided figure. If, starting from the site occupied by $\psi_{\max}$, we move in the V_- direction p steps and then in the T_- direction q steps [as exhibited by (17.28) and (17.29)], the boundary will continue with p steps in the U_+ direction, q steps in the V_+ direction, and again p steps in the T_+ direction, concluding with q steps in the U_- direction, as shown in Fig. 17.5. The boundary is thus symmetric under rotations of 120°.[1]

2. The boundary is symmetric under reflections in the y-axis.

3. There is one state at each site on the boundary, two states at each site on the next layer in, three states at the sites on the next layer, etc., until a triangular layer is reached, beyond which the multiplicity ceases to increase. Thus for "triangular" representations, characterized by $p = 0$ or $q = 0$, each site is only occupied once. The representation, in

[1] Provided we scale the axes so that the sites lie on equilateral triangles.

this graphical form, is conveniently pictured as a pyramid, with a six-sided base, each successive layer having each side shortened by one step until a triangular layer is reached, whereupon the pyramid is truncated. The multiplicity of states on each site on a given layer is now one. It is now easy to see that the triangle at the top of the pyramid will have $(p - q)$ steps on each side if $p \geq q$, and $(q - p)$ steps if $q > p$. We also see that there are altogether $q + 1$ layers if $p \geq q$, otherwise $(p + 1)$ layers. It is a simple exercise to count the number of states in an irreducible representation of this type, which we denote by (p, q). The triangle on top has

$$\sum_{l=1}^{p-q+1} l = \frac{1}{2}(p - q + 1)(p - q + 2)$$

sites and there are $(q + 1)$ triangles. The next outer layer adds $3(p - q + 2)$ points, and there are q such layers; the next one adds $3(p - q + 4)$ points, and there are $(q - 1)$ of them, etc. Hence the multiplicity is

$$\frac{1}{2}(q + 1)(p - q + 1)(p - q + 2) + 3\sum_{\nu=0}^{q}(q - \nu)(p - q + 2\nu + 2)$$

$$= \frac{1}{2}(p + 1)(q + 1)(p + q + 2) \quad (17.34)$$

4. There are, in general, several states for a given value of (t_3, y). We thus need another label to distinguish between them. We take advantage of the fact that the generators T_+, T_-, and T_3 satisfy commutation relations characteristic of the i-spin group, i.e., $SU(2)$. Thus, as long as we shift only along a $y =$ constant line, the operator

$$T^2 = \tfrac{1}{2}(T_+T_- + T_-T_+) + T_3^2 \quad (17.35)$$

is invariant, since

$$[T_\pm, T^2] = [T_3, T^2] = 0 \quad (17.36)$$

We can, therefore, label the states by the eigenvalue of T^2 in addition to t_3 and y. Consider, for example, the states on the top line of Fig. 17.5. They are all connected to the state on the extreme right by $T_\pm$ shift operators; they all therefore belong to an irreducible representation of $SU(2)$, i.e., they form an i-spin multiplet. Since there are $(p + 1)$ states, we can identify the value of the i-spin. It is

$$T = \tfrac{1}{2}p \quad (17.37)$$

and the value of t_3 at the site on the extreme right of that line is $t_3 = T = \tfrac{1}{2}p$.

When we go on to the next line, we may again start with the (unique) state on the extreme right, and generate from it by repeated application of T_-, an *i*-spin multiplet, which now has $(p + 2)$ members, and thus has

$$T = \tfrac{1}{2}(p + 1) \tag{17.38}$$

The second state on the next layer, going in, can be made orthogonal to the state belonging to $T = \frac{1}{2}(p + 1)$. It will satisfy

$$T_+\psi = 0 \tag{17.39}$$

This state may be used to generate the second *i*-spin multiplet on that line, and a count of points shows that here

$$T = \tfrac{1}{2}(p - 1) \tag{17.40}$$

In effect, the states on each layer of the pyramid can be chosen to form an *i*-spin multiplet. Incidentally, we can see in this way that the value of t_3 for the state of highest weight $\psi_{\max}$ is

$$(t_3)_{\max} = \tfrac{1}{2}(p + q) \tag{17.41}$$

This labeling is convenient, for it corresponds to nature's way of breaking the $SU(3)$ symmetry, but it is by no means unique. It is also possible to label the states by u_3, the eigenvalue of the operator U_3 defined in (17.23) (the analog of t_3), the quantity q, the eigenvalue of the operator

$$Q = T_3 + \tfrac{1}{2}Y \tag{17.42}$$

(the analog of Y), and the eigenvalue of

$$U^2 = \tfrac{1}{2}(U_+U_- + U_-U_+) + U_3^2 \tag{17.43}$$

There are, of course, only two additive quantum numbers, T_3 and Y, so that Q is not an independent labeling. However, because

$$[U_\pm, Q] = [U_3, Q] = 0 \tag{17.44}$$

it stands on the same footing as Y, provided we connect the states by $U_\pm$ shifting operators.

We may use the similarity between the u representation and the t representation to evaluate $(y)_{\max}$.[1] The u multiplet, to which $\psi_{\max}$ belongs and for which it is the lowest state, has

$$U = \tfrac{1}{2}q \tag{17.45}$$

so that

$$(\psi_{\max}, U_3, \psi_{\max}) \equiv (u_3)_{\max} = -\tfrac{1}{2}q$$

[1] The eigenvalue of Y for $\psi_{\max}$.

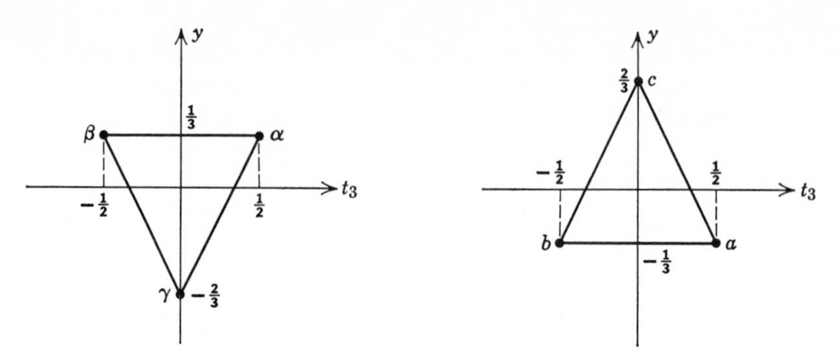

Fig. 17.6. The graphical representations of **3** and **3***.

Since, however

$$u_3 = \tfrac{3}{4}y - \tfrac{1}{2}t_3$$

we obtain

$$(y)_{\max} = \tfrac{1}{3}(p - q) \tag{17.46}$$

5. Aside from the trivial unitary-singlet representation (0, 0), the simplest representations are (1, 0) and (0, 1), denoted by **3** and **3***, respectively.[1] The graphical representation of these is given in Fig. 17.6. It should be pointed out that, in contrast to the situation in $SU(2)$, there are two *inequivalent* fundamental representations. In $SU(2)$, we found that the doublet which transformed according to

$$\chi' = U\chi$$

and the one which transformed according to

$$\bar{\chi}' = U^*\bar{\chi}$$

were equivalent, because U and U^* or, equivalently, τ_i and $-\tau_i^*$ were related by a similarity transformation. This statement is not true of the set λ_i and $-\lambda_i^*$ so that there is a difference between a triplet and its conjugate, as far as the transformation properties are concerned.[2]

We may use the commutation relations to determine the matrix elements of the shift operators between the states they connect. For example, we

[1] We shall denote a representation (p, q) by its dimensionality **N**, and the conjugate representation (q, p) by **N***, with the convention that **N** is associated with $p \geq q$. For example (3, 0) will be denoted by **10**, and (0, 3) by **10***.

[2] This can easily be understood in terms of the Sakata model; if we ignore baryon number, there is no difference between (p, n) and $(\bar{n}, \bar{p})$; there is, however, a difference between $(pn\Lambda^0)$ and $(\bar{n}\bar{p}\bar{\Lambda}^0)$, for we cannot ignore strangeness that is different in the two cases.

write for the states of **3** (or **3***), as labeled in Fig. 17.6

$$U_+ |\gamma\rangle = \lambda |\beta\rangle$$

Normalizing all states to unity, we find that

$$\lambda^2 = \langle\gamma| U_- U_+ |\gamma\rangle = \langle\gamma| [U_-, U_+] |\gamma\rangle$$
$$= \langle\gamma| T_3 - \tfrac{3}{2} Y |\gamma\rangle = 1$$

so that, with a definite choice of phase, we have $\lambda = 1$. It turns out that for the representation **3**, the product of the λ's for the three sides of the triangle is positive, and that all the λ's are equal in magnitude. We thus choose

$$\langle\beta| U_+ |\gamma\rangle = \langle\alpha| T_+ |\beta\rangle = \langle\gamma| V_- |\alpha\rangle = 1 \tag{17.47}$$

For the representations **3*** the product of the three λ's is negative. Because we want to keep the usual convention for the *i*-spin, we want the matrix element of T_+ to be positive; thus one other matrix element must be chosen to be positive and the other negative. We choose

$$\langle c| V_+ |b\rangle = \langle a| T_+ |b\rangle = -\langle c| U_+ |a\rangle = 1 \tag{17.48}$$

This is the choice of phases that corresponds to the convention used by de Swart[1] in his table of Wigner coefficients for $SU(3)$. The matrix elements (17.47) lead to the 3×3 matrices $\frac{1}{2}\lambda_i$ as representing the generators. The matrix elements (17.48) lead not to $-\frac{1}{2}\lambda_i^*$, as might be expected, but to the equivalent set

$$\tfrac{1}{2}\bar{\lambda}_i = -\tfrac{1}{2} W \lambda_i^* W^{-1}$$

with

$$W = \begin{pmatrix} 0 & 1 & 0 \\ -1 & 0 & 0 \\ 0 & 0 & -1 \end{pmatrix}$$

6. We may use the graphical technique to reduce products of two irreducible representations, a process analogous to the addition of angular momentum. In that procedure the additive nature of the quantum number T_3 played an important role in determining the sites occupied by the states in the product of two irreducible representations (Fig. 17.2). The same technique is applicable for $SU(3)$. The difference is that there are now two additive quantum numbers, T_3 and Y, so that the eigenvalues of T_3 and Y in a product $\psi(t_3, y)$, $\varphi(t_3', y')$ are $(t_3 + t_3')$ and $(y + y')$, respectively. In terms of graphs, the generalization of the procedure

[1] J. J. de Swart, *Rev. Mod. Phys.* **35**, 916 (1963).

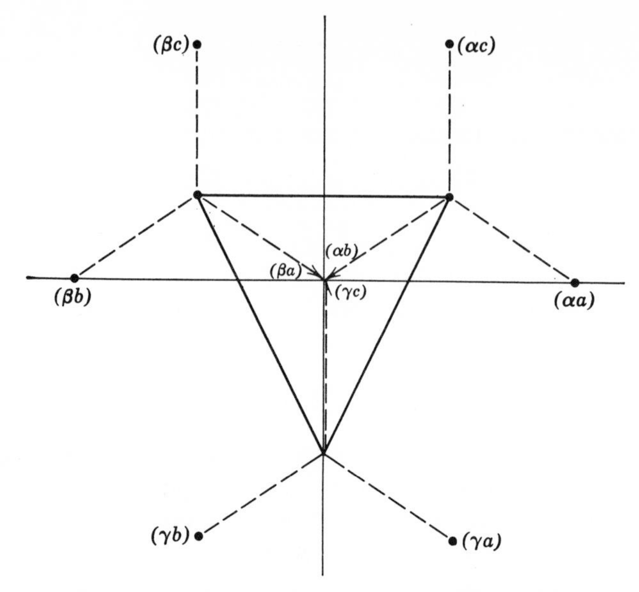

Fig. 17.7. The occupation sites of the product $\mathbf{3} \otimes \mathbf{3^*}$.

illustrated in Fig. 17.2 is a superposition on each state belonging to ψ, the whole graph representing φ, with the origin placed at the site of the states. The procedure is illustrated in Fig. 17.7 in which the multiplication $\mathbf{3} \times \mathbf{3^*}$ is worked out. We see from the figure that there are six boundary points and a triply occupied center site. From our general multiplicity rules we know that two of the center states belong with the six boundary points, forming an octet of states; the remaining state must be a singlet belonging to a different representation. We have thus derived

$$\mathbf{3} \otimes \mathbf{3^*} = \mathbf{8} \oplus \mathbf{1} \tag{17.49}$$

The graph shows that the octet represents the state (1, 1).

We can go further and determine the particular combinations of (α, β, γ) and (a, b, c) of (1, 0) and (0, 1) that make up the octet and the singlet. We write

$$\begin{aligned} |A_1\rangle &= |\alpha c\rangle \qquad & |A_2\rangle &= |\beta c\rangle \\ |C_1\rangle &= |\gamma a\rangle \qquad & |C_2\rangle &= |\gamma b\rangle \\ |B_1\rangle &= |\alpha a\rangle \qquad & |B_3\rangle &= |\beta b\rangle \end{aligned} \tag{17.50}$$

To find the state which belongs to the same i-spin multiplet as $|B_1\rangle$ and $|B_3\rangle$, we act on $|B_1\rangle$ with T_-:

$$T_- |B_1\rangle = |\beta a\rangle + |\alpha b\rangle \equiv \sqrt{2}\, |B_2\rangle$$

Here we have defined the normalized state

$$|B_2\rangle = \frac{1}{\sqrt{2}}(|\beta a\rangle + |\alpha b\rangle) \tag{17.51}$$

To find the *i*-spin singlet belonging to the octet, we apply V_- to the state $|A_1\rangle$. The resulting state is a superposition of $|B_2\rangle$ and the desired state $|D_0\rangle$. Orthogonalization leads to the normalized form

$$|D_0\rangle = \frac{1}{\sqrt{6}}(2\,|\gamma c\rangle + |\alpha b\rangle - |\beta a\rangle) \tag{17.52}$$

The reader can convince himself that

$$T_\pm\,|D_0\rangle = 0$$

The remaining combination of $|\gamma c\rangle$, $|\alpha b\rangle$, and $|\beta a\rangle$, orthogonal to both $|B_2\rangle$ and $|D_0\rangle$, must be the singlet. The normalized combination is

$$\frac{1}{\sqrt{3}}(-\,|\alpha b\rangle + |\beta a\rangle + |\gamma c\rangle) \tag{17.53}$$

The wavefunctions (17.50) through (17.52) may be used in the calculation of the matrix elements of the generators in the octet representation. Thus for example

$$\begin{aligned} \langle D_0|\, V_-\,|A_1\rangle &= \langle D_0|\, V_-\,|\alpha c\rangle = \langle D_0\,|\,\gamma c\rangle + \langle D_0\,|\,\alpha b\rangle \\ &= \frac{1}{\sqrt{6}}(2 + 1) = \sqrt{\frac{3}{2}} \end{aligned} \tag{17.54}$$

The results of this computation are easily represented on the octet graph in Fig. 17.8. These matrix elements are necessary for the construction of

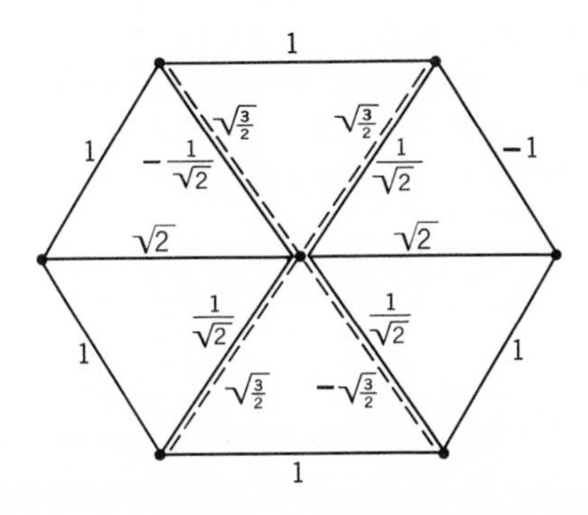

Fig 17.8. The matrix elements of the shift operators in the octet representation. The dotted lines are the ones joining the $Y = \pm 1$ states to the *i*-spin singlet D_0.

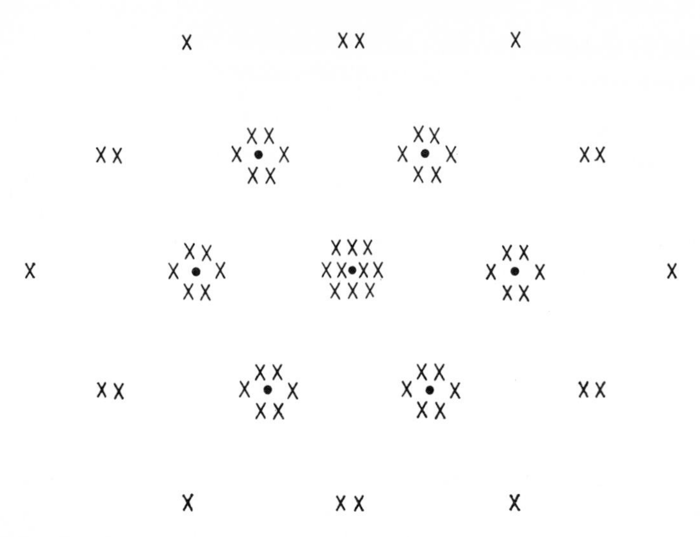

Fig. 17.9. The sites occupied by the states in $\mathbf{8} \otimes \mathbf{8}$ and their multiplicities.

the wavefunctions for the states in the reduction of

$$\mathbf{8} \otimes \mathbf{8} = \mathbf{1} \oplus \mathbf{8} \oplus \mathbf{8} \oplus \mathbf{10} \oplus \mathbf{10}^* \oplus \mathbf{27} \tag{17.55}$$

Incidentally, the reduction shown in (17.55) is easily obtained by the graphical technique used in the reduction of $\mathbf{3} \otimes \mathbf{3}^*$. Figure 17.9 shows the multiplicity of points at each site obtained by "multiplying" each point of an octet by another octet. With the help of the multiplicity rules illustrated in Fig. 17.5, and working from the outside in, the reader may easily convince himself of the correctness of (17.55).

The actual construction of the wavefunctions is a straightforward, though lengthy, procedure. The results may be found in the literature.[1]

7. There are other analogies with the theory of angular momentum. We may define "irreducible tensor operators," which are generalizations of the corresponding quantities in $SU(2)$, by means of their commutation relations with the generators. If these generators are denoted by F_i, rather than $T_+, T_-, \ldots, Y$, we have

$$[F_i, T_\alpha^{(\nu)}] = \sum_\beta T_\beta^{(\nu)} \langle \nu, \beta | F_i | \nu, \alpha \rangle \tag{17.56}$$

Here (ν) stands for (p, q) that label the irreducible representation and α stands for the collection of labels used to characterize the states within that representation, e.g. $\{t_3, y, T\}$. This is the analog of the corresponding

[1] See S. Gasiorowicz and S. L. Glashow, *loc. cit.*, for references.

definition in the theory of angular momentum[1]

$$[\mathbf{J}, T_M^J] = \sum_{M'} T_{M'}^J \langle JM' | \mathbf{J} | JM \rangle \tag{17.57}$$

The angular momentum operators themselves form irreducible tensors for the *vector* (i.e., angular momentum 1) representation. It is well known, as a consequence of the Wigner-Eckart theorem, that for matrix elements involving states of the same J the vector operator may be replaced by the generator. For example,

$$\langle JM' | \mathbf{r} | JM \rangle = \alpha(J) \langle JM' | \mathbf{J} | JM \rangle \tag{17.58}$$

In $SU(3)$, the generators F_i themselves again satisfy the commutation relations (17.56). In contrast to the situation in the theory of angular momentum, however, there is another set of operators, denoted by D_i, which also satisfy these relations. The existence of *two* sets of irreducible tensor operators in $SU(3)$ is connected with the fact that in the product of any representation $(p, q) \otimes (1, 1)$ (the octet), the representation (p, q) appears *twice*, unless $p = 0$ *or* $q = 0$ in which case it appears *once*.[2] This is in contrast with the theory of angular momentum where ψ_J appears but once in the product $\psi_J \otimes \psi_1$.

The operator D_i has the following form

$$D_i = \frac{2}{3} \sum_{jk} d_{ijk} F_j F_k \tag{17.59}$$

where the d_{ijk} were defined in (17.18). The proof that the D_i satisfy the required commutation relations

$$[D_i, F_j] = i f_{ijk} D_k \tag{17.60}$$

follows from the commutation relations for the F's once the identity

$$d_{ijk} f_{klm} + f_{klj} d_{ikm} = d_{kjm} f_{ikl} \tag{17.61}$$

is established. This identity is derived by inserting λ_i, λ_j, and λ_k into the general identity

$$[A, \{B, C\}] = \{[A, B], C\} + \{[A, C], B\} \tag{17.62}$$

and multiplying by a λ after which traces are taken. The generalization of (17.58) now reads

$$\langle \nu, \alpha' | 0_i^{(1,1)} | \nu, \alpha \rangle = C_1(\nu) \langle \nu, \alpha' | F_i | \nu, \alpha \rangle + C_2(\nu) \langle \nu, \alpha' | D_i | \nu, \alpha \rangle \tag{17.63}$$

[1] M. E. Rose, *op. cit.*, p. 82.

[2] This theorem can be proved with our graphical technique. It is, however, more easily proved using other techniques (S. Gasiorowicz and S. L. Glashow, *loc. cit.*).

The operators D_3 and D_8 are of particular interest. It is a matter of straightforward computation to show that

$$D_3 = \tfrac{1}{3}(V^2 - U^2) + \tfrac{2}{3}T_3Y \tag{17.64}$$

and

$$(2/\sqrt{3})D_8 = -\tfrac{2}{9}F^2 + \tfrac{2}{3}(T^2 - \tfrac{1}{4}Y^2) \tag{17-65}$$

Here $F^2 = \sum_{i=1}^{8} F_iF_i$ is an invariant operator that commutes with all the F_i. It can be written as

$$F^2 = \tfrac{1}{2}\{T_+, T_-\} + \tfrac{1}{2}\{U_+, U_-\} + \tfrac{1}{2}\{V_+, V_-\} + T_3^2 + \tfrac{3}{4}Y^2 \tag{17.66}$$

and its value for a representation (p, q) is easily obtained by calculating its matrix element for the "maximum state" for which we determined $t_3 = \tfrac{1}{2}(p + q)$ and $y = \tfrac{1}{3}(p - q)$. We find that

$$F^2 = \tfrac{1}{3}(p^2 + pq + q^2) + (p + q) \tag{17.67}$$

If a matrix representation of the F_i is sought, the coefficients on the right-hand side of the commutation relations (17.17), can be used for that purpose. Just like in $SU(2)$, where we found it possible to use

$$(T_i)_{jk} = -ie_{ijk} \tag{17.68}$$

so here we may use

$$(F_i)_{jk} = -if_{ijk} \tag{17.69}$$

Similarily, we can check that a possible matrix representation for the D_i is

$$(D_i)_{jk} = d_{ijk} \tag{17.70}$$

Equation (17.69) is proved by inserting λ_i, λ_j, λ_k into the Jacobi identity

$$[[A, B], C] + [[B, C], A] + [[C, A], B] = 0$$

multiplying by a λ and taking traces. We leave the check of (17.70) to the reader.

With these remarks we conclude this long mathematical preparation for a discussion of the evidence for, and the applications of unitary symmetry in, strong interaction physics. The length of the exposition is a reflection of a belief that most physicists are as yet not very much at home with group theory. For those who are, the tensor techniques involving Young tableaux are in many cases more direct.[1] A discussion of

[1] Such techniques are essential for the exploration of higher-rank groups, such as $SU(6)$, which have recently attracted much attention. An excellent exposition of these techniques may be found in M. Hamermesh, *Group Theory*, Addison-Wesley, Reading, Mass. (1963).

these techniques in the space available could only result in a series of recipes.

PROBLEMS

1. Use the graphical representation of the irreducible representations to show that in the product $(p, q) \otimes (1, 1)$ the representation (p, q) occurs twice, unless $pq = 0$. *Hint.* Draw the top right-hand corner of any representation (p, q) with the correct multiplicities. Multiply by the octet, and identify the points belonging to $(p + 1, q + 1)$, $(p + 2, q - 1)$, and $(p - 1, q + 2)$. What is left?
2. Follow the procedure of Messiah, *Quantum Mechanics*, Vol II, to prove the Wigner-Eckart theorem for $SU(3)$.
3. Show that the multiplicity of representations at a given occupation site one layer in from the boundary is 2, unless the boundary is triangular. *Hint.* If $|M\rangle$ is the state of maximum weight, $T_- |M\rangle$ and $U_+V_- |M\rangle$ are linearly independent, if there exists no θ such that $\cos \theta T_- |M\rangle + \sin \theta U_+V_- |M\rangle = 0$.
4. In the reduction of $\mathbf{8} \otimes \mathbf{8}$ the state of maximum weight is given by

$$|\mathbf{10},\ Y = 1,\ T = \tfrac{3}{2},\ T_3 = \tfrac{3}{2}\rangle = \frac{1}{\sqrt{2}}(\Sigma^+K^+ + p\pi^+)$$

   for the decuplet representation in which Σ^+ and π^+ are the $Y = 0$, $T = T_3 = 1$ members of the two octets and p, K^+ are the $Y = 1$, $T = T_3 = \frac{1}{2}$ members of the two octets. Use the matrix elements of the shift operators for the octet representation shown in Fig. 17.8 to work out the remaining terms in the decuplet representation.
5. Use the results of Problem 4 to show that the matrix elements of the shift operators in the $\mathbf{10}$ representation are given by

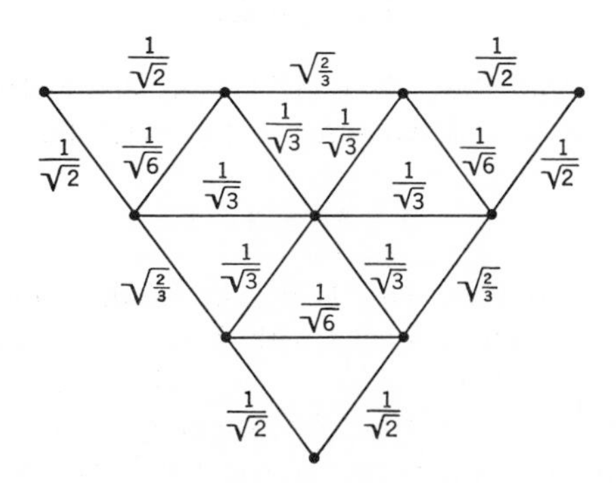

6. Work out $\mathbf{8} \otimes \mathbf{10}$. How many $\mathbf{1}$'s are there in $\mathbf{8} \otimes \mathbf{8} \otimes \mathbf{8} \otimes \mathbf{8}$?

18

The Eightfold Way

In our introduction to $SU(3)$ in Chapter 17, we pointed out that the origin of interest in that group was the symmetric Sakata model, with (p, n, Λ^0) forming the **3** representation, and the antiparticles forming the conjugate **3*** representation. One of the immediate successes of that model was that the nine possible baryon-antibaryon combinations could be divided into an octet and a singlet of meson states. If the spatial transformation properties of these "bound states" were appropriate (i.e., binding in the $^1S_0$ state), the octet could describe the (at that time) seven known mesons, π^+, π^0, π^-, K^+, K^0, $\bar{K}^0$, K^-, with a predicted eighth member. As our notation indicates, t_3 represents the third component of the i-spin; for the Sakata model the hypercharge can be written in the form

$$Y = y + \tfrac{2}{3}N_B \tag{18.1}$$

with N_B the baryon number. Thus (18.1) consistently describes the triplet and antitriplet. With this assignment, the octet that results from the multiplication

$$\mathbf{3} \otimes \mathbf{3^*} = \mathbf{8} \oplus \mathbf{1}$$

consists of an i-spin doublet with $Y = 1$, an i-spin doublet with $Y = -1$, a triplet with $Y = 0$, and a singlet with $Y = 0$, in exact correspondence with the known pseudoscalar mesons. In its classification of baryons, the Sakata model was less successful. By assigning the p, n, Λ^0 to the fundamental triplet, the model requires the baryons Σ and Ξ to belong to one or more multiplets contained in $\mathbf{3} \otimes \mathbf{3} \otimes \mathbf{3^*}$. In view of the similarities among the eight baryons, their relatively small mass differences, and the apparent absence of partners for the Σ and Ξ in $\mathbf{3} \otimes \mathbf{3} \otimes \mathbf{3^*} = \mathbf{15} \oplus \mathbf{6^*} \oplus \mathbf{3} \oplus \mathbf{3}$ (an examination of this product shows that only the **15** has a $Y = -1$ doublet state to accommodate the Ξ), the Sakata model was in some difficulties.

Gell-Mann[1] and independently Ne'eman[2] proposed that the baryons also belong to the regular octet representation. In this way the parallelism between the meson states and the baryon states, alluded to at the end of Chapter 16, is naturally explained. This assignment, generally known as the *eightfold way* requires a number of comments at the outset.

1. In order to associate the additive quantum numbers of $SU(3)$ with the i-spin and hypercharge, we must choose T_3 and Y, as defined in Chapter 17, to correspond to the physical quantities associated with these symbols. Thus the baryon number N_B no longer appears in any of the formulas.

2. All strongly interacting particles, discovered so far, have been made out of mesons and baryons, with other mesons and baryons (and anti-baryons) being produced alongside. Thus all the resonances so far observed, to be discussed later, must belong to representations contained in $\mathbf{8} \otimes \mathbf{8} \otimes \mathbf{8} \otimes \cdots$. All such particles have integral hypercharge. Since the hypercharge of the state of highest weight in any representation is $(y)_{\max} = \frac{1}{3}(p - q)$, this implies that only representations with $p = q$ (mod. 3) appear in the eightfold way.

3. It is still possible to consider the observed particles as being built up out of a fundamental triplet and a conjugate triplet. If the relation between the physical quantitites, T_3 and Y, and the diagonal quantum numbers is the same for the observed particles, so that t_3 and y for a state represent its z-component of i spin and hypercharge, the fundamental fields must carry hypercharge which is $\frac{1}{3}$ and $-\frac{2}{3}$ for members of the $\mathbf{3}$ and $\frac{2}{3}$ and $-\frac{1}{3}$ for members of the $\mathbf{3}^*$. This is the quark model of Gell-Mann[3] and Zweig.[4] If the fundamental fields also carry baryon number, the product $\mathbf{3} \times \mathbf{3}^*$ (quark-antiquark system) would carry no baryon number, as in the Sakata model. The baryons could be constructed out of the product

$$\mathbf{3} \otimes \mathbf{3} \otimes \mathbf{3} = \mathbf{10} \oplus \mathbf{8} \oplus \mathbf{8} \oplus \mathbf{1} \tag{18.2}$$

representing a bound state of three quarks. This would require that the quark fields carry baryon number $\frac{1}{3}$.[5] Field theory does not require that

[1] M. Gell-Mann, California Institute of Technology Synchrotron Lab Report No. CTSL-20, reprinted in M. Gell-Mann and Y. Ne'eman *The Eightfold Way*, W. A. Benjamin, New York (1964).
[2] Y. Ne'eman, *Nucl. Phys.* **26** 222 (1961).
[3] M. Gell-Mann, *Physics Letters* **8**, 214 (1964).
[4] G. Zweig (unpublished CERN report).
[5] This is not necessary, however; the triplets could carry no baryon number, and that attribute could be carried by an additional unitary-singlet spin-$\frac{1}{2}$ field which does carry the baryon number and whose presence in an octet could make the difference between a baryon and a meson.

there be a particle associated with each field.[1] If such quark particles existed, at least one of them would have to be stable, because baryon conservation would forbid the decay into "ordinary" particles. A search for new particles of this type has so far proved unsuccessful. In view of this failure and the difficulty of inventing a mechanism which would bind a quark and an antiquark, or three quarks, but not two quarks, for example, we will not discuss the quark model further.[2] The statement is also true of variations on this model, in which, by increasing the number of fundamental triplets, it is possible to avoid assignments of fractional hypercharge and baryon number to the fundamental fields.[3]

We begin our discussion by asking for the most general form of an $SU(3)$ invariant baryon-meson Yukawa coupling, the generalization of the charge independent form

$$\bar{\psi}_a(\tau_i)_{ab}\psi_b\phi_i \tag{18.3}$$

It is clear that if ψ_m is used to denote the eight baryon operators and ϕ_i the eight meson operators, the analog of (18.3) will be the form

$$\bar{\psi}_m(F_i)_{mn}\psi_n\phi_i = -if_{imn}\bar{\psi}_m\psi_n\phi_i \tag{18.4}$$

Here the matrix representation of F_i (17.69) has been used. It follows from the commutation relations (17.17) and the normalization

$$\operatorname{tr} \lambda_i\lambda_j = 2\,\delta_{ij} \tag{18.5}$$

that we may write

$$f_{imn} = \frac{1}{4i}\operatorname{tr}([\lambda_m, \lambda_n]\lambda_i) \tag{18.6}$$

Thus the coupling takes the form

$$-\tfrac{1}{4}\operatorname{tr}([\lambda_m\bar{\psi}_m, \lambda_n\psi_n]\lambda_i\phi_i) \tag{18.7}$$

If we define the 3×3 matrices

$$\bar{B} = \frac{1}{\sqrt{2}}\lambda_m\bar{\psi}_m \qquad B = \frac{1}{\sqrt{2}}\lambda_n\psi_n \qquad M = \frac{1}{\sqrt{2}}\lambda_i\phi_i \tag{18.8}$$

[1] However, if the field carries peculiar quantum numbers, such as baryon number $\frac{1}{3}$, a particle is unavoidable. If $q(x)$ represents a field of baryon number $\frac{1}{3}$, then $\bar{q}(x)\Psi_0$ creates a state of baryon number $\frac{1}{3}$, and the lightest of these must be stable, unless it can decay into a lower mass state of $-\frac{2}{3}$ and a baryon. At any rate some quark will have to be stable.

[2] The quark model, for which vector currents and axial vector currents (of interest in the weak interactions) may be taken to be $\frac{1}{2}\,\bar{q}(x)\gamma^\mu\lambda_i\,q(x)$ and $\frac{1}{2}\,\bar{q}(x)\gamma^\mu\gamma_5\lambda_i\,q(x)$ does lead to definite predictions for commutation relations among the currents. These relations are discussed at the end of Chapter 34.

[3] F. Gürsey, T. D. Lee, and M. Nauenberg, *Phys. Rev.*, **135,** B467 (1964).

we see that we can write the coupling in the form

$$-\text{tr}\,([\bar{B}, B]M) \tag{18.9}$$

In contrast to the charge-independent coupling, this is not the only Yukawa form possible. The existence of another set of octet operators D_i permits us to write a second invariant

$$\bar{\psi}_m(D_i)_{mn}\bar{\psi}_n\phi_i = d_{imn}\bar{\psi}_m\psi_n\phi_i \tag{18.10}$$

We may use (17.18) to derive

$$d_{imn} = \tfrac{1}{4}\,\text{tr}\,(\{\lambda_m, \lambda_n\}\lambda_i) \tag{18.11}$$

so that in terms of the matrices $\bar{B}$, B, and M, we obtain

$$\text{tr}\,(\{\bar{B}, B\}M) \tag{18.12}$$

The invariance of these couplings under infinitesimal transformations is most easily established by showing how the 3×3 matrices transform. We have for any field operator with octet transformation properties,

$$[F_i, \phi_j] = if_{ijk}\phi_k \tag{18.13}$$

Consequently, under an infinitesimal transformation

$$\begin{aligned}\phi_j' = e^{i\alpha_iF_i}\phi_je^{-i\alpha_iF_i} &\approx \phi_j + i\alpha_i[F_i, \phi_j]\\ &\cong \phi_j - \alpha_if_{ijk}\phi_k\end{aligned}$$

Since, however

$$\lambda_jf_{ijk} = -\frac{1}{2i}[\lambda_i, \lambda_k]$$

it follows that

$$\lambda_j\phi_j' \cong \lambda_j\phi_j - \frac{i}{2}\alpha_i[\lambda_i, \lambda_k\phi_k]$$

i.e.,

$$\begin{aligned}M' &\cong M - \frac{i}{2}\alpha_i[\lambda_i, M]\\ &\cong e^{-\frac{i}{2}\alpha_i\lambda_i}Me^{\frac{i}{2}\alpha_i\lambda_i}\end{aligned} \tag{18.14}$$

Thus the trace of a product of any number of such 3×3 octet matrices will be invariant.

It remains to express the operators $\phi_i, \ldots,$ in terms of field operators which create or annihilate states of fixed T_3 and Y, i.e., the particle states. If $\phi_{K^+}{}^+$ is the operator that, acting on the vacuum state, creates a K^+, it must satisfy the conditions

$$T_3\phi_{K^+}^+\Psi_0 = \tfrac{1}{2}\phi_{K^+}^+\Psi_0 \qquad Y\phi_{K^+}^+\Psi_0 = \phi_{K^+}^+\Psi_0$$

i.e.,

$$[T_3, \phi^+_{K^+}] = \tfrac{1}{2}\phi^+_{K^+} \qquad [Y, \phi^+_{K^+}] = \phi^+_{K^+} \tag{18.15}$$

Similarly

$$[T_3, \phi^+_{\pi^\pm}] = \pm\phi^+_{\pi^\pm} \qquad [Y, \phi^+_{\pi^\pm}] = 0$$

With the help of (18.13) we obtain the following commutation relations

$$\begin{aligned}
[T_3, \phi_1 \pm i\phi_2] &= \pm(\phi_1 \pm i\phi_2)\\
[T_3, \phi_4 \pm i\phi_5] &= \pm\tfrac{1}{2}(\phi_4 \pm i\phi_5)\\
[T_3, \phi_6 \pm i\phi_7] &= \mp\tfrac{1}{2}(\phi_6 \pm i\phi_7)\\
[Y, \phi_1 \pm i\phi_2] &= 0\\
[Y, \phi_4 \pm i\phi_5] &= \pm(\phi_4 \pm i\phi_5)\\
[Y, \phi_6 \pm i\phi_7] &= \mp(\phi_6 \pm i\phi_7)
\end{aligned} \tag{18.16}$$

These permit us to make the identification

$$\phi_{K^+} \leftrightarrow \frac{1}{\sqrt{2}}(\phi_4 + i\phi_5)^+ = \frac{1}{\sqrt{2}}(\phi_4 - i\phi_5)$$

$$\phi_{K^0} \leftrightarrow \frac{1}{\sqrt{2}}(\phi_6 - i\phi_7)$$

$$\phi_{\bar{K}^0} \leftrightarrow \frac{1}{\sqrt{2}}(\phi_6 + i\phi_7)$$

$$\phi_{\pi^\pm} \leftrightarrow \frac{1}{\sqrt{2}}(\phi_1 \mp i\phi_2)$$

We can also show, using the total *i*-spin operator, that

$$\phi_{\pi^0} \leftrightarrow \phi_3 \qquad \phi_{\eta^0} \leftrightarrow \phi_8$$

Thus the meson matrix has the form

$$M = \begin{pmatrix} \frac{1}{\sqrt{2}}\pi^0 + \frac{1}{\sqrt{6}}\eta^0 & \pi^+ & K^+ \\ \pi^- & -\frac{1}{\sqrt{2}}\pi^0 + \frac{1}{\sqrt{6}}\eta^0 & K^0 \\ K^- & \bar{K}^0 & -\frac{2}{\sqrt{6}}\eta^0 \end{pmatrix} \tag{18.17}$$

The abbreviations are: π^+ for ϕ_{π^+}, K^+ for ϕ_{K^+}, and so on.

The particular choice of phases for the operators in the matrix representation M was made so that under charge conjugation

$$CMC^{-1} = M^T \tag{18.18}$$

We write the baryon matrix in a similar form[1]

$$B = \begin{pmatrix} \frac{1}{\sqrt{2}}\Sigma^0 + \frac{1}{\sqrt{6}}\Lambda^0 & \Sigma^+ & p \\ \Sigma^- & -\frac{1}{\sqrt{2}}\Sigma^0 + \frac{1}{\sqrt{6}}\Lambda^0 & n \\ -\Xi^- & \Xi^0 & -\frac{2}{\sqrt{6}}\Lambda^0 \end{pmatrix} \tag{18.19}$$

and

$$\bar{B} = \begin{pmatrix} \frac{1}{\sqrt{2}}\bar{\Sigma}^0 + \frac{1}{\sqrt{6}}\bar{\Lambda}^0 & \bar{\Sigma}^- & -\bar{\Xi}^- \\ \bar{\Sigma}^+ & -\frac{1}{\sqrt{2}}\bar{\Sigma}^0 + \frac{1}{\sqrt{6}}\bar{\Lambda}^0 & \bar{\Xi}^0 \\ \bar{p} & \bar{n} & -\frac{2}{\sqrt{6}}\bar{\Lambda}^0 \end{pmatrix} \tag{18.20}$$

The most general form of the Yukawa coupling (omitting γ_5's) is α times the D-coupling plus $(1 - \alpha)$ times the F-coupling,

$$-(1 - \alpha)\,\mathrm{tr}\,([\bar{B}, B]M) + \alpha\,\mathrm{tr}\,(\{\bar{B}, B\}M) \tag{18.21}$$

When the traces are worked out and terms recombined, we obtain the following result:

$$\begin{aligned} \frac{1}{g}\mathcal{L}_I = {} & \bar{N}\boldsymbol{\tau}N \cdot \boldsymbol{\pi} + (1 - 2\alpha)\bar{\Xi}\boldsymbol{\tau}\Xi \cdot \boldsymbol{\pi} + \frac{3 - 4\alpha}{\sqrt{3}}\bar{N}N\eta^0 \\ & - \frac{3 - 2\alpha}{\sqrt{3}}\bar{\Xi}\Xi\eta - (1 - 2\alpha)\bar{N}\boldsymbol{\tau}K \cdot \boldsymbol{\Sigma} \\ & - \bar{\Xi}\boldsymbol{\tau}K^c \cdot \boldsymbol{\Sigma} - (1 - 2\alpha)\bar{\boldsymbol{\Sigma}} \cdot K^+\boldsymbol{\tau}N \\ & - \bar{\boldsymbol{\Sigma}} \cdot K^{c+}\boldsymbol{\tau}\Xi - \frac{3 - 2\alpha}{\sqrt{3}}\bar{N}\Lambda K \\ & - \frac{3 - 2\alpha}{\sqrt{3}}K^+\bar{\Lambda}N + \frac{3 - 4\alpha}{\sqrt{3}}\bar{\Xi}\Lambda K^c \\ & + \frac{3 - 4\alpha}{\sqrt{3}}K^{c+}\bar{\Lambda}\Xi + i(1 - \alpha\ \bar{\boldsymbol{\Sigma}} \times \boldsymbol{\Sigma} \cdot \boldsymbol{\pi} \\ & + \frac{2\alpha}{\sqrt{3}}\bar{\boldsymbol{\Sigma}}\Lambda \cdot \boldsymbol{\pi} + \frac{2\alpha}{\sqrt{3}}\bar{\Lambda}\boldsymbol{\Sigma} \cdot \boldsymbol{\pi} \\ & + \frac{2\alpha}{\sqrt{3}}\bar{\boldsymbol{\Sigma}} \cdot \boldsymbol{\Sigma}\eta^0 - \frac{2\alpha}{\sqrt{3}}\bar{\Lambda}\Lambda\eta^0 \end{aligned} \tag{18.22}$$

[1] With the standard convention for the Wigner coefficients, $(\phi_{\pi^+})^+ = (\phi_1 + i\phi_2)/\sqrt{2}$, when acting on the vacuum state, creates $-\Psi_{\pi^+}$. Also, since K^- is the antiparticle of K^+, (16.18) shows that $\phi_{K^-}^+$ acting on the vacuum must create the state $-\Psi_{K^-}$. For the baryons $\Psi_{\Sigma^+}^+$ will again create $-\Psi_{\Sigma^+}$; since Ξ is not the antiparticle of the proton, we must have $-\psi_{\Xi^-}^+$ to create the state $-\Psi_{\Xi^-}$.

Here we have used the *i*-spin doublets

$$N = \begin{pmatrix} p \\ n \end{pmatrix} \qquad \Xi = \begin{pmatrix} \Xi^0 \\ \Xi^- \end{pmatrix} \qquad K = \begin{pmatrix} K^+ \\ K^0 \end{pmatrix} \qquad K^c = \begin{pmatrix} \bar{K}^c \\ -K^- \end{pmatrix} \tag{18.23}$$

The F coupling is obtained by setting $\alpha = 0$, the D-coupling by setting $\alpha = 1$. The two couplings have different properties under the so-called R-symmetry, which replaces any matrix by its transpose. We shall see that both couplings seem to be present, so that there is no approximate symmetry under the R-reflections.

It goes without saying that this form of the coupling holds for any three octets. Under certain circumstances only one or the other of the F or D forms is possible. Consider the coupling of a vector meson to two pseudoscalar mesons P. Since the vector meson octet transforms oppositely to the meson under charge conjugation,[1] so that

$$CV_\mu C^{-1} = -V_\mu^T \tag{18.24}$$

the product $\text{tr}\,(V_\mu P\,\partial^\mu P)$ transforms according to

$$\begin{aligned} \text{tr}\,(V_\mu P\,\partial^\mu P) &\rightarrow -\text{tr}\,(V_\mu{}^T P^T\,\partial^\mu P^T) \\ &= -\text{tr}\,(V_\mu\,\partial^\mu P \cdot P) \end{aligned} \tag{18.25}$$

so that the symmetric combination (D-type) vanishes because of C invariance. We leave it to the reader to convince himself that the coupling of two vector mesons to a pseudoscalar meson must be of the D-type.

$$\epsilon_{\mu\nu\rho\sigma}\,\text{tr}\,(P\{\partial^\mu V^\nu, \partial^\rho V^\sigma\}) \tag{18.26}$$

with

$$\begin{aligned} \epsilon_{\mu\nu\rho\sigma} &= \begin{cases} \pm 1 \text{ if } (\mu, \nu, \rho, \sigma) \text{ is an even/odd permutation of } 0, 1, 2, 3. \\ 0 \text{ otherwise.} \end{cases} \end{aligned}$$

The invariance of a Lagrangian under $SU(3)$, like gauge invariance, implies the existence of conserved currents. If we write $\mathcal{L} = \mathcal{L}(\phi_\alpha, \partial_\mu\phi_\alpha)$ where ϕ_α stands for all field in $\mathcal{L}$ whatever their transformation properties, under an infinitesimal transformation

$$\phi_\alpha \rightarrow U\phi_\alpha U^{-1} \simeq \phi_\alpha + i\alpha_k(F_k)_{\alpha\beta}\phi_\beta$$

[1] A neutral vector meson is usually taken to have the same C transformation properties as the photon. If that neutral meson belongs to an octet, this modified (18.18) by a $-$ sign, if the phases for the charged members are suitably chosen.

and the change in the Lagrangian density is

$$\delta\mathfrak{L} = \frac{\partial\mathfrak{L}}{\partial\phi_\alpha}\,\delta\phi_\alpha + \frac{\partial\mathfrak{L}}{\partial(\partial_\mu\phi_\alpha)}\,\partial_\mu\,\delta\phi_\alpha$$
$$= \partial_\mu\left[\frac{\partial\mathfrak{L}}{\partial(\partial_\mu\phi_\alpha)}\,\delta\phi_\alpha\right]$$

when the equations of motion are used. Setting $\delta\mathfrak{L} = 0$, we obtain the result

$$i\alpha_k\,\partial_\mu\left[\frac{\partial\mathfrak{L}}{\partial(\partial_\mu\phi_\alpha)}\,(F_k)_{\alpha\beta}\phi_\beta\right] = 0$$

Thus the eight currents, defined by

$$\mathfrak{J}_\mu^{(k)} = i\,\frac{\partial\mathfrak{L}}{\partial(\partial^\mu\phi_\alpha)}\,(F_k)_{\alpha\beta}\phi_\beta \tag{18.27}$$

where it is understood that the F_k are represented by square matrices of the same dimensionality as the ϕ, satisfy the conservation law

$$\partial^\mu\mathfrak{J}_\mu^{(k)} = 0 \tag{18.28}$$

The coupling of a vector octet to two mesons has already been shown to be of F-type. The fundamental coupling of vector mesons to the baryon octet can also be hypothesized as an F-type coupling, but there is no reason to believe that phenomenological couplings of the D variety do not occur.[1]

One reason for our interest in the form of the couplings is that they greatly simplify the prediction of ratios of matrix elements of three-octet processes. If there should exist another heavy octet of particles of baryon number 1, these might be expected to decay into a baryon and a meson. From the form of the coupling we can read off a number of relations. For example, the following ratios for the squares of the matrix elements can be found

$$\left|\frac{M(\Xi_8^{0*} \to \Xi^- + \pi^+)}{M(N_8^{+*} \to p + \pi^0)}\right|^2 = 2(1 - 2\alpha)^2$$
$$\left|\frac{M(N_8^{+*} \to p + \eta^0)}{M(N_8^{+*} \to p + \pi^0)}\right|^2 = \frac{(3 - 4\alpha)^2}{3} \tag{18.29}$$

If the coupling is of F-type or D-type alone, one experimental width measurement determines all others. We shall see in Chapter 20 there is

[1] In fact, if we tried to construct a model of the strong interactions based only on vector mesons and baryons, with F-type couplings, the equations would have R symmetry; this is not observed in nature.

evidence for the existence of an octet of vectors mesons. These are the i-spin triplet ρ^+, ρ^0, ρ^-; the $Y = 1$ doublet K^{+*} and K^{0*}, with their $Y = -1$ antiparticles, and the i-spin singlet, denoted by ω_8. The expression (18.22) specialized to $\alpha = 0$ (for F coupling) leads to

$$|M(K^{+*} \to K^+ + \pi^0)/M(\rho^0 \to \pi^+ + \pi^-)|^2 = \tfrac{1}{4} \tag{18.30}$$

In an actual calculation of a decay rate, there are always kinematic factors coming from the dependence of the matrix element on the Q-value of the reaction and on the phase space, which also depends on the masses of the particles involved. Now it is a fact that $SU(3)$ is not an exact symmetry; the mass splitting among the members of the octets is quite large. Presumably the ratios of the coupling constants in the general trilinear forms (18.22) will also change. Although we will discuss some ideas about the breakdown of the mass degeneracy, there is little we can say about the departures of the coupling constants from the expected behavior. The experimental situation is far from clear; a hypothesis which is in reasonable accordance with the facts is that the most important way in which the breakdown of $SU(3)$ is to be taken into account is by putting the true physical masses in the kinematic factors.[1]

When interactions involving multiplets other than those belonging to octets are involved, we must, in general, calculate the Wigner coefficients. Some of these have by now been tabulated.[2] It is possible sometimes to use some tricks to obtain the wanted result. We shall consider a special example; in Chapter 29 we shall see that there is good evidence for a set of baryon-meson resonances belonging to the (3, 0), i.e., **10** representation of $SU(3)$.

The decuplet contains a $Y = 1$, $T = \frac{3}{2}$ state (whose members are denoted by Δ^{++}, Δ^+, Δ^0, Δ^-), a $Y = 0$, $T = 1$ state (Y_1^{*+}, Y_1^{*0}, Y_1^{*-}), a $Y = -1$, $T = \frac{1}{2}$ state (Ξ^{*0}, Ξ^{*-}) and a $Y = -2$, $T = 0$ state Ω^-. Within each i-spin multiplet we may calculate the ratios of matrix elements into two members of the octet by the usual methods. Thus if we want to compare the decays

$$\Delta^{++} \to p + \pi^+$$
$$\Delta^0 \to p + \pi^-$$

we must expand the $(p\pi^-)$ state in i-spin eigenstates; this yields, for example, the ratio

$$|M(\Delta^0 \to p + \pi^-)/M(\Delta^{++} \to p + \pi^+)|^2 = \tfrac{1}{3} \tag{18.31}$$

[1] For an attempt to estimate symmetry breaking effects on the coupling constants, see F. J. Ernst, R. L. Warnock and K. C. Wali, *Phys. Rev.* **141,** 1354 (1966).

[2] J. J. de Swart, *Rev. Mod. Phys.* **35,** 916 (1963); P. Mc Namee and F. Chilton, *Rev. Mod. Phys.* **36,** 1005 (1964).

It was pointed out in Chapter 17 that the states of a representation of $SU(3)$ could also be classified by their u-spin rather than the i-spin. In the decuplet, for example, Δ^-, Y_1^{-*}, Ξ^{-*}, Ω^- form a $U = \frac{3}{2}$ state, Δ^0, Y_1^{0*}, Ξ^{0*} a $U = 1$ state, etc. We may now use u-spin conservation to obtain a number of results relating different i-spin multiplets. Consider the decay

$$Y_1^{-*} \rightarrow \Lambda^0 + \pi^- \tag{18.32}$$

The initial state has $U = \frac{3}{2}$. In the final state, π^- belongs to $U = \frac{1}{2}$, as can be seen from Fig. 17.8 which represents the octet. The Λ^0 state is not a u-spin eigenstate. In fact, the triplet is obtained by applying U_- to the neutron state and normalizing, with the $U = 0$ state orthogonal to it. Thus

$$\Psi(U = 1, U_3 = 0) \propto U_-\Psi(n) = \tfrac{1}{2}\Psi(\Sigma^0) - \frac{\sqrt{3}}{2}\Psi(\Lambda^0) \tag{18.33}$$

and

$$\Psi(U = 0, U_3 = 0) = \frac{\sqrt{3}}{2}\Psi(\Sigma^0) + \tfrac{1}{2}\Psi(\Lambda^0) \tag{18.34}$$

We can therefore express the Λ^0 state in terms of $U = 0$ and 1 eigenstates

$$\Psi(\Lambda^0) = \tfrac{1}{2}\Psi(U = 0) - \frac{\sqrt{3}}{2}\Psi(U = 1) \tag{18.35}$$

In the decay under consideration only the $U = 1$ component contributes. Thus the matrix element for the decay (18.32) can be written as

$$M(Y_1^{-*} \rightarrow \Lambda^0 + \pi^-) = \frac{1}{\sqrt{2}} M_1 \tag{18.36}$$

since

$$\Psi(U = 1, U_3 = 0) \otimes \Psi(U = \tfrac{1}{2}, U_3 = \tfrac{1}{2})$$
$$= \sqrt{\tfrac{2}{3}}\,\Psi(U = \tfrac{3}{2}, U_3 = \tfrac{1}{2}) + \sqrt{\tfrac{1}{3}}\,\Psi(U = \tfrac{1}{2}, U_3 = \tfrac{1}{2}).$$

On the other hand, in the decay

$$\Delta^- \rightarrow n + \pi^-$$

the neutron is entirely a $U = \frac{3}{2}$ state, so that the matrix element here is

$$M(\Delta^- \rightarrow n + \pi^-) = M_1$$

By combining u-spin conservation and i-spin conservation,[1] we can carry

[1] The technique was developed by C. A. Levinson, H. J. Lipkin, and S. Meshkov, *Phys. Rev. Letters* **10**, 361 (1963).

out many calculations without explicitly requiring the form of the decuplet Wigner coefficients. We shall see later that there is some evidence for the belief that the photon may be treated as a $U = 0$ state; this technique is therefore particularly useful for comparing, say, photoproduction rates. We shall not go into more detail at this point because both the decay rates and the electromagnetic properties of strongly interacting particles are discussed later.

A much more difficult task than making predictions on the basis of $SU(3)$ is an assessment of how good the predictions should be. The first step along the path to the solution of this unsolved problem is to ask about the nature of the symmetry-breaking mechanism, and, in particular, about the transformation properties of that mechanism. To clarify what is meant, let us consider an example from atomic physics.

When an external magnetic field is applied to an atom, the rotational symmetry is destroyed. We would thus expect that the degeneracy of energy levels of a given angular momentum and a variety of J_z values, characteristic of the symmetry, will be removed. The fact that the perturbation is of the form

$$V = -\boldsymbol{\mu} \cdot \mathbf{H}_{\text{ext}}$$

where $\boldsymbol{\mu}$ is the magnetic dipole operator, has consequences which stem entirely from the fact that $\boldsymbol{\mu}$ transforms as a *vector* under rotations. Hence, to first order[1]

$$\Delta E = -(\Psi_{JM}, \boldsymbol{\mu}\Psi_{JM}) \cdot \mathbf{H}_{\text{ext}} = -\alpha_J(\Psi_{JM}, \mathbf{J}\Psi_{JM}) \cdot \mathbf{H}_{\text{ext}}$$

We can deduce, without knowing anything about the structure of the atom, that the energy shifts for the lines will be of the form ($\mathbf{H}$ in z-direction)

$$\Delta E = -\alpha_J H_{\text{ext}} M$$

i.e., that the split levels will be equally spaced. For stronger fields, when lowest order perturbation theory is inadequate, the transformation properties of the perturbation will be more complicated. The term quadratic in $\mathbf{H}$, for example, will transform as a combination of a scalar term and an irreducible tensor corresponding to $J = 2$, etc. The form of the splitting can again be obtained from the transformation properties alone, although there will be more parameters, undetermined except through a knowledge of the dynamics.

A similar situation obtains for the broken eightfold way. Let us consider the mass operator $\mathcal{M}$. In the limit of exact symmetry, we have

$$(\Psi_{\nu,\alpha}, \mathcal{M}\Psi_{\nu,\alpha}) = m_\nu \tag{18.37}$$

[1] The second step in this equation follows from the Wigner-Eckart theorem.

(Here again ν labels the representation and α the states within it.) In reality, $\mathcal{M}$ will contain a large term transforming as a unitary singlet, as well as any number of other irreducible tensors. The experimental fact that (ignoring electromagnetism) the states of fixed Y, belonging to a given *i*-spin multiplet, are still degenerate implies that in the expansion of $\mathcal{M}$ into irreducible tensor operators, only operators with $Y = 0$, $T = 0$ appear. Consider the masses of an octet. Since

$$\mathbf{8} \otimes \mathbf{8} = \mathbf{1} \oplus \mathbf{8} \oplus \mathbf{8} \oplus \mathbf{10} \oplus \mathbf{10^*} \oplus \mathbf{27}$$

the masses of the members of the octet can be written in terms of four constants: a singlet contribution, two octet contributions, and a contribution from the **27**. The decuplet tensors do not contribute, because they do not have $Y = T = 0$ members. If, for some reason, some of the tensors are absent, there will be more masses than constants, and we get a relation between the masses—a *sum rule*. In his fundamental paper on unitary symmetry, Gell-Mann proposed that the octet contribution to the mass operator be dominant (after the singlet contribution). This implies a mass relation. In analogy with the atomic physics example discussed above, we write in general

$$(\Psi_{\nu,\alpha}, \mathcal{M}\Psi_{\nu,\alpha}) = m_\nu + c_1(\nu)(\Psi_{\nu,\alpha}, F_8\Psi_{\nu,\alpha}) + c_2(\nu)(\Psi_{\nu,\alpha}, D_8\Psi_{\nu,\alpha}) \quad (18.38)$$

Using the identification $Y = (2/\sqrt{3})\, F_8$ and the expression (17.65) for D_8, we obtain the so-called Gell-Mann—Okubo mass formula[1]

$$(\Psi_{\nu,\alpha}, \mathcal{M}\Psi_{\nu,\alpha}) = M_1 + M_2 Y + M_3(T(T+1) - \tfrac{1}{4}Y^2) \quad (18.39)$$

In the special case in which the representation ν is triangular, i.e., $p = 0$ *or* $q = 0$, it is known that the octet appears but once in the decomposition of $\boldsymbol{\nu}^* \otimes \boldsymbol{\nu}$. In that case the mass formula reads

$$(\Psi_{\nu,\alpha}, \mathcal{M}\Psi_{\nu,\alpha}) = M_1 + M_2 Y \quad (18.40)$$

This could also have been obtained from (18.39) by noting that, in general, for a triangular representation, there is a linear relation between Y and T, of the form

$$T = \tfrac{1}{2}Y + \text{const.}$$

An alternate derivation of the octet formula, which follows from (18.39) and reads

$$\frac{M_\Xi + M_N}{2} = \frac{3M_\Lambda + M_\Sigma}{4} \quad (18.41)$$

[1] S. Okubo, *Progr. Theoret. Phys.*, **27**, 949 (1962).

is found by writing the most general "mass" term that transforms as an octet, in terms of the 3 × 3 matrices. This is

$$m_1 \operatorname{tr} (\bar{B}B\lambda_8) + m_2 \operatorname{tr} (\bar{B}\lambda_8 B) \tag{18.42}$$

which, when worked out and added to the singlet form

$$m_0 \operatorname{tr} \bar{B}B \tag{18.43}$$

yields the same formula.

The formula (18.41) agrees with experiment remarkably well. The corresponding formula for the pseudoscalar mesons is not as good, but when it is written for the squares of the masses

$$m_K^{\ 2} = \frac{3m_\eta^{\ 2} + m_\pi^{\ 2}}{4} \tag{18.44}$$

it also agrees with experiment. Even more dramatically, the equal spacing formula for the decuplet (18.40) is also satisfied; in fact, the properties of the Ξ^* and Ω^-, not to mention their existence, were predicted on the basis of $SU(3)$, and this is the most spectacular reason, but by no means the only one, for a growing belief in the validity of $SU(3)$ and in the octet nature of the symmetry breaking.

This concludes our discussion of $SU(3)$, but we shall return to its predictions again and again, in the proper places.

PROBLEMS

1. Prove that for any four 3 × 3 traceless matrices,

$$\operatorname{tr} (ABCD + ACBD + ABDC + ACDB + ADBC + ADCB)$$
$$= \operatorname{tr} AB \operatorname{tr} CD + \operatorname{tr} AC \operatorname{tr} BD + \operatorname{tr} AD \operatorname{tr} BC$$

2. Consider a phenomenological amplitude $\operatorname{tr} (\bar{\Psi}\lambda_i \Gamma \Psi M)$ where Ψ is a baryon octet matrix and M is a meson octet matrix. Show that (a) the amplitude has octet transformation properties, (b) the amplitude has the following transformation properties under charge conjugation

$$\operatorname{tr} (\bar{\Psi}\lambda_i \Gamma\Psi M) \to \operatorname{tr} (\bar{\Psi}\lambda_i^T(-C\Gamma^T C^{-1})\Psi M)$$

(Note that Γ is a Dirac matrix, e.g., 1, γ^α, γ_5, or the like.)

3. Show that there are only three CP invariant amplitudes linear in λ_6, and that they may be written in the form

$$if' \operatorname{tr} ([\bar{\Psi}, \Psi][\lambda_6, M]) + ig' \operatorname{tr} (\{\bar{\Psi}, \Psi\}[\lambda_6, M])$$
$$+ ih'(\operatorname{tr} (\bar{\Psi}M) \operatorname{tr} (\Psi\lambda_6) - \operatorname{tr} (\bar{\Psi}\lambda_6) \operatorname{tr} (\Psi M))$$

when the meson acts as a scalar meson (or, equivalently, when the meson is pseudoscalar but parity is not conserved). Show that there are four parity conserving amplitudes linear in λ_6, and show that they may be written in the form

$$ia \operatorname{tr} ([\bar{\Psi}, \gamma_5\Psi]\{\lambda_6, M\}) + ib \operatorname{tr} (\{\bar{\Psi}, \gamma_5\Psi\}[\lambda_6, M])$$
$$+ ic \operatorname{tr} (\bar{\Psi}\lambda_6\gamma_5\Psi M)$$
$$+ id \operatorname{tr} (\bar{\Psi}M\gamma_5\Psi\lambda_6)$$

19

Baryon Resonances

The eight baryons discussed in Chapter 14 share the property of being stable under the strong interactions. This property does not seem to be fundamental; a 5% increase in the mass of the Σ hyperon would make it strongly unstable, because it could then decay according to

$$\Sigma \rightarrow \Lambda^0 + \pi$$

The short lifetime may affect the independence of the decay characteristics on the mode of formation of the particle, but the production process and the decay should not be coupled too strongly if the state is to be classed as a particle. In practice it turns out that noticeable particles have a more or less defined mass, with the uncertainty or *width* and order of magnitude smaller than the mass. Furthermore, the particles appear to be characterized by sharp quantum numbers, such as spin, parity, strangeness, and *i*-spin, and many of them can be classified into $SU(3)$ supermultiplets. This chapter briefly discusses the known baryon resonances, how they were detected, and, as far as possible, their quantum numbers. All of these quantities are obtained from an examination of the decay products of the particles.

The mass of an unstable particle is determined by examining the energies and momenta of sets of particles created in a high energy reaction. If an unstable particle is frequently produced in the reaction, the distribution of the invariant mass

$$M^2 = (\sum E_i)^2 - (\sum \mathbf{p}_i)^2$$

for certain subsets of particles (the decay products) will show a significant peak. The location of the peak gives the mass of the particle, and its width gives information about the decay rate. As an example, Fig. 19.1 shows the (mass)^2 of the $(\Sigma^+\pi^-)$ system created in the reaction

$$\pi^- + p \rightarrow \Sigma^+ + K^0 + \pi^-$$

with $2 \cdot 2 - 2 \cdot 4$ BeV/c incident π^-.

There are three clear peaks above the background of uncorrelated events. All three peaks are associated with resonances which will be discussed later in the chapter.

In the case of reactions for which the final state consists of three particles, for example, in

$$K^- + p \rightarrow \Lambda^0 + \pi^+ + \pi^-$$

it is very useful to plot the events on a graph whose axes measure the energies of two of the particles. As was shown in Chapter 15, the density of points on such a *Dalitz plot* is proportional to the square of the matrix element as a function of the two energies, so that correlations will be very evident. Figure 19.2 illustrates this for the reaction

$$K^- + p \rightarrow \Lambda^0 + \pi^+ + \pi^-$$

Note that in the center of mass system

$$W = \omega_1 + \omega_2 + \omega_3$$
$$0 = \mathbf{p}_1 + \mathbf{p}_2 + \mathbf{p}_3$$

where $(\omega_i, \mathbf{p}_i) \cdots$ are the energies and momenta of the three particles in

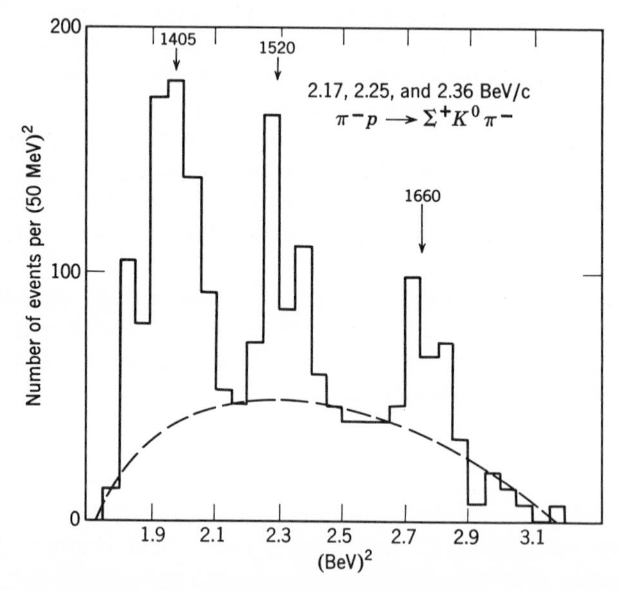

Fig. 19.1. Plot of the (mass)^2 of the $\Sigma^+\pi^-$ pair in the reaction $\pi^- + p \rightarrow \Sigma^+ + K^0 + \pi^-$. The dashed curve is the expected distribution for a constant matrix element and relativistic phase space.

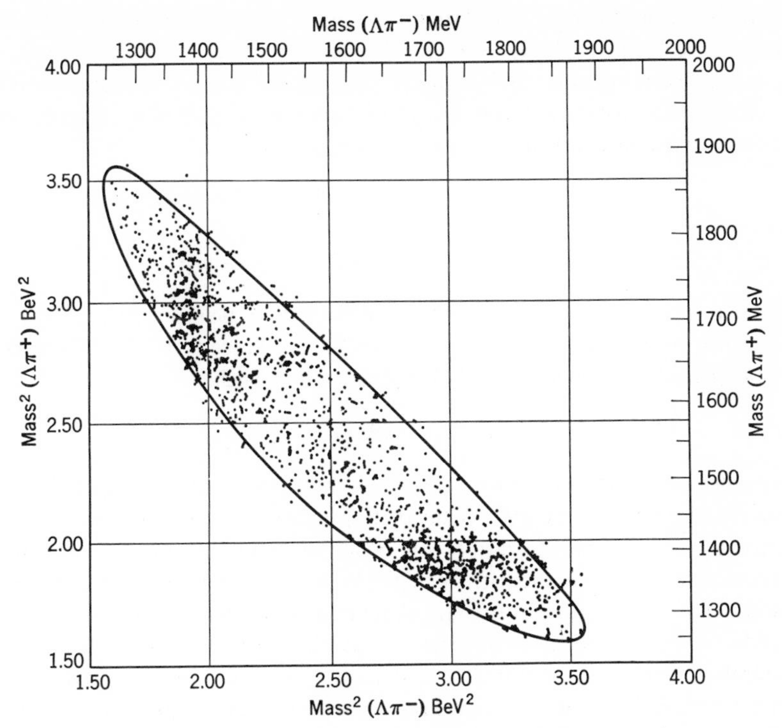

Fig. 19.2. Plot of the (mass)^2 of $\Lambda^0\pi^+$ and $\Lambda^0\pi^-$ in the reaction $K^- + p \rightarrow \Lambda^0 + \pi^+ + \pi^-$ at 1.51 BeV/c [from L. Alvarez et al., *Phys. Rev. Letters*, 10, 184 (1963)].

the final state, so that the invariant (mass)^2 of a pair

$$\begin{aligned}(\omega_2 + \omega_3)^2 - (\mathbf{p}_2 + \mathbf{p}_3)^2 &= (W - \omega_1)^2 - \mathbf{p}_1^{\,2} \\ &= W^2 + m_1^{\,2} - 2W\omega_1\end{aligned}$$

is proportional to the energy of the third particle. Thus the plot on Fig. 19.2 is equivalent to the type of Dalitz plot discussed in Chapter 15.

The usefulness of the Dalitz plot lies in that two resonances can often be looked at simultaneously, and distortions due to their interaction can at least be identified. Thus at lower energies, the two bands that appear in Fig. 19.2 cross. The events in the region belonging to both bands correspond to two pions resonating with the same Λ^0, very often, which can easily distort the shape of the resonance as well as shift its location.

In the experimental analysis of resonances, there are many possibilities of finding spurious peaks in the mass spectrum. Consider a simple example

$$K^- + p \rightarrow \Lambda^0 + \pi^+ + \pi^-$$

with the pions strongly correlated with the Λ^0 particles at a mass of 1385 MeV as seen in Fig. 19.2. *In a given experiment*, the $\pi^+\pi^-$ mass will show a strong peaking due to kinematic constraints alone! If the invariant (mass)^2 of the total system is s, and that of the (ij) pair is s_{ij}, energy and momentum conservation lead to the relation

$$s = s_{12} + s_{13} + s_{23} - m_1^2 - m_2^2 - m_3^2$$

so that

$$s_{12} = s - s_{13} - s_{23} + m_1^2 + m_2^2 + m_3^2$$

Thus peaking in s_{23} and s_{13} will be reflected in peaking in s_{12}. As an example, in the production of an $N^*(1518)$ resonance ($s = (1518)^2$), which subsequently decays into $N^*(1238)$ plus a pion, with

$$N^*(1238) \to N(938) + \pi(140)$$

a pion-pion correlation with "mass" $\sim$ 400 MeV may be expected, and has, in fact been observed. What distinguishes the spurious peak from a genuine one is that the latter will occur at the same mass in different experiments, whereas the former depends on the energy in the experiment. With four-particle final states the same complications can occur.

In our remarks so far we have used "unstable particles" and "resonances" interchangeably. Resonances are usually associated with a sharp peak in a cross section, but experience shows that the same peak (with the same quantum numbers) will occur in the final state of some complicated reaction as will occur in the elastic cross section for the scattering of the particles coming from the peak. Thus in the reaction

$$\pi^+ + p \to \pi^+ + \pi^- + \pi^+ + p$$

both (π^+p) and (π^-p) peak at 1238 MeV and a $(\pi^+\pi^-p)$ peak occurs near 1518 MeV. The same peaks appear in $\pi^+ + p \to \pi^+ + p$ and $\pi^- + p \to \pi^- + p$ cross sections. We do not, therefore, distinguish between these two qualitative descriptions of the phenomenon.

In addition to having a mass and a width, an unstable particle has other quantum numbers. Most important of these are the angular momentum and the parity. These are sometimes quite difficult to determine in practice, although sufficiently accurate experiments involving angular distributions, polarization correlations, and the like will always be able to yield this information. The internal quantum numbers, such as *i*-spin, can be determined either by looking for partners of different charge but the same mass or by looking at branching ratios for the decay into various channels. $SU(3)$ assignments are correspondingly harder to make, and here the Gell-Mann—Okubo mass formula (18.39) has played a major role in checking out potential partners of an incomplete set in a supermultiplet. The methods used are illustrated in our subsequent discussion of a number of baryon resonances.

Some $Y = 1$ Resonances

$N^*_{3/2}(1238)$.

This resonance,[1] also denoted by $\Delta(1238)$, was the first resonance to be discovered. Experiments carried out in the early 1950's showed a dramatic peak at 1238 MeV total center of mass energy, in the pion-nucleon scattering cross section. The width of the peak was approximately 120 MeV. It was observed both in the π^+p cross section and in the π^-p cross section, so that if we assume that the resonance is characterized by a definite *i*-spin, it must be in the $T = \frac{3}{2}$ state. The assignment was confirmed by the following considerations:

A decomposition of the pion-nucleon states into *i*-spin eigenstates yields[2]

$$\begin{aligned} \Psi_{\pi^+p} &= \Psi_{3/2,3/2} \\ \Psi_{\pi^0n} &= \sqrt{\tfrac{2}{3}}\Psi_{3/2,-1/2} + \sqrt{\tfrac{1}{3}}\Psi_{1/2,-1/2} \\ \Psi_{\pi^-p} &= \sqrt{\tfrac{1}{3}}\Psi_{3/2,-1/2} - \sqrt{\tfrac{2}{3}}\Psi_{1/2,-1/2} \end{aligned} \tag{19.1}$$

Thus the scattering amplitudes, assuming *i*-spin conservation are

$$\begin{aligned} (\Psi^{\text{out}}_{\pi^+p}, \Psi^{\text{in}}_{\pi^+p}) &= (\Psi^{\text{out}}_{3/2,3/2}, \Psi^{\text{in}}_{3/2,3/2}) \equiv A_{3/2} \\ (\Psi^{\text{out}}_{\pi^-p}, \Psi^{\text{in}}_{\pi^-p}) &= \tfrac{1}{3}A_{3/2} + \tfrac{2}{3}A_{1/2} \\ (\Psi^{\text{out}}_{\pi^0n}, \Psi^{\text{in}}_{\pi^-p}) &= \frac{\sqrt{2}}{3}A_{3/2} - \frac{\sqrt{2}}{3}A_{1/2} \end{aligned} \tag{19.2}$$

The total cross sections, at energies at which multiple meson production is insignificant, are

$$\begin{aligned} \sigma_{\pi^+} &= \rho\,|A_{3/2}|^2 \\ \sigma_{\pi^-} &= \sigma(\pi^-p \to \pi^-p) + \sigma(\pi^-p \to \pi^0n) = \rho(\tfrac{1}{3}\,|A_{3/2}|^2 + \tfrac{2}{3}\,|A_{1/2}|^2) \end{aligned} \tag{19.3}$$

[1] Until a standardized notation is accepted, we shall use N_T^* (mass) for a baryon resonance of *i*-spin T, hypercharge 1; for the resonances of hypercharge 0 and -1 respectively, the notation will be Y_T^* and Ξ_T^*. It has been proposed to use Δ_α for $T = \frac{3}{2}$, Σ_α for $T = 1$, N_α for $T = \frac{1}{2}$, and Λ_α for $T = 0$ states, with the subscript labeling the $SU(3)$ assignment.

[2] The *i*-spin state notation is Ψ_{T,T_3}.

The kinematical factors ρ are the same for both processes if we neglect the small mass differences. Thus the $T = \frac{3}{2}$ assignment predicts that at the resonance energy

$$\frac{\sigma_{\pi^+}}{\sigma_{\pi^-}} = 3 \tag{19.4}$$

which is well borne out by experiment. (See Fig. 19.3.)

A hint as to the angular momentum of the resonant state comes from the observation that the π^+p cross section at resonance is close to $8\pi/q_r^2$, where q_r is the center of mass momentum at the resonant energy. This value is the maximum value a cross section in the $J = \frac{3}{2}$ state can take. The angular distribution of the outgoing pion in the center of mass

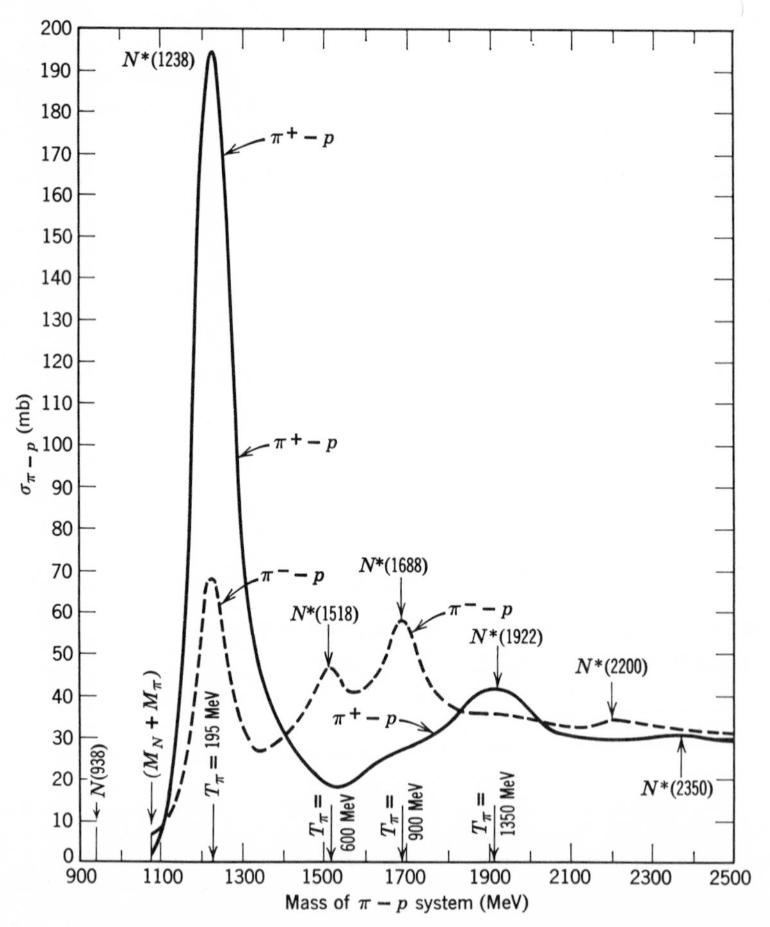

Fig. 19.3. Total cross sections for π^+p and π^-p scattering.

system relative to the initial pion direction depends on the angular momentum. The resonant state must be characterized by $J_z = \pm\frac{1}{2}$ with this choice of axis. In the decay into a pion and a spin-$\frac{1}{2}$ baryon, the final angular momentum states are

$$\begin{aligned}\Psi_J^{\prime 1/2} &= \sqrt{\frac{J+\frac{1}{2}}{2J}}\, Y^0_{J-1/2}(\theta,\phi)\chi^{1/2}_{1/2} + \sqrt{\frac{J-\frac{1}{2}}{2J}}\, Y^1_{J-1/2}(\theta,\phi)\chi^{-1/2}_{1/2} \\ \Psi_J^{\prime -1/2} &= \sqrt{\frac{J-\frac{1}{2}}{2J}}\, Y^{-1}_{J-1/2}(\theta,\phi)\chi^{1/2}_{1/2} + \sqrt{\frac{J+\frac{1}{2}}{2J}}\, Y^0_{J-1/2}(\theta,\phi)\chi^{-1/2}_{1/2}\end{aligned} \tag{19.5}$$

(where $Y_l^m(\theta, \phi)$ describe the orbital angular momentum in the decay), when the parity is such that $l = J - \frac{1}{2}$ is the orbital angular momentum, and

$$\begin{aligned}\Psi_J^{\prime 1/2} &= \sqrt{\frac{J+\frac{3}{2}}{2J+2}}\, Y^1_{J+1/2}(\theta,\phi)\chi^{-1/2}_{1/2} - \sqrt{\frac{J+\frac{1}{2}}{2J+2}}\, Y^0_{J+1/2}(\theta,\phi)\chi^{1/2}_{1/2} \\ \Psi_J^{\prime -1/2} &= \sqrt{\frac{J+\frac{1}{2}}{2J+2}}\, Y^0_{J+1/2}(\theta,\phi)\chi^{-1/2}_{1/2} - \sqrt{\frac{J+\frac{3}{2}}{2J+2}}\, Y^{-1}_{J+1/2}(\theta,\phi)\chi^{1/2}_{1/2}\end{aligned} \tag{19.6}$$

when the parity conservation requires $l = J + \frac{1}{2}$. For an unpolarized initial state, the angular distribution is determined by

$$I(\theta) \propto \tfrac{1}{2} \sum_{\substack{\text{nucleon}\\ \text{spins}}} \{|\Psi_J^{\prime 1/2}|^2 + |\Psi_J^{\prime -1/2}|^2\} \tag{19.7}$$

For $J = \frac{1}{2}$ we get $I(\theta) = \text{const}$; for $J = \frac{3}{2}$ we find

$$I(\theta) = \text{const}\,(1 + 3\cos^2\theta) \tag{19.8}$$

and the observed angular distribution at resonance agrees with this prediction. It should be noted that the angular distribution depends only on the value of J, and not on the parity, i.e., the orbital angular momentum. This is most easily seen in the helicity representation, where the concept of orbital angular momentum never enters.

Thus to determine the parity of the resonance more detailed measurements are required. The conclusion that the parity of the resonance is *even*, i.e., that the orbital angular momentum is 1,[1] was actually first reached by making a phase-shift analysis of the pion-nucleon scattering data and noting that the dominant phase shift (which could be either $P_{3/2}$ or $D_{3/2}$) behaved as q^3 rather than q^5 for small q. The conclusion was

[1] The parity of a resonance decaying into a pion and a nucleon with orbital angular momentum l is $(-1)^{l+1}$.

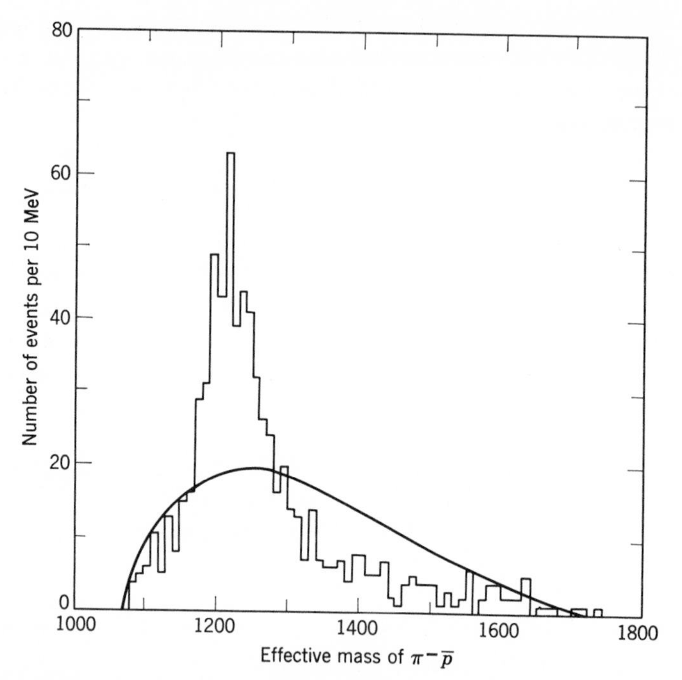

Fig. 19.4. Mass spectrum of $\pi^-\bar{p}$ in the reaction $p + \bar{p} \rightarrow p + \bar{p} + \pi^+ + \pi^-$ at 3.25 BeV/c. The spectrum expected on the basis of phase space alone is drawn in [from C. Baltay et al. "Nucleon Structure," Stanford Press, 1964].

borne out by experiments on the polarization of the outgoing nucleon in pion-nucleon scattering.

The $N^*_{3/2}(1238)$ has been seen in a large number of reactions, as has its antiparticle (Fig. 19.4) in the reaction

$$\bar{p} + p \rightarrow \bar{p} + p + \pi^+ + \pi^-$$

$N^*_{1/2}(1518)$.

The next peak in the pion-nucleon cross section curve shown on Fig. 19.3 occurs at 1518 MeV center of mass energy, in the π^-p cross section only. It is therefore a $T = \frac{1}{2}$ resonance; its width is again of the order of 120 MeV. In contrast to the so-called "3-3 resonance" discussed above, a phase-shift analysis of the angular distribution in the scattering is complicated by the somewhat larger number of phase shifts required at the higher energies as well as by the fact that the cross section for the inelastic processes such as

$$\pi^- + p \rightarrow \pi^+ + \pi^- + n$$

is not insignificant, so that complex phase shifts are required. In spite of that the phase-shift analysis has been carried out and supports the assignment $J = \frac{3}{2}$, parity *odd*, so that the decay is a D-state decay.[1] Actually, the quantum numbers were first determined through an analysis of the photoproduction data,

$$\gamma + p \rightarrow p + \pi^0$$

in which the resonance was first discovered.[2] The analysis is sufficiently involved so that we will merely outline the main steps, without deriving any of the formulas:[3]

The angular distribution of the photoproduced π^0 is well fitted by

$$\frac{d\sigma}{d\Omega} \propto (5 - 3\cos^2\theta) \tag{19.9}$$

characteristic of a $J = \frac{3}{2}$ resonance. If the parity is even, this would be a resonance having the same quantum numbers as the $T = \frac{3}{2}$ resonance at 1238 MeV. If we take the point of view that the principal features of the reaction

$$\gamma + p \rightarrow p + \pi^0$$

in the energy region between the two resonances are determined by contributions from them alone, the angular distribution should be that of (19.9). If the second resonance has odd parity, the angular distribution will be given by

$$\frac{d\sigma}{d\Omega} = \frac{1}{2}(|A|^2 + |B|^2)(5 - 3\cos^2\theta) - 2\,\mathrm{Re}\,AB^*\cos\theta \tag{19.10}$$

where A and B represent the energy dependent contributions from the two resonances. The data favors (19.9). Unfortunately, this is not decisive. It can be shown[4] that the phases of A and B are those of pion-nucleon scattering in the $P_{3/2}$ and $D_{3/2}$ states respectively. If the resonance is a $D_{3/2}$ resonance, then in the intermediate energy region the $P_{3/2}$ phase shift has gone through 90°, whereas the $D_{3/2}$ phase shift is approaching 90°, so that it is likely that A and B are close to 90° out of phase over much of the region, making the coefficient Re AB^* small. Sakurai[5] pointed out

[1] P. Auvil, C. Lovelace, A. Donnachie, and A. T. Lea, *Physics Letters* **12,** 76 (1964). This paper contains references to other work on this subject.

[2] M. Heinberg, W. M. McClelland, F. Turkot, W. M. Woodward, R. R. Wilson, and D. M. Zipoy, *Phys. Rev.* **110,** 1211 (1958); J. W. DeWire, H. E. Jackson, and R. Littauer, *Phys. Rev.* **110,** 1208 (1958); P. C. Stein and K. C. Rogers, *Phys. Rev.* **110,** 1209 (1958); F. P. Dixon and R. L. Walker, *Phys. Rev. Letters* **1,** 142, 458 (1958).

[3] R. F. Peierls, *Phys. Rev.* **118,** 325 (1960).

[4] See Chapter 26 page 449, for a proof which can easily be adapted to photoproduction.

[5] J. J. Sakurai, *Phys. Rev. Letters* **1,** 258 (1958).

that the polarization of the outgoing nucleon is given by

$$P = \frac{2 \operatorname{Im} AB^* \sin\theta}{d\sigma/d\Omega} \tag{19.11}$$

so that if the second resonance is of odd parity, appreciable polarization could be expected. This is borne out by measurements,[1] and the assignment $(\frac{3}{2})^-$ is generally accepted for the resonance.

The resonance $N^*_{3/2}(1518)$ decays primarily into $\pi + N$, but there is an approximately 20% branching ratio for the decay

$$N^*_{1/2} \rightarrow \pi + \pi + N$$

Some of the time, at least, the pion and the nucleon are correlated in the form of an $N^*_{3/2}(1238)$ state. It is an interesting phenomenon that in the decay

$$N^*_{1/2} \rightarrow p + \pi^+ + \pi^-$$

both $(p\pi^+)$ and $(p\pi^-)$ are products of an $N^*_{3/2}$ decay.[2]

$N^*_{1/2}(1688)$.

This resonance whose width is 100 MeV, was also discovered in the photoproduction of π^+ and π^0.[3] It appears as a peak in the π^-p cross section but not in the π^+p cross section, so that it must be a $T = \frac{1}{2}$ state. An analysis of the angular distribution in the center of mass system, in the form

$$\frac{d\sigma}{d\Omega} = \sum a_n \cos^n \theta \tag{19.12}$$

shows that there is little scattering of waves with $J = \frac{7}{2}$ or higher in the vicinity of the resonance,[4] while the photoproduction angular distribution is consistent with a $J = \frac{5}{2}$ assignment. The pion-nucleon phase-shift analysis of Auvil et al.[5] indicates that the F_{15} (F-wave, $T = \frac{1}{2}$, $J = \frac{5}{2}$) phase shift is rapidly rising at energies below the resonance, suggesting that $\frac{5}{2}^+$ is the appropriate assignment. The parity assignment is confirmed by measurements of the proton polarization in photoproduction in the

[1] P. C. Stein, *Phys. Rev. Letters* **2,** 473 (1959).

[2] J. Brown, G. Goldhaber, S. Goldhaber, J. Kadyk, T. O'Halloran, B. Shen, and G. Trilling, quoted in G. Goldhaber, Conference on Particle and High Energy Physics at Boulder, Colorado (1964).

[3] M. Heinberg et al., *loc. cit.*

[4] R. J. Cence, *Proceedings of the International Conference on Nucleon Structure at Stanford University*, Stanford University Press (1964).

[5] P Auvil et al., *loc cit.*

region between the 1518 and the 1688 resonance,[1] suggesting that these are of opposite parity, since the polarization is large.[2]

Some Higher Resonances. The data on pion-nucleon cross sections provides evidence for five more broad peaks. Such peaks, at these energies, are no more than departures from a completely monotonic behavior, but it must be remembered that at high energies, when many channels are open, a resonance in a particular state can only contribute to the cross section to a limited extent. At any rate the states are:

$N^*_{3/2}(1920)$ with width of the order of 200 MeV and spin-parity assignment $\frac{7}{2}^+$;

$N^*_{1/2}(2190)$ with width 200 MeV and a tentative assignment of $\frac{7}{2}^-$;

$N^*_{3/2}(2360)$ with width 200 MeV and an unconfirmed $\frac{9}{2}^-$ assignment;

$N^*_{1/2}(2650)$ with width 200 MeV and a tentative $\frac{9}{2}^+$ assignment;

$N^*_{3/2}(2825)$ with width 260 MeV and an unconfirmed spin-parity assignment of $\frac{11}{2}^+$.

References to the experiments in which these states were identified may be found in "Data on Particles and Resonant States"[3] which promises to be an annual publication and which should be considered an essential supplement to this chapter and the next one.

The $Y = 0$ Resonances

A very detailed treatment of the strange resonances, those with $Y = 0$ as well as $Y = -1$, appears in a review by Dalitz.[4] The treatment which we present is very much shorter, but necessary for completeness.

$Y_1^*(1385)$.

This resonance in the $\Lambda^0\pi$ system was first discovered in 1961 in the reaction[5]

$$K^- + p \rightarrow \Lambda^0 + \pi^+ + \pi^-$$

[1] C. Mencuccini, R. Querzoli, and G. Salvini, Proceedings of the Aix-en-Provence International Conference on Elementary Particles, C.E.N. Saclay (1961) p. 17.

[2] The $F_{5/2}$ assignment is further supported by the analysis of the differential π^-p cross section measurements reported by P. J. Duke, D. P. Jones, M. A. R. Kemp, P. G. Murphy, J. D. Prentice, J. J. Thresher, H. H. Atkinson, C. R. Cox, and K. S. Heard, *Phys. Rev. Letters* **15,** 468 (1965).

[3] A. H. Rosenfeld, A. Barbaro-Galtieri, W. H. Barkas, P. L. Bastien, J. Kirz, and M. Roos, *Rev. Mod. Phys.* **37,** 633 (1965).

[4] R. H. Dalitz, *Ann. Rev. Nucl. Sci.* **13,** Annual Reviews Inc., Palo Alto, Calif. (1963).

[5] M. H. Alston, L. W. Alvarez, P. Eberhard, M. L. Good, W. Graziano, H. K. Ticho, and S. G. Wojcicki, *Phys. Rev. Letters* **5,** 520 (1960).

The Dalitz plot in Fig. 19.2 shows clear evidence for the correlation, which has also been observed in the reactions[1]

$$\pi^- + p \to \Lambda^0 + \pi^- + K^+$$
$$\to \Lambda^0 + \pi^0 + K^0$$
$$\pi^+ + p \to \Lambda^0 + \pi^+ + K^+$$
$$K^- + d \to p + \Lambda^0 + \pi^-$$
$$K^- + \mathrm{He}^4 \to \mathrm{He}^3 + \Lambda^0 + \pi^-$$

All experiments agree on the mass and width

$$M_{Y_1^*} = 1385 \text{ MeV}$$
$$\Gamma = 50 \text{ MeV}$$

The antiparticle $\overline{Y_1^*}$ has also been observed in the reaction

$$\bar{p} + p \to \Lambda^0 + \bar{\Lambda}^0 + \pi^- + \pi^+$$

The decay mode $Y_1^* \to \Sigma + \pi$ is observed with

$$\Gamma(Y_1^* \to \Sigma + \pi) \simeq 0.1\, \Gamma(Y_1^* \to \Lambda^0 + \pi).$$

The method used for the determination of the spin and parity of the Y_1^* involves a study of the distribution of the polarization of the Λ^0 particle.[1] As will be shown in Chapter 33, the decay characteristics of the Λ^0 make a measurement of the Λ^0 polarization relatively easy. Thus the distribution of $\langle \mathbf{P} \rangle \cdot \mathbf{n}$ and $\langle \mathbf{P} \rangle \cdot \mathbf{m}$, where $\langle \mathbf{P} \rangle$ is the Λ^0 polarization and $\mathbf{n}$ and $\mathbf{m}$ are two directions defined in Fig. 19.5, can be measured. The

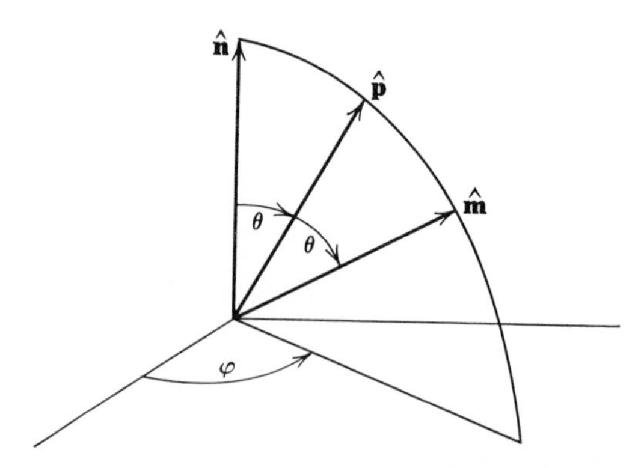

Fig. 19.5. The geometry for the decay of the Y_1^* · $\hat{\mathbf{n}}$ is the normal to the production plane, $\hat{\mathbf{p}}$ is the direction of the Λ^0 momentum, and $\hat{\mathbf{m}} = 2\hat{\mathbf{p}}(\hat{\mathbf{p}} \cdot \hat{\mathbf{n}}) - \hat{\mathbf{n}}$ is indicated in the figure.

[1] P. L. Bastien, M. Ferro-Luzzi, and A. H. Rosenfeld, *Phys. Rev. Letters* **6,** 702 (1961).

dependence of that distribution on the spin and parity of the Y_1^*, which is the source of the Λ^0's, is obtained as follows:[1]

We define $\hat{\mathbf{n}}$, the normal to the production plane, as quantization axis. The Y_1^* produced in a strong, parity-conserving reaction is in general described by a state vector

$$\Psi_\alpha = \sum_M C_M{}^\alpha \Psi_{JM} \tag{19.14}$$

($\alpha = J, J-1, J-2, \ldots, -J$), and the density matrix $\rho_{MM'}$ is given by

$$\rho_{MM'} = \sum_\alpha C_M{}^\alpha C_{M'}^{\alpha *} \tag{19.15}$$

In the case, however, in which the basis is chosen so that the z-axis coincides with the direction in which the spin of the Y_1^* *must* point (if polarized), then[2]

$$\begin{aligned}\langle J_3{}^k \rangle &= \sum_\alpha (\Psi_\alpha, J_3{}^k \Psi_\alpha) = \sum_\alpha \sum_{MM'} C_{M'}^{\alpha *} C_M{}^\alpha (\Psi_{JM'}, J_3^k \Psi_{JM}) \\ &= \sum_{MM'} \rho_{MM'} (J_3{}^k)_{MM'} = \sum_M \rho_{MM} M^k \end{aligned} \tag{19.16}$$

but any product of J_i which contains a J_i with $i \neq 3$ must have vanishing expectation value. From this we can show that $\rho_{MM'}$ must be diagonal, with the diagonal elements representing the probability that the z component of the Y_1^* spin has value M. The final $\pi\Lambda$ state may in general be written as[3]

$$\begin{aligned}\Phi^{\text{out}}_{\mathbf{p},\lambda,-\mathbf{p},0} &= \sum_\alpha \Psi_\alpha (\Psi_\alpha, \Phi^{\text{out}}_{\mathbf{p},\lambda;-\mathbf{p},0}) + \cdots \\ &= \sum_{MM'} \Psi_{JM} \rho_{MM'} \sqrt{\frac{2J+1}{4\pi}}\, \mathcal{S}_\lambda{}^*\, D^{(J)}_{M'\lambda}(\phi, \theta, 0)\end{aligned} \tag{19.17}$$

so that in general

$$\begin{aligned}\langle \boldsymbol{\sigma} \cdot \hat{\mathbf{n}} \rangle &= (\Phi_{\mathbf{p},\lambda',-\mathbf{p},0}, (\boldsymbol{\sigma} \cdot \hat{\mathbf{n}})_{\lambda'\lambda} \Phi_{\mathbf{p},\lambda,-\mathbf{p},0}) \\ &= \sum_{\lambda\lambda'} \sum_{MM'} (\boldsymbol{\sigma} \cdot \hat{\mathbf{n}})_{\lambda'\lambda} (\Phi_{\mathbf{p},\lambda',-\mathbf{p},0}, \Psi_{JM}) \\ &\qquad\qquad \times \rho_{MM'} \sqrt{\frac{2J+1}{4\pi}}\, \mathcal{S}_\lambda{}^*\, D^{(J)}_{M'\lambda}(\phi, \theta, 0) \\ &= \frac{2J+1}{4\pi} \sum_{\lambda\lambda'} \sum_{MM'} \mathcal{S}_{\lambda'} \mathcal{S}_\lambda{}^* (\boldsymbol{\sigma} \cdot \hat{\mathbf{n}})_{\lambda'\lambda} \rho_{MM'} \\ &\qquad\qquad \times D^{(J)}_{M'\lambda}(\phi, \theta, 0)\, D^{(J)*}_{M\lambda'}(\phi, \theta, 0)\end{aligned} \tag{19.18}$$

[1] This calculation is designed as a prototype of the general determination of spins and parities of resonances and that is why we are going into so much detail. An alternative approach (worked out for the general case) does not use the helicity formalism we favor. [N. Byers and S. Fenster, *Phys. Rev. Letters* **11,** 52 (1963).]

[2] Here $J_3{}^k$ denotes J_3 raised to the kth power.

[3] The terms left out in the expansion of the state vector in terms of a complete set is irrelevant. The second line in (19.17) makes use of material derived in Chapter 4, page 84.

In our case the Y_1^* density matrix is diagonal. We also have

$$(\boldsymbol{\sigma} \cdot \hat{\mathbf{m}})_{\lambda'\lambda} = ((\sigma_+ e^{-i\phi} + \sigma_- e^{i\phi}) \sin\theta + \sigma_3 \cos\theta)_{\lambda'\lambda} \tag{19.19}$$

with the angles relative to the direction **p** (since we are talking about helicity labels) appearing. Since $\sigma_\pm$ are raising and lowering operators we get

$$\begin{aligned}(\boldsymbol{\sigma} \cdot \hat{\mathbf{m}})_{\lambda'\lambda} &= 2\lambda \cos\theta\, \delta_{\lambda'\lambda} \\ &+ e^{-i\phi} \sin\theta \sqrt{(\tfrac{1}{2} - \lambda)(\tfrac{3}{2} + \lambda)}\, \delta_{\lambda',\lambda+1} \\ &+ e^{i\phi} \sin\theta \sqrt{(\tfrac{1}{2} + \lambda)(\tfrac{3}{2} - \lambda)}\, \delta_{\lambda',\lambda-1}\end{aligned} \tag{19.20}$$

The expression for $(\boldsymbol{\sigma} \cdot \hat{\mathbf{n}})_{\lambda'\lambda}$ is obtained by changing ϕ to $\pi + \phi$, i.e., changing the signs of the second and third terms. A little algebra yields the following result

$$\left\{\begin{matrix}\langle \boldsymbol{\sigma} \cdot \hat{\mathbf{m}} \rangle \\ \langle \boldsymbol{\sigma} \cdot \hat{\mathbf{n}} \rangle\end{matrix}\right\} = \sum_M P(M)\{|\mathcal{S}_{1/2}|^2 \cos\theta (d^J_{M,1/2}(\theta))^2 - |\mathcal{S}_{-1/2}|^2 \cos\theta (d^J_{M,-1/2}(\theta))^2 \\ \pm 2\,\mathrm{Re}(\mathcal{S}_{1/2}\mathcal{S}^*_{-1/2}) \sin\theta\, d^J_{M,1/2}(\theta)\, d^J_{M,-1/2}(\theta)\} \tag{19.21}$$

Parity conservation in the Y_1^* decay requires that

$$\mathcal{S}_{1/2} = \mathcal{S}_{-1/2} \tag{19.22}$$

so that we finally obtain

$$\left\{\begin{matrix}\langle \boldsymbol{\sigma} \cdot \hat{\mathbf{m}} \rangle \\ \langle \boldsymbol{\sigma} \cdot \hat{\mathbf{n}} \rangle\end{matrix}\right\} \propto \sum_M P(M)\{\cos\theta((d^J_{M,1/2}(\theta))^2 - (d^J_{M,-1/2}(\theta))^2) \\ \pm 2 \sin\theta\, d^J_{M,1/2}(\theta)\, d^J_{M,-1/2}(\theta)\} \tag{19.23}$$

The experimental distributions[1]

$$\begin{aligned}\langle \boldsymbol{\sigma} \cdot \hat{\mathbf{m}} \rangle &= 0.36 \pm 0.14 + (-3.5 \pm 1.05) \cos^2\theta + (4.0 \pm 1.3) \cos^4\theta \\ \langle \boldsymbol{\sigma} \cdot \hat{\mathbf{n}} \rangle &= -(0.39 \pm 0.15) + (1.36 \pm 0.30) \cos^2\theta\end{aligned} \tag{19.24}$$

fit the $\frac{3}{2}^+$ assignment well. This is the accepted result.

$Y_0^*(1405)$.

The existence of a $\Sigma^+\pi^-$ peak and a $\Sigma^-\pi^+$ peak, without the corresponding singly or doubly charged partners, was found in the reactions[2]

$$K^- + p \to \left\{\begin{matrix}\Sigma^\pm + \pi^\mp + \pi^+ + \pi^- \\ \Sigma^0 + \pi^0 + \pi^+ + \pi^-\end{matrix}\right.$$

[1] J. B. Shafer, J. J. Murray, and D. O. Huwe, *Phys. Rev. Letters* **10,** 176 (1963). The parity is obtained by observing that the angular dependence of $\langle \boldsymbol{\sigma} \cdot \hat{\mathbf{m}} \rangle$ for the decay of a particle $(J)^\pm$ is the same as the angular dependence of $\langle \boldsymbol{\sigma} \cdot \hat{\mathbf{n}} \rangle$ for a particle with quantum numbers $(J)^\mp$.

[2] M. H. Alston, L. W. Alvarez, M. Ferro-Luzzi, A. H. Rosenfeld, H. K. Ticho, and S. G. Wojcicki, *Proc. Intern. Conf. High Energy Physics, CERN*, Geneva (1962), p. 311, and R. H. Dalitz, *loc. cit.*

and

$$K^- + p \rightarrow \Sigma^\pm + \pi^\mp + \pi^0$$

Figure 19.1 shows the mass distribution from which we ascertain that

$$M_{Y_0^*} = 1405 \text{ MeV}$$
$$\Gamma = 35 \text{ MeV}$$

An analysis of the angular distribution and the polarization of the Σ^+ in

$$Y_0^* \rightarrow \Sigma^+ + \pi^-$$
$$\quad\quad \hookrightarrow p + \pi^0$$

indicates that $J = \frac{1}{2}$.[1] There is no evidence whether the parity is odd or even. It was suggested a long time ago by Dalitz and Tuan[2] that the $\bar{K} - N$ scattering data at low energies suggest that in the absence of other channels (such as $\pi\,\Sigma$) there would exist a $\bar{K} - N$ bound state with $T = 0$. If the $\Sigma\,\pi$ channel were available, such a bound state would no longer be stable and would appear as a resonance in the $\Sigma\,\pi$ system. If this resonance is indeed the bound state, the parity should be odd, since the $\bar{K}$ meson is pseudoscalar and $l = 0$.

$Y_0^*(1520)$.

The existence of a second $T = 0$ resonance at 1520 MeV may be deduced from data such as appear in Fig. 19.1. This resonance was also seen in the reaction[3]

$$K^- + p \rightarrow K^- + p + \pi^+ + \pi^-$$

where a plot of the K^-p mass shows a peak at 1520 MeV with a rather narrow width of

$$\Gamma \cong 16 \text{ MeV}$$

There are no charged counterparts to this resonance, hence the $T = 0$ assignment.

The $Y_0^*(1520)$ has the following decay modes[4]

$$\begin{aligned} Y_0^* &\rightarrow (\Sigma^+\,\pi^-) + (\Sigma^-\,\pi^+) + (\Sigma^0\,\pi^0) \qquad 55 \pm 7\% \\ &\rightarrow \bar{K}N \qquad 29 \pm 4\% \\ &\rightarrow \Lambda\pi\pi \qquad 16 \pm 2\% \end{aligned}$$

[1] A. Engler, H. E. Fisk, R. W. Kraemer, C. M. Meltzer, J. B. Westgard, T. C. Bacon, D. G. Hill, H. W. Hopkins, D. K. Robinson, and E. O. Salant, *Phys. Rev. Letters* **15,** 224 (1965).

[2] R. H. Dalitz and S. F. Tuan, *Ann. Phys.* (*N.Y.*), **3,** 307 (1960). See also J. K. Kim, *Phys. Rev. Letters* **14,** 29 (1965).

[3] S. P. Almeida and G. R. Lynch, *Physics Letters* **9,** 204 (1964).

[4] M. Watson, M. Ferro-Luzzi, and R. Tripp, *Phys. Rev.* **131,** B2248 (1963).

The spin and parity of the resonance are, roughly speaking, deduced as follows.[1]

The elastic K^-p scattering cross section may be fitted with the formula

$$\frac{d\sigma}{d\Omega} = \frac{1}{q^2}(A + B\cos\theta + C\cos^2\theta) \tag{19.25}$$

where B is small and C shows a sharp peak at the resonance. (See Fig. 19.6.)

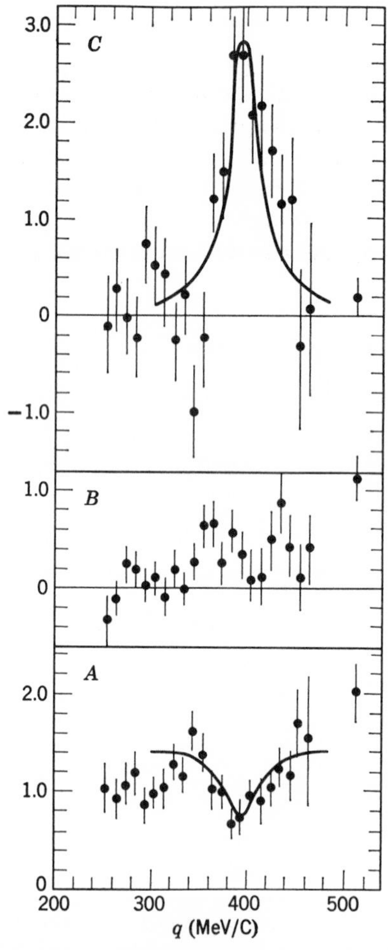

Fig. 19.6. Energy dependence of the coefficients A, B, C defined in equation (19.25). [From Ferro-Luzzi et al., *Phys. Rev. Letters* **8**, 28 (1962)].

[1] G. Snow, *Proc. Intern. Conf. on High Energy Physics, CERN*, Geneva (1962), p. 795.

At the lower energies it is presumably the $S_{1/2}$ scattering that dominates and in the resonant region, too, the $S_{1/2}$ amplitude is responsible for most of the nonresonant scattering (recall that 1500 MeV is fairly close to the threshold for $\bar{K}N$ scattering). The fact that no powers higher than $\cos^2\theta$ are needed to fit the angular distribution implies that $J = \frac{3}{2}$. The dip in the parameter A as well as the virtual absence of a $\cos\theta$ term imply that the resonance is a $D_{3/2}$ resonance, as the following expression shows

$$\frac{d\sigma}{d\Omega} = \rho\{|(S_{1/2} - D_{3/2}) + \cos\theta(2P_{3/2} + P_{1/2}) + 3D_{3/2}\cos^2\theta|^2 + \sin^2\theta\,|P_{1/2} - P_{3/2} + 3\cos\theta\, D_{3/2}|^2\} \quad (19.26)$$

This expression will actually be derived later (Chapter 23). It is correct on the assumption that in the partial waves $J \leq \frac{3}{2}$.

The data on the polarization of the Σ in the reaction

$$K^- + p \to \Sigma + \pi$$

have been used to show that only for odd relative $K\Sigma$ parity can we get a "reasonable" explanation of these data. Since the argument, which involves the properties of the resonance, is sufficiently involved, we merely mention the point here and refer the reader to the review of Dalitz for details.

$Y_1^*(1660)$.

This resonance, whose $T = 1$ assignment is established by its decay into $\Lambda\pi$, has a width of 45 ± 5 MeV. Its principal decay modes are[1]

$$\begin{aligned} Y_1^* \to \Lambda^0\pi &\sim 5\% \\ \Sigma\,\pi &\sim 30\% \\ \Lambda\pi\pi &\sim 20\% \\ \bar{K}N &\sim 15\% \\ Y_0^*(1405) + \pi &\sim 30\%^{2} \end{aligned}$$

The paper of Alvarez et al. contains some discussion of the angular distribution in the two-particle decay channels, which suggests that $J = \frac{3}{2}$ is a spin assignment to be preferred to $J = \frac{1}{2}$ or $J = \frac{5}{2}$. The evidence is, however, sufficiently weak so that no firm conclusion can be drawn.

[1] L. Alvarez, M. Alston, M. Ferro-Luzzi, D. Huwe, G. Kalbfleisch, D. Miller, J. Murray, A. H. Rosenfeld, J. B. Shafer, F. T. Solmitz, and S. G. Wojcicki, *Phys. Rev. Letters* **10,** 184 (1963).

[2] P. Eberhard, F. T. Shively, R. R. Ross, D. M. Siegel, J. R. Ficenec, R. I. Hulsizer, D. W. Mortara, M. Pripstein, and W. P. Swanson, *Phys. Rev. Letters* **14,** 466 (1965).

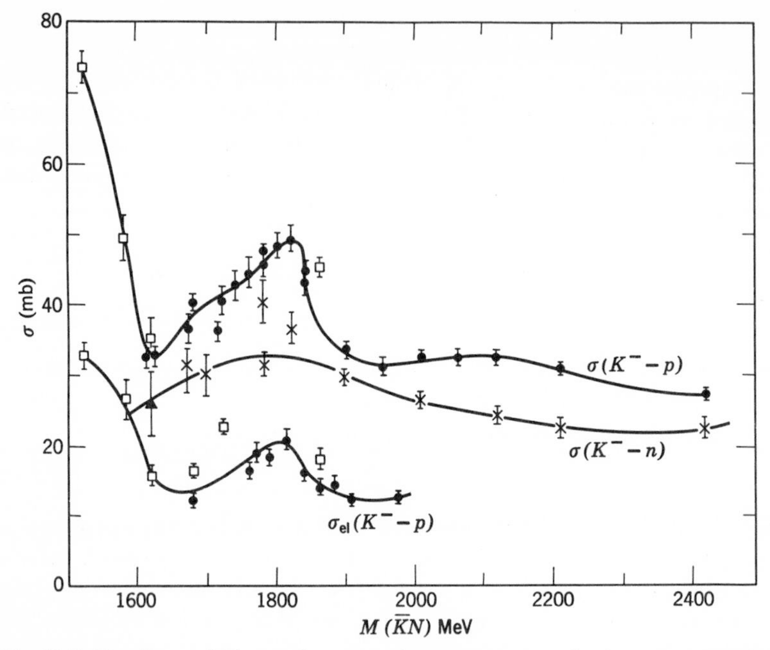

Fig. 19.7. The K^-p and K^-n total cross section measurements in the range 1520 MeV to 2500 MeV [from R. H. Dalitz, *Ann. Rev. Nucl. Sci.* **13** (1963)]

$Y_1^*(1765)$.

This resonance, whose width is of the order of 70 MeV has a spin-parity assignment of $\frac{5}{2}^-$. Its decay modes are[1]

$$Y_1^* \to \bar{K}N(60\,\%)$$
$$Y_1^* \to \Lambda^0\pi(16\,\%)$$
$$Y_1^* \to Y_1^*(1385) + \pi(10\,\%)$$
$$Y_1^* \to Y_0^*(1520) + \pi(10\,\%)$$
$$Y_1^* \to \Sigma\pi(<3\,\%)$$

$Y_0^*(1815)$.

The K^-p and K^-n total cross section measurements show a broad peak of width 50 MeV near 1815 MeV in the K^-p data, without a corresponding peak in the K^-n data, implying that the peak is associated with $T = 0$ (Fig. 19.7). The evidence from the angular distribution favors the

[1] R. B. Bell, R. W. Birge, Y. L. Pan, and R. T. Pu, *Phys. Rev. Letters* **16**, 203 (1966) where references to earlier work may be found.

$J = \frac{5}{2}^+$ assignment.[1] The decay $Y_0^* \to Y_1^*(1385) + \pi$ is observed with a 15% frequency.

The $Y = -1$ Resonances

$\Xi^*_{1/2}(1530)$.

A study of the reactions

$$K^- + p \to \begin{cases} K^+ + \Xi^- + \pi^0 & (a) \\ K^0 + \Xi^- + \pi^+ & (b) \\ K^0 + \Xi^0 + \pi^0 & (c) \end{cases}$$

revealed the existence of a resonance in the $\Xi\pi$ system with mass and width[2]

$$M = 1529 \pm 5 \text{ MeV}$$

$$\Gamma = 7.5 \pm 1.7 \text{ MeV}$$

Evidence for the $T = \frac{1}{2}$ assignment comes from a comparison of the production processes (a) and (b). If $T = \frac{3}{2}$, only the $T = 1$ component of the initial K^-p state can contribute to the production of the state, since the K has i-spin $\frac{1}{2}$. Thus writing

$$\Psi_{\Xi^-\pi^0} = \sqrt{\tfrac{2}{3}}\,\Psi_{3/2,-1/2} + \sqrt{\tfrac{1}{3}}\,\Psi_{1/2,-1/2}$$

$$\Psi_{\Xi^-\pi^+} = \sqrt{\tfrac{1}{3}}\,\Psi_{/2,1/2} + \sqrt{\tfrac{2}{3}}\,\Psi_{1/2,1/2}$$

and

$$\Psi_{3/2,-1/2} \otimes \Psi_{K^+} = \sqrt{\tfrac{1}{2}}\,\Psi_{2,0} - \sqrt{\tfrac{1}{2}}\,\Psi_{1,0}$$

$$\Psi_{3/2,1/2} \otimes \Psi_{K^0} = \sqrt{\tfrac{1}{2}}\,\Psi_{2\,0} + \sqrt{\tfrac{1}{2}}\,\Psi_{1,0}$$

we find that the prediction for the ratio of $\Xi^*_{1/2}$ production in the reaction (a) versus (b) is 2:1 for $T = \frac{3}{2}$. The experimental ratio is approximately 1:4, showing that $T = \frac{3}{2}$ is inadmissible. The spin and parity were

[1] O. Chamberlain, K. Crowe, D. Keefe, L. T. Kerth, A. Lemonick, T. Maung, and T. F. Zipf, *Phys. Rev.* **125,** 1696 (1962). V. Cook, B. Cork, T. F. Hoang, D. Keefe, L. T. Kerth, W. A. Wenzel, and T. F. Zipf, *Phys. Rev.* **123,** 320 (1961).

[2] G. M. Pjerrou, D. J. Prowse, P. E. Schlein, W. E. Slater, D. H. Stork, and H. K. Ticho, *Phys. Rev. Letters* **9,** 114 (1962). P. E. Schlein, D. D. Carmony, G. M. Pjerrou, W. E. Slater, D. H. Stork, and H. K. Ticho, *Phys. Rev. Letters* **11,** 167 (1963). L. Bertanza, V. Brisson, P. L. Connolly, E. L. Hart, I. S. Mitra, G. Moneti, R. R. Rau, N. P. Samios, I. Skillicorn, S. Yamamoto, M. G. Goldberg, L. Gray, J. Leitner, S. Lichtman, and S. Westgard, *Phys. Rev. Letters* **9,** 180 (1962).

studied by a variant of the method discussed in connection with the $Y_1^*(1385)$, and the favored assignment is $(\frac{3}{2})^+$.[1]

$\Xi^*_{1/2}(1820)$.

Another $\Xi\pi$ resonance was discovered at 1820 MeV with a width of the order of 16 MeV. The decay modes of the $\Xi^*_{1/2}(1820)$ are

$$\Xi^*_{1/2}(1820) \to \Lambda^0 \bar{K}(\sim 65\%)$$

which establishes the $T = \frac{1}{2}$ assignment, and

$$\begin{aligned}\Xi^*_{1/2}(1820) \to\ & \Xi\pi(\sim 5\%) \\ & \Xi^*_{1/2}(1530) + \pi(\sim 25\%) \\ & \Xi\pi\pi(\sim 5\%)\end{aligned}$$

A standard spin-parity analysis slightly favors the $(\frac{3}{2})^-$ assignment.

$\Xi^*_{1/2}(1933)$.

A resonance of width 140 MeV and tentative assignment $\frac{5}{2}^+$ has also been observed.

The Ω^- and $SU(3)$

In the limit in which $SU(3)$ is an exact symmetry, any unstable particle that decays into a member of the baryon octet and a member of the pseudoscalar meson octet must belong to one (or more) of the irreducible representations of $SU(3)$ contained in

$$\mathbf{8} \otimes \mathbf{8} = \mathbf{1} \oplus \mathbf{8} \oplus \mathbf{8} \oplus \mathbf{10} \oplus \mathbf{10^*} \oplus \mathbf{27}$$

If we now consider the $N^*_{3/2}(1238)$ and try to assign it to an irreducibly transforming supermultiplet, we find that, because of its *i*-spin, it can only belong to the **10** or the **27** representations. A number of people guessed[2] that the **10** was the correct assignment. The implications of the assignment were that

[1] P. E. Schlein et al., *loc. cit.* G. A. Smith, J. S. Lindsey, J. B. Shafer, J. J. Murray, *Phys. Rev. Letters* **14,** 25 (1965).
[2] R. E. Behrends, J. Dreitlein, C. Fronsdal, and B. Lee, *Rev. Mod. Phys.* **34,** 1 (1962); S. L. Glashow and J. J. Sakurai, *Nuovo Cimento* **26,** 622 (1962); M. Gell-Mann, *Proc. Intern. Ann. Conf. on High Energy Physics, CERN*, Geneva (1962), p. 805.

1. the $N^*_{3/2}$ should have the following partners:

a $Y = 0$ $\quad T = 1$ resonance of spin and parity $(\frac{3}{2})^+$

a $Y = -1$ $\quad T = \frac{1}{2}$ resonance of spin and parity $(\frac{3}{2})^+$

a $Y = -2$ $\quad T = 0$ resonance of spin and parity $(\frac{3}{2})^+$

as discussed in Chapter 18 (p. 284).

2. If the symmetry breaking transforms primarily as an octet, so that the Gell-Mann—Okubo mass formula is applicable, the separation between the levels should be constant.

The $Y_1^*(1385)$ fits into the multiplet, and, if this assignment to the decuplet is made, a $\Xi^*_{1/2}$ is predicted at 1529 MeV. Such a particle was indeed found, ($\Xi^*(1529)$) and the spin and parity assignment also seems to be borne out. Finally an *i*-spin singlet of negative charge and $Y = -2$ is predicted to exist with a mass of 1679 MeV. This mass is lower than the threshold for the decay

$$\Omega^- \to \Xi + \bar{K}$$

so that the Ω^- should be stable, decaying only through the weak interactions according to

$$\begin{aligned}\Omega^- \to\ & \Xi^- + \pi^0 \\ & \Xi^0 + \pi^- \\ & \Lambda^0 + K^- \\ & \Xi + 2\pi\end{aligned}$$

The particle, named Ω^-, has been found.[1] The mass, 1675 ± 8 MeV, has turned out to be in remarkable agreement with what has been predicted. The lifetime

$$\tau = (1.3 \pm 0.7)10^{-10} \text{ sec}$$

shows that the interaction is weak and apparently not different in magnitude from what would be expected.

Incidentally, if we assume that in the decays of the members of the **10** into a baryon and a pseudoscalar meson, the only violation of $SU(3)$ occurs in the masses then given the width of the $N^*_{3/2}(1238)$, the width of the Y_1^* and the $\Xi^*_{1/2}$ can be predicted. This comes about because $\mathbf{8} \otimes \mathbf{8}$ contains **10** only once, i.e., there is only one coupling constant. The

[1] V. E. Barnes et al., *Phys. Rev. Letters* **12,** 204 (1964), *Physics Letters* **12,** 134 (1964); G. S. Abrams et al., *Phys. Rev. Letters* **13,** 670 (1964).

expression for the width is[1]

$$\frac{\Gamma}{\mu} = \frac{f^2}{12\pi}\frac{m}{M}\left(\frac{p'}{\mu}\right)^3\left(\frac{E}{m} + 1\right) \tag{19.27}$$

Here μ is the pion mass, m the baryon mass, and M the mass of the decaying particle; p' is the center of mass momentum in the decay, and E is the energy of the baryon. It is the coupling constants f that are related by $SU(3)$. We may use the u-spin technique to relate the couplings for

$$\begin{aligned} M(N_{3/2}^{*++} \to p\pi^+) &= M(N_{3/2}^{*-} \to n\pi^-) \\ &= M(Y_1^{*-} \to \Lambda^0\pi^-) \\ &= M(\Xi_{1/2}^{*-} \to \pi^0\Xi^-) \\ &= M(\Xi_{1/2}^{*-} \to \pi^-\Sigma^0) \\ &= M(Y_1^{*-} \to \Sigma\,\pi) \end{aligned}$$

Since

$$\Psi_{\Lambda^0} = \frac{1}{2}\Psi_{U=0} - \frac{\sqrt{3}}{2}\Psi_{U=1}$$

and

$$\Psi_{\pi^0} = \frac{\sqrt{3}}{2}\Psi_{U=0} + \frac{1}{2}\Psi_{U=1}$$

we have, looking for terms which will contribute to the $U = \frac{3}{2}$ states only[2]

$$\Psi_{\Lambda^0}\Psi_{\pi^-} = -\frac{\sqrt{3}}{2}\Psi_1^0\Psi_{1/2}^{1/2} = -\sqrt{\frac{3}{4}}\left(\sqrt{\frac{2}{3}}\Psi_{3/2}^{1/2} + \cdots\right)$$

$$\Psi_{\Xi^0}\Psi_{\pi^-} = \Psi_1^{-1}\Psi_{1/2}^{1/2} = \frac{1}{\sqrt{3}}\Psi_{3/2}^{-1/2} + \cdots$$

$$\Psi_{\Xi^-}\Psi_{\pi^0} = \frac{1}{2}\Psi_{1/2}^{-1/2}\Psi_1^0 = \frac{1}{\sqrt{6}}\Psi_{3/2}^{-1/2} + \cdots$$

$$\Psi_{\Sigma^0}\Psi_{\pi^-} = \frac{1}{2}\Psi_1^0\Psi_{1/2}^{1/2} = \frac{1}{\sqrt{6}}\Psi_{3/2}^{1/2} + \cdots$$

$$\Psi_{\Sigma^-}\Psi_{\pi^0} = \frac{1}{2}\Psi_{1/2}^{1/2}\Psi_1^0 = \frac{1}{\sqrt{6}}\Psi_{3/2}^{1/2} + \cdots$$

[1] This formula is derived with the help of the phenomenological Lagrangian density

$$\mathfrak{L}_1 = \frac{f}{\mu}\bar{\psi}_\alpha(x)\psi_{\mu\alpha}(x)\,\partial^\mu\phi(x)$$

in which $\psi_{\mu\alpha}$ stands for the spin-$\frac{3}{2}$ field. See Appendix to Chapter 25, page 430.

[2] The labeling is $\Psi_U^{U_3}$.

This leads to the predictions

Process	Predicted width	Experiment
$Y_1^{*-} \rightarrow \Lambda\pi^-$	43 MeV	$\sim$ 44 MeV
$\Xi^{*-} \rightarrow \Xi\pi$	20 MeV	$\sim$ 7 MeV
$Y_1^{*-} \rightarrow \Sigma\pi$	6.8 MeV	$\sim$ 4.8 MeV

Comparison with experiment shows that unless the formula (19.27) enormously oversimplifies the dependence of the decay width on the masses of the particles involved, the predictions of $SU(3)$ for decays are not unreasonable, and that the conjecture that $SU(3)$ violations are concentrated in the masses is well borne out.

In conclusion, we note that the quantum numbers of the large number of resonances listed above are not yet sufficiently well known to allow us to group them in $SU(3)$ multiplets. It has been suggested that the $Y_0^*(1405)$ be a *unitary singlet*. In that case, the $Y_1^*(1660)$ must belong to an octet. If we take it to have spin $\frac{3}{2}$, as the data tentatively suggest, we may try to associate the $N_{\frac{1}{2}}^*(1518)$ and the $Y_0^*(1520)$ with it. The mass formula then predicts a $(\frac{3}{2})^-$ $\Xi_{\frac{1}{2}}^*$ of mass 1600 MeV. No such particle has been found, indicating most probably that the association is not correct.

It is not idle to ask about the appearance of resonances belonging to supermultiplets more complicated than the **8** and the **10**. There are some qualitative grounds for expecting the masses of the **27**, if it exists at all, to be high, as discussed in Chapter 25. Supermultiplets beyond that can only be produced in association with at least one other particle and so would not appear as bumps in the meson-nucleon cross section. There is a possibility that a $T = \frac{5}{2}$ N^* resonance has been seen at 1580 MeV.[1] Its decay mode appears to be

$$N_{5/2}^* \rightarrow N_{3/2}^*(1238) + \pi$$

In terms of $SU(3)$, it must belong to one of

$$\mathbf{8} \otimes \mathbf{10} = \mathbf{8} + \mathbf{10} + \mathbf{27} + \mathbf{35} \equiv (4, 1)$$

and the only representation containing $T = \frac{5}{2}$ is the **35**. If we attempt to treat $\pi - N^*$ scattering theoretically in the same way as π-N scattering, we do, in fact, predict a $T = J = \frac{5}{2}$ resonance. It will be interesting to see whether such a supermultiplet really exists.

[1] G. Goldhaber, S. Goldhaber, T. O'Halloran, and B. Shen, *Proceedings of the Second Coral Gables Conference on Symmetry Principles at High Energies*, *University of Miami*, 1965, W. H. Freeman and Co., San Francisco, Calif. G. Alexander, O. Benary, B. Reuter, A. Shapira, E. Simopoulou, and G. Yekutieli, *Phys. Rev. Letters* **15**, 207 (1965).

PROBLEMS

Use the following

$$J = \tfrac{3}{2}: \quad d_{\frac{3}{2}\frac{3}{2}}(\theta) = \frac{1 + \cos\theta}{2}\cos\frac{\theta}{2} \qquad d_{\frac{3}{2}\frac{1}{2}} = -\sqrt{3}\,\frac{1 + \cos\theta}{2}\sin\frac{\theta}{2}$$

$$d_{\frac{3}{2},-\frac{1}{2}} = \sqrt{3}\,\frac{1 - \cos\theta}{2}\cos\frac{\theta}{2} \qquad d_{\frac{3}{2},-\frac{3}{2}} = -\frac{1 - \cos\theta}{2}\sin\frac{\theta}{2}$$

$$d_{\frac{1}{2}\frac{1}{2}} = \frac{3\cos\theta - 1}{2}\cos\frac{\theta}{2} \qquad d_{\frac{1}{2},-\frac{1}{2}} = -\frac{1 + 3\cos\theta}{2}\sin\frac{\theta}{2}$$

$$J = \tfrac{5}{2} \quad d_{\frac{5}{2}\frac{5}{2}} = \left(\frac{1 + \cos\theta}{2}\right)^2\cos\frac{\theta}{2} \qquad d_{\frac{5}{2}\frac{3}{2}} = -\sqrt{5}\left(\frac{1 + \cos\theta}{2}\right)^2\sin\frac{\theta}{2}$$

$$d_{\frac{5}{2}\frac{1}{2}} = \frac{\sqrt{10}}{4}\sin^2\theta\cos\frac{\theta}{2} \qquad d_{\frac{5}{2},-\frac{1}{2}} = -\frac{\sqrt{10}}{4}\sin^2\theta\sin\frac{\theta}{2}$$

$$d_{\frac{5}{2},-\frac{3}{2}} = \sqrt{5}\left(\frac{1 - \cos\theta}{2}\right)^2\cos\frac{\theta}{2} \qquad d_{\frac{5}{2},-\frac{5}{2}} = -\left(\frac{1 + \cos\theta}{2}\right)^2\sin\frac{\theta}{2}$$

$$d_{\frac{3}{2}\frac{3}{2}} = \frac{5\cos\theta - 3}{2}\cos^3\frac{\theta}{2} \qquad d_{\frac{3}{2}\frac{1}{2}} = \frac{1 - 5\cos\theta}{\sqrt{2}}\cos^2\frac{\theta}{2}\sin\frac{\theta}{2}$$

$$d_{\frac{3}{2},-\frac{1}{2}} = \frac{1 + 5\cos\theta}{\sqrt{2}}\sin^2\frac{\theta}{2}\cos\frac{\theta}{2} \qquad d_{\frac{3}{2},-\frac{3}{2}} = -\frac{3 + 5\cos\theta}{2}\sin^3\frac{\theta}{2}$$

$$d_{\frac{1}{2}\frac{1}{2}} = \frac{5\cos^2\theta - 2\cos\theta - 1}{2}\cos\frac{\theta}{2} \qquad d_{\frac{1}{2},-\frac{1}{2}} = -\frac{5\cos^2\theta + 2\cos\theta - 1}{2}\sin\frac{\theta}{2}$$

to discuss the decay of a spin $\frac{5}{2}$ resonance into an unstable spin $\frac{3}{2}$ particle and a pion

$$N^{**}(\tfrac{5}{2}^+) \to N^*(\tfrac{3}{2}^+) + \pi^+$$
$$\hookrightarrow p + \pi^+$$

All observations must be expressed in terms of the final proton and π^+'s.

20

Boson Resonances

The Yukawa argument predicting the existence of mesons, responsible for the nuclear forces, was sufficiently qualitative that the spin and parity of the mesons could not be predicted. The nature of the pion was finally settled in 1951, but there remained room for other mesons. Thus Teller and collaborators[1] obtained a qualitative explanation of the strong spin-orbit term and the short-range repulsion in the nuclear forces on the basis of a vector meson exchange. Nambu, and also Frazer and Fulco,[2] pointed out that the existence of a boson state with vector quantum numbers would be helpful in understanding the detailed features of electron-nucleon scattering (see Chapter 25), and Sakurai,[3] in a very interesting paper which gave (almost) compelling reasons for the existence of vector mesons, explained in a qualitative way a large number of phenomena on the basis of perturbation arguments with vector mesons.

The questions posed by the theorists were beginning to be answered in 1961, when developments in rapid data analysis made a study of multiple pion correlations feasible. Not only were all the predicted vector mesons found, but many more mesons turned up. The attempt to fit the newly discovered particles into $SU(3)$ supermultiplets has been somewhat more successful with mesons than with baryons, and we shall present the data in this way. Needless to say, some of the assignments are still tentative.

[1] E. Teller and M. H. Johnson, *Phys. Rev.* **98,** 783 (1955); H. P. Duerr and E. Teller, *Phys. Rev.* **101,** 494 (1956).
[2] Y. Nambu, *Phys. Rev.* **106,** 1366 (1957); W. R. Frazer and J. R. Fulco, *Phys. Rev.* **117,** 1609 (1960).
[3] J. J. Sakurai, *Ann. Phys.* (*N.Y.*) **11,** 1 (1960).

A Nonet of Vector Mesons[1]

The ρ Meson. The first of the boson resonances was discovered in 1961 by a study of the invariant mass spectrum of pion pairs in the reaction

$$\begin{aligned}\pi^- + p &\rightarrow \pi^+ + \pi^- + n \\ &\rightarrow \pi^- + \pi^0 + p\end{aligned}$$

A typical mass plot is shown in Fig. 20.1. The mass of the particle is found to be

$$\begin{aligned} m_\rho &= 765 \text{ MeV} \\ \Gamma_\rho &= 125 \text{ MeV}\end{aligned}$$

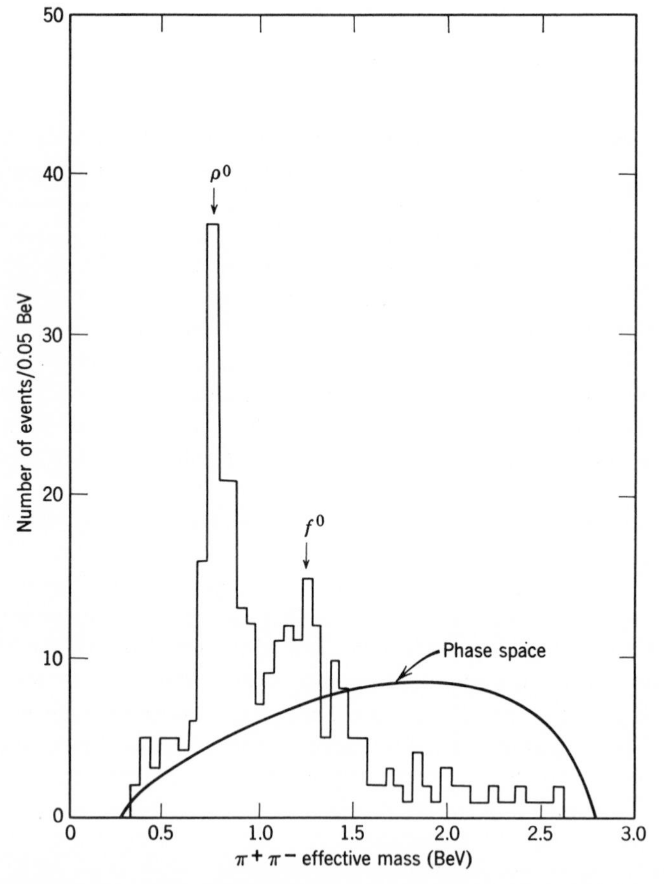

Fig. 20.1. The $\pi^+\pi^-$ mass distribution in $\pi^+ + p \rightarrow \pi^+ + \pi^- + (N^*)^{++}$ at 8 BeV/c. [From Aachen-Berlin-CERN, *Physics Letters* **12**, 356 (1964).]

[1] We are speaking of an octet and a singlet in the $SU(3)$ sense.

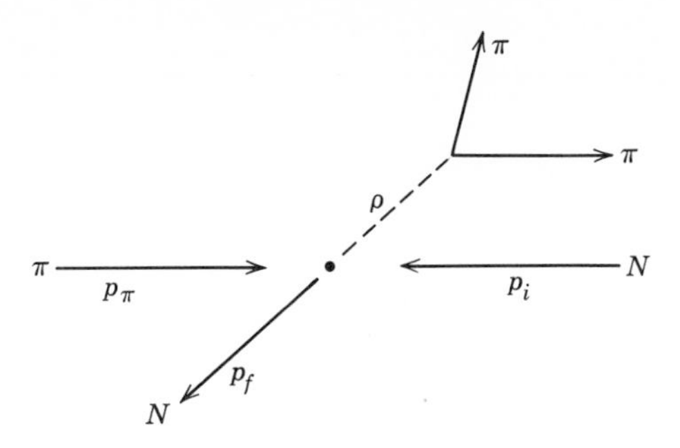

Fig. 20.2. Graphical representation of directions in ρ production.

The peak is found in the $\pi^+\pi^-$ and the $\pi^-\pi^0$ spectra.[1] In the reaction[2]

$$\pi^+ + p \rightarrow \begin{cases} \pi^+ + \pi^+ + n \\ \pi^+ + \pi^0 + p \end{cases}$$

the peak is found in the $\pi^+\pi^0$ spectrum, but not in the $\pi^+\pi^+$ spectrum, indicating that the peak is to be associated with $T = 1$. Since the resonance is broad, there is no question but that the decay is allowed by the strong interactions. We may thus use G-parity conservation to assign $G = +1$ to the resonance. From this we deduce that the ρ^0 must be *odd* under charge conjugation, since a $T = 1$ state changes sign under the transformation $e^{i\pi T_2}$ in

$$G = Ce^{i\pi T_2} \tag{20.1}$$

The two-pion state with i-spin $T = 1$ must be odd under the interchange of spatial coordinates, since it is odd under the interchange of internal variables. Thus the orbital angular momentum in the decay must be odd, hence the parity of the must be odd. Possible spin-parity assignments are therefore $1^-, 3^-, \ldots$. Information as to the spin assignment may be obtained from the angular distribution of the resonating pions in their center of mass system. Figure 20.2 shows a schematic graphical picture of the reaction. We choose a coordinate system in which the incident pion direction defines the z axis and

$$\hat{\mathbf{n}} = \frac{\mathbf{p}_f \times \mathbf{p}_\pi}{|\mathbf{p}_f \times \mathbf{p}_\pi|}$$

which is perpendicular to the plane of the paper (and coming out of it) represents the y axis. A ρ which is produced must have its spin aligned

[1] A. Erwin, R. March, W. Walker, and E. West, *Phys. Rev. Letters* **6,** 628 (1961).
[2] D. Stonehill, C. Baltay, H. Courant, W. Fickinger, E. Fowler, H. Kraybill, J. Sandweiss, J. Sanford, and H. Taft, *Phys. Rev. Letters* **6,** 624 (1961).

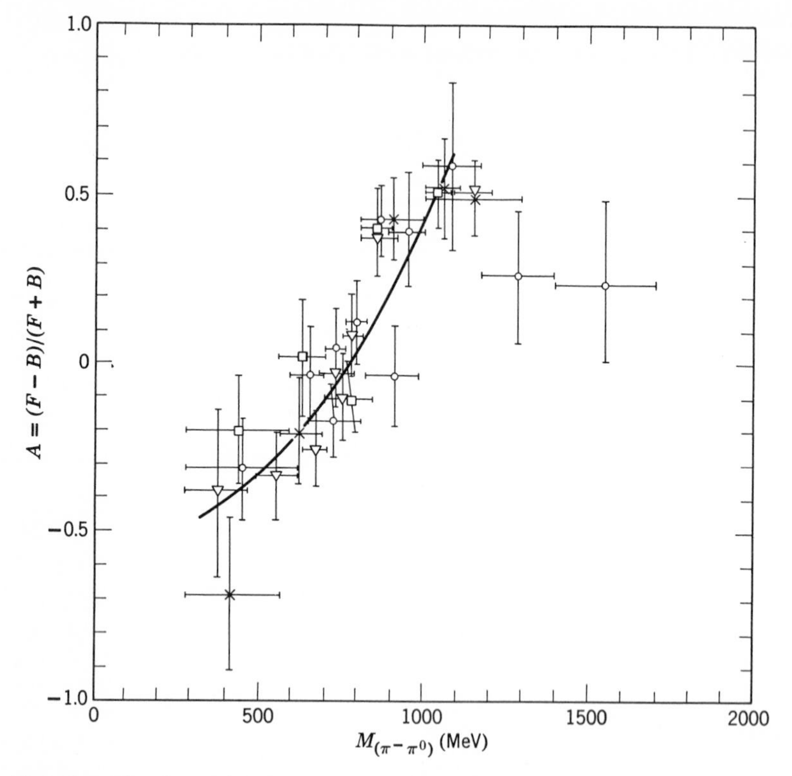

Fig. 20.3. The forward-backward asymmetry in the $\pi^-\pi^0$ angular distribution as a function of the $\pi^-\pi^0$ dipion mass, for low momentum transfers. The solid line represents the weighted mean values of results for the different experiments [from Saclay—Orsay—Bari—Bologna, *Nuovo Cimento* **35,** 713 (1965)].

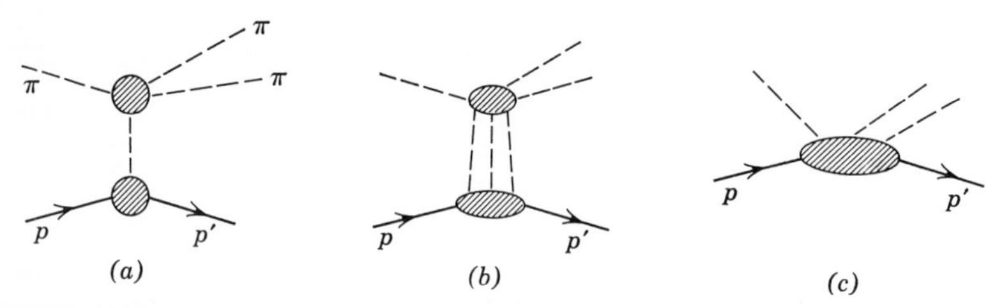

Fig. 20.4. Graphs contributing to production: (*a*) the one-pion exchange contribution, and (*b*), (*c*), some others. The graphs of type (*a*) are expected to dominate when the magnitude of $(p' - p)^2$ is small, i.e., of the order of a few pion masses squared.

perpendicular to the reaction plane. In the ρ rest frame its state is described by $\Sigma_m C_m \Psi_{Jm}$, and the angular distribution of the two pions relative to the z-axis will be of the form

$$\sum_m P(m)\, |\, Y_J{}^m(\theta, \phi)|^2$$

(with $\Sigma_m mP(m) = 0$). Thus the "maximum complexity" of the angular distribution gives a clue to the value of J. When a special class of ρ events, the "peripheral" events, is selected, the ρ may be viewed as an intermediate state in a reaction

$$\pi + \pi \rightarrow \rho \rightarrow \pi + \pi$$

In that case there is no z-component of the angular momentum in the initial state, and the angular distribution must be of the form

$$I(\theta) = |P_J(\theta)|^2 \tag{20.2}$$

The measured angular distribution for $\pi^-\pi^0$ pairs is of the form

$$I(\theta) = A + B \cos \theta + C \cos^2 \theta \tag{20.3}$$

which is interpreted as an angular distribution resulting from a coherent superposition of two amplitudes: (a) the amplitude corresponding to a $J = 1$ resonance, and (b) the amplitude to S-wave pion pair production (this necessarily in a $T = 2$ state since $J = 0$ is incompatible with $T = 1$). According to this interpretation, the angular distribution is

$$I(\theta) = |A_0 + A_1 \cos \theta|^2 = |A_0|^2 + 2 \operatorname{Re} A_0 A_1{}^* \cos \theta + |A_1|^2 \cos^2 \theta \tag{20.4}$$

The resonant amplitude A_1 becomes purely imaginary at the ρ mass, so that the interference term should change sign as it goes through the ρ mass. This is in fact observed (Fig. 20.3). The forward-backward asymmetry for the $\pi^+\pi^-$ has no such simple behavior and will be discussed below in connection with the ϵ-meson.

There is evidence, which will be discussed more fully in Chapter 27, that the ρ meson is produced peripherally, i.e., primarily through processes represented by the graph (*a*) in Fig. 20.4. Support for the peripheral model comes from the observed branching ratio

$$\frac{\sigma(\pi^- p \rightarrow \pi^+ \pi^- n)}{\sigma(\pi^- p \rightarrow \pi^- \pi^0 p)} = 1.8 \pm 0.3 \tag{20.5}$$

in the experiment of Erwin et al. In the $\pi\pi$ scattering part of graph (*a*) in Fig. 20.4, we use the *i*-spin decomposition

$$\begin{aligned} \Phi_{\pi^+\pi^-} &= \frac{1}{\sqrt{3}}\Phi_{0,0} + \frac{1}{\sqrt{2}}\Phi_{1,0} + \frac{1}{\sqrt{6}}\Phi_{2,0} \\ \Phi_{\pi^-\pi^0} &= \frac{1}{\sqrt{2}}\Phi_{1,-1} + \frac{1}{\sqrt{2}}\Phi_{2,-1} \end{aligned} \tag{20.6}$$

to predict, for a $T = 1$ resonance,

$$\frac{\sigma(\pi^+ + \pi^- \to \pi^+ + \pi^-)}{\sigma(\pi^- + \pi^0 \to \pi^- + \pi^0)} = 1 \tag{20.7}$$

In the first process, however, the exchanged pion is a π^+, and it is coupled to the nucleon with a strength $\sqrt{2}$ times larger than the π^0 to the nucleon (this from the *i*-spin conserving coupling $\bar{N}\boldsymbol{\tau}N \cdot \boldsymbol{\pi}$, for example). Thus the predicted ratio is 2:1, in good agreement with experiment. The ρ-meson has been seen in a large number of reactions, including $p\bar{p}$ annihilation.

The ω Meson. Another multipion resonance was discovered in the reaction[1]

$$p + \bar{p} \to \pi^+ + \pi^+ + \pi^0 + \pi^- + \pi^-$$

The mass distribution of all triplets of pions was plotted for each annihilation event. For triplets of charge 1 and 2 no structure was found beyond that expected from phase space, whereas, for the neutral mode $\pi^+\pi^0\pi^-$, a very sharp peak was found with mass 782 MeV and a width later determined to be[2]

$$\Gamma = 9.3 \pm 1.7 \text{ MeV}$$

(Fig. 20.5). This implies that the decay is strong and that the G parity is -1. Since $T = 0$, this implies that $C = -1$ for the neutral ω-meson.

The spin and parity were determined by an examination of the Dalitz plot (Fig. 20.6). If the momenta of the π^0, π^+, and π^- are $\mathbf{p}$, $\mathbf{q} - \frac{1}{2}\mathbf{p}$, and $-\mathbf{q} - \frac{1}{2}\mathbf{p}$ in the center of mass system, for $J = 0$ the matrix element for the decay will be a scalar constructed out of the vectors $\mathbf{p}$ and $\mathbf{q}$ only. (There is no polarization of the ω.) Furthermore, since $C = -1$, under the interchange of π^+ and π^-, i.e., under $\mathbf{q} \to -\mathbf{q}$, the wavefunction must

[1] B. C. Maglic, L. W. Alvarez, A. H. Rosenfeld and M. L. Stevenson, *Phys. Rev. Letters* **7**, 178 (1961).

[2] N. Gelfand, D. Miller, M. Nussbaum, J. Ratau, J. Schultz, J. Steinberger, T. H. Tan, L. Kirsch, and R. Plano, *Phys. Rev. Letters* **11**, 436 (1963).

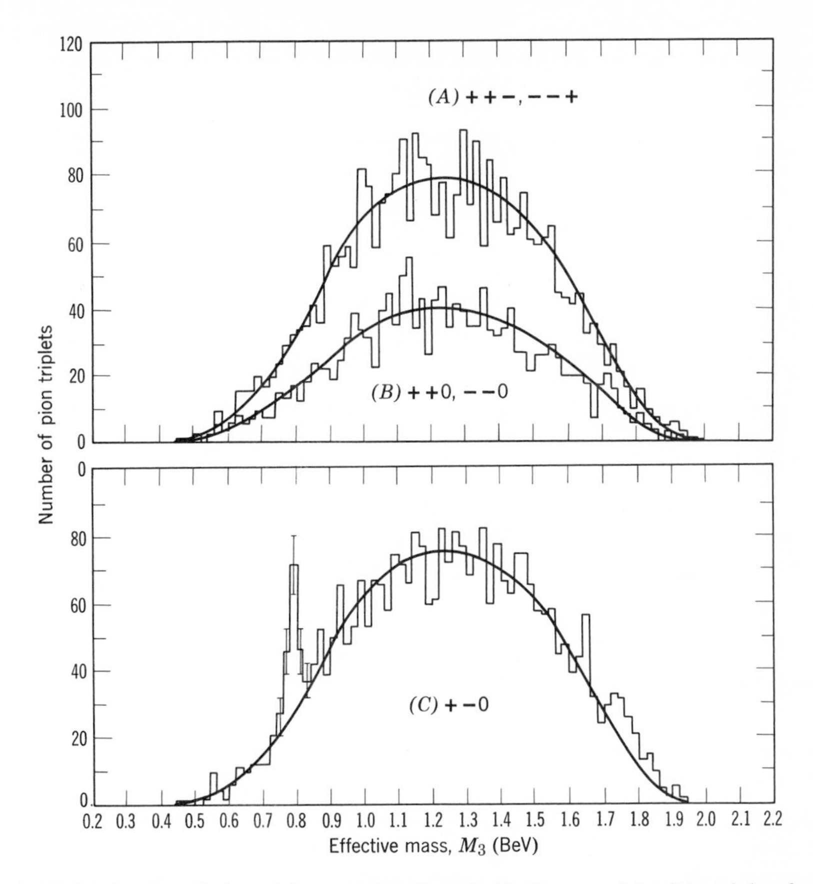

Fig. 20.5. Number of pion triplets as a function of effective mass M_3 of the triplets for the reaction $p + \bar{p} \to 2\pi^+ + 2\pi^- + \pi^0$ (*A*) triplets of $|Q| = 1$; (*B*) triplets of $|Q| = 2$; (*C*) triplets of $|Q| = 0$ [from B. Maglic et al., *Phys. Rev. Letters* **7,** 178 (1961)].

be antisymmetric for the final state. Thus the matrix element must be of the form

$$M_{\omega\to 3\pi} = \mathbf{p}\cdot\mathbf{q}f(\mathbf{p}^2, \mathbf{q}^2, (\mathbf{p}\cdot\mathbf{q})^2) \tag{20.8}$$

The energies of π^+ and π^- will be equal when

$$(\mathbf{q} + \tfrac{1}{2}\mathbf{p})^2 = (\mathbf{q} - \tfrac{1}{2}\mathbf{p})^2$$

i.e., when $\mathbf{p}\cdot\mathbf{q} = 0$. Thus there should be no decays when two pions have the same energy and $J = 0$. This implies that there should be a depletion of events in the region near the symmetry axes of the Dalitz plot. There is no evidence for that, so that $J = 0$ is excluded.

When the three-pion momenta are collinear, with the line making an angle $\hat{\boldsymbol{\theta}}$ with some quantization axis, the three-pion wavefunction is

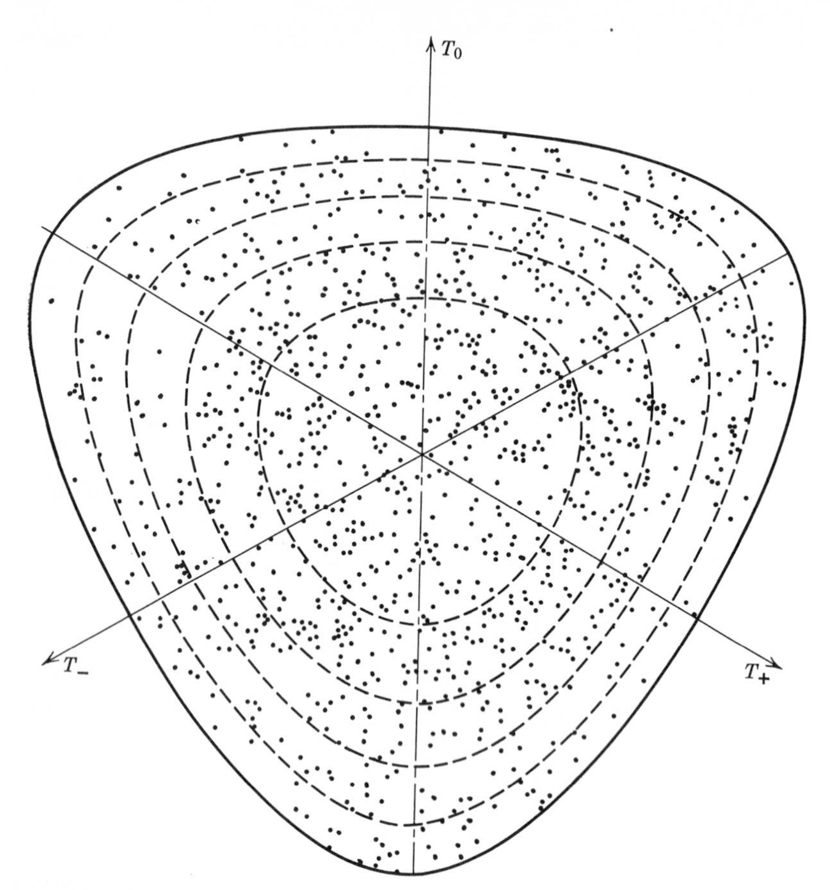

Fig. 20.6. Dalitz plot for the decay $\omega^0 \rightarrow \pi^+ + \pi_0 + \pi^-$ [from C. Alff et al., *Phys. Rev. Letters* **9,** 325 (1962)].

described by the angular factor $Y_J^{M}(\hat{\boldsymbol{\theta}})$. The parity of the state is $(-1)^3(-1)^J$, the first factor coming from the intrinsic parity of the three pions. Thus if the parity of the ω is $(-1)^J$ there will be no collinear events. These events lie on the boundary of the Dalitz plot. Such a depletion on the boundary is indeed observed, indicating that the possible assignments are $1^-, 2^+, 3^-, \ldots$.

We next present an argument, due to Dalitz, showing that the 2^+ assignment is inadmissible. Consider the configuration in which all three pions have the same energy. Momentum conservation requires that the momentum vectors all have the same length and be inclined at 120° to each other. It is clear that a rotation of 120° about the axis perpendicular to the decay plane is equivalent to two interchanges ($\pi^0 \leftrightarrow \pi^+$ followed by $\pi^+ \leftrightarrow \pi^-$). The matrix element is invariant under these interchanges, so that we must have

$$e^{2\pi i m/3} = 1 \tag{20.9}$$

where m is the value of the z-component of the ω angular momentum perpendicular to the decay plane. For $J < 3$, this implies $m = 0$. Now the operation $I_s R$, the product of a reflection in the origin followed by a rotation of 180° about the axis perpendicular to the decay plane, leads to the same configuration, so that the matrix element must be invariant under this operation, i.e.

$$1 = \eta_\omega(-1)^3 e^{im\pi}$$

Since $m = 0$, this implies $\eta_\omega = -1$, i.e., the ω has *odd parity*. If η_ω were $+1$, there would be a depletion of events in the center of the Dalitz plot where all pions have the same energy. No such depletion is observed, so that 2^+ is excluded. The accepted assignment is 1^-, so that the ω, like the ρ, is a vector meson.

The decay mode

$$\omega \rightarrow \pi^+ + \pi^-$$

violates G-parity conservation and can only take place if mediated by the electromagnetic interactions. As such, we might expect the rate to be a factor α^2 ($\alpha =$ the fine structure constant) smaller than the rate $\Gamma_{\omega\rightarrow 3\pi}$, since no real photons are emitted and therefore the matrix element involves at least two factors of e. It was pointed out by Glashow,[1] however, that the existence of a ρ^0 with a mass not so different from that of the ω acts to enhance the rate by a factor of roughly 100. When the electromagnetic interactions are taken into account, i-spin is no longer a good quantum number, and there is a possibility of "mixing" of the ρ^0 and ω states. If we write a phenomenological Lagrangian for the ρ^0 and ω, with an i-spin violating mixing term (λ of order α)

$$\begin{aligned}\mathcal{L} = -\tfrac{1}{2}(\partial_\mu \rho^0)^2 - \tfrac{1}{2}(\partial_\mu \omega)^2 - \tfrac{1}{2}\mu_\rho{}^2\rho^{02} - \tfrac{1}{2}\mu_\omega{}^2\omega^2 \\ - \tfrac{1}{2}\lambda(\mu_\rho{}^2 + \mu_\omega{}^2)\rho^0\omega - f\rho^0\pi^+\pi^- \end{aligned} \qquad (20.10)^2$$

then the mixing term λ may be eliminated if we define ρ', ω' by

$$\begin{aligned}\rho^0 &= \rho' \cos\theta + \omega' \sin\theta \\ \omega &= -\rho' \sin\theta + \omega' \cos\theta\end{aligned} \tag{20.11}$$

[1] S. L. Glashow, *Phys. Rev. Letters* **7**, 469 (1961). See also J. Bernstein and G. Feinberg, *Nuovo Cimento* **25**, 1343 (1962).

[2] The factor $\frac{1}{2}(\mu_\rho^{12} + \mu_\omega{}^2)$ is inserted to keep λ dimensionless.

and choose

$$\theta = \frac{1}{2}\tan^{-1}\lambda\frac{{\mu_\rho}^2 + {\mu_\omega}^2}{{\mu_\omega}^2 - {\mu_\rho}^2} \approx 20\lambda \tag{20.12}$$

Except for the decay term

$$-f\pi^+\pi^-(\cos\theta\rho' + \sin\theta\omega') \tag{20.13}$$

it is ρ' and ω' which are the free, decoupled, stable particles. According to (20.13), we expect the matrix element for $\omega' \to \pi^+\pi^-$ to be

$$M(\omega' \to \pi^+\pi^-) \approx \tan\theta M(\rho' \to \pi^+\pi^-) \tag{20.14}$$

The factor $\sin\theta \approx \theta \approx 20\lambda$, so that there is an enhancement of $\sim 10^2$.

Actually the situation is complicated by the fact that the amplitude for the appearance of a $\pi^+\pi^-$ pair is the sum of the amplitudes for the *i*-spin violating ω decay and the amplitude for the ρ^0 decaying into $\pi^+\pi^-$ at 782 MeV (the ρ^0 is broad enough to permit that). The possibility of interference complicates matters. The evidence is contradictory, two quoted values being[1]

$$\Gamma(\pi^+\pi^-)/\Gamma_{\text{tot}} \simeq 1.8^{+1.2}_{-0.6}\%$$

and[2]

$$\Gamma(\pi^+\pi^-)/\Gamma_{\text{tot}} < 0.8\%$$

The ϕ Meson. A study of the reactions

$$K^- + p \to \begin{cases} \Lambda^0 + K^+ + K^- \\ \Lambda^0 + K^0 + \bar{K}^0 \end{cases}$$

showed a peak in the K-$\bar{K}$ mass distribution at 1020 MeV with a width of $\Gamma = 3.1 \pm 0.6$ MeV.[3] The smallness of the width is not a reflection of weak coupling but rather of the small phase space available: the threshold for K^+K^- decay is 988 MeV, and for $K^0\bar{K}^0$ it is 995 MeV.

The *i*-spin of the ϕ is determined to be $T = 0$, since in the reaction

$$K^- + p \to \begin{cases} \Sigma^0 + K^+ + K^- \\ \Sigma^+ + K^- + K^0 \end{cases}$$

[1] W. D. Walker, J. Boyd, A. R. Erwin, P. H. Scatterblown, M. A. Thompson, and E. West, *Physics Letters* **8,** 208 (1964).

[2] G. Lütjens and J. Steinberger, *Phys. Rev. Letters* **12,** 517 (1964).

[3] L. Bertanza, V. Brisson, P. Connolly, E. Hart, I. Mittra, G. Moneti, R. Rau, N. Samios, S. Lichtman, I. Skillicorn, S. Yamamoto, L. Gray, M. Goldberg, J. Leitner, and J. Westgard, *Phys. Rev. Letters* **9,** 180 (1962); P. Connolly, E. Hart, K. Lai, G. London, G. Moneti, R. Rau, N. Samios, I. Skillicorn, S. Yamamoto, M. Goldberg, M. Gundzik, J. Leitner, and S. Lichtman, *Proceedings of the Sienna International Conference on Elementary Particles*, Italian Physical Society, Bologna (1963), p. 130.

a K^+K^- peak is seen, but there is no corresponding K^-K^0 peak. The transformation properties of the ϕ under charge conjugation depend on its spin. From the decay

$$\phi \to K^+ + K^-$$

we see that $C\phi C^{-1} = (-1)^J\phi$, and the parity is $(-1)^J$. Since $T = 0$, the G parity is $(-1)^J$. Now the observed branching ratio

$$\frac{\Gamma(\phi \to \pi^+\pi^-)}{\Gamma(\phi \to K + \bar{K})} < 0.08$$

can be understood if $G = -1$, which would imply that the spin is odd, as is the parity.[1] In order to distinguish between $J = 1$ and higher values of J, we consider the ratio

$$\alpha_J = \frac{\Gamma(\phi \to K^0 + \bar{K}^0)}{\Gamma(\phi \to K^+ + K^-)} = 0.8 \pm 0.1$$

Because of the mass difference between the charged and the neutral K's, this ratio is not unity, as expected from *i*-spin conservation, but is roughly proportional to

$$(p_0/p_+)^{2J+1}$$

where p_+ is the center of mass momentum in the charged K-decay, (152 MeV/c) and p_0 is the center of mass momentum in the neutral K-decay (107 MeV/c). The experimental ratio favors $J = 1$, which is the accepted assignment. It is perhaps worth mentioning that ϕ mesons are not produced copiously. In the reactions

$$K^- + p \to \begin{cases} \Lambda^0 + \phi \\ \Lambda^0 + \omega \end{cases}$$

with 2.24 BeV/c K^-, ω mesons are produced about four times as frequently as ϕ's.

[1] Another argument which anticipates some of the peculiarities of neutral K decays to be discussed in Chapter 32 goes as follows: under the combined transformation CP (P is parity), the ϕ must be even. If J is even, the two neutral particles which actually decay must both be even or both be odd under CP whereas for J odd, one must be even and the other odd. What is being used here is the fact that the K^0 and its antiparticle are not the states which are characterized by a definite lifetime, but it is (to a good approximation) the linear combinations even or odd under CP that decay, the former into $\pi^+\pi^-$ most of the time. In the ϕ decay only one of the K's decays into $\pi^+\pi^-$ so that J must be odd, J. J. Sakurai, *Phys. Rev. Letters* **9**, 472 (1962).

The K^* Meson. The K^* meson appears as a peak in the $K\pi$ mass distribution in reactions like[1]

$$K^- + p \rightarrow \begin{cases} p + \bar{K}^0 + \pi^- \\ p + K^- + \pi^0 \end{cases}$$

Its mass is 891 MeV, and the width is $\Gamma = 50 \pm 2$ MeV. The *i*-spin is determined to be $\frac{1}{2}$ from the observation of the branching ratio

$$R = \frac{\Gamma(K^{*-} \rightarrow \bar{K}^0 + \pi^-)}{\Gamma(K^{*-} \rightarrow K^- + \pi^0)} = 1.4 \pm 0.4$$

which is predicted to be 2 if $T = \frac{1}{2}$ and $\frac{1}{2}$ if $T = \frac{3}{2}$. This assignment is confirmed by the absence of a peak in the $\pi^- K^0$ spectrum in the reaction

$$\pi^- + p \rightarrow \Sigma^+ + \pi^- + K^0$$

Studies of angular distributions[2] similar to those carried on for the ρ meson lead to a spin and parity assignment of 1^-.

We have thus presented evidence for nine vector mesons, and it is pertinent to ask how these fit into the unitary symmetry scheme. The ρ-meson with $Y = 0$ and $T = 1$ and the K^*-mesons with $Y = \pm 1$ and $T = \frac{1}{2}$ seem like good candidates for an octet assignment with either the ω or the ϕ, both of which have $Y = T = 0$, with the remaining meson perhaps belonging to the singlet representation.

An application of the Gell-Mann—Okubo mass formula predicts that the mass of the $Y = T = 0$ member of the octet should be given by

$$m^2 = \frac{4m^2(K^*) - m^2(\rho)}{3} \cong (930 \text{ MeV})^2 \tag{20.15}$$

(if, as is customary for bosons, we use m^2 in the formula). A way to save the mass formula was suggested by Sakurai,[3] who pointed out that since $SU(3)$ is a quite badly violated symmetry, there will be, in fact, transitions between the unitary singlet eigenstate $\omega^{(0)}$, and ω_8 the *i*-spin singlet member of the octet. The transitions will be "medium strong" in that they violate $SU(3)$ but not *i*-spin conservation. Thus a phenomenological Lagrangian that includes the mixing (we leave off the kinetic energy terms) may be written in the form

$$\mathcal{L} = -m_8^2\omega_8^2 - m_1^2\omega^{(0)2} - \lambda^2\omega_8\omega^{(0)} \tag{20.16}$$

[1] M. Alston, L. Alvarez, P. Eberhard, M. L. Good, W. Graziano, H. Ticho, and S. Wojcicki, *Phys. Rev. Letters*, **6**, 300 (1961).
[2] G. Smith, J. Schwartz, D. Miller, G. Kalbfleisch, R. Huff, O. Dahl, and G. Alexander, *Phys. Rev. Letters* **10**, 138 (1963).
[3] J. J. Sakurai, *Phys. Rev. Letters* **9**, 472 (1962).

with $m_8 = 930$ MeV as predicted by the mass formula. The mass eigenstates, which are the physical particles ω and ϕ, are obtained as

$$\begin{aligned} \phi &= \cos\theta\,\omega_8 + \sin\theta\,\omega^{(0)} \\ \omega &= -\sin\theta\,\omega_8 + \cos\theta\,\omega^{(0)} \end{aligned} \tag{20.17}$$

with

$$\theta = \frac{1}{2}\tan^{-1}\frac{\lambda^2}{m_8^{\,2} - m_1^{\,2}}$$

diagonalizing (20.16). The physical masses now are

$$\begin{aligned} m^2(\phi) &= \tfrac{1}{2}(m_8^{\,2} + m_1^{\,2}) + \tfrac{1}{2}[(m_8^{\,2} - m_1^{\,2})^2 + \lambda^4]^{1/2} \\ m^2(\omega) &= \tfrac{1}{2}(m_8^{\,2} + m_1^{\,2}) - \tfrac{1}{2}[(m_8^{\,2} - m_1^{\,2})^2 + \lambda^4]^{1/2} \end{aligned} \tag{20.18}$$

If we take $m(\phi) = 1020$ MeV and $m(\omega) = 782$ MeV, we find

$$\begin{aligned} \lambda &= 650 \text{ MeV} \\ \cos\theta &= 0.77 \end{aligned} \tag{20.19}$$

The mixing theory makes some definite predictions. Consider the decay

$$\phi \to K\bar{K}$$

The unitary singlet component of ϕ must be coupled to $K\bar{K}$ in a P state, i.e., in a state antisymmetric under spatial exchange. Thus the internal symmetry state must also be antisymmetric. The only coupling possible, however, is of the form

$$\omega^{(0)}\,\mathrm{tr}\,(PP)$$

which is symmetric. Thus only the ω_8 component of the physical ϕ can decay into $K\bar{K}$. This coupling, however, is unique since only the antisymmetric coupling (F type) of the three octets satisfies the condition Thus the coupling must be

$$-i\,f_{ijk}\,V_\mu^{\,i}\,P^j\,\partial^\mu P^k \tag{20.20}$$

Hence the decay rate $\phi \to K\bar{K}$ can be related to the decay rate for $\rho^0 \to \pi^+\pi^-$, which involves the same octets. The matrix elements are related by the form of the coupling

$$g\,\mathrm{tr}\,(V_\mu[P, \partial^\mu P]) \tag{20.21}$$

The relevant terms are easily calculated to be

$$\begin{aligned} g[\sqrt{2}\rho_0^{\,\mu}(\pi^+\,\partial_\mu\pi^- - \pi^-\,\partial_\mu\pi^+) \\ + \sqrt{\tfrac{3}{2}}\omega_8^{\,\mu}(K^+\,\partial_\mu K^- - K^-\,\partial_\mu K^+ + K^0\,\partial_\mu\bar{K}^0 - \bar{K}^0\,\partial_\mu K^0)] \end{aligned} \tag{20.22}$$

Thus the ratio of the rates is

$$\frac{\Gamma(\phi \to K\bar{K})}{\Gamma(\rho^0 \to \pi^+\pi^-)} = \frac{3}{4}\cos^2\theta \tag{20.23}$$

provided the effect of mass splittings is ignored. The mass splitting cannot be ignored, particularly since the ϕ is so close to the $K\bar{K}$ threshold. Both the dependence of the phase space and the centrifugal factor p^J must take the masses into account.

We may use the Feynman rules to calculate the decay rate with the coupling of (20.22). In the rest frame of the ρ the matrix element is given by

$$\frac{g}{(2\pi)^{1/2}}\,\mathbf{e}\cdot(\mathbf{k}_+ - \mathbf{k}_-) = \frac{2g\mathbf{e}\cdot\mathbf{k}_+}{(2\pi)^{1/2}} \tag{20.24}$$

where $\mathbf{e}$ is the polarization vector and the $\mathbf{k}$'s represent the momenta of the outgoing pions. The square of this, when averaged over the polarization states of the ρ^0, is

$$\frac{1}{3}\sum_{\text{pol}}\frac{4g^2}{(2\pi)}(\mathbf{e}\cdot\mathbf{k}_+)^2 = \frac{2g^2}{3\pi}k_+^{\,2}$$

and the decay rate is

$$\begin{aligned}\Gamma &= \frac{1}{(2\pi)^4}\left(\frac{2}{3\pi}g^2k_+^{\,2}\right)\frac{1}{2m_\rho^{\,2}}(2\pi)^3 4\pi\int\omega_\pi\,d\omega_\pi k_+\,\delta(2\omega_\pi - m_\rho)\\ &= \frac{2g^2}{12\pi}\frac{k_+^{\,3}}{m_\rho^{\,2}}\end{aligned} \tag{20.25}$$

The same calculation can be carried out for the ϕ decay into K^+K^- and $K^0\bar{K}^0$ (or rather K_1K_2—the observed CP eigenstates). The result predicts[1]

$$\Gamma(\phi \to K^+K^-) = 1.6\text{ MeV} \qquad (\text{Exp } 1.9 \pm 0.4\text{ MeV})$$
$$\Gamma(\phi \to K_1K_2) = 1.0\text{ MeV} \qquad (\text{Exp } 1.2 \pm 0.3\text{ MeV})$$

when $\Gamma_\rho = 125$ MeV is used as input.

An interesting observation was made by Okubo.[2] He pointed out that if the unitary singlet is combined with the octet in the 3×3 matrix

$$V_9 = \begin{pmatrix} \frac{\omega_8}{\sqrt{6}} + \frac{\rho^0}{\sqrt{2}} + \frac{\omega^{(0)}}{\sqrt{3}} & \rho^+ & K^+ \\ \rho^- & \frac{\omega_8}{\sqrt{6}} - \frac{\rho^0}{\sqrt{2}} + \frac{\omega^{(0)}}{\sqrt{3}} & K^0 \\ K^- & \bar{K}^0 & -\frac{2\omega_8}{\sqrt{6}} + \frac{\omega^{(0)}}{\sqrt{3}} \end{pmatrix} \tag{20.26}$$

[1] In deriving these numbers we used (20.23) as applying to the ratio of the squares of the coupling constants.

[2] S. Okubo, *Physics Letters* **5**, 165 (1963).

and the two representations are always coupled together in that only V_9 appear in couplings, such as

$$\text{tr}\,(\{V_9, V_9\}P_8) \tag{20.27}$$

then (a) the unique D-type coupling of a vector meson to a vector meson and a pseudoscalar meson (20.27) forbids the decay

$$\phi \to \rho + \pi$$

in agreement with experiment; (b) the most general mass term in a phenomenological Lagrangian which *does not contain* tr V_9 in any form is

$$\begin{aligned} M^2\,\text{tr}\,(V_9V_9) - \mu^2\,\text{tr}\,(V_9V_9\lambda_8) &= (M^2 - \mu^2)\rho^+\rho + \left(M^2 + \frac{\mu^2}{2}\right)K^{*+}K^* \\ &\quad + (M^2 + \mu^2){\omega_8}^2 + M^2{\omega^{(0)}}^2 - 2\sqrt{2}\,\mu^2\omega_8\omega^{(0)} \end{aligned} \tag{20.28}$$

This predicts a definite mixing angle

$$\cos\theta = \frac{\sqrt{3}}{2}$$

and mass relations

$$m^2(\phi) = M^2 + 2\mu^2,\ m^2(\rho) = m^2(\omega) = M^2 - \mu^2: \tag{20.29}$$
$$m^2(K^*) = M^2 + \tfrac{1}{2}\mu^2$$

i.e.

$$m^2(\phi) + m^2(\rho) = 2m^2(K^*)$$
$$m^2(\rho) = m^2(\omega) \tag{20.30}$$

which are quite well satisfied. There is at this time no compelling reason for treating V_9 as a unit, without ever splitting off $\omega^{(0)}$ in the form of tr V_9, so that the observation of Okubo must be viewed as a curiosity. The reason for our scepticism about a fundamental role for the nonets in the form (20.26) is the resonance to be discussed next.

The Ninth Pseudoscalar Meson

The X^0-Meson. In a study of the reaction

$$K^- + p \to \Lambda^0 + \text{neutrals}$$

a new peak was found in the distribution of the (mass)^2 of the neutrals.[1]

[1] M. Goldberg, M. Gundzik, J. Leitner, M. Primer, P. L. Connolly, E. Hart, K. Lai, G. London, N. Samios, and S. Yamamoto, *Phys. Rev. Letters* **12,** 546 (1964); **13,** 249 (1964). G. Kalbfleisch, L. Alvarez, A. Barbaro-Galtieri, O. Dahl, P. Eberhard, W. Humphrey, J. Lindsey, D. Merrill, J. Murray, A. Rittenberg, R. Ross, J. B. Shafer, F. Shively, D. Siegel, G. Smith, and R. Tripp, *Phys. Rev. Letters* **12,** 527 (1964). G. Kalbfleisch, O. Dahl, and A. Rittenberg, *Phys. Rev. Letters* **13,** 349 (1964). P. Dauber, W. Slater, L. Smith, D. Stork, and H. Ticho, *Phys. Rev. Letters* **13,** 449 (1964).

The peak appears with

$$m(X^0) = 960 \pm 5 \text{ MeV}$$

The width is small, upper limits of 12 MeV or 4 MeV being quoted, but a much smaller width is possible. The principal decay mode of the X^0 is

$$X^0 \to \eta + 2\pi$$

If this decay is electromagnetic, as would be the case if the G parity of the X^0 were -1, the decay $X^0 \to 3\pi$ should be the dominant mode. There is no evidence for such a decay mode from which we conclude that $G = +1$ and that the $(\eta, 2\pi)$ mode is a strong interaction decay. Since

$$X^0 \to \eta^0 + 2\pi^0$$

is observed, $T = 0$ or 2. However, the production mechanism excludes $T = 2$, so that $T = 0$.

There is also some evidence for the decay

$$X^0 \to \pi^+ + \pi^- + \gamma$$

with a branching ratio of about 20%. From a Dalitz plot of these events as well as the $(\eta, 2\pi)$ events, the tentative conclusion that the spin-parity assignment is 0^- has been drawn.

It is clear from the success of the Gell-Mann—Okubo mass formula for the eight pseudoscalar mesons discussed in Chapter 15, that there is no appreciable $X^0 - \eta^0$ mixing of the type suggested in the $\omega^{(0)} - \omega_8$ system. Thus the construction of a nonet 3×3 matrix P_9 analogous to V_9 of (20.26) would not be appropriate. The nonet does reappear in the next class of resonances to be discussed.

A Nonet of 2^+ Resonances

The f^0-Meson. This resonance was first observed as a peak in the $\pi^+\pi^-$ spectrum in the reaction[1]

$$\pi^- + p \to n + \pi^+ + \pi^-$$

There is no corresponding peak in

$$\pi^- + p \to p + \pi^- + \pi^0$$

so that the resonance, mass 1250 MeV and width $\Gamma \approx 100$ MeV, must be a $T = 0$ state. Since this is a symmetric state, the spin must be even. This

[1] W. Selove, V. Hagopian, H. Brody, A. Baker, and E. Leboy, *Phys. Rev. Letters*, **9**, 272 (1963).

is confirmed by the observation of the decay mode[1]

$$f^0 \rightarrow 2\pi^0$$

The decay angular distribution requires a $\cos^4 \theta$ term,[2] from which we conclude that the spin and parity of the f^0 are 2^+.

The A_2-Meson. A study of the reactions[3]

$$\pi^+ + p \rightarrow \pi^+ + \pi^- + \pi^+ + p$$
$$\pi^- + p \rightarrow \pi^+ + \pi^- + \pi^- + p$$

at 3.65 BeV/c, shows that most of the time an $N^*_{3/2}(1238)$ is produced. In the first reaction $\rho^0 N^{*++}$, $\rho^0\pi^+p$, and $\pi^+\pi^-N^{*++}$ occur 30, 25, and 30% of the time, respectively. The ρ^0 produced with the N^{*++} tend to be produced "peripherally," i.e., with small momentum transfers. (The graph is the same as in Fig. 20.4a, except that the outgoing line labeled with p' is now an N^*.) On the other hand, the ρ^0 produced with a nonresonant π^+p do not seem to be peripheral. The mass plot of the $\pi^+\rho^0$ combination in these events shows a broad peak in the 1.0 to 1.4 BeV region.

Subsequently a study of the triplet of pions in[3]

$$\pi^- + p \rightarrow \pi^+ + \pi^- + \pi^- + p$$

at 3.2 BeV/c showed that the peak could be resolved into a lower one of mass 1090 MeV, called the A_1, and the higher one of mass 1320 MeV, called the A_2. The width of the A_2 is $\Gamma = 80$ MeV

The decay

$$A_2 \rightarrow \pi + \rho$$

implies that $T = 2, 1, 0$, but the observation of[4]

$$A_2 \rightarrow K^{\pm} + K^0$$

with a 10% branching ratio implies that $T = 1$. The observation of the A_2 peak in

$$\pi^- + p \rightarrow n + A_2^0$$
$$\hookrightarrow K_1^0 + K_1^0$$

[1] L. Sodickson, M. Wahlig, I. Mannelli, D. Frisch, and O. Fackler, *Phys. Rev. Letters* **12,** 485 (1964).

[2] Y. Lee, B. Roe, D. Sinclair, and J. Van der Velde, *Phys. Rev. Letters* **12,** 342 (1964).

[3] G. Goldhaber, J. Brown, S. Goldhaber, J. Kadyk, B. Shen, and G. Trilling, *Phys. Rev. Letters*, **12,** 336 (1964).

[4] S. U. Chung, O. Dahl, L. Hardy, R. Hess, G. Kalbfleisch, J. Kirz, D. Miller, and G. Smith, *Phys. Rev. Letters* **12,** 621 (1964).

(with both neutral mesons decaying into $\pi^+\pi^-$) implies that the spin and parity assignment must be $0^+, 2^+, 4^+, \ldots$, since the A_2 decays into two identical bosons. Since a 0^+ particle cannot decay into three pions (recall our discussion of the K quantum numbers in Chapter 15), we conclude that the lowest possible assignment is 2^+. The analysis of the Dalitz plot supports this assignment. There is no evidence yet for the permitted decay

$$A_2 \rightarrow X^0 + \pi$$

but the decay

$$A_2 \rightarrow \eta^0 + \pi$$

does occur about 20% of the time.

The $K^*(1410)$-Meson. Evidence for a $K\pi$ state with mass 1410 MeV and width $\Gamma \simeq 100$ MeV was found in the reaction[1]

$$K^- + p \rightarrow \bar{K}^0 + \pi^- + p$$

at 3.5 BeV/c and in

$$\pi^- + p \rightarrow \begin{cases} \Sigma^\pm \; \pi^\mp K^0 \\ \Sigma^- \; \pi^0 K^+ \\ \Lambda^0 \; \pi^0 K^0 \\ \Lambda^0 \; \pi^- K^+ \end{cases}$$

at 4 BeV/c.[2] The angular distribution favors the assignment 2^+. A discussion of the branching ratios is presented by Hardy et al., and the conclusion is drawn that $T = \frac{1}{2}$.

The f^*-Meson. In the study of K^-p interactions at 4.6–5.0 BeV/c with final states involving two K's, an enhancement has been observed in the K_1K_1 system[3] (both K's decaying into $\pi^+\pi^-$), at

$$m(f^*) = 1500 \text{ MeV}$$

with

$$\Gamma = 85 \text{ MeV}$$

[1] N. Haque *et al.*, *Physics Letters* **14**, 338 (1965).
[2] L. M. Hardy, S. Chung, O. Dahl, R. Hess, J. Kirz, and D. Miller, *Phys. Rev. Letters* **14**, 401 (1965). Further data is presented in S. Chung, O. Dahl, L. Hardy, R. Hess, L. Jacobs, J. Kirz, and D. Miller, *Phys. Rev. Letters* **15**, 325 (1965).
[3] V. E. Barnes, B. Culwick, P. Guidoni, G. Kalbfleisch, G. London, R. Palmer, D. Radojicic, D. Rahm, R. Rau, C. Richardson, N. Samois, J. Smith, B. Goz, N. Horwitz, T. Kikuchi, J. Leitner, and R. Wolfe, *Phys. Rev. Letters* **15**, 322 (1965).

The existence of the mode

$$f^* \to K_1 + K_1$$

implies that the f^* must have even spin and parity. The observation of the decay mode

$$f^* \to K + K^*$$

again excludes the 0^+ assignment, and the angular distribution appears to favor the 2^+ assignment. There is no evidence for a corresponding peak in $K^{\pm}K^0$ implying that $T = 0$.

An application of the mass formula to the $K^*(1410)$ and A_2 implies that in the form

$$M^2 = M_0^2 + \lambda(T(T+1) - \tfrac{1}{4}Y^2)$$

the values of the parameters M_0^2 and λ are

$$M_0^2 = (1440 \text{ MeV})^2 \qquad \lambda = -(425 \text{ MeV})^2$$

If $K^*(1410)$ and A_2 are members of an octet, the *i*-spin singlet should lie at 1440 MeV. This lies between the mass of f^0 and f^*, and the mixing theory must again be used to save the mass formula. The mixing angle turns out to be of the same order of magnitude as that required for the vector mesons.[1]

It is an interesting numerical fact that within the uncertainties of the masses of the 2^+-mesons,

$$m^2(K_{2^+}{}^*) - m^2(A_2) = m^2(K_{1^-}{}^*) - m^2(\rho) = m^2(K) - m^2(\pi) \quad (20.31)$$

This is the sort of relation we would expect if the various mesons were viewed as quark-antiquark bound states, with the mass differences coming about because of a splitting between the doublet and the singlet in the fundamental **3** (and **3***) representations.

Other Resonances

A perusal of the literature on resonances during any six-month period shows several new resonances appearing in various reactions. Some of these turn up in other experiments and finally become well established. Others never reappear. This section contains information on some which have this indefinite status either because they are new, or because their quantum numbers have not been determined.

[1] S. L. Glashow and R. H. Socolow, *Phys. Rev. Letters* **15**, 329 (1965).

The A_1 Peak. This peak, at 1090 MeV was already mentioned in connection with the A_2. The A_1 does decay into $\pi\rho$ (Fig. 20.7) but does not appear to decay into $K\bar{K}$ or $\pi\eta$.[1] On this basis the spin-parity assignments $0^+, 1^-, 2^+, \ldots$ are *excluded.* The favored assignment for the A_1 is that it be a pseudovector particle, with spin-parity 1^+. The question has been raised whether the A_1 is a bona fide resonance or merely a kinematic effect. The inconclusive considerations are too complicated to be summarized here, and the reader is referred to the literature.[2]

The ϵ-Meson. Our discussion of the ρ meson touched upon the angular distribution of the $\pi^{\pm}\pi^0$ pairs in the ρ center of mass system. The angular distribution could be interpreted as a coherent S-wave two-pion production and a ρ^+-decay. The forward-backward asymmetry in the $\pi^+\pi^-$ pairs (ρ^0) does not go through zero at the resonant energy (Fig. 20.8). If we again write the angular distribution in the form

$$I(\theta) = |A_0 + A_1 \cos\theta|^2 = |A_0|^2 + |A_1|^2 \cos^2\theta + 2 \operatorname{Re} A_0 A_1^* \cos\theta$$

we see that if the asymmetry term $2 \operatorname{Re} A_0 A_1^*$ does not vanish at the ρ mass

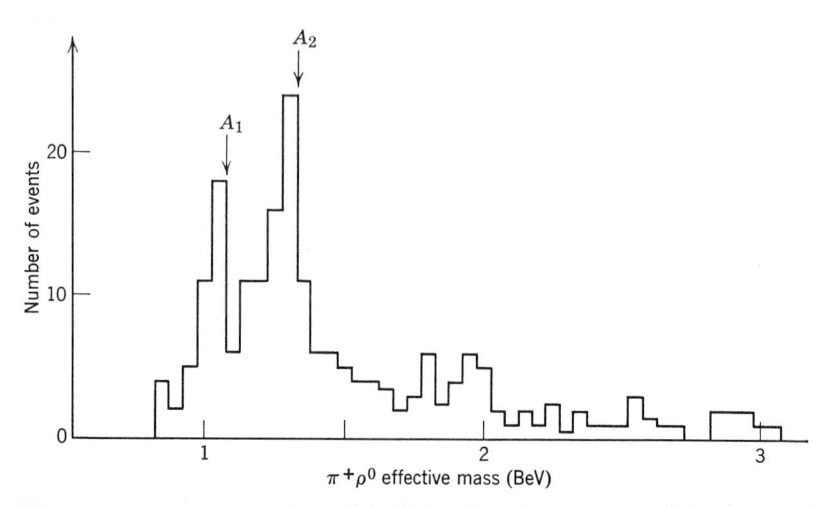

Fig. 20.7. $\pi^+\rho^0$ effective mass (in BeV) in the reaction $\pi^+ + p \rightarrow p + \pi^+ + \pi^+ + \pi^-$ at 8 BeV/c. The events are those for which there is no N^{*++} production. [From M. Deutschmann et al., *Physics Letters* **12,** 356 (1964).]

[1] M. Deutschmann, R. Schulte, H. Weber, W. Worsching, C. Grote, J. Klugow, S. Novak, S. Brandt, V. T. Cocconi, O. Czyzewski, P. Dalpiaz, G. Kellner, and D Morrison, *Phys. Rev. Letters* **12,** 356 (1964).

[2] R. T. Deck, *Phys. Rev. Letters* **13,** 169 (1964); U. Maor and T. A. O'Halloran Jr., *Physics Letters*, **15,** 281 (1965); N. P. Chang, *Phys. Rev. Letters* **14,** 806 (1965).

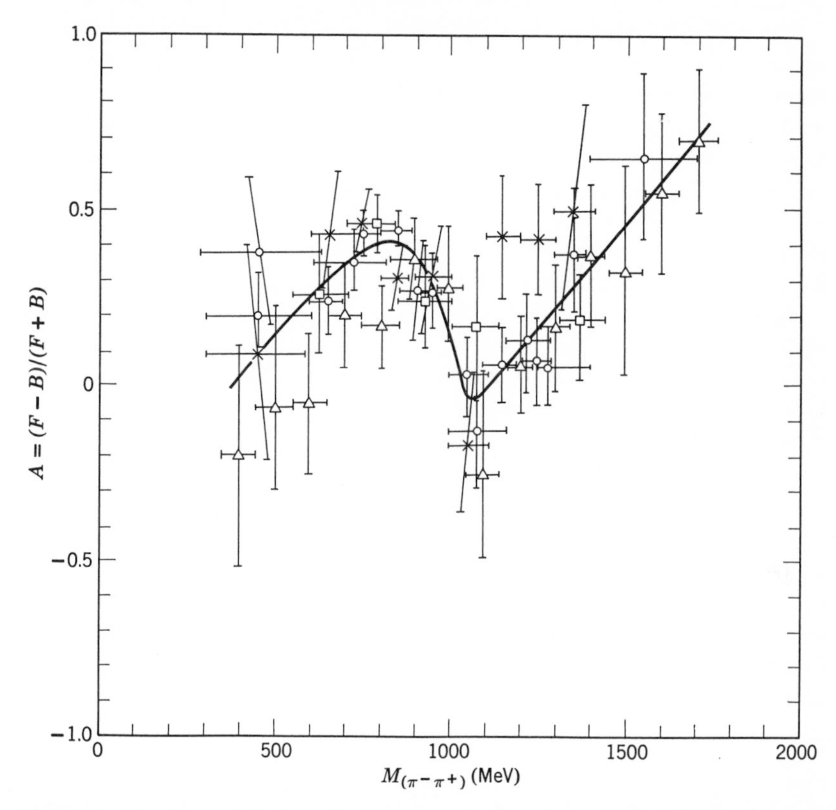

Fig. 20.8. The forward-backward asymmetry in the $\pi^+\pi^-$ angular distribution as a function of the $\pi^+\pi^-$ dipion mass. The solid line represents the weighted mean values of results for the different experiments [from Saclay et al., *Nuovo Cimento* **35,** 713 (1965)].

position where A_1 is purely imaginary, then this implies that A_0 too must have an appreciable imaginary component. Furthermore, the fact that the asymmetry is the same on both sides of the peak, where Re A_1 has opposite signs, implies that Re A_0 should have the same behavior. The asymmetry is most easily understood if there exists a $T = 0$, 0^+ particle, with mass and width close to that of the ρ.[1] Such a resonance has recently been found, with mass 720 MeV and width $\Gamma = 50$ MeV.[2]

[1] For detailed calculations, see L. Durand, III, and Y. T. Chiu, *Phys. Rev. Letters* **14,** 329 (1965).

[2] V. Hagopian, W. Selove, J. Alitti, J. P. Baton, and M. Neveu-Rene, *Phys. Rev. Letters* **14,** 1077 (1965). M. Feldman, W. Frati, J. Halpern, A. Kanofsky, M. Nussbaum, S. Richert, P. Yamin, A. Choudry, S. Devons, and J. Grunhaus, *Phys. Rev. Letters* **14,** 869 (1965).

The B Meson. In the reactions

$$\pi^{\pm} + p \rightarrow \pi^0 + \pi^+ + \pi^- + \pi^{\pm} + p$$

a peak in the four-pion invariant mass at 1220 MeV was found,[1] with width $\Gamma = 100$ MeV and with the neutral triplet of mesons $\pi^+\pi^0\pi^-$ strongly correlated at the ω mass. Thus the B was interpreted as a $\omega\pi$ resonance. The simple picture is, however, very much complicated by the following experimental observation.

As is known from our discussion of the ω-meson Dalitz plot, the three pions which come from the ω decay tend to cluster at the center of the Dalitz plot, i.e., they tend to share the energy equally. If the B events for which the triplet $\pi^+\pi^0\pi^-$ is so constrained are studied, it is found that the B peak is somewhat washed out, whereas if the triplet of mesons from the border regions of the Dalitz plot (where they are less likely to come from an ω), are studied the B appears very sharply. It seems, therefore, that the B is a four-pion resonance, but not primarily a $\pi\omega$ resonance.[2] Since the B does not decay into 2π or $K\bar{K}$, the quantum numbers cannot be 1^- or 3^-. Clearly more data are needed to clear up this situation.

In spite of the present state of confusion, there is hope that there will soon be enough structure apparent in the mesonic spectrum to suggest a simple view of the ever-growing number of states.

PROBLEMS

1. Consider the multiplicative quantum number A (Bronzan-Low) assigned to the bosons as follows:

J^P	$A = +1$	$A = -1$
0^-		η^0, π, K, X^0
1^-	ϕ, ρ, K^*, γ	ω
2^+	f	f^*, A_2, $K^*(1410)$
1^+		A_1

Make a table of allowed and forbidden reactions (not forbidden for reasons other than A-parity conservation) and compare with the experimental facts as outlined in this chapter.

2. Discuss the symmetric Dalitz plot population for the decay of a 1^+ meson into three pions for the cases $(a)T = 0$, $(b)T = 1$, and (c) $T = 2$.

[1] M. Abolins, R. L. Lander, W. Melhop, N. Xuong, and P. Yager, *Phys. Rev. Letters* **11,** 381 (1963).

[2] G. Goldhaber, S. Goldhaber, J. Kadyk, and B. Shen, *Phys. Rev. Letters* **15,** 118 (1965).

21

Properties of S-Matrix Elements I: Unitarity

Our contact with strong interaction physics has, until now, been limited to areas which could be discussed without a knowledge of the dynamics of the strong interactions. It is remarkable that these areas are as large as they are, and undoubtedly an understanding of the origin of the symmetries will bring us many more relationships. Nevertheless, sooner or later a cross section, an angular distribution, or an excitation function must be faced and an attempt made to construct a model that predicts to some degree of accuracy the experimental results of interest. In quantum electrodynamics the first model to be tried, the quantum treatment of a relativistically invariant interaction between the electron field and the photon field, has worked better than anyone had any right to expect.

In the treatment of the interaction among the strongly interacting particles there has been no such success. The absence of a correspondence principle has made the construction of a satisfactory Lagrangian difficult; even the number of fields to be dealt with at a fundamental level is not known, although simplicity always favors the minimum number, as in the quark model of Gell-Mann and Zweig. Whatever the fields, the strong coupling has prevented meaningful calculations, and over the last decade attention has turned to less ambitious attempts, in which the search is more concentrated on relationships between matrix elements. The approach, popularly known as the dispersion-relations approach or "*S*-matrix theory," deals with the properties of scattering matrix elements which follow form the structure of relativistic field theory.[1] In this chapter, we study a set of relations between these matrix elements which follow from the unitarity of the *S* matrix, or equivalently from the completeness of the set of all physical states.

[1] There is a large group of theorists exploring the possibility that only the *S* matrix, and not the underlying field theory, is meaningful. Whether we believe in field theory, or only abstract those properties from it that apply to the *S* matrix, does not, at this stage, affect the computational efforts.

The elements of the scattering matrix can be expressed as Fourier transforms of matrix elements of field operators [e.g. (7.24)], and the unitarity relations that will be derived do not depend on the equations of motion, so that the problem of what the fundamental fields are, and how they interact, can be bypassed at this stage. We note, parenthetically, that this happens because field theory is very flexible in some ways and allows the construction of a local field corresponding to any stable particle.[1] Thus we can speak of pion fields or nucleon fields, even though the ultimate subatomic constituents may be described by fields with entirely different properties and quantum numbers.

To develop the unitarity conditions, we shall consider a specific problem —pion-nucleon scattering:

$$N_p + \pi_q \rightarrow N_{p'} + \pi_{q'}$$

The indices p, q, p', q' stand not only for the momenta, but also for the spin states and i-spin states. The unitarity of the S-matrix

$$SS^+ = S^+S = 1 \tag{21.1}$$

implies that for the operator R, defined by

$$R = S - 1 \tag{21.2}$$

the following relation holds

$$R + R^+ = -R^+R \tag{21.3}$$

Thus the matrix elements satisfy

$$(\Psi^{\text{in}}_{p'q'}, R\Psi^{\text{in}}_{pq}) + (\Psi^{\text{in}}_{p'q'}, R^+\Psi^{\text{in}}_{pq})$$
$$= -\sum_n (\Psi^{\text{in}}_{p'q'}, R^+\Psi^{\text{in}}_{n})(\Psi^{\text{in}}_{n}, R\Psi^{\text{in}}_{pq}) \tag{21.4}$$

On the right-hand side the summation is over a complete set of physical states which span the Hilbert space. In terms of the T matrix defined in (9.10) by

$$(\Psi^{\text{in}}_{k}, R\Psi^{\text{in}}_{l}) = -(2\pi)^4 i\, \delta(P_k - P_l)\, T(k, l) \tag{21.5}$$

the above relation takes the form

$$i(T^*(p, q; p', q') - T(p', q'; p, q))$$
$$= -(2\pi)^4 \sum_n \delta(P_n - p - q)\, T^*(n; p', q')\, T(n; p, q) \tag{21.6}$$

[1] R. Haag, *Phys. Rev.* **112,** 669 (1958); K. Nishijima, *Phys. Rev.* **112,** 995 (1958); W. Zimmermann, *Nuovo Cimento* **10,** 567 (1958). When dealing with the strong interactions, we usually assume that the electromagnetic and weak interactions can be "turned off" in some way, so that baryons and pseudoscalar mesons are stable.

The left-hand side of this equation is closely related to the imaginary part of the scattering amplitude, and (21.6) relates it to a sum over products of matrix elements connecting the initial and final states to all physical states with the same energy-momentum as the initial and final states. These "intermediate states" must be such that there exists a transition amplitude from them to the initial and final states. Thus the states n must have the same quantum numbers as the initial state. For pion-nucleon scattering, for example, we only need to consider intermediate states that have baryon number 1, odd-half integral spin, i-spin $\frac{1}{2}$ or $\frac{3}{2}$, and the same charge as the initial state. At low energies, below the threshold for pion production, e.g., when

$$(M + m_\pi)^2 \leq (p + q)^2 \leq (M + 2m_\pi)^2 \tag{21.7}$$

the sum is over one-nucleon—one-pion states only. There we have, quite explicitly

$$\begin{aligned} i(T^*(p, q; p', q') - T(p', q'; p, q)) \\ = -(2\pi)^4 \sum_{\text{spin}} \int \frac{d^3p''}{(2\pi)^3} \int \frac{d^3q''}{(2\pi)^3} \frac{1}{\rho_N(p'')\rho_\pi(q'')} \\ \delta(p + q - p'' - q'')\, T^*(p'', q''; p', q')\, T(p'', q''; p, q) \end{aligned} \tag{21.8}$$

The sum is over the two spin states of the nucleon in the intermediate state, and the density factors

$$\begin{aligned} \rho_N(p'') &= \frac{1}{(2\pi)^3} \frac{E_{p''}}{M} \\ \rho_\pi(q'') &= \frac{2\omega_{q''}}{(2\pi)^3} \end{aligned} \tag{21.9}$$

appear because of our choice of normalization of states. In contrast to the "virtual states" which appear in Feynman graphs, the particles in the intermediate states are on the physical mass shell, so that $E_{p''} = (\mathbf{p}''^2 + M^2)^{1/2}$ and $\omega_{q''} = (\mathbf{q}''^2 + m_\pi^2)^{1/2}$.

In the special case of *forward scattering*, we have $p' = p$ and $q' = q$, so that the unitarity condition reads

$$\begin{aligned} 2 \operatorname{Im} T(p, q; p, q) &= -(2\pi)^4 \sum_n \delta(P_n - p - q)\, |T(n; p, q)|^2 \\ &= -(2\pi)^4 \sum_n \int \frac{d^3p_1}{(2\pi)^3} \cdots \frac{d^3p_n}{(2\pi)^3} \\ &\quad \times \frac{\delta(p + q - p_1 - \cdots - p_n)}{\rho(p_1) \cdots \rho(p_n)} |T(p_1 \cdots p_n; p, q)|^2 \end{aligned} \tag{21.10}$$

Let us now look back on (9.16): the expression for the cross section $\sigma(p + q \to n)$ is, in the center of mass system

$$\sigma(p + q \to n) = \frac{(2\pi)^6 M}{2\omega_q E_q\left(\frac{q}{\omega_q} + \frac{q}{E_q}\right)}\left[(2\pi)^4 \int \frac{d^3 p_1}{(2\pi)^3} \cdots \frac{d^3 p_n}{(2\pi)^3}\right.$$
$$\left.\times \frac{(\delta p_1 + \cdots + p_n - p - q)}{\rho(p_1) \cdots \rho(p_n)} |T(p_1, p_n, \ldots p_n; p, q)|^2\right]$$

Comparison with (21.10) shows that we get[1]

$$\operatorname{Im} T(p, q; p, q) = -\frac{qW}{(2\pi)^6 M} \sigma_{\text{tot}}(W) \tag{21.11}$$

This is the well-known *optical theorem* that relates the imaginary part of the scattering amplitude to the total cross section. If we note that the conventional, elastic, scattering amplitude $f(W, \theta, \phi)$, in terms of which the differential cross section for elastic scattering at a center of mass energy W is

$$\frac{d\sigma_{\text{el}}}{d\Omega} = \sum_{\text{spin}} |f(W, \theta, \phi)|^2 \tag{21.12}$$

bears the relation

$$f(W, \theta, \phi) = -\frac{16\pi^5 M}{W} T(p', q'; p, q) \tag{21.13}$$

to the transition matrix element for boson-fermion scattering, we recover the conventional form

$$\operatorname{Im} f(W, 0) = \frac{q}{4\pi} \sigma_{\text{tot}}(W) \tag{21.14}$$

Note that (21.13) takes the form

$$f(W, \theta, \phi) = -\frac{8\pi^5}{W} T(p', q'; p, q) \tag{21.15}$$

for boson-boson scattering and

$$f(W, \theta, \phi) = -\frac{32\pi^5 M^2}{W} T(p', q'; p, q) \tag{21.16}$$

[1] We leave it to the reader to convince himself that the manifestly covariant expression corresponding to (21.11) is

$$\operatorname{Im} T(p, q; p, q) = -\frac{[(p \cdot q)^2 - p^2 q^2]^{1/2}}{(2\pi)^6 M} \sigma_{\text{tot}}((p + q)^2)^{1/2})$$

for fermion-fermion scattering, the difference arising from the difference in our normalization of one-fermion states and one-boson states.[1] The form of the optical theorem (21.14) remains unchanged, however.

We shall find the optical theorem very useful in later applications. The remarkable relationship between the elastic amplitude and the total cross section is more easily understood in terms of the wave description of scattering.[2] In such terms, any reduction in the intensity of the incident beam of particles can only come about as a result of an interference that takes place between the incoming beam and the coherent scattered wave in the forward direction. This explains the linear appearance of the forward scattering amplitude rather than its square.

The unitarity condition (21.6) was derived without any reference to field theory or even relativistic invariance. We will show now that a generalization of this relation emerges from a field theoretic treatment. For this purpose, we consider the expression for the scattering matrix element derived in Chapter 7 (7.14)

$$(\Psi^{\text{in}}_{p'q'}, R\Psi^{\text{in}}_{pq}) = i^2 \iint dx\, dy \frac{e^{iq'y}}{(2\pi)^{3/2}} \vec{K}_y \times (\Psi_{p'}, \theta(y_0 - x_0)[\phi_\beta(y), \phi_\alpha(x)]\Psi_p)\overleftarrow{K}_x \frac{e^{-iqx}}{(2\pi)^{3/2}} \tag{21.17}$$

From this it follows that

$$(\Psi^{\text{in}}_{pq}, R\Psi^{\text{in}}_{p'q'})^* = i^2 \iint dx\, dy \frac{e^{iq'y}}{(2\pi)^{3/2}} \vec{K}_y \times (\Psi_{p'}, \theta(x_0 - y_0)[\phi_\beta(y), \phi_\alpha(x)]\Psi_p)\overleftarrow{K}_x \frac{e^{-iqx}}{(2\pi)^{3/2}} \tag{21.18}$$

Thus

$$\begin{aligned}(\Psi^{\text{in}}_{pq}, R\Psi^{\text{in}}_{p'q'})^* &+ (\Psi^{\text{in}}_{p'q'}, R\Psi_{pq}) \\ &= -(2\pi)^4 i\, \delta(p + q - p' - q')(T(p'q'; pq) - T^*(pq, p'q')) \\ &= -\frac{1}{(2\pi)^3} \iint dx\, dy e^{iq'y} \vec{K}_y (\Psi_{p'}, [\phi_\beta(y), \phi_\alpha(x)]\Psi_p)\overleftarrow{K}_x e^{-iqx}\end{aligned} \tag{21.19}$$

[1] The relations (21.13), (21.15), and (21.16) are easily derived using the so-called elastic-unitarity condition (21.8), which must read, in the center oi mass system

$$\text{Im}\, f(\mathbf{k}', \mathbf{k}) = \frac{k}{4\pi} \int d\Omega_{\hat{\mathbf{k}}''} f^*(\mathbf{k}', \mathbf{k}'') f(\mathbf{k}'', \mathbf{k})$$

in terms of f.

[2] L. I. Schiff, *Progr. Theoret. Phys.* (*Kyoto*) **11**, 288 (1954).

If we introduce the variables ξ and z defined by

$$y = \xi + \tfrac{1}{2}z \qquad (21.20)$$
$$x = \xi - \tfrac{1}{2}z$$

and use

$$K_y\, \phi_\beta(y) \equiv j_\beta(y) = e^{iP_\mu\xi^\mu} j_\beta(\tfrac{1}{2}z) e^{-iP_\mu\xi^\mu} \qquad (21.21)$$

and

$$K_y\, \phi_\alpha(x) \equiv j_\alpha(x) = e^{iP_\mu\xi^\mu} j_\alpha(-\tfrac{1}{2}z) e^{-iP_\mu\xi^\mu} \qquad (21.22)$$

then the right-hand side of (21.19) becomes

$$-(2\pi)^4\,\delta(p + q - p' - q')\frac{1}{(2\pi)^3}\int dz e^{i(q+q')z/2}\left(\Psi_{p'}, \left[j_\beta\left(\frac{z}{2}\right), j_\alpha\left(-\frac{z}{2}\right)\right]\Psi_p\right) \qquad (21.23)$$

From this we get

$$i[T(p', q'; p, q) - T^*(p, q; p', q')]$$
$$= \frac{1}{(2\pi)^3}\int dz e^{i(q+q')z/2}\left(\Psi_{p'}, \left[j_\beta\left(\frac{z}{2}\right), j_\alpha\left(-\frac{z}{2}\right)\right]\Psi_p\right) \qquad (21.24)$$

The integration over z now yields delta functions, and the result is

$$i(T(p', q'; p, q) - T^*(p, q; p', q'))$$
$$= (2\pi)\sum_n \{\delta(P_n - p - q)(\Psi_{p'}, j_\beta(0)\Psi_n)(\Psi_n, j_\alpha(0)\Psi_p)$$
$$- \delta(P_n - p + q')(\Psi_{p'}, j_\alpha(0)\Psi_n)(\Psi_n, j_\beta(0)\Psi_p)\} \qquad (21.25)$$

With the help of the relation

$$(\Psi_n^{\text{out}}, j_\alpha(0)\Psi_p) = -(2\pi)^{3/2}\,T(n; p, q) \qquad (21.26)$$

obtained from (6.28), we see that the first term in (21.25) reproduces (21.6).[1] Thus, for consistency we must show that the second term vanishes in the physical region, i.e., in the region in which the variables p and q are physical momenta for a nucleon and a pion. In the center of mass system, we have with $p = (E, \mathbf{p})$ and $p' = (E, \mathbf{p}')$ that

$$P_n^2 = (p + q)^2 = (E_p + \omega_p)^2 \geq (M + m_\pi)^2 \qquad (21.27)$$

in the first term. On the other hand, in the second term

$$P_n^2 = (p - q')^2 = (E_p - \omega_p)^2 - (\mathbf{p} + \mathbf{p}')^2$$

and it is easily checked that for real momenta this is always less than $(M + m_\pi)^2$, so that there will be no contribution from the second term and no conflict with (21.6).

[1] We choose for the complete set of physical states Ψ_n in (21.25) the "out" states.

The second term will, however, contribute when the momenta take on unphysical values (corresponding to imaginary scattering angles, for example), such that

$$(p - q')^2 > (M + m_\pi)^2 \tag{21.28}$$

It should be stressed that the scattering matrix, which relates physical states, is only defined in the physical region.[1] We shall, however, find that the nonphysical region also has significance in field theory. Thus when we speak of a matrix element like $T(p', q'; p, q)$ in the nonphysical region, we assume that there is a well-defined mathematical procedure of arriving at the value of that function; specifically, we take it for granted that some sort of analytic continuation into the nonphysical region is possible. To any order in perturbation theory there is an explicit expression for $T(p', q'; p, q)$, and this presumably can be checked. This topic will be discussed in Chapter 22.

In order to understand the presence of the additional term in the unitarity relation, let us represent the scattering amplitude by the shaded box in Fig. 21.1.

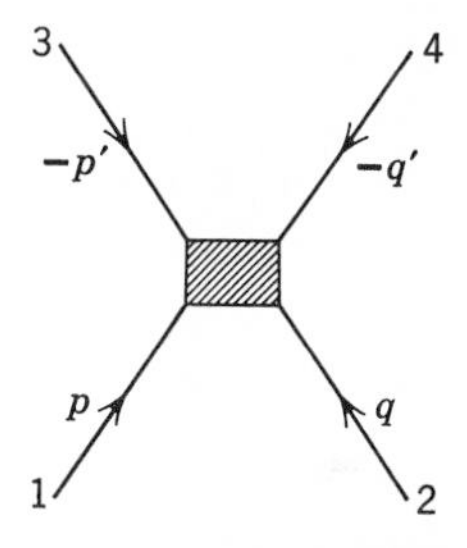

Fig. 21.1. The sum of all the graphs for pion-nucleon scattering. In the physical region particles 1 and 2 enter, 3 and 4 leave. The arrows correspond to the flow of momentum and not charge.

With the notations

$$\begin{aligned} s &= (p + q)^2 = (p' + q')^2 \\ t &= (p' - p)^2 = (q' - q)^2 \\ u &= (p - q')^2 = (p' - q)^2 \end{aligned} \tag{21.29}$$

which satisfy

$$s + t + u = 2M^2 + 2m_\pi^2 \tag{21.30}$$

we see that the physical region corresponds to $s \geq (M + m_\pi)^2$ and the region in which the second term contributes is $u > (M + m_\pi)^2$. We notice that when particles 1 and 2 are incident and particles 3 and 4 leave, the center of mass energy of the scattering process represented in Fig. 21.1 is W, with $W^2 = s$. The physical region will be called the s-channel. If we now consider particles 1 and 4 incident and 2 and 3 outgoing, which implies that $q_0' < 0$ and $q_0 < 0$, the center of mass energy W satisfies

$$W^2 = u \tag{21.31}$$

Thus what we called the nonphysical region is such only for the process "1" + "2" → "3" + "4." For the crossed process it is a perfectly good

[1] The physical region is defined as that domain in which all variables take on physical values, i.e., all momenta are real, as are all angles.

physical region, with, according to (21.25), a perfectly good unitarity condition. If the initial process represents

$$\pi^+ + p \rightarrow \pi^+ + p$$

for instance, the *crossed process* will represent

$$\pi^- + p \rightarrow \pi^- + p$$

In perturbation theory, the two processes are simply related term by term. This symmetry was first pointed out by Gell-Mann and Goldberger, who made the simple but profound observation that in relativistic field theory for each graph there is a graph with the external lines crossed (for example, Fig. 8.3). We shall often make use of *crossing symmetry*.[1]

We might expect that the t-channel, in which the two pions are incident and particles 1 and 4 leave, which describes the process

$$\pi^+ + \pi^- \rightarrow p + \bar{p}$$

should also have a unitarity condition, with terms contributing when the momenta are such that the t-channel is "physical."[2] That there is such a unitarity condition can be seen by considering a different representation of the pion-nucleon scattering amplitude. We start with the reduction formula (7.19), which, after the separation of the center of mass motion yields

$$T(p', q'; p, q) = -\frac{i}{(2\pi)^3}\int dz e^{i(p+q')z/2}\theta(z_0)\left(\Psi_{p'}, \left[j_\beta\left(\frac{z}{2}\right), f\left(-\frac{z}{2}\right)\right]\Psi_q\right) u(p) \tag{21.32}$$

Yet another form, derived by the same techniques, is

$$T(p', q'; p, q) = -\frac{i}{(2\pi)^3}\int dz e^{i(p'+q)z/2}\bar{u}(p')\theta(z_0)\left(\Psi_{q'}, \left[f\left(\frac{z}{2}\right), j_\alpha\left(-\frac{z}{2}\right)\right]\Psi_p\right) \tag{21.33}$$

which yields

$$T^*(p, q; p', q') = \frac{i}{(2\pi)^3}\int dz e^{i(p+q')z/2}\theta(-z_0)\left(\Psi_{p'}, \left[j_\beta\left(\frac{z}{2}\right), \bar{f}\left(-\frac{z}{2}\right)\right]\Psi_q\right) u(p) \tag{21.34}$$

[1] The relation between crossed processes was used to simplify electrodynamic calculations in J. M. Jauch and F. Rohrlich, *The Theory of Photons and Electrons*, who used the name "substitution rule" for the crossing relationship.

[2] This channel is obtained from the s-channel by "crossing" the outgoing nucleon line with the incoming pion line. Also, in the t-channel, symmetry under the interchange of the two pion lines is a consequence of the Bose statistics of the two-pion state.

Thus we find

$$
\begin{aligned}
&i[T(p', q'; p, q) - T^*(p, q; p', q')] \\
&\quad = \frac{1}{(2\pi)^3}\int dz e^{i(p+q')z/2}\left(\Psi_{p'}, \left[j_\beta\left(\frac{z}{2}\right), \bar{f}\left(-\frac{z}{2}\right)\right]\Psi_q\right) u(p) \\
&\quad = (2\pi)\sum_n \{\delta(P_n - p - q)(\Psi_{p'}, j_\beta(0)\Psi_n)(\Psi_n, \bar{f}(0)\Psi_q)\, u(p) \\
&\qquad - \delta(P_n - p + p')(\Psi_{p'}, \bar{f}(0)\Psi_n)u(p)(\Psi_n, j_\beta(0)\Psi_q)\}
\end{aligned} \tag{21.35}
$$

We may use (6.30) to convince ourselves that the first term yields the s-channel unitarity condition as before. The second term involves the matrix element $(\Psi_n, j_\beta(0)\Psi_q)$, which aside from kinematical factors describes the process

$$\pi + \pi \rightarrow \text{“}n\text{”}$$

Similarly $(\Psi_{p'}, \bar{f}(0)\Psi_n)u(p)$ describes the process

$$\text{“}n\text{”} \rightarrow p + \bar{p}$$

It might be expected that the region in which the second term of (21.36) contributes is where

$$t \geq (2M)^2 \tag{21.36}$$

since only then can the $N\bar{N}$ state be “physical.” Actually, the term will contribute whenever the t-channel center of mass energy is such that any physical state n can be created. The lowest mass state that can be made by $\pi + \pi$ is the two-pion state, so that the t-channel unitarity contributes when

$$t \geq (2m_\pi)^2 \tag{21.37}$$

In that region $T(p + \bar{p} \rightarrow n)$ must be defined by an analytic continuation of the amplitude for

$$p + \bar{p} \rightarrow \text{“}n\text{”}$$

For the s and u channels there are also such nonphysical contributions coming from the hitherto neglected *one nucleon* contributions to the sums in (21.25). These only contribute to the unitarity condition for initial pion-nucleon states with $J = \frac{1}{2}$ and $T = \frac{1}{2}$. Let us now evaluate the contribution in the s-channel. We get

$$
\begin{aligned}
&i(T(p', q'; p, q) - T^*(p, q; p', q'))_s \\
&\quad = (2\pi)\int \frac{d^3p''}{E_{p''}/M}\,\delta(p'' - p - q)(\Psi_{p'}, j_\beta(0)\Psi_{p''})(\Psi_{p''}, j_\alpha(0)\Psi_p) \\
&\quad = 4\pi M\int dp''\,\delta(p''^2 - M^2)\,\delta(p'' - p - q)(\Psi_{p'}, j_\beta(0)\Psi_{p''})(\Psi_{p''}, j_\alpha(0)\Psi_p)
\end{aligned} \tag{21.38}
$$

It is clear that this cannot give a contribution in the physical region. In fact, the second delta function shows us that the contribution is at $s = M^2$.

To obtain a more explicit form, we must examine the matrix element $(\Psi_{p''}, j_\alpha(0)\Psi_p)$ more closely. This is a matrix element of a pseudoscalar quantity. Furthermore, an application of reduction techniques would ultimately yield an expression of the form

$$(\Psi_{p''}, j_\alpha(0)\Psi_p) \propto \bar{u}(p'')\left(\Psi_0, \iint dx\, dy e^{ip'y - ipx}\, 0(x, y)\Psi_0\right) u(p)$$

If we also note that the matrix element must transform as a vector in i-spin space, we are led to the most general form for

$$(\Psi_{p''}, j_\alpha(0)\Psi_p) = i\, \bar{u}(p'')\gamma_5\tau_\alpha\, u(p)\, F(p^2, p''^2, (p'' - p)^2) \quad (21.39)$$

F is a function, unknown to us, that can be formed out of the invariants p''^2, p^2, and $(p'' - p)^2$. As we see from (21.38),

$$\begin{aligned} (p'' - p)^2 &= m_\pi^2 \\ p^2 &= M^2 \\ p''^2 &= M^2 \end{aligned} \quad (21.40)$$

so that F is just a constant. We write it conventionally thus

$$F(M^2, M^2, m_\pi^2) = \frac{g}{(2\pi)^3} \quad (21.41)$$

with g real, since $j_\alpha(0)$ is a hermitian operator. In lowest order perturbation g coincides with g_0, the coupling constant in the interaction lagrangian

$$\mathcal{L}_I = ig_0\, \bar{\psi}(x)\gamma_5\tau_\alpha\, \psi(x)\, \phi_\alpha(x) \quad (21.42)$$

and it is therefore often called the renormalized coupling constant. The existence of a g does not depend, however, on a particular model lagrangian. It is just a number that sets the scale for the matrix element connecting a nucleon to a pion-nucleon state with the same quantum numbers. The right-hand side of (21.38) may now be worked out. We get

$$\begin{aligned} 4\pi M\, \delta(s - M^2) \int dp''\, \delta(p'' - p - q)\left(-\frac{g^2}{(2\pi)^6}\right) \\ \times \sum_{\substack{\text{spin} \\ r}} \bar{u}(p')\gamma_5\tau_\beta\, u^{(r)}(p'')\, \bar{u}^{(r)}(p'')\gamma_5\tau_\alpha\, u(p) \\ = -\frac{g^2}{(2\pi)^5}\, \delta(s - M^2)\, \bar{u}(p')\tau_\beta\tau_\alpha\gamma_5(M + \not{p} + \not{q})\gamma_5\, u(p) \\ = \frac{g^2}{(2\pi)^5}\, \delta(s - M^2)\, \bar{u}(p')\tau_\beta\tau_\alpha \frac{\not{q} + \not{q}'}{2}\, u(p) \end{aligned} \quad (21.43)$$

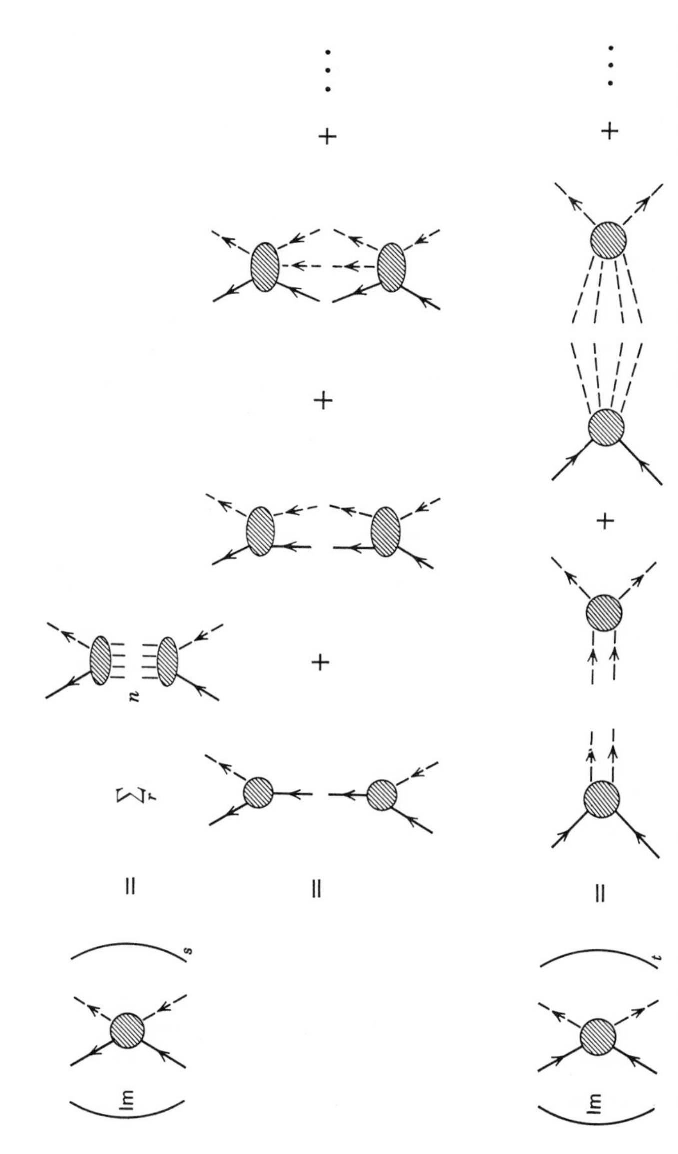

Fig. 21.2. Graphical representation of the absorptive part of the scattering amplitude in the s- and t-channels. There are no three-pion terms in the second series because of the G-parity selection rule, which forbids the conversion of an odd number of pions into an even number of pions.

Similarly, the u-channel one-nucleon term contribution is

$$-\frac{g^2}{(2\pi)^5}\,\delta(u-M^2)\,\bar{u}(p')\tau_\alpha\tau_\beta\frac{\not{q}+\not{q}'}{2}\,u(p) \tag{21.44}$$

We leave it to the reader to show that if the pion were scalar, (21.43) would be changed to

$$-\frac{g^2}{(2\pi)^5}\,\delta(s-M^2)\,\bar{u}(p')\tau_\beta\tau_\alpha\left(2M+\frac{\not{q}+\not{q}'}{2}\right)u(p) \tag{21.45}$$

The one-particle terms are the only place in which more or less specific statements are made about the interaction and the particles. They represent the only model dependence of the unitarity relations.

To summarize: we have found with the help of the reduction formulas of field theory that the absorptive part of the scattering amplitude may be written as a sum of terms involving matrix elements connecting the initial and final states to a complete set of intermediate states with the same quantum numbers, energy, and momentum. Such contributions arise not only in the physical region, but also in regions that are physical when some of the outgoing particles are viewed as incoming ones and vice versa. Furthermore, we have found that, given the transformation properties involved, the one-particle contributions to the unitarity sums, whenever there are any, can be written down explicitly. The results are presented graphically in Fig. 21.2.

PROBLEMS

1. Show that the matrix element $(\Psi_0, A(x)\,B(y)\Psi_0)$, with $A(x)$ and $B(y)$ representing *scalar fields*, may be written in the form

$$(\Psi_0, A(x)\,B(y)\Psi_0) = i\int_0^\infty d\mu^2\,\rho_{AB}(\mu^2)\,\Delta^{(+)}(x-y,\mu^2)$$

where

$$\Delta^{(+)}(x-y,\mu^2) = -i\int\frac{dp}{(2\pi)^3}\,e^{-ip(x-y)}\,\theta(p_0)\,\delta(p^2-\mu^2)$$

2. If $(\Psi_0, B(y)\,A(x)\Psi_0)$ is represented by

$$(\Psi_0, B(y)\,A(x)\Psi_0) = i\int d\mu^2\,\rho_{BA}(\mu^2)\,\Delta^{(+)}(y-x,\mu^2)$$

what is the relation between $\rho_{AB}(\mu^2)$ and $\rho_{BA}(\mu^2)$ implied by local commutativity

$$[A(x), B(y)] = 0 \qquad (x-y)^2 < 0$$

3. Use the results of Problem 1 and 2 to write down a representation for

$$(\Psi_0, [A(x), B(y)]\Psi_0)$$

Suppose $\{A(x), B(y)\} = 0$ for $(x - y)^2 < 0$ were assumed in Problem 2; how would the matrix element $(\Psi_0, [A(x), B(y)]\Psi_0)$ behave outside the light cone, i.e., for $(x - y)^2 < 0$?

4. Obtain representations for $(\Psi_0, \psi_\alpha(x)\, \bar{\psi}_\beta(y)\, \Psi_0)$ and $(\Psi_0, \bar{\psi}_\beta(y)\, \psi_\alpha(x)\, \Psi_0)$ using the most general forms

$$(2\pi)^3 \sum_{P^{(n)}=p} (\Psi_0, \psi_\alpha(0)\Psi_n)(\Psi_n, \bar{\psi}_\beta(0)\Psi_0)$$

$$= \int dm^2\, \delta(p^2 - m^2)\theta(p_0)[\delta_{\alpha\beta}\, \rho^{(1)}_{\psi\bar{\psi}}(m^2) + (\not{p})_{\alpha\beta}\, \rho^{(2)}_{\psi\bar{\psi}}(m^2)$$

$$+ (\gamma_5)_{\alpha\beta}\, \rho^{(3)}_{\psi\bar{\psi}}(m^2) + (\gamma_5 \not{p})_{\alpha\beta}\, \rho^{(4)}_{\psi\bar{\psi}}(m^2)]$$

. . .

Write down the representation for $(\Psi_0, \{\psi_\alpha(x), \bar{\psi}_\beta(y)\}\Psi_0)$ using the property that $\{\psi_\alpha(x), \bar{\psi}_\beta(y)\} = 0$ for $(x - y)^2 < 0$. What simplification occurs when parity is conserved?

5. Given the representation

$$(\Psi_0, [A(x), A(y)]\Psi_0) = i \int d\mu^2 \rho(\mu^2)\, \Delta(x - y, \mu^2)$$

and the fact that $A(x)$ is a scalar field satisfying the field equation

$$(\Box + m^2)\, A(x) = g\, A^2(x)$$

calculate $\rho(\mu^2)$ to order g^2.

6. Obtain an expression for Z_2, as defined in (7.35) in terms of the functions $\rho^{(1)}(\mu^2)$ that appear in Problem 4 with parity conservation.

22

Properties of S-Matrix Elements II: Analyticity

The unitarity of the S-matrix yields information about the imaginary parts of S-matrix elements. In this chapter we will show that the structure of the scattering amplitude yields significant information about the real part as well. The far-reaching developments to be described below started with a paper by Gell-Mann, Goldberger, and Thirring,[1] who observed that the form of the reduction formula for the scattering amplitude permits an analytic continuation of the amplitude to complex values of the energy. We shall soon illustrate the importance of this esoteric result.

The argument of Gell-Mann, Goldberger, and Thirring proceeded along the following lines.

Consider the forward scattering amplitude for photon-proton scattering, say. The part of the amplitude that need concern us is given by

$$T_{\lambda\lambda}(p, q; p, q) = -\frac{i}{(2\pi)^3}\int dz e^{iqz}\theta(z_0)\left(\Psi_p, \left[j_\lambda\left(\frac{z}{2}\right), j_\lambda\left(-\frac{z}{2}\right)\right]\Psi_p\right) \tag{22.1}$$

which is quite similar to (21.17), with $q = q'$, the photon four momenta equal for forward scattering. The local commutativity condition that we have associated with microscopic causality reads

$$\left[j_\lambda\left(\frac{z}{2}\right), j_\lambda\left(-\frac{z}{2}\right)\right] = 0 \qquad z^2 < 0 \tag{22.2}$$

This condition, together with the step function $\theta(z_0)$, implies that the amplitude is a Fourier transform of a function which vanishes outside the forward light cone. This has definite implications for the dependence of the Fourier transform on the energy variable. In the laboratory system

[1] M. Gell-Mann, M. L. Goldberger, and W. Thirring, *Phys. Rev.* **95,** 1612 (1954) M. Goldberger, *Phys. Rev.* **99,** 979 (1955).

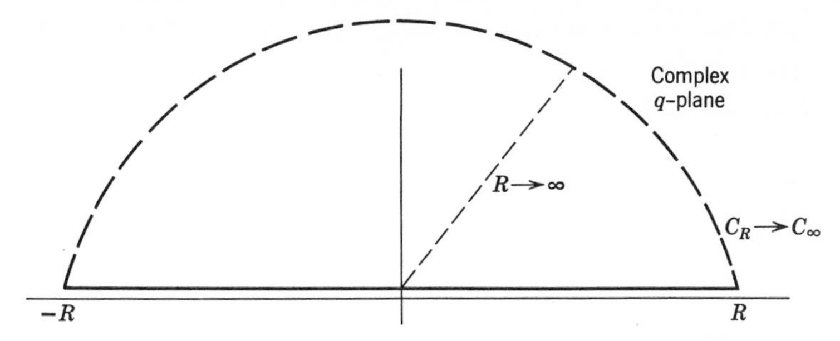

Fig. 22.1. Contour used in the representation of $f(q)$ in (22.5).

the scattering amplitude is of the form

$$-\frac{i}{(2\pi)^3}\int_0^\infty dz_0 \int d^3z e^{iq(z_0-\hat{\mathbf{q}}\cdot\mathbf{z})}\, F_{\lambda\lambda}(\mathbf{z}, z_0) \tag{22.3}$$

and since

$$F_{\lambda\lambda}(\mathbf{z}, z_0) = 0 \qquad z_0 < |\mathbf{z}| \tag{22.4}$$

the integration is over a region such that $z_0 - \hat{\mathbf{q}}\cdot\mathbf{z} > 0$. If we now let q, the photon energy, become complex, with $\operatorname{Im} q > 0$, the integral is still well defined, so that we may use the representation (22.1) to define the analytic continuation of the forward scattering amplitude to the upper half of the complex plane for q. If we write $f(q)$ for $T_{\lambda\lambda}(p, q; p, q)$ in the laboratory system, our conclusion concerning the analyticity of $f(q)$ in the upper half of the complex q-plane allows us to write a representation of $f(q)$ that makes explicit the analyticity properties and is useful for further manipulations. This representation is just the Cauchy formula

$$f(q) = \frac{1}{2\pi i}\int_C \frac{f(\zeta)}{\zeta - q}\, d\zeta \tag{22.5}$$

where the contour consists of the real axis from $-\infty$ to $+\infty$ and a semicircle in the upper half plane to close it (Fig. 22.1). Thus we have

$$f(q) = \frac{1}{2\pi i}\int_{-\infty}^{\infty} dq' \frac{f(q')}{q' - q} + \frac{1}{2\pi i}\int_{C_\infty} d\zeta\, \frac{f(\zeta)}{\zeta - q} \tag{22.6}$$

The representation (22.6) breaks down if $f(q)$ is not bounded as $q \to \infty$ in the upper half plane. We will proceed on the assumption that $f(q) \to 0$ as $q \to \infty$, so that the second term in (22.6) does not contribute, and discuss other cases later. We do not, in fact, know the behavior of $f(q)$ at infinity. The behavior of $f(q)$ at infinity reflects the behavior of the matrix element

$$\left(\Psi_p, \left[j_\lambda\left(\frac{z}{2}\right), j_\lambda\left(-\frac{z}{2}\right)\right]\Psi_p\right)_{z^2\to 0}$$

and this may be quite singular. It is generally postulated that the mathematical nature of field operators is such that the singularity can be no worse than a finite derivative of a delta function, or equivalently that $f(q)$ is bounded by a polynomial in q as $q \to \infty$.

With $f(q)$ vanishing at infinity, we have

$$f(q) = \frac{1}{2\pi i}\int_{-\infty}^{\infty} dq' \frac{f(q')}{q' - q}$$

and as q approaches the real axis from above, we have[1]

$$\begin{aligned} f(q) &= \lim_{\epsilon\to 0} \frac{1}{2\pi i}\int_{-\infty}^{\infty} dq' \frac{f(q')}{q' - q - i\epsilon} \\ &= \frac{1}{2\pi i} P\int_{-\infty}^{\infty} dq' \frac{f(q')}{q' - q} + \frac{1}{2\pi i}\int_{-\infty}^{\infty} dq' f(q')(\pi i\, \delta(q - q')) \end{aligned}$$

so that[2]

$$f(q) = \frac{1}{\pi i} P\int_{-\infty}^{\infty} dq' \frac{f(q')}{q' - q} \tag{22.7}$$

Taking the real part of both sides, we get

$$\operatorname{Re} f(q) = \frac{1}{\pi} P\int_{-\infty}^{\infty} dq' \frac{\operatorname{Im} f(q')}{q' - q} \tag{22.8}$$

We may easily show that[3]

$$\operatorname{Im} f(-q') = -\operatorname{Im} f(q') \tag{22.9}$$

This allows us to rewrite (22.8) in the following manner:

$$\begin{aligned} \operatorname{Re} f(q) &= \frac{1}{\pi} P\int_{-\infty}^{0} dq' \frac{\operatorname{Im} f(q')}{q' - q} + \frac{1}{\pi}\int_{0}^{\infty} dq' \frac{\operatorname{Im} f(q')}{q' - q} \\ &= \frac{1}{\pi} P\int_{0}^{\infty} dq'' \frac{\operatorname{Im} f(q'')}{q'' + q} + \frac{1}{\pi} P\int_{0}^{\infty} dq' \frac{\operatorname{Im} f(q')}{q' - q} \\ &= \frac{1}{\pi} P\int_{0}^{\infty} 2q'\, dq' \frac{\operatorname{Im} f(q')}{q'^2 - q^2} \end{aligned} \tag{22.10}$$

[1] The scattering amplitude must be defined as this kind of a boundary value to give meaning to the integral in (22.3); the integral has the given structure because of the appearance of $\theta(z_0)$ in (22.1).

[2] P in the integral stands for the instruction that the principal value be taken.

[3] From (22.1) it is easily shown that $f(-q)$ is given by the same integral with $-\theta(-z_0)$ appearing in it; this, however, is just the representation for $f^*(q)$.

This remarkable relation, expressing the real part of the forward scattering amplitude in terms of the imaginary part is a direct consequence of the microscopic causality condition, and it is thus completely model independent. If we now use the optical theorem

$$\operatorname{Im} f(q') = \frac{q'}{4\pi}\,\sigma_{\text{tot}}(q') \tag{22.11}$$

we obtain a form of the *forward scattering dispersion relation*

$$\operatorname{Re} f(q) = \frac{1}{2\pi^2}\,P\int_0^\infty dq'\,\frac{q'^2\sigma_{\text{tot}}(q')}{q'^2 - q^2} \tag{22.12}$$

The amplitude for zero frequency is just the Thomson amplitude[1]

$$f(0) = -\frac{\alpha}{M} \tag{22.13}$$

This implies that

$$-\frac{\alpha}{M} = \frac{1}{2\pi^2}\int_0^\infty \sigma_{\text{tot}}(q')\,dq' \tag{22.14}$$

which is obviously wrong, since the right-hand side is positive.

The error lies in our assumption that $f(q)$ vanishes as $q \to \infty$. If $f(q)$ does not vanish but $f(q)/q$ does, we go through the same steps as before, obtaining

$$\frac{f(q)}{q} = -\frac{\alpha/M}{q} + \frac{1}{2\pi i}\lim_{\epsilon\to 0}\int dq'\,\frac{f(q')}{q'(q' - q - i\epsilon)} \tag{22.15}$$

There is an additional term present, coming from the pole of $f(q)/q$ at the origin. If we convert the integral to one over positive frequencies only, we obtain

$$\operatorname{Re} f(q) = -\frac{\alpha}{M} + \frac{q^2}{\pi}\,P\int_0^\infty 2q'\,dq'\,\frac{\operatorname{Im} f(q')}{q'^2(q'^2 - q^2)}$$

[1] The amplitude at zero frequency can only depend on the charge and the mass of the target particle and must equal the classical Thomson amplitude. All structure effects and radiative corrections can have only one effect at zero frequency, and that is to renormalize the charge and mass to their physical values. An explicit proof may be found in W. Thirring, *Phil. Mag*. **41,** 1193 (1950), M. Gell-Mann and M. Goldberger, *Phys. Rev*. **96,** 1428 (1954), or F. Low, *Phys. Rev*. **96,** 1433 (1954).

This may be rewritten in the form

$$\operatorname{Re} f(q) = -\frac{\alpha}{M} + \frac{q^2}{2\pi^2} P \int_0^\infty dq' \frac{\sigma_{tot}(q')}{q'^2 - q^2} \tag{22.16}$$

This formula could also be obtained by subtracting from $\operatorname{Re} f(q)$, as defined in (22.12), the expression for $\operatorname{Re} f(0)$ in terms of the integral in (22.12).[1] If the behavior at infinity is more divergent, an expression of the type

$$\operatorname{Re} f(q) = \sum_{n=1}^{N} C_n(q^2)^{n-1} + \frac{(q^2)^N}{\pi} P \int_0^\infty dq'^2 \frac{\operatorname{Im} f(q')}{(q'^2)^N(q'^2 - q^2)} \tag{22.17}$$

can be derived. Actually, since cross sections are not expected to grow indefinitely with energy, the relation (22.16) is presumably correct, and comparison with experiment may be used to check the validity of the assumptions that went into its derivation.

The example of forward photon scattering is unfortunately somewhat misleading. Proofs of forward dispersion relations for other systems are either very complicated or do not exist. The difficulty occurs as soon as the particle that is scattered has mass; the exponent in (22.3), as a function of the frequency ω of the particle, is now

$$e^{i\omega z_0 - i\sqrt{\omega^2 - m^2}\hat{\mathbf{q}}\cdot\mathbf{z}}$$

and the simple representation (22.1) does not, in fact, exist for complex ω; thus rather complicated methods must be used to establish that the forward scattering amplitude for pion-nucleon scattering has analyticity properties which allow us to derive a dispersion relation.[2] For *K*-nucleon scattering or *N*-*N* scattering, no proofs of a general nature have been found.

In the absence of proofs, supporting evidence for analyticity properties of scattering (and other) amplitudes may be found in the properties of terms in the perturbation series.[3] Even though certain analyticity properties

[1] It should be noted that with each subtraction more information must be put in; in the case of photon scattering $f(0)$ can be derived separately, but in meson theory, where there are no threshold theorems, $f(0)$ must be measured experimentally.

[2] See, for example, G. Källén, *Elementary Particle Physics*, Addison-Wesley Publishing Co., Inc., Reading, Mass. (1964), Chapter 5.

[3] For a detailed discussion of the derivation of analyticity properties from the Feynman rules, see J. D. Bjorken and S. D. Drell, *Relativistic Quantum Fields*, McGraw-Hill Book Co., Inc., New York (1965).

true for each term in a series are not necessarily true for the sum—especially since it is far from clear that the series converges—nevertheless, it has been a working hypothesis to believe in the general validity of results suggested by the Feynman graphs.

Proofs of analyticity properties for arbitrary Feynman graph contributions are very complicated. We shall illustrate the procedure by a simple example. Consider the one-particle matrix element of a scalar current (cf. Chapter 13) $(\Psi_{p'}, j(0)\Psi_p)$. The simplest nontrivial graph contributing to this matrix element is shown in Fig. 22.2. The particle whose initial momentum is p, and whose final momentum is p', is assumed to have a mass M. It emits a particle of mass μ and turns into a particle of mass M_1. Since all the particles are scalar, the contribution of this graph is proportional to

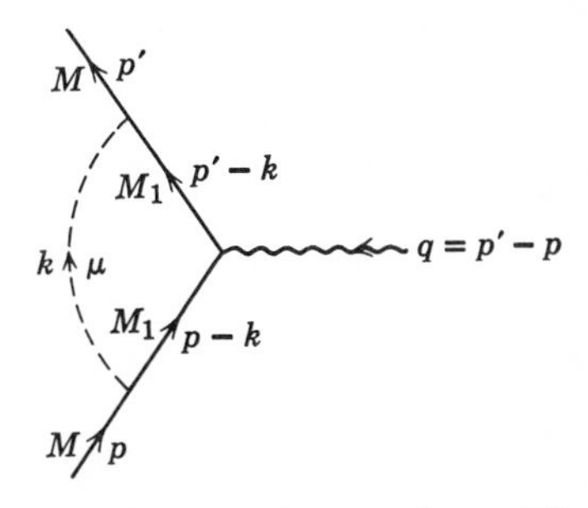

Fig. 22.2. Simplest graph contributing to current matrix element.

$$\int d^4k \frac{1}{k^2 - \mu^2 + i\epsilon} \frac{1}{(p' - k)^2 - M_1{}^2 + i\epsilon} \frac{1}{(p - k)^2 - M_1{}^2 + i\epsilon} \tag{22.18}$$

We use an old technique of Feynman to work out the integral. We use the formula

$$\frac{1}{A_1 A_2 \cdots A_n} = (n - 1)! \int_0^1 d\alpha \cdots \int_0^1 d\alpha_n \frac{\delta(\alpha_1 + \alpha_2 + \cdots + \alpha_n - 1)}{[\alpha_1 A_1 + \cdots + \alpha_n A_n]^n} \tag{22.19}$$

which is easily derived by induction from

$$\frac{1}{A_1 A_2} = \int_0^1 d\alpha \frac{1}{[A_1\alpha + A_2(1 - \alpha)]^2}$$

to write the integral in the form

$$2\int_0^1 d\alpha_1 \int_0^1 d\alpha_2 \int_0^1 d\alpha_3\, \delta(\alpha_1 + \alpha_2 + \alpha_3 - 1) \int d^4k$$

$$\times \frac{1}{[k^2 - 2k(\alpha_1 p + \alpha_2 p') + \alpha_1(p^2 - M_1{}^2) + \alpha_2(p'^2 - M_1{}^2) - \mu^2\alpha_3 + i\epsilon]^3}$$

$$= 2\int_0^1 d\alpha_1 \int_0^1 d\alpha_2 \int_0^1 d\alpha_3\, \delta(\alpha_1 + \alpha_2 + \alpha_3 - 1) \int d^4k$$

$$\times \frac{1}{[(k - \alpha_1 p - \alpha_2 p')^2 - (\alpha_1 p + \alpha_2 p')^2 + (M^2 - M_1{}^2)(\alpha_1 + \alpha_2) - \mu^2\alpha_3 + i\epsilon]^3} \tag{22.20}$$

With the help of

$$2\int d^4k \frac{1}{[k^2 - L + i\epsilon]^3} = -\frac{i\pi^2}{L} \tag{22.21}$$

we obtain[1]

$$-i\pi^2\int_0^1 d\alpha_1\int_0^1 d\alpha_2\int_0^1 d\alpha_3$$

$$\times \frac{\delta(\alpha_1 + \alpha_2 + \alpha_3 - 1)}{\alpha_1^2M^2 + \alpha_2^2M^2 + 2\alpha_1\alpha_2 p\cdot p' + (M_1^2 - M^2)(\alpha_1 + \alpha_2) + \mu^2\alpha_3} \tag{22.22}$$

Singularities can only occur when the denominator vanishes. In terms of the variable

$$q^2 = (p' - p)^2 = 2M^2 - 2p\cdot p'$$

which will be allowed to range over the whole complex plane, the denominator has the form

$$D(\alpha_i, q^2) = (\alpha_1 + \alpha_2)^2M^2 + (\alpha_1 + \alpha_2)(M_1^2 - M^2) + \alpha_3\mu^2 - \alpha_1\alpha_2q^2 \tag{22.23}$$

It is clear that for complex q^2 the denominator can never vanish. Thus all singularities must lie on the real axis in the q^2-plane. If we use the relation $\Sigma_{i=1}^3 \alpha_i = 1$ we may write the denominator function as

$$D(\alpha_i, q^2) = -M^2\alpha_3(1 - \alpha_3) + M_1^2(1 - \alpha_3) + \alpha_3\mu^2 - \alpha_1\alpha_2q^2 \tag{22.24}$$

Noting that the external particles of mass M are stable, so that

$$M < M_1 + \mu$$

we see that

$$D(\alpha_i, q^2) > [M_1(1 - \alpha_3) - \mu\alpha_3]^2 - \alpha_1\alpha_2q^2$$

Hence the singularities must lie in the region $q^2 > 0$, at

$$q^2 = \frac{M_1^2(1 - \alpha_3) - M^2\alpha_3(1 - \alpha_3) + \mu^2\alpha_3}{\alpha_1\alpha_2} \tag{22.25}$$

with the subsidiary condition $\Sigma_{i=1}^3 \alpha_i = 1$. The minimum value of the right-hand side is obtained by using the standard technique of Lagrange multipliers. We write

$$\frac{M_1^2(1 - \alpha_3) - M^2\alpha_3(1 - \alpha_3) + \mu^2\alpha_3}{\alpha_1\alpha_2}\left(-\frac{d\alpha_1}{\alpha_1} - \frac{d\alpha_2}{\alpha_2}\right)$$
$$+ \frac{d\alpha_3}{\alpha_1\alpha_2}(-M_1^2 - M^2(1 + 2\alpha_3) + \mu^2) = 0$$

[1] If the particles are spinors, certain γ-matrices and terms like $(M_1 - \not p + \not k)$ appear in the numerator of the integrand. These may make the integral divergent, but they do not change the analyticity properties of the integral, which depend only on the properties of the denominator.

and

$$\lambda(d\alpha_1 + d\alpha_2 + d\alpha_3) = 0$$

We add these, set the coefficients of the differentials equal to zero, and obtain, after some simple manipulation, the result that the minimum occurs when

$$\mathring{\alpha}_1 = \mathring{\alpha}_2 = \frac{\mu^2}{M^2 + \mu^2 - M_1{}^2} \qquad \mathring{\alpha}_3 = \frac{M^2 - M_1{}^2 - \mu^2}{M^2 + \mu^2 - M_1{}^2} \tag{22.26}$$

If $M^2 < \mu^2 + M_1{}^2$, as is the case for the nucleon-photon vertex, say, (where $M_1 = M$), we get $\mathring{\alpha}_1 = \mathring{\alpha}_2 \geq \frac{1}{2}$, which violates the subsidiary condition. Thus the minimum must occur at $\mathring{\alpha}_1 = \mathring{\alpha}_2 = \frac{1}{2}$; $\mathring{\alpha}_3 = 0$. This, in turn, implies that the singularities start on the positive q^2-axis at the threshold point

$$q^2_{\min} = 4M_1{}^2 \tag{22.27}$$

If $M^2 > \mu^2 + M_1{}^2$, then the singularities start at

$$q^2_{\min} = \frac{1}{\mu^2}\,[M_1{}^2 - (M - \mu)^2][(M + \mu)^2 - M_1{}^2] \tag{22.28}$$

which is smaller than $4M_1{}^2$ but still positive.[1] Thus, aside from subtractions, determined by the behavior at infinity, we obtain the representation

$$(\Psi_{p'}, j(0)\Psi_p) = F((p' - p)^2) = F(q^2) = \frac{1}{\pi}\int_{q^2_{\min}}^{\infty} dq'^2 \frac{\operatorname{Im} F(q'^2)}{q'^2 - q^2 - i\epsilon} \tag{22.29}$$

It has been shown that the analyticity properties exhibited by this representation are characteristic of each term in the perturbation expansion of the current matrix element.

We will now show that this particular set of singularities is dictated by the unitarity condition. Let us apply the reduction technique to

$$(\Psi_{p'}, j(0)\Psi_p)$$

We may obtain two equivalent forms

$$(\Psi_{p'}, j(0)\Psi_p) = \frac{i}{(2\pi)^{3/2}}\int dx e^{ip'x}\theta(x_0)(\Psi_0, [f(x), j(0)]\Psi_p) \tag{22.30}$$

and

$$(\Psi_{p'}, j(0)\Psi_p) = \frac{i}{(2\pi)^{3/2}}\int dx e^{-ipx}\theta(-x_0)(\Psi_{p'}, [j(0), f^+(x)]\Psi_0) \tag{22.31}$$

[1] For the deuteron form factor (Fig. 22.3), $M_1 = \mu = \frac{1}{2}(M_d + E_B)$ where E_B is the binding energy. Thus $q^2_{\min} = 8M_dE_B$.

from which we derive

$$(\Psi_p, j(0)\Psi_{p'})^* = -\frac{i}{(2\pi)^{3/2}}\int dx e^{ip'x}\theta(-x_0)(\Psi_0, [f(x), j(0)]\Psi_p)$$

Hence

$$(\Psi_{p'}, j(0)\Psi_p) - (\Psi_p, j(0)\Psi_{p'})^*$$
$$= \frac{i}{(2\pi)^{3/2}}\int dx e^{ip'x}(\Psi_0, [f(x), j(0)]\Psi_p) \quad (22.32)$$

We insert a complete set of states and carry out the x-integration in usual fashion to obtain

$$\begin{aligned}(\Psi_{p'}, j(0)\Psi_p) &- (\Psi_p, j(0)\Psi_{p'})^* \\ &= 2i \operatorname{Im} F(q^2) \\ &= (2\pi)^{5/2} i \sum_n \{\delta(P_n - p')(\Psi_0, f(0)\Psi_n)(\Psi_n, j(0)\Psi_p) \\ &\qquad - \delta(P_n - p + p')(\Psi_0, j(0)\Psi_n)(\Psi_n, f(0)\Psi_p)\} \quad (22.33)\end{aligned}$$

The first term on the right-hand side vanishes. The reason for this is that the only state Ψ_n, whose energy and momentum is p', must be a one-particle state, and we have already shown in Chapter 13 that the condition that the particle, whose energy momentum is p', with $p'^2 = M^2$ have renormalized mass M, is

$$(\Psi_0, f(0)\Psi_{p'}) = 0 \quad (22.34)$$

The second term does give a contribution in the unphysical region[1] where $P_n^2 = q^2 > 0$. In fact, the contribution comes for all values of q^2 larger than the square of the mass of the lowest state Ψ_n for which $(\Psi_0, j(0)\Psi_n)$ does not vanish. In the case of the electromagnetic current matrix element, the lowest such state is a two-pion state, so that the imaginary part is nonvanishing for $q^2 > 4m_\pi^2$. For the process described in Fig. 22.2, the current (which we took to be scalar for simplicity) "creates" a pair of particles of mass M_1, so that the threshold is

$$q^2 = 4M_1^2 \quad (22.35)$$

[1] Since the current $j(x)$ was taken to be a hermitian operator, it might appear that $F(q^2) = (\Psi_{p'}, j(0)\Psi_p)$ has to be real, in contradiction with (22.33). This is, in fact, so in the physical region in which $q^2 = (p' - p)^2$ and the momenta $\mathbf{p}'$ and $\mathbf{p}$ are real. The right-hand side of (22.33) is nonvanishing for $q^2 > 0$, and there $F(q^2)$ should be viewed as an analytic continuation of the current matrix element to the unphysical region. The preceding considerations have shown that such an analytic continuation is possible.

in agreement with (22.27). In the case that $M^2 > M_1^2 + \mu^2$, the situation is a little more delicate. As shown by Mandelstam,[1] to whose paper we must refer the reader for details, the branch cut still goes from $4M_1^2$ to infinity, but the structure of the matrix element $(\Psi_n, f(0)\Psi_p)$ forces the deformation of the cut back along the real axis to the value quoted in (22.28) and then on to $+\infty$.

We note, parenthetically, that the starting points of the line of singularities are in accord with physical intuition.[2] For the case $M^2 < M_1^2 + \mu^2$, which arises in the nucleon form factor, the extension of the particle structure is governed by the mass of the lightest particle which can be emitted by the nucleon, and reabsorbed after interacting with the electromagnetic field. The radius is thus given by

$$r_0 \approx \frac{1}{2M_1}$$

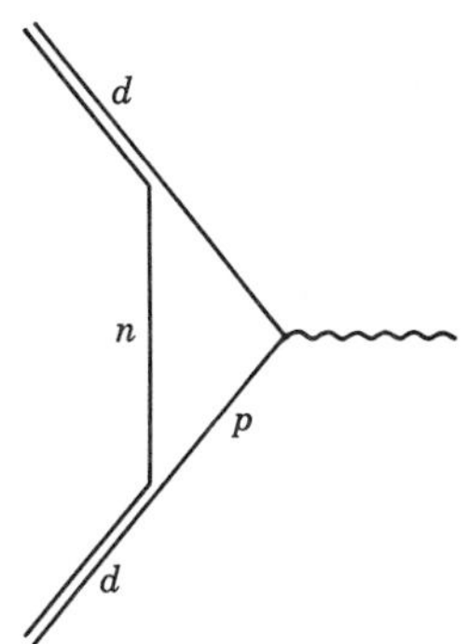

Fig. 22.3. Perturbation graph for deuteron electromagnetic structure.

When $M^2 > M_1^2 + \mu^2$, as would be the case for the deuteron vertex, for example, (Fig. 22.3) the size of the structure is that of the deuteron, so that one would expect that[3]

$$r_0 \approx \frac{1}{4\sqrt{M_1 E_B}} \tag{22.36}$$

In our discussion we have tacitly taken $p^2 = p'^2 = M^2$, so that (22.29) really reads

$$F(p^2, p'^2, q^2) = \frac{1}{\pi}\int_{q^2_{\min}}^{\infty} dq'^2 \, \frac{G(p^2, p'^2, q'^2)}{q'^2 - q^2} \tag{22.37}$$

with

$$G(p^2, p'^2, q'^2) = \frac{1}{2i}\lim_{\epsilon\to 0}\,(F(p^2, p'^2, q'^2 + i\epsilon) - F(p^2, p'^2, q'^2 - i\epsilon)) \tag{22.38}$$

the discontinuity of $F(p^2, p'^2, q^2)$ across the cut in the complex q^2-plane, as shown in Fig. 22.4. If p^2 and/or p'^2 are allowed to vary, new singularities

[1] S. Mandelstam, *Phys. Rev. Letters* **4,** 84 (1960). See also R. Blankenbecler and L. Cook, *Phys. Rev.* **119,** 1745 (1960).

[2] Y. Nambu, *Nuovo Cimento* **9,** 610 (1958).

[3] A loosely bound system of two particles with mass M_1 and binding energy E has a wavefunction that behaves as $e^{-\alpha r}$, with $\alpha = \sqrt{M_1 E_B}$. Hence the charge distribution behaves like $e^{-2\alpha r}$. Thus $\langle r\rangle \sim 1/2\alpha$ and the mean distance to the center of mass system is one half of that, i.e., $1/4\alpha$.

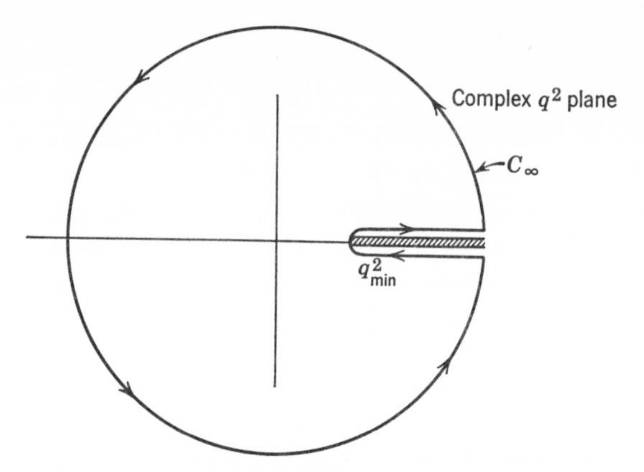

Fig. 22.4. Contour in complex q^2-plane leading to the representation (22.37).

may appear in the q^2-plane, either in the movement of a singularity causing the prolongation of the cut (as mentioned above) or in a more complicated fashion, destroying the simple representation. It is only for restricted values of p^2 and p'^2 that the validity of (22.29) has been proved for every graph contributing to the vertex.[1]

When scattering amplitudes are considered, matters get more complicated, because even with the external particles on the mass shell, the amplitude depends on two variables, such as the incident energy and the scattering angle. In many cases the analyticity properties of the scattering amplitude in both variables are of interest. An analysis of a large class of Feynman graphs as well as nonrelativistic scattering theory supports the conjecture of Mandelstam[2] regarding the behavior of the amplitude for the general two-particle process $A + B \rightarrow C + D$. Let us assign to these particles masses $m_A, \ldots, m_D$ and four momenta $p_A, \ldots, p_D$. The process is represented graphically in Fig. 22.5*a*. With the usual variables

$$\begin{aligned} s &= (p_A + p_B)^2 \\ t &= (p_A - p_C)^2 \\ u &= (p_A - p_D)^2 \end{aligned} \tag{22.39}$$

related by

$$s + t + u = m_A{}^2 + m_B{}^2 + m_C{}^2 + m_D{}^2 \tag{22.40}$$

only two of the variables are independent. In the center of mass system for the reaction $A + B \rightarrow C + D$ we find that

$$s = W^2$$

[1] Y. Nambu, *Nuovo Cimento* **6,** 1064 (1957).
[2] S. Mandelstam, *Phys. Rev.* **112,** 1334 (1958).

where W is the center of mass total energy. The variable t is called the momentum transfer variable, since, if the particle C has the same mass as the particle A and the particle D the same mass as B (as in pion-nucleon scattering, for example), $-t$ is the square of the momentum transfer for $A + B \rightarrow C + D$. The Mandelstam conjecture may be stated as follows: with any "four-line process" (as shown in Fig. 22.5) there is associated a function $\Phi(s, t, u)$, with s, t, u related by (22.40). This function is analytic in the cut planes in the s, t, and u variables, with the cuts determined by the generalized unitarity conditions. This function has certain boundary values when the variables approach the cuts. Thus when s approaches (from above) the cut in the s-plane which extends from some s_0 to $+\infty$, the boundary value is the scattering amplitude for the s-channel, i.e., for the process $A + B \rightarrow C + D$. If t approaches the cut in the t-plane from above, the function Φ becomes the amplitude for the process $A + \bar{C} \rightarrow \bar{B} + D$, and likewise for the u-channel.

As we shall see, a "four-line" amplitude can in general be written in the form of a sum of products of certain kinematical factors, e.g., spinors and γ-matrices, with what we call *invariant functions*. The Mandelstam conjecture for such invariant functions may be expressed in the form of the so-called Mandelstam representation

$$\begin{aligned}\Phi(s, t, u) &= \frac{1}{\pi}\int_{s_0}^{\infty} ds' \, \frac{\varphi_1(s')}{s' - s} + \frac{1}{\pi}\int_{t_0}^{\infty} dt' \, \frac{\varphi_2(t')}{t' - t} + \frac{1}{\pi}\int_{u_0}^{\infty} du' \, \frac{\varphi_3(u')}{u' - u} \\ &\quad + \frac{1}{\pi^2}\iint_{s_0 t_0}^{\infty} ds' \, dt' \, \frac{\rho_{12}(s', t')}{(s' - s)(t' - t)} + \frac{1}{\pi^2}\iint_{t_0 u_0}^{\infty} dt' \, du' \, \frac{\rho_{23}(u', t')}{(t' - t)(u' - u)} \\ &\quad + \frac{1}{\pi^2}\iint_{s_0 u_0}^{\infty} ds' \, du' \, \frac{\rho_{13}(s', u')}{(s' - s)(u' - u)}\end{aligned} \tag{22.41}$$

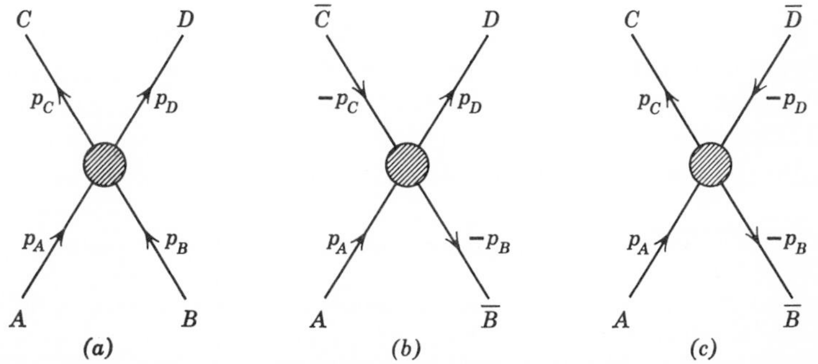

Fig. 22.5. (*a*) The kinematics for the process $A + B \rightarrow C + D$; (*b*) the kinematics for the process $A + \bar{C} \rightarrow \bar{B} + D$; (*c*) the kinematics for the process $A + \bar{D} \rightarrow \bar{B} + C$.

The thresholds for the s, t, and u-cuts (s_0, t_0, u_0) are determined by the generalized unitarity conditions.

The Mandelstam representation (22.39) is incomplete as it stands, since the behavior of the "spectral functions" $\rho_{ij}(x, y)$ is not known for large values of the arguments, so that (22.39) may actually have to be rewritten with a number of subtractions, analogous to the modifications represented in (22.17). It should also be noted that the functions $\varphi_i(s') \cdots$ could contain delta-function singularities of the type $g_1^2\, \delta(s' - \mu_1^2)$ so that the function $\Phi(s, t, u)$ would then contain poles of the type $g_1^2/(\mu_1^2 - s)$. We shall expand on the implications of the Mandelstam conjecture in the next chapter, when pion-nucleon scattering will be discussed. At this point we merely want to point out that in spite of its simplicity, the Mandelstam representation does not follow from the analyticity properties in *one variable* when the other is held fixed in the physical region. For example, it is *not true* that the analyticity properties of the vertex function in two variables are such that we may write

$$F(p^2, p'^2, q^2) = \frac{1}{\pi^2} \iint dx\, dy \frac{f(p^2, x, y)}{(x - p'^2)(y - q^2)}$$

The Mandelstam representation has been shown to hold for potential scattering, for potentials of the form[1]

$$V(r) = \int_\mu^\infty d\sigma \frac{e^{-\sigma r}}{r}$$

and for a large class of Feynman graphs.[2]

For processes involving more than four lines, e.g., production amplitudes, the number of variables increases very rapidly. The analysis of even the simplest graphs is so complicated that not even conjectures, regarding the behavior in more than one variable, have been made. There are indications that if we take one variable at a time, with all other variables in the physical range (so that energies and momenta are real and positive—except for the one variable taken to be complex—and scattering angles are real too), the total amplitude is again analytic in that particular variable, with only the cuts due to unitarity coming in. As an illustration consider the amplitude

[1] See M. L. Goldberger and K. M. Watson, *Collision Theory*, John Wiley and Sons, Inc. New York (1964).

[2] There are graphs, even to fourth order, in perturbation theory for which the Mandelstam representation breaks down. These are graphs in which there are anomalous thresholds, similar to the ones discussed for the vertex function. For a simple guide to anomalous thresholds, see S. Coleman and R. Norton, *Nuovo Cimento* **38**, 438 (1965), where references to the literature may also be found.

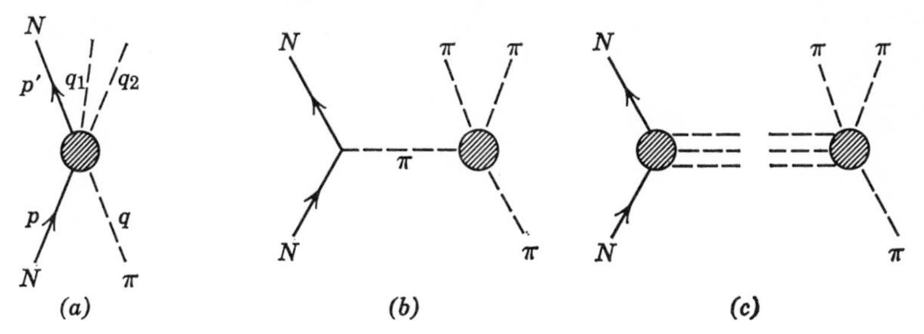

Fig. 22.6. (*a*) The process $\pi + N \rightarrow N + \pi + \pi$; (*b*) the one-particle term in the absorptive part of the amplitude in the t-channel; (*c*) a multiparticle term in the absorptive part in the t-channel.

for the process

$$\pi + N \rightarrow N + \pi + \pi$$

shown in Fig. 22.6*a*. We may safely assume that if the variables $s = (p + q)^2$, $s_{11} = (p' + q_1)^2$, $s_{12} = (p' + q_2)^2, \ldots$ are in the physical range, the amplitude is analytic in the one variable $t = (p' - p)^2$, with the cuts determined by unitarity, as shown in Fig. 22.5*b* and *c*. Thus we obtain the following representation for $A(s_1, s_{11}, s_{12}, \ldots; t)$

$$\begin{aligned} A(s, s_{11}, \ldots; t) &= \frac{1}{\pi}\int_{t_0}^{\infty} dt' \frac{B(s, s_{11}, \ldots; t')}{t' - t} \\ &= \frac{1}{\pi}\int_{t_0}^{\infty} dt' \frac{C\, \delta(t' - \mu^2)\, A_{\pi\pi}(\ldots; t')}{t' - t} \\ &\qquad + \frac{1}{\pi}\int_{t_0}^{\infty} dt' \frac{\tilde{B}(\ldots; t')}{t' - t} \end{aligned} \tag{22.42}$$

In this equation, B stands for the absorptive part in the t-channel. As shown in Fig. 22.6*b* and *c*, this consists of two parts: a one-particle part, which consists of a product of a pion-nucleon vertex, and a four-pion function, with the mass of one of the pions being t' and a multiparticle part, arising from the interchange of three pions (two are forbidden by G parity) and other particles. The one-pion exchange term, upon integration of the delta function yields the product of the pion-nucleon coupling constant and the true pion-pion scattering amplitude as residues of a pole at $t = \mu^2$. We shall see in Chapter 27 that there is some evidence that in experiments involving production with low negative values of t the one-particle exchange terms tend to dominate the total amplitude, and indirect knowledge of, say, the pion-pion cross section is obtainable. Crucial to a justification of this type of analysis are representations of the type shown in (22.42) and that is why we mention them at this point.

PROBLEMS

1. Consider the box graph

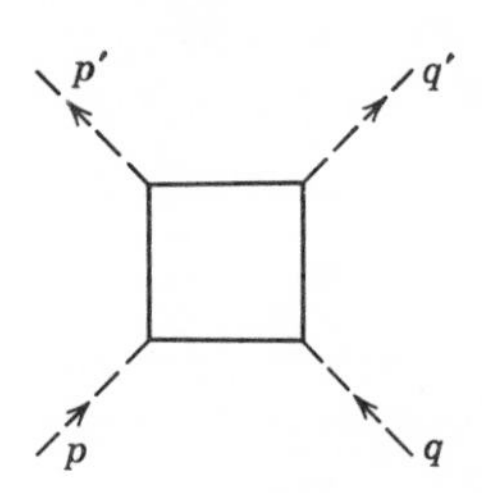

Write down an expression for the amplitude represented by the graph in terms of $s = (p + q)^2$ and $t = (q' - q)^2$, involving only integrals over the Feynman parameters [analogous to (22.22)]. Discuss the analyticity properties of the amplitude as a function of s, when $t \leq 0$.

2. Use the representation for the vacuum expectation value of a commutator obtained in Problem 3, Chapter 21 to obtain a representation for

$$(\Psi_0, \theta(x_0 - y_0)[A(x), B(y)]\Psi_0)$$

What are the analyticity properties of the Fourier transform of this matrix element? What determines the location of the singularities?

23

Pion-Nucleon Scattering and the Forward Scattering Dispersion Relations

In the 1950's, before the existence of internal symmetries became part of the lore of particle physics, pions and nucleons had a rather special position in the family of particles, and pion-nucleon scattering was widely regarded as *the* dynamical problem. The study of the analyticity properties of scattering amplitudes that was stimulated by the work of Gell-Mann, Goldberger, and Thirring immediately led to simple relations between the real part of the pion-nucleon scattering amplitude and the cross section,[1] and the successful confrontation of these relations with experiment marked a turning point in the theoretical approach to strong interaction dynamics. Although the derivation and testing of forward scattering dispersion relations preceded the study of the analyticity properties of the scattering amplitude as a function of both the center of mass energy and the cosine of the scattering angle, we shall invert the historical development and take the opportunity to range a little more widely over the subject of pion-nucleon scattering. Thus we shall start with the Mandelstam conjecture to obtain the forward scattering relations, even though the former is only conjectured and the latter have been proved independent of perturbation theory.[2]

Before we can write down a Mandelstam representation, we must work through some kinematical preliminaries, which, while tedious, are necessary here and useful elsewhere. To begin with, we observe that if, in the amplitude

$$T_{\beta\alpha}(p', q'; p, q) = -\frac{i}{(2\pi)^3}\int dz\, e^{i\frac{q+q'}{2}z}\,\theta(z_0)\left(\Psi_{p'}, \left[j_\beta\left(\frac{z}{2}\right), j_\alpha\left(-\frac{z}{2}\right)\right]\Psi_p\right) \tag{23.1}$$

[1] R. Karplus and M. Ruderman, *Phys. Rev.* **98,** 771 (1955); M. L. Goldberger, *Phys. Rev.*, **99,** 979 (1955).
[2] K. Symanzik, *Phys. Rev.* **105,** 743 (1957); N. N. Bogoliubov, B. V. Medvedev, and M. Polivanov, *Fortsch. Physik* **6,** 169 (1958).

further contractions of the two nucleon states are carried out, an expression of the form

$$T_{\beta\alpha}(p', q'; p, q) = \frac{1}{16\pi^5} \bar{u}(p')\mathcal{T}_{\beta\alpha} u(p) \tag{23.2}$$

is obtained. Here $\mathcal{T}_{\beta\alpha}(p', q'; p, q)$ is a 4×4 matrix standing between the spinors $\bar{u}(p')$ and $u(p)$. The most general form of $\mathcal{T}_{\beta\alpha}(p', q'; p, q)$ is

$$-A + B^\mu \gamma_\mu + C^{\mu\nu}[\gamma_\mu, \gamma_\nu] + D^\mu \gamma_\mu \gamma_5 + E\gamma_5 \tag{23.3}$$

with the scalar, vector, . . . functions constructed out of the momenta p, p' of the nucleons and q, q' of the mesons in the initial and final state respectively. The last two terms will not appear because the scattering amplitude must be a scalar;[1] parity is conserved in the strong interactions. The tensor terms must be of the form

$$C^{\mu\nu} = C_1 p^\mu p'^\nu + C_2 p^\mu q^\nu + \cdots$$

Since these stand between the spinors $\bar{u}(p')$ and $u(p)$, the condition

$$(p - M)\, u(p) = \bar{u}(p')(p' - M) = 0$$

may be used to eliminate these. Similar arguments show that there is only one vector term and that we may rewrite (23.2) in the form

$$T_{\beta\alpha}(p', q'; p, q) = \frac{1}{16\pi^5} \bar{u}(p')\left[-A_{\beta\alpha}(s, t, u) + \frac{q + q'}{2} B_{\beta\alpha}(s, t, u)\right]u(p) \tag{23.4}$$

The functions $A_{\beta\alpha}(s, t, u)$ and $B_{\beta\alpha}(s, t, u)$ are functions of the scalar variables

$$s = (p + q)^2$$
$$t = (q' - q)^2$$
$$u = (p' - q)^2$$

with

$$s + t + u = 2M^2 + 2m_\pi^2 \tag{23.5}$$

As far as the *i*-spin dependence is concerned, we note that the spinors $\bar{u}(p')$ and $u(p)$ have a suppressed label indicating that they are really eight-component spinors, representing the nucleonic state, proton or neutron. Thus the functions $A_{\beta\alpha}$ and $B_{\beta\alpha}$ are still 2×2 matrices in *i*-spin space. The most general form which we can write is

$$\begin{aligned} A_{\beta\alpha} &= \delta_{\beta\alpha} A^{(+)} + \tfrac{1}{2}[\tau_\beta, \tau_\alpha] A^{(-)} \\ B_{\beta\alpha} &= \delta_{\beta\alpha} B^{(+)} + \tfrac{1}{2}[\tau_\beta, \tau_\alpha] B^{(-)} \end{aligned} \tag{23.6}$$

[1] It is not possible to construct a pseudoscalar with three independent four-vectors.

Thus for a given energy and scattering angle there are four numbers that describe pion-nucleon scattering. This is as expected: for each *i*-spin $T = \frac{1}{2}$ and $\frac{3}{2}$ there are two amplitudes, one describing the scattering in which the nucleon spin remains unchanged in direction, the other in which it flips.

The dynamics of pion-nucleon scattering is described by the functions $A^{(\pm)}(s, t, u)$ and $B^{(\pm)}(s, t, u)$. It is for these "invariant amplitudes" that we postulate the cut plane analyticity properties conjectured by Mandelstam. The invariant amplitudes will be represented by expressions like that shown in (22.41). The only additional terms come from the single particle terms whose imaginary parts were worked out in (21.43) and (21.44). There we found that

$$i(T_{\beta\alpha}(p', q'; p, q) - T_{\alpha\beta}{}^*(p, q; p', q'))$$
$$= \frac{g^2}{(2\pi)^5}\, \bar{u}(p')\frac{\not{q} + \not{q}'}{2}\{\tau_\beta\tau_\alpha\, \delta(s - M^2) - \tau_\alpha\tau_\beta\, \delta(u - M^2)\}u(p)$$
$$+ \text{ (multiple particle contributions)} \quad (23.7)$$

This implies that we can write

$$A^{(\pm)}(s, t, u) = \frac{1}{\pi^2}\int_{(M+m_\pi)^2}^{\infty} ds' \int_{4m_\pi{}^2}^{\infty} dt' \frac{\rho_{12}^{(\pm)}(s', t')}{(s' - s)(t' - t)}$$
$$+ \frac{1}{\pi^2}\int_{(M+m_\pi)^2}^{\infty} du' \int_{4m_\pi{}^2}^{\infty} dt' \,\frac{\rho_{23}^{(\pm)}(u', t')}{(u' - u)(t' - t)}$$
$$+ \frac{1}{\pi^2}\int_{(M+m_\pi)^2}^{\infty} ds' \int_{(M+m_\pi)^2}^{\infty} du' \,\frac{\rho_{31}^{(\pm)}(s', u')}{(s' - s)(u' - u)} \quad (23.8)$$

since the one-particle terms only appear associated with $\frac{\not{q} + \not{q}'}{2}$, i.e., with the $B^{(\pm)}$ terms. Note that the lower limits on the integrals above are determined by the onset of the one-nucleon one-pion thresholds in the *s*- and *u*-channels and the two-pion state in the *t*-channel.

Equation (23.7) can be shown, after a little algebra, to imply that

$$i\delta_{\beta\alpha}(B^{(+)}(s, t, u) - B^{(+)*}(s, t, u)) + \frac{i}{2}\,[\tau_\beta, \tau_\alpha](B^{(-)}(s, t, u) - B^{(-)*}(s, t, u))$$
$$= \tfrac{1}{2}\, g^2(\tau_\beta\tau_\alpha\, \delta(s - M^2) - \tau_\alpha\tau_\beta\, \delta(u - M^2))$$
$$+ \text{ (multiple particle contributions)} \quad (23.9)$$

i.e.

$$\text{Im } B^{(\pm)}(s, t, u) = -\tfrac{1}{4}\, g^2(\delta(s - M^2) \mp \delta(u - M^2))$$
$$+ \text{ (multiple particle contributions)} \quad (23.10)$$

These one-particle terms appear in the Mandelstam representation for $B^{(\pm)}(s, t, u)$ in the following form

$$
\begin{aligned}
B^{(\pm)}(s, t, u) = &-\frac{g^2}{4\pi}\left(\frac{1}{M^2 - s} \mp \frac{1}{M^2 - u}\right) \\
&+ \frac{1}{\pi^2}\int_{(M+m_\pi)^2}^{\infty} ds' \int_{4m_\pi{}^2}^{\infty} dt' \, \frac{\sigma_{12}^{(\pm)}(s', t')}{(s' - s)(t' - t)} \\
&+ \frac{1}{\pi^2}\int_{(M+m_\pi)^2}^{\infty} du' \int_{4m_\pi{}^2}^{\infty} dt' \, \frac{\sigma_{23}^{(\pm)}(u', t')}{(u' - u)(t' - t)} \\
&+ \frac{1}{\pi^2}\int_{(M+m_\pi)^2}^{\infty} ds' \int_{(M+m_\pi)^2}^{\infty} du' \, \frac{\sigma_{31}^{(\pm)}(s', u')}{(s' - s)(u' - u)}
\end{aligned}
\tag{23.11}
$$

This is so, because the scattering amplitude in a given channel, the s-channel for instance, is defined as the boundary value of the appropriate combination of $A^{(+)}(s, t, u), \ldots B^{(-)}(s, t, u)$ as s approaches the real axis from above in the physical region $(M + m_\pi)^2 \leq s \leq \infty$ with t and u in their physical regions. Thus the imaginary part of any one of the amplitudes, $A^{(+)}, \ldots, B^{(-)}$ is given by

$$
\lim_{\epsilon \to 0+} \frac{F(s + i\epsilon, t, u) - F(s - i\epsilon, t, u)}{2i}
$$

If we use

$$
\frac{1}{M^2 - s - i\epsilon} = P\frac{1}{M^2 - s} + i\pi\,\delta(s - M^2)
$$

we see that the single particle terms in (23.11) are appropriately chosen.

The physical region for the t variable is best determined by considering the center of mass system. There we write

$$
\begin{aligned}
q &= (\omega, \mathbf{q}) \qquad & q' &= (\omega, \mathbf{q}') \\
p &= (E, -\mathbf{q}) \qquad & p' &= (E, -\mathbf{q}')
\end{aligned}
$$

and find that

$$
s = (E + \omega)^2 \equiv W^2 \tag{23.12}
$$

and

$$
\begin{aligned}
t &= -(\mathbf{q}' - \mathbf{q})^2 \\
&= -2q^2(1 - \cos\theta)
\end{aligned}
\tag{23.13}
$$

Thus the physical region for t in the s channel is

$$
-4q^2 \leq t \leq 0 \tag{23.14}
$$

Incidentally, it is just a matter of some computations to show that if s is in the physical region and $\cos\theta$ is bounded by 1, then $u < (M + m_\pi)^2$;

in fact

$$u \leq \frac{(M^2 - m_\pi^{\ 2})^2}{s} < (M - m_\pi)^2$$

If we look at (23.8) and (23.11) and note that we can write

$$\frac{1}{(s' - s)(u' - u)} = \left(\frac{1}{s' - s} + \frac{1}{u' - u}\right)\frac{1}{u' + s' + t - 2M^2 - 2m_\pi^{\ 2}}$$

we see that the single fixed-momentum transfer dispersion relations

$$A^{(\pm)}(s, t) = \frac{1}{\pi}\int_{(M+m_\pi)^2}^{\infty} ds' \frac{\alpha_1^{(\pm)}(s', t)}{s' - s} + \frac{1}{\pi}\int_{(M+m_\pi)^2}^{\infty} du' \frac{\alpha_2^{(\pm)}(u', t)}{u' - u}$$

$$B^{(\pm)}(s, t) = -\frac{g^2}{4\pi}\left(\frac{1}{M^2 - s} \mp \frac{1}{M^2 - u}\right)$$
$$+ \frac{1}{\pi}\int_{(M+m_\pi)^2}^{\infty} ds' \frac{\beta_1^{(\pm)}(s', t)}{s' - s} + \frac{1}{\pi}\int_{(M+m_\pi)^2}^{\infty} du' \frac{\beta_2^{(\pm)}(u', t)}{u' - u} \tag{23.15}$$

follow immediately. These dispersion relations have actually been proved independent of perturbation theory for small values of t.[1]

It can be seen from (23.1) that the following formal identity holds

$$T^*_{\beta\alpha}(p, -q'; p', -q) = T_{\beta\alpha}(p', q'; p, q) \tag{23.16}$$

This is an expression of crossing symmetry. It follows from this identity that

$$\left((\bar{u}(p)\left(-A^{\beta\alpha}(u, t, s) - \frac{\not{q} + \not{q}'}{2} B^{\beta\alpha}(u, t, s)\right)u(p')\right)^*$$
$$= \bar{u}(p')\left(-A^{\beta\alpha}(u, t, s)^* - \frac{\not{q} + \not{q}'}{2} B^{\beta\alpha}(u, t, s)^*\right)u(p)$$
$$= \bar{u}(p')\left(-A^{\beta\alpha}(s, t, u) + \frac{\not{q} + \not{q}'}{2} B^{\beta\alpha}(s, t, u)\right)u(p)$$

i.e.

$$A^{(\pm)}(u, t, s)^* = \pm A^{(\pm)}(s, t, u)$$
$$B^{(\pm)}(u, t, s)^* = \mp B^{(\pm)}(s, t, u) \tag{23.17}$$

[1] A discussion of the nonperturbative proofs may be found in the lectures of H. Lehmann, *Suppl. Nuovo Cimento*, **14,** 153 (1959). A proof that for small $t < 0$ the single dispersion relations hold to each order in perturbation theory is given by K. Symanzik, *Progr. Theoret. Phys.* (*Kyoto*) **20,** 690 (1958).

This means that (23.15) can be rewritten as follows

$$A^{(\pm)}(s, t, u) = \frac{1}{\pi}\int_{(M+m_\pi)^2}^{\infty} dx\alpha^{(\pm)}(x, t)\left(\frac{1}{x-s} \pm \frac{1}{x-u}\right)$$

$$B^{(\pm)}(s, t, u) = -\frac{g^2}{4\pi}\left(\frac{1}{M^2-s} \mp \frac{1}{M^2-u}\right) + \frac{1}{\pi}\int_{(M+m_\pi)^2}^{\infty} dx\beta^{(\pm)}(x, t)\left(\frac{1}{x-s} \mp \frac{1}{x-u}\right)$$

In order to obtain the correct signs it must be remembered that in the denominators the variable s has a small positive imaginary part and the variable u, given by $u = 2M^2 + 2m^2 - t - s$ actually has a small negative imaginary part when s is in the physical region. If the notation

$$\begin{aligned} P &= \tfrac{1}{2}(p' + p) \\ Q &= \tfrac{1}{2}(q' + q) \\ \Delta &= p' - p = q - q' \\ t &= \Delta^2; \;\; \nu = P \cdot Q \end{aligned} \tag{23.18}$$

is introduced, these equations may be written in the more symmetric-looking form

$$A^{(\pm)}(\nu, t) = \frac{1}{\pi}\int_{Mm_\pi + \frac{1}{4}t}^{\infty} d\nu' \alpha^{(\pm)}(\nu', t)\left(\frac{1}{\nu'-\nu} \pm \frac{1}{\nu'+\nu}\right)$$

$$B^{(\pm)}(\nu, t) = \frac{g^2}{8\pi}\left(\frac{1}{\frac{1}{2}m_\pi^2 + \frac{1}{4}t + \nu} \mp \frac{1}{\frac{1}{2}m_\pi^2 + \frac{1}{4}t - \nu}\right) + \frac{1}{\pi}\int_{Mm_\pi + \frac{1}{4}t}^{\infty} d\nu' \beta^{(\pm)}(\nu', t)\left(\frac{1}{\nu'-\nu} \mp \frac{1}{\nu'+\nu}\right) \tag{23.19}$$

Before going on to a discussion of forward scattering we will work out the phase shift analysis of pion-nucleon scattering so that we will later be able to express the real parts of $A^{(\pm)}$ and $B^{(\pm)}$ in terms of the phase shifts. First of all, it should be noted that the phase shifts describing the reaction

$$\pi^- + p \rightarrow \pi^- + p$$

will actually be complex, since in addition to the above elastic process, the inelastic process

$$\pi^- + p \rightarrow \pi^0 + n$$

is also possible. We can, however, get real phase shifts by taking advantage of the fact that i-spin is conserved, so that at low energies, the scattering of the pion by a nucleon in the total i-spin $\frac{1}{2}$ or $\frac{3}{2}$ states is truly elastic. The general amplitude which is of the form

$$F_{\beta\alpha} = \delta_{\beta\alpha}F^{(+)} + \tfrac{1}{2}[\tau_\beta, \tau_\alpha]F^{(-)} \tag{23.20}$$

may be decomposed into *i*-spin amplitudes with the help of a projection operator technique. The projection operators for $T = \frac{3}{2}$ and $T = \frac{1}{2}$ are obtained with the help of the relation

$$\mathbf{T} = \tfrac{1}{2}\boldsymbol{\tau} + \mathbf{t}_\pi$$

This implies that

$$\mathbf{t}_\pi \cdot \boldsymbol{\tau} = T(T + 1) - 2 - \tfrac{3}{4} \tag{23.21}$$

Now for the meson *i*-spin operator we use the representation (16.23) so that the projection operators

$$\Pi_{\frac{1}{2}} = \frac{1 - \mathbf{t}_\pi \cdot \boldsymbol{\tau}}{3} \qquad \Pi_{\frac{3}{2}} = \frac{2 + \mathbf{t}_\pi \cdot \boldsymbol{\tau}}{3} \tag{23.22}$$

constructed with the help of (23.21), take the form

$$\begin{aligned}(\Pi_{\frac{1}{2}})_{\beta\alpha} &= \tfrac{1}{3}\,\delta_{\beta\alpha} + \tfrac{1}{3}ie_{\beta\alpha i}\tau_i = \tfrac{1}{3}(\delta_{\beta\alpha} + \tfrac{1}{2}[\tau_\beta, \tau_\alpha])\\ (\Pi_{\frac{3}{2}})_{\beta\alpha} &= \tfrac{2}{3}\,\delta_{\beta\alpha} - \tfrac{1}{3}ie_{\beta\alpha i}\tau_i = \tfrac{1}{3}(2\,\delta_{\beta\alpha} - \tfrac{1}{2}[\tau_\beta, \tau_\alpha])\end{aligned} \tag{23.23}$$

with the indices denoting that these are 2×2 matrices in the nucleon isospinor space, suppressed. Thus (23.20) may be rewritten in the form

$$\begin{aligned}F_{\beta\alpha} &= F^{(+)}(\Pi_{\frac{1}{2}} + \Pi_{\frac{3}{2}})_{\beta\alpha} + F^{(-)}(2\Pi_{\frac{1}{2}} - \Pi_{\frac{3}{2}})_{\beta\alpha}\\ &= (\Pi_{\frac{1}{2}})_{\beta\alpha}(F^{(+)} + 2F^{(-)}) + (\Pi_{\frac{3}{2}})_{\beta\alpha}(F^{(+)} - F^{(-)})\end{aligned} \tag{23.24}$$

Let us now consider the scattering amplitude F in a particular *i*-spin state. Whereas for spin-0 particles we would write the conventional expansion

$$F(W, \cos\theta) = \sum_{l=0}^{\infty} (2l + 1) f_l(W) P_l(\cos\theta)$$

we cannot do this here, because the nucleon has spin $\frac{1}{2}$. Thus the scattering amplitude $F(W, \cos\theta)$ is really a 2×2 matrix in the nucleon spin space. There are actually only two independent amplitudes if parity is conserved, and there is invariance under time reversal.[1] This may be restated as follows:

The interaction in a given orbital angular momentum l contributes to scattering in the states of total angular momentum $J = l \pm \frac{1}{2}$, and these amplitudes are in general different. We write

$$F(W, \cos\theta) = \sum_{l=0}^{\infty} (2l + 1)(f_{l-}(W)\Pi_{J=l-\frac{1}{2}} + f_{l+}\Pi_{J=l+\frac{1}{2}})\, P_l(\cos\theta) \tag{23.25}$$

Here $\Pi_{J=l-\frac{1}{2}}$ and $\Pi_{J=l+\frac{1}{2}}$ are again projection operators. They are constructed with the help of the identity

$$J(J + 1) = l(l + 1) + \tfrac{3}{4} + \mathbf{L} \cdot \boldsymbol{\sigma} \tag{23.26}$$

[1] This can be seen most easily in the helicity representation, Chapter 4.

and have the form

$$\Pi_{J=l-\frac{1}{2}} = \frac{l - \boldsymbol{\sigma} \cdot \mathbf{L}}{2l + 1}$$

$$\Pi_{J=l+\frac{1}{2}} = \frac{l + 1 + \boldsymbol{\sigma} \cdot \mathbf{L}}{2l + 1} \tag{23.27}$$

We use

$$\begin{aligned}\boldsymbol{\sigma} \cdot \mathbf{L} P_l(\cos\theta) &= \boldsymbol{\sigma} \cdot \mathbf{L} P_l(\hat{\mathbf{q}}' \cdot \hat{\mathbf{q}}) = -\boldsymbol{\sigma} \cdot \mathbf{q}' \times i\nabla_{\mathbf{q}'} P_l(\hat{\mathbf{q}}' \cdot \hat{\mathbf{q}}) \\ &= -i\boldsymbol{\sigma} \cdot \hat{\mathbf{q}}' \times \hat{\mathbf{q}} P_l'(\cos\theta)\end{aligned}$$

to obtain the form

$$\begin{aligned}F(W, \cos\theta) &= \sum_{l=0}^{\infty} (l f_{l-}(W) + (l+1) f_{l+}(W))\, P_l(\cos\theta) \\ &\quad - i\boldsymbol{\sigma} \cdot \hat{\mathbf{q}} \times \hat{\mathbf{q}}' \sum_{l=0}^{\infty} (f_{l-}(W) - f_{l+}(W))\, P_l'(\cos\theta) \end{aligned} \tag{23.28}$$

for each i-spin scattering amplitude.

To connect this with the form (23.4), we must evaluate the latter in the center of mass system. We use the explicit representations of the spinors, given in (2.47) to calculate

$$\begin{aligned}\bar{u}(p')\, u(p) &= \frac{1}{2M(M+E)} (M + \rho_3 E + i\rho_2 \boldsymbol{\sigma} \cdot \mathbf{q}')(M + \rho_3 E + i\rho_2 \boldsymbol{\sigma} \cdot \mathbf{q}) \\ &= \frac{(M+E) - (E-M)\cos\theta}{2M} - i\boldsymbol{\sigma} \cdot \hat{\mathbf{q}} \times \hat{\mathbf{q}}' \frac{E-M}{2M}\end{aligned}$$

and

$$\begin{aligned}\bar{u}(p') \frac{\not{q} + \not{q}'}{2} u(p) &= \frac{(E+M)(W-M)}{2M} + \frac{(E-M)(W+M)}{2M} \cos\theta \\ &\quad - i\, \frac{(E-M)(W+M)}{2M} \boldsymbol{\sigma} \cdot \hat{\mathbf{q}} \times \hat{\mathbf{q}}'\end{aligned}$$

respectively.

Thus we obtain with the help of (21.13), which states that

$$F(W, \cos\theta) = -\frac{16\pi^5 M}{W} T(W, \cos\theta)$$

the relations

$$\begin{aligned}\frac{E+M}{2W} (A(W, \cos\theta) - (W - M)B(W, \cos\theta)) & \\ - \frac{E-M}{2W} \cos\theta (A + (W+M)B) & \\ = \sum_l (l f_{l-} + (l+1) f_{l+}) P_l(\cos\theta) & \\ \frac{E-M}{2W} (A + (W+M)B) = \sum_i (f_{l-} - f_{l+}) P_l'(\cos\theta) & \end{aligned} \tag{23.29}$$

With the help of the formulas

$$(2l + 1)\cos\theta P_l(\cos\theta) = lP_{l-1}(\cos\theta) + (l + 1)P_{l+1}(\cos\theta)$$
$$(2l + 1)P_l(\cos\theta) = P'_{l+1}(\cos\theta) - P'_{l-1}(\cos\theta)$$
$$lP_l(\cos\theta) = \cos\theta P_l'(\cos\theta) - P'_{l-1}(\cos\theta)$$

the reader may convince himself of the following result[1]

$$f_{l\pm}(W) = \frac{1}{2}\int_{-1}^{1} d(\cos\theta)\left\{\left[\frac{E + M}{2W}(A - (W - M)B)\right]P_l(\cos\theta) - \left[\frac{E - M}{2W}(A + (W + M)B)\right]P_{l\pm1}(\cos\theta)\right\} \tag{23.30}$$

With this long preparation, we are at last ready to discuss the analysis of pion-nucleon scattering data. We shall restrict ourselves to fairly low energies, at which inelastic processes, such as pion production, are not important. In that energy region, the partial-wave amplitudes may be written in terms of *real* phase shifts

$$f_{l\pm}(W) = \frac{e^{i\delta_{l\pm}(W)}\sin\delta_{l\pm}(W)}{q} \qquad (23.31)^2$$

In the region under consideration only S- and P-waves are likely to be important. The scattering amplitude is thus given by

$$F(W, \cos\theta) \approx f_{S_{1/2}} + (f_{P_{1/2}} + 2f_{P_{3/2}})\cos\theta - i\boldsymbol{\sigma}\cdot\hat{\mathbf{n}}\sin\theta(f_{P_{1/2}} - f_{P_{3/2}}) \tag{23.32}$$

In this expression we have used spectroscopic notation and defined the normal to the reaction plane by

$$\hat{\mathbf{n}} = \frac{\hat{\mathbf{q}} \times \hat{\mathbf{q}}'}{|\hat{\mathbf{q}} \times \hat{\mathbf{q}}'|}$$

where $\hat{\mathbf{q}}'$ and $\hat{\mathbf{q}}$ are unit vectors in the direction of the final and initial meson momenta. The differential cross section thus is

$$\frac{d\sigma}{d\Omega} = |f_{S_{1/2}} + (f_{P_{1/2}} + 2f_{P_{3/2}})\cos\theta|^2 + \sin^2\theta\,|f_{P_{1/2}} - f_{P_{3/2}}|^2 \tag{23.33}$$

[1] If we note that A and B depend on W^2 only and that $E = (W^2 + M^2 - m_\pi^2)/2W$, we can readily prove the so-called MacDowell symmetry relation

$$f_{l+}(-W) = -f_{l+1-}(W)$$

[2] This can be shown when (23.28) is inserted into (21.8): the condition

$$\mathrm{Im}\, f_{l\pm}(W) = q\,|\,f_{l\pm}(W)|^2$$

in the c.m. system follows.

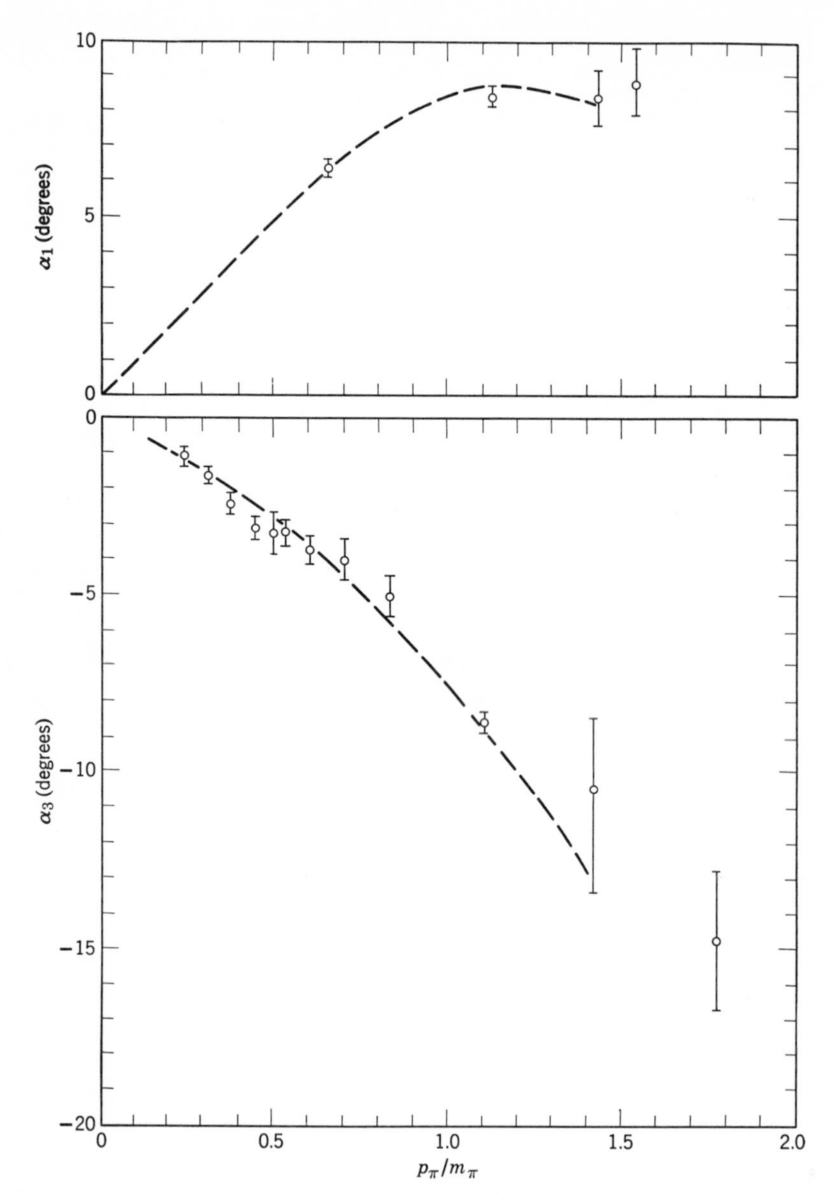

Fig. 23.1. The pion-nucleon *S*-wave phase shifts for $T = \frac{1}{2}(\alpha_1)$ and for $T = \frac{3}{2}(\alpha_3)$ as a function of the pion center of mass momentum. [From J. Hamilton, *Strong Interactions and High Energy Physics*, Oliver and Boyd Ltd. (1964).]

and the polarization of the nucleon in the final state may easily be shown to be given by

$$\langle \boldsymbol{\sigma} \cdot \hat{\mathbf{n}} \rangle \frac{d\sigma}{d\Omega} = 2 \sin \theta \operatorname{Im} [(f^*_{P_{1/2}} - f^*_{P_{3/2}})(f_{S_{1/2}} + f_{P_{1/2}} \cos \theta + 2 f_{P_{3/2}} \cos \theta)] \tag{23.34}$$

It was with the help of an analysis of the angular distribution in terms of phase shifts that the quantum numbers of the resonance $N^*(1238)$ were determined;[1] subsequent polarization measurements confirmed that the choice of phase shifts (among a set of possible ones) which made the $J = T = \frac{3}{2}$ state resonant was indeed the correct one. Figure 23.1 shows the S-wave phase shifts for $T = \frac{1}{2}$ and $\frac{3}{2}$. These are conventionally denoted by α_1 and α_3 respectively (the notation for the P-wave phase shifts is $\alpha_{2T,2J}$), and we can determine the scattering lengths from the data. The values

$$a_1 = \lim_{q \to 0} \frac{\alpha_1}{q/m_\pi} = 0.178 \pm 0.005$$

$$a_3 = \lim_{q \to 0} \frac{\alpha_3}{q/m_\pi} = -0.107 \pm 0.003 \tag{23.35}$$

where q is the pion center of mass momentum, will be of particular interest to us.

At higher energies, when inelastic processes become possible, the form of the partial wave scattering amplitude is

$$f_{l\pm}(W) = \frac{\eta_{l\pm} e^{2i\delta_{l\pm}} - 1}{2iq} = \frac{\eta_{l\pm} \sin 2\,\delta_{l\pm}}{2q} + i\,\frac{1 - \eta_{l\pm} \cos 2\,\delta_{l\pm}}{2q} \tag{23.36}$$

with

$$0 \leq \eta_l \leq 1 \tag{23.37}$$

The effective doubling of parameters at each energy makes a determination of them much more difficult, given the fairly large experimental errors. Nevertheless, such phase-shift analyses have been carried out.[2]

For forward scattering there are no spin-flip terms. The first term of (23.28) yields

$$F(W, \theta = 0) = \sum_{l=0}^{\infty} (l f_{l-}(W) + (l+1) f_{l+}(W))$$

$$\operatorname{Re} F(W, \theta = 0) = \frac{1}{2q} \sum_{l=0}^{\infty} (l \sin 2\,\delta_{l-} + (l+1) \sin 2\,\delta_{l+}) \tag{23.38}$$

[1] For a discussion of the ambiguities in a phase-shift analysis, see H. Bethe and F. de Hoffmann, *Mesons and Fields II*, Harper and Row, Inc; New York (1955).

[2] P. Auvil, C. Lovelace, A, Donnachie, and A. T. Lea, *Physics Letters* **12**, 76 (1964) and references cited therein.

We shall be interested in the real part of the scattering amplitude in the laboratory system, because the dispersion relations that we will soon discuss take the simplest form there. To find the relation between the scattering amplitude in the center of mass system and the amplitude in the laboratory system, it is quickest to make use of the fact that the optical theorem

$$\operatorname{Im} F(W, \theta = 0) = \frac{q}{4\pi}\,\sigma_{\text{tot}}(W) \tag{23.39}$$

reads the same in the two reference frames. Thus we have[1]

$$\frac{F_{\text{c.m.}}(s)}{F_{\text{lab}}(s)} = \frac{q}{q_L}\,\frac{\sigma_{\text{tot}}(s)}{\sigma_{\text{tot}}(s)} = \frac{q}{q_L} \tag{23.40}$$

It is now a matter of just a little algebra to find that in terms of the invariant quantity s,

$$q = \left[\frac{(\sqrt{s}+M+m_\pi)(\sqrt{s}+M-m_\pi)(\sqrt{s}-M+m_\pi)(\sqrt{s}-M-m_\pi)}{4s}\right]^{1/2} \tag{23.41}$$

On the other hand in the laboratory frame

$$s = (\omega_{q_L} + M)^2 - q_L{}^2 = M^2 + m_\pi{}^2 + 2M\omega_{q_L}$$

which yields after a little computation

$$q_L = \left[\frac{(\sqrt{s}+M+m_\pi)(\sqrt{s}+M-m_\pi)(\sqrt{s}-M+m_\pi)(\sqrt{s}-M-m_\pi)}{4M^2}\right]^{1/2} \tag{23.42}$$

Thus the real part of the forward scattering amplitude is easily determined in the laboratory frame, once the phase shifts are given.

In the forward direction $t = 0$. Also, if ω_L is the pion energy in the laboratory system, then $\nu = M\omega_L$ and the relations (23.19) may be written in the form

$$A^{(\pm)}(\omega_L) = \frac{1}{\pi}\int_{m_\pi}^{\infty} d\omega'\,\alpha^{(\pm)}(\omega')\left(\frac{1}{\omega'-\omega_L} \pm \frac{1}{\omega'+\omega_L}\right)$$

$$B^{(\pm)}(\omega_L) = \frac{g^2}{8\pi M}\left(\frac{1}{(m_\pi^2/2M)+\omega_L} \mp \frac{1}{(m_\pi^2/2M)-\omega_L}\right)$$

$$+\frac{1}{\pi}\int_{m_\pi}^{\infty} d\omega'\,\beta^{(\pm)}(\omega')\left(\frac{1}{\omega'-\omega_L} \mp \frac{1}{\omega'+\omega_L}\right) \tag{23.43}$$

[1] Here q_L denotes the pion momentum in the laboratory frame.

Since $q = q'$, it follows that

$$T_{\beta\alpha}(p, q; p, q) = \frac{1}{16\pi^5}(-A_{\beta\alpha}(\omega_L) + B_{\beta\alpha}(\omega_L)\bar{u}(p)\not{q}u(p))$$
$$= \frac{1}{16\pi^5}(-A_{\beta\alpha}(\omega_L) + \omega_L B_{\beta\alpha}(\omega_L)) \tag{23.44}$$

Since the scattering amplitude is given by

$$F_{\text{lab}}(\omega_L) = \frac{q_L}{q} F_{\text{c.m.}} = \frac{W}{M} F_{\text{c.m.}} = \frac{W}{M}\left(-\frac{16\pi^5 M}{W}\right) T$$
$$= -16\pi^5 T(p, q; p, q) \tag{23.45}$$

it follows that

$$F_{\text{lab}}^{(\pm)}(\omega_L) = A^{(\pm)}(\omega_L) - \omega_L B^{(\pm)}(\omega_L) \tag{23.46}$$

In order to derive relations for quantities most directly related to experimentally measured quantities, and to bring out certain symmetries in the dispersion relations to be written down, let us consider the scattering amplitudes for $\pi^\pm$ by protons. It is clear that since $F_{\pi^+} = F(T = \frac{3}{2})$, we have

$$F_{\pi^+} = F^{(+)} - F^{(-)} \tag{23.47}$$

Also

$$F_{\pi^-} = \tfrac{1}{3}F(T = \tfrac{3}{2}) + \tfrac{2}{3}F(T = \tfrac{1}{2})$$
$$= \tfrac{1}{3}(F^{(+)} - F^{(-)}) + \tfrac{2}{3}(F^{(+)} + 2F^{(-)})$$
$$= F^{(+)} + F^{(-)} \tag{23.48}$$

Thus we may write two forms of the dispersion relations. On one hand we have

$$\operatorname{Re} F^{(-)}(\omega_L) = \frac{1}{2}\operatorname{Re}(F_{\pi^-}(\omega_L) - F_{\pi^+}(\omega_L))$$
$$= -\frac{g^2\omega_L}{8\pi M^2}\frac{m_\pi^2}{(m_\pi^2/2M)^2 - \omega_L}$$
$$+ \frac{2\omega_L}{\pi} P\int_{m\pi}^{\infty} d\omega' \frac{\alpha^{(-)}(\omega')}{\omega'^2 - \omega_L^2} + \frac{\omega_L}{\pi} P\int_{m\pi}^{\infty} d\omega' \frac{2\omega'\beta^{(-)}(\omega')}{\omega'^2 - \omega_L^2}$$
$$= -\frac{1}{2}\frac{g^2\omega_L m_\pi^2/4\pi M^2}{\left(\dfrac{m_\pi^2}{2M}\right)^2 - \omega_L^2} + \frac{2\omega_L}{\pi} P\int_{m\pi}^{\infty} d\omega'$$
$$\times \frac{\alpha^{(-)}(\omega') + \omega'\beta^{(-)}(\omega')}{\omega'^2 - \omega_L^2}$$

With

$$\text{Im}\, F^{(-)}(\omega_L) = \alpha^{(-)}(\omega_L) + \omega_L \beta^{(-)}(\omega_L) = \frac{q_L}{4\pi} \frac{\sigma_{\pi^-}^{\text{tot}}(\omega_L) - \sigma_{\pi^+}^{\text{tot}}(\omega_L)}{2}$$

we thus get

$$\text{Re}\,(F_{\pi^-}(\omega_L) - F_{\pi^+}(\omega_L)) = \frac{g^2 \omega_L m_\pi^2/4\pi M}{\omega_L^2 - (m_\pi^2/2M)^2} + \frac{\omega_L}{2\pi^2} P \int_{m_\pi}^{\infty} \frac{q'\, d\omega'}{\omega'^2 - \omega_L^2} (\sigma_{\pi^-}^{\text{tot}}(\omega') - \sigma_{\pi^+}^{\text{tot}}(\omega')) \quad (23.49)$$

On the other hand, a little computation leads to the form

$$\begin{aligned} \text{Re}\, F^{(+)}(\omega_L) &= \frac{1}{2} \text{Re}\,(F_{\pi^-}(\omega_L) + F_{\pi^+}(\omega_L)) \\ &= -\frac{g^2}{4\pi M} - \frac{2}{\pi} \int_{m_\pi}^{\infty} d\omega' \beta^+(\omega') \\ &\quad + \frac{2g^2}{8\pi M} \frac{(m_\pi^2/2M)^2}{(m_\pi^2/2M)^2 - \omega_L^2} \\ &\quad + \frac{1}{\pi} P \int_{m_\pi}^{\infty} \frac{2\omega'\, d\omega'}{\omega'^2 - \omega_L^2} \frac{q'}{8\pi} (\sigma_{\pi^-}^{\text{tot}}(\omega') + \sigma_{\pi^+}^{\text{tot}}(\omega')) \end{aligned} \quad (23.50)$$

The total $\pi^{\pm}p$ cross section at high energies can be fitted[1] to

$$\begin{aligned} \sigma^{(+)} &= \frac{1}{2}(\sigma_{\pi^-} + \sigma_{\pi^+}) = \sigma_\infty + \frac{b^{(+)}}{m_\pi^2}\left(\frac{q_L}{m_\pi}\right)^{-\alpha} \\ \sigma^{(-)} &= \frac{1}{2}(\sigma_{\pi^-} - \sigma_{\pi^+}) = \frac{b^{(-)}}{m_\pi^2}\left(\frac{q_L}{m_\pi}\right)^{-\alpha} \end{aligned} \quad (23.51)$$

above 5 BeV, with the parameters listed in the table below

α	σ_∞	$b^{(+)}$	$b^{(-)}$
0.5	$1.050 m_\pi^{-2}$	2.21	0.34
0.7	$1.125 m_\pi^{-2}$	3.54	0.77

Thus, roughly speaking, the $\sigma_{\pi^\pm}$ approach a common value of about 22 mb. These results imply that the relation (23.50) needs a subtraction. When

[1] These parameters are taken from G. Hohler, G. Ebel, and J. Giesecke, *Z. Physik* **180,** 430 (1964). See however the remarks on page 385.

this is done, we get the following expression

$$\begin{aligned}\operatorname{Re}(F_{\pi^-}(\omega_L) + F_{\pi^+}(\omega_L)) &= F_{\pi^-}(m_\pi) + F_{\pi^+}(m_\pi) \\ &+ \frac{g^2 q_L^2}{8\pi M}\left(\frac{m_\pi}{M}\right)^2 \frac{1}{(m_\pi^2/2M)^2 - \omega_L^2} \\ &+ \frac{q_L^2}{4\pi^2}\int_{m_\pi}^{\infty} \frac{\omega'\, d\omega'}{q'(\omega'^2 - \omega_L^2)} (\sigma_{\pi^-}^{\text{tot}}(\omega') + \sigma_{\pi^+}^{\text{tot}}(\omega'))\end{aligned} \tag{23.52}$$

where a m_π^2/M^2 correction to the one-particle term has been dropped. If we make the (unnecessary) corresponding subtraction in (23.49) and then combine that with (23.52), we can write the more symmetric relations

$$\begin{aligned}\operatorname{Re} F_{\pi^\pm}(\omega_L) &= \frac{1}{2}\left(1 + \frac{\omega_L}{m_\pi}\right) F_{\pi^\pm}(m_\pi) + \frac{1}{2}\left(1 - \frac{\omega_L}{m_\pi}\right) F_{\pi^\pm}(m_\pi) \\ &\pm \frac{2f^2\omega_L^2}{m_\pi^2} \frac{1}{\omega_L \pm \dfrac{m_\pi^2}{2M}} \frac{1}{1 - \dfrac{m_\pi^2}{4M^2}} \\ &+ \frac{q_L^2}{4\pi^2} P\int_{m_\pi}^{\infty} \frac{d\omega'}{q'}\left(\frac{\sigma_{\pi\pm}^{\text{tot}}(\omega')}{\omega' - \omega_L} + \frac{\sigma_{\pi\mp}^{\text{tot}}(\omega')}{\omega' + \omega_L}\right)\end{aligned} \tag{23.53}$$

where

$$f^2 = \frac{g^2}{4\pi}\left(\frac{m_\pi}{2M}\right)^2 \tag{23.54}$$

The real part of the scattering amplitude is given by[1]

$$\begin{aligned}\operatorname{Re} F_{\pi^+}(\theta = 0) &= \left(1 + \frac{2\omega_L}{M} + \frac{m_\pi^2}{M^2}\right)\frac{1}{2q_L} \\ &\times \sum_l (l \sin 2\delta_{l-}^{(T=3/2)} + (l+1)\sin 2\delta_{l+}^{(T=3/2)}) \\ \operatorname{Re} F_{\pi^-}(\theta = 0) &= \frac{1}{2}\left(1 + \frac{2\omega_L}{M} + \frac{m_\pi^2}{M^2}\right)\frac{1}{3q_L}\sum_l (2l \sin 2\delta_{l-}^{(T=1/2)} + l \sin 2\delta_{l-}^{(T=3/2)} \\ &+ (2l+2)\sin 2\delta_{l+}^{(T=1/2)} + (l+1)\sin 2\delta_{l+}^{(T=3/2)})\end{aligned} \tag{23.55}$$

At low energies, where only S and P phase shifts are important, these quantities are known from phase-shift analyses. Thus there is only one

[1]These relations are derived by noting that

$$\begin{aligned}\operatorname{Re} F_L^T = \frac{q_L}{q}\operatorname{Re} F_{\text{c.m.}}^T &= \left(\frac{q_L}{q}\right)^2 \frac{q}{q_L}\operatorname{Re} F_{\text{c.m.}}^T \\ &= (W/M)^2 \frac{1}{2q_L}\sum_l (l \sin 2\delta_{l-}^T + (l+1)\sin 2\delta_{l+}^T)\end{aligned}$$

unknown parameter, $g^2/4\pi$ in the dispersion relations. This parameter may be evaluated, and it is found that[1]

$$f^2 = 0.081 \pm 0.002$$

i.e.

$$\frac{g^2}{4\pi} = 14.4 \pm 0.4 \tag{23.56}$$

The significance of this determination depends, of course, on the validity of the dispersion relations in general. There are a number of papers in the literature in which the comparison between theory and experiment is discussed. Figure 23.2 gives an illustration of the kind of agreement

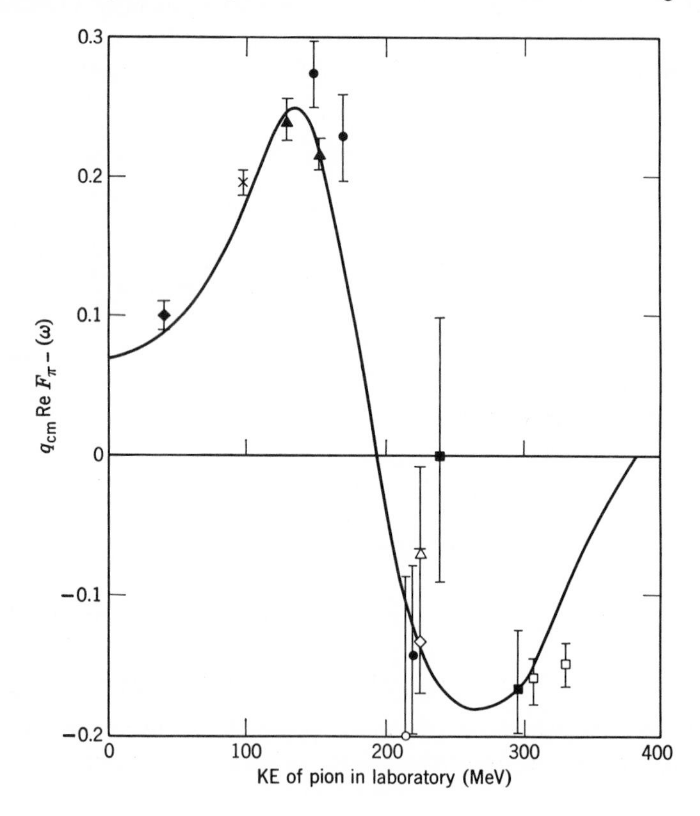

Fig. 23.2. Comparison of the prediction of the forward scattering dispersion relations for $\pi^- + p \to \pi^- + p$ with experiment. [From J. Deahl et al. *Phys. Rev.* **124,** 1987 (1961).]

[1] See J. Hamilton and W. S. Woolcock, *Rev. Mod. Phys.* **35,** 737 (1963) for references and critical discussion of various evaluations of $g^2/4\pi$.

found between the predicted real part of the scattering amplitude and the same quantity determined experimentally.

The dispersion relation (23.49) does not contain any subtraction terms. When ω is set equal to m_π, we obtain a sum rule. Noting that

$$F_{\pi^-}(m_\pi) - F_{\pi^+}(m_\pi) = \frac{2}{3}(F^{1/2}(m_\pi) - F^{3/2}(m_\pi))$$
$$= \frac{2}{3}\left(1 + \frac{m_\pi}{M}\right) \lim_{q \to 0} \frac{\sin 2\alpha_1 - \sin 2\alpha_3}{q}$$
$$= \frac{2}{3m_\pi}\left(1 + \frac{m_\pi}{M}\right)(a_1 - a_3)$$

we obtain

$$\frac{2}{3}\left(1 + \frac{m_\pi}{M}\right)(a_1 - a_3) \simeq 4f^2 + \frac{m_\pi^{\ 2}}{2\pi^2}\int_{m_\pi}^{\infty} \frac{d\omega'}{q'}\,(\sigma_{\pi^-}^{\text{tot}}(\omega') - \sigma_{\pi^+}^{\text{tot}}(\omega')) \tag{23.57}$$

When the experimental values of the scattering lengths (23.35) are inserted, we find that

$$f^2 = 0.082 \pm 0.008$$

in excellent agreement with the value quoted in (23.56). This may be taken as experimental evidence that

$$\sigma_{\pi^-}^{\text{tot}} - \sigma_{\pi^+}^{\text{tot}} \to 0$$

fast enough to make (23.49) valid.

If we are willing to abandon complete model-independence, we can, of course, obtain independent information about the subtraction constant $(F_{\pi^-}(m_\pi) + F_{\pi^+}(m_\pi))$. Although there are no general threshold theorems of the type that determine the subtraction constant for photon scattering (22.13), there is a class of models for which a type of threshold theorem exists. This is the class of models in which the pion field $\phi_\alpha(x)$ can be written as the divergence of an axial vector current[1]

$$m_\pi^2\phi_\alpha(x) = \lambda\partial_\mu J_\alpha^{A\mu}(x) \tag{23.58}$$

If that is the case, then

$$\begin{aligned} j_\alpha(x) &= (\Box + m_\pi^2)\phi_\alpha(x) \\ &= \partial_\mu(\partial^\mu\phi_\alpha(x) + \lambda J_\alpha^{A\mu}(x)) \\ &\equiv \partial_\mu K_\alpha^\mu(x) \end{aligned} \tag{23.59}$$

[1] Thus theory will be discussed in detail in Chapter 34 in connection with the weak interactions.

and it follows that the transition amplitude is of the form

$$T_{\beta\alpha}(p', q'; p, q) = -\frac{1}{(2\pi)^{3/2}}(\Psi^{\text{out}}_{p'q'}, j_\alpha(0)\Psi_p)$$
$$= -\frac{iq_\mu}{(2\pi)^{3/2}}(\Psi^{\text{out}}_{p'q'}, K^\mu_\alpha(0)\Psi_p) \tag{23.60}$$

Thus the scattering amplitude vanishes in the limit in which the pion *four-momentum* vanishes. The only difficulty which could arise is that the matrix element of $K^\mu_\alpha(0)$ has a singularity as $q \to 0$. If we consider $\pi^\pm p$ scattering, such a singularity could only occur because of the pole due to the intermediate neutron state. If we keep the neutron mass M_n different from the proton mass M_p then a denominator of the type

$$\frac{1}{(p+q)^2 - M_n^2} = \frac{1}{M_p^2 - M_n^2 + 2p \cdot q + q^2}$$

gives no trouble at $q = 0$. This threshold theorem applies to the *scattering amplitude with the mass of the external pion extrapolated to zero*, and its utility depends on the assumption that this extrapolation does not change the cross section integrals in any significant way. The scattering amplitudes do not actually vanish at threshold, and it is off-hand difficult to say whether the amplitudes $F_{\pi^\pm}(m_\pi)$ are "small". It is nevertheless possible to test the validity of the threshold theorem by a consistency condition which follows from it.[1] The condition is obtained by the following method:

When $M_n \neq M_p$ and q^2, q'^2 are not equal to $m_\pi{}^2$, then the dispersion relations for

$$A^{(+)}(\nu, t, q^2, q'^2) = \tfrac{1}{2}(A_{\pi^+p}(\nu, t, q^2, q'^2) + A_{\pi^-p}(\nu, t, q^2, q'^2))$$
$$B^{(+)}(\nu, t, q^2, q'^2) = \tfrac{1}{2}(B_{\pi^+p}(\nu, t, q^2, q'^2) + B_{\pi^-p}(\nu, t, q^2, q'^2))$$

take the form

$$A^{(+)}(\nu, t, q^2, q'^2) = -\frac{g^2}{4\pi}\frac{\nu_B{}^2\Delta}{\nu_B{}^2 - \nu^2} + \frac{2}{\pi}\int_{\nu_0}^{\infty}\frac{\nu'\,d\nu'}{\nu'^2 - \nu^2}\,\text{Im}\,A^{(+)}(\nu', t, q^2, q'^2)$$

$$B^{(+)}(\nu, t, q^2, q'^2) = -\frac{g^2}{4\pi}\frac{\nu}{\nu_B{}^2 - \nu^2} + \frac{2\nu}{\pi}\int_{\nu_0}^{\infty}\frac{d\nu'}{\nu'^2 - \nu^2}\,\text{Im}\,B^{(+)}(\nu', t, q^2, q'^2)$$

where

$$\Delta = M_p - M_n$$
$$\nu_B = \tfrac{1}{2}(M_p^2 - M_n^2) + \tfrac{1}{4}(q^2 + q'^2 - t)$$
$$\nu_0 = Mm_\pi + \tfrac{1}{2}m_\pi^2 - \tfrac{1}{4}(q^2 + q'^2 - t)$$

[1] S. L. Adler, *Phys. Rev.* **137,** B1022 (1965); **139,** B1638 (1965). We follow a method due to N. Fuchs.

Hence

$$A^{(+)}(\nu, t, q^2, q'^2) - \frac{\nu}{M_p} B^{(+)}(\nu, t, q^2, q'^2) = -\frac{g^2}{4\pi M_p}\frac{\Delta\nu_B M_p - \nu^2}{\nu_B^2 - \nu^2} + \frac{2}{\pi}\int_{\nu_0}^{\infty}\frac{d\nu'}{\nu'}\operatorname{Im} A^{(+)}(\nu', t, q^2, q'^2)$$

$$+ \frac{2\nu^2}{\pi}\int_{\nu_0}^{\infty}\frac{d\nu'}{\nu'}\frac{1}{\nu'^2 - \nu^2}\left(\operatorname{Im} A^{(+)}(\nu', t, q^2, q'^2) - \frac{\nu'}{M_p}\operatorname{Im} B^{(+)}(\nu', t, q^2, q'^2)\right) \tag{23.61}$$

When $t = 0$ and $q^2 = q'^2$, the above quantity is the forward scattering amplitude in the laboratory frame (equation (23.46)), and at $\nu = M_p\omega_L = 0$ we get the requirement that

$$\frac{g^2}{4\pi M_p} = \frac{2}{\pi}\int_0^{\infty}\frac{d\nu'}{\nu'}\operatorname{Im} A^{(+)}(\nu', 0, 0, 0) \tag{23.62}$$

This requirement has been found to be satisfied to an accuracy of 10% lending support to the class of theories characterized by (23.58).

It is worth pointing out that the forward dispersion relations could have been used to determine the parity of the pion. We noted in (21.45) that the pole term, coming from the one-particle contribution to the unitarity sum, would have quite a different form for scalar mesons; in particular, the sign of the pole term would be different, making agreement with experiment impossible. This distinction in the sign of the residue of the pole comes about because the one-nucleon intermediate state may be viewed as a "bound state" of the initial pion and nucleon in the *P*-state when the pion is pseudoscalar, and in the *S*-state when the pion is scalar.

In view of this result, it has been suggested that the *K*-meson parity could perhaps be established by comparing *K*-nucleon forward scattering dispersion relations with experiment. Unfortunately, the *K*-meson dispersion relations do not share the simplicity of the pion-nucleon relations. In addition to pole terms, arising from nonvanishing matrix elements for

$$K^- + p \to \Lambda^0$$

and

$$K^- + p \to \Sigma^0$$

there is an as yet uncharted "unphysical" region in which the processes

$$K^- + p \to \begin{cases} \Lambda^0 + \pi^0 \\ \Lambda^0 + 2\pi \\ \Sigma + \pi \end{cases}$$

for laboratory energies $M_{\Lambda^0} + m_\pi - M_N \leq \omega_L \leq m_K$ (i.e., *unphysical K* momenta) contribute to the imaginary part of the forward scattering amplitude. When $\omega_L \geq m_K$ the optical theorem may be used, but without a specific model we cannot evaluate the contribution to the unphysical continuum to the forward scattering *K-N* amplitude.

The forward scattering dispersion relations have become a subject of new interest because of recent measurements of high energy $\pi^\pm p$ elastic scattering for very small momentum transfers, where it is possible to detect the Coulomb interference contribution.[1] As (23.28) shows, the scattering amplitude, when Coulomb effects are ignored, consists of two terms, one of which is called the nonspin-flip amplitude, and the other, proportional to

$$\boldsymbol{\sigma} \cdot \hat{\mathbf{q}} \times \hat{\mathbf{q}}' = \boldsymbol{\sigma} \cdot \hat{\mathbf{n}} \sin \theta$$

called the spin-flip amplitude. At very small angles, the contribution of the second term becomes much smaller because of the sin θ factor. (This statement does not take into account the extremely improbable case that the coefficient is anomalously large.) Thus the scattering is described by only one amplitude. If we define

$$A_{\pi^\pm} = \frac{\sqrt{\pi}}{q} F_{\pi^\pm} \tag{23.63}$$

then the elastic differential cross section $\dfrac{d\sigma}{dt}$ is given by

$$\frac{d\sigma}{dt} = |A_{\pi^\pm}|^2$$

When Coulomb effects are included, this is changed to

$$\frac{d\sigma_{\pi^\pm}}{dt} = |A_c(t) + \text{Re}\, A_{\pi^\pm} + i\, \text{Im}\, A_\pm|^2 \tag{23.64}$$

The Coulomb amplitude[2] behaves as $1/|t|$. It is negative for π^+p scattering, corresponding to the repulsion, and positive for π^-p scattering. If the charge independent part of the amplitude is written in the form

$$A_{\pi^\pm} = (\alpha + i)e^{a+bt} \tag{23.65}$$

then the total cross section determines a by the optical theorem

$$\text{Im}\, A_{\pi^\pm}(t = 0) = e^a = \frac{\sqrt{\pi}}{q}\frac{q}{4\pi}\sigma^{\text{tot}}_{\pi^\pm} \tag{23.66}$$

and α, b may be varied to give a fit to the data. Figures (23.3) and (23.4)

[1] K. J. Foley, R. S. Gilmore, R. S. Jones, and S. J. Lindenbaum, W. A. Love, S. Ozaki, E. H. Willen, R. Yamada, and L. C. L. Yuan, *Phys. Rev. Letters* **14**, 862 (1965).
[2] The form usually used is that of H. Bethe, *Ann. Phys. (N.Y.)* **3**, 190 (1958).

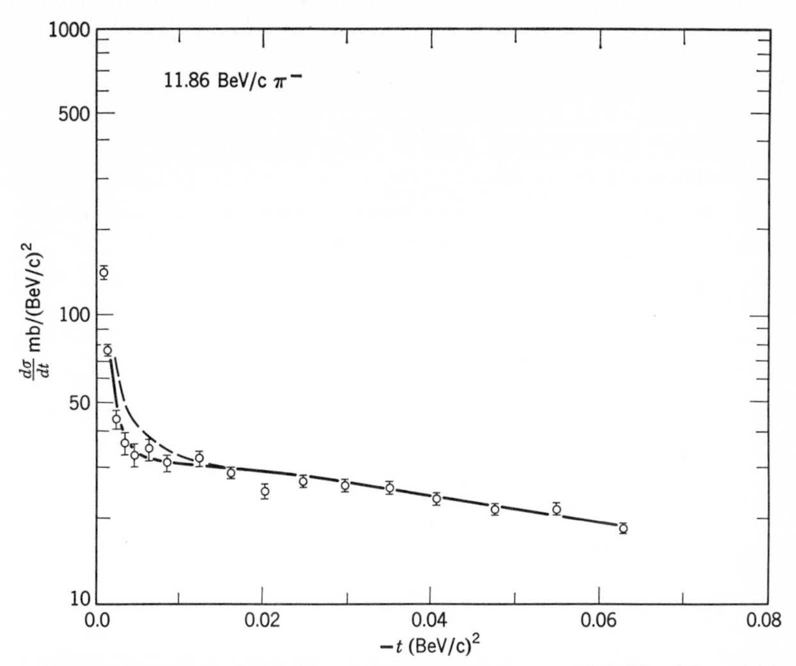

Fig. 23.3. Small angle π^-p elastic scattering cross section, at 12 BeV/c. The solid line represents the best fit, varying α and b. The dotted line is the best fit for $\alpha = 0$. [From K. J. Foley et al., *Phys. Rev. Letters* **14,** 862 (1965).]

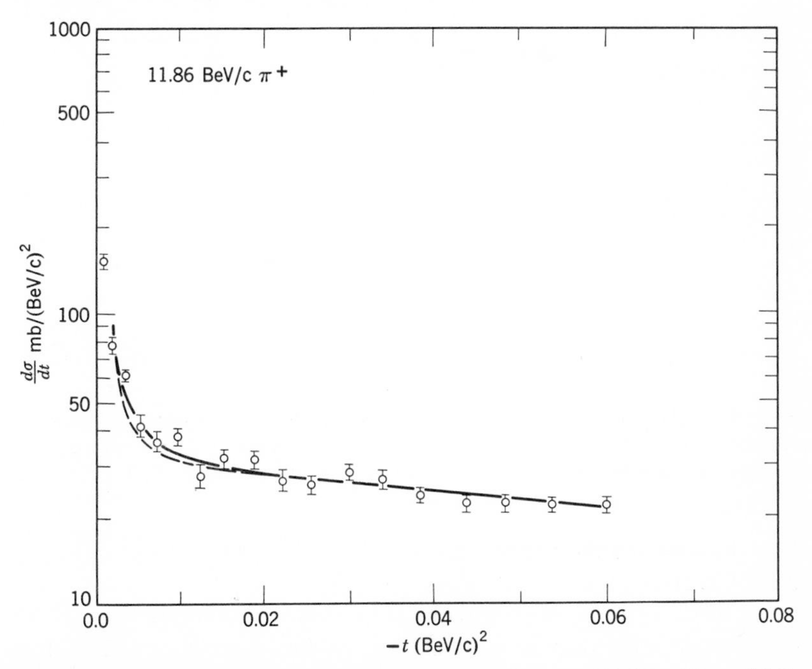

Fig. 23.4. Small angle π^+p elastic scattering cross section at 12 BeV/c. The solid line represents the best fit, varying α and b. The dotted line is the best fit with $\alpha = 0$. (From K. J. Foley et al., *loc. cit.*)

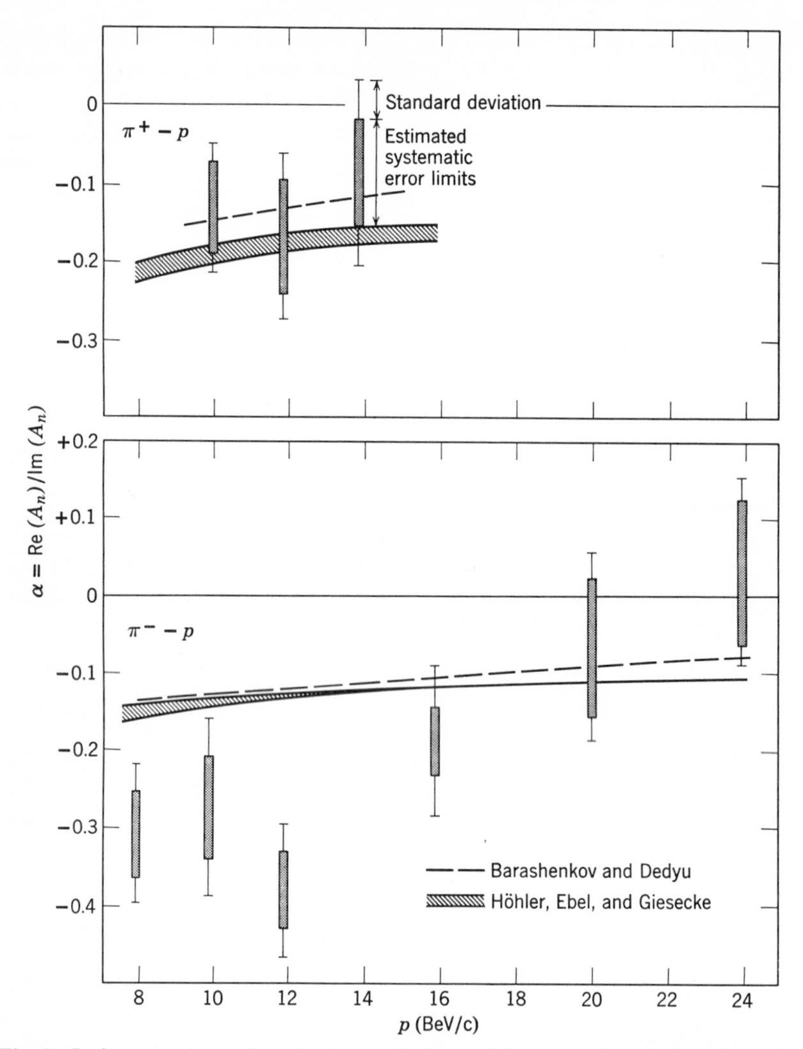

Fig. 23.5. A comparison of α with the predictions of the forward scattering dispersion relations. The solid rectangles represent the estimated limits of the systematic errors; the flags add one standard deviation of the statistical errors. (From K. J. Foley et al., *loc. cit.*)

show the data and the need for a nonvanishing α. The value of b is found to be $\sim 5\ (\text{BeV}/c)^{-2}$.

The value of Re $F_{\pi^\pm}$ may now be compared with that predicted by the forward scattering dispersion relations.[1] If the assumptions (23.51) are

[1] V. S. Barashenkov and V. I. Dedyu, *Nuclear Physics* **64,** 636 (1965); G. Höhler, G. Ebel, and J. Giesecke, *Z. Physik* **180,** 430 (1964).

made about the total cross section at high energies, there is no agreement, as Fig. 23.5 shows. Agreement with the dispersion relations is possible *provided* $(\sigma_{\pi^-}^{\text{tot}} - \sigma_{\pi^+}^{\text{tot}})$ *changes sign above 20 BeV.*[1]

It is clear that since the dispersion relations can be proved on the basis of some very general assumptions which are model independent and fundamental to relativistic field theory, clear evidence of their failure would be of the greatest importance. Whether they are indeed saved by an oscillation in the difference $(\sigma_{\pi^-} - \sigma_{\pi^+})$, and, if so, the reasons for such an oscillation at very high energies will undoubtedly be topics of intensive study, both experimental and theoretical, in coming years.

PROBLEMS

1. Use the fact that

$$\mathbf{t}^{(1)} \cdot \mathbf{t}^{(2)} = \tfrac{1}{2}T(T+1) - 2$$

and the representation

$$(t_a^{(k)})_{bc} = -i\epsilon_{abc}$$

to construct projection operators for two-pion states of $T = 2$, 1, and 0.

2. The most general form of the pion-pion scattering amplitude is of the form

$$T_{\gamma\delta,\alpha\beta}(s, t, u) = \delta_{\alpha\gamma}\delta_{\beta\delta}A + \delta_{\alpha\delta}\delta_{\beta\gamma}B + \delta_{\alpha\beta}\delta_{\gamma\delta}C$$

Use the projection operators constructed above to relate A, B, and C to scattering amplitudes for definite *i*-spin states.

3. Work out the forward scattering dispersion relations for pion-pion scattering in a form that incorporates the optical theorem and crossing symmetry so that integrals over positive frequencies only appear. If the total cross sections become constant at high energies, how many arbitrary constants have to be introduced?

4. What is the generalization of (23.4) for pion-nucleon scattering when parity conservation is not assumed? Discuss the process as far as possible.

[1] B. Lautrup, P. Møller Nielsen and P. Olesen, *Phys. Rev.* **140,** B984 (1965).

24

Properties of Partial Wave Amplitudes

The derivation of the forward scattering dispersion relations and their successful application led to an immediate interest in the analyticity properties of the scattering amplitudes in nonforward directions. It was soon discovered that one could derive dispersion relations for the scattering amplitude at a fixed nonvanishing momentum transfer.[1] Such relations, of the general form

$$\operatorname{Re} F(s, t) = \frac{1}{\pi} P \int_{4m_\pi^2}^{\infty} dx \operatorname{Im} F(x, t)\left(\frac{1}{x - s} \pm \frac{1}{x - u}\right) \tag{24.1}$$

[with $s = 4(q^2 + m_\pi^2)$, $u = -2q^2(1 + \cos\theta)$, and $t = -2q^2(1 - \cos\theta)$ for pion-pion scattering in the center of mass system] could potentially lead to a set of equations connecting different partial wave amplitudes, obtained by the insertion of the partial wave expansion in terms of $\cos\theta = 1 + 2t/(s - 4m_\pi^2)$

$$F(s, t) = \sum_l (2l + 1)a_l(s)P_l\left(1 + \frac{2t}{s - 4m_\pi^2}\right) \tag{24.2}$$

into the dispersion relation. This set of equations could be cut off at some value of l and provide an approximation scheme different from perturbation theory. There were two difficulties with the program: (a) the series (24.2) does not converge when the argument of the Legendre polynomial exceeds a certain value; there is a certain range of values of x where such an expansion could not be used on the right-hand side of (24.1); (b) if (24.1) had to be written in subtracted form, the subtraction "constant" would now be a completely unknown function of t. These difficulties were in part removed by the conjecture of Mandelstam regarding the explicit

[1] A. Salam, *Nuovo Cimento* **3,** 424 (1956); R. Capps and G. Takeda, *Phys. Rev.* **103,** 1877 (1956); G. F. Chew, M. L. Goldberger, F. Low, and Y. Nambu, *Phys. Rev.* **106,** 1337 (1957). This last paper known in the trade as CGLN has been enormously useful for the detailed understanding of low energy pion-nucleon scattering.

t-dependence of the scattering amplitude. The Mandelstam representation made possible a direct study of partial wave amplitudes, a matter of great interest, since these amplitudes are severely restricted by the unitarity condition. In the "elastic" domain they must be of the form

$$f_l(q) = e^{i\delta_l}\frac{\sin\delta_l}{q} \qquad \delta_l \text{ real} \tag{24.3}$$

and above those energies,

$$f_l(q) = \frac{1}{2iq}(\eta_l e^{2i\delta_l} - 1) \qquad 0 \leq \eta_l \leq 1$$

still satisfies

$$|qf_l(q)| \leq 1 \tag{24.4}$$

Since resonances are characterized by definite angular momentum and parity quantum numbers, the possibility exists that their location and nature could most easily emerge from a study of partial wave amplitudes. In this chapter we shall discuss some properties of partial wave amplitudes without actually discussing the real resonances. For pedagogic purposes we will make our main remarks with reference to a theory involving the scattering of two neutral pseudoscalar mesons, of the kind which could be described by an interaction Lagrangian like

$$L_1 = \lambda \int d^3x :\phi^4(x): \tag{24.5}$$

Our interest is in the elastic scattering amplitude, so that we shall study the T matrix, $T(q_3, q_4; q_1, q_2)$, in terms of which the scattering amplitude is given by

$$F(W, \cos\theta) = -\frac{8\pi^5}{W}T(q_3, q_4; q_1, q_2) \tag{24.6}$$

with W the center of mass energy. We assume that T has analyticity properties represented by the Mandelstam form

$$T(s, t) = \frac{1}{\pi^2}\iint_{4m_\pi^2}^{\infty} dx\, dy \rho(x, y) \times \left[\frac{1}{(x-s)(y-t)} + \frac{1}{(x-t)(y-u)} + \frac{1}{(x-u)(y-s)}\right] \tag{24.7}$$

with

$$s = (q_1 + q_2)^2 = (q_3 + q_4)^2$$
$$t = (q_3 - q_1)^2 = (q_2 - q_4)^2$$
$$u = (q_4 - q_1)^2 = (q_2 - q_3)^2$$

In the center of mass system we have $q_1 = (\omega, \mathbf{q})$, $q_2 = (\omega, -\mathbf{q})$, $q_3 = (\omega, \mathbf{q}')$, and $q_4 = (\omega, -\mathbf{q}')$, so that

$$\begin{aligned} s &= W^2 = (2\omega)^2 = 4(q^2 + m_\pi^{\ 2}) \\ t &= -(\mathbf{q}' - \mathbf{q})^2 = -2q^2(1 - \cos\theta) \\ u &= -(\mathbf{q}' + \mathbf{q})^2 = -2q^2(1 + \cos\theta) \end{aligned} \tag{24.8}$$

The partial wave amplitude $f_l(W)$, defined by

$$f_l(W) = \frac{1}{2}\int_{-1}^{1} d(\cos\theta) P_l(\cos\theta)\left(-\frac{8\pi^5}{W}\,T(s, t)\right) \tag{24.9}$$

may be obtained from (24.7). We write (24.7) in the form

$$\begin{aligned} T(s, t) = \frac{1}{\pi^2}\iint\limits_{4m_\pi^2}^{\infty} dx\,dy\,\rho(x, y)\Bigg[&\frac{1}{(x - s)}\,\frac{1}{(y + 2q^2 - 2q^2\cos\theta)} \\ &+ \frac{1}{(y - s)}\,\frac{1}{(x + 2q^2 + 2q^2\cos\theta)} + \frac{1}{(x + y - u - t)} \\ &\times\left(\frac{1}{x + 2q^2 - 2q^2\cos\theta} + \frac{1}{y + 2q^2 + 2q^2\cos\theta}\right)\Bigg] \end{aligned}$$

and then utilize

$$\frac{1}{2}\int_{-1}^{1} d(\cos\theta)\,\frac{P_l(\cos\theta)}{\zeta - \cos\theta} = Q_l(\zeta) \tag{24.10}$$

Since $u + t = -4q^2$, we obtain, after a little manipulation

$$\begin{aligned} f_l(W) = -\frac{8\pi^3}{2q^2\sqrt{s}}\iint\limits_{4m_\pi^2}^{\infty} dx\,dy\,\rho(x, y)(1 + (-1)^l) \\ \times\left(\frac{1}{x - s} + \frac{1}{x + y + 4q^2}\right) Q_l\left(1 + \frac{y}{2q^2}\right) \end{aligned} \tag{24.11}$$

Incidentally, there is only one spectral function $\rho(x, y)$ in the problem, because the physical process is always

$$\pi^0 + \pi^0 \rightarrow \pi^0 + \pi^0$$

in all three channels, s, t, and u, so that $T(s, t, u)$ is unchanged by the permutation of all the variables. Furthermore, since the mesons are bosons, the amplitude must be symmetric under the interchange of q_3 and

q_4 say, so that we must have

$$\rho(x, y) = \rho(y, x) \tag{24.12}$$

We might try to insert (24.11) into the relation

$$\text{Im}\, f_l(W) = q\,|f_l(W)|^2 \tag{24.13}$$

which follows from

$$f_l(W) = \frac{e^{i\delta_l(W)} \sin \delta_l(W)}{q} \tag{24.14}$$

and thus get a complicated, but nevertheless explicit, equation for $\rho(x, y)$. This program, formulated much more elegantly,[1] has been much discussed in recent years.[2] In addition to the sheer numerical complication of an equation of this type, there are special problems connected with the behavior of the spectral functions $\rho(x, y)$ as the arguments become large. Since nothing of practical interest has emerged from this approach so far,[3] we shall not embark on a discussion of this ambitious program, but rather see what can be said about the partial wave amplitudes without taking into account their coupling through $\rho(x, y)$.

The function

$$t_l(\nu) \equiv \sqrt{s}\, f_l(\nu) \tag{24.15}$$

where $\nu = q^2$, vanishes for odd l; this follows from the factor $(1 + (-1)^l)$ in (24.11) and is a direct consequence of the boson nature of the particle being discussed. In the similar problem of pion-pion scattering, which differs from the one under consideration only in that three i-spin states $T = 0, 1, 2$ are involved, we similarly find that for $T = 0, 2$ only even angular momentum states scatter, whereas for $T = 1$ only odd angular momentum states scatter. The analyticity properties of $t_l(\nu)$ may be read off from the representation (24.11). The singularities can only occur for real ν; they occur when $4m_\pi^2 \leq s < \infty$, i.e., for $q^2 = \nu \geq 0$ (from the

[1] The unitarity condition is written in the form

$$\text{Im}\, T(s, t) = -2\pi^4 q/W \int d\Omega_{\hat{q}''} T^*(s, -(\mathbf{q}' - \mathbf{q}'')^2) T(s, -(\mathbf{q}'' - \mathbf{q})^2)$$

into which (24.7) is then substituted.

[2] A very clear discussion of this program (known as "strip approximation") is found in S. C. Frautschi, "Regge Poles and S-Matrix Theory," W. A. Benjamin Inc., New York (1963). The use of the Mandelstam representation in potential scattering is discussed in M. L. Goldberger and K. M. Watson, *Collision Theory*, John Wiley and Sons, Inc., New York (1964) Chapter 10.

[3] The impact of the program has nevertheless been large; it has led to the notions of "the bootstrap" (Chapter 25) and "Regge Poles" (Chapter 28).

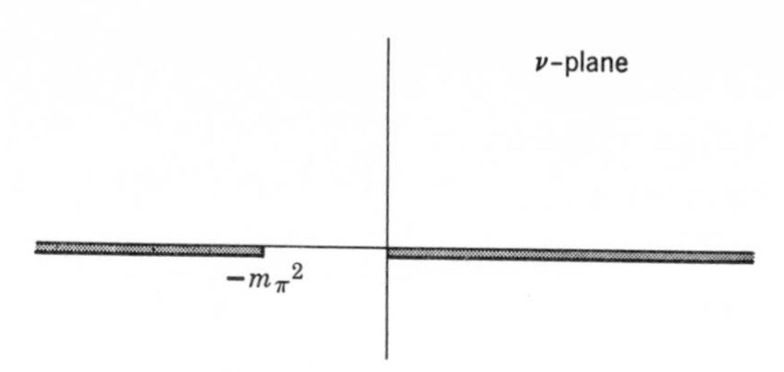

Fig. 24.1. Analyticity domain of $t_l(\nu)$.

first denominator) and when $-2m_\pi^2 \geq \nu > -\infty$ from the second denominator. The Legendre function of the second kind,[1] $Q_l(x)$ is defined outside a cut from $x = 1$ to $x = -1$, so that it gives rise to a cut in the ν plane from $-m_\pi^2$ to $-\infty$. A first glance at (24.11) might suggest that there is a pole at $q^2 = 0$. However, as $q^2 \to 0$

$$Q_l\left(1 + \frac{y}{2q^2}\right) \sim \left(\frac{q^2}{y}\right)^{l+1}$$

so that there is no singularity. In fact, the threshold behavior

$$t_l(\nu) \sim \nu^l \tag{24.16}$$

is just the familiar $\delta_l(q) \sim q^{2l+1}$ behavior of low energy potential scattering theory.[2]

The result of these considerations is that the function $t_l(\nu)$ is analytic in the cut ν-plane, with cuts from $-\infty$ to $-m_\pi^2$ and from 0 to $+\infty$ (Fig. 24.1). We may thus write a representation for this function, which incorporates the unitarity condition that[3]

$$\text{Im } t_l(\nu) = \frac{q}{\sqrt{s}} |t_l(\nu)|^2 \tag{24.17}$$

with

$$0 \leq \nu \leq 3m_\pi^2 \tag{24.18}$$

For $\nu \geq 3m_\pi^2$, we have

$$\text{Im } t_l(\nu) = \frac{q}{\sqrt{s}} |t_l(\nu)|^2 + \frac{\sqrt{s}}{q} \frac{1 - \eta_l^2}{4}$$

because above the threshold for inelastic processes (namely the process

[1] See, for example, W. Magnus and F. Oberhettinger, *Special Functions of Mathematical Physics*, Chelsa Publishing Co., New York (1949).

[2] The derivation of the threshold behavior given here implies that $\int dy\, \rho(x, y) y^{-l-1}$ converges (cf. (24.11)). The result is more general.

[3] Note that $t_l(\nu) \equiv (-8\pi^5)^{\frac{1}{2}} \int_{-1}^{1} d(\cos\theta) P_l(\cos\theta) T(\nu, \cos\theta)$; hence the absence of numerical factors in (24.17).

$(2\pi^0 \to 4\pi^0)$ the scattering amplitude is given by

$$f_l(\nu) = \frac{\eta_l e^{2i\delta_l(\nu)} - 1}{2iq}$$

Writing, in general

$$\text{Im}\, t_l(\nu) = \rho_l(\nu)\, |t_l(\nu)|^2 \tag{24.19}$$

we get

$$t_l(\nu) = \frac{1}{\pi}\int_0^{\infty} d\nu' \frac{\rho_l(\nu')\,|t_l(\nu')|^2}{\nu' - \nu - i\epsilon} + \frac{1}{\pi}\int_{-\infty}^{-m_\pi^2} d\nu' \frac{\phi_l(\nu')}{\nu' - \nu - i\epsilon} \tag{24.20}$$

where $\phi_l(\nu)$ is the unknown discontinuity in $t_l(\nu)$ across the left-hand cut. Note that had we been discussing the scattering of scalar mesons, for which the process

$$\pi^0 + \pi^0 \to \pi^0$$

is allowed, pole terms corresponding to single particle terms in the unitarity sum would appear. We would thus have additional terms coming from

$$T_1(s, t, u) = \frac{g^2}{(2\pi)^6}\left(\frac{1}{s - m_\pi^2} + \frac{1}{u - m_\pi^2} + \frac{1}{t - m_\pi^2}\right) \tag{24.21}$$

The partial wave projections of the second and third terms can be absorbed in the last term of (24.20), the only difference being that the cut extends to $-\frac{1}{4}m_\pi^2$ from the left-hand side. The first term yields a contribution

$$-\frac{g^2}{8\pi}\frac{\delta_{l0}}{4\nu + 3m_\pi^2}$$

i.e., a pole term which appears in the S-state only, to $t_l(\nu)$.

We found in (24.16) that, contrary to appearances, the function

$$h_l(\nu) \equiv [t_l(\nu)]/\nu^l \tag{24.22}$$

does not have an lth order pole at $\nu = 0$. We may thus write a representation analogous to (24.20) for the function $h_l(\nu)$. This function vanishes much more strongly than $t_l(\nu)$ as $\nu \to \infty$ when $l \geq 1$. For $l = 0$ there is no change. Since it is unlikely that $(1 - \eta_l(\nu)) \to 0$ when $\nu \to \infty$, it appears that for ν on the real axis Im $t_l(\nu) \to$ const when $\nu \to \infty$, so that (24.20) may need a subtraction (for $l = 0$). Thus we should write

$$t_0(\nu) = a_0 + \frac{\nu}{\pi}\int_0^{\infty} \frac{d\nu'}{\nu'}\frac{\text{Im}\, t_0(\nu')}{\nu' - \nu - i\epsilon} + \frac{\nu}{\pi}\int_{-\infty}^{-m_\pi^2} d\nu' \frac{\phi_0(\nu')}{\nu'(\nu' - \nu)} \tag{24.23}$$

The presence of an arbitrary constant in the S-wave scattering amplitude may be associated with the freedom of choosing the coupling constant λ in (24.5).

An equation of the type (24.23) is a rather complicated integral equation for $t_l(\nu)$, which depends on the unknown function $\phi_l(\nu)$, as well as on inelastic processes. Even if the latter are neglected, i.e., we only take into account "elastic unitarity" as given in (24.17), we cannot do much without knowledge of the function $\phi_l(\nu)$.[1]

The expression in (24.20) viewed as a *representation* of a unitary partial wave amplitude is nevertheless very useful, because in a limited domain the left-hand cut contributions may be represented simply in terms of a few arbitrary parameters that can be "fitted" to experiments and yield an amplitude which approximates the true one over a limited energy range. We shall now illustrate this with a few examples.

The Effective Range Approximation. With the help of the relation

$$\operatorname{Im} \frac{\nu^l}{t_l(\nu)} = -\frac{q\nu^l}{\sqrt{s}} \qquad 0 \leq \nu \leq 3m_\pi^2 \tag{24.24}$$

which follows directly from (24.17), we note that the function

$$g_l(\nu) \equiv \frac{\nu^l}{t_l(\nu)} + \frac{iq\nu^l}{\sqrt{s}} \tag{24.25}$$

has the following analyticity properties.

1. It has poles at the zeros of the partial wave amplitude.

2. It has a cut along the positive real axis *extending from* $3m_\pi^2$ *to* $+\infty$, provided we define $q = +\sqrt{\nu}$ when ν approaches the positive real axis from above, and $q = -\sqrt{\nu}$ when ν approaches the positive real axis from below, and draw the cut from the branch point at $\nu = -4m_\pi^2$, (coming from the square root in $\sqrt{s}$) to $-\infty$.

3. It has a left-hand cut extending from $-m_\pi^2$ (or possibly $-\frac{1}{4}m_\pi^2$) to $-\infty$. Thus, if it is assumed that there are no complex zeros of $t_l(\nu)$ close to the origin, we may expand the regular function $g_l(\nu)$ in a Taylor series about the origin. We thus have

$$g_l(\nu) \cong a_l + b_l\nu \tag{24.26}$$

or, equivalently

$$\nu^l \operatorname{Re} \frac{1}{t_l(\nu)} \cong a_l + b_l\nu \tag{24.27}$$

[1] In principle the function is obtainable, because it involves the crossed channels which in this problem also represent $\pi^0\pi^0$ scattering. In fact, however, this function is unknown. In spite of much work nobody has succeeded in solving (24.20) or equivalently, in obtaining an amplitude that satisfies the required analyticity properties, elastic unitarity, and crossing. In the Mandelstam form it is the unitarity condition that is unmanageable; in the partial-wave form it is crossing that is unmanageable.

We may write this in terms of the phase shifts as[1]

$$\frac{q^{2l+1}}{\sqrt{s}} \cot \delta_l(\nu) \simeq a_l + b_l \nu \tag{24.28}$$

This is the well-known *effective range formula* with relativistic kinematics, which guarantees a two-parameter fit of low energy scattering. The range of validity of the formula depends on the range of validity of the two-term expansion on the right-hand side of (24.26), and this depends on the influence of the left-hand cut. If the dominant part of the left-hand cut lies far away (this will be seen to correspond roughly to a short-range potential), the validity of (24.28) may extend over a considerable range. The corresponding scattering amplitude

$$f_l(\nu) = \frac{1}{q \cot \delta_l(\nu) - iq}$$

has the form

$$f_l(\nu) = \frac{1}{\dfrac{\sqrt{s}}{\nu^l}(a_l + b_l \nu) - iq} = \frac{\nu^l / b_l \sqrt{s}}{\nu + \dfrac{a_l}{b_l} - i \dfrac{\nu^l}{b_l \sqrt{s}} q} \tag{24.29}$$

If $a_l > 0$ and $b_l < 0$, then $g_l(\nu)$ has a zero at $\nu = -a_l/b_l \equiv \nu_R > 0$ and in the vicinity of that value of $\nu, f_l(\nu)$ is approximately given by

$$f_l(\nu) = \frac{1}{q} \frac{-\frac{1}{2}\Gamma}{(\nu - \nu_R + \frac{1}{2} i \, \Gamma)} \tag{24.30}$$

with $\Gamma = \dfrac{2\nu_R{}^l q_R}{|b_l| \sqrt{s_R}}$. This is a resonance form; near $\nu = \nu_R$ the phase shift is increasing and crosses $\pi/2$ (mod. π) at $\nu = \nu_R$. If a_l is decreased, until it becomes negative, while b_l stays negative, the denominator will vanish for some value of $\nu = -\nu_B$, with $q = i\sqrt{\nu_B}$. The root will occur when

$$-\nu_B + \frac{a_l}{b_l} - (-1)^l \frac{\nu_B^{l+\frac{1}{2}}}{|b_l| \sqrt{s_B}} = 0$$

For a a_l/b_l small, the root is at $\nu_B \approx \dfrac{a_l}{b_l}$, and the residue at the pole of $f_l(\nu)$ is

$$\approx [(-1)^{l+1} \nu_B{}^l]/(|b_l| \sqrt{s_B})$$

Such a pole corresponds to a bound state,[2] and as mentioned in our discussion of pion-nucleon scattering (in connection with the pole terms for scalar pions), the sign of the residue alternates with l.

[1] $(-a_l)^{-1}$ is generally called the scattering length and $2b_l$ the effective range.
[2] See Chapter 23, especially discussion on page 381.

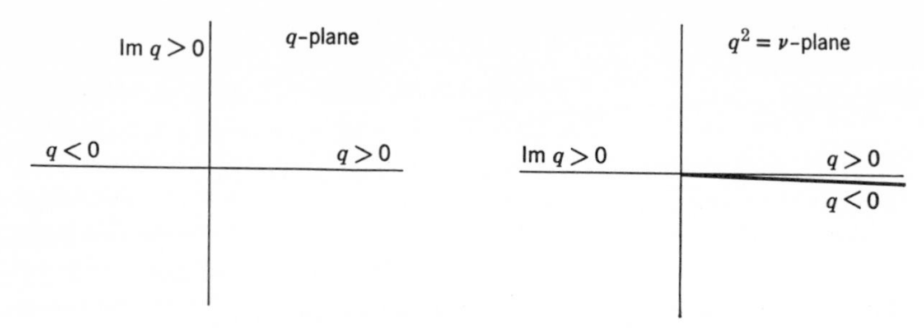

Fig. 24.2. Relation between the q-plane and the ν-plane.

It is worth pointing out that the Breit-Wigner form (24.30) is deceptive in that it suggests that $f_l(\nu)$ has a complex pole at

$$\nu = \nu_R - \frac{i}{2}\Gamma,$$

in contradiction with the postulated analyticity properties. Actually the position of the singularity must be investigated in the form (24.29). If we write $f_l(\nu)$ in the form

$$f_l(\nu) \approx \frac{-\nu_R{}^l/|b_l|\sqrt{s_R}}{q^2 - \nu_R + iq\nu_R{}^l/|b_l|\sqrt{s_R}}$$

which is correct when

$$\beta \equiv \frac{\nu_R{}^l}{|b_l|\sqrt{s_R}} \ll \frac{\nu_R}{q_R} = q_R$$

then

$$f_l(\nu) \approx \frac{-\beta}{(q + \frac{1}{2}i\beta + \sqrt{\nu_R - \frac{1}{4}\beta^2})(q + \frac{1}{2}i\beta - \sqrt{\nu_R - \frac{1}{4}\beta^2})}$$

has its poles in the lower half of the complex q-plane. With our definition of the relation between q and ν stated before and shown on Fig. 24.2, we see that in the ν-plane the singularities actually lie on the second Riemann sheet.

Equation with a Given Left-hand Cut Discontinuity. If the discontinuity in $t_l(\nu)$ across the left-hand cut, $\phi_l(\nu)$, is known, then $t_l(\nu)$ may be found, provided we can solve the nonlinear singular integral equation (24.20). A method to bypass this difficulty has been devised by Chew and Mandelstam[1] and much studied in recent years.[2] It consists of writing the

[1] G. F. Chew and S. Mandelstam, *Phys. Rev.* **119**, 467 (1960).
[2] The literature may be traced from A. W. Martin and J. L. Uretsky, *Phys. Rev.* **135**, B803 (1964).

scattering amplitude as a ratio of functions: the numerator function is defined to have no discontinuities across the right-hand cut, and the denominator function is to have no discontinuities across the left-hand cut.[1] Actually, to ensure the correct threshold properties of the function $t_l(\nu)$, namely that it behave as ν^l as $\nu \to 0$, we consider the function

$$\tilde{h}_l(\nu) = \frac{s^l}{\nu^l} t_l(\nu) \tag{24.31}$$

The s^l is inserted so as not to change the behavior of the function at infinity.[2] We now write

$$\tilde{h}_l(\nu) = \frac{N_l(\nu)}{D_l(\nu)} \tag{24.32}$$

By definition, $N_l(\nu)$ has no right-hand singularities, so that for $\nu > 0$,

$$\begin{aligned} D_l(\nu + i\epsilon) - D_l(\nu - i\epsilon) &= N_l(\nu)(\tilde{h}_l^{-1}(\nu + i\epsilon) - \tilde{h}_l^{-1}(\nu - i\epsilon)) \\ &= -2i\left(\frac{\nu}{s}\right)^{l+½} N_l(\nu) \equiv -2i\rho_l(\nu)N_l(\nu) \end{aligned} \tag{24.33}$$

which allows us to write

$$D_l(\nu) = 1 - \frac{\nu - \nu_0}{\pi}\int_0^\infty d\nu' \frac{\rho_l(\nu')N_l(\nu')}{(\nu' - \nu_0)(\nu' - \nu - i\epsilon)} \tag{24.34}$$

Similarly, for $\nu < -m_\pi^2$ we have

$$N_l(\nu + i\epsilon) - N_l(\nu - i\epsilon) = 2i\phi_l(\nu)D_l(\nu)$$

i.e.

$$N_l(\nu) = \frac{1}{\pi}\int_{-\infty}^{-m_\pi^2} d\nu' \frac{\phi_l(\nu')\, D_l(\nu')}{\nu' - \nu} \tag{24.35}$$

Note that we have normalized D_l to unity at $\nu = \nu_0$; we can always do this by scaling both N_l and D_l. We can now insert (24.35) into (24.33) to obtain the integral equation

$$D_l(\nu) = 1 - \frac{\nu - \nu_0}{\pi^2}\int_0^\infty d\nu' \int_{-\infty}^{-m_\pi^2} d\nu'' \frac{\rho_l(\nu')\phi_l(\nu'')\, D_l(\nu'')}{(\nu' - \nu_0)(\nu' - \nu)(\nu'' - \nu')} \tag{24.36}$$

Note that with the definition

$$K_l(\nu, \nu'', \nu_0) = \frac{\nu - \nu_0}{\pi}\int_0^\infty \frac{d\nu' \rho_l(\nu')}{(\nu' - \nu_0)(\nu' - \nu)(\nu' - \nu'')} \tag{24.37}$$

[1] When the masses of the particles are not equal (as in πN scattering), there are left-hand cuts not only on the negative axis, but also in the complex plane. These are, however, always separated from the right-hand unitarity cut; the notion of left-hand cut must be enlarged to include all left-hand singularities.

[2] See A. W. Martin and J. L. Uretsky, *loc. cit.*

this may be rearranged to have the form

$$D_l(\nu) = 1 + \frac{1}{\pi}\int_{-\infty}^{-m_\pi^2} d\nu'' K_l(\nu, \nu'', \nu_0)\phi_l(\nu'')\, D_l(\nu'')$$

If it is $L_l(\nu)$, defined by

$$L_l(\nu) = \frac{1}{\pi}\int_{-\infty}^{-m_\pi^2} d\nu' \, \frac{\phi_l(\nu')}{\nu' - \nu} \tag{24.38}$$

that is given directly, the integral equation may be cast in another form. This is obtained by noting that the function $N_l(\nu)$ may be written as

$$N_l(\nu) = L_l(\nu)\, D_l(\nu) + \frac{\nu - \nu_0}{\pi}\int_0^\infty d\nu' \rho_l(\nu') \, \frac{L_l(\nu')N_l(\nu')}{(\nu' - \nu_0)(\nu' - \nu)}$$

To check the correctness of this, it is merely necessary to observe that for $\nu > 0$, Im $N_l(\nu) = 0$ and that for $\nu < -m_\pi^2$, $N_l(\nu)$ has the correct left-hand cut discontinuity. Insertion of the form for $D_l(\nu)$ into the above relation yields the integral equation

$$N_l(\nu) = L_l(\nu) + \frac{\nu - \nu_0}{\pi}\int_0^\infty d\nu' \, \frac{\rho_l(\nu')N_l(\nu')}{\nu' - \nu_0} \, \frac{L_l(\nu') - L_l(\nu)}{\nu' - \nu} \tag{24.39}$$

This integral equation is clearly nonsingular, but it cannot be solved analytically unless $L_l(\nu)$ is particularly simple. In many practical applications it has been customary to take for $L_l(\nu)$ the Born approximation to the scattering amplitude, so that for weak coupling

$$\tilde{h}_l(\nu) \approx \frac{L_l(\nu) + 0(g^4)}{1 - 0(g^2)} \approx L_l(\nu) = \tilde{h}_l^{\mathrm{B.A}}(\nu)$$

If the Born approximation $\left(\text{the } l\text{th partial wave projection of } \dfrac{g^2}{t - m^2}\right)$ is substituted for $L_l(\nu)$, a rather complicated equation emerges. In order to get some insight into the nature of the solutions of such equations, Pagels[1] has recently suggested that $K(\nu, \nu', \nu_0)$ be approximated by a series of pole terms of the form ($s = 4(\nu + m_\pi^2)$)

$$\frac{1}{\nu - \nu_0} K_l(\nu, \nu', \nu_0) = \frac{s^2 C_l}{(s - s')(s - s_0)(s - a_l)} + \left\{\begin{matrix}\text{cycl. perm. in} \\ s', s_0\end{matrix}\right\} \tag{29.40}$$

Limitations of space forbid the further discussion of this approximation scheme which promises to simplify the numerical work involved in the solution of the equations. We shall be satisfied with considering the simple

[1] H. Pagels, *Phys. Rev.* **140,** B 1599 (1965).

case of the S-wave amplitude, with

$$L_0(\nu) = \lambda = \text{constant}$$

in order to make some comments of a general nature.

We choose $\nu_0 = 0$ for simplicity, and immediately see that $N(\nu) = \lambda$. Thus

$$D(\nu) = 1 - \frac{\lambda\nu}{\pi}\int_0^\infty \frac{d\nu'}{\sqrt{\nu's'}}\frac{1}{\nu' - \nu} \tag{24.41}$$

With the change of variables,

$$\nu' = (y - m_\pi^{\,2})^2/4y$$

the integral is easily converted into the form

$$D(\nu) = 1 - \frac{2\nu\lambda}{\pi}\int_0^\infty \frac{dy}{y^2 - 4\nu(y + m_\pi^{\,2})} \tag{24.42}$$

The interesting property of this expression is that *when* $\lambda < 0$, $D(\nu)$ *vanishes for some value of* $\nu < 0$. This implies that the scattering amplitude

$$t(\nu) = \frac{\lambda}{D(\nu)} \tag{24.43}$$

has a pole for some value of $\nu < 0$. If the root of (24.42) is denoted by $-\nu_B$, then in the vicinity of that root

$$D(\nu) = (\nu + \nu_B)\left(\frac{dD}{d\nu}\right)_{\nu=-\nu_B} + 0((\nu + \nu_B)^2)$$

so that

$$t(\nu) \cong \frac{\lambda}{\nu + \nu_B}\left[-\frac{2\lambda}{\pi}\int_0^\infty \frac{y^2\,dy}{[y^2 + 4\nu_B(y + m_\pi^{\,2})]^2}\right]^{-1} \tag{24.44}$$

Such a pole, with negative residue, corresponds to a bound state (cf. (24.21) and following discussion). The result that a bound state is generated for $\lambda < 0$ is not specific to our choice of $L(\nu)$;[1] that function plays a role in the partial wave integral equation analogous to that of a potential in the Schrödinger equation, and, with the right kind of potential, a bound state can always develop. It is reassuring that in dispersion theory, too, the right "interaction" can lead to a bound state. Once such a bound state pole has been found on the negative real axis, there is no need to make it part of the given $L_0(\nu)$ and recalculate. If we introduce

[1] What is peculiar to this choice of $L(\nu)$ is that only one bound state is possible, and that there will exist a bound state no matter how small λ is, provided $\lambda < 0$.

the notation

$$I(\nu) \equiv \frac{\nu}{\pi}\int_0^\infty \frac{d\nu'}{\sqrt{\nu' s'}}\frac{1}{\nu' - \nu}$$

so that [(24.43), (24.41)]

$$t(\nu) = \frac{\lambda}{1 - \lambda I(\nu)}$$

and

$$1 - \lambda I(-\nu_B) = 0$$

then the modified $\bar{N}(\nu)$ will be given by

$$\bar{N}(\nu) = \lambda + \frac{\kappa}{\nu + \nu_B}$$

(with κ such that the pole in the N/D calculation has the right residue for $\nu = -\nu_B$). This leads to a modified $\bar{D}(\nu)$ given by

$$\begin{aligned}\bar{D}(\nu) &= 1 - \frac{\nu}{\pi}\int_0^\infty \frac{d\nu'}{\sqrt{\nu' s'}}\frac{1}{\nu' - \nu}\left(\lambda + \frac{\kappa}{\nu' + \nu_B}\right)\\ &= 1 - \lambda I(\nu) - \frac{\kappa}{\nu + \nu_B}(I(\nu) - I(-\nu_B))\\ &= (1 - \lambda I(\nu))\left(1 + \frac{1}{\lambda}\frac{\kappa}{\nu + \nu_B}\right)\end{aligned}$$

so that

$$\bar{t}(\nu) = \frac{\bar{N}(\nu)}{\bar{D}(\nu)} = \frac{\lambda}{1 - \lambda I(\nu)} = \frac{N(\nu)}{D(\nu)} = t(\nu) \qquad \text{QED}$$

The general expression for $t_l(\nu)$ is given by

$$t_l(\nu) = \frac{N_l(\nu)}{1 - \dfrac{\nu}{\pi} P\displaystyle\int_0^\infty \frac{d\nu'}{\sqrt{\nu' s'}}\frac{N_l(\nu')}{\nu' - \nu} - i\dfrac{q}{\sqrt{s}} N_l(\nu)} \tag{24.45}$$

We now note that if for some value $\nu = \nu_R$, Re $D_l(\nu_R) = 0$, i.e.

$$1 - \frac{\nu_R}{\pi} P\int_0^\infty \frac{d\nu'}{\sqrt{\nu' s'}}\frac{N_l(\nu')}{\nu' - \nu_R} = 0$$

then at $\nu = \nu_R$, $\delta_l(\nu_R) = \pi/2$. If the phase shift is rising when it crosses $\pi/2$, we speak of a resonance. In the vicinity of ν_R, we have

$$\text{Re } D_l(\nu) \cong (\nu - \nu_R)\left(\frac{d \text{ Re } D_l(\nu)}{d\nu}\right)_{\nu_R}$$

so that

$$t_l(\nu) \cong \frac{1}{q} \frac{q_R N_l(\nu_R) \Big/ \left(\frac{d}{d\nu} \operatorname{Re} D_l(\nu)\right)_{\nu_R}}{\nu - \nu_R - \frac{iq_R}{\sqrt{s_R}} N_l(\nu_R) \Big/ \left(\frac{d}{d\nu} \operatorname{Re} D_l(\nu)\right)_{\nu_R}} \tag{24.46}$$

Comparison with the familiar Breit-Wigner form for a resonance shows that we may call

$$\gamma = -2q_R N_l(\nu_R)\sqrt{s_R} \Big/ \left(\frac{d}{d\nu} \operatorname{Re} D_l(\nu)\right)_{\nu_R} \tag{24.47}$$

the width of the resonance.

The N/D solution of the partial wave dispersion equation with a given left-hand "input" is not unique. This is most directly seen by noting that the replacement

$$D_l(\nu) \to D_l(\nu) + \sum_i \frac{\alpha_i}{\nu - \nu_i} \tag{24.48}$$

does not violate the analyticity conditions required of $t_l(\nu)$ provided the signs of the α_i are so chosen that the poles which are generated in $t_l(\nu)$ lie on the second sheet and not on the physical sheet. This can best be clarified by the consideration of a single term $\alpha/(\nu - \nu_0)$ with α very small. The amplitude now has the form

$$t(\nu) = \frac{N(q^2)}{D_R(q^2) - \frac{iq}{\sqrt{S}} N(q^2) + \alpha/(q^2 - q_0^2)} \tag{24.49}$$

and it has a zero at $q^2 = q_0^2$. In general $N(q^2)$ and $D(q^2)$ will vary slowly in the vicinity of $q^2 = q_0^2$, so that we will treat these functions as constant. The denominator will vanish at

$$q = \pm q_0 + \eta$$

and when α is very small, so that η is small too, we have approximately

$$\pm 2q_0\eta\left(D_R \mp \frac{iq_0}{\sqrt{s_0}} N\right) + \alpha = 0$$

i.e.

$$\eta\left(\pm 2q_0 D_R - \frac{2iq_0^2}{\sqrt{s_0}} N\right) = -\alpha$$

Thus

$$\eta = \mp \frac{\alpha}{2q_0} \frac{D_R \pm iq_0 N/\sqrt{s_0}}{D_R^2 + (q_0 N/\sqrt{s_0})^2}$$

and

$$\mathrm{Im}\,\eta = -\frac{\alpha N/2\sqrt{s_0}}{D_R^{\,2} + q_0^{\,2}N^2/s_0}$$

The pole will lie in the lower half q-plane, provided

$$\alpha N(q_0^{\,2}) > 0 \tag{24.50}$$

This so-called CDD[1] ambiguity in the solution of the partial wave dispersion equations has the consequence of introducing additional resonances. They differ from what might be called "dynamical" resonances, which arise when the CDD are absent but Re $D(\nu) = 0$, in that they can apparently be introduced at will. Simple models show that this ambiguity has a counterpart in a Hamiltonian approach to the problem of the scattering; they correspond to the possibility of the existence of additional particles that are unstable and can decay into the particles whose scattering is being studied. We leave it to the reader to convince himself that a "dynamical" resonance in the scattering amplitude may be separated out and exhibited as a CDD term without any change in the amplitude, just as was done with the "dynamical" bound state earlier.

A more realistic model than the one that we have just worked out might correspond to the replacement of $\phi_l(\nu)$ by a series of delta functions. This corresponds to a choice

$$L_l(\nu) = \sum \frac{\lambda_i}{\nu + \alpha_i} \tag{24.51}$$

If we limit ourselves to S waves and keep only one term,

$$\phi_0(s) = \pi\lambda_0\,\delta(\nu + \alpha_0)$$

then

$$N(\nu) = -\frac{\lambda_0\,D(-\alpha_0)}{\nu + \alpha_0} \tag{24.52}$$

and

$$D(\nu) = 1 + \frac{\lambda_0\,D(-\alpha_0)}{\pi}(\nu - \nu_0)\int_0^\infty d\nu'\,\frac{q'}{\sqrt{s'}}\,\frac{1}{\nu' - \nu_0}\,\frac{1}{\nu' - \nu}\,\frac{1}{\nu' + \alpha_0} \tag{24.53}$$

It is simpler to work with nonrelativistic kinematics in which case $\sqrt{s'}$ is dropped. We shall also choose $\nu_0 = -\infty$. Then

$$\begin{aligned} D(\nu) &= 1 + \frac{\lambda_0\,D(-\alpha_0)}{\pi}\int_0^\infty d\nu' q' \frac{1}{(\nu' - \nu)(\nu' + \alpha_0)} \\ &= 1 + \frac{\lambda_0\,D(-\alpha_0)}{\sqrt{\alpha_0} - i\sqrt{\nu}} \end{aligned}$$

[1] L. Castillejo, R. H. Dalitz, and F. J. Dyson, *Phys. Rev.* **101**, 453 (1956).

and

$$S = 1 + 2i\sqrt{\nu}\, f_0(\nu)$$

can easily be seen to have the form

$$S = \frac{q^2 - iqR + \alpha_0 + R\sqrt{\alpha_0}}{q^2 + iqR + \alpha_0 + R\sqrt{\alpha_0}} \tag{24.54}$$

where $q = \sqrt{\nu}$ and $R \equiv \lambda_0\, D(-\alpha_0)$. The interesting aspect of this amplitude is that (provided $\lambda_0 < 2\sqrt{\alpha_0}$) it is just the S-wave amplitude generated by the solution of the Schrödinger equation with the potential[1]

$$V(r) = 4\lambda_0\sqrt{\alpha_0}\,\frac{e^{-2r\sqrt{\alpha_0}}}{\left(1 - \dfrac{\lambda_0}{2\sqrt{\alpha_0}}\, e^{-2r\sqrt{\alpha_0}}\right)^2} \tag{24.55}$$

This observation supports the remark made earlier that "distant" left-hand contributions correspond to contributions from short-range potentials.

In general, the only way to treat the effect of inelastic processes is to introduce their influence through the function $\eta_l(\nu)$ in

$$t_l(\nu) = \sqrt{s}\,\frac{\eta_l(\nu)e^{2i\delta_l(\nu)} - 1}{2iq}$$

If this function is known, either from experiment or from a model, the N/D method can be generalized to take this into account.[2]

Define the function

$$R(\nu) \equiv \exp\left[-\frac{iq}{\pi}\int_{\nu_{\text{inel}}}^{\infty} d\nu'\,\frac{\log \eta(\nu')}{q'(\nu' - \nu)}\right] \tag{24.56}$$

with ν_{inel} the threshold for inelastic processes when η departs from unity. This may be written in the form[3]

$$\begin{aligned} R(\nu) &= e^{i\phi(\nu)} && \nu < \nu_{\text{inel}} \\ &= \eta(\nu)e^{i\phi(\nu)} && \nu > \nu_{\text{inel}} \end{aligned} \tag{24.57}$$

with

$$\phi(\nu) = -\frac{q}{\pi}\, P\int_{\nu_{\text{inel}}}^{\infty} d\nu'\,\frac{\log \eta(\nu')}{q'(\nu' - \nu)} \tag{24.58}$$

[1] V. Bargmann, *Rev. Mod. Phys.* **21,** 488 (1949); C. Eckart, *Phys. Rev.* **35,** 1303 (1930).

[2] G. Frye and R. L. Warnock, *Phys. Rev.* **130,** 478 (1963). This paper contains a very complete discussion of partial wave dispersion relations. Our discussion follows that of M. Froissart, *Nuovo Cimento* **22,** 191 (1961).

[3] This $\phi(\nu)$ has nothing to do with the $\phi_l(\nu)$ which appeared earlier in the chapter.

We may thus write

$$t(\nu) = \frac{\eta(\nu)e^{2i\delta(\nu)} - 1}{2iq/\sqrt{s}} = \frac{R(\nu)e^{-i\phi(\nu)}e^{2i\delta(\nu)} - 1}{2iq/\sqrt{s}}$$
$$= \frac{R(\nu)e^{2i\alpha(\nu)} - 1}{2iq/\sqrt{s}} \tag{24.59}$$

where we have introduced

$$\alpha(\nu) = \delta(\nu) - \tfrac{1}{2}\phi(\nu) \tag{24.60}$$

Now if we define

$$a(\nu) = \frac{e^{2i\alpha(\nu)} - 1}{2iq/\sqrt{s}} \tag{24.61}$$

we see that

$$a(\nu) = \frac{t(\nu)}{R(\nu)} + \frac{1}{2iq/\sqrt{s}}\left(\frac{1}{R(\nu)} - 1\right) \tag{24.62}$$

has the following properties: (a) it has the same analyticity properties as $t(\nu)$, since $R(\nu)$, as defined by (24.56), does not vanish anywhere in the finite plane, and the branch cut, caused by the presence of q in its definition, can be taken along the positive real axis; (b) it satisfies the elastic unitarity condition on the right-hand cut; (c) the discontinuity across the left-hand cut is known when $\eta(\nu)$ and the left-hand discontinuity of $t(\nu)$ are known. We may thus find $a(\nu)$ by a standard N/D method.[1]

It is clear from (24.60) which reads

$$\delta(\nu) = \alpha(\nu) + \frac{q}{4\pi} P\int_{\nu_{\rm inel}}^{\infty} \frac{d\nu'}{q'} \frac{\log 1/\eta(\nu')}{\nu' - \nu} \tag{24.63}$$

that rapid changes in the inelasticity will have a large effect on $\delta(\nu)$. It has been argued by Ball and Frazer[2] that a sharp drop in η could be responsible for some of the observed resonances, such as the $D_{3/2}$ πN resonance. More detailed calculations have been somewhat discouraging, and we do not pursue this subject further.

A special case in which inelastic processes can be studied in detail is when the inelasticity is caused by the presence of other two-particle

[1] It has recently been pointed out by M. Bander, P. W. Coulter and G. L. Shaw, *Phys. Rev. Letters* **14,** 270 (1965), that the solutions for the scattering amplitude in a multi-channel problem and the solutions obtained with an equivalent inelasticity factor do not always coincide. The latter can develop CDD singularities.

[2] J. Ball and W. Frazer, *Phys. Rev. Letters* **7,** 204 (1961).

channels only, e.g., in

$$\begin{aligned}\pi^- + p \rightarrow\ & \pi^- + p \\ & \pi^0 + n \\ & K^0 + \Lambda^0 \\ & K^0 + \Sigma^0 \\ & K^+ + \Sigma^- \\ & \rho^- + p \\ & \cdot \\ & \cdot \\ & \cdot\end{aligned}$$

The range of problems that can be treated in this way is quite large if we are willing to treat resonances as particles. There is a considerable literature on the subject of multichannel processes.[1] We must limit ourselves to a very brief account.

The properties most useful for representations over a limited energy region are, as for the single channel case, analyticity and unitarity. We consider the latter property first, in the s-channel only.

Let us consider a number of two-particle states, denoted by the letters $a, b, c, \ldots$. If the particles in a given initial state have four-momenta p, q (and masses M_a, m_a), and those in the final state have four-momenta p', q' (and masses M_b, m_b), the transition matrix will be denoted by $T_{ba}(p', q'; p, q)$. The generalization of the unitarity condition (21.8) reads

$$i(T_{ba}(p', q'; p, q) - T_{ab}{}^*(p, q; p', q')) = (2\pi)^4 \sum_{\text{spins}} \sum_c \int \frac{d^3p_c}{(2\pi)^3} \int \frac{d^3q_c}{(2\pi)^3}$$
$$\times \frac{\delta(p_c + q_c - p - q)}{\rho_c^{(1)}(p_c)\rho_c^{(2)}(q_c)} T_{cb}{}^*(p_c, q_c; p', q')T_{ca}(p_c, q_c; p, q)$$

Here $\rho_c(p_c) = E_c/(2\pi)^3 M_c$ for fermions and $2\omega_c/(2\pi)^3$ for bosons. In the center of mass system this becomes

$$i(T_{ba}(p', q'; p, q) - T_{ab}{}^*(p, q; p', q')) = (2\pi)^4 \sum_{\text{spins}} \sum_c \int d\Omega_{\hat{p}_c}$$
$$\times \int p_c{}^2\, dp_c \frac{\delta(\sqrt{p_c{}^2 + M_c{}^2} + \sqrt{p_c{}^2 + m_c{}^2} - \sqrt{s})}{(2\pi)^6 \rho_c^{(1)}(p_c)\rho_c^{(2)}(p_c)} T_{cb}{}^* T_{ca} \quad (24.64)$$

With the change of variables

$$E = \sqrt{p_c{}^2 + M_c{}^2} + \sqrt{p_c{}^2 + m_c{}^2} \quad (24.65)$$

[1] The literature may, in part, be traced from the article of R. H. Dalitz, *Ann. Rev. Nuc. Sci.* **13**, p. 339, Annual Reviews, Inc. Palo Alto (1963).

we obtain

$$\frac{p_c^{\,2}\,dp_c\,\delta(\sqrt{p_c^{\,2}+M_c^{\,2}}+\sqrt{p_c^{\,2}+m_c^{\,2}}-\sqrt{s})}{(2\pi)^6\rho_c^{(1)}(p_c)\rho_c^{(2)}(p_c)} = \frac{p_c}{4W}\,n_c\,dE\,\delta(E-\sqrt{s}) \tag{24.66}$$

with

$$n_c = \begin{cases} 4M_c m_c & \text{if } c \text{ is a two-fermion state} \\ 2M_c & \text{if } c \text{ is a boson-fermion state} \\ 1 & \text{if } c \text{ is a boson-boson state} \end{cases}$$

Thus the unitarity condition simplifies to

$$i(T_{ba} - T_{ab}{}^*) = (2\pi)^4 \sum_{\text{spins}} \sum_c \int d\Omega_{\hat{p}_c} T_{cb}{}^* \frac{p_c n_c}{4W} T_{ca} \tag{24.67}$$

Now it is generally possible to expand $T_{ba}(p', q'; p, q)$ in the center of mass system in a complete set of eigenfunctions of the total angular momentum J and the parity ϵ, so that

$$(T_{ba}(p', q'; p, q))_{\text{cm}} = -\sum_{J,\epsilon} t_{ba}^{(J,\epsilon)}(s)\Phi_{J,\epsilon}(\hat{\mathbf{p}}', \hat{\mathbf{p}}) \tag{24.68}$$

The functions $\Phi_{J,\epsilon}$ may be chosen to satisfy the reality condition

$$\Phi^*_{J,\epsilon}(\hat{\mathbf{p}}, \hat{\mathbf{p}}') = \Phi_{J,\epsilon}(\hat{\mathbf{p}}', \hat{\mathbf{p}}) \tag{24.69}$$

and the orthonormal property

$$\frac{1}{4\pi}\sum_{\text{spins}} \int d\Omega_{\hat{p}''} \Phi^*_{J'\epsilon'}(\hat{\mathbf{p}}'', \hat{\mathbf{p}}')\Phi_{J,\epsilon}(\hat{\mathbf{p}}'', \hat{\mathbf{p}}) = \delta_{JJ'}\,\delta_{\epsilon\epsilon'}\Phi_{J,\epsilon}(\hat{\mathbf{p}}', \hat{\mathbf{p}}) \tag{24.70}$$

For a pair of spinless particles, the Legendre polynomials form such a set, and we also found it in our decomposition of the pion-nucleon scattering matrix. This expansion may be used to obtain the unitarity condition for the $t_{ba}^{(J,\epsilon)}(s)$; it is

$$i(t_{ba}^{(J,\epsilon)}(s) - t_{ab}^{(J,\epsilon)*}(s)) = -\sum_c t_{cb}^{(J,\epsilon)*}(s)\,\frac{16\pi^5 p_c\,n_c}{W}\,t_{ca}^{(J,\epsilon)}(s) \tag{24.71}$$

The generalization of the one-channel relation

$$(t(\omega - i\epsilon))^* = t(\omega + i\epsilon) \tag{24.72}$$

with the help of which we relate $\text{Im}\, t(\omega)$ to $[t(\omega + i\epsilon) - t(\omega - i\epsilon)]/2i$ is the matrix condition

$$(t(s - i\epsilon))^+ = t(s + i\epsilon) \tag{24.73}$$

so that the left-hand side of (24.71) stands for

$$i(t_{ba}^{(J,\epsilon)}(s + i\epsilon) - t_{ba}^{(J,\epsilon)}(s - i\epsilon))$$

It should be stressed that for each particle pair transition $a \to b$, there is a submatrix of amplitudes and not just one amplitude. The reason, of course, is that the angular momentum and parity are not enough to determine the state of a two-particle system at rest. For example, in the transition

$$\pi + N \to \rho + N^*$$

in the state $\frac{1}{2}^-$, say, there are three amplitudes, leading to the possible final states ${}^6D_{1/2}$, ${}^4D_{1/2}$, and ${}^2S_{1/2}$.[1] We shall henceforth understand the labels $a, b, \ldots$ to denote not only the particle pair (in the representation in which i-spin is diagonal) but also the possible orbital angular momentum and spin states.

For the purpose of constructing an effective range formula, analogous to (24.28), we must know something about the threshold behavior of the matrix elements. When the masses are different, the momentum transfer variable in the center of mass system is

$$t = (p' - p)^2 = (E_{p'}^{(1)} - E_p^{(2)})^2 + 2\,|\mathbf{p}|\,|\mathbf{p}'| \cos\theta - \mathbf{p}^2 - \mathbf{p}'^2$$

and the argument of the Q_l function appearing in a term like that of (24.11) is $\dfrac{1}{2pp'}(y - m_1^2 - m_2^2 + 2E_p^{(1)}E_p^{(2)})$. Thus as $p \to 0$, the amplitude behaves as p^{l_a} where l_a is the orbital angular momentum of the initial state. Similarly, at the threshold of the outgoing particle pair, the matrix element varies as $(p')^{l_b}$. Thus we have

$$t_{ba}^{(J,\epsilon)}(s) = p'^{l_b} A_{ba}(s) p^{l_a} \tag{24.74}$$

where the amplitude $A_{ba}(s)$ has no threshold zeros. The unitarity condition now reads

$$i(A_{ba}(s + i\epsilon) - A_{ba}(s - i\epsilon)) = -\sum_c A_{bc}(s - i\epsilon)\frac{16\pi^5 n_c p_c^{2l_c+1}}{W} A_{ca}(s + i\epsilon)$$

so that we can write the matrix equation

$$i[A^{-1}(s - i\epsilon) - A^{-1}(s + i\epsilon)] = 2R \tag{24.75}$$

where the diagonal matrix

$$R_{c'c} = -\frac{8\pi^5 n_c}{W} p_c^{2l_c+1}\,\delta_{cc'} \tag{24.76}$$

has been introduced. Of course, p_c must be real.

[1] Equivalently, each submatrix is of the form $f_J(\lambda_1'\lambda_2'; \lambda_1\lambda_2)$ with the λ's denoting the helicities.

The discussion of the analyticity properties is tedious because when the masses are unequal, the Mandelstam representation leads to a more complicated left-hand cut structure.[1] We will not go into details, but merely state the result that there will always be a left-hand cut structure (denoted by L) and separate from it a right-hand cut generated by the branch points at $s_a = (M_a + m_a)^2, s_b, \ldots$.

It follows that the matrix

$$M(s) \equiv A^{-1}(s) - iR(s) \tag{24.77}$$

has no right-hand cut; its singularities are the cuts L and single poles wherever det $A(s)$ vanishes. A *zero range* approximation consists of replacing $M(s)$ by a hermitian constant matrix, which must be chosen so that the matrix $t(s)$ is symmetric. Furthermore, bound states and resonances of the multichannel system, for which

$$A(s) = [M(s) + iR(s)]^{-1} \tag{24.78}$$

occur at the values of s for which

$$\det\,(M(s) + iR(s)) = 0 \tag{24.79}$$

We can illustrate this by considering the coupled $\pi\pi$ and KK system in an S state. We then obtain

$$t^{-1}(s) = M(s) - \frac{8i\pi^5}{\sqrt{s}}\begin{pmatrix} q_\pi & 0 \\ 0 & q_K \end{pmatrix}$$

In terms of the center of mass energy of one of the pions, ω, this reads[2]

$$t^{-1}(s) = \begin{pmatrix} M_{11} - \dfrac{8i\pi^5}{\omega}\sqrt{\omega^2 - m_\pi^{\,2}} & M_{12} \\ M_{12} & M_{22} - \dfrac{8i\pi^5}{\omega}\sqrt{\omega^2 - m_K^{\,2}} \end{pmatrix} \tag{24.80}$$

The condition (24.79) now reads

$$\left(K_{11} - i\sqrt{1 - \frac{m_\pi^{\,2}}{\omega^2}}\right)\left(K_{22} - i\sqrt{1 - \frac{m_K^{\,2}}{\omega^2}}\right) - K_{12}^{\,2} = 0 \tag{24.81}$$

[1] S. MacDowell, *Phys. Rev.* **116,** 774 (1960).

[2] Although the unitarity condition in the physical region requires that q_π and q_K be real, one may use (24.80) to define $t^{-1}(s)$ and therefore $t(s)$ for all q_π and q_K. When ω becomes less than m_K, q_K is defined as $i\,|q_K|$.

provided we write

$$M_{ij} = 4\pi^5 K_{ij} \tag{24.82}$$

The matrix (24.80) is easily inverted, and we get

$$t(s) = \frac{1/4\pi^5}{\left(K_{11} - i\sqrt{1 - \frac{m_\pi^2}{\omega^2}}\right)\left(K_{22} - i\sqrt{1 - \frac{m_K^2}{\omega^2}}\right) - K_{12}^2} \times \begin{pmatrix} K_{22} - i\sqrt{1 - \frac{m_K^2}{\omega^2}} & -K_{12} \\ -K_{12} & K_{11} - i\sqrt{1 - \frac{m_\pi^2}{\omega^2}} \end{pmatrix} \tag{24.83}$$

Several properties of multichannel problems, exhibited in this simple example, are worth pointing out.

1. If the condition (24.81) is satisfied, it can only be so for $\omega < m_\pi$.[1] Under these circumstances all the four amplitudes have a pole at $\omega = \omega_B$. An example of this is the Λ^0 pole in the amplitudes for $K^-p \to K^-p$, $\pi^\pm\Sigma^\mp \to \pi^\pm\Sigma^\mp$ as well as $K^-P \to \pi^\pm\Sigma^\mp$.

2. For values $m_\pi < \omega < m_K$ the denominator can be written as

$$\left(K_{11} - i\sqrt{1 - \frac{m_\pi^2}{\omega^2}}\right)\left(K_{22} + \sqrt{\frac{m_K^2}{\omega^2} - 1}\right) - K_{12}^2 \tag{24.84}$$

and there may be a value $\omega = \omega_R$ at which the *real part* of the denominator vanishes; at $\omega = \omega_R$ the $\pi\pi$ scattering amplitude is purely imaginary, a behavior characteristic of a resonance. An example of this is the Y_1^* which is a resonance in the $\pi\Lambda^0$ system but is not seen directly in $KN \to KN$ or $KN \to \pi\Lambda^0$.

3. Consider the case $K_{12} = 0$, when the $\pi\pi$ and the KK channels are decoupled. If for $\omega = \omega_0$ the KK amplitude has a pole, which in this example corresponds to the vanishing of $\left(K_{22} - \sqrt{\frac{m_K^2}{\omega_0^2} - 1}\right)$, as K_{12} is slowly increased, the real part of the denominator will vanish in the vicinity of the former location of the KK bound state, and give rise to a resonance in the $\pi\pi$ system. The $Y_0^*(1405)$ resonance may be a reflection of a KN "bound state" although the strong coupling between all channels makes such an identification problematical.

[1] A solution for $\omega > m_K$ is only possible for an accidental special relation between the π and K masses.

4. It goes without saying that it is of course possible for the real part of the denominator to vanish for $\omega > m_K$, in which case all channels show a resonant behavior.

In summary, the following can be said: The Mandelstam representation has made possible the study of dispersion relations for partial wave amplitudes. The unitarity condition can be satisfied only below the production threshold, so that, aside from a phenomenological representation of absorption, the method outlined in this chapter is limited to two-particle states. In this sense, the dispersion relations resemble the nonrelativistic Schrödinger equation, in the solution of which two-particle unitarity is directly incorporated, and in which absorption must be represented through an optical potential. The analogy goes further in that the dynamical properties of the partial wave scattering amplitude are determined by a given function $L_l(\nu)$, just as the nonrelativistic scattering is determined by the potential. With this analogy before us, it is easy to make a critical assessment of the inadequacies of the dispersion approach to scattering:

1. The N/D approach does not incorporate the distinguishing feature of high-energy physics, namely multiple-particle processes.

2. It does not exhibit the relation between different partial wave amplitudes. Even when all the $L_l(\nu)$ are derived from a single Born term, for example, there is no obvious connection between the equations for different l values.

3. In contrast to the solutions of the Schrödinger equation, the N/D solutions have a number of undesirable features; they suffer from the CDD ambiguity; even aside from this, it is not clear that they are unique solutions; the solutions may depend on the value of the subtraction point ν_0 at which $D(\nu)$ is normalized to unity; finally, with the conventional choice of a Born approximation calculation to determine the left-hand cut discontinuity, the integrals do not usually converge, and a cut-off parameter must be introduced.

The justification for taking so much space for what is undeniably an nadequate approach is the following:

1. The dispersion approach to partial wave amplitudes does show that a number of phenomena, familiar from nonrelativistic potential scattering theory may, on the basis of a very few assumptions, also characterize low energy particle physics phenomena. Among these are the existence of an effective range approximation, the manifestation of bound states as poles in the scattering amplitude, the correlation of resonances with poles on the nonphysical sheet, the manifestation of a resonance in all channels of a many-channel process, etc.

2. It may just turn out that much of the structure of the low-energy particle spectrum is determined in the two-particle domain, with the contribution of many-particle states represented by a few adjustable parameters. Thus, although the approach outlined above has proved to be inadequate for a complete understanding of low energy dynamics, it may still provide a much more meaningful phenomenological framework than a phase shift analysis, for example. We shall see in the next chapter that, for reasons not well understood, the interaction generated by the exchange of single particle does predict, in the effective range framework (or the N/D framework) the existence of resonances with the observed quantum numbers.

3. The experience with this simplified approach is necessary before more complicated approaches involving three-particle states and/or off-the-mass-shell amplitudes can be studied.

PROBLEMS

1. Consider the two-channel problem discussed in the chapter. If there is a resonance above the K-threshold, show that near the resonance

$$\frac{1}{4\pi}\sigma_{\pi\pi\to\pi\pi} = \frac{\frac{1}{4}\Gamma_1^2}{(\omega-\omega_R)^2+\frac{1}{4}(\Gamma_1+\Gamma_2)^2}$$

$$\frac{1}{4\pi}\sigma_{KK\to\pi\pi} = \frac{\Gamma_2}{\Gamma_1}\cdot\frac{1}{4\pi}\sigma_{\pi\pi\to\pi\pi}$$

$$\frac{1}{4\pi}\sigma_{KK\to KK} = \left(\frac{\Gamma_2}{\Gamma_1}\right)^2\cdot\frac{1}{4\pi}\sigma_{\pi\pi\to\pi\pi}$$

where

$$\tfrac{1}{2}\Gamma_1 = \left(\frac{q_K}{\omega}\right)_R\sqrt{K_{11}^2+\left(\frac{q_\pi}{\omega}\right)_R^2} \qquad \tfrac{1}{2}\Gamma_2 = \left(\frac{q_\pi}{\omega}\right)_R\sqrt{K_{22}^2+\left(\frac{q_K}{\omega}\right)_R^2}$$

2. Consider a one-channel partial wave dispersion relation with elastic unitarity. If $L(\nu)$, as given in (24.37) say, is of the form

$$L(\nu) = \frac{\lambda}{\nu+\alpha}$$

what can you say about $(\delta(\infty)-\delta(0))$, where $\delta(\nu)$ is the phase shift? How does this change when there is a bound state? How many bound states can there be with this $L(\nu)$?

3. If

$$L(\nu) = \sum_{n=1}^{N}\frac{\lambda_n}{\nu+\alpha_n}$$

what is the maximum number of bound states possible? What can we say about $(\delta(\infty)-\delta(0))$?

4. Show that the following figure depicts the form of the scattering amplitude

$$T_e \equiv \frac{\eta e^{2i\delta} - 1}{2i}$$

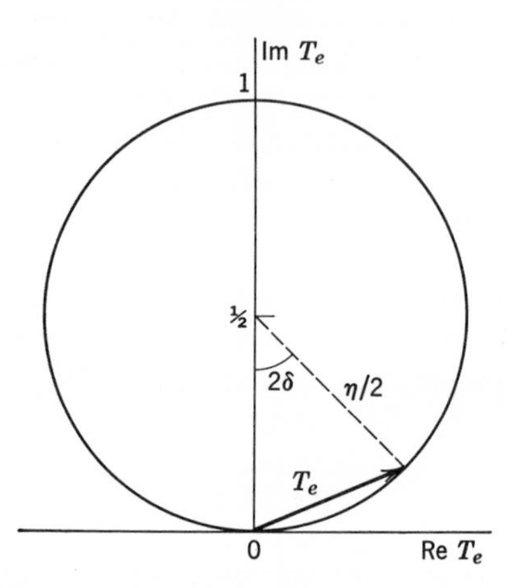

5. Show that the figures below describe resonant trajectories for the following cases

(*a*) resonance with elasticity greater than 0.5
(*b*) resonance with elasticity smaller than 0.5
(*c*) resonance with an attractive nonresonant background of the same quantum numbers (spin and parity)
(*d*) resonance with a repulsive nonresonant background of the same spin and parity.

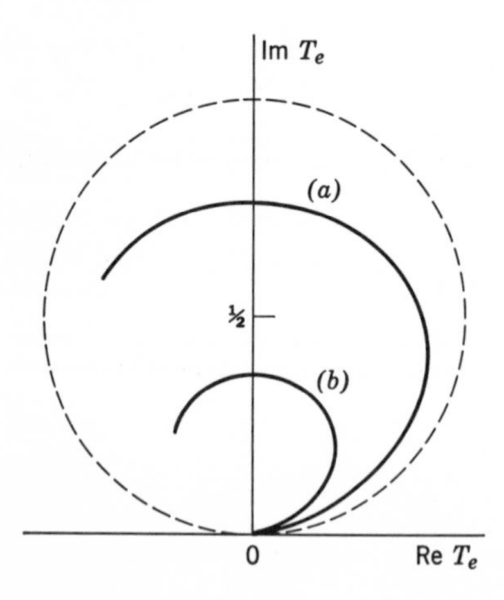

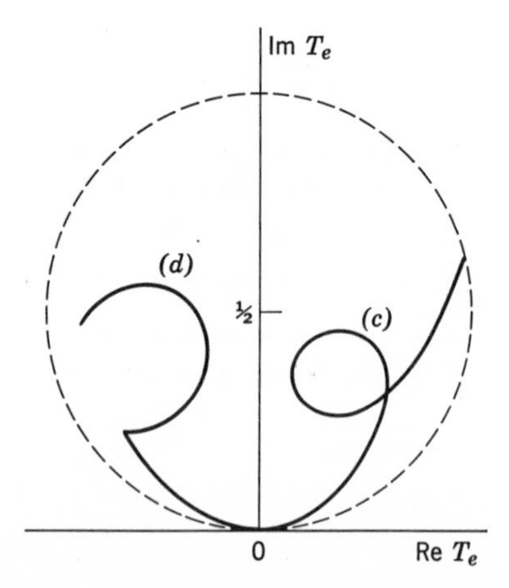

25

The Dynamical Origin of Resonances

During the last five years it has become clear that the major task of strong interaction physics is to explain the spectrum of resonances that is slowly being unveiled in the laboratory. Because of the intractability of field theory when the coupling is strong, and because of the lack of any hint as to what the "fundamental" constituents of matter are, no analog to the Schrödinger equation with a well-defined set of quasi-eigenstates has been developed. Rather, the direction of research in this area has been determined by the manifestation of bound states and resonances as singularities in scattering matrix elements. The point of view which has guided work on this problem is roughly the following:

1. The detailed nature of the low energy part of the spectrum is determined by the long-range part of the "potential" acting between pairs of particles, with the shorter range contributions representable by a small number of adjustable parameters.

2. The "potential" between pairs of particles is generated by the exchange of particles. In contrast to the atomic problem, in which photons give rise to the potential, or to the nuclear problem in which mesons give rise to the potential between nucleons, no particles are distinguished as "potential-producing." Thus in meson-nucleon interactions, the same mesons and the same nucleons may be exchanged. If there are excited baryon states, they too can be exchanged, giving rise to a potential responsible for their existence. The spectrum which emerges must therefore require a set of self-consistency constraints on the masses and widths (or coupling constants) of the members of the spectrum.

3. The tool to be used in this approach is a combination of analyticity, which gives representations of scattering amplitudes, crossing symmetry which provides the connection between the process being studied and the "potential" which determines it, and unitarity which provides a basic constraint on the amplitudes.

The implementation of this program, generally known as the "bootstrap" theory, would, strictly speaking, require the calculation of a scattering amplitude which satisfies the crossing conditions. Even approximate incorporation of crossing requires rather complicated numerical work. The best that we can do in the limited space of this chapter is to give a qualitative description of the procedure which has been followed. Although a commitment to the bootstrap point of view requires a simultaneous treatment of all particles,[1] all calculations so far have been restricted to the mutual coupling of a small subset of them. We shall discuss the generation of the $N^*(1238)$ resonance ($T = J = \frac{3}{2}$) in the pion-nucleon interaction.

The procedure, roughly speaking, is the following: partial wave dispersion relations of the form

$$f_l(W) = \frac{1}{\pi}\int_{\text{right-hand cut}} dW' \frac{\rho(W')\,|f_l(W')|^2}{W' - W - i\epsilon} + \frac{1}{\pi}\int_{\text{left-hand cut}} dW' \frac{\phi_l(W')}{W' - W} \tag{25.1}$$

are written down, with the neglect of multipion processes responsible for the simple form of the first integral. The function $\phi_l(W')$ is determined by a simple approximation and the equation is then solved by the N/D method, with zeros of Re $D(W)$ determining the position of the resonance. The function $\phi_l(W)$ should, of course, contain the effect of the resonance determined by our solution. A guide to the construction of the function $\phi_l(W)$ comes from crossing symmetry. As has been pointed out in connection with the Mandelstam representation, the T-matrix, as a function of the variables s, t, and u, describes a number of processes. For pion-nucleon scattering, for example, the dependence on the t-variable is such that when $t > 4m_\pi^2$ the T-matrix describes the process $\pi + \pi \rightarrow N + \bar{N}$ and when $u \geq (M + m_\pi)^2$ it describes pion-nucleon scattering. These properties are described pictorially in Fig. 25.1. Figure 25.1*a* and *b* show the relationship between the processes in the s- and u-channels and s- and t-channels respectively. Figure 25.1*c*, *d*, *e*, . . . , show the one-particle (including the N^* resonance) contributions to these channels.

Since the left-hand cut contributions to a partial wave amplitude come from the projecting out of that partial wave from the terms which depend on $\cos\theta$, i.e., from the terms which depend on u and t, we can put the discontinuities across the left-hand cut in correspondence with the intermediate states in the u and t channels. By analogy with nonrelativistic scattering, the mass of the exchanged system is related to the reciprocal

[1] It must be admitted in principle that partial "bootstrap" equations still have solutions, albeit with not quite correct masses and widths, because otherwise no solution could exist without the inclusion of *all* states.

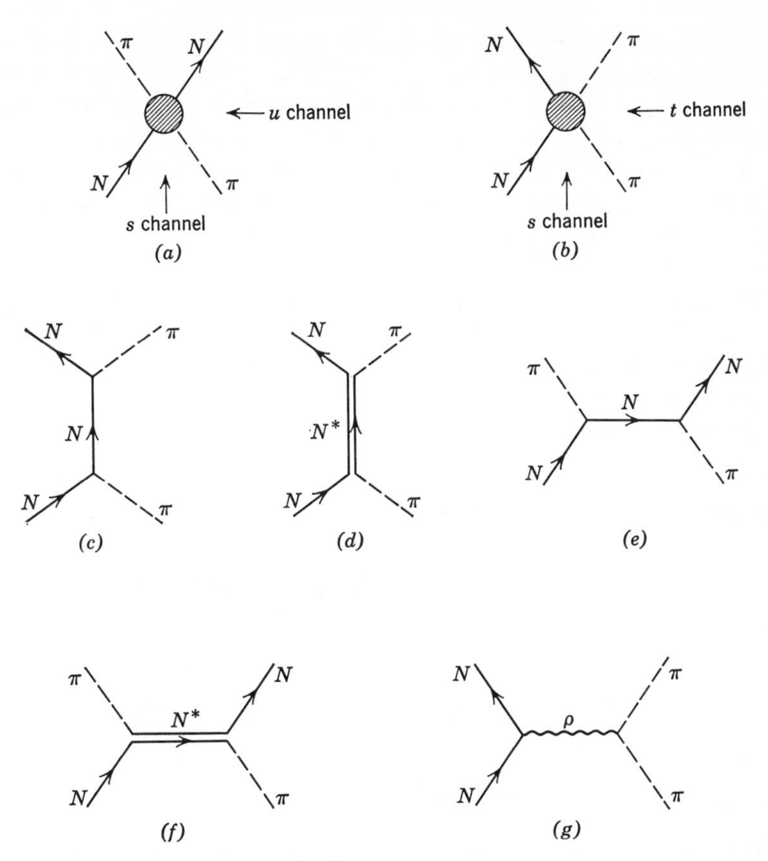

Fig. 25.1. (*a*) Relation between *s*- and *u*-channels in pion-nucleon scattering; (*b*) relation between *s*- and *t*-channels; (*c*) one-nucleon intermediate state in *s*-channel; (*d*) one-N^* intermediate state in *s*-channel; (*e*) one-nucleon "exchange" in *u*-channel; (*f*) one-N^* "exchange" in *u*-channel; (*g*) one ρ-meson "exchange" in *t*-channel.

of the range of the "potential." Thus a term like

$$\int_{4m_\pi^2}^{\infty} dt' \int ds' \, \frac{\rho(s't')}{(s'-s)(t'-t)} = \int_{4m_\pi^2}^{\infty} dt' \, \frac{a(t',s)}{t' + (\mathbf{p}-\mathbf{p}')^2}$$

has a Fourier transform

$$V(r) = 4\pi \int_{4m_\pi^2}^{\infty} dt' \, a(t',s) e^{-\sqrt{t'}r}$$

indicating a potential or range $1/2m_\pi$. A term like

$$\int_{(M+m_\pi)^2}^{\infty} du' \, b(u',s) \frac{1}{u'-u}$$

acts like a very complicated energy dependent potential.

The approximation we are led to comes from the reasonable view that the long-range part of the potential (corresponding to the exchange of low mass states) should play the major role in the low energy structure. There can only be resonances if the approximate "potential" is attractive. This approximation need not be good for the determination of the S-wave scattering amplitude, since this amplitude is very sensitive to short-range effects; there is no centrifugal barrier to keep the S-wave out from the central region.

We shall begin by showing that the one-nucleon exchange in the u-channel gives rise to an attractive potential in just the right state to give rise to a resonance in the $T = J = \frac{3}{2}$ state. We work in the limit of large nucleon mass, the "static limit."[1]

From (23.30) we have

$$f_{l\pm}(W) = \frac{E+M}{2W}(A_l - (W-M)B_l) + \frac{E-M}{2W}(-A_{l\pm 1} - (W+M)B_{l\pm 1}) \quad (25.2)$$

with

$$\begin{Bmatrix} A_l \\ B_l \end{Bmatrix} = \frac{1}{2}\int_{-1}^{1} d(\cos\theta)\, P_l(\cos\theta) \begin{Bmatrix} A(W, \cos\theta) \\ B(W, \cos\theta) \end{Bmatrix} \quad (25.3)$$

The one-particle terms come from

$$A^{(\pm)}(W, \cos\theta) = 0$$

$$B^{(\pm)}(W, \cos\theta) = -\frac{g^2}{4\pi}\left(\frac{1}{M^2 - s} \mp \frac{1}{M^2 - (2M^2 + 2m_\pi^2 - s - t)}\right) \quad (25.4)$$

The first term will only contribute to B_0. If we just wanted to calculate the pion-nucleon scattering amplitude in some approximation, we would keep it. It contributes small repulsive terms to the f_{0+} and f_{1-} amplitudes. Both of these amplitudes are not easily calculated: the former because S-waves are strongly affected by short-range forces, the latter, because they turn out to be very small, and therefore very sensitive, to small inaccuracies. In our case, it is in the spirit of the bootstrap approach not to put poles in the s-channel, since these should presumably come out of

[1] The results obtained will be the same as those of the pioneering paper of G. F. Chew and F. E. Low, *Phys. Rev.* **107**, 1570 (1956), which has served as a prototype of all calculations of this type.

the calculation. With this omission we get

$$
\begin{aligned}
B_L^{(\pm)} &= \pm \frac{g^2}{4\pi}\frac{1}{2}\int_{-1}^{1} d(\cos\theta) P_L(\cos\theta)\frac{1}{s - M^2 - 2m_\pi{}^2 - 2p^2(1-\cos\theta)} \\
&= \pm \frac{g^2}{4\pi}\frac{1}{2p^2}\frac{1}{2}\int_{-1}^{1} d(\cos\theta) P_L(\cos\theta) \\
&\quad \times \frac{1}{(s - M^2 - 2m_\pi{}^2 - 2p^2)/2p^2 + \cos\theta} \\
&= \pm \frac{g^2}{4\pi}\frac{(-1)^L}{2p^2} Q_L\left(\frac{s - M^2 - 2\omega^2}{2p^2}\right)
\end{aligned}
\tag{25.5}
$$

where p is the center of mass momentum and ω the pion energy. When M, the nucleon mass, is taken to be very large, so that nucleon recoil can be neglected, then

$$s \approx (M + \omega)^2 \approx M^2 + 2M\omega$$

and

$$\frac{s - M^2 - 2\omega^2}{2p^2} \approx \frac{M\omega}{p^2}$$

In the large M limit, we use

$$Q_L\left(\frac{M\omega}{p^2}\right) \approx \frac{\sqrt{\pi}\,\Gamma(L+1)}{2^{L+1}\Gamma(L+\frac{3}{2})}\left(\frac{p^2}{M\omega}\right)^{L+1} \tag{25.6}$$

to obtain the following results

$$
\begin{aligned}
B_0^{(\pm)} &= \pm \frac{g^2}{4\pi}\frac{1}{2M\omega} \\
B_1^{(\pm)} &= \mp \frac{g^2}{4\pi}\frac{p^2}{6M^2\omega^2} \\
B_2^{(\pm)} &= 0\left(\frac{1}{M^3}\right)
\end{aligned}
\tag{25.7}
$$

If we use $E - M \cong \dfrac{p^2}{2M}$ in (25.2) and $W - M \approx \omega$, we get after some calculations

$$
\begin{aligned}
f_{0+}^{(\pm)} &= \mp \frac{g^2}{8\pi M} \\
f_{1-}^{(\pm)} &= \pm \frac{g^2}{4\pi}\frac{p^2}{12M^2\omega} \\
f_{1+}^{(\pm)} &= \mp \frac{g^2}{4\pi}\frac{p^2}{6M^2\omega}
\end{aligned}
\tag{25.8}
$$

so that in terms of the *i*-spin amplitudes, with

$$\begin{aligned} f^{(3/2)} &= f^{(+)} - f^{(-)} \\ f^{(1/2)} &= f^{(+)} + 2f^{(-)} \end{aligned} \tag{25.9}$$

the final result is

$$\begin{aligned} f_{0+}^{(1/2)} &= \frac{2Mf^2}{m_\pi^2} & f_{0+}^{(3/2)} &= -\frac{2Mf^2}{m_\pi^2} \\ f_{1-}^{(1/2)} &= \frac{f^2p^2}{3m_\pi^2\omega} & f_{1-}^{(3/2)} &= -\frac{2f^2p^2}{3m_\pi^2\omega} \\ f_{1+}^{(1/2)} &= -\frac{2f^2p^2}{3m_\pi^2\omega} & f_{1+}^{(3/2)} &= \frac{4f^2p^2}{3m_\pi^2\omega} \end{aligned} \tag{25.10}$$

in terms of the coupling constant

$$f^2 = \frac{g^2}{4\pi}\left(\frac{m_\pi}{2M}\right)^2$$

A positive sign in the Born approximation amplitude reflects an attractive potential, and a negative sign a repulsive potential. We observe an attraction in the S state for $T = \frac{1}{2}$ which we discount; an attraction in the $T = \frac{1}{2}$, $J^P = \frac{1}{2}^+$ state (the nucleon state) and a much stronger (by a factor 4) attraction in the $T = \frac{3}{2}$, $J = \frac{3}{2}^+$ state, the N^* state. If the Born approximation is used for $N(\omega)$ in an N/D representation of the amplitude, so that

$$f_{l\pm}^{(T)}(\omega) = \frac{(f_{l\pm}^{(T)}(\omega))_{B.A.}}{1 - \dfrac{\omega}{\pi}\displaystyle\int_{m_\pi}^{\infty} \frac{d\omega'}{\omega'}\, p'\, \frac{(f_{l\pm}^{(T)}(\omega'))_{B.A.}}{\omega' - \omega - i\epsilon}} \tag{25.11}$$

then a positive $(f_{l\pm}^{(T)}(\omega))_{B.A.}$ will lead to a resonance. Actually in

$$f_{+1}^{(3/2)} = \frac{\dfrac{4f^2p^2}{3m_\pi^2\omega}}{1 - \dfrac{4f^2\omega}{3\pi m_\pi^2} P\displaystyle\int_{m_\pi}^{\infty} d\omega' \frac{p'^3}{\omega'(\omega' - \omega)} - \frac{4ip^3f^2}{3\omega m_\pi^2}} \tag{25.12}$$

the integral in D diverges. In this case the divergence comes from the nonrelativistic approximations made, but when the spin of the exchanged particle is ≥ 1, the divergence is intrinsic to the problem. We must therefore cut off the integral. The magnitude of the cut-off is chosen to fit the position of the resonance. We write

$$f_{+1}^{(3/2)} \approx \frac{4f^2p^2/3\omega m_\pi^2}{1 - (\omega/\omega_R) - 4if^2p^3/3\omega m_\pi^2} \tag{25.13}$$

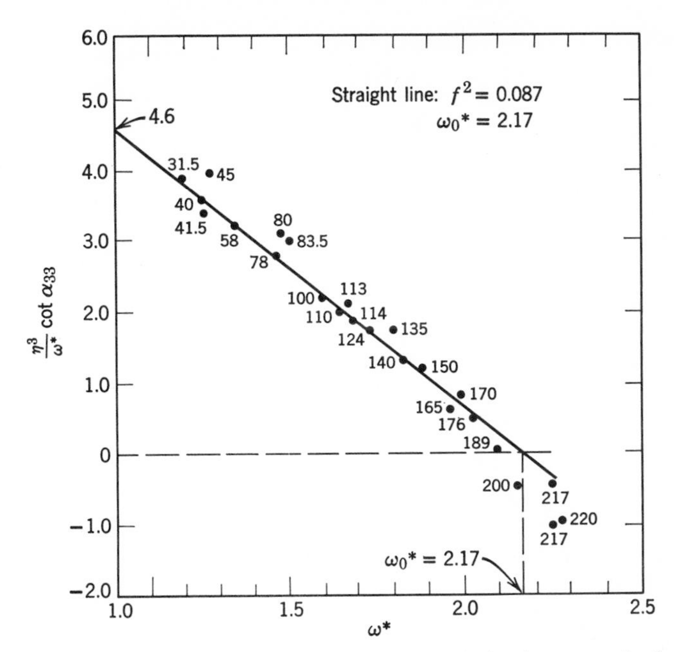

Fig. 25.2. A plot of $\eta^3 \cot \alpha_{33}/\omega^*$, with $\eta = p_\pi/m_\pi$ and ω^* the pion energy in the center of mass system, as a function of the center of mass energy. [From Barnes et al., *Phys. Rev.* **117,** 225 (1960).]

This can be rewritten in the form

$$\frac{p^3}{\omega m_\pi^{\ 2}} \cot \alpha_{33} = \frac{1}{4f^2/3}\left(1 - \frac{\omega}{\omega_R}\right) \tag{25.14}$$

Interestingly enough, the data on pion-nucleon scattering fits the linear approximation to Re $D(\omega)$ reasonably well. Figure 25.2 shows the data, with a predicted value of $f^2 \simeq 0.08$, in good agreement with the value obtained from the forward scattering dispersion relations.

In the spirit of the bootstrap approach, we should now consider the contribution of the exchange of the N^* and see whether it modifies the result, and whether the nucleon is also "predicted." In the partial bootstrap approach, which we are discussing, there is no question of "predicting" the pion. Its structure must be determined separately, perhaps in a joint $\pi\rho\omega$ problem. To see how the scattering amplitude in different states in the s-channel is affected by the exchange of a particular particle (T, J^P) in the u-channel, we discuss the so-called *crossing matrix*. In general, the exchange of any particle in the u-channel gives rise to scattering in all angular momentum states in the s-channel. We can only hope for simplicity in the static limit, which we treat, and hope that this limit gives a qualitatively correct description of the dominant effect.[1]

The fixed momentum transfer dispersion relations (23.19) take a simpler form in the static limit, in which

$$\nu \simeq M\omega \tag{25.15}$$

With the change of variable of integration $\omega' = \nu'/M$, the $M \to \infty$ limit of these relations is

$$A^{(\pm)}(\omega, t) = \frac{1}{\pi}\int_{m\pi}^{\infty} d\omega' \alpha^{(\pm)}(\omega', t)\left(\frac{1}{\omega' - \omega} \pm \frac{1}{\omega' + \omega}\right)$$

$$B^{(\pm)}(\omega, t) = \frac{g^2}{8\pi M}\left(\frac{1}{\omega} \pm \frac{1}{\omega}\right) + \frac{1}{\pi}\int_{m\pi}^{\infty} d\omega' \beta^{(\pm)}(\omega', f)\left(\frac{1}{\omega' - \omega} \mp \frac{1}{\omega' + \omega}\right) \tag{25.16}$$

Thus in the static limit crossing symmetry implies invariance under the simultaneous interchange $\omega \leftrightarrow -\omega$, $\hat{\mathbf{q}} \leftrightarrow -\hat{\mathbf{q}}'$ and $\alpha \leftrightarrow \beta$. If we write (23.28) in the form

$$F^{(\pm)}(\omega, \cos\theta) = \sum_l \left\{ f_{l+}^{(\pm)}(\omega)\left(l + 1 - i\boldsymbol{\sigma}\cdot\hat{\mathbf{q}}' \times \hat{\mathbf{q}}\frac{d}{d(\cos\theta)}\right)P_l(\cos\theta) + f_{l-}^{(\pm)}\left(l + i\boldsymbol{\sigma}\cdot\hat{\mathbf{q}}' \times \hat{\mathbf{q}}\frac{d}{d(\cos\theta)}\right)P_l(\cos\theta)\right. \tag{25.17}$$

then crossing symmetry implies that

$$\begin{aligned} F^{(\pm)}(\omega, \cos\theta) &= \pm F^{(\pm)}(-\omega, \cos\theta) \\ &= \pm\sum_l \left\{ f_{l+}^{(\pm)}(-\omega)\left(l + 1 + i\boldsymbol{\sigma}\cdot\hat{\mathbf{q}}' \times \hat{\mathbf{q}}\frac{d}{d(\cos\theta)}\right)P_l(\cos\theta) \right. \\ &\qquad \left. + f_{l-}^{(\pm)}(-\omega)\left(l - i\boldsymbol{\sigma}\cdot\hat{\mathbf{q}}' \times \hat{\mathbf{q}}\frac{d}{d(\cos\theta)}\right)P_l(\cos\theta)\right\} \end{aligned}$$

This allows us to make the identification

$$\begin{aligned} f_{l+}^{(\pm)}(\omega) &= \pm\left(\frac{1}{2l+1} f_{l+}^{(\pm)}(-\omega) + \frac{2l}{2l+1} f_{l-}^{(\pm)}(-\omega)\right) \\ f_{l-}^{(\pm)}(\omega) &= \pm\left(\frac{2l+2}{2l+1} f_{l+}^{(\pm)}(-\omega) - \frac{1}{2l+1} f_{l-}^{(\pm)}(-\omega)\right) \end{aligned} \tag{25.18}$$

Equation (25.18) may be written in the form

$$\begin{pmatrix} f_{l+}^{(\pm)}(\omega) \\ f_{l-}^{(\pm)}(\omega) \end{pmatrix} = \frac{1}{2l+1}\begin{pmatrix} 1 & 2l \\ 2l+2 & 1 \end{pmatrix}\begin{pmatrix} \pm f_{l+}^{(\pm)}(-\omega) \\ \pm f_{l-}^{(\pm)}(-\omega) \end{pmatrix} \tag{25.19}$$

[1] That this is so has been claimed by P. Carruthers, *Phys. Rev.* **133**, B497 (1964).

The matrix is denoted by L and is called the static angular momentum crossing matrix. Its significance will be explained in a moment.

With the relations

$$\begin{aligned} f^{(3/2)} &= f^{(+)} - f^{(-)} \\ f^{(1/2)} &= f^{(+)} + 2f^{(-)} \end{aligned} \tag{25.20}$$

we may immediately obtain

$$\begin{aligned} f^{(+)}(\omega) - f^{(-)}(\omega) = f^{(3/2)}(\omega) &= Lf^{(+)}(-\omega) + Lf^{(-)}(-\omega) \\ &= L(\tfrac{2}{3}f^{(1/2)}(-\omega) + \tfrac{1}{3}f^{(3/2)}(-\omega)) \end{aligned}$$

and

$$\begin{aligned} f^{(+)}(\omega) + 2f^{(-)}(\omega) = f^{(1/2)}(\omega) &= Lf^{(+)}(-\omega) - 2Lf^{(-)}(-\omega) \\ &= L(\tfrac{4}{3}f^{(3/2)}(-\omega) - \tfrac{1}{3}f^{(1/2)}(-\omega)) \end{aligned}$$

Thus the crossing relation may be written in a way to include the *i*-spin crossing properties. We obtain

$$\begin{bmatrix} f_{l+}^{(3/2)}(\omega) \\ f_{l-}^{(3/2)}(\omega) \\ f_{l+}^{(1/2)}(\omega) \\ f_{l-}^{(1/2)}(\omega) \end{bmatrix} = \frac{1}{3(2l+1)} \begin{bmatrix} 1 & 2l & 2 & 4l \\ 2(l+1) & -1 & 4(l+1) & -2 \\ 4 & 8l & -1 & -2l \\ 8(l+1) & -4 & -2(l+1) & 1 \end{bmatrix} \begin{bmatrix} f_{l+}^{(3/2)}(-\omega) \\ f_{l-}^{(3/2)}(-\omega) \\ f_{l+}^{(1/2)}(-\omega) \\ f_{l-}^{(1/2)}(-\omega) \end{bmatrix} \tag{25.21}$$

Note that the crossing matrix M, obtained above, as well as the crossing matrix L satisfy the condition

$$M^2 = \mathbf{1} \tag{25.22}$$

To illustrate the significance of the crossing matrix, let us consider pion-nucleon scattering and assume that a particle with a given angular momentum and parity J^P and *i*-spin T is exchanged. This implies that in case the u-channel is physical, i.e., $u \geq (M + m_\pi)^2$, or in the static limit when $\omega \leq -m_\pi$, there will only be scattering in the state (J^P, T) so that only $(f^{(T)}(-\omega))_{J^P}$ will be nonvanishing. The crossing matrix then shows in which states there will be scattering in the s-channel when *it* is physical, i.e., when $\omega \geq m_\pi$. A positive element of the scattering matrix will indicate an attractive potential.

Consider, for example, the exchange of a nucleon. In the crossed channel the contribution to the amplitude is only that which comes from the pole term, which gives[1]

$$f_{1-}^{(3/2)}(-\omega) = 0 \qquad f_{1-}^{(1/2)}(-\omega) = \frac{3p^2f^2}{m_\pi^{\ 2}\omega}$$

When this form for $f_{-1}^{(1/2)}(-\omega)$ is inserted into the right-hand side of (25.21) we obtain the results seen before, namely

$$f_{1+}^{(3/2)}(\omega) = \frac{4p^2f^2}{3m_\pi^{\ 2}\omega} \qquad f_{1+}^{(1/2)}(\omega) = f_{1-}^{(3/2)}(\omega) = -\frac{2p^2f^2}{3m_\pi^{\ 2}\omega} \qquad f_{1-}^{(1/2)}(\omega) = \frac{p^2f^2}{3m_\pi^{\ 2}\omega} \tag{25.23}$$

It is easy to see that the exchange of an N^* leads to the strongest attraction in the state having the quantum numbers of the nucleon.[2] Thus if we ask for the condition that the N and the N^* "bootstrap" each other,[3] we must find for what f^2 and f'^2 the condition (with $l = 1$)

$$\frac{p^2}{m_\pi^{\ 2}\omega}\begin{pmatrix} f'^2 \\ 0 \\ 0 \\ 3f^2 \end{pmatrix} = \frac{1}{9}\begin{pmatrix} 1 & 2 & 2 & 4 \\ 4 & -1 & 8 & -2 \\ 4 & 8 & -1 & -2 \\ 16 & -4 & -4 & 1 \end{pmatrix}\frac{p^2}{m_\pi^{\ 2}\omega}\begin{pmatrix} f'^2 \\ 0 \\ 0 \\ 3f^2 \end{pmatrix} \tag{25.24}$$

is satisfied. It is easily seen that

$$f'^2 \equiv f^2_{N^*N\pi} = \tfrac{3}{2}f^2 \tag{25.25}$$

This oversimplified calculation illustrates how a self-consistency requirement can be imposed. It may also say something about the origin of the excited states of the nucleon. Carruthers[4] has argued that there is reason to believe that other pion-nucleon resonances act cooperatively to generate each other. We saw above how the P_{11} and P_{33} states[5] generate each other;

[1] We see in (25.4) that the direct pole term in the s-channel is

$$-(g^2/4\pi)(M^2 - s)^{-1} \sim g^2/8\pi M\omega$$

Crossing leads to

$$B_0^{(\pm)} \simeq -g^2/8\pi M\omega$$

i.e.

$$f_{0+}^{(\pm)}(-\omega) \simeq g^2/8\pi M$$

and

$$f_{1-}^{(\pm)}(-\omega) \simeq \frac{p^2f^2}{m_\pi^{\ 2}\omega}$$

We ignore the $l = 0$ terms as they are of no interest in this connection.

[2] The working out of N^* exchange without assuming the static limit is done by assuming that it acts as a particle of spin $\frac{3}{2}$, which may be treated with the help of the formalism briefly outlined in the appendix to this chapter. See E. Abers and C. Zemach, *Phys. Rev.* **131,** 2305 (1963).

[3] G. F. Chew, *Phys. Rev. Letters* **9,** 233 (1962).

[4] P. Carruthers, *Phys. Rev.* **133,** B497 (1964).

[5] We use the spectroscopic notation $(l)_{2T,2J}$.

Carruthers argues that the nucleon (P_{11}) exchange acts to lead to an attraction in the F_{37} state, whereas the isobar (P_{33}) exchange leads to an attraction in the F_{15} state; furthermore, the F_{15} and F_{37} states "bootstrap" each other, as can be seen from the static crossing matrix (with $l = 3$). Again H_{19} and $H_{3,11}$ reinforce each other and are further bound by the exchange of the lower excited states of the nucleon. Thus a sequence of resonances, corresponding to some of the resonances discussed in Chapter 19, is predicted. Incidentally, the D states and the G states ($l = 2, 4$) do not seem to belong to this mutually cooperating set.

How much credence should we put into these admittedly crude calculations? Supporting evidence for the point of view that the nucleon spectrum is determined by low mass states comes from calculations of Donnachie, Hamilton, and Lea.[1] These authors study partial wave dispersion relations for $f_{l\pm}^{(T)}(W)/q^{2l}$. The left-hand cut discontinuities include (a) nucleon exchange, whose scale is determined by the pion-nucleon coupling constant f^2; (b) N^* exchange, with the experimentally determined mass and width; (c) ρ-meson exchange. ρ-meson exchange is a way of representing the projection of the $T = J = 1$ amplitude for $\pi\pi \rightarrow N\bar{N}$ in the t-channel (see Fig. 25.1), and it involves the $\rho\pi\pi$ coupling (determined by the ρ width) as well as the ρNN coupling. The latter is estimated with the help of the anomalous vector moment of the nucleons (see Chapter 26) and a relation between the ρ-nucleon coupling and the S-wave pion-nucleon phase shifts. In addition the projection of the $T = J = 0$ amplitude for $\pi\pi \rightarrow N\bar{N}$ in the t-channel, which is also estimated from the S-wave pion-nucleon phase shifts, is used, together with some subtraction constants (the P-wave pion-nucleon scattering lengths) which are introduced to make the partial wave dispersion relations more convergent and thereby decrease the influence of higher mass (i.e., shorter range) contributions which are not well understood. With this input the authors find that (a) the amplitudes P_{33}, D_{13}, F_{15}, and F_{37} can resonate; (b) the P_{33} resonance is satisfactorily explained; (c) the D_{13} and F_{15} amplitudes should resonate below 810 MeV and 1150 MeV respectively (the actual values are 600 MeV and 900 NeV); (d) it is plausible that the F_{37} resonates in the vicinity of the observed position (1350 MeV); (e) phase shifts calculated at 98 MeV, 120 MeV, 224 MeV, and 310 MeV are by and large in agreement with the phase shifts obtained from phase-shift analyses of the data on angular distributions and, in some cases, polarization data for the recoil proton. The details of these calculations are unfortunately too complicated to be exhibited in a meaningful way, but the amount that different parts of the input contribute is shown in Fig. 25.3.

[1] A. Donnachie, J. Hamilton, and A. T. Lea, *Phys. Rev.* **135,** B515 (1964). See also A. Donnachie and J. Hamilton, *Ann. Phys.* (*N.Y.*) **31,** 410 (1965).

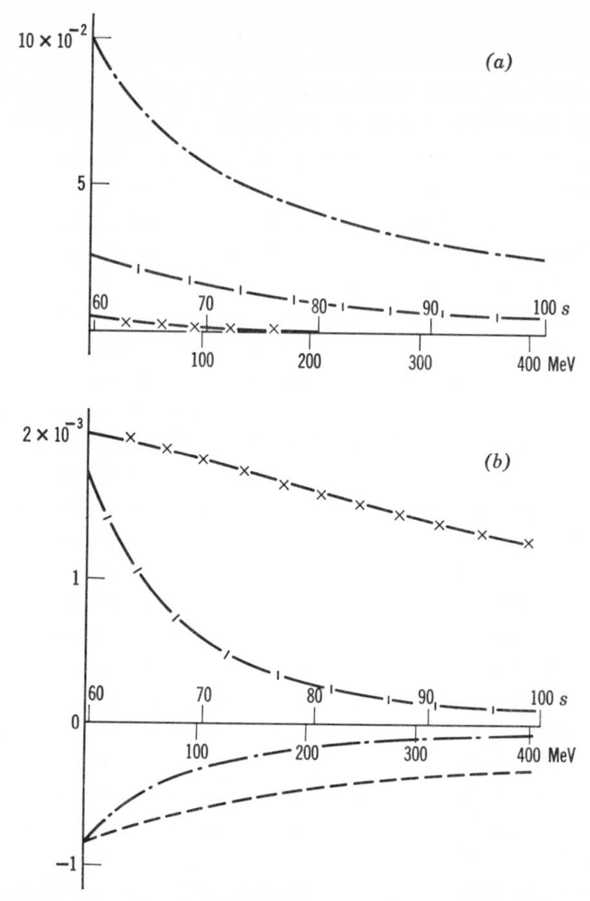

Fig. 25.3. Contributions of various left-hand cuts to the amplitudes (*a*) P_{33} (the crossed N^* cut is too small to show); (*b*) the D_{13} amplitude; (*c*) the F_{15} amplitude, and (*d*) the F_{37} amplitude (here the solid line shows their sum). –·–·–·– indicates Born terms in the *s*- and *u*-channel, – – – – indicates the contribution from the crossed cut (N^*-exchange), –|– |–|–stands for the contribution of the $T = J = 0\pi\pi$ exchange in the *t*-channel, and –×–×–×– represents the $T = J = 1\pi\pi$ exchange in the *t*-channel. [From Donnachie et al., *Phys. Rev.* **135,** B555 (1965).]

Since the nucleons and pions are now believed to be members of $SU(3)$ octets, it is a matter of some interest to see whether the resonant state, caused by the one-baryon exchange in the *u*-channel, is predicted to be a decuplet. In a static limit, the only modification of the calculation done at the beginning of this chapter consists of augmenting the *i*-spin part of the crossing matrix to relate various $SU(3)$ amplitudes. From a practical point of view there are, of course, other differences. Whereas we might, for the sake of simplicity, ignore nucleon recoil in pion-nucleon scattering,

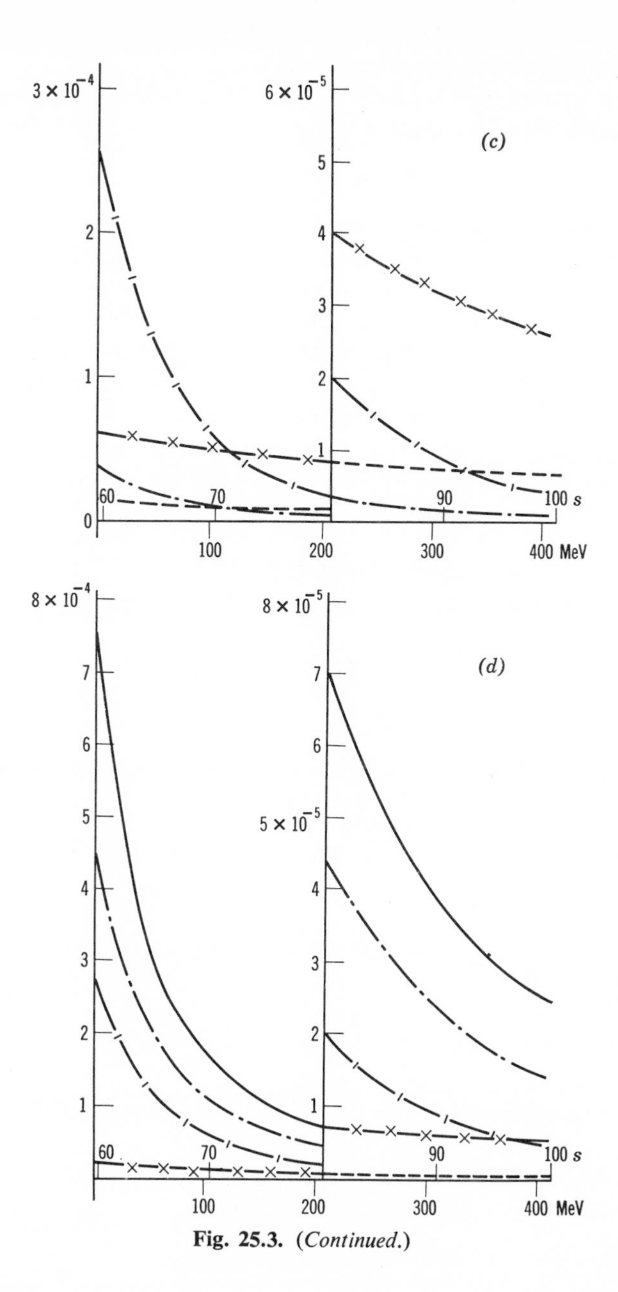

Fig. 25.3. (*Continued.*)

such an approximation becomes ridiculous in K^-p scattering. It is only because detailed calculations bear out the qualitative arguments to be given[1] that we feel free to present them.

In the scattering of two octets, the initial and final states can belong to the supermultiplets contained in

$$\mathbf{8} \otimes \mathbf{8} = \mathbf{1} \oplus \mathbf{8}_a \oplus \mathbf{8}_s \oplus \mathbf{10} \oplus \mathbf{10}^* \oplus \mathbf{27}$$

Thus all amplitudes are linear combinations of the following ones $F(\mathbf{1})$, $F(\mathbf{8}_s \leftrightarrow \mathbf{8}_s)$, $F(\mathbf{8}_s \leftrightarrow \mathbf{8}_a)$, $F(\mathbf{8}_a \leftrightarrow \mathbf{8}_a)$, $F(\mathbf{10})$, $F(\mathbf{10}^*)$, and $F(\mathbf{27})$. In order to obtain the crossing properties of these amplitudes we shall express them in terms of amplitudes whose crossing properties are manifest. For example, the amplitude for the process $\pi^+ + p \to \pi^+ + p$ in the u-channel becomes the amplitude for the process $\pi^- + p \to \pi^- + p$ in the s-channel. When both of these amplitudes are expressed in terms of the F's listed above, information is obtained about the crossing properties of the F's. The following expressions are most easily worked out with the help of tables of $SU(3)$ Wigner coefficients[2]

$$\begin{aligned}
F(\pi^+ + p \to \pi^+ + p) &= \tfrac{1}{2}F(\mathbf{27}) + \tfrac{1}{2}F(\mathbf{10}) \\
F(\pi^- + p \to \pi^- + p) &= \tfrac{1}{5}F(\mathbf{27}) + \tfrac{1}{6}F(\mathbf{10}) + \tfrac{1}{6}F(\mathbf{10}^*) + \tfrac{3}{10}F(\mathbf{8}_{ss}) \\
&\quad + \tfrac{1}{6}F(\mathbf{8}_{aa}) + \frac{1}{\sqrt{5}}\,F(\mathbf{8}_{as}) \\
F(K^+ + p \to K^+ + p) &= F(\mathbf{27}) \\
F(K^- + p \to K^- + p) &= \tfrac{7}{40}F(\mathbf{27}) + \tfrac{1}{12}F(\mathbf{10}) + \tfrac{1}{12}F(\mathbf{10}^*) + \tfrac{1}{5}F(\mathbf{8}_{ss}) \\
&\quad + \tfrac{1}{3}F(\mathbf{8}_{aa}) + \tfrac{1}{8}F(\mathbf{1}) \\
F(\Lambda^0 + \eta^0 \to \Lambda^0 + \eta^0) &= \tfrac{27}{40}F(\mathbf{27}) + \tfrac{1}{5}F(\mathbf{8}_{ss}) + \tfrac{1}{8}F(\mathbf{1}) \qquad (25.26) \\
F(\Lambda^0 + \pi^0 \to \Lambda^0 + \pi^0) &= \tfrac{3}{10}F(\mathbf{27}) + \tfrac{1}{4}F(\mathbf{10}) + \tfrac{1}{4}F(\mathbf{10}^*) + \tfrac{1}{2}F(\mathbf{8}_{ss}) \\
\frac{1}{\sqrt{3}}F(\pi^0 + p \to K^+ + \Lambda) &= \tfrac{1}{20}F(\mathbf{27}) + \tfrac{1}{12}F(\mathbf{10}^*) - \tfrac{1}{20}F(\mathbf{8}_{ss}) \\
&\quad - \tfrac{1}{12}F(\mathbf{8}_{aa}) - \frac{1}{3\sqrt{5}}\,F(\mathbf{8}_{as}) \\
\frac{1}{\sqrt{3}}F(K^- + p \to \pi^0 + \Lambda^0) &= -\tfrac{1}{10}F(\mathbf{27}) - \tfrac{1}{12}F(\mathbf{10}) + \tfrac{1}{12}F(\mathbf{10}^*) \\
&\quad + \tfrac{1}{10}F(\mathbf{8}_{ss}) - \frac{1}{6\sqrt{5}}\,F(\mathbf{8}_{as})
\end{aligned}$$

[1] A. W. Martin and K. C. Wali, *Phys. Rev.* **130,** 2455 (1963).

[2] J. J. de Swart, *Rev. Mod. Phys.* **35,** 916 (1963). P. McNamee and F. Chilton, *Rev. Mod. Phys.* **36,** 1005 (1964).

These can be inverted. We obtain, for example

$$F(\mathbf{27}) = F(K^+ + p \to K^+ + p)$$

$$F(\mathbf{10}) = 2F(\pi^+ + p \to \pi^+ + p) - F(K^+ + p \to K^+ + p)$$

etc. Thus under crossing,

$$F(\mathbf{27}) = F(K^+ + p \to K^+ + p) \to F(K^- + p \to K^- + p)$$

which is known in terms of the $SU(3)$ F's from (25.26). In this way the $SU(3)$ crossing matrix may be constructed. The result is[1]

	27	**10**	**10***	$\mathbf{8}_{ss}$	$\mathbf{8}_{aa}$	$\mathbf{8}_{as}$	**1**
27	$\frac{7}{40}$	$\frac{1}{12}$	$\frac{1}{12}$	$\frac{1}{5}$	$\frac{1}{3}$	0	$\frac{1}{8}$
10	$\frac{9}{40}$	$\frac{1}{4}$	$\frac{1}{4}$	$\frac{2}{5}$	0	$\sqrt{\frac{4}{5}}$	$-\frac{1}{8}$
10*	$\frac{9}{40}$	$\frac{1}{4}$	$\frac{1}{4}$	$\frac{2}{5}$	0	$-\sqrt{\frac{4}{5}}$	$-\frac{1}{8}$
$\mathbf{8}_{ss}$	$\frac{27}{40}$	$\frac{1}{2}$	$\frac{1}{2}$	$-\frac{3}{10}$	$-\frac{1}{2}$	0	$\frac{1}{8}$
$\mathbf{8}_{aa}$	$\frac{9}{8}$	0	0	$-\frac{1}{2}$	$\frac{1}{2}$	0	$-\frac{1}{8}$
$\mathbf{8}_{as}$	0	$\sqrt{\frac{5}{4}}$	$-\sqrt{\frac{5}{4}}$	0	0	0	0
1	$\frac{27}{8}$	$-\frac{5}{4}$	$-\frac{5}{4}$	1	-1	0	$\frac{1}{8}$

(25.27)

Let us now consider baryon exchange. The magnitude of this term depends on the D/F ratio in the Yukawa baryon meson coupling. If the coupling is

$$g(\alpha \bar{B} D_i B + (1 - \alpha)\bar{B} F_i B)P_i \tag{25.28}$$

then the amplitude in the u-channel will be of the form

$$\frac{g^2}{M^2 - u}(\alpha^2 D_j D_i + (1 - \alpha)^2 F_j F_i + \alpha(1 - \alpha)(F_j D_i + D_j F_i)) \tag{25.29}$$

This must be some combination of $F(\mathbf{8}_{ss})$, $F(\mathbf{8}_{aa})$, and $F(\mathbf{8}_{as})$. We know that if the coupling were pure F, ($\alpha = 0$) the amplitude would be $F(\mathbf{8}_{aa})$;

[1] See also J. J. de Swart, *Nuovo Cimento* **31**, 420 (1964).

if the coupling were pure D, the amplitude would be $F(\mathbf{8}_{ss})$. Hence a product of the 8×8 matrices F_jF_i must be proportional to an $\mathbf{8}_{aa}$ projection operator. The proportionality coefficient in

$$(\Pi(\mathbf{8}_{aa}))_{ji} = \lambda F_jF_i \tag{25.30}$$

is determined by the requirement that

$$(\Pi(\mathbf{8}_{aa}))_{kj}(\Pi(\mathbf{8}_{aa}))_{ji} = (\Pi(\mathbf{8}_{aa}))_{ki}$$

i.e.,

$$\lambda^2F^2F_kF_i = \lambda F_kF_i$$

Now F^2 can be calculated with the help of the representation

$$(F_i)_{jk} = -if_{ijk} \tag{25.31}$$

with the f_{ijk} tabulated in Chapter 17. It turns out that $F^2 = 3$, so that

$$F_jF_i = 3(\Pi(\mathbf{8}_{aa}))_{ji} \tag{25.32}$$

Similarly, using the fact that a calculation gives $D^2 = \frac{5}{3}$, we obtain

$$D_jD_i = \tfrac{5}{3}(\Pi(\mathbf{8}_{ss}))_{ji} \tag{25.33}$$

The operator D_jF_i may be viewed as a "symmetry-flip" operator, in that it acts on an antisymmetric octet to give a symmetric octet. To see this, we note that

$$(\tfrac{3}{5}D_kD_j)(D_jF_i)\,|\mathbf{8}_a\rangle = D_kF_i\,|\mathbf{8}_a\rangle$$
$$(\tfrac{1}{3}F_kF_j)(D_jF_i)\,|\mathbf{8}_a\rangle = 0$$

The second relation follows from

$$(D_jF_j)_{ab} = -i\,d_{jac}f_{jab} = 0$$

which vanishes because of the symmetry properties of the d and f. If we write

$$DF\,|\mathbf{8}_a\rangle = \xi\,|\mathbf{8}_a\rangle$$

we can obtain the coefficient ξ from

$$\xi^2\langle\mathbf{8}_a\,|\,\mathbf{8}_a\rangle = \langle\mathbf{8}_a|\,FDDF\,|\mathbf{8}_a\rangle$$
$$= 5\langle\mathbf{8}_a\,|\,\mathbf{8}_a\rangle$$

The operator F_jD_i can similarly be shown to act as an exchange operator, changing a symmetric state to an antisymmetric one. Thus we may write (using the notation X for exchange operator)

$$D_jF_i + F_jD_i = \sqrt{5}(X^{as} + X^{sa})_{ji} \tag{25.34}$$

The one-baryon octet exchange may, therefore, be represented by the column

$$\begin{pmatrix} 0 \\ 0 \\ 0 \\ 5\alpha^2/3 \\ 3(1-\alpha)^2 \\ \sqrt{5}\,\alpha(1-\alpha) \\ \sqrt{5}\,\alpha(1-\alpha) \\ 0 \end{pmatrix} \begin{matrix} \mathbf{27} \\ \mathbf{10} \\ \mathbf{10^*} \\ \mathbf{8}_{ss} \\ \mathbf{8}_{aa} \\ \mathbf{8}_{as} \\ \mathbf{8}_{sa} \\ \mathbf{1} \end{matrix}$$

or, if we treat (*as*) and (*sa*) as indistinguishable, as we already did in the crossing matrix, by the column

$$\begin{pmatrix} 0 \\ 0 \\ 0 \\ 5\alpha^2/3 \\ 3(1-\alpha)^2 \\ 2\sqrt{5}\,\alpha(1-\alpha) \\ 0 \end{pmatrix} \tag{25.35}$$

This column vector, when multiplied by the crossing matrix, yields the amplitudes in the s-channel. The value of α determines which states will resonate, i.e., for which the potential generated by the one-baryon exchange will be attractive. In the calculation of Martin and Wali, the best results were obtained with $\alpha \approx \frac{2}{3}$, which can be seen to make the **10** the most likely to bind. The spatial part of the calculation is the same as for the pion-nucleon problem (in the limit of exact $SU(3)$ with all the baryon masses and all the meson masses degenerate), so that the spin and parity will again be $\frac{3}{2}^+$.

It is interesting that a value of $\alpha \approx \frac{2}{3}$ can be obtained from the requirement that the decuplet exchange and the baryon exchange form a "reciprocal bootstrap" system.[1] The baryon-exchange amplitude may be written as

$$\frac{g^2}{M^2-u}\left[\frac{5}{3}\alpha^2 F(\mathbf{8}_{ss}) + 3(1-\alpha)^2 F(\mathbf{8}_{aa}) + \sqrt{5}\,\alpha(1-\alpha)(F(\mathbf{8}_{as}) + F(\mathbf{8}_{sa}))\right] \tag{25.36}$$

[1] R. E. Cutkosky, *Ann. Phys.* (*N.Y.*) **23,** 415 (1963).

or, in matrix form (rows and columns labeled by $\mathbf{8}_a$ and $\mathbf{8}_s$), as

$$\frac{g^2}{M^2 - u}\begin{pmatrix} 3(1-\alpha)^2 & \sqrt{5}\,\alpha(1-\alpha) \\ \sqrt{5}\,\alpha(1-\alpha) & \frac{5}{3}\alpha^2 \end{pmatrix} \begin{matrix} a \\ s \end{matrix} \quad \text{(columns } a, s\text{)} \tag{25.37}$$

The eigenvalues of this matrix are easily seen to be 0 and $\frac{5}{3}\alpha^2 + 3(1-\alpha)^2$, and the eigenstate corresponding to the nonvanishing eigenvalue is the baryon state

$$\Phi_B \propto 3(1-\alpha)\Phi_{\mathbf{8}_a} + \sqrt{5}\,\alpha\Phi_{\mathbf{8}_s} \tag{25.38}$$

If a decuplet is exchanged, the column to be multiplied by the crossing matrix is

$$\frac{\gamma^2}{M^2 - u}\begin{pmatrix} 0 \\ 1 \\ 0 \\ 0 \\ 0 \\ 0 \\ 0 \end{pmatrix} \tag{25.39}$$

This leads to the following octet amplitudes in the s-channel

$$F(\mathbf{8}_{ss}) = \frac{\frac{1}{2}\gamma^2}{M^2 - s} \qquad F(\mathbf{8}_{aa}) = 0 \qquad F(\mathbf{8}_{as}) = \frac{\sqrt{5}}{4}\frac{\gamma^2}{M^2 - s}$$

and they can again be written in matrix form

$$\frac{\gamma^2}{M^2 - s}\begin{pmatrix} 0 & \sqrt{5}/4 \\ \sqrt{5}/4 & \frac{1}{2} \end{pmatrix} \tag{25.40}$$

The eigenvalues are $\frac{1}{4}(1 \pm \sqrt{6})$, and the eigenstate corresponding to the attractive amplitude is easily calculated to be

$$\Phi_B \propto \sqrt{5}\,\Phi_{\mathbf{8}_a} + (1 + \sqrt{6})\Phi_{\mathbf{8}_s} \tag{25.41}$$

Comparison with (25.38) shows that self-consistency requires

$$\alpha = \frac{3}{2 + \sqrt{6}} \simeq 0.67 \tag{25.42}$$

The considerations outlined above form a basis for some optimism that a qualitative understanding of the mechanism of resonance formation has been found. Thus encouraged, a number of people[1] have undertaken

[1] See bibliography.

detailed, relativistic calculations of a bootstrap system for mesons and for baryons. The results have not been in quantitative agreement with the known spectrum, and the calculations have been marred by a number of ambiguities. The procedure has consisted of solving the pair of equations (24.34), (24.35), in which $\phi_l(\nu)$ is calculated from the Born approximation corresponding to the exchange of all the single particles which are supposed to participate in the bootstrap. The coupling constants and the masses are taken as free parameters. Zeros of $D_l(\nu)$ (or its real part) are looked for, and the $N_l(\nu)$ function is calculated at the position of the zero in the denominator. The free parameters are varied until the zeros in the denominator occur at exactly the mass values which determine $\phi_l(\nu)$, and the residue at the pole has the value which the pole in the t- or u-channel has. The ambiguities arise in several connections:

1. The form of the amplitude for which the dispersion relations are written down: a way of ensuring the proper threshold behavior for the amplitude is to treat the related amplitude $h_l(\nu)$, where

$$t_l(\nu) = q^{2l}\, h_l(\nu)$$

This, however, leads to trouble at infinity. A choice which does not do violence to the problem at infinity is

$$t_l(\nu) = \frac{q^{2l}}{W^{2l}}\, h_l(\nu)$$

but this will, in general, lead to poles at the origin for the scattering amplitude, and there is no reason why they should be there. Furthermore, in a problem involving particles of unequal mass, the last form violates the threshold condition for the crossed process. A "best" form for the related amplitude has been suggested by Martin and Uretsky.[1]

2. It turns out that, in the approximate calculations, the results depend on the arbitrary subtraction point ν_0 at which the denominator function is normalized to unity. There are more sophisticated versions of the N/D procedure which do not suffer from this difficulty.[2]

3. In a multichannel N/D approximation, the transition matrix frequently turns out to be nonsymmetric, in violation of the condition imposed by time reversal invariance.[3]

4. When the spin of the exchanged particle is equal to or larger than one, the integral equation for N requires a cut-off in the integral term. This appears to be a fundamental difficulty in the approach which treats

[1] A. Martin and J. L. Uretsky, *Phys. Rev.* **135,** B803 (1964).
[2] G. L. Shaw, *Phys. Rev. Letters* **12,** 345 (1964).
[3] J. D. Bjorken and M. Nauenberg, *Phys. Rev.* **121,** 1250 (1961).

resonant states as particles, and it is presumably a counterpart of the nonrenormalizability of Lagrangian theories involving particles of spin ≥ 1. The cut-off thus becomes an additional parameter in the theory.

5. In the numerical treatment of the equations, poles in the scattering amplitude appear, the sign of whose residue is opposite to that demanded by their identification as bound states. Such poles are said to represent "ghost states."[1] Such states violate unitarity, and their appearance is very likely a consequence of the violation of crossing symmetry.[2]

Even when all of these difficulties are coped with in one way or another, the numerical results of the calculations are not particularly encouraging, in that the widths of the resonances turn out to be 2 to 3 times too large.[3] The conclusion which one is led to at this time is that (*a*) the notion that low-energy phenomena are significantly determined by the properties of low-mass states, with only a few parameters required to represent the unknown "short-range" effects, has had qualitative and quantitative success; (*b*) the bootstrap explanation of the resonance parameters, such as mass and width, requires for implementation a much more sophisticated approach than the N/D method with simple one-particle exchange. It is, of course, entirely possible that the resonance parameters are particularly sensitive to "short-range" effects, in which case self-consistency conditions among low-mass states alone will not give the right answers.

An Extended Footnote on Perturbation Theory with Spin $\frac{3}{2}$ Particles

A free spin $\frac{3}{2}$ particle may be described by the quantity $\psi_{\alpha\mu}(x)$ which transforms as a four-component spinor on the α index and as a vector on the μ index.[4] It satisfies the equation

$$(i\gamma^{\mu}\partial_{\mu} - M)_{\alpha\beta}\psi_{\beta\nu}(x) = 0 \qquad \nu = 0, 1, 2, 3$$

and the subsidiary conditions

$$(\gamma^{\mu})_{\alpha\beta}\psi_{\beta\mu}(x) = 0 \qquad \partial^{\mu}\psi_{\alpha\mu}(x) = 0$$

which can easily be seen, in the nonrelativistic limit, to ensure that $\psi_{\alpha\mu}$

[1] G. Källén and W. Pauli, *Kgl. Danske Vidensk. Selsk. Mat.-Fys. Medd.* **30,** No. 7 (1955).

[2] M. A. Ruderman and S. Gasiorowicz, *Nuovo Cimento* **8,** 861 (1958).

[3] J. R. Fulco, G. L. Shaw and D. Y. Wong, *Phys. Rev.* **137,** B1242 (1965).

[4] W. Rarita and J. Schwinger, *Phys. Rev.* **60,** 61 (1941); see also H. Umezawa, *Quantum Field Theory*, Amsterdam, North Holland Publishing Co., 1956. The field $\psi_{\alpha\mu}$ transforms under the reducible $[(\frac{1}{2}, 0) \oplus (0, \frac{1}{2})] \otimes (\frac{1}{2}, \frac{1}{2})$ representation of the homogeneous Lorentz group.

reduces to the spin $\frac{3}{2}$ part of a product of a three-vector and a two-component spinor. The coupling of a spin $\frac{3}{2}$ particle to the nucleon and the pion is usually taken to be

$$\frac{G}{m}\,\bar{\psi}(x)\,\psi_\mu(x)\,\partial^\mu\,\phi(x) + \text{h.c.}$$

for a positive parity spin $\frac{3}{2}$ particle. Calculations require the projection operator

$$\Lambda_{\mu\nu}(p) = \sum_{r=-3/2}^{3/2} u_\mu^{(r)}(p)\,\bar{u}_\nu^{(r)}(p)$$

If this is written in the form

$$\Lambda_{\mu\nu}(p) = \frac{\not{p} + M}{2M}(ag_{\mu\nu} + b\gamma_\mu\gamma_\nu + cp_\mu p_\nu + d\gamma_\mu p_\nu + ep_\mu\gamma_\nu)$$

and the subsidiary conditions in momentum space are applied, then a properly normalized projection operator is found to be

$$\Lambda_{\mu\nu} = \frac{\not{p} + M}{2M}\left(g_{\mu\nu} - \frac{1}{3}\gamma_\mu\gamma_\nu - \frac{2p_\mu p_\nu}{3M^2} + \frac{p_\mu\gamma_\nu - p_\nu\gamma_\mu}{3M}\right)$$

The propagator may, in analogy with the spin $\frac{1}{2}$ case, be taken to be

$$\frac{2M\Lambda_{\mu\nu}(p)}{p^2 - M^2 + i\epsilon}$$

but this is not an unambiguous choice. If only N^*- exchange graphs are to be treated, then one may treat the exchange graph as representing a one-particle intermediate state, for the discussion of which only $\Lambda_{\mu\nu}$ is really needed. For these graphs the above propagator is the correct one.

PROBLEMS

1. Consider a static model for $\rho\omega\pi$ coupling, with the vector mesons ρ, ω treated as static sources. The vector mesons will only interact through the operator properties of the spin matrix **S**. With a *P*-wave coupling, $\mathbf{S}\cdot\mathbf{q}$ (where $\mathbf{q}$ is the pion momentum), show that a $\pi\omega$ resonance of spin and parity 2^- is predicted if the dominant contribution to the scattering comes from ρ-exchange.
2. Calculate the pion-nucleon *S*-wave phase shifts in Born approximation, on the assumption that the *S*-wave scattering is entirely due to ρ-meson exchange. If the coupling of the ρ to the pions and nucleons is determined by (*a*) the $\rho \to 2\pi$ width and (*b*) a "universal" coupling of ρ to the *i*-spin current

$$\mathcal{L}_I = f\rho_\mu^{\alpha}(\phi T^\alpha\,\partial^\mu\phi + \tfrac{1}{2}\bar{\psi}\gamma^\mu\tau^\alpha\psi + \cdots)$$

what are the predicted scattering lengths? If the ρ-exchange model is only used for the ratio of the scattering lengths, and $(a_1 - a_3) = 0.285$ as given by the forward scattering dispersion relations, what are a_1 and a_3?

3. Consider the scattering of a pion by a $T = J = \frac{3}{2}$ particle in the static limit with P-wave coupling. Show that a resonance with $T = J = \frac{5}{2}$ is predicted.

4. The Rarita-Schwinger theory describes a particle of spin 2 by a field $\phi_{\mu\nu}(x)$, symmetric in μ, ν, and satisfying the conditions

$$\partial^{\mu}\phi_{\mu\nu}(x) = 0; \qquad g^{\mu\nu}\phi_{\mu\nu}(x) = 0$$

Show that such a field has the right transformation properties to describe a particle of spin 2. If the field is expanded in plane wave states, with $h_{\mu\nu}(\lambda)$ $(\lambda = 1, 2, \ldots, 5)$ playing the role of $e_\mu(\lambda)$ in the vector field case, show that

$$\sum_{\lambda} h_{\mu\nu}(\lambda)h_{\rho\sigma}(\lambda) = \tfrac{1}{2}P_{\mu\rho}P_{\nu\sigma} + \tfrac{1}{2}P_{\mu\sigma}P_{\nu\rho} - \tfrac{1}{3}P_{\mu\nu}P_{\rho\sigma}$$

where

$$P_{\mu\nu} = g_{\mu\nu} - \frac{q_\mu q_\nu}{m^2}$$

and the particle has mass m and four-momentum q.

26

Form Factors

The study of scattering cross sections for the strongly interacting particles is not the only way of obtaining information about their structure. In field theoretical language the matrix elements which are studied in pion-nucleon collisions are $(\Psi_n, j_\pi(0)\Psi_N)$, with n representing a pion-nucleon state in the case of elastic scattering. An entirely different aspect of the structure of particles is obtained when $(\Psi_n, j_\mu(0)\Psi_N)$ is measured, with $j_\mu(x)$ the electric current operator. Such matrix elements may either be measured with real photons, as in the case of photoproduction

$$\gamma + N \rightarrow N + \pi\,(+\cdots)$$

which is easily seen to be determined by

$$(2\pi)^4 i\,\frac{e^\mu(q,\lambda)}{(2\pi)^{3/2}}(\Psi_n, j_\mu(0)\Psi_N) \tag{26.1}$$

or with virtual photons. The source of the virtual photons is most conveniently taken to be an electron or a muon, since these particles have no strong interactions, so that the electromagnetic interaction can be isolated. Examples of measurements with virtual photons are electron-proton scattering, in which $(\Psi_{p'}, j_\mu(0)\Psi_p)$ is measured, or electroproduction, e.g., $e + p \rightarrow e + p + \pi$, in which $(\Psi_{p\pi^0}, j_\mu(0)\Psi_p)$ is measured, with $q^2 \neq 0$ for the photon (Fig. 26.1). Limitations of space prevent us from discussing any topic other than the electron-nucleon scattering.[1]

Because the fundamental interaction of the electromagnetic field with charged matter is weak, the contributions to the cross section for electron-proton scattering, which come from the two-photon exchange, are expected

[1] For an excellent discussion of photoproduction see P. Stichel, *Fortschr. Physik* **13,** 73 (1965).

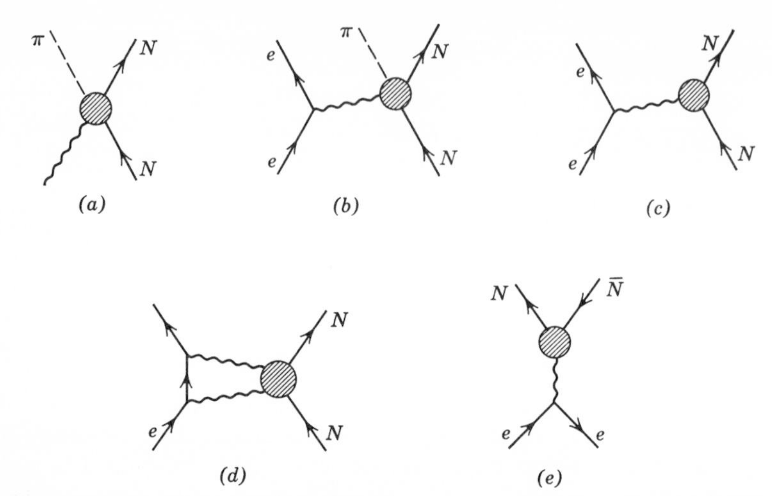

Fig. 26.1. Processes used in the measurement of current matrix elements. (*a*) Photoproduction, (*b*) electroproduction, (*c*) electron-proton scattering (*d*) with two-photon exchange, and (*e*) pair production in electron-positron collisions.

to be small. Detailed calculations[1] as well as experiments[2] support this, and we shall deal only with the one-photon exchange terms in what follows. The one-photon terms should also give the dominant effect in processes like

$$e^+ + e^- \rightarrow p + \bar{p}$$

Until now energy limitations have prevented the carrying out of experiments of this type, which will become feasible with the completion of storage rings for high energy electrons.

If the proton were just a heavy positron, the matrix element for the contribution of the graph in Fig. 26.1*c* would be

$$T = \frac{1}{(2\pi)^6} \bar{u}_e(k')(-e\gamma^\mu)\, u_e(k) \frac{-g_{\mu\nu}}{(k' - k)^2 + i\epsilon} \bar{u}(p')e\gamma^\nu u(p) \qquad (26.2)$$

with k, k' representing the initial and final electron four-momenta and

[1] S. D. Drell and M. A. Ruderman, *Phys. Rev.* **106**, 561 (1957); S. D. Drell and S. Fubini, *Phys. Rev.* **113**, 741 (1959).

[2] Experiments showing that $[\sigma(e^+p) - \sigma(e^-p)]/[\sigma(e^+p) + \sigma(e^-p)] \leq 0.02$ are reported by A. Browman and J. Pine, *Proceedings of the International Conference on Nucleon Structure at Stanford University*, R. Hofstadter and L. I. Schiff (Ed.), Stanford University Press, 1964. The difference between the two cross sections is due to the interference between the one-photon and the two-photon exchange terms which differ in sign for the two processes.

p, p' representing the corresponding quantities for the proton. With the current defined by

$$j_\mu(x) = e\bar{\psi}_p(x)\gamma_\mu\psi_p(x) + \cdots \tag{26.3}$$

T may be written in the form

$$T = \frac{e}{(2\pi)^3}\,\bar{u}_e(k')\gamma^\mu\, u_e(k)\,\frac{1}{(k'-k)^2 + i\epsilon}\,(\Psi_{p'}, j_\mu(0)\Psi_p) \tag{26.4}$$

Because of the strong interactions of the nucleon with other particles (such as charged mesons), which also interact with the electromagnetic field, the current matrix element is much more complicated than the corresponding matrix element for the electron. In fact, the most general form of the current matrix element[1]

$$(\Psi_{p'}, j_\mu(0)\Psi_p) = \frac{e}{(2\pi)^3}\,\bar{u}(p')[\gamma_\mu F_1{}^p((p'-p)^2) \\ + i\sigma_{\mu\nu}(p'-p)^\nu F_2{}^p((p'-p)^2)]u(p) \tag{26.5}$$

will be involved. Here $F_1{}^p$ and $F_2{}^p$ are completely unknown proton form factors. With the help of

$$\begin{aligned}\bar{u}(p')i\sigma_{\mu\nu}(p'-p)^\nu\, u(p) &= -\tfrac{1}{2}\bar{u}(p')[\gamma_\mu, \not{p}' - \not{p}]\, u(p) \\ &= -\tfrac{1}{2}\bar{u}(p')(\gamma_\mu(\not{p}' - M) - (M - \not{p})\gamma_\mu)\, u(p) \\ &= \bar{u}(p')(2M\gamma_\mu - (p'+p)_\mu)\, u(p)\end{aligned}$$

the matrix element may be rewritten in a form which is more convenient for squaring and trace-taking, namely

$$\frac{e}{(2\pi)^3}\,\bar{u}(p')[\gamma_\mu(F_1{}^p + 2MF_2{}^p) - (p_\mu' + p_\mu)F_2{}^p]u(p) \tag{26.6}$$

The functions $F_1{}^p$ and $F_2{}^p$ depend on

$$t = (p'-p)^2 = (k'-k)^2$$

The threshold values of these functions may be determined from the behavior of a nonrelativistic proton in an external electromagnetic field.

[1] If account is taken of the fact that the spinors $u(p')$ and $u(p)$ obey the Dirac equation, it is easily seen that the most general form can only involve γ_μ, $(p'-p)_\mu$ and $i\sigma_{\mu\nu}(p'-p)^\nu$ multiplied by general functions of p^2, p'^2 and $(p'-p)^2$, standing between the spinors. The current conservation condition, which reads $i(p'-p)_\mu(\Psi_{p'}, j^\mu(0)\Psi_p) = 0$ eliminates the second of the terms. The hermiticity of the current operator requires that the F_i be real.

The interaction energy is

$$\begin{aligned} H_1 &= \int d^3x\, A^\mu_{\text{ext}}(x)(\Psi_{p'}, j_\mu(x)\Psi_p) \\ &= \int d^3x\, A^\mu_{\text{ext}}(x)e^{i(p'-p)x}(\Psi_{p'}, j_\mu(0)\Psi_p) \\ &= \int d^3x e^{-i(\mathbf{p}'-\mathbf{p})\cdot\mathbf{x}} A^\mu_{\text{ext}}(x)(\Psi_{p'}, j_\mu(0)\Psi_p) \end{aligned} \tag{26.7}$$

in the frame in which the electron and the proton do not charge energies. With a scalar potential we get, using

$$\bar{u}(p')\gamma_0\, u(p) = u^+(p')\, u(p) \approx 1$$

and

$$\bar{u}(p')\gamma_i\, u(p) = \frac{p_i + p_i'}{2M} + i\,\frac{(\boldsymbol{\sigma} \times (\mathbf{p}' - \mathbf{p}))_i}{2M}$$

the result

$$H_1 = e\, F_1{}^p(0)\,\frac{1}{(2\pi)^3}\int d^3x e^{-i(\mathbf{p}'-\mathbf{p})\cdot\mathbf{x}} A^0_{\text{ext}}(x)$$

from which we deduce that

$$F_1{}^p(0) = 1 \tag{26.8}$$

since $eF_1{}^p(0)$ is the effective charge of the proton. The reader may recall that similar arguments were used to identify the charge in the calculation of the radiative corrections to Coulomb scattering in Chapter 13.

The interaction with a vector potential yields

$$H_1 = -\left\{\frac{ie}{2M}(F_1{}^p(0) + 2M\, F_2{}^p(0))\boldsymbol{\sigma} \times (\mathbf{p}' - \mathbf{p}) + \frac{e}{2M}(\mathbf{p} + \mathbf{p}')\right\} \cdot \frac{1}{(2\pi)^3}\int d^3x e^{-i(\mathbf{p}'-\mathbf{p})\cdot\mathbf{x}}\mathbf{A}^{\text{ext}}(x) \tag{26.9}$$

The second term is of the form $e\langle\mathbf{v}\rangle \cdot \mathbf{A}$ and represents the interaction of a moving charge with a vector potential. The first term may be rewritten in the form

$$-\frac{e}{2M}(1 + 2M\, F_2{}^p(0))\,\frac{1}{(2\pi)^3}\int d_3x e^{-i(\mathbf{p}'-\mathbf{p})\cdot\mathbf{x}}\,\boldsymbol{\sigma}\cdot\nabla \times \mathbf{A}^{\text{ext}}(x)$$

which is just the expression for the interaction of a magnetic dipole with an external magnetic field. The magnetic moment is given by

$$\frac{e}{2M}(1 + 2M\, F_2{}^p(0))$$

which implies that the anomalous moment, beyond that expected of a particle obeying the Dirac equation, is $eF_2^p(0)$. Thus

$$F_2^p(0) = \frac{\mu_p}{2M} \tag{26.10}$$

with μ_p the anomalous moment in nuclear magnetons.

The cross section for electron proton scattering, which follows from (26.4) with the form (26.6) for the proton current, is easily calculated in standard fashion. The result is usually quoted in the laboratory frame. With E the electron energy and θ the electron scattering angle, we find that

$$\begin{aligned}\frac{d\sigma}{d\Omega} = &\frac{\alpha^2}{4E^2 \sin^4 \theta/2} \frac{\cos^2 \theta/2}{1 + (2E/M) \sin^2 \theta/2} \\ &\times \left\{(F_1^p(t))^2 - \frac{t}{4M^2}\left(4M^2(F_2^p(t))^2 + 2(F_1^p(t) + 2M\, F_2^p(t))^2 \tan^2 \frac{\theta}{2}\right)\right\}\end{aligned} \tag{26.11}$$

It is convenient to define the quantities

$$\begin{aligned} G_E^p(t) &= F_1^p(t) + \frac{t}{2M} F_2^p(t) \\ G_M^p(t) &= F_1^p(t) + 2M\, F_2^p(t) \end{aligned} \tag{26.12}$$

in terms of which the cross section has the form[1]

$$\begin{aligned} \frac{d\sigma}{d\Omega} &= \left(\frac{d\sigma}{d\Omega}\right)_{\text{Mott}} \left\{\frac{(G_E^p(t))^2 - \dfrac{t}{4M^2}(G_M^p(t))^2}{1 - t/4M^2} - \frac{t}{2M^2}(G_M^p(t))^2 \tan^2 \frac{\theta}{2}\right\} \\ \left(\frac{d\sigma}{d\Omega}\right)_{\text{Mott}} &= \frac{\alpha^2}{4E^2 \sin^4 \theta/2} \frac{\cos^2 \theta/2}{1 + (2E/M) \sin^2 \theta/2} \end{aligned} \tag{26.13}$$

with $(d\sigma/d\Omega)_{\text{Mott}}$ representing the cross section for a structureless proton. Note that the ratio $(d\sigma/d\Omega)/(d\sigma/d\Omega)_{\text{Mott}}$, when evaluated for a fixed momentum transfer $-t$, yields a straight line when plotted against $\tan^2 \theta/2$, with the slope and intercept yielding information about the form factors. The rectilinear behavior is a consequence of the one-photon exchange mechanism, and has been checked to good accuracy for $0 < -t < 1.8(\text{BeV}/c)^2$.[2]

[1] Observe that there is no interference between G_E and G_M terms in the cross section.

[2] References may be found in the talk on "The Rosenbluth Formula" by D. R. Yennie, Stanford Conference, *loc. cit.* The quantity t is there expressed in $(\text{fermi})^{-2}$. We shall use $(\text{BeV}/c)^2 = 25\ (\text{fermi})^{-2}$.

It is important to obtain similar information about the neutron. The neutron form factors can be extracted from the data on electron-deuteron scattering, in particular from the reaction

$$e + d \rightarrow n + p + e$$

Lack of space and the availability of an excellent review of this subject excuses us from a discussion of this fairly intricate analysis.[1] Suffice it to say that for values of $-t \lesssim 1.8$ (BeV/c)^2 reasonably reliable values of the neutron form factors are available. Figures 26.2 and 26.3 summarize the data for the nucleonic form factors. It follows from (26.8), (26.10) and (26.12) that

$$G_E{}^p(0) = 1$$
$$G_M{}^p(0) = 1 + \mu_p \cong 2.79 \tag{26.14}$$

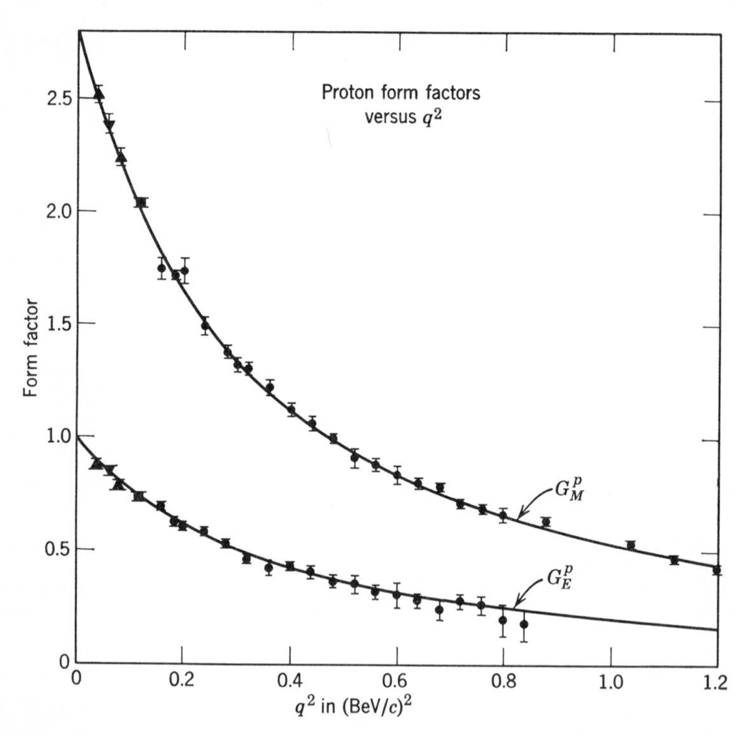

Fig. 26.2. The proton form factors as a function of $q^2 = -t$ in (BeV/c)^2. [From Hughes et al., *Phys. Rev.* **139,** B458 (1965).]

[1] S. D. Drell and F. Zachariasen, *Electromagnetic Structure of Nucleons*, Oxford University Press (1961). For a more recent review, see L. I. Schiff and T. Griffy in *High Energy Physics*, Academic Press (1966).

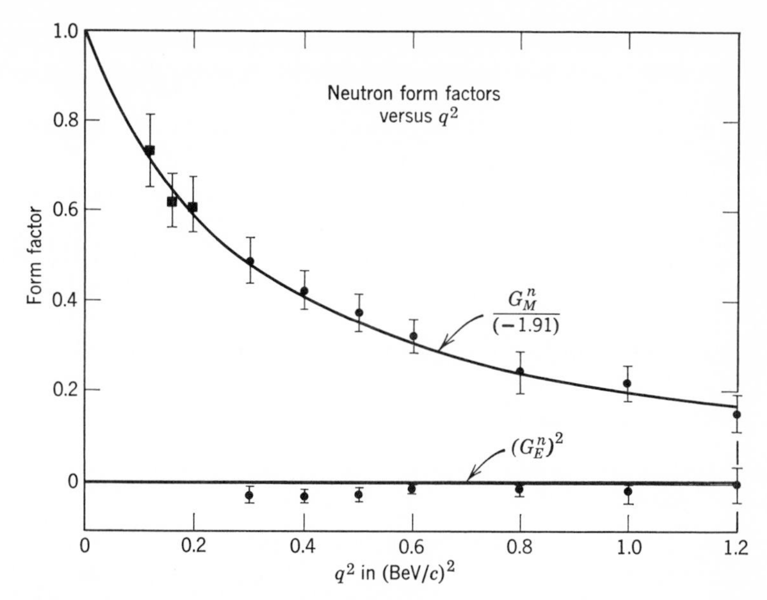

Fig. 26.3. The neutron form factors as a function of $q^2 = -t$ in $(\text{BeV}/c)^2$. (From Hughes et al., *loc. cit.*)

Similarly, for the neutron form factors

$$G_E{}^n(0) = 0$$

$$G_M{}^n(0) = \mu_n \cong -1.91 \tag{26.15}$$

An easy way to summarize the data is by means of a so-called "dipole fit"[1]

$$\begin{aligned} G_E{}^p(t) &\cong \frac{G_M{}^p(t)}{1 + \mu_p} \cong \frac{G_M{}^n(t)}{\mu_n} \\ &\cong -\frac{4M^2}{t}\frac{G_E{}^n(t)}{\mu_n} \cong \frac{1}{(1 - t/(0.71)^2)^2} \end{aligned} \tag{26.16}$$

The last term shows that

$$\left(-\frac{dG_E{}^n(t)}{dt}\right)_{t=0} \cong \frac{\mu_n}{4M^2}$$

i.e.,

$$\left(\frac{dF_1{}^n(t)}{dt}\right)_{t=0} \cong 0 \tag{26.17}$$

Any fit to the data must satisfy (26.17) in addition to (26.14) and (26.15). This condition is a consequence of the experimental fact that measurements

[1] J. R. Dunning Jr., K. W. Chen, A. A. Cone, G. Hartwig, N. F. Ramsey, J. K. Walker and R. Wilson, *Phys. Rev.* **141,** 1286 (1966). This fit is good to $q^2 = 4(\text{BeV}/c)^2$.

of the electron-neutron interaction by the atomic scattering of slow neutrons can be completely explained by the magnetic moment of the neutron.[1]

The dipole fit is by no means a unique fit; it is, in fact, difficult to interpret the shape factor $(1 - t/a^2)^{-2}$.

A more rational approach to the *shape* factor (as opposed to the threshold values of the G's which normalize the form factors) comes from a combined use of analyticity and unitarity[2]. We shall assume that the form factors F_1 and F_2 are real analytic functions of t, with singularities determined by unitarity. Since most of the structure of the nucleons is determined by the strong interactions, it is useful to arrange the calculation in such a way that advantage can be taken of i-spin conservation. As the form of the electric current operator

$$j_\mu(x) = e\bar{\psi}(x)\gamma_\mu \frac{1 + \tau_3}{2} \psi(x) + \cdots \tag{26.18}$$

shows, the current has mixed transformation properties under rotations in i-spin space: part of it transforms as a scalar and part as a third component of an isovector. This fact will be seen to have a natural interpretation within the framework of $SU(3)$.

As a consequence of these mixed transformation properties, the proton and neutron current matrix elements may be expressed in terms of the matrix elements of an isovector and an isoscalar current. We may therefore write

$$\begin{aligned} j_\mu{}^p(x) &= j_\mu{}^s(x) + j_\mu{}^v(x) \\ j_\mu{}^n(x) &= j_\mu{}^s(x) - j_\mu{}^v(x) \end{aligned} \tag{26.19}$$

Corresponding isovector and isoscalar form factors $F_1{}^v(t), \ldots, F_2{}^s(t)$, may be defined. The threshold values are

$$\begin{aligned} F_1{}^v(0) = G_E{}^v(0) = F_1{}^s(0) = G_E{}^s(0) &= \tfrac{1}{2} \\ 2MF_2{}^v(0) = \tfrac{1}{2}(\mu_p - \mu_n) \equiv \mu^v &\cong 1.85 \\ 2MF_2{}^s(0) = \tfrac{1}{2}(\mu_p + \mu_n) \equiv \mu^s &\cong -0.06 \end{aligned} \tag{26.20}$$

The form factors $F_i^{v,s}(t)$ will be assumed to be given by representations of the type

$$F_i(t) = \frac{1}{\pi}\int_{t_i}^{\infty} dt' \,\frac{\varphi_i(t')}{t' - t} \tag{26.21}$$

[1] L. L. Foldy, *Phys. Rev.* **87,** 688 (1952).

[2] The dispersion approach to the form factor calculations was first tried by G. F. Chew, R. Karplus, S. Gasiorowicz and F. Zachariasen, *Phys. Rev.* **110,** 265 (1958) and P. Federbush, M. L. Goldberger and S. B. Treiman, *Phys. Rev.* **112,** 642 (1958).

The lower limit on the integral, as well as $\varphi_i(t')$, are to be determined by unitarity. This is most conveniently used not for the matrix element $(\Psi_{p'}, j_\mu(0)\Psi_p)$ but rather for $(\Psi_{p\bar{p}}, j_\mu(0)\Psi_0)$ which describes the process

$$\gamma \to N + \bar{N}$$

This process is simply related to the one we are interested in; graphically we merely interchange a photon line with a nucleon line and the nucleon pair production process is described by the same form factors analytically continued to a different region. Application of the reduction formalism allows us to write

$$\begin{aligned}(\Psi_{\bar{p}p}^{\text{out}}, j_\mu(0)\Psi_0) &= \frac{e}{(2\pi)^3}\,\bar{u}(p)\{\gamma_\mu(F_1((p+\bar{p})^2) + 2MF_2((p+\bar{p})^2)) \\ &\qquad + (\bar{p}_\mu - p_\mu)F_2((p+\bar{p})^2)\}v(\bar{p}) \\ &= \frac{i}{(2\pi)^{3/2}}\int dx e^{i\bar{p}x}(\Psi_p, \theta(x_0)[\bar{f}(x), j_\mu(0)]\Psi_0)v(\bar{p})\end{aligned} \tag{26.22}$$

For the inverse process we write

$$\begin{aligned}(\Psi_0, j_\mu(0)\Psi_{\bar{p}p}^{\text{in}}) &= \frac{e}{(2\pi)^3}\,\bar{v}(\bar{p})\{\gamma_\mu(F_1((p+\bar{p})^2) + 2MF_2((p+\bar{p})^2)) \\ &\qquad + (\bar{p}_\mu - p_\mu)F_2((p+\bar{p})^2)\}u(p) \\ &= \frac{i}{(2\pi)^{3/2}}\int dx e^{-i\bar{p}x}\,\bar{v}(\bar{p})(\Psi_0, \theta(-x_0)[j_\mu(0), f(x)]\Psi_p)\end{aligned} \tag{26.23}$$

Subtraction of the complex conjugate of this from (26.22) yields, after some manipulation,

$$\begin{aligned}&\frac{e}{(2\pi)^3}\,\bar{u}(p)\{\gamma_\mu \operatorname{Im}(F_1(t) + 2MF_2(t)) + (\bar{p}_\mu - p_\mu)\operatorname{Im} F_2(t)\}v(\bar{p}) \\ &\quad = \frac{1}{2}\cdot\frac{1}{(2\pi)^{3/2}}\int dx e^{i\bar{p}x}(\Psi_p, [\bar{f}(x), j_\mu(0)]\Psi_0)v(\bar{p}) \\ &\quad = \frac{(2\pi)^{5/2}}{2}\sum_n \delta(P_n - p - \bar{p})(\Psi_p, \bar{f}(0)\Psi_n)(\Psi_n, j_\mu(0)\Psi_0)\end{aligned} \tag{26.24}$$ [1]

There will be contributions from the right-hand side when $t = (p + \bar{p})^2$ is large enough. The onset of the imaginary part of $F_i(t)$ depends on the mass of the lowest state Ψ_n which yields a nonvanishing $(\Psi_n, j_\mu(0)\Psi_0)$. It is here that the separation of the current into isoscalar and isovector terms becomes convenient. For the isoscalar current, which is odd under charge

[1] The other term in the commutator contains $\delta(P_n + \bar{p})(\Psi_n, \bar{f}(0)\Psi_0)$. Thus Ψ_n must be a one-nucleon state, for which $(\Psi_{\bar{N}}, \bar{f}(0)\Psi_0) = 0$ (mass renormalization condition of Chapter 13).

conjugation and which does not change under rotation in *i*-spin space

$$Gj_\mu{}^s(0)G^{-1} = -j_\mu{}^s(0) \tag{26.25}$$

Hence the lowest state is a three-pion state, whose lowest (mass)^2, which determines the lower limit on the isoscalar form factor representation integral, is $(3m_\pi)^2$. For the isovector current

$$Gj_\mu{}^v(0)G^{-1} = j_\mu{}^v(0) \tag{26.26}$$

and the lower limit on the integral is $(2m_\pi)^2$.

Before a meaningful representation can be written down, we must know something about the behavior of the form factors at infinity. We unfortunately know nothing. Perturbation theory suggests that the dispersion relations for $F_1(t)$ require one subtraction, and those for $F_2(t)$ do not require any. This is consistent with the general feeling that the charge of the proton is something not readily understood in terms of the extended structure of that particle (the equality with the electron charge in spite of the great difference in structure of the two particles argues against this), whereas the anomalous moment can, qualitatively at least, be understood in terms of the pion cloud. Thus we write

$$\begin{aligned} F_1{}^s(t) &= \frac{1}{2} + \frac{t}{\pi}\int_{9m_\pi^2}^{\infty} dt' \, \frac{\varphi_1{}^s(t')}{t'(t'-t-i\epsilon)} \\ F_1{}^v(t) &= \frac{1}{2} + \frac{t}{\pi}\int_{4m_\pi^2}^{\infty} dt' \, \frac{\varphi_1{}^v(t')}{t'(t'-t-i\epsilon)} \end{aligned} \tag{26.27}$$

and

$$\begin{aligned} F_2{}^s(t) &= \frac{1}{\pi}\int_{9m_\pi^2}^{\infty} dt' \, \frac{\varphi_2{}^s(t')}{t'-t-i\epsilon} \\ F_2{}^v(t) &= \frac{1}{\pi}\int_{4m_\pi^2} dt' \, \frac{\varphi_2{}^v(t')}{t'-t-i\epsilon} \end{aligned} \tag{26.28}$$

If we are interested only in the shape of the form factors, we may want to write a subtracted version of (26.28).

In the calculation of the imaginary part of the form factors, it is interesting to test the assumption that the states n can be well represented by the vector meson resonances, ρ^0 for the isovector form factor, and ω, ϕ for the isoscalar form factor. The graphical representation of this approximation is shown in Fig. 26.4. For the interaction of the vector meson of energy momentum Q with the nucleon, we write a phenomenological matrix element

$$(\Psi_p, \bar{f}(0)\Psi_Q)v(\bar{p}) = \frac{e^\lambda(\eta, Q)}{(2\pi)^3}\,\bar{u}(p)\{\gamma_\lambda F_{VNN}^{(1)} + i\sigma_{\lambda\tau}(p+\bar{p})^\tau F_{VNN}^{(2)}\}v(\bar{p}) \tag{26.29}$$

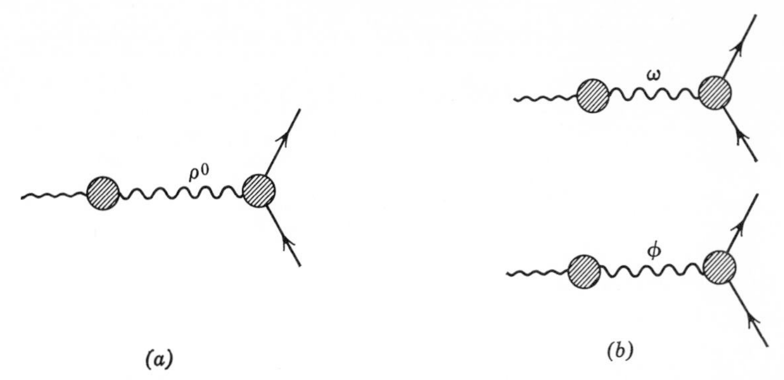

Fig. 26.4. Graphs dominating (*a*) the isovector form factor, (*b*) the isoscalar form factor.

with $e^\lambda(\eta, Q)$ describing the vector meson polarization state. The coupling of the vector meson to the nucleon is also described by form factors, but in the calculation of the imaginary part of F_i only their value at $t = m_V^2$ enter into (26.24). We similarly describe the matrix element for the conversion of a photon into a vector meson by a constant

$$(\Psi_Q, j_\mu(0)\Psi_0) = -\frac{1}{(2\pi)^{3/2}}\, e_\mu(\eta, Q)\, \frac{em_V^2}{2\gamma_V} \tag{26.30}$$

This matrix element cannot be a constant away from $t = m_V^2$. It must, in fact, vanish when $t \to 0$. Perhaps the easiest way to see this is to note that a matrix element of this form leads to a "mixing" of the photon and the vector meson. Such mixing invariably displaces the masses of the particles; the only way to insure that the photon remains massless (gauge invariance) is that the mixing disappear for $t = 0$. With these forms we obtain

$$\begin{aligned}
\operatorname{Im} F_1(t) &= \sum_{\substack{\text{vector}\\ \text{mesons}}} \frac{\pi m_V^{\,2} F_{VNN}^{(1)}}{2\gamma_V}\, \delta(t - m_V^{\,2})\\
\operatorname{Im} F_2(t) &= \sum_{\substack{\text{vector}\\ \text{mesons}}} \frac{\pi m_V^{\,2} F_{VNN}^{(2)}}{2\gamma_V}\, \delta(t - m_V^{\,2})
\end{aligned} \tag{26.31}$$

when the one-particle phase space and the sum over polarizations are worked out in (26.24). The sum extends over all vector mesons (labeled with V, so far) which contribute. We thus obtain for the isovector form factors

$$\begin{aligned}
F_1^{\,v}(t) &\approx \frac{1}{2} + \frac{1}{2}\,\frac{F_{\rho NN}^{(1)}}{\gamma_\rho}\,\frac{t}{m_\rho^{\,2} - t}\\
F_2^{\,v}(t) &\approx \frac{F_{\rho NN}^{(2)}}{2\gamma_\rho}\,\frac{m_\rho^{\,2}}{m_\rho^{\,2} - t}
\end{aligned} \tag{26.32}$$

and for the isoscalar form factors

$$F_1^s(t) = \frac{1}{2} + \frac{1}{2}\frac{F^{(1)}_{\omega NN}}{\gamma_\omega}\frac{t}{m_\omega^2 - t} + \frac{1}{2}\frac{F^{(1)}_{\phi NN}}{\gamma_\phi}\frac{t}{m_\phi^2 - t}$$

$$F_2^s(t) = \frac{F^{(2)}_{\omega NN}}{2\gamma_\omega}\frac{m_\omega^2}{m_\omega^2 - t} + \frac{F^{(2)}_{\phi NN}}{2\gamma_\phi}\frac{m_\phi^2}{m_\phi^2 - t} \tag{26.33}$$

In terms of the $G_E^v(t), \ldots$, we have the result

$$\begin{aligned} G_E^v(t) &= \frac{1}{2} + \frac{1}{2}\frac{F^{(1)}_{\rho NN}}{\gamma_\rho}\frac{t}{m_\rho^2 - t} + \frac{F^{(2)}_{\rho NN}}{2\gamma_\rho}\frac{m_\rho^2}{2M}\frac{t}{m_\rho^2 - t} \\ &= \frac{1}{2}\left(1 - a + \frac{a}{1 - t/m_\rho^2}\right) \end{aligned} \tag{26.34}$$

Similarly

$$\begin{aligned} G_E^s(t) &= \frac{1}{2}\left(1 - b - c + \frac{b}{1 - t/m_\omega^2} + \frac{c}{1 - t/m_\phi^2}\right) \\ G_M^v(t) &= G_M^v(0)\left(1 - a' + \frac{a'}{1 - t/m_\rho^2}\right) \\ G_M^s(t) &= G_M^s(0)\left(1 - b' - c' + \frac{b'}{1 - t/m_\omega^2} + \frac{c'}{1 - t/m_\phi^2}\right) \end{aligned} \tag{26.35}$$

A reasonably good fit to the data is given by[1]

$$\begin{aligned} G_E^v(t) &= \frac{1}{2}\left[+1.23 \pm 0.71 + \frac{3.28 \pm 0.07}{1 - t/(0.75)^2} - \frac{3.51 \pm 0.70}{1 - t/(1.22)^2}\right] \\ G_E^s(t) &= \frac{1}{2}\left[-0.21 \pm 0.12 + \frac{1.93 \pm 0.06}{1 - t/(0.79)^2} - \frac{0.72 \pm 0.10}{1 - t/(1.03)^2}\right] \\ G_M^v(t) &= 2.353\left[+0.31 \pm 0.05 + \frac{2.23 \pm 0.02}{1 - t/(0.75)^2} - \frac{1.54 \pm 0.04}{1 - t/(1.22)^2}\right] \\ G_M^s(t) &= 0.44\left[-0.19 \pm 0.14 + \frac{2.10 \pm 0.08}{1 - t/(0.79)^2} - \frac{0.91 \pm 0.11}{1 - t/(1.03)^2}\right] \end{aligned} \tag{26.36}$$

This shows that in the region $-t \leqslant 1.2(\text{BeV}/c)^2$ the notion of vector meson dominance has some merit. Whether one wants to take the data as implying that there exists another $T = 1$ vector meson of mass 1.2 BeV,

[1] See, for example, E. B. Hughes, T. A. Griffy, M. R. Yearian, and R. Hofstadter, *Phys. Rev.* **139,** B458 (1965).

is a matter of taste. The above model is too crude to allow us to deduce significant information about vector meson-baryon couplings.[1]

Some relations between the isoscalar and isovector parameters emerge from an application of *unitary symmetry*, which is the next topic to be discussed. The electromagnetic current, which contains contributions from all charged hadrons, with a universal coupling (so that all charge form factors for charged particles have magnitude 1 for zero momentum transfer), is known to transform in part as the third component of the *i*-spin current, and in part as an isoscalar current. The Gell-Mann—Nishijima formula, written in the form

$$Q = T_3 + \tfrac{1}{2}Y \tag{26.37}$$

with Y the hypercharge, suggested, soon after attention was drawn to $SU(3)$, that the electric current transforms as a combination of the currents $\mathcal{F}_\mu^{(3)}$ and $\frac{1}{\sqrt{3}}\,\mathcal{F}_\mu^{(8)}$ (see (17.21)).[2] If these appear with equal coefficients, as in the formula (26.37), the current, which we may write as $\mathcal{F}_\mu^{(3)} + \frac{1}{\sqrt{3}}\,\mathcal{F}_\mu^{(8)}$, transforms as the $U = 0$ member of the octet representation.

In terms of vector mesons the $U = 0$ state is the one orthogonal to

$$U\_\Psi_{K^{0*}} = -\sqrt{\frac{1}{2}}\,\Psi_{\rho_0} + \sqrt{\frac{3}{2}}\,\Psi_{\omega_8}$$

Thus the electric current may be considered to be coupled to the combination $\sqrt{3}\,\rho_\mu{}^0 + \omega_{8\mu}$ only. This implies that

$$\frac{1}{\gamma_\rho}:\frac{1}{\gamma_\omega} = \sqrt{3}:1 \tag{26.38}$$

It should be noted that, with our assumption, the unitary singlet vector meson $\omega^{(0)}$ is not coupled to the electromagnetic current. If $\omega^{(0)} - \omega_8$ mixing is taken into account (see Chapter 20), both the physical ω and the

[1] For the current state of the art, see M. Gell-Mann and F. Zachariasen, *Phys. Rev.* **124,** 953 (1961); M. Gell-Mann, D. Sharp, and W. Wagner, *Phys. Rev. Letters* **8,** 261 (1962); D. A. Geffen, *Phys. Rev.* **128,** 374 (1962); R. Dashen and D. Sharp, *Phys. Rev.* **133,** B1585 (1964).

[2] S. Coleman and S. L. Glashow, *Phys. Rev. Letters* **6,** 1423 (1961); N. Cabibbo and R. Gatto, *Nuovo Cimento* **21,** 872 (1961). These transformation properties emerge quite naturally from a quark model of the type proposed by Gell-Mann and Zweig. With other models, additional unitary singlet contributions could appear [M. Nauenberg, *Phys. Rev.* **135,** B1047 (1964)].

physical ϕ are coupled to the electromagnetic current, and we have

$$\frac{1}{\gamma_\rho}:\frac{1}{\gamma_\omega}:\frac{1}{\gamma_\phi} = \sqrt{3}:-\sin\theta:\cos\theta$$
$$\cong 1.73:-0.62:0.78 \qquad (26.39)$$

Further predictions emerge only if we assume pure F-coupling of the vector mesons to the baryons. Although this is a valid assumption at zero-momentum transfer, where the conserved currents may in effect be treated by perturbation theory, it is doubtful whether a quantitative statement can be made in the experimental region.

Unitary symmetry does give a number of predictions, which follow from the assignment of octet transformation properties to the electric current. For example, since the electric current is just the $U = 0$ generator of $SU(3)$, it follows that for any product of current densities their expectation value for all members belonging to a U-spin multiplet is the same. In particular, this means that for the electromagnetic self-energies (in the limit of unitary symmetry, except for electromagnetic effects), the following relations follow

$$\delta m_p = \delta m_{\Sigma^+} \qquad (U = \tfrac{1}{2}\text{ multiplet})$$
$$\delta m_n = \delta m_{\Xi^0} = \frac{1}{4}\,\delta m_{\Sigma^0} + \frac{3}{4}\,\delta m_{\Lambda^0} - \frac{\sqrt{3}}{2}\,\delta_{\Sigma^0\Lambda^0} \qquad (U = 1\text{ multiplet})$$
$$\delta m_{\Sigma^-} = \delta m_{\Xi^-} \qquad (U = \tfrac{1}{2}\text{ multiplet})$$
$$\sqrt{3}\,\delta m_{\Lambda^0} - \sqrt{3}\,\delta m_{\Sigma^0} + 2\,\delta_{\Sigma^0\Lambda^0} = 0 \qquad (U = 0\text{ state}) \qquad (26.40)$$

Here $\delta_{\Sigma^0\Lambda^0}$ is a "mixing mass," which would appear in the form

$$\delta m_{\Sigma^+}\bar{\Sigma}^+\Sigma^+ + \cdots + \delta_{\Sigma^0\Lambda^0}\bar{\Sigma}^0\Lambda^0 + \cdots$$

in an effective Lagrangian. From (26.40) there results the remarkable formula

$$(m_{\Xi^-} - m_{\Xi^0}) = (m_p - m_n) + (m_{\Sigma^-} - m_{\Sigma^+}) \qquad (26.41)$$

The three terms are 6.5 ± 1.5 MeV, -1.3 MeV and 7.7 ± 0.2 MeV, showing that the formula is in surprising agreement with experiment. The formula

$$\delta_{\Sigma^0\Lambda^0} = \frac{1}{\sqrt{3}}(m_p - m_{\Lambda^0} - m_{\Sigma^+} + m_{\Sigma^0}) \cong 1.5\text{ MeV} \qquad (26.42)$$

has found a somewhat esoteric confirmation in the difference in binding energies of the hypernuclei ${}_\Lambda He^4$ and ${}_\Lambda H^4$. Because of the $\delta_{\Sigma^0\Lambda^0}$ term, the physical Σ^0 and Λ^0 particles are no longer pure i-spin states, nor are π^0 and η^0, since there is now mixing, with a mixing angle estimated from

(26.42) to be 0.02 radians. This leads to a $\Lambda^0\Lambda^0\pi^0$ coupling (forbidden by *i*-spin conservation) which is -0.05 times the $\Sigma^0\Lambda^0\pi^0$ coupling,[1] and leads to a violation of charge symmetry in hypernuclei which is in rough agreement with what is observed.

Another prediction of this general type is that

$$(\Psi_0, j_{\mu_1} \cdots j_{\mu_n} \Psi_{\pi^0}) - \sqrt{3}(\Psi_0, j_{\mu_1} \cdots j_{\mu_n} \Psi_{\eta^0}) = 0 \tag{26.43}$$

i.e.,

$$M(\eta^0 \to 2\gamma) = \frac{1}{\sqrt{3}} M(\pi^0 \to 2\gamma)$$

The transformation properties of the current lead to definite predictions regarding relations between photoproduction amplitudes. With the process[2]

$$\gamma + N \to \Delta$$

there is only one coupling, since **10** appears but once in **8** $\otimes$ **8**. Thus the amplitude for

$$\gamma + p \to N^{*+}$$

determines, except for kinematic factors, the amplitudes for decays like

$$Y_1^{0*} \to \Lambda^0 + \gamma$$

Most interesting are the predictions of form factors, hence magnetic moments of the baryons. Since $(\Psi_{B'}, j_\mu(0)\Psi_B)$ contains both F- and D-couplings, we expect to be able to express all baryon form factors in terms of two of them, for example, those of the neutron and the proton. This is most easily done in terms of the 3×3 matrix representation of the octets. The electric current is now represented by the matrix

$$Q = \lambda_3 + \frac{1}{\sqrt{3}}\lambda_8 = \frac{1}{3}\begin{pmatrix} 2 & 0 & 0 \\ 0 & -1 & 0 \\ 0 & 0 & -1 \end{pmatrix} \tag{26.44}$$

The form factors are given by

$$\begin{aligned} &\tilde{F}_1(t)\,\mathrm{tr}\,(\bar{B}'\gamma_\mu[Q, B]) + \tilde{D}_1(t)\,\mathrm{tr}\,(\bar{B}'\gamma_\mu\{Q, B\}) \\ &\qquad + i\tilde{F}_2(t)\,\mathrm{tr}\,(\bar{B}'\sigma_{\mu\nu}q^\nu[Q, B]) + i\tilde{D}_2(t)\,\mathrm{tr}\,(\bar{B}'\sigma_{\mu\nu}q^\nu\{Q, B\}) \end{aligned} \tag{26.45}$$

[1] R. H. Dalitz and F. von Hippel, *Physics Letters* **10**, 155 (1964).

[2] As shown by M. Gourdin and Ph. Salin, *Nuovo Cimento* **27**, 193 (1963), this process is adequately described by the phenomenological interaction

$$\mathfrak{L}_1 = \frac{e\gamma_1}{m_\pi} F^{\mu\nu}(x)(\bar{\psi}_\nu(x)\gamma_\mu\gamma_5\psi(x) + \text{h.c.})$$

with $\gamma_1 \simeq 0.37$, $\psi_\nu(x)$ the spin $\frac{3}{2}$ field and $F^{\mu\nu}(x)$ the electromagnetic field.

Here the condition that at zero momentum transfer the coupling be of pure F-type requires that

$$\tilde{F}_1(0) = 1$$
$$\tilde{D}_1(0) = 0 \qquad (26.46)$$

From (26.45) follow, among others,

$$\mu_{\Sigma^+} = \mu_p$$
$$\mu_{\Lambda^0} = \tfrac{1}{2}\mu_n$$
$$\mu_{\Sigma\Lambda} = -\frac{\sqrt{3}}{2}\mu_n \qquad (26.47)$$

Here $\mu_{\Sigma\Lambda}$ is the mixed moment instrumental in the decay $\Sigma^0 \rightarrow \Lambda^0 + \gamma$. The value of these predictions depends on whether the mass splitting is significant in the dynamics of the magnetic moments. If these moments are determined by the low energy structure, the π-K mass difference is quite significant, and the departure from the predictions listed may be very large. The only data so far[1] are

$$\mu_{\Lambda^0} = -0.73 \pm 0.17 \qquad (26.48)$$

and

$$1 + \mu_{\Sigma^+} = 4.3 \pm 1.5 \qquad (26.49)$$

consistent with the predictions of $SU(3)$.

We conclude this chapter with a brief discussion on the status of the theoretical understanding of the magnitude of the anomalous moments of the nucleons. The methods which have been used so far involve dispersion relations and the use of unitarity, as it appears in (26.24). For the purposes of this calculation we do not treat the vector mesons as stable particles. Thus the low-mass states, which are hopefully the most important ones in (26.24), are the two-pion states for the isovector form factors and the three-pion states for the isoscalar form factors. Since the shape of the isoscalar form factor involves both the ω and the ϕ mesons, the $K\bar{K}$ system is presumably also involved.

It is clear that a matrix element like $(\Psi_p, \bar{f}(0)\Psi_n)$ which represents the process

$$\text{“}n\text{”} \rightarrow N + \bar{N}$$

is unmanageable at this time, unless n is a two-particle state. We could discuss the $K\bar{K}$ contribution to the isoscalar form factor in this way, but since we know that the ω, whose main decay mode is $\omega \rightarrow 3\pi$, is just as

[1] D. A. Hill, K. K. Li, E. W. Jenkins, T. F. Kycia, and H. Ruderman, *Phys. Rev. Letters* **15,** 85 (1965). (This letter contains references to previous measurements); A. D. McInturff and C. E. Roos, *Phys. Rev. Letters* **13,** 246 (1964).

important we must abandon the attempt to calculate μ^s in the conventional way. For the calculation of μ^v only two-pion intermediate states are considered in (26.24). Thus the quantity of interest is

$$\frac{(2\pi)^{5/2}}{2}\sum_n \delta(P_n - p - \bar{p})(\Psi_p, f(0)\Psi_{2\pi})v(\bar{p})(\Psi_{2\pi}, j_\mu(0)\Psi_0)$$

$$= \frac{(2\pi)^{5/2}}{2}\int\frac{d^3k_1}{2\omega_{k_1}}\int\frac{d^3k_2}{2\omega_{k_2}}\,\delta(k_1 + k_2 - p - \bar{p})$$

$$\times\,(\Psi_p, f(0)\Psi^{\text{out}}_{k_1k_2})\,v(\bar{p})(\Psi^{\text{out}}_{k_1k_2}, j_\mu(0)\Psi_0) \quad (26.50)$$

First, let us consider the matrix element $(\Psi^{\text{out}}_{k_1k_2}, j_\mu(0)\Psi_0)$. This is related to the pion electromagnetic form factor, defined by[1]

$$(\Psi^{\text{out}}_{k_1k_2}, j_\mu(0)\Psi_0) = \frac{ie}{(2\pi)^{3/2}}\,e_{3\alpha\beta}(k_1 - k_2)_\mu F_\pi(t) \quad (26.51)$$

The kinematic factors are just those that remain in lowest-order perturbation theory when $F_\pi(t) = 1$. Again the absence of charge renormalization is equivalent to the condition

$$F_\pi(0) = 1 \quad (26.52)$$

The pion form factor $F_\pi(t)$ is evidently also an analytic function of t in the complex t-plane cut from $4m_\pi^2$ to $+\infty$. If we consider only two-pion intermediate states in the unitarity condition, a form for $F_\pi(t)$ can be obtained rather quickly. It is important to remember that the two-pion state in $(\Psi^{\text{out}}_{k_1k_2}, j_\mu(0)\Psi_0)$ must be a state characterized by $T = J = 1$.

We shall first prove that the phase of $F_\pi(t)$ is that of the pion-pion scattering amplitude in the $T = J = 1$ state. Time-reversal invariance implies that

$$(\Psi^{\text{in}}_{2\pi}, j_0(0\Psi)_0) = (\Psi^{\text{in}}_{2\pi}, j_0(0)\Psi_0)'$$
$$= (\Psi^{\text{out}}_{2\pi}, j_0(0)\Psi_0)^* \quad (26.53)$$

(In an angular momentum representation, motion reversal will not change $\Psi_{2\pi}$, except for the "in" $\leftrightarrow$ "out"). Now

$$(\Psi^{\text{out}}_{2\pi}, j_0(0)\Psi_0) = (\Psi^{\text{in}}_{2\pi}, Sj_0(0)\Psi_0) = e^{2i\delta_{\pi\pi}(T=J=1)}(\Psi^{\text{in}}_{2\pi}, j_0(0)\Psi_0)$$
$$= e^{2i\delta_{\pi\pi}}(\Psi^{\text{out}}_{2\pi}, j_0(0)\Psi_0)^*$$

so that

$$F_\pi(t) = e^{2i\delta_{\pi\pi}}F_\pi^*(t) \quad (26.54)$$

which establishes the result.[2] Now, if the $\pi\pi$ scattering amplitude in the

[1] The pion i-spin component labels α and β were omitted from (26.50) for the sake of brevity. The sum in the unitarity expression always extends over all relevant quantum numbers.

[2] This is a special case of what is usually called the Fermi-Watson-Aidzu theorem.

$T = J = 1$ state is given by $N(t)/D(t)$, where $D(t)$ only has the right-hand cut and $N(t)$ only the left-hand cut (as in Chapter 24), it is clear that

$$F_\pi(t) = \frac{D(0)}{D(t)} \tag{26.55}$$

satisfies the phase condition, the analyticity properties, and the normalization condition (26.52).

The reader can easily check that

$$(\Psi^{\text{out}}_{k_1k_2}, \Psi^{\text{in}}_{\bar{p}p}) = -(2\pi)^{5/2} i\, \delta(k_1 + k_2 - p - \bar{p})\, \bar{v}(\bar{p})(\Psi^{\text{out}}_{k_1k_2}, f(0)\Psi_p) \tag{26.56}$$

so that

$$\bar{v}(\bar{p})(\Psi^{\text{out}}_{k_1k_2}, f(0)\Psi_p) = (2\pi)^{3/2}\, T(k_1k_2, p\bar{p}) \tag{26.57}$$

the T-matrix for the process $N + \bar{N} \to 2\pi$. This is of the form

$$T_{\beta\alpha}(k_1k_2, p\bar{p}) = \frac{1}{16\pi^5} \bar{v}(\bar{p}) \left[-A_{\beta\alpha}(s, t, u) + \frac{k_2 - k_1}{2} B_{\beta\alpha}(s, t, u) \right] u(p) \tag{26.58}$$

where $A_{\beta\alpha}$ and $B_{\beta\alpha}$ are known explicitly, if we assume that they are represented by the one-nucleon terms and by an isobar term with the $N^*_{3/2 3/2}$ treated as a stable particle. There is one major additional effect, however. As illustrated in Fig. 26.5, the two pions in the final state interact, and in the $T = J = 1$ state they interact strongly. This needs to be taken into account, and can be done so in each partial wave. We leave it to the reader to convince himself that, if the amplitude for $N + \bar{N} \to 2\pi$ in the $T = J = 1$ state is of the form[1]

$$f_{N\bar{N}\to\pi\pi}(t) = \frac{1}{\pi} \int_{\text{left-hand cut}} dt' \frac{\Delta_L f(t')}{t' - t} \tag{26.59}$$

in the absence of the $\pi\pi$ interaction, then in the presence of the $\pi\pi$ interaction the amplitude is[2]

$$f_{N\bar{N}\to\pi\pi}(t) = \frac{1}{D(t)} \frac{1}{\pi} \int_{\text{left-hand cut}} dt' \frac{\Delta_L f(t') \cdot D(t')}{t' - t} \tag{26.60}$$

For a resonance in the $\pi\pi$ amplitude, $D(t)$ will look like a Breit-Wigner resonance denominator. If $D(t')$, which varies slowly away from $t = m_\rho^2$, can be neglected inside of (26.60), the net result is just the dispersion

[1] Recall that we are only discussing pole terms in $N\bar{N} \to \pi\pi$, and thus there is no right-hand cut in the t-channel without final state interaction.

[2] See W. R. Frazer and J. R. Fulco, *Phys. Rev.* **117**, 1603 (1960); this solution, as that in (26.55), is subject to the usual *CDD* ambiguity mentioned in Chapter 24. Again $f_{N\bar{N}\to\pi\pi}(t)$ must have the same phase as $f_{\pi\pi}(t)$ in the same (J, T) state.

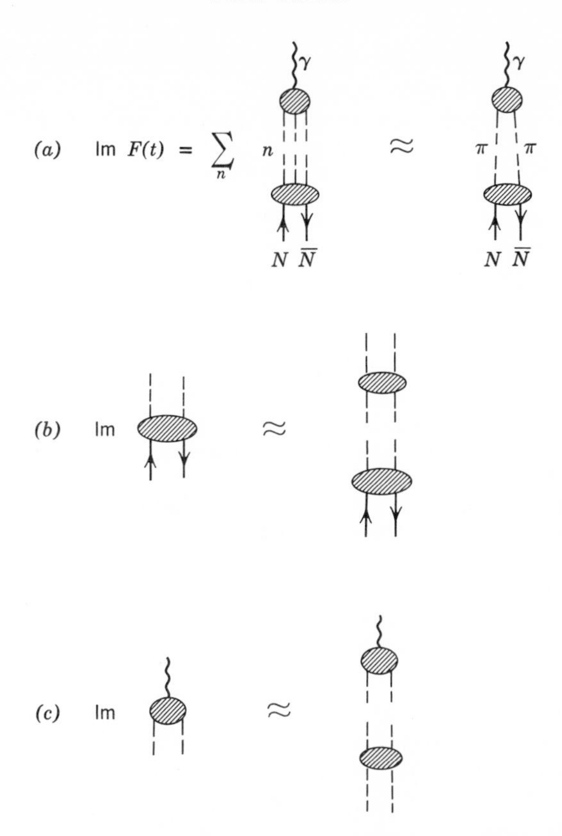

Fig. 26.5. Schematic representation of the isovector form factor calculation.

calculation of the graph shown in Fig. 26.6, with the form factor multiplied by $|D(0)/D(t)|^2$, which is the ρ-meson propagator in the limit of a very sharp resonance. This factor does not affect the calculation of the vector magnetic moment very much.[1] For the rest, the isobar exchange leads to

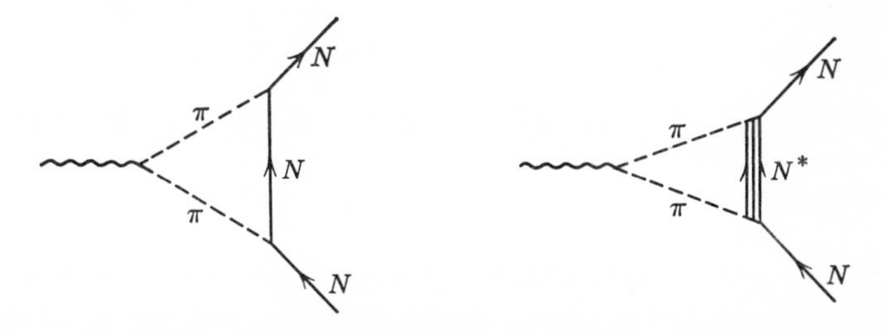

Fig. 26.6. Perturbation calculation of the form factor.

[1] J. S. Ball and D. Y. Wong, *Phys. Rev.* **130,** 2112 (1963).

a divergent integral, and it is necessary to cut it off. If this is done at $4M^2$, reasonable agreement with experiment is obtained. It is difficult to take seriously the numerical value obtained from the calculation[1]

$$\mu^v = 1.90$$

(to be compared with the experimental value 1.85). Nevertheless, the result is encouraging from the point of view that the magnetic moments can be calculated with the inclusion of low-lying mass states alone.[2]

PROBLEMS

1. Use lowest-order perturbation in (26.58), with neglect of the $\pi\pi$ interaction, to calculate the right-hand side of (26.50) and hence μ^v to this order. Use $g^2/4\pi \simeq 15$ to obtain a numerical estimate.
2. Write down a form for an invariant, parity conserving $\omega\rho\pi$ coupling. If the decay $\omega \to 3\pi$ is determined by the graph

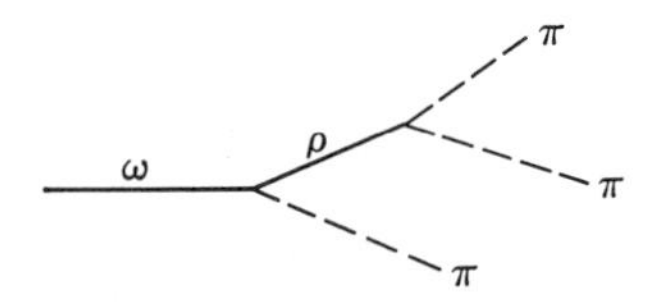

(with the pions properly symmetrized), estimate the strength of the coupling, given that $\Gamma_\omega \simeq 10$ MeV, and using the $\rho \to 2\pi$ decay as described by

$$\mathcal{L}_I = f_\rho \phi T^\alpha \partial_\mu \phi \rho_\alpha{}^\mu$$

with $f_\rho{}^2/4\pi \simeq 2.4$

3. If we assume that the π^0 decay is determined by the graph

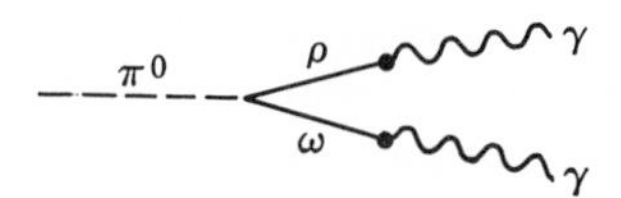

with the vector meson-photon coupling given by (26.30) and the $SU(3)$ relation

$$\frac{1}{\gamma_\rho} : \frac{1}{\gamma_\omega} = \sqrt{3} : 1$$

(which neglects $\omega - \phi$ mixing), find γ_ρ from the known lifetime of the π^0 and the $\omega\rho\pi$ coupling determined in Problem 2.

4. Calculate the rate for

$$\omega \to e^+ + e^-$$

[1] P. Federbush, M. L. Goldberger, and S. B. Treiman, *loc. cit.*

[2] More recently calculations relating the anomalous moments to the photomeson production amplitudes have provided similar encouragement. S. D. Drell and H. R. Pagels, *Phys. Rev.* **140,** B397 (1965). S. Fubini, G. Furlan and C. Rossetti, *Nuovo Cimento*, **43,** 161 (1966).

27

One-Particle Exchange Mechanism

With the availability of high energy particles from existing accelerators, much information has been obtained on inelastic processes, such as

$$\pi^+ + p \rightarrow \pi^+ + \pi^- + \pi^+ + p$$

for example. At first sight it might appear that what we have discussed up to this point leaves us quite unprepared to cope with such processes. Indeed, if one insists on treating the final state as a multiparticle state, the large number of variables make even a representation of a matrix element for such a process unmanageable. Fortunately, it turns out that when correlations between different particles are explored, pairs and sometimes triplets of particles appear to be decay products of resonances. Thus, as a first approximation, a process like the one we started with may, in a certain domain of incident energy and momentum transfer, be well described as[1]

$$\pi^+ + p \rightarrow \rho^0 + N^{*++}$$

with the ρ^0 and N^* in the final state subsequently decaying into their usual end-products. It is an experimental fact that in most inelastic collisions involving 3–5 particles in the final state, in the 2–10 BeV energy region, "resonance production" is very common, so that, in effect, the final states involve only two particles. As another example, in the process

$$K^+ + p \rightarrow K^0 + \pi^+ + p$$

at 3 BeV/c, it turns out that roughly half of the time the reaction is really

$$K^+ + p \rightarrow K^{*+}(891) + p$$

and half the time it is

$$K^+ + p \rightarrow K^0 + N^{*++}(1238)$$

with very few nonresonant events, as the Dalitz plot (Fig. 27.1) shows.

[1] This has been observed to be the case for 2–3 BeV/c, 4 BeV/c, and 8 BeV/c incident π^+. Much of the information quoted in this chapter comes from the excellent review article by J. D. Jackson, *Rev. Mod. Phys.* **37**, 484 (1965), which also contains many references.

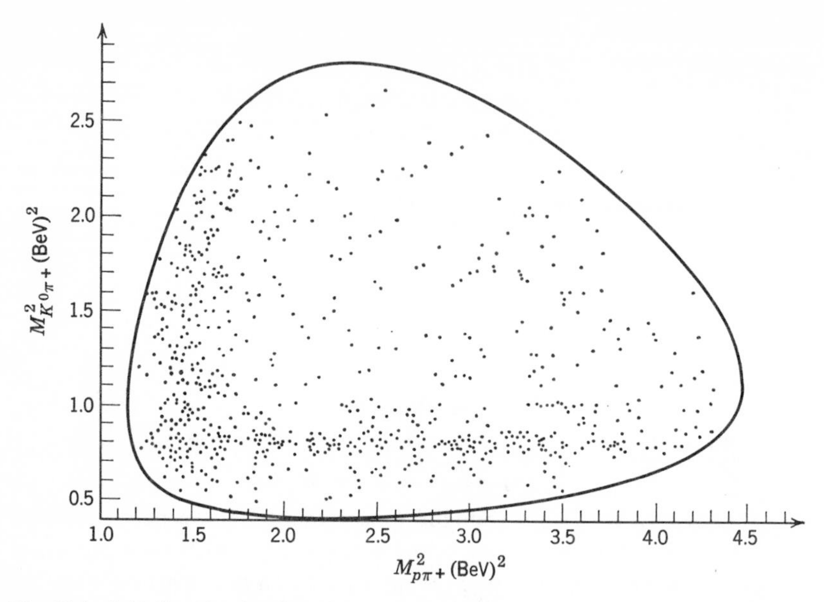

Fig. 27.1. A Dalitz plot for the $K^+ + p \to K^0 + p + \pi^+$ events at 3 BeV/c. Note that most events lie in one or the other of the N^* or K^* bands. [From Ferro-Luzzi et al., *Nuovo Cimento* **36,** 1101 (1965).]

Even with these simplifying features we are not well equipped to study high-energy events, except for one class of processes. These are the processes characterized by a low-momentum transfer. For such events—and these, in fact, turn out to be the majority of events—the value of t is still close to the unphysical value $t = m_\pi^2$, at which the one-pion exchange graph (Fig. 27.2) would give the dominant (in fact, infinite) contribution. If we make the approximation that the process is dominated by the one-pion exchange term, even in the physical region $|t| \approx 0.1$–0.2 (BeV/c)^2, we can actually calculate the cross section for the process in terms of low-energy matrix elements. For example, the matrix element for the process $\pi^+ + p \to \rho^0 + N^{*++}$ is given by

$$T(N^*\rho;\, \pi N) = (\Psi_\rho, j_\pi(0)\Psi_\pi)\frac{1}{t - m_\pi^{\;2} + i\epsilon}(\Psi_{N^*}, j_\pi(0)\Psi_N) \tag{27.1}$$

The matrix element $(\Psi_\rho, j_\pi(0)\Psi_\pi)$ is closely related to the matrix element for the decay

$$\rho^0 \to \pi^+ + \pi^-$$

except that the mass of one of the pions is not m_π^2, but rather t. If we assume that the matrix element is not very sensitive to variations in the mass of the meson, we can make the replacement

$$(\Psi_{\rho}, j_{\pi}(0)\Psi_{\pi}) = \frac{1}{(2\pi)^3} 2f_{\rho\pi\pi}\epsilon^{\mu}(\rho)q_{\mu} \tag{27.2}$$

with q the incident pion four-momentum and $\epsilon^{\mu}(\rho)$ the ρ-meson polarization vector, which satisfies the condition

$$\epsilon^{\mu}(\rho)q_{\mu}' = 0 \tag{27.3}$$

If we make the same approximation for the πNN^* vertex, we can write[1]

$$(\Psi_{N^*}, j_{\pi}(0)\Psi_{N}) = \frac{f_{N^*N\pi}}{(2\pi)^3}\, \bar{u}_{\mu}(p')\, u(p)(q^{\mu} - q'^{\mu}) \tag{27.4}$$

The approximation used in relating the current matrix elements to physical current matrix elements is the one which was used with some success in high-energy electrodynamics and which is known as the Weizsäcker-Williams approximation (Chapter 12).

If the process is

$$\pi + N \rightarrow \rho + N$$

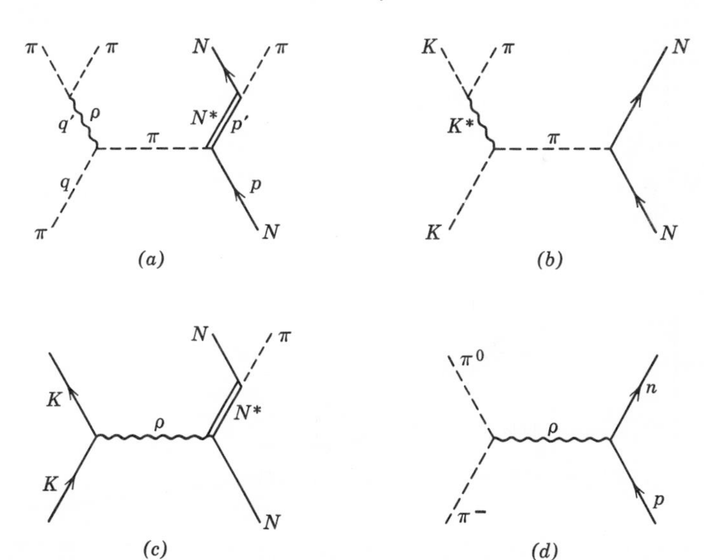

Fig. 27.2. One-particle exchange graphs for the processes (*a*) $\pi + p \rightarrow \rho + N^*$; (*b*) $K^+ + N \rightarrow K^* + N$; (*c*) $K^+ + N \rightarrow K^0 + N^*$; (*d*) $\pi^- + p \rightarrow \pi^0 + n$.

[1] The Rarita-Schwinger formalism for the description of spin $\frac{3}{2}$ particles is briefly described in the appendix to Chapter 25. See also the footnote on page 310.

in (27.1) we replace $(\Psi_{N^*}, j_\pi(0)\Psi_N)$ with $(\Psi_N, j_\pi(0)\Psi_N)$ and then use

$$(\Psi_N, j_\pi(0)\Psi_N) = \frac{ig}{(2\pi)^3}\,\bar{u}(p')\gamma_5\, u(p) \tag{27.5}$$

This is the matrix element (with obvious modifications) that would be used for the graph shown in Fig. 27.2*b*. The graph shown in Fig. 27.2*c* involves the exchange of a ρ-meson, and will be discussed later.

We can now calculate the cross section for the process

$$\pi^+ + p \rightarrow \rho^+ + p$$

for example, with the help of (27.1), (27.2), and (27.5). The result of a

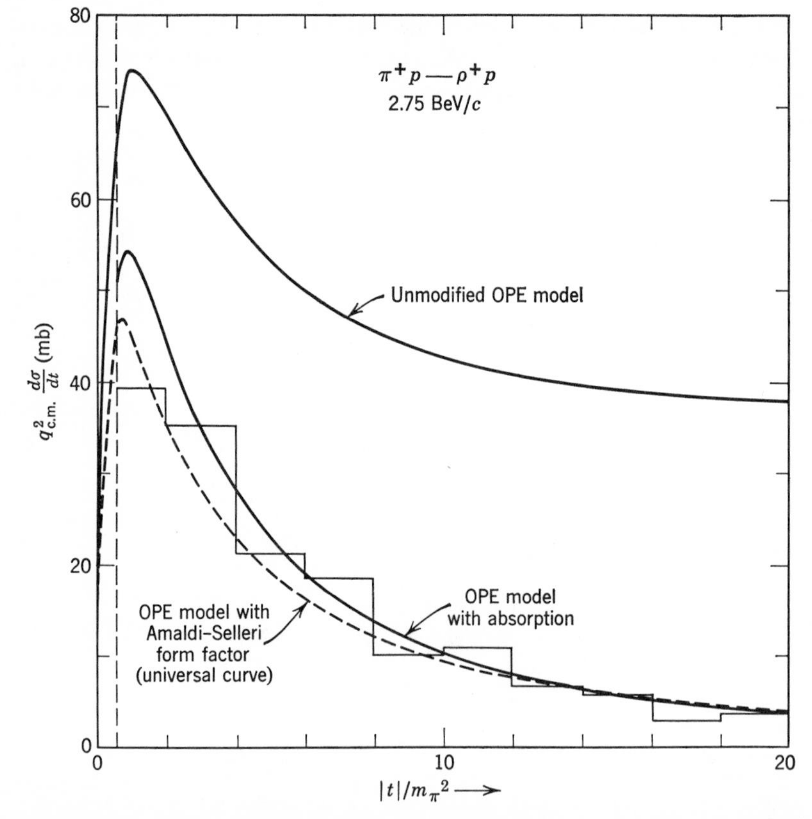

Fig. 27.3. Plot of the differential cross section for $\pi^+ + p \rightarrow \rho^+ + p$ at 2.75 BeV/*c*. The histogram represents the data, the upper curve the result of (27.6), and the lower solid curve the result of a one-pion exchange calculation with absorption. [From J. D. Jackson, *Rev. Mod, Phys.* **37**, 484 (1965).]

straightforward perturbation calculation is

$$\frac{d\sigma}{dt} = \frac{\pi t}{sq_{\text{cm}}^2}\frac{f_{\rho\pi\pi}^2}{4\pi}\frac{g^2}{4\pi}\frac{(t-(m_\rho-m_\pi)^2)(t-(m_\rho+m_\pi)^2)}{4m_\rho^{\,2}(t-m_\pi^{\,2})^2} \tag{27.6}$$

Figure 27.3 shows the comparison with experiment. It is clear that, although the forward peaking is qualitatively reproduced, the experimental peaking is much stronger.

It is not obvious that the failure does not lie in the restriction to a one-pion exchange. Treiman and Yang[1] gave a necessary condition for the validity of the OPE model. They pointed out that one consequence of an exchange of a spinless particle is that the plane formed by the pion momentum and the ρ-momentum on one hand and that formed by the initial and final momentum of the nucleon are uncorrelated. This comes about because a spinless particle cannot transmit any information involving a direction. The Treiman-Yang test is generally satisfied for low-momentum transfers, but not for higher ones, which is encouraging for the model (Fig. 27.4).

More detailed tests were suggested by Gottfried and Jackson.[2] As an illustration let us consider ρ^0-meson production. A convenient coordinate system in the ρ^0 rest frame is defined by taking the incident pion direction $\mathbf{p}_\pi$ as z-axis, and $\mathbf{p}_f \times \mathbf{p}_\pi$, where $\mathbf{p}_f$ is the direction of the outgoing nucleon or N^*, as the direction of the y-axis, as shown in Fig. 20.2. The ρ^0-meson state vector is of the form $\Psi_\alpha = \sum_{m=1}^{3} C_M{}^\alpha \Psi_{JM}$ $(J = 1; \alpha = 1, 2, 3)$, and the decay amplitude is given by

$$(\Phi_{p1,0;-p1,0}, S\Psi_\alpha) = \mathcal{S}_0 \sum_M C_M{}^\alpha\, D_{M0}^{(1)*}(R_{p_1,p_\pi}) \tag{27.7}$$

where $\mathbf{p}_1$ is the momentum of one of the mesons from $\rho^0 \to \pi^+ + \pi^-$ in the ρ^0 rest frame. The angular distribution of the pions is given by

$$\begin{aligned}\sum_\alpha |(\Phi_{p1,0;-p1,0}, S\Psi_\alpha)|^2 &= |\mathcal{S}_0|^2 \sum_\alpha \sum_{MM'} C_M{}^\alpha C_{M'}^{\alpha *}\, D_{M0}^{(1)}(R_{p_1,p_\pi})\, D_{M'0}^{(1)}(R_{p_1,p_\pi}) \\ &= |\mathcal{S}_0|^2 \sum_{MM'} \rho_{MM'} e^{i(M-M')\phi}\, d_{M0}^{(1)}(\theta)\, d_{M'0}^{(1)}(\theta)\end{aligned} \tag{27.8}$$

where

$$\rho_{MM'} = \sum_\alpha C_M{}^\alpha C_{M'}^{\alpha *} \tag{27.9}$$

is the density matrix of the ρ-meson. We leave it to the reader to convince himself that parity conservation in the decay implies that

$$\rho_{-M,-M'} = (-1)^{M-M'} \rho_{MM'} \tag{27.10}$$

[1] S. B. Treiman and C. N. Yang, *Phys. Rev. Letters* **8,** 140 (1962).

[2] K. Gottfried and J. D. Jackson, *Nuovo Cimento* **33,** 309 (1964).

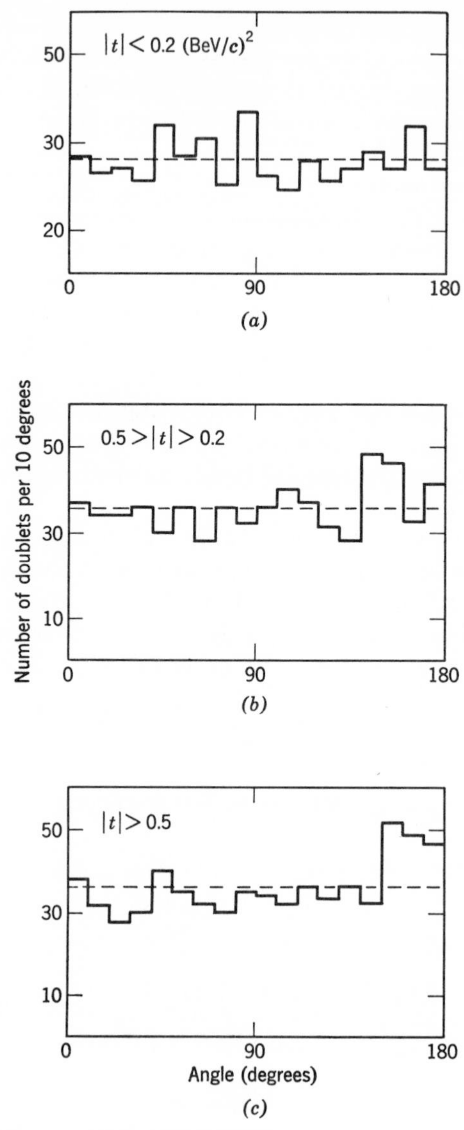

Fig. 27.4. Distribution of Treiman-Yang angle for various regions of $|t|$, the square of the momentum transfer to π^+p in the reaction $p + \bar{p} \rightarrow p + \bar{p} + \pi^+ + \pi^-$. [From T. Ferbel et al., Phys. Rev. **138**, B1528 (1965).]

With the help of this condition, and

$$\rho_{MM'} = \rho^*_{M'M} \tag{27.11}$$

and using

$$d^{(1)}_{00}(\theta) = \cos\theta, \qquad d^{(1)}_{11}(\theta) = \tfrac{1}{2}(1 + \cos\theta)$$
$$d^{(1)}_{10}(\theta) = -d^{(1)}_{-1,0}(\theta) = d^{(1)}_{0,-1}(\theta) = -\frac{1}{\sqrt{2}}\sin\theta \tag{27.12}$$

the angular distribution of the pions will in general be of the form

$$W(\theta, \phi) = \rho_{00}\cos^2\theta + \rho_{11}\sin^2\theta - \rho_{1,-1}\sin^2\theta\cos 2\phi - \sqrt{2}\,\mathrm{Re}\,\rho_{10}\sin 2\theta\cos\phi \tag{27.13}$$

Now for one-pion exchange the initial state leading to the production of the meson is a two-pion state. In the coordinate system chosen above, the initial state must have $l = 1$ and $M = 0$, since there are no spins, and the orbital angular momentum cannot have nonvanishing M-values along the quantization axis which we have chosen. Thus *only* ρ_{00} *can be nonvanishing for pure one-pion exchange.*[1] In many reactions this seems to be the case (see Fig. 27.5).

The "angular" evidence thus favors the one-pion exchange model, and the source of the difficulty must be sought elsewhere. Ferrari and Selleri[2] have tried to improve the theory within the framework of the one-pion exchange model by allowing for some t-dependence in the pion propagator, as well as in the two-pion current matrix elements. The result of this modification is that the cross section in (27.6) is multiplied by $|F(t)|^2$, where $F(0) = 1$ and $F(t) \to 0$ as $t \to \infty$. Although such a form factor was found, and it works over a reasonably large range of energies, the actual structure of the form factor

$$F(t) \sim \frac{A}{t - (2 \cdot 4m_\pi)^2} \tag{27.14}$$

cannot be reconciled with the qualitative notion that such form factors should be associated with some resonance that plays a role in the structure of the vertex. Here the "resonance mass," $2.4m_\pi$, is just too small.

[1] When the particle exchanged is a vector meson (e.g., a ω), the production process $\omega + \pi \to \rho$ in the ρ rest frame can only occur in the P-orbital angular momentum state, so that angular momentum and parity can be conserved. The production matrix element is thus proportional to $C(1, 1, 1; M, 0, M)$. This vanishes for $M = 0$, so that for vector-meson exchange the angular distribution must be of the form

$$\sin^2\theta(\rho_{11} - \rho_{1,-1}\cos 2\phi).$$

[2] E. Ferrari and F. Selleri, *Suppl. Nuovo Cimento* **24,** 453 (1962). This paper contains references to a large number of studies of the one-pion exchange effect.

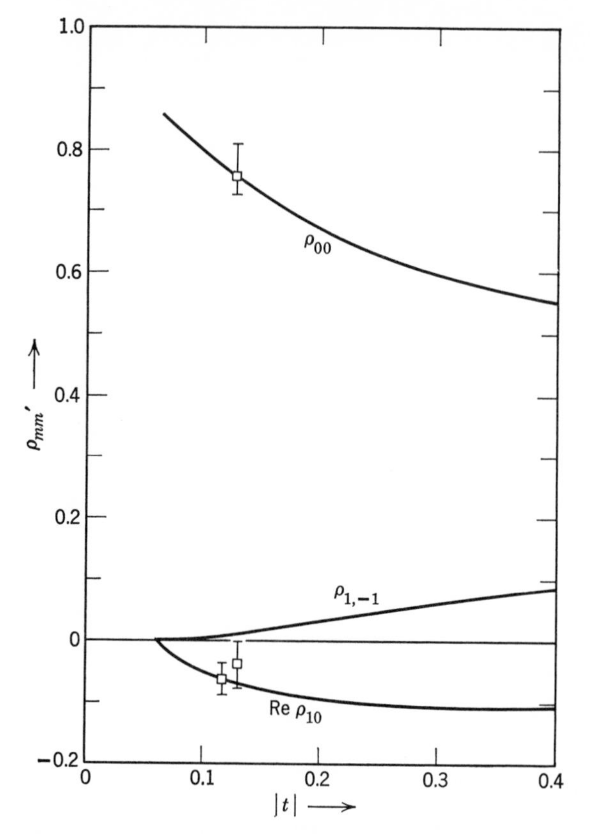

Fig. 27.5. Comparison of measured and calculated values of the elements of the ρ^0-meson density matrix, in the reaction $\pi^+ + p \to \rho^0 + N^*$ at 4 BeV/c. A dominant ρ_{00} component suggests the exchange of a spinless particle. The deviations from $\rho_{00} = 1$ are due to absorption. (From J. D. Jackson, *loc. cit.*)

A different way of explaining the much more drastic peaking in the forward direction than is accounted for by (27.6) is the following:

When use is made of the relation

$$Wq_{\text{cm}} = Mq_L \tag{27.15}$$

where q_L is the momentum of the incident pion in the laboratory system, we see that

$$q_L^{\ 2} \frac{d\sigma}{dt} = \frac{\pi t}{4M^2 m_\rho^{\ 2}} \frac{f_{\rho\pi\pi}^2}{4\pi} \frac{g^2}{4\pi} \frac{(t - (m_\rho - m_\pi)^2)(t - (m_\rho + m_\pi)^2)}{(t - m_\pi^{\ 2})^2} \tag{27.16}$$

is a universal form, independent of energy. The total cross section for the reaction is

$$\sigma = \int_{t_{\min}}^{t_{\max}} dt \frac{d\sigma}{dt}$$

The limits of integration are given by $\cos \theta_L = \pm 1$ in the expression

$$t = m_\rho^2 + m_\pi^2 - 2\omega_\pi(q_L)\, \omega_\rho(q_L') + 2q_L q_L' \cos \theta_L \qquad (27.17)$$

Thus at high energies

$$\sigma \sim \sigma_0 q_L'^2 \sim W^2 \qquad (27.18)$$

a behavior which shows that unitarity is being violated.[1] A decomposition of the cross section (27.6) into partial wave contributions confirms this. The manner in which unitarity is being violated can be guessed on physical grounds. Our peripheral model treats the longest-range force (exchange of the lightest particle) only. This may be adequate for the high angular momentum partial waves, which, because of the centrifugal barrier, never penetrate the short-distance domain. The low partial waves which can, so to speak, get close to the core of the nucleon, can react in other ways. Unitarity implies that when, for a given partial wave, a large number of reaction channels are open, the amplitude in any one of them is reduced. Thus the inclusion of absorption effects for the low partial waves will leave the strongly forward-peaked high partial wave contributions to our reaction unchanged, but it will cut down the part of the amplitude that varies slowly with angle. This is just what is observed experimentally.

The implementation of the program is not easy. The most popular procedure has been:[2]

1. The decomposition of the reaction amplitude into helicity amplitudes (see Chapter 4, pages 82 and 83); this is more general than the partial wave decomposition, which has to be worked out differently for each final state.

2. The modification of each helicity amplitude by a factor which takes into account the reduction in the amplitude due to open channels for the two initial particles, as well as by a factor which does the same for the two particles in the final state.

[1] Limitations of space prevent us from showing that, if we assume the validity of the Mandelstam representation, then $|\sigma/\log W| <$ const. follows from unitarity. [M. Froissart, *Phys. Rev.* **123,** 1053 (1961).]

[2] N. Sopkovitch, *Nuovo Cimento* **26,** 186 (1962); L. Durand III and Y. Chiu, *Phys. Rev Letters* **12,** 399 (1964); M. Ross and G. Shaw, *Phys. Rev. Letters* **12,** 627 (1964); A. Dar and W. Tobocman, *Phys. Rev. Letters* **12,** 511 (1964); K. Gottfried and J. D. Jackson, *Nuovo Cimento* **34,** 735 (1964).

We shall illustrate the procedure for the process $\pi^+ + p \to \rho^+ + p$. The Born approximation for this process is given by

$$\frac{2igf_{\rho\pi\pi}}{(2\pi)^6}\frac{\bar{u}^{(r)}(p')\gamma_5\, u^{(s)}(p)}{t - m_\pi^{\ 2}}\, e_\mu(\kappa, q')q^\mu \tag{27.19}$$

Here (r), (s), and (κ) represent the values of the z-components of the spin of the particles involved. The corresponding amplitude for the process involving particles with helicities (λ_2), (λ_1), and (λ) is clearly

$$\frac{2igf_{\rho\pi\pi}}{(2\pi)^6}\frac{1}{t - m_\pi^{\ 2}}\,\bar{u}(p', \lambda_2)\gamma_5\, u(p, \lambda_1)\, e_\mu(\lambda, q')q^\mu \tag{27.20}$$

Our first task is, therefore, to construct the helicity wave functions.

A vector meson at rest has as its polarization vector

$$e_\mu(+1, 0) = -\frac{1}{\sqrt{2}}\begin{pmatrix} 0 \\ 1 \\ i \\ 0 \end{pmatrix} \qquad e_\mu(0, 0) = \begin{pmatrix} 0 \\ 0 \\ 0 \\ 1 \end{pmatrix} \qquad e_\mu(-1, 0) = \frac{1}{\sqrt{2}}\begin{pmatrix} 0 \\ 1 \\ -i \\ 0 \end{pmatrix} \tag{27.21}$$

A Lorentz transformation, which gives the meson momentum q' in the z-direction, yields

$$e_\mu(\lambda, q'\hat{\mathbf{i}}_z) = L_\mu^{\ \nu}(\hat{\mathbf{i}}_z q')\, \epsilon_\nu(\lambda, 0) \tag{27.22}$$

i.e.,

$$e_\mu(\pm 1, q'\hat{\mathbf{i}}_z) = \mp\frac{1}{\sqrt{2}}\begin{pmatrix} 0 \\ 1 \\ \pm i \\ 0 \end{pmatrix} \qquad e_\mu(0, \hat{\mathbf{i}}_z q') = \begin{pmatrix} q'/m_\rho \\ 0 \\ 0 \\ \omega'/m_\rho \end{pmatrix} \tag{27.23}$$

This operation must be followed by a rotation which takes the z-axis into the direction $\hat{\mathbf{q}}'$, which in our coordinate system is characterized by the spherical angles $(\theta, 0)$. Since for such a rotation

$$R(0, \theta, 0) = e^{-i\theta J_2} \tag{27.24}$$

we obtain the helicity polarization vectors

$$e_\mu(\pm 1, q') = \mp\frac{1}{\sqrt{2}}\begin{pmatrix} 0 \\ \cos\theta \\ \pm i \\ -\sin\theta \end{pmatrix} \qquad e_\mu(0, q') = \begin{pmatrix} q'/m_\rho \\ \dfrac{\omega'}{m_\rho}\sin\theta \\ 0 \\ \dfrac{\omega'}{m_\rho}\cos\theta \end{pmatrix} \tag{27.25}$$

and

$$e_\mu(\lambda, q')q^\mu = \begin{cases} -\dfrac{q}{\sqrt{2}} \sin\theta & \lambda = +1 \\ \dfrac{q}{m_\rho}(q' - \omega' \cos\theta) & \lambda = 0 \\ \dfrac{q}{\sqrt{2}} \sin\theta & \lambda = -1 \end{cases} \tag{27.26}$$

The helicity spinors are similarly constructed. The spinor representing a spin $\frac{1}{2}$ particle moving with momentum p in the z-direction, with helicity λ_1 is given by

$$u(\hat{\mathbf{i}}_z p, \lambda_1) = \frac{1}{\sqrt{2M(E+M)}}\begin{pmatrix} E+M \\ \sigma_3 p \end{pmatrix}\chi_{\lambda_1} = \frac{1}{\sqrt{2M(E+M)}}\begin{pmatrix} E+M \\ 2\lambda_1 p \end{pmatrix}\chi_{\lambda_1}$$

$$\chi_{1/2} = \begin{pmatrix} 1 \\ 0 \end{pmatrix} \qquad \chi_{-1/2} = \begin{pmatrix} 0 \\ 1 \end{pmatrix} \tag{27.27}$$

The spinor moving in the opposite direction is given by (4.80), i.e.

$$\begin{aligned} u(p, \lambda_1) &= (-1)^{1/2-\lambda_1} e^{-i\pi\sigma_2/2} u(\hat{\mathbf{i}}_z p, \lambda_1) \\ &= \frac{(-1)^{1/2-\lambda_1}}{\sqrt{2M(E+M)}}\begin{pmatrix} E+M \\ 2\lambda_1 p \end{pmatrix} \begin{cases} \begin{pmatrix} 0 \\ 1 \end{pmatrix} & \lambda_1 = +\frac{1}{2} \\ \begin{pmatrix} -1 \\ 0 \end{pmatrix} & \lambda_1 = -\frac{1}{2} \end{cases} \end{aligned} \tag{27.28}$$

i.e.

$$u(p, +\tfrac{1}{2}) = \frac{1}{\sqrt{2M(E+M)}}\begin{pmatrix} E+M \\ p \end{pmatrix}\begin{pmatrix} 0 \\ 1 \end{pmatrix}$$

$$u(p, -\tfrac{1}{2}) = \frac{1}{\sqrt{2M(E+M)}}\begin{pmatrix} E+M \\ -p \end{pmatrix}\begin{pmatrix} 1 \\ 0 \end{pmatrix}$$

The spinor representing a particle with helicity λ_2 moving in the direction $\mathbf{p}'$ (which makes an angle θ w.r.t. $\mathbf{p}$) is given by

$$\begin{aligned} u(p', \lambda_2) &= e^{-i\theta\sigma_2/2} u(p, \lambda_2) \\ &= \frac{1}{\sqrt{2M(E'+M)}}\begin{pmatrix} E'+M \\ 2\lambda_2 p' \end{pmatrix}\begin{pmatrix} \cos\theta/2 & -\sin\theta/2 \\ \sin\theta/2 & \cos\theta/2 \end{pmatrix}\chi_{\lambda_2} \end{aligned} \tag{27.29}$$

so that

$$\begin{aligned} \bar{u}(p', \tfrac{1}{2}) &= \frac{1}{\sqrt{2M(E'+M)}}(E'+M, -p')(-\sin\theta/2, \cos\theta/2) \\ \bar{u}(p', -\tfrac{1}{2}) &= \frac{1}{\sqrt{2M(E'+M)}}(E'+M, p')(\cos\theta/2\,, \sin\theta/2) \end{aligned} \tag{27.30}$$

Consequently, with

$$\gamma_5 = \rho_1 \tag{27.31}$$

we obtain[1]

$$\begin{aligned}
&\bar{u}(p', \tfrac{1}{2})\gamma_5\, u(p, \tfrac{1}{2}) \\
&\quad = \left(\frac{1}{4M^2(E'+M)(E+M)}\right)^{1/2} (E'+M, -p') \begin{pmatrix} 0 & 1 \\ 1 & 0 \end{pmatrix} \begin{pmatrix} E+M \\ p \end{pmatrix} \\
&\qquad \times (-\sin\theta/2\,, \cos\theta/2) \begin{pmatrix} 0 \\ 1 \end{pmatrix} \\
&\quad = \left(\frac{(E+M)(E'+M)}{4M^2}\right)^{1/2} \left(\frac{p}{E+M} - \frac{p'}{E'+M}\right) \cos\frac{\theta}{2}
\end{aligned} \tag{27.32a}$$

Similarly

$$\bar{u}(p', \tfrac{1}{2})\gamma_5\, u(p, -\tfrac{1}{2}) = \left(\frac{(E+M)(E'+M)}{4M^2}\right)^{1/2} \left(\frac{p}{E+M} + \frac{p'}{E'+M}\right) \sin\frac{\theta}{2} \tag{27.32b}$$

$$\bar{u}(p', -\tfrac{1}{2})\gamma_5\, u(p, \tfrac{1}{2}) = \left(\frac{(E+M)(E'+M)}{4M^2}\right)^{1/2} \left(\frac{p}{E+M} + \frac{p'}{E'+M}\right) \sin\frac{\theta}{2} \tag{27.32c}$$

$$\bar{u}(p', -\tfrac{1}{2})\gamma_5\, u(p, -\tfrac{1}{2}) = -\left(\frac{(E+M)(E'+M)}{4M^2}\right)^{1/2} \left(\frac{p}{E+M} - \frac{p'}{E'+M}\right) \cos\frac{\theta}{2} \tag{27.32d}$$

Thus, with the notation

$$\begin{aligned}
C &= \frac{2igf_{\rho\pi\pi}}{(2\pi)^6}\, q \left(\frac{(E+M)(E'+M)}{4M^2}\right)^{1/2} \\
\eta_{\pm} &= \frac{p}{E+M} \pm \frac{p'}{E'+M}
\end{aligned} \tag{27.33}$$

[1] Incidentally, the vector and spinor helicity wave functions constructed above may be used to make up spin $\frac{3}{2}$ helicity wave functions of the form

$$w_\mu(\mathbf{p}, \lambda) = \sum_{\lambda'} C(1, \tfrac{1}{2}, \tfrac{3}{2}, \lambda - \lambda', \lambda', \lambda)\, u(\mathbf{p}, \lambda')\, e_\mu(\lambda - \lambda', \mathbf{p})$$

as done by G. Ebel, (unpublished DESY report).

the various helicity amplitudes may be written as[1]

$$A\left(W, \theta; 1, \frac{1}{2}, \frac{1}{2}\right) = A\left(W, \theta; -1, -\frac{1}{2}, -\frac{1}{2}\right)$$

$$= -A\left(W, \theta; 1, -\frac{1}{2}, -\frac{1}{2}\right)$$

$$= -A\left(W, \theta; -1, \frac{1}{2}, \frac{1}{2}\right) = \frac{C\eta_-}{t - m_\pi^2}\left(-\frac{1}{\sqrt{2}} \sin\theta \cos\frac{\theta}{2}\right)$$

$$= -\frac{\sqrt{2}\, C\eta_-}{t - m_\pi^2} \sin\frac{\theta}{2}\left(1 - \sin^2\frac{\theta}{2}\right)$$

$$A\left(W, \theta; 1, \frac{1}{2}, -\frac{1}{2}\right) = A\left(W, \theta; 1, -\frac{1}{2}, \frac{1}{2}\right) = -A\left(W, \theta; -1, \frac{1}{2}, -\frac{1}{2}\right)$$

$$= -A\left(W, \theta; -1, -\frac{1}{2}, \frac{1}{2}\right)$$

$$= -\frac{\sqrt{2}\, C\eta_+}{t - m_\pi^2} \cos\frac{\theta}{2} \sin^2\frac{\theta}{2}$$

$$A\left(W, \theta; 0, \frac{1}{2}, \frac{1}{2}\right) = -A\left(W, \theta; 0, -\frac{1}{2}, -\frac{1}{2}\right)$$

$$= \frac{C\eta_-}{t - m_\pi^2} \cos\frac{\theta}{2}\left(\frac{q'}{m_\rho} - \frac{\omega'}{m_\rho}\cos\theta\right)$$

$$A\left(W, \theta; 0, \frac{1}{2}, -\frac{1}{2}\right) = A\left(W, \theta; 0, -\frac{1}{2}, \frac{1}{2}\right)$$

$$= \frac{C\eta_+}{t - m_\pi^2} \sin\frac{\theta}{2}\left(\frac{q'}{m_\rho} - \frac{\omega'}{m_\rho}\cos\theta\right) \tag{27.34}$$

Following our discussion in Chapter 4 (eq. 4.98), each of these amplitudes may be decomposed into angular momentum amplitudes according to

$$A(W, \theta, \lambda, \lambda_2, \lambda_1) = \frac{1}{2\pi}\sum_J \left(J + \frac{1}{2}\right) a_J(W; \lambda, \lambda_2, \lambda_1)\, d^{(J)}_{\lambda_1, \lambda - \lambda_2}(\theta) \tag{27.35}$$

It is each of the amplitudes $a_J(W; \lambda, \lambda_2, \lambda_1)$ that is to be modified by

[1] It is easily checked that the amplitudes satisfy the condition

$$A(W, \theta, -\lambda_\rho, -\lambda', -\lambda) = (-1)^{\lambda_\rho + \lambda + \lambda'} A(W, \theta; \lambda_\rho, \lambda', \lambda)$$

required by parity conservation.

factors that take into account the absorption in the initial and final states. The conventional (but by no means well-established) procedure is to replace $a_J(W; \lambda, \lambda_2, \lambda_1)$ by

$$e^{i\delta_f^J(W)}\, a_J(W; \lambda, \lambda_2, \lambda_1) e^{i\delta_i^J(W)} \tag{27.36}$$

where δ_i^J and δ_f^J are the *complex* phase shifts in the initial and final states. In the practical calculations of $A(W, \theta; \lambda, \lambda_2, \lambda_1)$ for small angles, the sum over J is replaced by an integral, and the approximate expression

$$d^J_{\lambda_1, \lambda-\lambda_2}(\theta) \cong J_{\lambda-\lambda_1-\lambda_2}\left((2J+1)\sin\frac{\theta}{2}\right) \tag{27.37}$$

is used, so that

$$A(W, \theta; \lambda, \lambda_2, \lambda_1) \cong \frac{1}{2\pi}\int_{x_{\min}}^{\infty} x\, dx J_{\lambda-\lambda_1-\lambda_2}\left(2x \sin\frac{\theta}{2}\right) a(W, x; \lambda, \lambda_2, \lambda_1) \tag{27.38}$$

The amplitudes $a(W, x, \lambda; \lambda_2, \lambda_1)$ are obtained by noting that (*a*) with

$$\begin{aligned} t &= (\omega_\rho - \omega_\pi)^2 - (\mathbf{q}' - \mathbf{q})^2 \\ &= m_\rho^{\ 2} + m_\pi^{\ 2} - 2\omega_\rho\omega_\pi + 2qq'\left(1 - 2\sin^2\frac{\theta}{2}\right) \end{aligned} \tag{27.39}$$

all the amplitudes $A(W, \theta; \lambda, \lambda_2, \lambda_1)$ are of the form

$$\sum_{n=\lambda-\lambda_1-\lambda_2} c_n \frac{w^n}{\epsilon^2 + w^2} \qquad w = 2\sin\frac{\theta}{2} \tag{27.40}$$

and (*b*)

$$\frac{w^n}{\epsilon^2 + w^2} = \epsilon^n \int_0^\infty x\, dx J_n(wx)\, K_n(\epsilon x) \tag{27.41}$$

Thus if we have

$$A(W, \theta; \lambda, \lambda_2, \lambda_1) \cong \sum_{n=\lambda-\lambda_1-\lambda_2} \frac{c_n(W; \lambda, \lambda_2, \lambda_1) w^n}{\epsilon^2 + w^2} \tag{27.42}$$

then

$$a(W, x; \lambda, \lambda_2, \lambda_1) \cong 2\pi \sum_n c_n(W; \lambda, \lambda_2, \lambda_1)\epsilon^n K_n(\epsilon x) \tag{27.43}$$

A crude way to take into account absorption is to multiply $a(W, x; \lambda, \lambda_2, \lambda_1)$ by a factor of the form

$$1 - C_1\, e^{-\gamma_1(x-\frac{1}{2})^2} \tag{27.44}$$

When $C_1 \cong 1$, this form represents total absorption for low partial waves ($x \leftrightarrow J + \frac{1}{2}$). The Gaussian form of the fall-off of the absorption with increasing J is a reflection of the observed dependence of the elastic scattering amplitude

$$f(q, t) \cong \frac{iq}{4\pi} \sigma_{\text{tot}} e^{bt} \tag{27.45}$$

on momentum transfer.[1] A reasonable fit for $\pi^- p$ elastic scattering at 4 BeV/c lead Gottfried and Jackson to a choice

$$C_1 \cong 0.76 \qquad \gamma_1 \cong 0.04 \tag{27.46}$$

for the process under consideration.

A large number of calculations have been performed in the last two years, and the following random comments illustrate the situation at this time:[2]

1. The one-pion exchange model with absorption works well for the reactions

$$\pi^+ + p \rightarrow \rho^+ + p$$

over a range of 1.6 BeV/c–8 BeV/c. The factor (27.44), which really should be written as

$$(1 - C_1 e^{-\gamma_1(x-\frac{1}{2})^2})^{\frac{1}{2}}(1 - C_2 e^{-\gamma_2(x-\frac{1}{2})^2})^{\frac{1}{2}}$$

to treat the initial (1) channel, which is known from elastic scattering, differently from the final (2) channel, can be modified somewhat without changing the overall agreement with experiment. A favored choice for the ρ-nucleon channel parameters is

$$C_2 \cong 1$$

$$\gamma_2 \cong \tfrac{3}{4}\gamma_1$$

An example of the accuracy of the model is obtained from a comparison of the experimental and theoretical values of the ρ density matrix. The theoretical values,[3] averaged over an angular interval $1 \geq \cos\theta \geq 0.7$, for the ρ^+ production at 4 BeV/c are

$$\langle\rho_{00}\rangle = 0.65 \qquad \langle\rho_{1,-1}\rangle = 0.06 \qquad \langle\rho_{10}\rangle = -0.18$$

in adequate agreement with the experimental values

$$\langle\rho_{00}\rangle = 0.70 \pm 0.08 \qquad \langle\rho_{1,-1}\rangle = 0.17 \pm 0.14 \qquad \langle\rho_{10}\rangle = -0.07 \pm 0.07$$

2. The reaction

$$\pi^+ + p \rightarrow \rho + N^*$$

[1] K. Gottfried and J. D. Jackson, *Nuovo Cimento* **34,** 735 (1964). See also M. Baker and R. Blankenbecler, *Phys. Rev.* **128,** 415 (1962).

[2] J. D. Jackson and J. T. Donohue, K. Gottfried, R. Keyser, and B. E. Y. Svenson, *Phys. Rev.* **139,** B428 (1965).

[3] Limitations of space prevent us from exhibiting this fairly straightforward calculation.

at 4 BeV/c can also be understood in terms of one-pion exchange, with the choice of parameters

$$C_2 \cong 1$$
$$\gamma_2 \cong \tfrac{1}{2}\gamma_1$$

as is shown in Fig. 27.6.

The fit in the shape of the angular distribution is not matched by a corresponding success in the total cross section, the theoretical cross

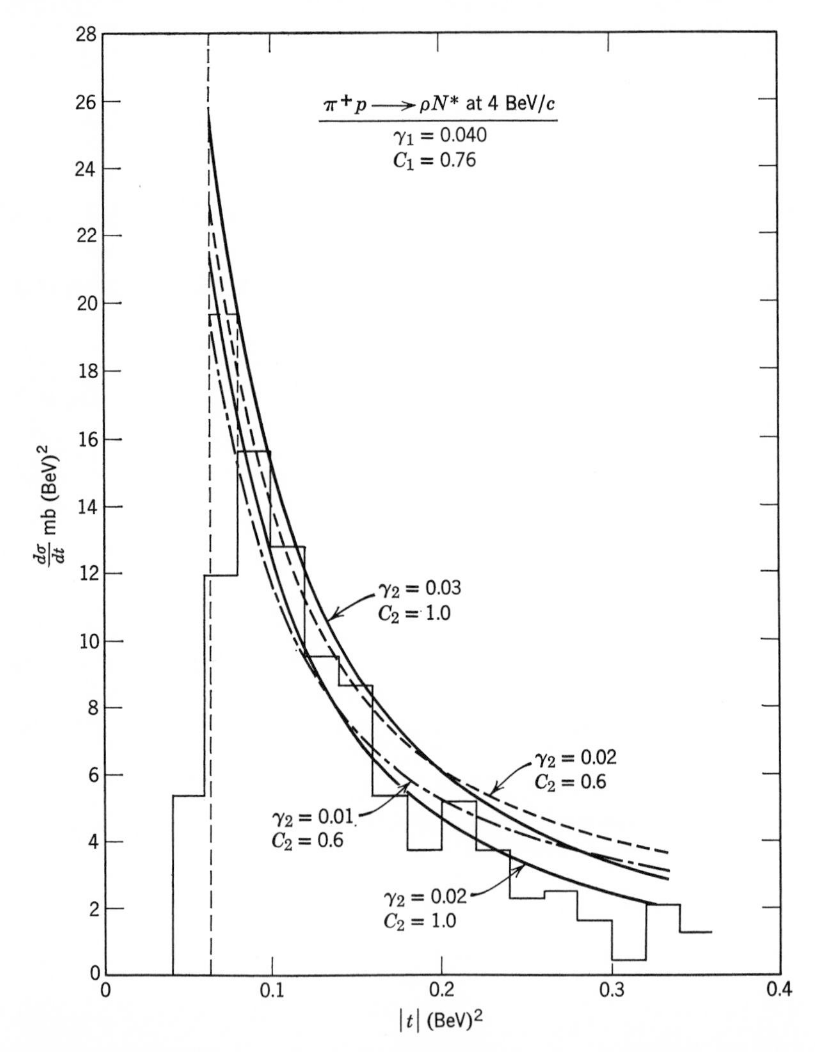

Fig. 27.6. Differential cross section for the reaction $\pi^- + p \rightarrow \rho + N^*$ at 4 BeV/c, for different values of the parameters C_2 and γ_2. The normalization of the data is arbitrary. (From J. D. Jackson, *loc. cit.*)

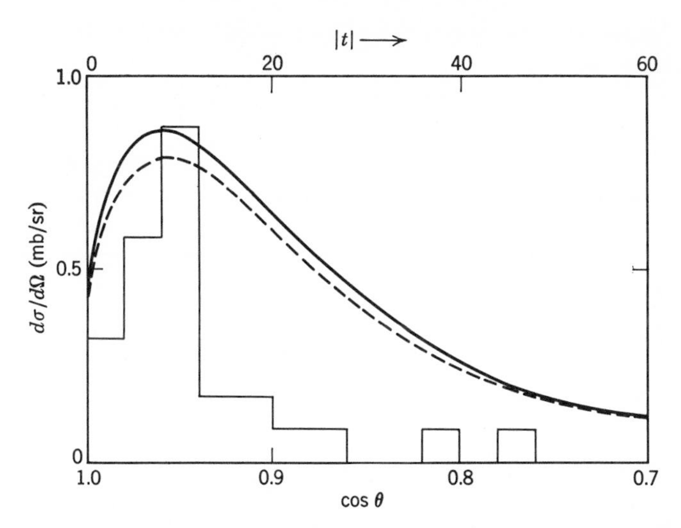

Fig. 27.7. The differential cross section for $K^+ + p \to K^{*+} + p$ at 5 BeV/c calculated with the coupling constants determined to give a fit at 3 BeV/c. Both pion exchange and vector-meson exchange are included. The two theoretical curves correspond to two different choices of vector meson couplings. [From J. D. Jackson et al., *Phys. Rev.* **139,** B428 (1965).]

sections being a factor 2–3 too large. This discrepancy disappears at 8 BeV/c, where the two agree

$$\sigma_{\text{exp}} \cong 1 \text{ mb}$$
$$\sigma_{\text{theo}} \cong 1.2 \text{ mb}$$

3. The data on K^* production in the reaction

$$K^+ + p \to K^* + p$$

at 3 BeV/c cannot be fitted by the one-pion mechanism. It can be explained if vector-meson exchange, first suggested in another context by Sakurai and Stodolsky,[1] is included. When the parameters describing the vector-meson couplings are determined at the above energy, it is found that the theoretical cross section at large momentum transfers is much too large (Fig. 27.7). This is a consequence of an s-dependence in the Born cross section compared with that obtained for pion exchange as in (27.6).

[1] L. Stodolsky and J. J. Sakurai, *Phys. Rev. Letters* **11,** 90 (1963); L. Stodolsky, *Phys. Rev.* **134,** B1099 (1964). These authors were able to explain the data for $KN \to KN^*$, where no pion can be exchanged, provided the ρ-meson couples to NN^* analogously to the γNN^* coupling observed in photoproduction. In terms of fields this is

$$\frac{G}{m_\pi}(\partial^\mu \rho^\nu - \partial^\nu \rho^\mu)(\bar{\psi}_\nu \gamma_\mu \gamma_5 \psi + \text{h.c.}).$$

4. The ρ-exchange model definitely fails to explain the data on $\pi^- p$ charge exchange (Fig. 27.8).[1]

In summary, it is fair to say that to a large extent the details of resonance production reactions with small momentum transfer are understandable in terms of a one-pion exchange model, provided care is taken that the most obvious violations of unitarity are avoided. The procedure for doing this is ambiguous, and other uncertainties do not allow a critical test of the currently used method, which consists of taking into account absorption

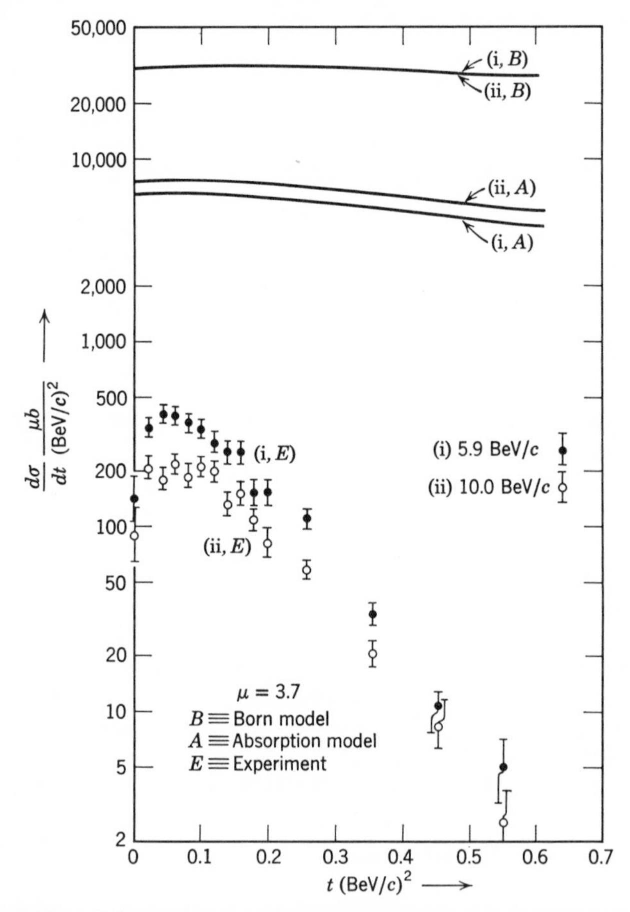

Fig. 27.8. Differential $\pi^- p$ charge exchange cross sections at laboratory momenta of 5.9 and 10.0 BeV/c. The lines labeled B represent the Born approximation, and those labeled A represent the results of calculations which include absorption. [From V. Barger and M. Ebel, *Phys. Rev.* **138,** B1148 (1965).]

[1] V. Barger and M. Ebel, *Phys. Rev.* **138,** B1148 (1965).

of low partial waves alone. For processes in which the exchange of higher spin particles takes place, the model is less successful. It may be that a treatment of higher spin multipion resonances as "particles," though successful in the direct channel, needs modification when these resonances are exchanged and are far off the mass shell.

PROBLEMS

1. Consider the process shown in the graph below

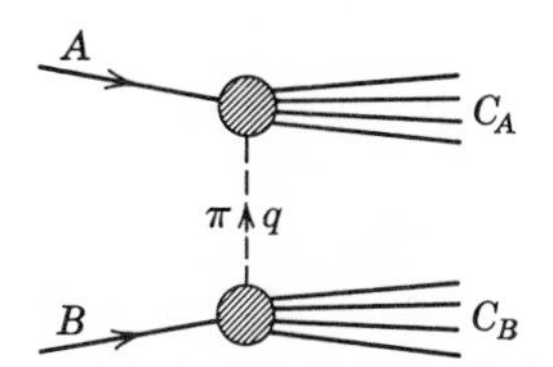

Use relations like $(\Psi_{C_A}, j_\pi \Psi_A) = (2\pi)^{3/2}\, T(\pi + A \to C_A)$ to obtain the expression

$$d\sigma(A + B \to C_A + C_B) = \frac{4}{(2\pi)^4}\int \frac{d^4q}{(q^2 - m_\pi^2)^2} \times \left[\frac{((q \cdot p_A)^2 - q^2 p_A^2)((q \cdot p_B)^2 - q^2 p_B)^2}{(p_A \cdot p_B)^2 - p_A^2 p_B^2}\right]^{1/2} \times d\sigma(A + \pi \to C_A)\, d\sigma(B + \pi \to C_B)$$

Show that with a change of variables this may be written in the form

$$d\sigma(A + B \to C_A + C_B) = \frac{1}{4\pi^3 p^2 W^2}\int \frac{dt}{(t - m_\pi^2)^2}\int dW_{A\pi} p_{A\pi} W_{A\pi}^2 \times \int dW_{B\pi} p_{B\pi} W_{B\pi}^2\, d\sigma(A + \pi \to C_A)\, d\sigma(B + \pi \to C_B)$$

where $t = q^2$, W is the total energy and p the momentum in the center of mass system, and $W_{A\pi}$, $p_{A\pi}$ are the total energy and the momentum of particle A in the $A\pi$ center of mass system (and similarly for $W_{B\pi}$, $p_{B\pi}$).

2. Show that if the amplitude for $\pi + N \to \rho + N$ is denoted by $T_{\lambda_N'\lambda_\rho'\lambda_N}(s, t)$, then the density matrix for the ρ-meson is given by

$$\rho_{\lambda_\rho'\lambda_\rho} = N \sum_{\lambda_N \lambda_N'} T_{\lambda_N'\lambda_\rho',\lambda_N} T^*_{\lambda_N'\lambda_\rho,\lambda_N}$$

which satisfies tr $\rho = 1$ if N is appropriately chosen. Find the density matrix in the (J, M) representation. Calculate the density matrix for the ρ-meson in the OPE model.

28

Elastic Scattering at High Energies

A very small subset of high energy reactions—the peripheral inelastic two-body processes discussed in Chapter 27—can be reasonably well understood in terms of simple concepts of field theory or S-matrix theory (analyticity, crossing, unitarity). The model of one-particle exchange dominance, corrected by a phenomenological term describing absorption of low partial waves, was seen to give a good description of these processes over a large energy range. No such theory exists for other high energy processes. Our aim in this chapter is to provide a phenomenological framework for the description of high energy processes in terms of the smallest number of arbitrary parameters. We shall concentrate on the elastic scattering data.

The most easily measured scattering matrix element is the imaginary part of the scattering amplitude in the forward direction (momentum transfer $t = 0$), since it is related to the total cross section by the optical theorem. If we use the amplitude defined in (23.63)[1]

$$A(s, t) = \frac{\sqrt{\pi}}{q_{\mathrm{cm}}} f_{\mathrm{cm}}(s, t) \tag{28.1}$$

so that the elastic differential cross section has the simple form

$$\frac{d\sigma_{\mathrm{el}}(s, t)}{dt} = |A(s, t)|^2 \tag{28.2}$$

then the optical theorem yields

$$\mathrm{Im}\, A(s, 0) = \frac{\sigma_{\mathrm{tot}}(s)}{\sqrt{16\pi}} \tag{28.3}$$

[1] We remind the reader once more that the amplitude that is most convenient to use when we discuss analycity is $T(s, t, u)$, which is related to the center of mass scattering amplitude $f_{\mathrm{cm}}(s, t)$ by $T(s, t, u) = -\sqrt{s} f_{\mathrm{cm}}(s, t)/8\pi^5$ (if the states are normalized as if they were all bosons). When talking about unitarity, the form $A(s, t)$ is more compact. Note that the quantity denoted by f here, was denoted by F in Ch. 23.

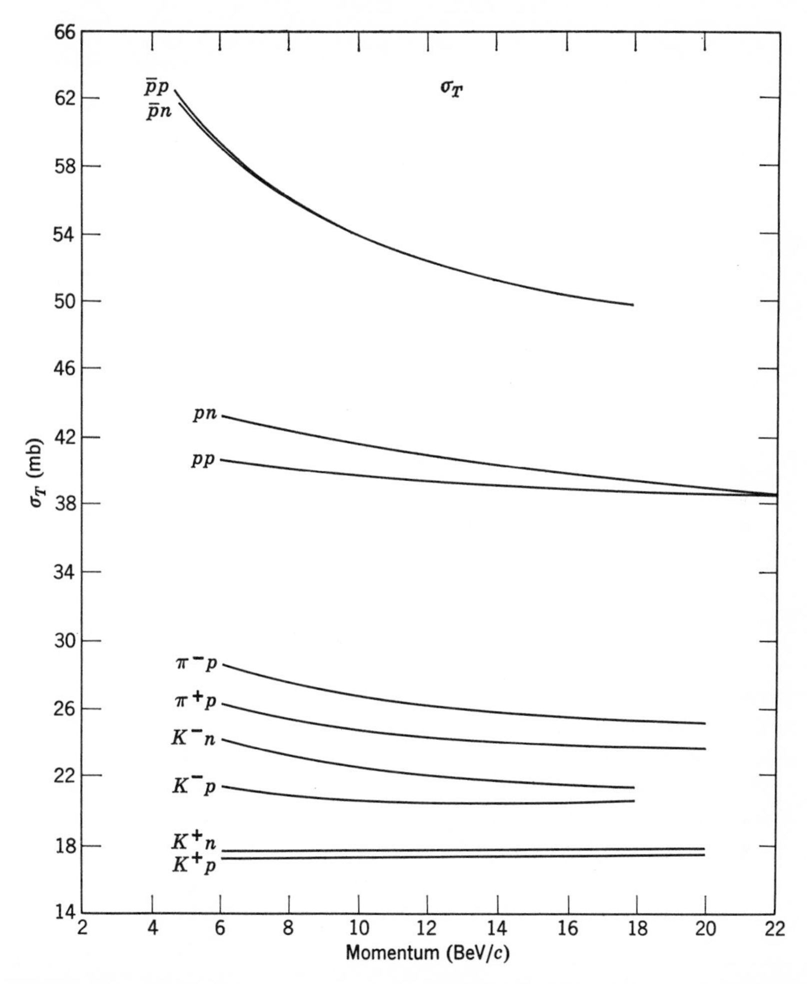

Fig. 28.1. Schematic representation for the high energy total cross sections for nucleons, antinucleons and mesons on nucleons, as functions of the laboratory momentum of the incident particle. [(Compiled from data in W. Galbraith et al., *Phys. Rev.* **138,** B913 (1965).]

The total cross sections for nucleons and mesons are shown in Fig. 28.1. The striking aspect of the curves shown there is the absence of visible structure. If the cross sections do tend to zero, they do so remarkably slowly. A more plausible interpretation of the data is that they tend to constant values. There is nothing unphysical about constant cross sections; a simple classical model, the "black-sphere" model, yields constant cross sections at high energies.

If we ignore spin, the scattering amplitude may be written in the form

$$A(s, t) = \frac{\sqrt{\pi}}{q} \sum_l (2l + 1) f_l(s) P_l(\cos\theta)$$
$$= \frac{\sqrt{\pi}}{q} \sum_l (2l + 1) \frac{\eta_l(s) e^{2i\delta_l(s)} - 1}{2iq} P_l(\cos\theta) \tag{28.4}$$

Here δ_l are real phase shifts, and η_l represents absorption. When there is no absorption, $\eta_l = 1$; for total absorption $\eta_l = 0$. The black-sphere model assumes total absorption for all partial waves up to some maximum value of l determined by the classical value corresponding to some effective interaction radius R

$$\eta_l = 0 \qquad l \leqslant qR \tag{28.5}$$

and no scattering at all for the higher partial waves. Thus

$$A(s, 0) \approx \frac{i\sqrt{\pi}}{2q^2} \sum_{l=0}^{qR} (2l + 1)$$
$$= i\sqrt{\pi}\, R^2/2 \tag{28.6}$$

and

$$\sigma_{\text{tot}}(s) = 4\sqrt{\pi}\, \text{Im}\, A(s, 0) = 2\pi R^2 \tag{28.7}$$

It should be pointed out that this simple model is certainly not correct as far as a description of the differential cross section is concerned. The scattering amplitude for small angles is

$$A(s, t) \approx \frac{i\sqrt{\pi}}{2q^2} \sum_{l=0}^{qR} (2l + 1) P_l(\cos\theta)$$
$$\approx \frac{i\sqrt{\pi}}{2q^2} \int_0^{qR} 2x\, dx\, J_0(x\theta) \cong \frac{i\sqrt{\pi}\, R}{q\theta} J_1(q\theta R)$$
$$= \frac{i\sqrt{\pi}\, R}{\sqrt{-t}} J_1(R\sqrt{-t}) \tag{28.8}$$

where we have written $-t = 4q^2 \sin^2 \theta/2 \approx (q\theta)^2$ and used the small-angle approximation

$$P_l(\cos\theta) \approx J_0((l + \tfrac{1}{2})\theta) \tag{28.9}$$

This disagrees with the data that have been parametrized in the form

$$\frac{d\sigma}{dt} = \left(\frac{d\sigma}{dt}\right)_{t=0} e^{at+bt^2}$$

If cross sections become constant at high energies, and do so sufficiently rapidly, the cross section for the scattering of particles by a given target must be the same as the cross section for the antiparticle by the same target,[1] i.e.

$$\lim_{s\to\infty} (\sigma_{\text{tot}}^{AB}(s) - \sigma_{\text{tot}}^{\bar{A}B}(s)) = 0 \tag{28.10}$$

We shall outline the argument for the case of the total cross sections σ_{π^-p} and σ_{π^+p}. As can be seen from (23.49), a dispersion relation written down on the assumption that $\sigma_{\pi^-p} - \sigma_{\pi^+p} \to 0$ as $s \to \infty$, the assumption that

$$\sigma_{\pi^-p}(s) - \sigma_{\pi^+p}(s) \to \Delta\sigma \tag{28.11}$$

implies that we can write a once-subtracted dispersion relation for[2]

$$g(\omega^2) \equiv \frac{f_{\pi^-}(\omega) - f_{\pi^+}(\omega)}{2\omega} = \frac{f_{\pi^-}(\omega) - f_{\pi^-}(-\omega)}{2\omega} \tag{28.12}$$

namely

$$\begin{aligned}\operatorname{Re}(g(\omega^2) - g(0)) = \frac{2Mf^2}{\omega_B^{\,2}}\frac{\omega^2}{\omega^2 - \omega_B^{\,2}} + \frac{\omega^2}{4\pi^2}\\ \times P\int_{m_\pi}^{\infty} \frac{q'\,d\omega'}{\omega'^2(\omega'^2 - \omega^2)}(\sigma_{\pi^-p}(\omega') - \sigma_{\pi^+p}(\omega'))\end{aligned} \tag{28.13}$$

Thus at high energies

$$\operatorname{Re} g(\omega^2) \approx \frac{\Delta\sigma}{8\pi^2}\log\frac{\omega^2}{m_\pi^{\,2}} + \text{const} \tag{28.14}$$

In terms of center of mass amplitudes the result is that

$$\operatorname{Im} f_{\pi^-p}(s, 0) - \operatorname{Im} f_{\pi^+p}(s, 0) \to \frac{\sqrt{s}}{8\pi}\Delta\sigma$$

$$\operatorname{Re} f_{\pi^-p}(s, 0) - \operatorname{Re} f_{\pi^+p}(s, 0) \to \frac{\sqrt{s}}{4\pi^2}\Delta\sigma\log\frac{s}{m_\pi^{\,2}}$$

and each individual $\operatorname{Im} f_{\pi^\pm p}(s, 0) \to \sqrt{s} \times (\text{const})$. The result implies that at high energies the elastic scattering amplitude is relatively real. This goes against all experience with models supporting the intuitive idea that the imaginary part of the scattering amplitude, which gets positive contributions from every inelastic process, must be larger than the real part, which gets contributions from a large number of terms with random signs. This contradiction can only be removed if $\Delta\sigma = 0$. The Pomeranchuk theorem has been proved under somewhat weaker assumptions by Weinberg[3] who showed that, provided $\sigma_{\pi^-p}(s) - \sigma_{\pi^+p}(s)$ does not change

[1] I. Ia. Pomeranchuk, *Soviet Phys. JETP* **34**(7), 499 (1958).
[2] $f_{\pi^-}(-\omega) = f_{\pi^+}(\omega)$ by crossing symmetry.
[3] S. Weinberg, *Phys. Rev.* **124,** 2049 (1961).

sign an infinite number of times, then (*a*) if $g(\omega^2)$ is bounded, then

$$(\sigma_{\pi^- p}(s) - \sigma_{\pi^+ p}(s)) \log s \to 0 \qquad \text{as } s \to \infty$$

(*b*) if $g(\omega^2)$ is not bounded, but $g(\omega^2) = 0(\log \omega)$, then

$$\frac{\sigma_{\pi^- p}(s)}{\sigma_{\pi^+ p}(s)} \to 1$$

The theorem has been generalized by Pomeranchuk and Okun[1] to the statement that in the forward direction at high energies exchange scattering amplitudes are always dominated by the elastic amplitudes. For example, for charge exchange scattering,

$$\frac{\left(\dfrac{d\sigma(\pi^- p \to \pi^0 n)}{dt}\right)_{t=0}}{\left(\dfrac{d\sigma(\pi^- p \to \pi^- p)}{dt}\right)_{t=0}} \to 0 \qquad \text{as } s \to \infty \tag{28.15}$$

and for strangeness exchange scattering

$$\frac{\left(\dfrac{d\sigma(K^- p \to \Sigma^\pm \pi^\mp)}{dt}\right)_{t=0}}{\left(\dfrac{d\sigma(K^- p \to K^- p)}{dt}\right)_{t=0}} \to 0 \qquad \text{as } s \to \infty \tag{28.16}$$

The elastic scattering processes differ from the (inelastic) exchange scattering processes in the possible quantum numbers of what is exchanged between the two particles. As is illustrated in Fig. 28.2, what is exchanged in any process is determined by the initial and final states *in the t-channel*. In the elastic process $\pi^- p \to \pi^- p$, for example, the intermediate state in the t-channel can be characterized by charge $Q = 0$, i-spin states $T =$ 0, 1, 2. In contrast, for $\pi^- p \to \pi^0 n$ (exchange scattering), the intermediate state has $Q = -1$, and only $T = 1, 2$ is possible. The reader can readily convince himself that only for elastic scattering will there be an exchange of a system with the quantum numbers of the vacuum ($Q = 0$; $T = 0$; etc). We shall prove the Okun-Pomeranchuk theorem by showing that, if one assumes that at high energies some particular exchange dominates all others, and if the exchange amplitude at high energies is not real for $t = 0$, then the dominant exchange must be the one associated with the vacuum quantum numbers.

[1] I. Ia. Pomeranchuk, *loc. cit.*; L. B. Okun and I. Ia. Pomeranchuk, *Soviet Phys. JETP* **3,** 307 (1956).

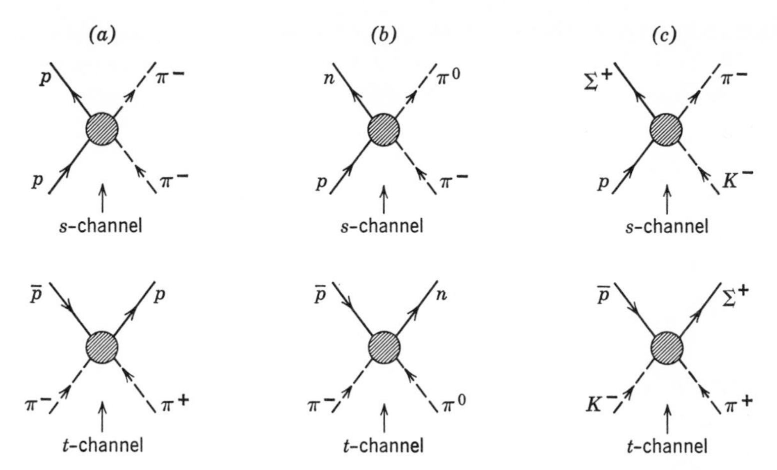

Fig. 28.2. The graphical representation of elastic and exchange scattering and the relation to processes under crossing from the s- to the t-channel. (a) Elastic scattering, (b) the charge exchange reaction $\pi^- p \rightarrow \pi^0 n$, ($c$) the strangeness exchange reaction $K^- p \rightarrow \Sigma^+ \pi^-$.

We shall consider pseudoscalar meson-baryon scattering. These will be treated as octets and denoted by $\{a\}$ and $\{b\}$, respectively. Our tools will be the optical theorem and crossing symmetry. The optical theorem implies that the imaginary part of the forward elastic scattering amplitude for the scattering of a particular meson a by a particular baryon b is given by

$$\operatorname{Im} \langle ab|\, A(s, 0)\, |ab\rangle = \frac{\sigma_{\text{tot}}^{ab}(s)}{\sqrt{16\pi}} > 0 \tag{28.17}$$

Crossing symmetry relates this amplitude to the amplitude for the process $a + \bar{a} \rightarrow b + \bar{b}$, analytically continued to what in the t-channel is an unphysical region, namely $s \rightarrow \infty$, $t = 0$. If the amplitude in the t-channel is denoted by B, then crossing symmetry requires that

$$\langle ab|\, A(s, 0)\, |ab\rangle = \langle \bar{b}b|\, B(0, s)\, |a\bar{a}\rangle \tag{28.18}$$

As we saw in our discussion of unitary symmetry, the states $|a\bar{a}\rangle$ and $|b\bar{b}\rangle$ may be expanded in terms of a complete set of states which transform irreducibly under $SU(3)$. We write

$$|a\bar{a}\rangle = \sum_{\lambda,\mu} |\lambda, \mu\rangle\langle\lambda, \mu \mid a\bar{a}\rangle \tag{28.19}$$

Here the notation $\langle\lambda\mu \mid a\bar{a}\rangle$ is used for the $SU(3)$ Wigner coefficients, λ labels the irreducible representation (for octet-octet scattering, $\lambda =$ **1**, **8**, **8′**, **10**, **10***, and **27**), and μ is the i-spin and hypercharge labeling. μ

is also supposed to distinguish between a decuplet, for example, obtained from reducing out two-boson states and one obtained from reducing out two-fermion states or a boson-fermion state. We may rewrite (28.18) as follows

$$\langle ab|\, A(s, 0)\, |ab\rangle = \sum_{\lambda\mu} \sum_{\lambda'\mu'} \langle \bar{b}b|\, \lambda'\mu'\rangle\langle\lambda'\mu'|\, B(0, s)\, |\lambda\mu\rangle\langle\lambda\mu \mid a\bar{a}\rangle \tag{28.20}$$

We shall assume exact unitary symmetry. This is not necessary for the proof, and the more general case is in fact discussed by Amati et al.[1] whose method of proof we follow, but the condition

$$\langle\lambda'\mu'|\, B(0, s)\, |\lambda, \mu\rangle = \delta_{\lambda\lambda'}\, \delta_{\mu\mu'}\, B_\lambda(0, s) \tag{28.21}$$

simplifies the argument somewhat.

It is generally true that for all irreducible representations the identity representation **1** is contained in the reduction of $\{a\} \times \{\bar{a}\}$ exactly once. Thus we may write

$$|\mathbf{1}\rangle = N_A \sum_a |a\bar{a}\rangle \tag{28.22}$$

where N_A is a normalization coefficient. From this it follows that

$$\sum_a \langle\lambda\mu \mid a\bar{a}\rangle = \begin{cases} = N_A^{-1} & \text{if } \lambda \text{ is the identity representation} \\ = 0 & \text{otherwise.} \end{cases}$$

We write this as

$$\begin{aligned} \sum_a \langle\lambda\mu \mid a\bar{a}\rangle &= \frac{1}{N_A}\, \delta(\lambda, \mathbf{1}) \\ \sum_b \langle b\bar{b} \mid \lambda'\mu'\rangle &= \frac{1}{N_B}\, \delta(\lambda', \mathbf{1}) \end{aligned} \tag{28.23}$$

These relations imply that when λ is not the identity representation, then some of the coefficients $\langle\lambda\mu \mid a\bar{a}\rangle$ and $\langle b\bar{b} \mid \lambda'\mu'\rangle$ must be negative, while others are positive, as otherwise the cancellation in

$$\sum_a \langle\lambda\mu \mid a\bar{a}\rangle = 0 \qquad \lambda \neq \mathbf{1}$$

would not be possible. For $\lambda = \mathbf{1}$ they can all be chosen positive.

Let us now assume that a particular λ dominates in the exchange. This means that at high energies, when λ-exchange is dominant,

$$\begin{aligned} \text{Re}\, \langle ab|\, A(s, 0)\, |ab\rangle &= \text{Re}\, B_\lambda(0, s) \sum_\mu \langle\bar{b}b \mid \lambda\mu\rangle\langle\lambda\mu \mid a\bar{a}\rangle \\ \text{Im}\, \langle ab|\, A(s, 0)\, |ab\rangle &= \text{Im}\, B_\lambda(0, s) \sum_\mu \langle\bar{b}b \mid \lambda\mu\rangle\langle\lambda\mu \mid a\bar{a}\rangle \end{aligned} \tag{28.24}$$

[1] D. Amati, L. L. Foldy, A. Stanghellini, and L. Van Hove, *Nuovo Cimento* **32,** 1685 (1964).

It is clear from the second of these relations that if $\lambda \neq \mathbf{1}$ then for some mesons in the set $\{a\}$ and some baryons in the set $\{b\}$ the combination of the three factors in the second of the relations (28.24) will be negative, i.e., the inequality will be violated. Hence for the *imaginary part* the dominant exchange amplitude must be the one associated with the identity representation (quantum numbers of the vacuum). For exchange reactions

$$\begin{aligned} \text{Re}\,\langle a'b'\,|A(s,0)|\,ab\rangle &= \sum_{\mu\lambda} \langle b'\bar{b} \mid \lambda\mu\rangle\langle\lambda\mu \mid a\bar{a}'\rangle\, \text{Re}\, B_\lambda(0,s) \\ \text{Im}\,\langle a'b'\,|A(s,0)|\,ab\rangle &= \sum_{\mu\lambda} \langle b'\bar{b} \mid \lambda\mu\rangle\langle\lambda\mu \mid a\bar{a}'\rangle\, \text{Im}\, B_\lambda(0,s) \end{aligned} \tag{28.25}$$

there is no contribution from the exchange of $\lambda = \mathbf{1}$. Thus

$$(d\sigma(a + b \rightarrow a' + b')/dt)_{t=0}$$

can only be comparable with $(d\sigma(a + b \rightarrow a + b)/dt)_{t=0}$ if for some $\lambda \neq \mathbf{1}$,

$$\text{Re}\, B_\lambda(0,s) \gtrsim \text{Im}\, B_\mathbf{1}(0,s) \gg \text{Im}\, B_\lambda(0,s) \tag{28.26}$$

This, however, implies that the exchange amplitude is real at $t = 0$ at high energies, contradicting the initial assumption. The experimental data is in accord with the predictions of Pomeranchuk and Okun. (Figure 28.3).

If we now turn to elastic scattering with small but nonvanishing momentum transfer ($0 < -t \leqslant 1$ (BeV/c)^2), we find that the differential scattering cross section for a large number of elastic processes, $pp \rightarrow pp$, $\bar{p}p \rightarrow \bar{p}p$, $\pi^+p \rightarrow \pi^+p$, $\pi^-p \rightarrow \pi^-p$, $K^+p \rightarrow K^+p$, and $K^-p \rightarrow K^-p$ can all be fitted by an almost universal shape curve[1]

$$\frac{d\sigma}{dt} = \left(\frac{d\sigma}{dt}\right)_{t=0} e^{at+bt^2} \tag{28.27}$$

with the parameters a and b listed in the Table 28-1.[2] These parameters are, to a good approximation, independent of energy, and

$$a \approx 10(\text{BeV}/c)^{-2} \qquad 0.01 < \frac{b}{a^2} < 0.04 \tag{28.28}$$

As was pointed out earlier, this form is not in agreement with the predictions of the black-sphere model. To see what kind of a model will

[1] K. J. Foley, S. J. Lindenbaum, W. A. Love, S. Ozaki, J. J. Russell, and L. C. Yuan, *Phys. Rev. Letters* **10,** 376, 543 (1963); **11,** 425, 503 (1963).

[2] These values have been taken from the table in R. Serber, *Phys. Rev. Letters* **13,** 32 (1964).

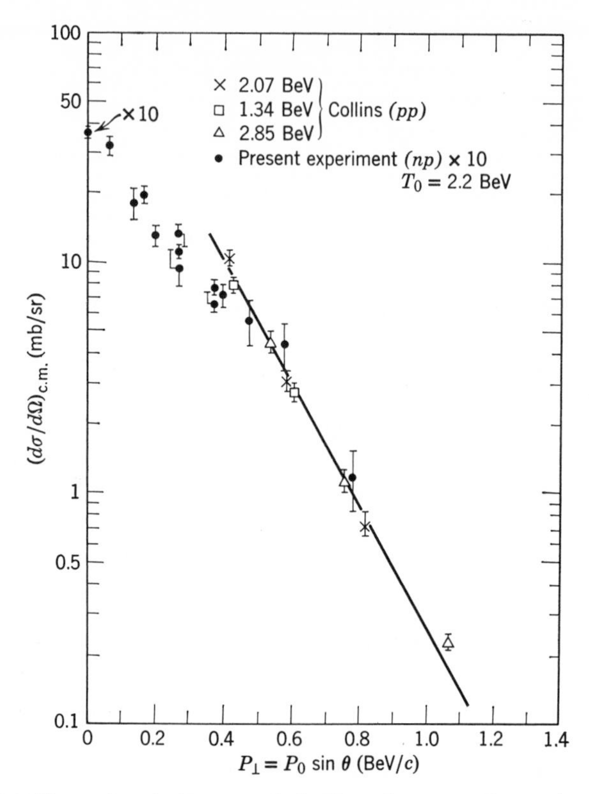

Fig. 28.3. Data illustrating the Pomeranchuk-Okun theorem. The *np* charge-exchange cross section is compared with the *pp* elastic cross section multiplied by 0.1 in the graph. [From J. L. Friedes et al., *Phys. Rev. Letters* **15,** 38 (1965).]

work, we write

$$\begin{aligned} A(s, t) &= \frac{\sqrt{\pi}}{q} \sum_{l=0}^{\infty} (2l + 1) f_l(s)\, P_l(\cos\theta) \\ &\approx \frac{\sqrt{\pi}}{q} \int 2x\, dx\, f_x(s) J_0(\theta x) \\ &\approx 2q\sqrt{\pi} \int d\rho\, \rho\, f(\rho, s)\, J_0(\rho\sqrt{-t}) \end{aligned} \tag{28.29}$$

This is called an *impact parameter representation*,[1] since the variable of

[1] The most thorough modern discussion of this representation may be found in R. Blankenbecler and M. L. Goldberger, *Phys. Rev.* **126,** 766 (1962). See also R. Glauber, *Lectures in Theoretical Physics*, Interscience Publishers, Inc., New York (1958).

Table 28-1

Values of the Parameters a and b

	Incident mom. in lab. (Bev/c)	$(d\sigma/dt)_{t=0}$ $mb/(\text{BeV}/c)^2$	a $(\text{BeV}/c)^{-2}$	b $(\text{BeV}/c)^{-4}$
pp	6.8	106 ± 2	9.78 ± 0.21	3.28 ± 0.31
	8.8	106 ± 2	9.62 ± 0.22	2.30 ± 0.31
	12.8	104 ± 4	10.03 ± 0.28	2.19 ± 0.38
	16.7	92 ± 5	9.79 ± 0.40	1.48 ± 0.59
	19.6	97 ± 7	10.48 ± 0.43	2.25 ± 0.56
$\bar{p}p$	7.2	181 ± 16	13.15 ± 0.47	—
	8.9	178 ± 8	12.98 ± 0.61	0.40 ± 1.49
	12.0	140 ± 10	11.67 ± 0.98	−2.77 ± 2.50
π^+p	6.8	43.7 ± 1.6	8.58 ± 0.23	2.24 ± 0.29
	8.8	41.0 ± 1.2	8.79 ± 0.23	2.38 ± 0.32
	12.8	37.8 ± 1.8	8.93 ± 0.27	2.36 ± 0.34
	16.7	32.6 ± 1.9	8.94 ± 0.35	2.82 ± 0.43
π^-p	7.0	42.6 ± 1.7	9.45 ± 0.25	2.75 ± 0.34
	8.9	41.4 ± 1.0	9.22 ± 0.21	2.54 ± 0.30
	13.0	42.4 ± 1.6	9.71 ± 0.26	3.02 ± 0.35
	17.0	35.2 ± 1.8	9.07 ± 0.37	2.12 ± 0.53
K^+p	6.8	19.7 ± 4.3	6.13 ± 0.88	1.36 ± 0.77
	9.8	19.7 ± 1.5	6.23 ± 0.45	0.99 ± 0.55
	12.8	22.0 ± 1.4	6.93 ± 0.38	1.18 ± 0.45
K^-p	7.2	38.9 ± 11	10.2 ± 1.2	3.97 ± 0.92
	9.0	37 ± 11	10.5 ± 1.2	4.2 ± 1.0

integration, $\rho \approx (l + \frac{1}{2})/q$ has this interpretation in the classical limit. Equation 28.29 is a Fourier-Bessel tranform, and it may be inverted to give

$$f(\rho, s) = \frac{1}{2q\sqrt{\pi}} \int_0^\infty d(\sqrt{-t})(\sqrt{-t})\, A(s, t)\, J_0(\rho\sqrt{-t}) \tag{28.30}$$

If we take

$$A(s, t) \approx i \operatorname{Im} A(s, 0)e^{at/2} = \frac{i\sigma_{\text{tot}}}{4\sqrt{\pi}}\, e^{at/2} \tag{28.31}$$

which yields

$$\frac{d\sigma}{dt} = \frac{\sigma_{\text{tot}}^2}{16\pi}\, e^{at} \tag{28.32}$$

we find that

$$f(\rho, s) = \frac{i\sigma_{\text{tot}}(s)}{8\pi q}\int_0^\infty d(\sqrt{-t})\sqrt{-t}\, J_0(\rho\sqrt{-t})e^{-\frac{a}{2}(\sqrt{-t})^2}$$
$$= \frac{i\sigma_{\text{tot}}}{8\pi q a} e^{-\rho^2/2a} \tag{28.33}$$

For total absorption of the low partial waves (small impact parameter) we would expect that

$$f(\rho, s) \approx \frac{i}{2q}$$

obtained by setting $\eta_l = 0$ in (28.4). We thus predict that

$$a \lessapprox \frac{\sigma_{\text{tot}}}{4\pi} \tag{28.34}$$

For *pp* scattering it turns out that $4\pi a \approx 50$ mb, whereas $\sigma_{\text{tot}} \approx 40$ mb. Thus the model corresponds to a central, strongly absorbing region, with a Gaussian tail and a "width" of $\sqrt{2a} \approx 0.9$ fermis.

A somewhat less phenomenological interpretation of an exponential t-dependence in the scattering amplitude emerged from a theory, which for a short time (1961–63) excited a great deal of interest and which may still play a very important role in elementary particle physics. Space limitations allow us only the briefest description of the theory of *Regge Poles.*[1]

In an attempt to prove the validity of the Mandelstam representation for potential scattering, Regge[2] took the unconventional approach of working with the partial wave expansion

$$f(q^2, \cos\theta) = \sum_{l=0}^{\infty} (2l+1) f_l(q^2)\, P_l(\cos\theta) \tag{28.35}$$

In general, such a series does not allow a very large extension of the analyticity domain in $\cos\theta$ by letting $\cos\theta$ become complex in $P_l(\cos\theta)$, because such a series only converges in a small domain outside $-1 \leq \cos\theta \leq 1$, namely the ellipse, with foci at ± 1, extending to the closest singularity of $f(s, \cos\theta)$ in the complex $\cos\theta$ plane. To expand the domain, Regge studied the partial wave scattering amplitudes as a function of complex l, and found justification for the following steps:

[1] A simple introduction to the theory of Regge Poles and their connection with the Mandelstam representation may be found in S. C. Frautschi, *Regge Poles and S Matrix Theory*, W. A. Benjamin Inc., New York (1963). A more extensive account, which shows how complicated the subject had become in a very short time, is that of R. Oehme in *Strong Interactions and High Energy Physics* (Scottish Universities' Summer School, 1963), R. G. Moorhouse, Ed., Plenum Press, New York (1964).

[2] T. Regge, *Nuovo Cimento* **14,** 951 (1959).

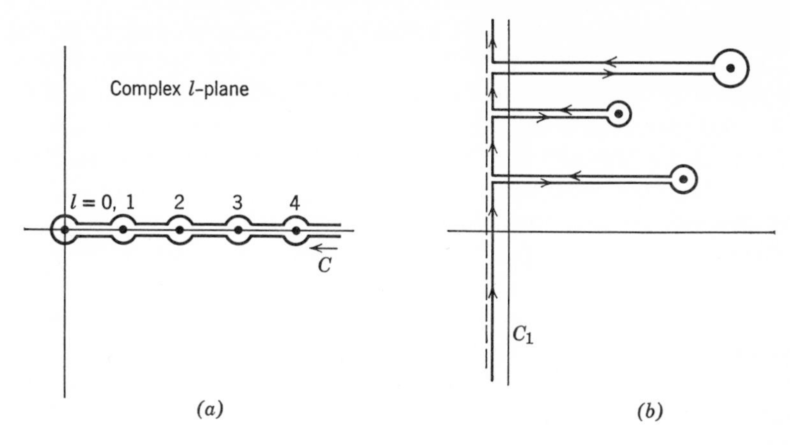

Fig. 28.4. (*a*) Original contour of integration in the complex angular momentum plane; (*b*) deformed contour in the complex angular momentum plane.

1. The sum in (28.35) may be rewritten in the form

$$f(q^2, \cos\theta) = \frac{i}{2}\int_C dl(2l + 1)f(l, q^2)\frac{P_l(-\cos\theta)}{\sin \pi l} \tag{28.36}$$

with the contour shown in Fig. 28.4*a*. This is completely equivalent to (28.35), because $1/(\sin \pi l)$ has poles at $l = 0, \pm 1, \pm 2, \ldots$ with residue $(-1)^l/\pi$.

2. The contour may be deformed to the one shown in Fig. 28.4*b*. In addition to knowing something about $f(l, q^2)$, which is the hard part, it is necessary to know the properties of the Legendre functions $P_l(\cos\theta)$ in unfamiliar regions,[1] to establish this fact.

What Fig. 28.4*b* shows is that to the right of the line $\operatorname{Re} l \geq -\frac{1}{2}$, the function $f(l, q^2)$ only has isolated poles.

If $f(l, q^2)$ at the poles, whose position in general depends on q^2, is written in the form

$$f(l, q^2) \approx \frac{\beta_n(q^2)}{l - \alpha_n(q^2)} \tag{28.37}$$

then the scattering amplitude takes the form

$$f(q^2, \cos\theta) = \frac{i}{2}\int_{C_1} dl\,\frac{(2l + 1)}{\sin \pi l}f(l, q^2)P_l(-\cos\theta) - \pi\sum_n \frac{(2\alpha_n(q^2) + 1)\beta_n(q^2)}{\sin \pi\alpha_n(q^2)}P_{\alpha_n(q^2)}(-\cos\theta) \tag{28.38}$$

[1] W. Magnus and F. Oberhettinger, *Special Functions of Mathematical Physics*, Chelsea Publishing Co., New York (1949).

The first term is generally called the "background integral" and the second is the contribution from the "Regge Poles." In this form we have a representation in which we may let $\cos\theta$ become complex without destroying the validity of the representation, and Regge was able to accomplish his aim and establish the Mandelstam representation for a certain class of potentials (superposition of Yukawa potentials) in nonrelativistic potential scattering theory.

The by-products turned out to be vastly more interesting. First of all, Regge was able to show that the Mandelstam representation for potential scattering needed only a finite number of subtractions in the t-variable, a result not obtained by anybody else. This comes from the fact that as $z \to \infty$, $P_l(z) \sim |z|^l$. Thus the Regge pole farthest to the right in the complex l-plane determines the asymptotic behavior of $f(q^2, \cos\theta)$ as $\cos\theta$ becomes large. Next, Regge showed that the functions $\alpha_n(q^2)$, which determine the position of the poles in the complex l-plane, have the following properties: (*a*) they lie on the real axis when $q^2 < 0$, i.e., below threshold; (*b*) they move to the right, starting at $l \leq -1$ when $q^2 = -\infty$, leave the real axis at $q^2 = 0$, and then travel in the upper half-plane, returning to the line $l = -1$ (Fig. 28.5).

If an $\alpha(q^2)$, coming from the left, crosses $l = 0$ at some value of $q^2 < 0$, we see from the sum in (28.38) that the scattering amplitude has a pole.

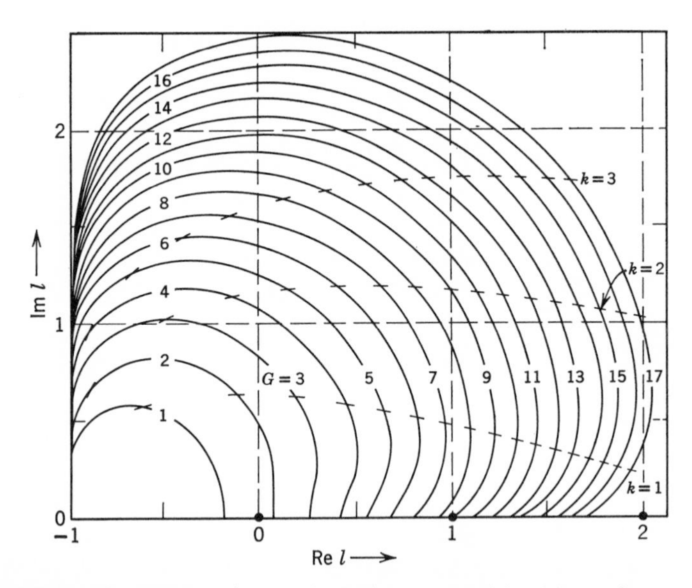

Fig. 28.5. Leading Regge trajectory for Yukawa potentials of increasing strength (as measured by G) as seen in the complex l-plane. (From C. Lovelace and D. Masson, *CERN Proceedings*, 1962.)

In the vicinity of that pole the scattering amplitude has the form

$$f(q^2, \cos\theta) \approx \frac{\gamma}{q^2 - q_0^{\,2}} P_0(\cos\theta) \qquad (q_0^{\,2} < 0)$$

characteristic of a bound state for zero angular momentum. As Fig. 28.5 shows, a minimum potential strength is required to reach $l = 0$; otherwise the trajectory, the path of the Regge pole in the complex l-plane parametrized by q^2), turns over before reaching $l = 0$. If the potential is too weak, there are no bound states! If, on the other hand, the potential is strong, the trajectory may cross the point $l = 1$ before leaving the axis; there the scattering amplitude would have the form characteristic of a bound state of angular momentum 1. In the meantime, a second Regge pole could appear and cross $l = 0$, giving rise to a second S-wave bound state. Thus the Regge poles are intimately connected with the bound states of a system. If, for $q^2 > 0$ a trajectory passes close to an integral value of l with $\operatorname{Im} \alpha_n(q^2) > 0$, then we can easily see that the particular partial wave scattering amplitude has a resonant form. To show this, consider $f(q^2, \cos\theta)$ in the region where a Regge pole dominates, so that

$$f(q^2, \cos\theta) \approx \frac{\gamma_n(q^2)}{\sin \pi\alpha_n(q^2)} P_{\alpha_n}(-\cos\theta) \tag{28.39}$$

then[1]

$$\begin{aligned} f_l(q^2) &= \frac{1}{2}\int_{-1}^{1} d(\cos\theta)\, f(q^2, \cos\theta)\, P_l(\cos\theta) \\ &= \gamma_n(q^2) \frac{1}{\pi(\alpha_n(q^2) - l)(\alpha_n(q^2) + l + 1)} \end{aligned} \tag{28.40}$$

When, at q_r^2 $\operatorname{Re} \alpha_n(q^2) = l$, the partial wave scattering amplitude has the form

$$\frac{\gamma_n(q_r^2)}{2l+1} \frac{1}{(q^2 - q_r^2)\left(\dfrac{d \operatorname{Re}\alpha_n(q^2)}{dq^2}\right)_{q_r^2} + i \operatorname{Im} \alpha_n(q_r^2)} \tag{28.41}$$

characteristic of a resonance.

The impact of the Regge theory on particle physics came from the conjecture by Mandelstam[2] and others that in relativistic theory, too, the

[1] We use the relation

$$\frac{1}{2}\int_{-1}^{1} dz P_l(z)\, P_\alpha(-z) = \frac{\sin \pi\alpha}{\pi(\alpha - l)(\alpha + l + 1)}.$$

[2] G. F. Chew, S. C. Frautschi, and S. Mandelstam, *Phys. Rev.* **126,** 1202 (1962); S. C. Frautschi, M. Gell-Mann, and F. Zachariasen, *Phys. Rev.* **126,** 2204 (1962); R. Blankenbecler and M. L. Goldberger, *Phys. Rev.* **126,** 766 (1962); V. N. Gribov, *Soviet Phys. JETP* **14,** 478, 1395 (1962).

partial wave scattering amplitudes are functions analytic in the complex l-plane (to the right of some line Re $l < 0$), except for poles, the location of these poles $\alpha_n(s)$ again defining a trajectory whose path is associated with the existence of particles, stable or unstable, as in the nonrelativistic case. This conjecture could solve an unpleasant problem associated with the existence of particles of spin ≥ 1. Suppose we have a particle of spin J acting as a bound state in the s-channel of some system (e.g., the f^0, with $J = 2$ in the $\pi\pi$ system). The contribution of this particle to the scattering in the s-channel is given by

$$T(s, t) \sim \frac{g^2}{s - s_B} P_J\left(1 + \frac{2t}{s - s_0}\right)$$

By crossing symmetry, the same term will contribute to the scattering amplitude in the t-channel, and there such a term implies that, as the center of mass energy in that channel ($W_t \sim \sqrt{t}$) becomes large, the scattering amplitude increases as t^J. A detailed examination of the Mandelstam representation shows that there is no obvious way of canceling this singular contribution. The Regge mechanism will, however, do it. The pole contribution will be

$$\frac{g^2}{s - s_B} P_{\alpha(s)}\left(1 + \frac{2t}{s - s_0}\right)$$

with $\alpha(s_B) = J$. When we consider the t-channel, however, s is a momentum transfer variable, so that in the physical region $s < 0$, and there is no reason to believe that $\alpha(s)$, $s < 0$ could not be < 1. This remains the principal reason for our interest in the Regge mechanism.

The Regge theory can be taken over into relativistic theory with only minor modifications. One of them is that, because of the presence of exchange potentials, a trajectory will be characterized by a label, the "signature" that distinguishes whether the physical states will have even l or odd l.[1] In effect, the Regge pole contributions appear in the form

$$\frac{1}{2} \frac{\gamma_n(s)}{\sin \pi\alpha_n(s)} (P_{\alpha_n(s)}(-\cos\theta) \pm P_{\alpha_n}(\cos\theta))$$

This implies that a trajectory, which gives rise to an S-wave bound state, can only give rise to a D-wave bound state after that. This ensures that in the scattering of two identical pions, no odd angular momentum bound states can appear.

As in nonrelativistic theory, the trajectory of a Regge pole determines the asymptotic behavior of the amplitude when $\cos\theta$, i.e., $2t/(s - s_0) \to \infty$.

[1] See Appendix in S. C. Frautschi, M. Gell-Mann, and F. Zachariasen, *loc. cit.*

Because of crossing symmetry, this is no longer an academic statement. It implies that for $s \to \infty$ and $t < 0$, the behavior of the scattering amplitude is

$$T(s, t) \sim \sum_n \left(\frac{s}{s_0}\right)^{\alpha_n(t)} \beta_n(t)$$

where $\alpha_n(t)$ are the locations of the Regge poles which determine the bound states and resonances *in the t-channel*, and s_0 is some parameter with the dimensions of $(\text{mass})^2$. We thus speak of an "exchange" of Regge poles, which is a generalization of the exchange of single particles. What can we say about the trajectories $\alpha_n(t)$? The fact that for forward scattering $T(s, 0) \sim s$ implies that at $t = 0$ the leading trajectory must be characterized by $\alpha_P(0) = 1$. This trajectory has been called the Pomeranchuk trajectory; if we associate with it the quantum numbers of the vacuum, the Pomeranchuk theorems follow immediately.

In a plot of Re $\alpha_P(t)$ versus t, the slope near $t = 0$ must be positive. To see this, let us write, for small t,

$$\alpha_P(t) \cong 1 + \epsilon_P t \tag{28.42}$$

then

$$\frac{d\sigma}{dt} = \left(\frac{d\sigma}{dt}\right)_{t=0} e^{2(\alpha_P(t)-1)\log(s/s_0)} \tag{28.43}$$

where s_0 is some fixed constant with the dimensions of $(\text{mass})^2$. Since the differential cross section falls as $-t$ increases, ϵ_P must be positive.

The form predicted by (28.43) is the same as (28.27) for small t, except that the Regge theory predicts that the constant a increases with s logarithmically. There is no evidence for such a shrinkage of the diffraction peak with increasing energy,[1] and it is now believed that, if the Regge theory is correct in the relativistic quantum mechanics domain, the energies now available in the laboratory are presumably not high enough to reveal the simple Regge pole structure. We shall, therefore, not pursue this theory further, except to mention that attempts have been made to associate particles in "Regge recurrence" families. Particles belonging to a family must have the same internal quantum numbers and differ in spin by 2. It is possible to view some baryon resonances [e.g., N and $N^*(1688)$; $N^*(1238)$, and $N^*(1920)$; Λ^0 and $Y_0{}^*(1815)$, etc.] as connected in this way.

In view of the inapplicability of the Regge theory at the energies now being studied, it is of interest to ask what other models might explain the shape form shown in (28.27). Van Hove[2] has studied elastic scattering as

[1] Except for pp elastic scattering; for $p\bar{p}$ scattering there is anti-shrinkage.
[2] L. Van Hove, *Nuovo Cimento* **28,** 798 (1963).

the shadow scattering of inelastic processes and then made some simple assumptions about the nature of the inelastic processes at high energies, which do yield a form like (28.27). To see this, let us consider the unitarity condition for the elastic scattering amplitude (21.6).[1]

$$\begin{aligned} i[T(p', q'; p, q) - T^*(p, q; p', q')] \\ &= (2\pi)^4 \sum_n \delta(P_n - p - q) T^*(n; p', q') T(n; p, q) \\ &= \sum_n 2\pi\, \delta(P_n - p - q)(\Psi_{p'}, j(0)\, \Psi_n^{\text{out}})(\Psi_n^{\text{out}}, j(0)\, \Psi_p) \qquad (28.44) \end{aligned}$$

The right-hand side may be decomposed into an elastic part and an inelastic one

$$\begin{aligned} (2\pi)^4 \sum_{\text{inel}} \delta(P_n - p - q) T^*(n; p'. q') T(n; p, q) \\ + (2\pi)^4 \int \frac{d^3p''}{(2\pi)^3 \rho(p'')} \int \frac{d^3q''}{(2\pi)^3 \rho(q'')} \\ \times \delta(p'' + q'' - p - q) T^*(p'', q''; p', q') T(p'', q''; p, q) \qquad (28.45) \end{aligned}$$

As things stand, $T(p', q'; p, q)$ contains spin dependent terms as well as terms that do not depend on the spin. It is generally believed that at high energies the former are much smaller than the latter. As justification of this statement consider, for example, (23.28), which represents the pion-nucleon elastic scattering amplitude. We may write it in the form

$$\begin{aligned} f(W, \cos\theta) = \frac{1}{2iq} \sum_l (-1 + l\eta_{l-} e^{2i\delta_{l-}} + (l+1)\eta_{l+} e^{2i\delta_{l+}}) P_l(\cos\theta) \\ + \frac{\boldsymbol{\sigma} \cdot \hat{\mathbf{n}}}{2q} \sin\theta \sum_l (\eta_{l-} e^{2i\delta_{l-}} - \eta_{l+} e^{2i\delta_{l+}}) P_l'(\cos\theta) \qquad (28.46) \end{aligned}$$

and argue that when absorption is large, the first term dominates.

The unitarity condition (28.44) is thus supposed to be written for the spin independent terms, with the spin-dependent ones appearing on the right-hand side with the inelastic contribution.

Consider the state vector Φ_{pq} defined by

$$\Phi_{pq} = \sum_{\substack{n \\ (\text{inel})}} \Psi_n^{\text{out}} (\Psi_n^{\text{out}}, j(0) \Psi_p) \qquad (28.47)$$

This state vector is orthogonal to any state $\Psi_{p'q'}^{\text{out}}$, reached from Ψ_{pq}^{in} by an

[1] We use the T-matrix here, rather than the amplitude A, because the unitarity condition in Chapter 21 is worked out in terms of T.

elastic scattering. We associate it with the final inelastic state. We may write

$$(2\pi)\sum_{\text{inel}}(\Psi_n^{\text{out}}, j(0)\Psi_{p'})^*\,\delta(P_n - p - q)(\Psi_n^{\text{out}}, j(0)\Psi_p)$$
$$= (2\pi)\sum_{\substack{m,n\\ \text{inel}}}(\Psi_m^{\text{out}}, j(0)\Psi_{p'})^*(\Psi_m^{\text{out}}, \delta(P - p - q)\Psi_n^{\text{out}})$$
$$\times(\Psi_n^{\text{out}}, j(0)\Psi_p)$$
$$= 2\pi(\Phi_{p'q'}\,\delta(P - p - q)\Phi_{pq}) \tag{28.48}$$

where P is now an operator. In general, the labels (p, q) and (p', q') on the Φ only label the origin of the inelastic states. If we now make use of the experimetal information that at high energies the inelastic processes consist primarily of "jets" of particles moving (with small transverse momentum 0.3–0.5 BeV/c) in the forward and backward direction in the center of mass system, we may ignore all correlations other than with the directions determined by (p, q) and (p', q') in the center of mass system. Roughly speaking, we may write

$$(\Phi_{p'q'}, \delta(P - p - q)\Phi_{pq}) = \chi(\hat{\mathbf{p}}\cdot\hat{\mathbf{p}}', s) \tag{28.49}$$

in the center of mass system[1]. The function measures the overlap between the two inelastic uncorrelated jets, which differ only in the direction defined by the initial state. If R represents the rotation which takes the direction $\hat{\mathbf{p}}$ into the direction $\hat{\mathbf{p}}'$, then

$$R\Phi_{pq} = \sum_{\text{inel}} R\Psi_n^{\text{out}}(R\Psi_n^{\text{out}}, Rj(0)\Psi_p)$$
$$= \sum_{\text{inel}} \Psi_n'^{\text{out}}(\Psi_n'^{\text{out}}, j(0)\,\Psi_{p'}) = \Phi_{p'q'}$$

Hence[2]

$$\chi^*(\hat{\mathbf{p}}\cdot\hat{\mathbf{p}}', s) = (\Phi_{pq}, \delta(P - q - p)R\,\Phi_{pq}) = (R^{-1}\,\Phi_{pq}, \delta(P - q - p)\,\Phi_{pq})$$
$$= \chi(\hat{\mathbf{p}}\cdot\hat{\mathbf{p}}', s)$$

i.e., $\chi(\hat{\mathbf{p}}\cdot\hat{\mathbf{p}}', s)$ is real, as it of course must be. If we now write

$$T(p', q'; p, q)_{\text{c.m.}} = -\frac{\sqrt{s}}{8\pi^5}\sum(2l + 1)f_l(s)\,P_l(\cos\theta) \tag{28.50}$$

and

$$\chi(\hat{\mathbf{p}}\cdot\hat{\mathbf{p}}', s) = \frac{\sqrt{s}}{8\pi^5 q}\sum_l(2l + 1)\,\chi_l(s)P_l(\cos\theta) \tag{28.51}$$

[1] A detailed discussion may be found in L. Van Hove, *Rev. Mod. Phys.* **36**, 655 (1964).

[2] If $\hat{p}$ is chosen as z-axis, and $\hat{p}'$ makes an angle $(\theta, 0)$ with it, then $R^{-1}\Phi_{pq}$ will be associated with a jet in the direction (θ, π). We assume that there is no dependence on ϕ, so that the last step follows.

then we obtain

$$\mathrm{Im}\, f_l(s) = q\, |f_l(s)|^2 + \frac{1}{2q}\chi_l(s) \tag{28.52}$$

or, in terms of $\hat{f}_l = 2qf_l$

$$\mathrm{Im}\, \hat{f}_l(s) = |\hat{f}_l(s)|^2/2 + \chi_l(s) \tag{28.53}$$

If the elastic scattering amplitude is assumed to be imaginary, then

$$\tfrac{1}{2}\, |\mathrm{Im}\, \hat{f}_l(s)|^2 - \mathrm{Im}\, \hat{f}_l(s) + \chi_l(s) = 0$$

i.e.,

$$\mathrm{Im}\, \hat{f}_l(s) = 1 - \sqrt{1 - 2\chi_l(s)} \tag{28.54}$$

The particular choice of root is dictated by the fact that for sufficiently large values of l, when $l \gg qR$, where R is the dimension of the system, both f_l and χ_l should approach zero.

Van Hove shows that for an uncorrelated jet model

$$\begin{aligned} \chi(\hat{\mathbf{p}} \cdot \hat{\mathbf{p}}', s) &= \chi(0, s)e^{At} \\ t &= -2q^2(1 - \cos\theta) \approx -(q\theta)^2 \end{aligned} \tag{28.55}$$

Using the same kind of calculations which lead to (28.33) one finds that

$$\chi_l(s) = \chi_0 e^{-l^2/4Aq^2} \tag{28.56}$$

The total cross section is given by

$$\begin{aligned} \sigma_{\text{tot}} &= \sum_l (2l + 1)\frac{4\pi}{2q^2}\, \mathrm{Im}\, \hat{f}_l \\ &\approx \int_0^\infty x\, dx\, \frac{4\pi}{q^2}\left(1 - \sqrt{1 - 2\chi_0 e^{-x^2/4Aq^2}}\right) \\ &\approx 2\pi \int_0^\infty dy \left(1 - \sqrt{1 - 2\chi_0 e^{-y/4A}}\right) \\ &= 16\pi A\left(1 - \sqrt{1 - 2\chi_0} + \log \frac{1 + \sqrt{1 - 2\chi_0}}{2}\right) \end{aligned} \tag{28.57}$$

The elastic cross section is given by

$$\sigma_{\text{el}} = \frac{\pi}{q^2} \sum (2l + 1)\, |\hat{f}_l|^2 \tag{28.58}$$

so that the inelastic cross section is

$$\begin{aligned} \sigma_{\text{incl}} &= \sigma_{\text{tot}} - \sigma_{\text{el}} \\ &= \frac{2\pi}{q^2} \sum_l (2l + 1)\left\{\mathrm{Im}\, \hat{f}_l - \frac{1}{2}\, |\hat{f}_l|^2\right\} \\ &= \frac{2\pi}{q^2} \sum_l (2l + 1)\chi_l \end{aligned} \tag{28.59}$$

This is easily calculated to be

$$\sigma_{\text{incl}} \cong \frac{2\pi}{q^2}\int_0^\infty 2x\,dx\chi_0 e^{-x^2/4Aq^2}$$
$$= 8\pi A\chi_0 \tag{28.60}$$

Thus

$$\frac{\sigma_{\text{el}}}{\sigma_{\text{tot}}} = 1 - \frac{\sigma_{\text{incl}}}{\sigma_{\text{tot}}} = 1 - \frac{\chi_0}{2\left(1 - \sqrt{1-2\chi_0} + \log\dfrac{1+\sqrt{1-2\chi_0}}{2}\right)} \tag{28.61}$$

is independent of A. It may therefore be used to determine χ_0. Incidentally, we see from (28.54) that $0 \leq \chi_0 \leq \frac{1}{2}$. This implies, as a little computation shows, that

$$0 \leq \frac{\sigma_{\text{el}}}{\sigma_{\text{tot}}} \leq 0.185 \tag{28.62}$$

The experimental values of this ratio for $\pi^- p$-scattering at 17.6 BeV is 0.160 ± 0.013, so that the consistency condition is satisfied for this process. For pp-scattering, the ratio at 19.6 BeV is 0.244 ± 0.012, so that a modification of the treatment is required.[1]

The differential elastic scattering cross section is given by

$$\frac{d\sigma_{\text{el}}}{dt} = \pi\left|\sum_l (2l+1)\frac{\hat{f}_l}{2q^2}P_l(\cos\theta)\right|^2 \tag{28.63}$$

This involves calculating

$$\sum(2l+1)\frac{\hat{f}_l}{2q^2}P_l(\cos\theta) \approx \int_0^\infty \frac{x\,dx}{q^2}\left[1 - \sqrt{1-2\chi_0 e^{-x^2/4Aq^2}}\right]J_0(x\theta)$$
$$\approx \int_0^\infty b\,db\left[1-\sqrt{1-2\chi_0 e^{-b^2/4A}}\right]J_0(b\sqrt{-t}) \tag{28.64}$$

Van Hove[2] shows that this leads to the form

$$\frac{d\sigma_{\text{el}}}{dt} = \frac{\sigma_{\text{tot}}^2}{16\pi}e^{at+bt^2} \tag{28.65}$$

[1] W. N. Cottingham and R. F. Peierls, *Phys. Rev.* **137,** B147 (1965), have studied both processes quantitatively, and they obtain good agreement with experiment provided a real part of the elastic scattering amplitude (in the form of a difference of two Gaussian functions) is included for pp-scattering.

[2] L. Van Hove, *Rev. Mod. Phys., loc. cit.*

where

$$a = 2a_1 A \qquad b = \frac{a_2}{{a_1}^2}$$

The constants a_1 and a_2 depend on χ_0. For $\chi_0 = 0$, $a_1 = 1$, $a_2 = 0$; for $\chi_0 = \frac{1}{2}$, $a_1 = 0.86$, $a_2 = 0.03$, so that b/a^2 is always very small.

The Gaussian shape for $\chi(\hat{\mathbf{p}} \cdot \hat{\mathbf{p}}', s)$ of (28.55) also appears in the multiperipheral model,[1] in which the inelastic processes are described by a series of graphs of the type shown in Fig. 28.6. Limitations of space prevents us from discussing this interesting model, in which the origin of the Gaussian shape is again a certain decoupling between the initial state and the final inelastic state. The main point to be made in this connection is that there is some general justification for the form (28.55), and that form serves to explain the main features of the diffraction part of the scattering.

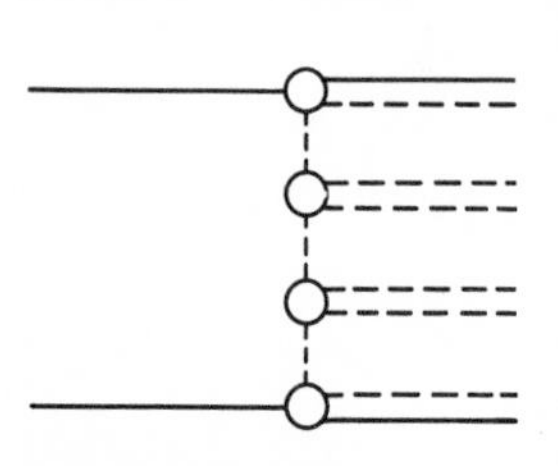

Fig. 28.6. The type of graphs summed up in the multiperipheral calculation of inelastic scattering.

The Van Hove model presented above makes definite predictions for the angular distribution of elastic scattering for large momentum transfers. The simple calculation of (28.64) for large $|t|$ may be found in the paper of Cottingham and Peierls. The result is the prediction that for large momentum transfers

$$\frac{d\sigma_{\mathrm{el}}}{dt} \sim \frac{e^{-\alpha\sqrt{-t}}}{t^2} \tag{28.66}$$

Actually the data (pp elastic scattering) are fitted much better by an empirical fit[2]

$$\frac{d\sigma}{d\Omega} = Ae^{-p_\perp/p_0} \tag{28.67}$$

where $p_\perp = p \sin \theta$, with p the center of mass momentum, $p_0 = 0.151$ BeV/c and $A = 34$ mb/ster (Fig. 28.7). There is no theoretical reason known at this time for the appearance of the transverse momentum in such an expression. Wu and Yang[3] have speculated that the particular dependence of (28.67) may be a reflection of a difficulty of accelerating different parts of a nucleon without breaking it up. If this is indeed the reason then a number of predictions follow for fixed-angle processes

[1] D. Amati, S. Fubini, and A. Stanghellini, *Nuovo Cimento* **26,** 896 (1962).
[2] J. Orear, *Phys. Rev. Letters,* **12,** 112 (1964).
[3] T. T. Wu and C. N. Yang, *Phys. Rev.* **137,** B708 (1965).

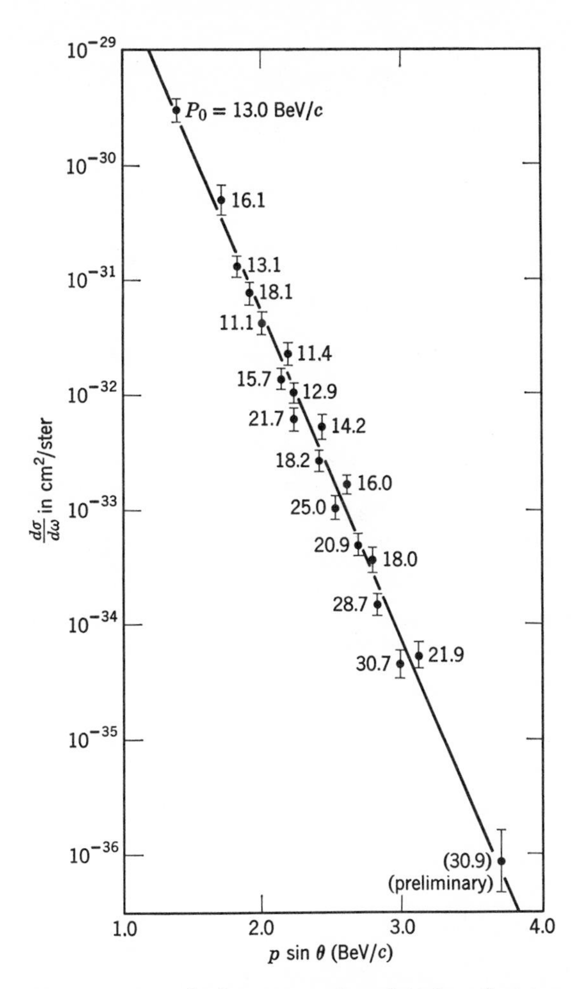

Fig. 28.7. Proton-proton elastic cross section plotted against transverse momentum transfer. The line is the exponential $Ae^{-p \sin \theta / p_0}$ with $A = 34$ mb/ster and $p_0 = 0.151$ BeV/c. [From J. Orear, *Phys. Rev. Letters* **12,** 112 (1964).]

($\theta \neq 0$,) as the energy becomes very large, namely)

$$\log \frac{d\sigma}{d\Omega}(\theta, pn \to pn) \Big/ \log \frac{d\sigma}{d\Omega}(\theta, pp \to pp) \to 1$$
$$\log \frac{d\sigma}{d\Omega}(\theta, pp \to pN^*) \Big/ \log \frac{d\sigma}{d\Omega}(\theta, pp \to pp) \to 1 \tag{28.68}$$

and others. Clearly, with the present rate of accumulation of data, a new major area of elementary particle physics is being opened up.

PROBLEMS

1. Show that (28.65) follows from (28.64). Use

$$\int_0^\infty x\, dx\, J_0(yx) e^{-\frac{1}{2}x^2} = e^{-\frac{1}{2}y^2}$$

2. The so-called Regge trajectories $\alpha_n(t)$ associated with the low-mass particles, have the property that

$$\alpha_n(m_n^2) = J_n$$

where m_n is the mass of the particle and J_n its spin. If it is assumed that

$$\alpha_n(t) \simeq \alpha_n(0) + \epsilon t$$

with a "universal" ϵ, and the $N_{\frac{1}{2}}(940)$ (the nucleon) and the $N_{\frac{1}{2}}(1512)$ lie on the same trajectory, find ϵ, and

(*a*) Find the mass of the 2^+-particle that lies on the same trajectory as the "Pomeranchuk," for which $\alpha_p(0) = 1$.

(*b*) Find the mass of the 3^--particle, which is the "Regge recurrence" of the ρ-meson. At what value of J, does the ρ-meson trajectory intersect the $t = 0$ axis?

(*c*) Consider the baryon octet and its $\frac{5}{2}^+$ Regge recurrences. Check the Gell-Mann—Okubo formula for this predicted octet.

PART IV

Introduction

The weak interactions, like the electromagnetic ones, play a dual role in elementary particle physics. On one hand, they manifest themselves in the interactions of leptons, where they can be studied without contamination by the problematical strong-interaction renormalization effects. On the other hand, they serve as tools for probing certain fundamental quantities—the matrix elements of currents involving the strongly interacting particles.

A very exciting development, not quite a decade old, is the discovery of the nature of the weak interaction of the leptons. This development was the result of a new class of experiments, which originated in the revolutionary paper of Lee and Yang on the question of parity conservation in the weak interactions. Within a very short time (but still anticipated by theoretical speculations), a form for the weak interaction Hamiltonian, for leptons at least, was determined by a series of unusually inventive experiments. Insofar one could "see through" the strong renormalization effects, the weak interactions appeared to share with electromagnetism a certain kind of universality. It appears likely that the fundamental constituents of matter, whatever they are, enter into the weak interactions in a very one-sided way: only one of the two helicity states appears to couple, so that the field operator for the spin $\frac{1}{2}$ baryon (or quark) appears in the weak interaction in the form $(1 - \gamma_5)\, \psi(x)$. Such, at least, is the most symmetric form of the theory that one can formulate. To check whether this is so, and to find a way of formulating the notion of universality in some general way, is the object of present studies of the weak interactions.

The discovery that $SU(3)$ appears to be the underlying symmetry of the strong interactions had an important impact on the understanding of the weak ones. There is mounting evidence to support the conjecture that the hadronic[1] quantities, which play a role in the weak interactions, are vector currents and axial currents, and that these are $SU(3)$ currents. Here again a similarity with electromagnetism appears: the quantities which interact with an electromagnetic field are also currents (only vector currents), and these too are $SU(3)$ currents.[2]

In Part IV of this book an attempt is made to present the phenomenology, and the data which have given us the considerable insight recently gained into the weak interactions. After a brief review of the status of the weak interactions just before the overthrow of parity, the question of tests of parity, charge conjugation and time reversal is investigated. Whereas parity conservation tests are easily visualized, this cannot be said of the other symmetries, which, we show, are not quite independent, since Lorentz invariance alone requires the validity of the combined operation CPT. In Chapter 30 the tests for all three symmetries are discussed in a formal but comprehensive manner.

Chapter 31 is devoted to a brief description of the series of experiments that within a year (1957) completely clarified the nature of the weak interactions at low energies, at least as far as non-strange particles are concerned. In the subsequent two chapters we discuss the weak interactions of the strange particles. These can be divided into two classes. The more fundamental ones are the leptonic interactions, in which a strongly interacting particle decays with the emission of a lepton pair. Because of our newly gained understanding of the way in which leptons interact with other systems, these interactions are analogous to electromagnetic interactions of hadrons. It is true that the study of decays yields a very limited amount of information. Nevertheless, even before experiments with neutrino beams provide us with the same kind of flexibility that electron beams provide for the study of form factors, enough can be learned about weak currents to specify their transformation properties. The other class of processes consists of nonleptonic processes, which may be viewed as an analog of the Coulomb scattering of two protons, for example. It is clear that there the strong interaction processes are much more effective in masking the basic interactions, and this field therefore shares with other strong interactions the difficulty of a lack of workable theory for strong couplings. We, therefore, limit ourselves to a phenomenological discussion of some of the nonleptonic processes.

In the final chapter, more detailed information about the currents is discussed. Evidence that the weak currents are part of an octet of $SU(3)$ currents is presented, and some very attractive conjectures about their

[1] We remind the reader that "hadrons" are the strongly interacting particles.

[2] Specifically, the generating current discussed in Chapter 26 on page 445.

properties are discussed: (*a*) that the axial currents, although not conserved, differ from conserved currents only slightly, and (*b*) that the commutation relations between currents are simple and involve the symmetry of the strong interactions in a fundamental way. The last is a point of view which has been adopted by Gell–Mann, and support for it is presented in a recent calculation of the strong renormalization of the axial current—a calculation in excellent agreement with experiment.

Our optimistic view of the progress made should not blind us to the fact that there is an enormous amount that we do not understand. Why are the weak interactions so simple? There is no apparent equivalent of gauge invariance which, at least formally, expresses the universality of the electromagnetic interactions. There is no counterpart to the photon, at least not with mass less than 2 BeV. The weak interaction of the leptons is very singular, and although energies of several hundred BeV in the center of mass system would be required to bring the theory to a point of breakdown, we know that the theory, as formulated, cannot be right. Finally, there is a feeling that progress in this area can only be made in conjunction with an understanding of the vast difference between hadrons and leptons—a field in which ideas have been singularly scarce. If these matters are not discussed in the succeeding chapters, it is only because no light has been shed on them.

The whole field of weak interactions has been happily served by a number of excellent reviews. Some are mentioned in relevant places in the text. Among the others are the review articles on the weak interactions in the *Annual Reviews of Nuclear Science*,[1] the very useful lectures by J. D. Jackson at the 1962 Brandeis Summer Institute,[2] and a very comprehensive collection of reprints edited by P. K. Kabir.[3] In addition, the proceedings of the conferences on Weak Interactions at Brookhaven (1963) and at Argonne (1965) are an indispensable source of the latest information, frequently presented in very digestible form. Many long derivations could be omitted because of the availability of the excellent textbook by G. Källén.[4] Finally, we would like to recommend the very un-pedestrian treatment of beta decay by H. Lipkin.[5]

[1] E. J. Konopinski, *Ann. Rev. Nucl. Sci.* **9** (1959); G. Feinberg and L. M. Lederman, *Ann. Rev. Nucl. Sci.* **13** (1963); T. D. Lee and C. S. Wu, *Ann. Rev. Nucl. Sci.* **15** (1965).
[2] J. D. Jackson in *Elementary Particles Physics and Field Theory*, W. A. Benjamin, New York (1963).
[3] *The Development of Weak Interaction Theory*, P. K. Kabir (Ed.), Gordon and Breach Science Publishers, New York. (1963).
[4] G. Källén, *Elementary Particle Physics*, Addison-Wesley Publishing Co., Reading, Massachusetts (1964).
[5] H. J. Lipkin, *Beta Decay for Pedestrians*, North-Holland Publishing Co., Amsterdam (1962).

29

"Classical Theory" of Beta Decay

The theory of β-decay, i.e., the theory describing the process

$$n \to p + e^- + \bar{\nu}$$

dates from 1934, and is due to Fermi.[1] In 1931 Pauli postulated the existence of a new, highly penetrating particle, the neutrino, in order to explain the continuous energy distribution of the electrons observed in nuclear decays. Subsequently, Fermi pointed out that, in analogy with quantum electrodynamics, a coupling of the form

$$H_I = C_V \int d^3x (\bar{\psi}_p(x)\gamma^\mu \, \psi_n(x))(\bar{\psi}_e(x)\gamma_\mu \, \psi_\nu(x)) \tag{29.1}$$

could be used to describe the process. We shall presently see that the coupling constant C_V, whose dimensions are $[M]^{-2}$ when $\hbar = c = 1$, is very small. Thus our perturbation treatment, which follows, is justified at energies such that $\langle E \rangle^2 C_V \ll 1$, where E is the largest energy appearing in the process, e.g., the nucleon mass, in beta decay.

One of the earliest observations about the Fermi type of coupling was that it is not the most general coupling possible. Even if we ignore the possibility of derivative couplings, Lorentz invariance, together with invariance under time reversal and space inversions, allows for five couplings, contained in H_w below[2]

$$H_w = \int d^3x \Big[C_S(\bar{\psi}_p(x) \, \psi_n(x))(\bar{\psi}_e(x) \, \psi_\nu(x)) + C_V(\bar{\psi}_p(x)\gamma^\alpha \psi_n(x))(\bar{\psi}_e(x)\gamma_\alpha \, \psi_\nu(x))$$

[1] E. Fermi, *Nuovo Cimento* **11,** 1 (1934); *Z. Phys.* **88,** 161 (1934). Pauli's neutrino hypothesis was never published. For a very interesting historical review (which also contains much useful material) see C. S. Wu, "The Neutrino," in *Theoretical Physics in the Twentieth Century*, M. Fierz and V. F. Weisskopf (Eds.), Interscience Publishers, New York (1960).

[2] Because of our choice of metric, the number which is denoted by C_A in (29.2) is opposite in sign to the conventional C_A as in M. E. Rose "*Beta and Gamma-Ray Spectroscopy*" (Interscience Publishers Inc. New York, 1955), for example.

$$+ \frac{1}{2} C_T(\bar{\psi}_p(x)\sigma^{\alpha\beta}\psi_n(x))(\bar{\psi}_e(x)\sigma_{\alpha\beta}\,\psi_\nu(x))$$
$$+ C_A(\bar{\psi}_p(x)\gamma^\alpha\gamma_5\psi_n(x))(\bar{\psi}_e(x)\gamma_\alpha\gamma_5\psi_\nu(x))$$
$$+ C_P(\bar{\psi}_p(x)i\gamma_5\psi_n(x))(\bar{\psi}_e(x)i\gamma_5\psi_\nu(x))\Big] \tag{29.2}$$

where $\sigma^{\alpha\beta} = \frac{i}{2}[\gamma^\alpha, \gamma^\beta]$ and the C_i are real constants. The β-decay amplitude, to lowest order in the coupling is[1]

$$(\Psi_f, H_w\Psi_i) = \sum_{i=S}^{P} C_i \int_{x_0=0} d^3x(\Psi_f, (\bar{\psi}_p\Gamma^i\psi_n)(\bar{\psi}_e\Gamma_i\psi_\nu)\Psi_i)$$
$$= \sum_i C_i \int_{x_0=0} d^3x(\tilde{\Psi}_f, \bar{\psi}_p(x)\Gamma^i\psi_n(x)\tilde{\Psi}_i)$$
$$\times (\Psi_e, \bar{\psi}_e(x)\Psi_0)\Gamma_i \frac{1}{(2\pi)^{3/2}}\sqrt{2m_\nu}\, v(p_\nu)e^{-i\mathbf{p}_\nu\cdot\mathbf{x}} \tag{29.3}$$

The factor $\sqrt{2m_\nu}$ accompanying the neutrino spinor has the effect of making the normalization factor in the rate identical with that of a photon. This makes taking the limit $m_\nu \to 0$ a little easier. The quantity $(\Psi_e, \bar{\psi}_e(x)\Psi_0)$ is just the electron wave function; if the charge of the nucleus is small, so that the Coulomb effects can be neglected, we write

$$(\Psi_e, \bar{\psi}_e(x)\Psi_0) = e^{-i\mathbf{p}_e\cdot\mathbf{x}} \frac{\bar{u}(p_e)}{(2\pi)^{3/2}} \tag{29.4}$$

Otherwise the wave function for the electron in the nuclear Coulomb field must be used. For the purposes of this chapter (29.4) is adequate.

The nuclear matrix elements

$$\int d^3x e^{-i(\mathbf{p}_e+\mathbf{p}_\nu)\cdot\mathbf{x}}(\tilde{\Psi}_f, \bar{\psi}_p(x)\Gamma_i\,\psi_n(x)\tilde{\Psi}_i) \tag{29.5}$$

involve the operators

$$\begin{aligned}
\bar{\psi}_p(x)\,\psi_n(x) &\cong \psi_p^{+}(x)\,\psi_n(x) \\
\bar{\psi}_p(x)\gamma^k\,\psi_n(x) &\cong \psi_p^{+}(x)\rho_1\sigma^k\,\psi_n(x) \cong 0 \\
\bar{\psi}_p(x)\gamma^0\,\psi_n(x) &= \psi_p^{+}(x)\,\psi_n(x) \\
\bar{\psi}_p(x)\sigma^{ij}\,\psi_n(x) &\cong \psi_p^{+}(x)\sigma^{ij}\,\psi_n(x) = e_{ijk}\,\psi_p^{+}(x)\sigma_k\,\psi_n(x) \\
\bar{\psi}_p(x)\sigma^{0k}\,\psi_n(x) &= -\psi_p^{+}(x)\rho_2\sigma^k\,\psi_n(x) \cong 0 \\
\bar{\psi}_p(x)\gamma^0\gamma_5\,\psi_n(x) &= \psi_p^{+}(x)\rho_1\,\psi_n(x) \cong 0 \\
\bar{\psi}_p(x)\gamma^k\gamma_5\,\psi_n(x) &\cong \psi_p^{+}(x)\sigma^k\,\psi_n(x) \\
\bar{\psi}_p(x)i\gamma_5\,\psi_n(x) &= -\psi_p^{+}(x)\rho_2\,\psi_n(x) \cong 0
\end{aligned} \tag{29.6}$$

[1] We use $\tilde{\Psi}_f$ and $\tilde{\Psi}_i$ to denote the final and initial states of the *nucleus*.

In these expressions, the terms on the extreme right represent the nonrelativistic approximation. It is reasonable to make this approximation, since the momenta of nucleons in a nucleus are usually quite small ($v \lesssim \frac{1}{10}$). With this approximation there are only two nuclear matrix elements which need to be considered

$$\int_{x_0=0} d^3x e^{-i\boldsymbol{\Delta}\cdot\mathbf{x}} (\tilde{\Psi}_f, \psi_p^+(x)\, \psi_n(x)\Psi_i) \tag{29.7}$$

and

$$\int_{x_0=0} d^3x e^{-i\boldsymbol{\Delta}\cdot\mathbf{x}} (\tilde{\Psi}_f, \psi_p^+(x)\boldsymbol{\sigma}\, \psi_n(x)\tilde{\Psi}_i) \tag{29.8}$$

In the expressions above

$$\boldsymbol{\Delta} = \mathbf{p}_e + \mathbf{p}_\nu = \mathbf{p}_i - \mathbf{p}_f$$

is the momentum transfer in the decay of the nucleus. In general, unless angular momentum or parity selection rules intervene, the principal contribution to the matrix element comes from unity in the expansion

$$e^{-i\boldsymbol{\Delta}\cdot\mathbf{x}} = 1 - i\,\boldsymbol{\Delta}\cdot\mathbf{x} - \tfrac{1}{2}(\boldsymbol{\Delta}\cdot\mathbf{x})^2 + \cdots$$

since the momentum transfers are such that $|\boldsymbol{\Delta}|\, R$ (with R a length of the order of a nuclear radius) is small.

From the point of view of nuclear physics, the corrections of order $|\boldsymbol{\Delta}|\, R$, etc., are very interesting as they give us information about the nuclear states before and after the decay; for our purposes, the leading term in $e^{-i\boldsymbol{\Delta}\cdot\mathbf{x}}$ yields the most direct information about the β-interaction. If the initial and final states differ in parity, or if they differ by more than one unit of angular momentum, relativistic corrections and higher terms in $e^{-i\boldsymbol{\Delta}\cdot\mathbf{x}}$ will give rise to such "forbidden" decays. We shall limit our inquiry to allowed transitions. In what follows we shall use the notation

$$\begin{aligned} \int d^3x(\tilde{\Psi}_f, \psi_p^+(x)\, \psi_n(x)\tilde{\Psi}_i) &= \langle \mathbf{1} \rangle \\ \int d^3x(\tilde{\Psi}_f, \psi_p^+(x)\boldsymbol{\sigma}\, \psi_n(x)\tilde{\Psi}_i) &= \langle \boldsymbol{\sigma} \rangle \end{aligned} \tag{29.9}$$

Electromagnetic corrections, arising principally from the distortion of the electron wave function by the nuclear Coulomb field, will be neglected.

The transition matrix element for the allowed transitions is $\mathcal{M}$, where

$$S = 1 - 2\pi i\, \delta(E_f + E_e + E_\nu - E_i)\mathcal{M}$$

and

$$\mathcal{M} = \frac{\sqrt{2m_\nu}}{(2\pi)^3}[\langle \mathbf{1}\rangle(C_S\,\bar{u}(p_e)\,v(p_\nu) + C_V\,\bar{u}(p_e)\gamma_0\,v(p_\nu)) + \langle\sigma^k\rangle(\tfrac{1}{2}C_T\,\bar{u}(p_e)\sigma_k\,v(p_\nu) + C_A\,\bar{u}(p_e)\gamma_k\gamma_5\,v(p_\nu))] \quad (29.10)$$

To find the decay rate, we must obtain the absolute value of the square of (29.10). We shall also sum over the observed neutrino polarization states, making use of the fact that when the neutrino mass is zero

$$\lim_{m_\nu\to 0} 2m_\nu \sum_{\text{spins}} v(p_\nu)\,\bar{v}(p_\nu) = (\not{p}_\nu - m_\nu) = \not{p}_\nu \quad (29.11)$$

If we also sum over the electron polarization states, and only deal with experiments in which the decaying nucleus is unpolarized, so that the average over initial nuclear polarization states leads to

$$\overline{\langle\sigma^k\rangle\langle\mathbf{1}\rangle} = 0$$
$$\overline{\langle\sigma^k\rangle\langle\sigma^l\rangle} = \tfrac{1}{3}\,\delta_{kl}\,|\langle\boldsymbol{\sigma}\rangle|^2 \quad (29.12)$$

we are left with the calculation of

$$\frac{1}{(2\pi)^6}\frac{1}{2m_e}\Big[|C_S|^2\,|\langle\mathbf{1}\rangle|^2\,\mathrm{tr}\,(\not{p}_\nu\not{p}_e) + C_V^{\,2}\,|\langle\mathbf{1}\rangle|^2\,\mathrm{tr}\,(\gamma_0\not{p}_\nu\gamma_0\not{p}_e) + 2C_SC_V\,|\langle\mathbf{1}\rangle|^2\,\mathrm{tr}\,(\not{p}_\nu\gamma_0 m_e) + \frac{1}{12}\,C_T^{\,2}\,|\langle\boldsymbol{\sigma}\rangle|^2\,\mathrm{tr}\,(\sigma_k\not{p}_\nu\sigma_k\not{p}_e) + \frac{1}{3}\,C_A^{\,2}\,|\langle\boldsymbol{\sigma}\rangle|^2\,\mathrm{tr}\,(\gamma_k\gamma_5\not{p}_\nu\gamma_5\gamma_k\not{p}_e) + \frac{1}{3}\,m_eC_AC_T\,|\langle\boldsymbol{\sigma}\rangle|^2\,\mathrm{tr}\,(\sigma_k\not{p}_\nu\gamma_5\gamma^k)\Big]$$

A straightforward evaluation of the traces yields the following result

$$\frac{2E_eE_\nu}{m_e(2\pi)^6}\Big[C_S^{\,2}\,|\langle\mathbf{1}\rangle|^2\Big(1 - \frac{\mathbf{p}_e\cdot\mathbf{p}_\nu}{E_eE_\nu}\Big) + \frac{2m_e}{E_e}\,C_SC_V\,|\langle\mathbf{1}\rangle|^2 + C_V^{\,2}\,|\langle\mathbf{1}\rangle|^2\Big(1 + \frac{\mathbf{p}_e\cdot\mathbf{p}_\nu}{E_eE_\nu}\Big) + \frac{1}{4}\,C_T^{\,2}\,|\langle\boldsymbol{\sigma}\rangle|^2\Big(1 + \frac{1}{3}\frac{\mathbf{p}_e\cdot\mathbf{p}_\nu}{E_eE_\nu}\Big) - \frac{m_e}{E_e}\,C_AC_T\,|\langle\boldsymbol{\sigma}\rangle|^2 + C_A^{\,2}\,|\langle\boldsymbol{\sigma}\rangle|^2\Big(1 - \frac{1}{3}\frac{\mathbf{p}_e\cdot\mathbf{p}_\nu}{E_eE_\nu}\Big) \quad (29.13)$$

In the search for the form of the β-interactions before the events of 1957, the following properties of the above expressions were of interest:

1. The interference terms between S and V, and those between T and A, have a marked energy dependence. No such dependence on the electron energy is observed in the experimental spectrum. From the absence of these "Fierz interference terms" it was deduced that[1]

$$\begin{aligned} C_S C_V &\approx 0 \\ C_T C_A &\approx 0 \end{aligned} \tag{29.14}$$

2. Transitions, for which the change in the nuclear state involves no angular momentum change and no parity change, can take place via both the "Fermi transitions," for which $|\langle \mathbf{1} \rangle|^2 \neq 0$ and the "Gamow-Teller transitions,[2] for which $|\langle \boldsymbol{\sigma} \rangle|^2 \neq 0$. An exception are 0–0 transitions, which cannot occur via A or T, since $\langle J = 0 | \boldsymbol{\sigma} | J = 0 \rangle = 0$. Thus if we wish to study the magnitude of C_S or C_V we must look for 0–0, no parity change, transitions. In such transitions the S and V spectra differ only in the *electron-neutrino correlations*: for the scalar interaction, the electron and the neutrino prefer to come out of the nucleus in opposite directions, whereas the vector interaction leads to a preference for parallel electron and neutrino momenta.

3. The transitions which involve $|\Delta \mathbf{J}| = 1$, no parity change, in the nuclear states can only go via the Gamow-Teller transitions, i.e., they involve T and A. These can again be distinguished by the electron-neutrino correlation, with the correlation somewhat washed out by the factor $\frac{1}{3}$ which comes from the average over nuclear polarization states.

The experimental situation in 1956 was confused: because of the great difficulty of correlation experiments, there were conflicting results. Thus the decay

$$He^6 \rightarrow Li^6 + e^- + \nu$$

involving a change from $J = 0$ to $J = 1$ (no nuclear parity change), indicated a negative correlation in one experiment, suggesting $C_A \neq 0$, and showed a positive correlation, i.e., a tensor interaction, in another. Similarly, the decay

$$Ne^{19} \rightarrow F^{19} + e^+ + \nu$$

with a spin change from $J = \frac{1}{2}$ to $J = \frac{1}{2}$ (no parity change), indicated the preponderance of S and T, with rather large errors, whereas the decay

$$A^{35} \rightarrow Cl^{35} + e^+ + \nu$$

rather clearly indicated the dominance of V.

[1] M. Fierz, *Z. Phys.* **104,** 553 (1937).
[2] G. Gamow and E. Teller, *Phys. Rev.* **49,** 895 (1936).

Under the impact of the developments of 1956–57, the A^{35} experiment and the He^6 experiment were redone, and they showed that V and A were indeed dominant.[1]

The β-decay interaction, because it involves nonrelativistic nucleons, yields no information about the presence and magnitude of C_P. Information about this term came somewhat indirectly in connection with the idea of a *Universal Fermi Interaction.* With the quantitative analysis of the weak interactions of the μ-meson (muon), it became clear that the decay

$$\mu^{\pm} \rightarrow e^{\pm} + \nu + \nu$$

and the nuclear capture

$$\mu^{-} + Z_A \rightarrow (Z - 1)_A + \nu$$

could properly be described by a four-fermion interaction similar to that responsible for the β-decay, and that the strength of the interaction was roughly the same for these processes as for the β-interaction.[2] The proposal that all of these be represented by the *same* four-fermion interaction[3]

$$H_w = \sum C_i(\bar{\psi}_1\Gamma^i\psi_2)(\bar{\psi}_3\Gamma_i\psi_4) \tag{29.15}$$

was very attractive.[4] One of its consequences was that the pion should

[1] For a review of this class of experiments, a description of the apparatus, and further references, see J. S. Allen, *Rev. Mod. Phys.* **31,** 791 (1959).

[2] In view of the fact that the neutron has a lifetime of 10^3 sec and the muon has a lifetime of 10^{-6} sec, this statement may seem a little strange. Since the phase space varies as the fifth power of the maximum electron energy, the phase-space factor can account for this factor of 10^9. A very simple discussion of the μ-capture may be found in E. Fermi, *Elementary Particles*, Yale University Press, New Haven (1951).

[3] G. Puppi, *Nuovo Cimento* **5,** 587 (1948); O. Klein, *Nature* **161,** 897 (1948); J. Tiomno and J. A. Wheeler, *Rev. Mod. Phys.* **21,** 144 (1949); T. D. Lee, M. Rosenbluth, and C. N. Yang, *Phys. Rev.* **75,** 905 (1949).

[4] The ambiguity between $\Sigma\, C_i(\bar{\psi}_1\Gamma_i\psi_2)(\bar{\psi}_3\Gamma_i\psi_4)$ and the form in which ψ_2 and ψ_4 are interchanged is not fundamental. It leads to a linear transformation of the C_i. The equivalence of the two forms implies that

$$\sum C_i(\Gamma^i)_{ab}(\Gamma^i)_{cd} = \sum \tilde{C}_i(\Gamma^i)_{ad}(\Gamma^i)_{cb}$$

which, with appropriately normalized Γ^i, leads to a relation of the type

$$\tilde{C}_j = \sum_i C_i\, \tfrac{1}{4}\, \mathrm{tr}\,(\Gamma^i\Gamma^j\Gamma^i\Gamma^j).$$

This is known as a Fierz transformation. See, for example, R. H. Good, *Rev. Mod. Phys.* **27,** 187 (1955).

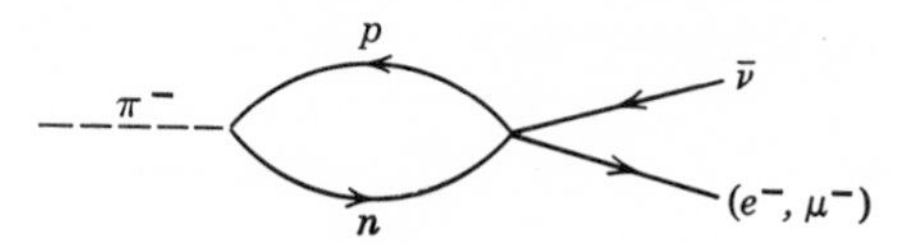

Fig. 29.1. A possible graph leading to the decay $\pi^- \to (\mu^-, e^-) + \bar{\nu}$.

decay weakly

$$\pi \to \mu + \nu$$
$$\pi \to e + \nu$$

through a mechanism, of which the one shown in Fig. 29.1 is typical. The decay is obtained from the matrix element

$$(\Psi_{e^-\bar{\nu}}, H_w \Psi_\pi) = \sum_i C_i(\Psi_{e^-\bar{\nu}}, \bar{\psi}_e \Gamma^i \psi_\nu \Psi_0)(\Psi_0, \bar{\psi}_p \Gamma_i \psi_n \Psi_\pi) \qquad (29.16)$$

Parity and angular momentum conservation imply that only the axial vector and the pseudoscalar terms in $(\Psi_0, \bar{\psi}_p \Gamma_i \psi_n \Psi_{\pi^-})$ do not vanish. We thus get

$$\frac{\sqrt{2m_\nu}}{(2\pi)^3}\{C_A\, \bar{u}(p_e)\gamma_\mu\gamma_5\, v(p_\nu)(\Psi_0, \bar{\psi}_p\gamma^\mu\gamma_5\psi_n\Psi_{\pi^-})$$
$$- C_P\, \bar{u}(p_e)\gamma_5\, v(p_\nu)(\Psi_0, \bar{\psi}_p\gamma_5\psi_n\Psi_{\pi^-})\} \qquad (29.17)$$

Lorentz invariance requires that the form of the matrix elements involving the strongly interacting particles be such that

$$(2\pi)^{3/2} C_A(\Psi_0, \bar{\psi}_p\gamma^\mu\gamma_5\psi_n\Psi_{\pi^-}) = i\frac{q^\mu}{m_\pi} f_A(q^2) = i\frac{q^\mu}{m_\pi} f_A(m_\pi^2)$$
$$(2\pi)^{3/2} C_P(\Psi_0, \bar{\psi}_p i\gamma_5\psi_n\Psi_{\pi^-}) = if_P(q^2) = if_P(m_\pi^2) \qquad (29.18)$$

We use $q = p_e + p_\nu$ and the Dirac equation for the leptons to obtain

$$T = -\frac{\sqrt{2m_\nu}}{(2\pi)^{9/2}}\left(i\frac{f_A}{m_\pi}\,\bar{u}(p_e)(\not{p}_e + \not{p}_\nu)\gamma_5\, v(p_\nu) + if_P\,\bar{u}(p_e)\gamma_5\, v(p_\nu)\right)$$
$$= -i\frac{\sqrt{2m_\nu}}{(2\pi)^{9/2}}\left(\frac{m_e}{m_\pi}f_A + f_P\right)\bar{u}(p_e)\gamma_5\, v(p_\nu) \qquad (29.19)$$

The square of the matrix element, summed over lepton spins, is

$$\sum_{\text{spins}} |T|^2 = \frac{1}{(2\pi)^9}\left(\frac{m_e}{m_\pi}f_A + f_P\right)^2 2m_\nu \sum_{\text{spins}} \bar{u}\gamma_5\, v(-\bar{v}\gamma_5\, u)$$
$$= \frac{1}{(2\pi)^9}\left(\frac{m_e}{m_\pi}f_A + f_P\right)^2 \frac{1}{2m_e}\,\text{tr}\,(\gamma_5\not{p}_\nu\gamma_5(-\not{p}_e))$$
$$= \frac{1}{(2\pi)^9}\left(\frac{m_e}{m_\pi}f_A + f_P\right)^2 \frac{2p_e\cdot p_\nu}{m_e} \qquad (29.20)$$

The decay rate is easily calculated. We have

$$\begin{aligned}\Gamma &= \frac{(2\pi)^4\,|T|^2}{2m_\pi/(2\pi)^3}\int \frac{d^3p_\nu}{2E_\nu}\,\frac{d^3p_e}{E_e/m_e}\,\delta(p_\nu + p_e - q)\\ &= \frac{(2\pi)^7}{2m_\pi}\,\frac{1}{(2\pi)^9}\left(\frac{m_e}{m_\pi}f_A + f_P\right)^2\left(\frac{m_\pi^{\,2} - m_e^{\,2}}{m_e}\right)4\pi m_e\\ &\quad\times\int_0^\infty \frac{p_e\,dp_e}{2E_e}\,\delta(p_e + E_e - m_\pi)\\ &= \frac{m_\pi}{8\pi}\left(1 - \frac{m_e^{\,2}}{m_\pi^{\,2}}\right)^2\left(\frac{m_e}{m_\pi}f_A + f_P\right)^2 \end{aligned}\tag{29.21}$$

The ratio of the decay rates into electron-neutrino to muon-neutrino modes is given by[1]

$$R = \frac{\Gamma(\pi \to e + \nu)}{\Gamma(\pi \to \mu + \nu)} = \left(\frac{m_\pi^{\,2} - m_e^{\,2}}{m_\pi^{\,2} - m_\mu^{\,2}}\right)^2 \frac{\left(\dfrac{m_e}{m_\pi}f_A + f_P\right)^2}{\left(\dfrac{m_\mu}{m_\pi}f_A + f_P\right)^2}\tag{29.22}$$

This quantity is very sensitive to a small admixture of f_P. The experimentally observed ratio,[2] which is very sensitive to the value of f_P/f_A

$$R = (1.25 \pm 0.03) \times 10^{-4}$$

is obtained with the choice

$$f_P = 0\tag{29.23}$$

which implies that there can be no pseudoscalar interaction. The lifetime of the pion implies that[3]

$$f_A = 2.12 \times 10^{-7}\tag{29.24}$$

a number which will be kept for future reference.

The pion decay yields no information about the tensor interaction, because

$$(\Psi_0, \bar\psi_p\sigma_{\mu\nu}\psi_n\Psi_\pi) = 0$$

This is most easily seen from the impossibility of constructing an antisymmetric tensor out of a single four vector q^μ. The tensor interaction does contribute to the decay

$$\pi^\pm \to e^\pm + \nu + \gamma$$

[1] M. A. Ruderman and R. Finkelstein, *Phys. Rev.* **76,** 1458 (1949).

[2] E. di Capua, R. Garland, L. Pondrom, and A. Strelzoff, *Phys. Rev.* **133B,** 1333 (1964), where references to earlier work may be found.

[3] Compare $f_A^{\,2}/4\pi = 3.5 \times 10^{-15}$ with, for example, $f_{\rho\pi\pi}^2/4\pi \simeq 2.4$ to get an idea of how weak the weak interactions are.

and a rough perturbation type calculation, with a reasonable cut-off, yields[1]

$$R_\gamma = \frac{\Gamma(\pi \to e + \nu + \gamma)}{\Gamma(\pi \to \mu + \nu)} \geqslant 0.02$$

in complete disagreement with the experimental upper limit[2]

$$R_\gamma < 10^{-4}$$

Since this result was in contradiction with the He^6 experiment, and, furthermore, since an incorrect experiment had yielded

$$\Gamma(\pi \to e\nu)/\Gamma(\pi \to \mu\nu) < 10^{-5}$$

in 1956, the situation in the weak interactions of the pion was also confused. The confusion was soon to be removed in an unprecedented burst of experimental activity stimulated by a paper of T. D. Lee and C. N. Yang, entitled "Question of Parity Conservation in Weak Interactions."

[1] Kjwata, S. Ogawa, H. Okoni, B. Sakita, and S. Oneda, *Progr. Theoret. Phys.* **13,** 19 (1955).

[2] J. M. Cassels, M. Rigby, A. M. Wetherell, and J. R. Wormwald, *Proc. Phys. Soc. A.* (*London*) **70,** 729 (1957).

30

Parity, Charge Conjugation and Time Reversal, and Tests for Their Validity

In 1956, in an attempt to solve the $\tau - \theta$ puzzle, Lee and Yang[1] asked the question of what the evidence for parity conservation in the weak interactions really was. They pointed out that, although there was good evidence that parity was conserved in the strong and electromagnetic interactions, no experiments performed until that time had tested the parity conservation in the weak interactions. In the case of electromagnetic interactions, the empirical rule of Laporte that in dipole radiation "even" states only combine with "odd" states is equivalent to parity conservation.[2] The Laporte rules have been tested to a high degree of accuracy, so that one may conclude that if the transition amplitude is

$$T = T_{\text{scalar}} + T_{\text{pseudoscalar}}$$

then $|T_P/T_S| < 10^{-6}$. In the strong interactions the evidence for parity conservation is equally strong. For example, Tanner[3] has pointed out that the decay of the 13.2 MeV excited state $\text{Ne}^{20*}(1^+)$ into 0^{16} (with spin-parity 0^+), and an α-particle is forbidden by parity conservation. If parity were violated, evidence for this decay would appear in a resonance in the α-spectrum in the reaction

$$p + \text{F}^{19} \rightarrow 0^{16} + \alpha$$

The absence of such a phenomenon leads to the conclusion that for the strong interactions[4]

$$\left|\frac{T_P}{T_S}\right| < 10^{-4}$$

[1] T. D. Lee and C. N. Yang, *Phys. Rev.* **104,** 254 (1956).

[2] For example, see E. P. Wigner, *Group Theory and its Applications to the Quantum Mechanics of Atomic Spectra,* Academic Press, New York (1959).

[3] N. Tanner, *Phys. Rev.* **107,** 1203 (1957).

[4] In a recent experiment F. Boehm and E. Kankeleit, *Phys. Rev. Letters* **14,** 312 (1965), observed the circular polarization of the order of 10^{-4} (i.e., a preference of one helicity state over the other) of a γ-transition in Ta^{181}. They find $|T_P/T_S| \sim 10^{-6}$ in agreement with an estimate based on parity violation by the weak interactions.

In the weak interactions, all the classical β-decay experiments measuring quantities such as the shape of the electron spectra, the electron-neutrino correlations, $\beta - \gamma$ correlations, forbidden spectra, etc., deal with *scalar* quantities. If we again write a transition matrix in the form

$$T = T_S + T_P$$

then the term in the absolute square of this, which indicates parity nonconservation, must be the interference term. Such a term is pseudoscalar, and rate measurements are such as to average this term away. In order to test parity conservation, a pseudoscalar quantity must be measured, and Lee and Yang, after making this crucial observation, proceeded to list experiments which might reveal whether parity was conserved in the weak interactions.

One of the experiments was a measurement of the correlation between the momentum of the electron in the β-decay of a polarized nucleus and the direction of polarization of the nucleus. Such a correlation experiment measures the coefficient of the quantity $\mathbf{p}_e \cdot \langle \mathbf{J} \rangle$ in the matrix element, and this quantity is clearly pseudoscalar, since under spatial inversion $\mathbf{p}_e \rightarrow -\mathbf{p}_e$ but $\mathbf{J} \rightarrow \mathbf{J}$. Another experiment testing parity conservation would be the measurement of the longitudinal polarization of the electron. Only when parity is conserved is the expectation value of $\langle \boldsymbol{\sigma}_e \cdot \mathbf{p}_e \rangle$ guaranteed to be zero.

In the weak decay of hyperons, the Λ^0, for example, parity nonconservation might manifest itself in a correlation between the direction of the Λ^0-polarization (which will be perpendicular to the production plane) and the pion momentum in

$$\Lambda^0 \rightarrow p + \pi^-$$

Such a correlation would manifest itself in an up-down asymmetry in the number of pions relative to the production plane.

The experiments suggested by Lee and Yang, and many others, were soon carried out, with the startling result that parity was *not* conserved in the weak interactions. We shall discuss some of these experiments later, but we must first turn to some other questions that were raised in conjunction with this one. These questions had to do with a general theorem of quantum field theory, first noted by Lüders[1] but also discussed by Schwinger and Pauli[2] (who in a very timely fashion had drawn attention to its general nature in 1955), which states that every possible state of a system of particles is a possible state of the corresponding system of

[1] For a simple exposition of his original work, see G. Lüders, *Ann. Phy. N.Y.* **2,** 1 (1957).
[2] W. Pauli, *Niels Bohr and the Development of Physics*, McGraw-Hill Book Co., New York (1955).

antiparticles with space and time inverted. More formally, this theorem states that under very general conditions there is invariance under the product of the operations C (charge conjugation), P (parity) and T (time reversal). In order to prove this *CPT theorem*, it is actually sufficient to assume Lorentz invariance, the existence of a vacuum state and local commutativity. As a matter of fact, the last statement can be replaced by a weaker one.[1] We shall explore some of the consequences of this theorem later in the chapter, but we shall first illustrate it for a four-fermion interaction, so that we may obtain a clear grasp of what is involved.[2]

We consider the interaction Hamiltonian

$$H_w = \int_{x_0=0} d^3x\, \mathcal{H}_w(x) \tag{30.1}$$

with

$$\begin{aligned}\mathcal{H}_w(x) = {} & \sum_i C_i(\bar{\psi}_1\Gamma_i\psi_2)(\bar{\psi}_3\Gamma_i\psi_4) \\ & + \sum_i C_i^*(\bar{\psi}_2\gamma_0\Gamma_i^+\gamma_0\psi_1)(\bar{\psi}_4\gamma_0\Gamma_i^+\gamma_0\psi_3) \\ & + \sum_i C_i'(\bar{\psi}_1\Gamma_i\psi_2)(\bar{\psi}_3\Gamma_i\gamma_5\psi_4) \\ & + \sum_i C_i'^*(\psi_2\gamma_0\Gamma_i^+\gamma_0\psi_1)(\bar{\psi}_4\gamma_0\gamma_5\Gamma_i^+\gamma_0\psi_3)\end{aligned} \tag{30.2}$$

Here the second and fourth terms are the Hermitian conjugate terms; H_w must be Hermitian even though we do not necessarily assume that the C_i are real. The terms with the C_i' are of opposite parity to those with the C_i-coupling strengths. The Γ_i are the complete set of Dirac matrices, 1, γ^μ, $\frac{1}{\sqrt{2}}\,\sigma^{\mu\nu}$, $\gamma^\mu\gamma_5$, and $i\gamma_5$. With this choice

$$\begin{aligned}\gamma_0\gamma^{\mu+}\gamma_0 &= \gamma^\mu \\ \gamma_0\sigma^{\mu\nu+}\gamma_0 &= -\frac{i}{2}\,\gamma_0[\gamma^{\nu+}, \gamma^{\mu+}]\gamma_0 = \frac{i}{2}\,[\gamma^\mu, \gamma^\nu] = \sigma^{\mu\nu} \\ \gamma_0(\gamma^\mu\gamma^5)^+\gamma_0 &= \gamma_0\gamma_5\gamma^{\mu+}\gamma_0 = \gamma^\mu\gamma^5 \\ \gamma_0(i\gamma_5)^+\gamma_0 &= i\gamma_5\end{aligned} \tag{30.3}$$

The behavior of H_w under *charge conjugation* is obtained by noting that the transformation corresponds to the replacement

$$\begin{aligned}\psi_\alpha &\to \bar{\psi}_\rho\mathcal{C}_{\rho\alpha} \\ \bar{\psi}_\beta &\to -(\mathcal{C}^{-1})_{\beta\sigma}\psi_\sigma\end{aligned} \tag{30.4}$$

[1] R. Jost, *Helv. Phys. Acta* **30,** 409 (1957).
[2] The approach is that of T. D. Lee, R. Oehme, and C. N. Yang, *Phys. Rev.* **106,** 340 (1957).

with

$$\mathcal{C}^{-1}\gamma^\mu\mathcal{C} = -\gamma^{\mu T} \tag{30.5}$$

as shown in Chapter 2. This implies that

$$\begin{aligned}\bar{\psi}_\alpha(1)\Gamma_{\alpha\beta}\,\psi_\beta(2) &\rightarrow -\mathcal{C}_{\alpha\rho}{}^{-1}\psi_\rho(1)\Gamma_{\alpha\beta}\,\bar{\psi}_\sigma(2)\mathcal{C}_{\sigma\beta}\\ &= -\psi_\rho(1)\,\bar{\psi}_\sigma(2)(\mathcal{C}\Gamma^T\mathcal{C}^{-1})_{\sigma\rho}\\ &= \bar{\psi}(2)(\mathcal{C}\Gamma^T\mathcal{C}^{-1})\,\psi(1)\end{aligned} \tag{30.6}$$

the change in sign coming from the interchange in order of the anticommuting spinor operators. The relations

$$\begin{aligned}\mathcal{C}(\gamma^\mu)^T\mathcal{C}^{-1} &= -\gamma^\mu\\ \mathcal{C}(\sigma^{\mu\nu})^T\mathcal{C}^{-1} &= -\sigma^{\mu\nu}\\ \mathcal{C}(\gamma^\mu\gamma_5)^T\mathcal{C}^{-1} &= \gamma^\mu\gamma_5\\ \mathcal{C}\gamma_5{}^T\mathcal{C}^{-1} &= \gamma^5\\ \mathcal{C}(\sigma^{\mu\nu}\gamma_5)^T\mathcal{C}^{-1} &= -\sigma^{\mu\nu}\gamma_5\end{aligned} \tag{30.7}$$

imply that the overall effect of charge conjugation is to interchange

$$\sum C_i(\bar{\psi}_1\Gamma_i\psi_2)(\bar{\psi}_3\Gamma_i\psi_4) \rightarrow \sum C_i(\bar{\psi}_2\Gamma_i\psi_1)(\bar{\psi}_4\Gamma_i\psi_3)$$

since a left over minus sign in one of the parentheses will be matched by a similar minus sign in the other. Thus for the even couplings, charge conjugation is equivalent to the interchange

$$C_i \leftrightarrow C_i{}^* \tag{30.8}$$

An examination of the odd terms shows that for these charge conjugation is equivalent to the interchange

$$C_i' \leftrightarrow -C_i'{}^* \tag{30.9}$$

Actually, a more general expression, allowing for an arbitrary phase factor in (30.4), leads to the effective transformation rule

$$C\{C_i, C_i'\}C^{-1} = \{\eta_C C_i{}^*, -\eta_C C_i'{}^*\} \tag{30.10}$$

with

$$\eta_C = \eta_{C1}{}^*\eta_{C2}\eta_{C3}{}^*\eta_{C4} \tag{30.11}$$

still a phase factor.

It should be quite obvious that under *space inversion*, the terms with the C_i transform with a sign opposite to those with the C_i', so that we have

$$P\{C_i, C_i'\}P^{-1} = \{\eta_P C_i, -\eta_P C_i'\} \tag{30.12}$$

with the phase factor

$$\eta_P = \overset{*}{\eta}_{P1}\eta_{P2}\overset{*}{\eta}_{P3}\eta_{P4} \tag{30.13}$$

Thus under the combined transformation CP, we get

$$CP\{C_i, C_i'\}P^{-1}C^{-1} = \{\eta_C\eta_P C_i^*, \eta_C\eta_P C_i'^*\} \tag{30.14}$$

Thus a theory will be invariant under the combined CP operation only if, in effect, all the C_i and the C_i' are real. For parity conservation we can choose the overall phase factor η_P in such a way that H_w is unaltered only if either all the C_i' or all the C_i vanish.

Next, let us consider time reversal. The operation changes ψ and $\bar{\psi}$ according to

$$\begin{aligned}\psi_\alpha &\rightarrow \bar{\psi}_\rho B_{\rho\alpha}\\ \bar{\psi}_\beta &\rightarrow B^{-1}{}_{\rho\sigma}\psi_\sigma\end{aligned} \tag{30.15}$$

where

$$B(\gamma^\mu)^*B^{-1} = \gamma^\mu \tag{30.16}$$

but it also reverses the order of factors in any product of operators.[1] Thus we get

$$\begin{aligned}\bar{\psi}_\alpha(1)(\Gamma_i)_{\alpha\beta}\psi_\beta(2) &\rightarrow (\psi_\sigma(1)B^{-1}_{\alpha\sigma}\Gamma_{i\alpha\beta}\bar{\psi}_\lambda(2)B_{\lambda\beta})_{\text{reversed}}\\ &= \bar{\psi}(2)B\Gamma_i^T B^{-1}\psi(1)\end{aligned}$$

Also

$$\begin{aligned}\bar{\psi}(3)\Gamma_i\gamma_5\psi(4) &\rightarrow -\bar{\psi}(4)\gamma_5 B\Gamma_i^T B^{-1}\psi(3)\\ &= \bar{\psi}(4)\gamma_0\gamma_5\gamma_0 B\Gamma_i^T B^{-1}\psi(3)\end{aligned}$$

With the help of these we deduce that under time reversal we effectively make the interchange

$$\begin{aligned}C_i &\leftrightarrow \eta_T C_i^*\\ C_i' &\leftrightarrow \eta_T C_i'^*\end{aligned} \tag{30.17}$$

in H_w, with η_T an over-all phase factor

$$\eta_T = \eta_{T1}^*\eta_{T2}\eta_{T3}^*\eta_{T4} \tag{30.18}$$

We have thus proved that with

$$\Theta \equiv TPC \tag{30.19}$$

we get

$$\Theta H_w(C_i, C_i')\Theta^{-1} = \eta_{CPT}\, H_w(C_i, C_i') \tag{30.20}$$

The first consequence which we deduce from the CPT theorem is that, if any one of the symmetries C, P, T is violated, then at least one other symmetry must be violated. The second result which will now be proved is that the equality of the masses of stable particles and their antiparticles is a consequence of the CPT theorem alone.

[1] Note that the reversal in the order of operators is to be carried out without regard to the statistics obeyed by the field.

Let us describe the particle A at rest by $\Psi_A(s, \sigma)$ where s denotes its spin and σ its z-component. Now

$$M_A = (\Psi_A(s, \sigma), H\Psi_A(s, \sigma)) \tag{30.21}$$

From

$$H = \Theta H \Theta^{-1} \tag{30.22}$$

it follows that[1]

$$\begin{aligned} M_A &= (\Psi_A(s, \sigma), C^{-1}P^{-1}T^{-1}HTPC\Psi_A(s, \sigma)) \\ &= (P\Psi_{\bar{A}}(s, \sigma), T^{-1}HTP\Psi_{\bar{A}}(s, \sigma)) \end{aligned} \tag{30.23}$$

where $\bar{A}$ is the antiparticle of A. Since the particle is at rest

$$P\Psi_{\bar{A}}(s, \sigma) = \eta\Psi_{\bar{A}}(s, \sigma)$$

so that the parity transformation has no effect on the expectation value. Finally, we rewrite this as

$$\begin{aligned} M_A &= (T\Psi_{\bar{A}}(s, \sigma), HT\Psi_{\bar{A}}(s, \sigma))^* \\ &= (\Psi_{\bar{A}}(s, -\sigma), H^+\Psi_{\bar{A}}(s, -\sigma)) \\ &= (\Psi_{\bar{A}}(s, -\sigma), H\Psi_{\bar{A}}(s, -\sigma)) \end{aligned} \tag{30.24}$$

Since the mass is independent of the value of the z-component of the spin, we have

$$M_A = M_{\bar{A}} \tag{30.25}$$

Another consequence of the CPT theorem is the equality of the lifetimes of particle and antiparticles,[2] at least to first order in H_w. We write

$$H = H_s + H_w \tag{30.26}$$

Here H_s is the strong and electromagnetic part of the Hamiltonian (including, of course, the free Hamiltonian), which satisfies

$$\begin{aligned} [H_s, P] &= 0 \\ [H_s, C] &= 0 \end{aligned} \tag{30.27}$$

We shall use H_s to define the states and let H_w give rise to transitions between them. The states are Ψ_A for the particle A which would be stable under H_s alone, and $\Psi_{\bar{A}} \equiv C\Psi_A$ for the antiparticle $\bar{A}$. Similarly, the

[1] For brevity, we write for the P, C, T operators in Hilbert space the symbols P, C, T rather than $U(P)$, $U(C)$, $U(T)K$.

[2] G. Lüders and B. Zumino, *Phys. Rev.* **106,** 345 (1957), give a general formal proof of this result to all order in H_w. The paper does not, however, show how the decaying particle is to be described; it does show that, if the Hamiltonian can be used, according to some prescription, to determine the lifetime, then the result follows.

final states are Ψ_B and $\Psi_{\bar{B}} \equiv C\Psi_B$. We specifically assume that

$$(\Psi_{\bar{B}}, \Psi_B) = 0$$

In general H_w need not conserve parity; we may, therefore, write it as a sum of terms even and odd, respectively, under P

$$H_w = H_w^{(+)} + H_w^{(-)} \tag{30.28}$$

The rate for the decay

$$A \to B$$

is given by

$$\Gamma_{AB} = \frac{1}{2s_A + 1} \sum_{\sigma} \sum_{\substack{\text{final} \\ \text{spins}}} |(\Psi_B, (H_w^{(+)} + H_w^{(-)}) \Psi_A(s, \sigma))|^2$$

$$= \frac{1}{2s_A + 1} \sum_{\sigma} \sum_{\substack{\text{final} \\ \text{spins}}} \{|(\Psi_B, H_w^{(+)}\Psi_A)|^2 + |(\Psi_B, H_w^{(-)}\Psi_A)|^2\} \tag{30.29}$$

since a rate, which is a scalar, does not measure the pseudoscalar interference term.

It follows from the *CPT* theorem that

$$\begin{aligned}(\Psi_B, H_w\Psi_A) &= (\Psi_B, C^{-1}T^{-1}P^{-1}H_wPTC\Psi_A) \\ &= (\Psi_{\bar{B}}, T^{-1}(H_w^{(+)} - H_w^{(-)})T\Psi_{\bar{A}}) \\ &= (\tilde{\Psi}_{\bar{B}}, (H_w^{(+)} - H_w^{(-)})\tilde{\Psi}_{\bar{A}})^* \\ &= (\tilde{\Psi}_{\bar{A}}, (H_w^{(+)} - H_w^{(-)})\tilde{\Psi}_{\bar{B}})\end{aligned} \tag{30.30}$$

In the last line, the $\sim$ denotes the state with reversed motion; in particular, in an angular momentum representation the reversed states are just states with z-components of the spins having signs opposite to what they had before. We may now use the form (30.30) to calculate Γ_{AB}. It is

$$\Gamma_{AB} = \frac{1}{(2s_A + 1)} \sum_{\sigma} \sum_{\substack{\text{final} \\ \text{spins}}} \{|(\tilde{\Psi}_{\bar{A}}, H_w^{(+)}\tilde{\Psi}_{\bar{B}})|^2 + |(\tilde{\Psi}_{\bar{A}}, H_w^{(-)}\tilde{\Psi}_{\bar{B}})|^2\}$$

This, however, is just the rate for the decay $\bar{A} \to \bar{B}$, so that we have shown that

$$\Gamma_{AB} = \Gamma_{\bar{A}\bar{B}} \tag{30.31}$$

These predictions of the *CPT* theorem are in excellent agreement with experiment. As long as one believed that charge conjugation invariance was a property of all interactions, these results did not appear to be connected with the space-time properties of the interactions; with C violated, as we shall see, the result follows from *CPT* that is a direct consequence of Lorentz invariance, provided we formulate this property field-theoretically. Because of this last proviso the *CPT* theorem is not quite on the same footing as the well-established energy conservation, for example, and we must devise tests of it. In this chapter, however, we shall assume that there is *CPT* invariance, and turn to the question of tests for *C*, *P* and *T*.

It was already pointed out that the observation of a pseudoscalar quantity, such as $\mathbf{J} \cdot \mathbf{p}$ or, in a four-body decay, a nonvanishing expectation value for $\mathbf{p}_1 \cdot \mathbf{p}_2 \times \mathbf{p}_3$, is direct evidence for parity nonconservation. Since under time reversal both spins and momenta are reversed, we might expect that a nonvanishing expectation value of a quantity like $\mathbf{J} \cdot \mathbf{p}_1 \times \mathbf{p}_2$ would be evidence for a *T* violation. This is clearly false, since in pion-nucleon scattering such a term is present, and there is no question about the validity of *T* in the strong interactions.[1] A more careful analysis is evidently needed; furthermore, it is difficult offhand to think of tests for charge conjugation invariance.

In our analysis[2] we shall assume that the interaction Hamiltonian responsible for the decays that we will be considering, H_w, is of the form

$$H_w = \sum_i (C_i H_i^{(+)} + C_i' H_i^{(-)}) \tag{30.32}$$

with $H_i^{(\pm)}$ so arranged that

$$T H_w T^{-1} = \sum (C_i^* H_i^{(+)} + C_i'^* H_i^{(-)}) \tag{30.33}$$

An example of such a structure, which is quite general, is shown in (30.2). In this way, *T* invariance will be reflected in the reality of the C_i and the C_i'. By the same token, since $H_i^{(\pm)}$ are defined to have the property that

$$P H_i^{(\pm)} P^{-1} = \pm H_i^{(\pm)} \tag{30.34}$$

parity conservation implies $C_i' = 0$, and, as in (30.10), charge conjugation invariance requires that the C_i be real and the C_i' pure imaginary. Our next task is to express certain observables in terms of the C_i and C_i'.

Consider the decay of a particle *A* into a final state *B*. Both Ψ_A and Ψ_B are eigenstates of the strong (+ electromagnetic) Hamiltonian H_s. We shall characterize the final state *B* by spins and momenta; it also has an

[1] Time-reversal invariance has been checked by a test of detailed balance in the reactions $p + t \to d + d$ [L. Rosen and J. E. Brolley, *Phys. Rev. Letters* **2,** 98 (1959)] and $C^{12} + \alpha \to N^{14} + d$ [D. Bodansky, S. F. Eccles, G. W. Farwell, M. E. Rickey, and P. C. Robinson, *Phys. Rev. Letters* **2,** 101 (1959)]. An analysis of *p-p* scattering by A. Abashian and E. M. Haffer, *Phys. Rev. Letters* **1,** 255 (1958) and P. Hillman, A. Johansson and G. Tibell, *Phys. Rev.* **110,** 1218 (1958) leads to the same limit on *T*-violating terms in the amplitude, viz. something of the order of 3%. This is not good enough to check that *T* is not violated in the electromagnetic interactions of strongly interacting particles.

[2] This analysis follows that given by T. D. Lee in some unpublished lecture notes.

"out" label, since it is a state which asymptotically, in the future, has the plane wave character described by the variables $\{\mathbf{p}, s\}$. The expectation value of any observable O, such as $\mathbf{J} \cdot \mathbf{p}$, $\mathbf{p}_1 \cdot \mathbf{p}_2 \times \mathbf{p}_3$, etc., is given by

$$\langle O \rangle = \frac{\dfrac{1}{2s_A + 1} \sum_\sigma \sum_{ps} O(p, s, \sigma) \, |(\Psi_B^{\text{out}}(p, s), H_w \Psi_{A\sigma})|^2}{\dfrac{1}{2s_A + 1} \sum_\sigma \sum_{ps} |(\Psi_B^{\text{out}}(p, s), H_w \Psi_{A\sigma})|^2} \tag{30.35}$$

The denominator merely serves as a normalizing factor, and we shall ignore it in what follows.

If the final state B contains particles which interact either electromagnetically, or strongly, as in

$$n \rightarrow p + e^- + \bar{\nu}$$

or

$$\Lambda^0 \rightarrow p + \pi^-$$

then the final state interaction makes $(\Psi_B^{\text{out}}, H_w \Psi_A)$ complex. We have, in fact

$$(\Psi_B^{\text{out}}(p, s), H_w \Psi_{A\sigma}) = \sum_{Jm} (\Psi_B^{\text{out}}(p, s), \Psi_B^{\text{out}}(J, m))(\Psi_B^{\text{out}}(J, m), H_w \Psi_{A\sigma})$$

and using the same arguments which led to the result in (26.36) we can prove that the phase of $(\Psi_B^{\text{out}}(J, m), H_w \Psi_{A\sigma})$ is just the scattering phase shift for the strongly interacting particles in the final state. Thus we may write

$$\begin{aligned}(\Psi_B^{\text{out}}(p, s), H_w \Psi_{A\sigma}) &= \sum_{Jm} (\Psi_B^{\text{out}}(p, s), \Psi_B^{\text{out}}(J, m)) e^{i\delta_J} \\ &\qquad \times |(\Psi_B^{\text{out}}(J, m), H_w \Psi_{A\sigma})| \\ &\equiv \sum_J e^{i\delta_J} \, M(J, p, s, \sigma)\end{aligned} \tag{30.36}$$

Hence

$$\langle O \rangle \propto \sum_{ps\sigma} \sum_J \sum_{J'} O(p, s, \sigma) e^{i\delta_J} \, M(J, p, s, \sigma) e^{-i\delta_{J'}} \, M^*('J\, , p, s, \sigma)$$

The decomposition (30.32) implies that we can write

$$M(J, p, s, \sigma) = \sum_i C_i \, M_i^{(+)}(J, p, s, \sigma) + \sum_i C_i' \, M_i^{(-)}(J, p, s, \sigma)$$

so that

$$\begin{aligned}\langle O \rangle \propto \sum_{ps\sigma} \sum_{JJ'} \sum_{ij} O(p, s, \sigma) e^{i(\delta_J - \delta_{J'})} \{ & C_i C_j^* \, M_i^{(+)}(J, p, s, \sigma) \, M_j^{(+)*}(J', p, s, \sigma) \\ & + C_i C_j'^* \, M_i^{(+)}(J, p, s, \sigma) \, M_j^{(-)*}(J', p, s, \sigma) \\ & + C_i' C_j^* \, M_i^{(-)}(J, p, s, \sigma) \, M_j^{(+)*}(J', p, s, \sigma) \\ & + C_i' C_j'^* \, M_i^{(-)}(J, p, s, \sigma) \, M_j^{(-)*}(J', p, s, \sigma) \}\end{aligned} \tag{30.37}$$

There are different kinds of correlations $O(p, s, \sigma)$ which can be measured. They may be even or odd under motion reversal, i.e., under a reversal of sign of the momentum and the spin vector, and they may be even of odd under space inversion, i.e., under a reversal of sign of the momentum alone. We shall denote them by

$$\begin{aligned} E^{(+)} &: \mathbf{p}_1 \cdot \mathbf{p}_2, \mathbf{S}_1 \cdot \mathbf{S}_2 \\ E^{(-)} &: \mathbf{S} \cdot \mathbf{p} \\ O^{(+)} &= \mathbf{S} \cdot \mathbf{p}_1 \times \mathbf{p}_2, \mathbf{S}_1 \cdot \mathbf{S}_2 \times \mathbf{S}_3 \\ O^{(-)} &= \mathbf{p}_1 \cdot \mathbf{p}_2 \times \mathbf{p}_3 \end{aligned} \tag{30.38}$$

the superscript referring to the behavior under the parity transformation. We will now show that the quantities $M_i^{(\pm)}(J, p, S, \sigma)$ have symmetry properties such that $\langle E^{(\pm)} \rangle$, $\langle O^{(\pm)} \rangle$ pick out different, well-defined bilinear combinations of the C_i. One property, which is self-evident, is that

$$M_i^{(\pm)}(J, p, s, \sigma) = \pm M_i^{(\pm)}(J, -p, s, \sigma) \tag{30.39}$$

The other one, which states

$$M_i^{(\pm)*}(J, p, s, \sigma) = M_i^{(\pm)}(J, -p, -s, -\sigma) \tag{30.40}$$

follows from the time-reversal invariance of $H_i^{(\pm)}$. We have

$$\begin{aligned} \sum_J e^{i\delta_J} M_i^{(\pm)}(J, p, s, \sigma) \\ &= (\Psi_B^{\text{out}}(p, s), H_i^{(\pm)} \Psi_{A\sigma}) \\ &= (\Psi_B^{\text{out}}(p, s), T^{-1} H_i^{(\pm)} T \Psi_{A\sigma}) \\ &= (T\Psi_B^{\text{out}}(p, s), H_i^{(\pm)} T \Psi_{A\sigma})^* \\ &= (\Psi_B{}^{\text{in}}(-p, -s), H_i^{(\pm)} \Psi_{A,-\sigma})^* \\ &= \sum_{Jm} (\Psi_B{}^{\text{in}}(-p, -s), \Psi_B{}^{\text{in}}(J, m))^* (S\Psi_B^{\text{out}}(J, m), H_i^{(\pm)} \Psi_{A,-\sigma})^* \\ &= \sum_{Jm} e^{2i\delta_J} (\Psi_B(-p, -s), \Psi_B(J, m))^* (\Psi_B^{\text{out}}(J, m), H_i^{(\pm)} \Psi_{A,-\sigma})^* \end{aligned} \tag{30.41}$$

which leads to the required result. The two symmetry properties may be used to reformulate (30.37). We get

$$\begin{aligned} \langle O \rangle \propto \sum_{JJ'} \sum_{ij} e^{i(\delta_J - \delta_{J'})} \\ \times \Big\{ & C_i C_j^* \sum_{ps\sigma} O(p, s, \sigma)\, M_i^{(+)}(J, p, s, \sigma)\, M_j^{(+)}(J', -p, -s, -\sigma) \\ &+ C_i C_j'^* \sum_{ps\sigma} O(p, s, \sigma)\, M_i^{(+)}(J, p, s, \sigma)\, M_j^{(-)}(J', -p, -s, -\sigma) \\ &+ C_i' C_j^* \sum_{ps\sigma} O(p, s, \sigma)\, M_i^{(-)}(J, p, s, \sigma)\, M_j^{(+)}(J', -p, -s, -\sigma) \\ &+ C_i' C_j'^* \sum_{ps\sigma} O(p, s, \sigma)\, M_i^{(-)}(J, p, s, \sigma)\, M_j^{(-)}(J', -p, -s, -\sigma) \Big\} \end{aligned} \tag{30.42}$$

Consider now an observable of the type $O^{(+)}$, such as $\mathbf{S}\cdot\mathbf{p}_1\times\mathbf{p}_2$. It has the property that

$$O(-p, -s, -\sigma) = -O(p, s, \sigma) \tag{30.43}$$

From this it follows:

1. In the second and third terms, the sum over p involves three terms, two of which are even when $\mathbf{p}\leftrightarrow-\mathbf{p}$, and one of which is odd. Thus the coefficients of the terms containing an interference term $C_iC_j'^*$ (or complex conjugate) *vanish*. Therefore, a measurement of an $O^{(+)}$-correlation yields no information about parity conservation, as already pointed out by Lee and Yang.

2. The remaining two terms may be rewritten, with the help of (30.43) in the form

$$\begin{aligned}\langle O^{(+)}\rangle \propto \sum_{JJ'} e^{i(\delta_J-\delta_{J'})}\Bigg\{&\sum_{ij} C_iC_j^* \times \frac{1}{2}\sum_{ps\sigma} O^{(+)}(p, s, \sigma)\\ &\times (M_i^{(+)}(J, p, s, \sigma)\, M_j^{(+)}(J', -p, -s, -\sigma)\\ &- M_i^{(+)}(J, -p, -s, -\sigma)\, M_j^{(+)}(J', p, s, \sigma) + \sum_{ij} C_i'C_j'^* \cdots\Bigg\}\\ \propto \frac{1}{4}\sum_{JJ'}\sum_{ij}&(\cos(\delta_J-\delta_{J'}) + i\sin(\delta_J-\delta_{J'}))\\ &\times [(C_iC_j^* + C_i^*C_j) + (C_iC_j^* - C_i^*C_j)]\sum_{ps\sigma} O^{(+)}(p, s, \sigma)\\ &\times \{M_i^{(+(}(J, p, s, \sigma)\, M_j^{(+)}(J', -p, -s, -\sigma)\\ &- M_i^{(+)}(J, -p, -s, -\sigma)\, M_j^{(+)}(J', p, s, \sigma)\} + \cdots\end{aligned}$$

The important thing to notice about this expression is that the $\sum_{ps\sigma}$ is antisymmetric under the simultaneous interchange $i\leftrightarrow j$, $J\leftrightarrow J'$. Consequently, the coefficients of that sum are

$$\cos(\delta_J-\delta_{J'})(C_iC_j^* - C_i^*C_j)$$

and

$$\sin(\delta_J-\delta_{J'})(C_iC_j^* + C_i^*C_j)$$

Thus an observation of a correlation like $\mathbf{S}\cdot\mathbf{p}_1\times\mathbf{p}_2$ implies either that time reversal invariance is violated, $(C_iC_j^* - C_i^*C_j)$ nonvanishing, *or* that there is no violation of time reversal, but $(\delta_J-\delta_{J'})$ is nonvanishing, i.e., there is a final state interaction. In pion-nucleon scattering, where such a correlation is observed, it is the latter conclusion that must be drawn. If, on the other hand, such a correlation were observed in the decay

$$K^+ \to \pi^0 + \mu^+ + \nu$$

then one would have to conclude that T is violated.

Other observables can be treated in a similar fashion. We summarize the results in a Table 30-1, in which the entries indicate which symmetry is violated, as far as one can tell from an observation of the correlation at the head of the column. The two rows indicate whether the observation depends on the presence of a final state interaction or not.

Table 30-1

	$E^{(+)}$	$E^{(-)}$	$O^{(+)}$	$O^{(-)}$
$\cos(\delta_J - \delta_{J'})$ (present even in absence of final state interaction)	—	P, C	T, C	P, T
$\sin(\delta_J - \delta_{J'})$ (depends on the strength of the final state interaction)	T, C	P, T	—	P, C

This discussion covers all the tests which have been carried out, except for those based on the curious behavior of the K^0 and $\overline{K^0}$ under decay; this subject will be covered in Chapter 32. We must now turn to the results of the experiments suggested by Lee and Yang, and others following their lead.

31

The Form of the Beta Interaction

The first experiment designed to test parity conservation in the weak interactions, technically the most difficult one, involved an examination of the correlation between the direction of the electron momentum and the polarization of the decaying nucleus, i.e., a term of the type $\langle \mathbf{J} \rangle \cdot \mathbf{p}$. The experiment was carried out at the National Bureau of Standards by Wu, Ambler, Hayward, Hoppes, and Hudson,[1] who polarized Co^{60} nuclei and measured the correlation of the electron momentum and the magnetic field causing the polarization. Because of the small magnetic moments of nuclei $\sim(e\hbar/M_pc)$ it takes enormous fields to orient the spin of the nucleus. Such fields are created by lining up the electrons in the unfilled electronic shell of paramagnetic salts. The need for low temperatures (0.01°K) adds to the difficulties of this experiment.

Co^{60}, whose use was suggested by Lee and Yang, decays to an excited state of Ni^{60}. The latter reaches its ground state via two successive γ-ray transitions ($4^+ \rightarrow 2^+ \rightarrow 0^+$). For an oriented nucleus, the angular distribution of the γ-ray relative to the polarization direction is given by an expression of the type

$$W(\theta) = A + B\cos^2\theta + C\cos^4\theta \tag{31.1}$$

Parity is conserved in the electromagnetic interactions, so that only even powers of $\cos\theta$ appear. The value of

$$a = \frac{W(\pi/2) - W(0)}{W(\pi/2)}$$

may be used to measure the magnitude of the nuclear polarization. The correlation between the polarization of the Co^{60} nucleus and the electron

[1] C. S. Wu, E. Ambler, R. Hayward, D. Hoppes, and R. Hudson, *Phys. Rev.* **105,** 1413 (1957) and **106,** 1361 (1957).

momentum may be obtained from

$$\mathcal{M} = \frac{\sqrt{2m_\nu}}{(2\pi)^3} \langle\sigma^k\rangle[\tfrac{1}{2} C_T\, \bar{u}(p_e)\sigma_k\, v(p_\nu) + \tfrac{1}{2} C_T'\, \bar{u}(p_e)\sigma_k\gamma_5\, v(p_\nu)$$
$$+ C_A\, \bar{u}(p_e)\gamma_k\gamma_5\, v(p_\nu) + C_A'\, \bar{u}(p_e)\gamma_k\, v(p_\nu)] \quad (31.2)$$

which is just the form of (29.10) with possible pseudoscalar terms added, so that

$$C_T\sigma_k \rightarrow C_T\sigma_k + C_T'\sigma_k\gamma_5$$
$$C_A\gamma_k\gamma_5 \rightarrow C_A\gamma_k\gamma_5 + C_A'\gamma_k$$

and with the Fermi terms omitted, since the β-decay, Co^{60}–Ni^{60*} is a $5^+ \rightarrow 4^+$ transition. A straightforward calculation, which is too long to be carried out here, reveals that (*a*) the energy dependent ($\sim m_e/E_e$) Fierz interference terms are now proportional to Re $(C_A'C_T'^* - C_AC_T^*)$, and it is this coefficient which is, within experimental errors, zero; (*b*) The angular distribution of the electrons relative to the polarization is given by[1]

$$W_e(\theta) = 1 + \alpha \cos\theta \quad (31.3)$$

where θ is the angle between the spin direction of the nucleus and the momentum of the electron ($\cos\theta = \hat{\mathbf{p}}_e \cdot \langle\mathbf{J}\rangle/J$), and[2]

$$\alpha = -v_eP\,\frac{2\,\text{Re}\,(C_TC_T'^* - C_AC_A'^*)}{|C_T|^2 + |C_T'|^2 + |C_A|^2 + |C_A'|^2} \quad (31.4)$$

v_e is the electron velocity and P is the magnitude of the polarization of the Co^{60}.

In the experiment under consideration the electron angular distribution was measured in the presence of an orienting magnetic field, which was also reversed. The results, shown in Fig. 31.1, indicate that the γ-anisotropy decayed in about 5 minutes because of depolarization caused by thermal effects, and the electron asymmetry coefficient behaved very similarly, with the electron distribution following the direction of the magnetic field.

Since in this experiment $v_e \approx 0.6$ and the magnitude of the polarization is also $\approx$60%, the experimentally measured value

$$\alpha \approx -0.4$$

[1] The complete computation may be found in G. Källén, *Elementary Particle Physics*, Addison-Wesley Publishing Co., Reading, Mass. (1964), Chapter 13. The metric and the representation of the γ-matrices used there differ from ours.

[2] Because of our choice of metric the coefficients C_i' are related as follows to the coefficients defined by Lee and Yang: $C_S' = -(C_S')_{LY}$, $C_V' = -(C_V')_{LY}$, $C_T' = -(C_T')_{LY}$, $C_A' = (C_A')_{LY}$, $C_P' = -(C_P')_{LY}$. Note that $C_A = -(C_A)_{LY}$ with the other unprimed C's the same.

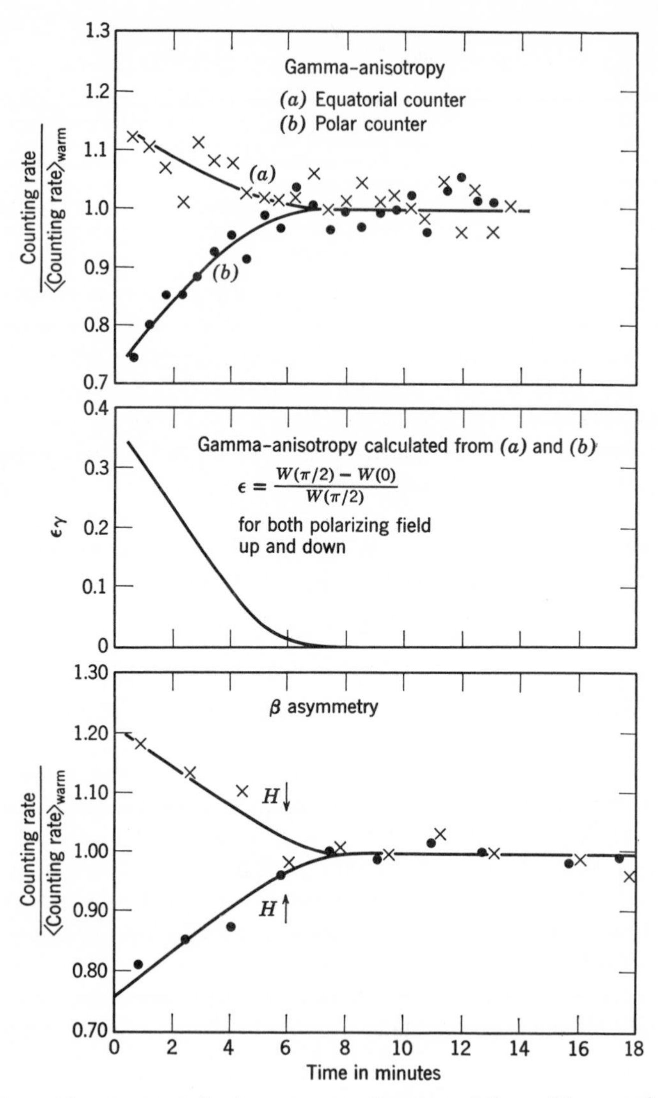

Fig. 31.1. γ-anisotropy and β-anisotropy as a function of time. The warming-up of the sample of Co^{60} leads to a depolarization in approximately 5 minutes. [From Wu *et al.*, *Phys. Rev.* **105,** 1413 (1957).]

implies that

$$\frac{2\,\mathrm{Re}\,(C_T C_T'^* - C_A C_A'^*)}{|C_T|^2 + |C_T'|^2 + |C_A| + |C_A'|^2} \approx 1 \tag{31.5}$$

In the expression written down in (31.4) we did not include the term that comes in because of the final state interaction, here the Coulomb interaction. Since the effect observed is so large, ($|\alpha|$ is ≈ 0.4 instead of $\approx 0.4Z/137$) it clearly does not depend on the final state interaction. We therefore deduce that *P and C* are violated (see Table 30-1). This large violation obscures the presence of a Z-dependent term that could give us information about T violation.

If we assume that there is invariance under T, then the coupling constants are real, and it is possible to explain both the absence of the Fierz terms and the results shown in (31.5) with the choice

$$\begin{aligned} C_A &= C_A' = 0 \\ C_T &= C_T' \end{aligned} \tag{31.6}$$

or

$$\begin{aligned} C_T &= C_T' = 0 \\ C_A &= -C_A' \end{aligned} \tag{31.7}$$

so that only one of the terms is present in the Gamow-Teller part of the β-interaction

$$\tfrac{1}{2}C_T(\bar{\psi}_p\sigma^{\alpha\beta}\psi_n)(\bar{\psi}_e\sigma_{\alpha\beta}(1+\gamma_5)\psi_\nu) + C_A(\bar{\psi}_p\gamma^\alpha\gamma_5\psi_n)(\bar{\psi}_e\gamma_\alpha(1-\gamma_5)\psi_\nu) \tag{31.8}$$

The motivation for making the rather big leap from experiment to these special forms is theoretical. The matrices $1 \pm \gamma_5$ have rather special properties. Since

$$\left(\frac{(1 \pm \gamma_5)}{2}\right)^2 = \frac{(1 \pm \gamma_5)}{2} \tag{31.9}$$

they are projection operators. If we look back at Chapter 2 (2.48) to (2.51), we see that $\frac{1}{2}(1 \pm \gamma_5)$ is a projection operator for positive/negative helicity states of a massless particle. The forms (31.6) and (31.7) therefore suggest that neutrinos of one helicity only enter into the weak interactions. This *two-component neutrino theory*[1] requires that parity not be conserved, because a space reflection changes the sign of the helicity.

Equation 31.8 may be rewritten as

$$\tfrac{1}{2}C_T(\bar{\psi}_p\sigma^{\alpha\beta}\psi_n)(\bar{\psi}_e(1+\gamma_5)\sigma_{\alpha\beta}\psi_\nu) + C_A(\bar{\psi}_p\gamma^\alpha\gamma_5\psi_n)(\bar{\psi}_e(1+\gamma_5)\gamma_\alpha\psi_\nu) \tag{31.10}$$

[1] A normal spin $\frac{1}{2}$ particle obeying the Dirac equation is described by a four-component spinor.

In both terms the electron operator appears as the adjoint of $(1 - \gamma_5)\,\psi_e(x)$. The appearance of this operator $(1 - \gamma_5)$ implies that the electron in the decay will be polarized. The expectation value of the longitudinal polarization for the electron is

$$\langle \boldsymbol{\sigma} \cdot \hat{\mathbf{p}} \rangle = \frac{\sum_r u^{(r)+}(p)(1 - \gamma_5)\boldsymbol{\sigma} \cdot \hat{\mathbf{p}}(1 - \gamma_5)\, u^{(r)}(p)}{\sum_r u^{(r)+}(p)(1 - \gamma_5)(1 - \gamma_5)\, u^{(r)}(p)}$$
$$= \frac{\operatorname{tr}\{(1 - \gamma_5)\boldsymbol{\sigma} \cdot \hat{\mathbf{p}}(1 - \gamma_5)[(\not{p} + m)/2m]\gamma^0\}}{\operatorname{tr}\{(1 - \gamma_5)(1 - \gamma_5)[(\not{p} + m)/2m]\gamma^0\}}$$
$$= -\frac{4\mathbf{p} \cdot \hat{\mathbf{p}}/m}{4E/m} = -\frac{v}{c} \tag{31.11}$$

The two possibilities (31.6) and (31.7) for the Co^{60} experiment can be depicted pictorially as in Fig. 31.2 (p. 526).

The figure shows in each case the observed fact that the electron momentum tends to point in a direction opposite to that of the Co^{60} spin. In the decay $(5^+ \rightarrow 4^+)$ one unit of angular momentum must be carried off by the two leptons. With electron having a preference for negative helicity, it is clear that angular momentum conservation requires that, if the neutrino comes off in the same direction as the electron (as preferred by the correlation $(1 + \frac{1}{3}\mathbf{p}_e \cdot \mathbf{p}_\nu/E_e E_\nu)$ for the tensor interaction), it must have negative helicity, whereas in the case of the axial interaction, where the preference is for antiparallel neutrino-electron momenta, the neutrino which participates in this reaction will have to have positive helicity.

The prediction that electrons in the β-decay should have maximum helicity, $-v/c$, was first checked in Co^{60}-decay by Frauenfelder et al.[1] The relatively slow electrons (in the 100 keV range) were deflected through a little more than 90° by an electrostatic field,[2] as a consequence of which the beam that originally was longitudinally polarized ended up transversely polarized. This transverse polarization was then measured by scattering the electrons in the backward hemisphere from a high Z-material.

There are other methods for measuring the longitudinal polarization of an electron, which do not involve converting the polarization to a transverse one. As we pointed out in the section on electrodynamics, the longitudinal polarization can be measured by Møller scattering off a

[1] H. Frauenfelder, R. Bobone, E. Von Goeler, N. Levine, H. R. Lewis, R. Peacock, A. Rossi, and G. De Pasquali, *Phys. Rev.* **106,** 386 (1957); see also P. E. Cavanagh, J. Turner, C. Coleman, G. Gard, and B. Ridley, *Phil. Mag.* **21,** 1105 (1957).

[2] A magnetic field would also rotate the spin.

target containing electrons polarized in a direction along, or opposite, to the line of collision. This technique was used to confirm the magnitude and sign of the electron helicity in other Gamow-Teller transitions. Similar results were obtained with the use of an experimental arrangement which takes advantage of the fact that the bremsstrahlung of a longitudinally polarized electron has the same longitudinal polarization (Chapter 12). The photon helicity can be measured by scattering them off polarized electrons.

It should be noted that the appearance of the electron operator in the form $(1 - \gamma_5)\,\psi_e(x)$ implies that in β^+-decay, the positrons will have helicity $+\,v/c$, as can be shown by replacing $u(p)$ by $v(p)$ in (31.11). This was confirmed in a very ingenious experiment by Page and Heinberg[1] who studied the decay of Na^{22} in a $(3^+ \to 2^+)$ transition. A pictorial representation like that of Fig. 31.2 now shows the positron with positive helicity, so that the neutrino emitted in positron decay must have helicity opposite to that of the neutrino emitted in electron decay. This is relevant if it should turn out that only one of the terms in (31.8) is correct. In that case we could consider the neutrino as being described by a two-component spinor (only one helicity state) and the antineutrino by the two-component spinor with the opposite helicity state.

Let us now turn to the Fermi transitions. One experiment which has a bearing on this is the $0^+ \to 0^+$ decay $Ga^{66} \xrightarrow{e^+} Zn^{66}$. The photons coming from the annihilation of the positrons in flight were scattered off polarized electrons, and in this way it was deduced that the positron helicity was consistent with the value $+\,v/c$. This value suggests that here too the electron operator appears in the combination $(1 - \gamma_5)\psi_e$ in the Fermi part of the weak interaction. This would imply that the suggestive combination

$$C_S(\bar{\psi}_p\psi_n)(\bar{\psi}_e(1 + \gamma_5)\psi_\nu) + C_V(\bar{\psi}_p\gamma^\alpha\psi_n)(\bar{\psi}_e\gamma_\alpha(1 - \gamma_5)\psi_\nu) \tag{31.12}$$

is to be considered, with the two possible *two-component neutrino* alternatives

$$C_S = 0 \tag{31.13}$$

or

$$C_V = 0 \tag{31.14}$$

If we pursue the idea of a two-component neutrino interaction, we must check whether the same helicity neutrino appears in the Fermi transitions as appears in the Gamow-Teller transitions. A way of testing this is to examine a transition that involves both Fermi and Gamow-Teller terms and ask whether there is any interference. Clearly, if the two neutrinos

[1] L. A. Page and M. Heinberg, *Phys. Rev.* **106**, 394 (1957).

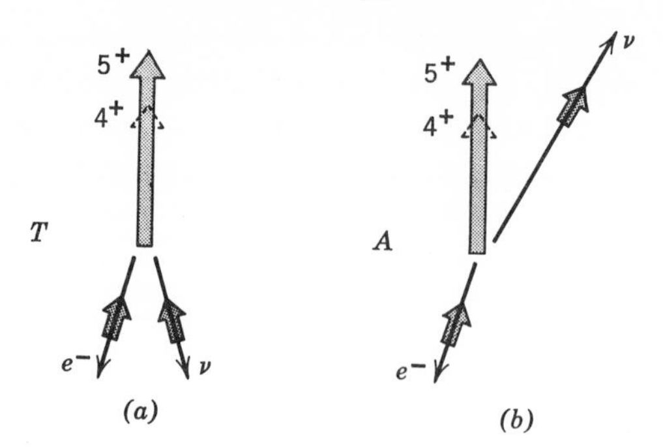

Fig. 31.2. Schematic picture of Co^{60} decay for (*a*) tensor interaction and (*b*) axial vector interaction.

have opposite helicity, the S and/or V amplitude cannot interfere with the T and/or A amplitude. If we draw a schematic diagram analogous to that of Fig. 31.2, indicating the helicities and the electron-neutrino correlations (Fig. 31.3), we see that there will be interference only if S and T *both* appear and/or V and A *both* appear. Transitions which should be examined are such that $\Delta \mathbf{J} = 0$ $(0 \nrightarrow 0)$ with no parity change. Under those circumstances the correlation of the electron momentum with the nuclear polarization contains a term in addition to the one in (31.4) which describes pure Gamow-Teller transitions. This term will contain $C_S C_T'$ and $C_V C_A'$. Since few beta-decaying nuclei are readily polarized, an alternate experimental approach is possible.[1] If the nucleus is initially unpolarized, after emitting the β-ray it will be polarized, and it is then undergoes a γ-decay, a measurement of the helicity of the photon in effect measures this polarization. The correlation between the photon spin $\mathbf{S}$ and the polarization $\langle \mathbf{J} \rangle$ is consistent with parity conservation in the radiative decay, since $\mathbf{S} \cdot \mathbf{J}$ is a scalar. In effect, a measurement of $\langle \mathbf{p}_e \cdot \mathbf{S} \rangle$ is

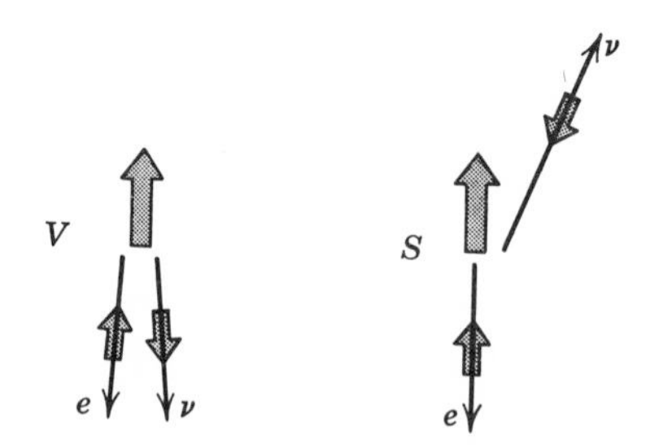

Fig. 31.3. Helicities in the Fermi transitions.

[1] J. D. Jackson, S. Treiman, and H. Wyld, *Phys. Rev.* **106,** 517 (1957); K. Alder, B. Stech, and A. Winther, *Phys. Rev.* **107,** 728 (1957).

equivalent to a measurement of $\langle \mathbf{p}_e \cdot \mathbf{J} \rangle$. The study of the decay Sc^{46} $(4^+ \to 4^+)$ followed by the radiative de-excitation $(4^+ \to 2^+ \to 0^+)$ led to the conclusion that there is Fermi-Gamow-Teller interference.[1] On the basis of the two-component neutrino theory, we have the possibilities

$$S \text{ and } T$$

or

$$V \text{ and } A$$

As mentioned in Chapter 29, the electron-neutrino correlation experiments showed that the second alternative is the correct one.

A definitive experiment in which the neutrino helicity was actually measured confirmed the *V-A* form of the beta interaction.[2] The principle of the experiment is the following: Consider electron capture from the *K*-shell

$$e^-(K) + A(0) \to B^*(1) + \nu$$

in which the initial nucleus has spin 0, and the final nucleus is left in an excited state of spin 1. The nucleus $B^*(1)$ and the neutrino will travel in opposite directions, and conservation of angular momentum demands that their helicities be the same. If now the excited nucleus decays radiatively

$$B^*(1) \to B(0) + \gamma$$

then the photon will be circularly polarized, and if it is emitted in a direction opposite to that of the neutrino, then it will have the helicity of the neutrino. Whereas photon helicity measurements are relatively easy, the direction of the neutrino momentum is not readily found. What was done, was in effect a measurement of the direction of motion of the recoiling nucleus, $B^*(1)$. If the energy difference between $B^*(1)$ and $B(0)$ is ΔE, the momentum of $B^*(1)$ is p, and the photon is emitted with an angle θ relative to the direction of motion, then energy and momentum conservation show that the photon energy is given by

$$E_\gamma = \Delta E + \frac{p\,\Delta E \cos\theta}{M_B} - \frac{(\Delta E)^2}{2M_B} \tag{31.15}$$

If a photon of this energy strikes a stationary $B(0)$ nucleus, it will be resonantly absorbed, by exciting $B(0)$ to the state $B^*(1)$, provided that the photon energy, less the recoil energy of $B(0)$, i.e.

$$E_\gamma - \frac{p_\gamma^2}{2M_B} \cong \Delta E + \frac{p\,\Delta E \cos\theta}{M_B} - \frac{(\Delta E)^2}{M_B}$$

[1] F. Boehm and A. Wapstra, *Phys. Rev.* **107,** 1202 (1957).

[2] M. Goldhaber, L. Grodzins, and A. Sunyar, *Phys. Rev.* **109,** 1015 (1958).

is equal to ΔE again. Thus the condition for resonant absorption or resonant scattering is that

$$p \cos \theta = \Delta E \tag{31.16}$$

If p, which is also the neutrino energy, is equal to the photon energy, the condition for resonant scattering is that the photon move in a direction opposite to that of the neutrino.

Goldhaber et al. rather miraculously found a nucleus, $Eu^{152}(0^-)$, which gives an excited state $Sm^{152}(1^-)$ upon K capture; the latter decays to the 0^+ ground state. The neutrino energy is $\approx$900 keV and $\Delta E \approx 960$ keV, so that the photon circular polarization is not quite 100%. Resonant scattering was found to take place and the photon helicity was measured, with the conclusion that the *neutrino emitted had negative helicity*, so that the β-interaction is axial. If we accept the two-component neutrino theory, the interaction is some combination of *vector* and *axial* only.

Before we discuss the relative magnitude and phase of C_V and C_A, we turn to corroborating evidence for this theory obtained from a study of the decay chain

$$\begin{array}{l} \pi \rightarrow \mu + \nu \\ \qquad \hookrightarrow e + \nu + \nu \end{array}$$

It was suggested by Lee and Yang that parity nonconservation in the π-decay will give rise to a longitudinal polarization of the muon, and parity nonconservation in the muon decay will give rise to a correlation between the electron direction and the muon polarization. On the basis of the two-component neutrino theory the magnitude of the muon

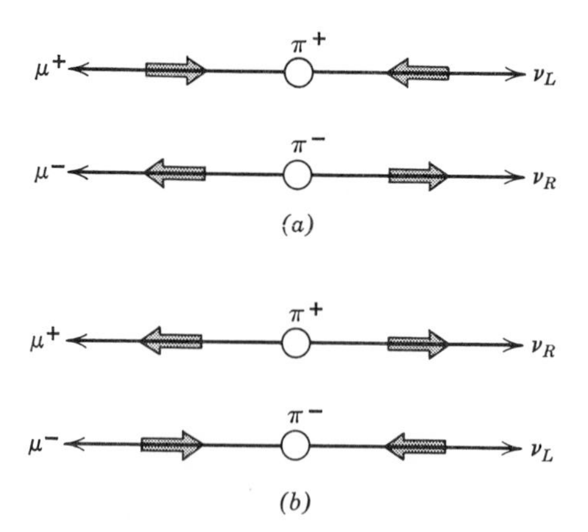

Fig. 31.4. Two possible schemes for $\pi^\pm$ decay into $\mu^\pm$ and ν. The helicities must always be the same to conserve angular momentum, and the relation between π^+ and π^- follows from *CPT*.

polarization can be seen to be $\pm v/c$.[1] In pion decay there are two alternatives, described pictorially in Fig. 31.4. If the μ^+ is coupled to the same kind of neutrino as the e^+ is in β-decay, then scheme (*a*) is correct. The helicity of the μ^- from π^- decay has been measured by utilizing the spin dependence of μ^--electron scattering (with polarized electrons) or the spin dependence of Coulomb scattering of the muons.[2] The result is that the μ^- is *right-handed,* so that scheme (*a*) is correct.[3]

This particular result is consistent with the idea of *lepton conservation.* If we call the ν_L accompanying e^+ β-decay a lepton, and the e^+ an antilepton, the the ν_R which appears in the decay of the neutron ($n \rightarrow p + e^- + \bar{\nu}$) will be called an antineutrino. The pion, having no lepton number, must decay into a lepton and an antilepton. Since ν_L is a lepton, μ^+, like e^+ must be an antilepton.

The test of this idea is the decay

$$\mu^+ \rightarrow e^+ + \nu + \nu$$

According to the two-component neutrino theory with lepton conservation, one of the two neutrinos must be left-handed and the other right-handed. A qualitative suggestion of how this might be tested is obtained from a pictorial representation (Fig. 31.5), which makes use of the experimental fact that in μ^+-decay the e^+ are preferentially emitted in a direction, opposite to the μ^+-direction.[4] What the picture shows is that for lepton nonconservation it is difficult to conserve angular momentum in the configuration in which the electron and the two neutrinos are collinear. This configuration corresponds to the top of the electron spectrum, and if two ν_R's or two ν_L's should be emitted, there would be a dip in the spectrum at high energies. No such dip is observed (Fig. 31.6).

Incidentally, the experiment of Garwin et al. referred to above is worth describing because of its ingenuity and by-products. In this experiment,

[1] For the two-component neutrino theory predictions, see A. Salam, *Nuovo Cimento* **5,** 299 (1957); L. Landau, *Nucl. Phys.* **3,** 127 (1957); T. D. Lee and C. N. Yang, *Phys. Rev.* **105,** 1671 (1957).

[2] A. T. Alikanov, Tu. V. Galaktionov, Yu. V. Gorodkov, G. P. Eliseyev, and V. A. Lubimov, *Transl. Soviet Phys. JETP* **11,** 1380 (1960); G. Backenstoss, B. Hyams, G. Knop, P. Marin, and U. Stierlin, *Phys. Rev. Letters* **6,** 415 (1961).

[3] The muon operator $\psi(x)$ could appear in the interaction in either form $(1 \pm \gamma_5)\,\psi(x)$ and still emerge from the pion decay with a fixed helicity sign required by the neutrino helicity and angular momentum conservation. If it were massless, the muon coupled with $(1 - \gamma_5)\,\psi(x)$ would have to have a definite helicity. The electron in $\pi^\pm \rightarrow e^\pm + \nu$ is almost massless: the combination $(1 - \gamma_5)$ requires the e^- to be left-handed, whereas the ν helicity requires it to be right-handed. The decay can only proceed because the e is *not* massless and the rate is proportional to m_e^2.

[4] J. I. Friedman and V. L. Telegdi, *Phys. Rev.* **105,** 1681 (1957); R. L. Garwin, L. M. Lederman, and M. Weinrich, *Phys. Rev.* **105,** 1415 (1957).

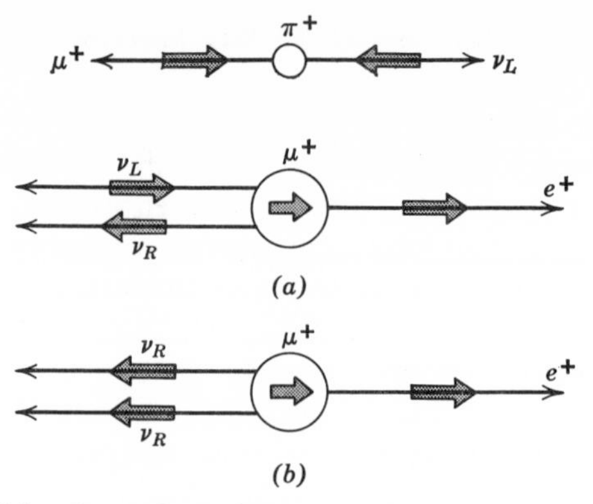

Fig. 31.5. Two possible schemes for μ-decay: (*a*) with lepton conservation (*b*) without lepton conservation. The μ's are labeled with their spins.

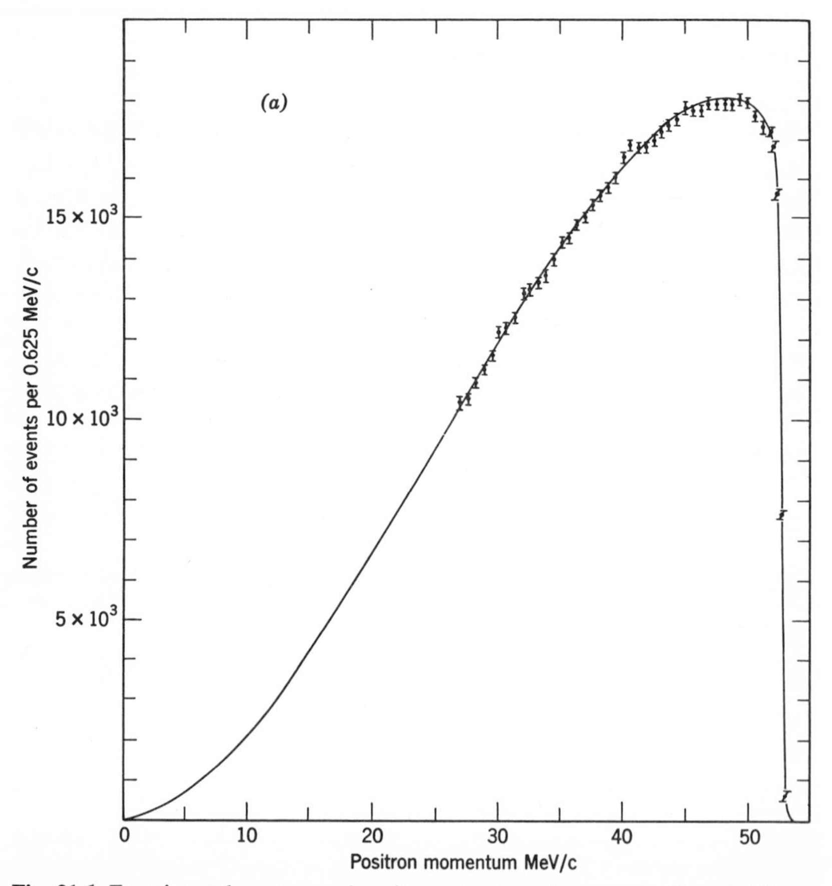

Fig. 31.6. Experimental spectrum of positrons in $\mu^+ \rightarrow e^+ + \nu + \bar{\nu}$. The solid line is the theoretically predicted spectrum based on the interaction (31.18) corrected for electromagnetic effects. [From M. Bardon et al. *Phys. Rev. Letters* **14,** 449 (1965).]

the muons emitted in the forward direction from pions decaying in flight were stopped in graphite. A magnetic field, perpendicular to the line of flight, precesses the muons with a frequency characterized by their magnetic moment. If there is muon polarization (which implies parity nonconservation in the pion decay) and a correlation between the positron momentum and the polarization direction (which implies parity nonconservation in the muon decay) then the positron spectrum should show the effect of this periodic precession. One would expect the positron spectrum intensity, at a given point, to have the following dependence

$$I_{e^+}(t) = e^{-\lambda t}[1 + \alpha r \cos(\omega t + \phi)] \tag{31.17}$$

where $\omega = geB/2m_\mu c$ (with g the gyromagnetic ratio of the muon) is the precession frequency, ϕ some fixed phase, constant for the experiment, r (with $r \leq 1$) a depolarization factor, and α an asymmetry factor characteristic of the muon decay. Garwin et al. did not study the intensity as a function of time, but rather set their counters for a fixed time and varied the magnetic field. The variation of the positron intensity at a fixed time after the muon arrival in the graphite varied sinusoidally with the strength of the magnetic field. The period was fitted with $g \simeq 2$. The interesting by-product of this experiment was, in fact, the first determination of g for the muon. That measurement, and subsequent improvements, show that as far as the magnetic moment is concerned, there is no sign that the muon has interactions other than weak and electromagnetic. The muon thus is just a heavy electron. The depolarization of μ^+ in graphite is relatively small, and the parameter α, characteristic of the muon decay, turned out to be close to $-\frac{1}{3}$. Lack of space prevents us from exhibiting the detailed calculation terms which yield the electron energy spectrum and angular distribution relative to the muon polarization direction.[1] These are quite straightforward, and they show that the value of $\alpha \cong -\frac{1}{3}$ implies that in the most general form of the interaction consistent with a two-component neutrino interaction

$$\tilde{C}_S(\bar{\psi}_\nu(1 + \gamma_5)\psi_\mu)(\bar{\psi}_e(1 - \gamma_5)\psi_\nu) + \tilde{C}_V(\bar{\psi}_\nu\gamma^\alpha(1 - \gamma_5)\psi_\mu)(\bar{\psi}_e\gamma_\alpha(1 - \gamma_5)\psi_\nu)$$

$\tilde{C}_S$ must be very small. We thus write[2]

$$\mathcal{H}_\mu(x) = \frac{1}{\sqrt{2}} G_\mu(\bar{\psi}_\nu\gamma^\alpha(1 - \gamma_5)\psi_\mu)(\bar{\psi}_e\gamma_\alpha(1 - \gamma_5)\psi_\nu) \tag{31.18}$$

[1] The calculations are worked out in complete detail in G. Källén, *Elementary Particle Physics*, Addison-Wesley Publishing Co., Reading, Mass. (1964).

[2] The factor $1/\sqrt{2}$ is inserted to compensate for a factor 2 which comes from squaring $\bar{\psi}_e\gamma_\alpha(1 - \gamma_5)\psi_\nu$.

This is to be compared with the beta interaction, now written as

$$\frac{1}{\sqrt{2}}(C_V\bar{\psi}_p\gamma^\alpha\psi_n + C_A\bar{\psi}_p\gamma^\alpha\gamma_5\psi_n)(\bar{\psi}_e\gamma_\alpha(1-\gamma_5)\psi_\nu) \tag{31.19}$$

The similarity is suggestive, and it remains to determine the numerical values of the three constants which appear in the expressions in (31.18) and (31.19).

The magnitude of G_μ is determined from the lifetime of the muon. The rate is given by

$$\Gamma = (2\pi)^4 \frac{(2\pi)^3}{E_\mu/m_\mu} \iiint \frac{d^3p_e}{E_e/m_e}\frac{d^3p_{\nu_1}}{2p^0_{\nu_1}}\frac{d^3p_{\nu_2}}{2p^0_{\nu_2}}\,\delta(P_\mu - p_e - p_{\nu_1} - p_{\nu_2}) \times \frac{1}{2}\sum_{\text{spins}} |\mathcal{H}_\mu|^2$$

$$= \frac{(2\pi)^7}{E_\mu/m_\mu} \iiint \frac{d^3p_e}{E_e/m_e}\frac{d^3p_{\nu_1}}{2p^0_{\nu_1}}\frac{d^3p_{\nu_2}}{2p^0_{\nu_2}}\,\delta(P_\mu - p_e - p_{\nu_1} - p_{\nu_2}) \times \frac{1}{(2\pi)^{12}}\frac{G_\mu^2}{4m_em_\mu}\,\mathrm{tr}\,((1+\gamma_5)\gamma^\alpha \not{P}_\mu\gamma^\beta\not{p}_{\nu_2})\,\mathrm{tr}\,((1+\gamma_5)\gamma_\alpha\not{p}_{\nu_1}\gamma_\beta\not{p}_e)$$

when the electron mass m_e is set equal to zero in the matrix element. The trace calculation is easily done, and we get

$$\frac{G_\mu^2}{4E_\mu\pi^5}\iiint \frac{d^3p_e}{2E_e}\frac{d^3p_{\nu_1}}{2p^0_{\nu_1}}\frac{d^3p_{\nu_2}}{2p^0_{\nu_2}}\,\delta(P_\mu - p_e - p_{\nu_1} - p_{\nu_2})(P_\mu\cdot p_{\nu_1})(p_e\cdot p_{\nu_2})$$

The calculation is shortened by replacing $d^3p_{\nu_2}/2p^0_{\nu_2}$ by $d^4p_{\nu_2}\theta(p^0_{\nu_2})\,\delta(p^2_{\nu_2})$ and eliminating p_{ν_2}. If we go to the muon rest frame we get, upon expanding out the argument of the delta function

$$\frac{G_\mu^2}{2m_\mu\pi^5}\int\frac{d^3p_e}{2E_e}\int\frac{d^3p_{\nu_1}}{2p_{\nu_1}}\,\delta\left(\frac{1}{2}m_\mu^{\ 2} - m_\mu E_e - m_\mu p_{\nu_1} + E_ep_{\nu_1} - p_ep_{\nu_1}\cos\theta_{e\nu}\right) \times m_\mu p_{\nu_1}\left(\frac{1}{2}m_\mu^{\ 2} - m_\mu p_{\nu_1}\right)$$

The integration over the angle between the electron and neutrino directions yields limits on the p_{ν_1} and p_e integrations

$$\tfrac{1}{2}m_\mu - E_e \le p \le \tfrac{1}{2}m_\mu \qquad 0 \le p_e \cong E_e \le \tfrac{1}{2}m_\mu$$

The result, with M standing for the proton mass, is

$$\tau_\mu = \frac{1}{\Gamma_\mu} = \frac{192\pi^3}{G_\mu^{\ 2}m_\mu^{\ 5}} = \frac{192\pi^3}{(G_\mu M^2)^2}\left(\frac{M}{m_\mu}\right)^4\frac{1}{m_\mu} \tag{31.20}$$

From the numerical value

$$\tau_\mu = 2.2 \times 10^{-6} \text{ sec}$$

we obtain

$$G_\mu M^2 = 1.02 \times 10^{-5} \tag{31.21}$$

We determine the value of C_V and C_A in the beta interaction by considering first the decay rate for a pure Fermi transition, which gives C_V. The best nucleus for this purpose is O^{14} which decays to N^{14*}. The transition is $0^+ \to 0^+$, and O^{14}, N^{14*} form an *i*-spin triplet together with C^{14}. The energy released in nuclear beta decay is always so small that the nuclear recoil can be completely neglected in the kinematics. The matrix element for the transition is

$$\mathcal{M} = \frac{\sqrt{2m_\nu}}{(2\pi)^6} \frac{C_V}{\sqrt{2}} \bar{u}(p_\nu)\, \gamma_0(1 - \gamma_5)\, v(p_e)(\Psi_{N_{14}^*}, a_n^+ a_p \Psi_{O^{14}})$$

so that

$$\sum_{\text{spins}} |\mathcal{M}|^2 = \frac{\frac{1}{2}C_V^2}{(2\pi)^{12}} |(\Psi_{N_{14}^*}, a_n^+ a_p \Psi_{O_{14}})|^2 \frac{4E_e}{m_e} p_\nu(1 - \cos\theta_{e\nu})$$

is obtained after a simple trace computation. We can now calculate

$$\Gamma = (2\pi)^7 \int \frac{d^3p_e}{E_e/m_e} \int \frac{d^3p_\nu}{2p_\nu} \delta(E_e + p_\nu - \Delta E) \sum_{\text{spins}} |\mathcal{M}|^2$$

The momentum conservation delta function gets rid of the integration over the recoil momentum, and the recoil energy is too small to appear in the energy conservation delta function. Thus[1]

$$\Gamma = \frac{1}{\tau} = \frac{C_V^2}{2\pi^3} |(\Psi_{N_{14}^*}, a_n^+ a_p \Psi_{O_{14}})|^2 \int_0^{p_{\max}} p_e^2\, dp_e(\Delta E - E_e)$$

$$\equiv \frac{C_V^2}{2\pi^3} |(\Psi_{N_{14}^*}, a_n^+ a_p \Psi_{O_{14}})|^2 m_e^5 F(\eta_0) \tag{31.22}$$

Note that the product $F(\eta_0)\tau$ yields a measure of the strength of the coupling. For the O^{14} decay, the value of ΔE is 2.32 MeV and the measured value is[2]

$$F(\eta_0)\tau \log 2 = 3075 \pm 10 \text{ sec}$$

[1] A straightforward integration yields

$$F(\eta_0) = -\tfrac{1}{4}\eta_0 - \tfrac{1}{12}\eta_0^3 + \tfrac{1}{30}\eta_0^5 + \tfrac{1}{4}\sqrt{1 + \eta_0^2} \log(\eta_0 + \sqrt{\eta_0^2 + 1})$$

where $\eta_0 = \sqrt{(\Delta E/m_e)^2 - 1}$.

[2] Conventionally the half-life is used in the so-called *ft*-values. Since $T_{1/2} = \tau \log 2$, we get the $F\tau \log 2$ factor.

To determine C_V^2 we still need $|(\psi_{N^*_{14}}, a_n{}^+ a_p \Psi_{O_{14}})|^2$. Now the combination of operators $a_n{}^+ a_p$ is just a charge lowering operator; it is, in fact, equivalent to[1]

$$a_N^+ \tau_- a_N = \tfrac{1}{2}\, a_N{}^+(\tau_1 - i\tau_2) a_N = (T_1 - iT_2)_N = (T_-)_N \approx T_-$$

Now, assuming charge independence in the nuclear transition, we have

$$(\Psi_{N^*_{14}}, T_- \Psi_{O_{14}}) = \langle T = 1, T_3 = 0 \,|T_-|\, T = 1, T_3 = 1\rangle = \sqrt{2}$$

Hence we can find

$$(C_V M^2)^2 = \pi^3 \left(\frac{M}{m_e}\right)^4 \frac{h}{m_e c^2} \frac{1}{F(\eta_0)\tau \log 2}$$

or

$$C_V M^2 \cong 1.01 \times 10^{-5} \tag{31.23}$$

The equality of C_V and G_μ is quite remarkable. When electromagnetic radiative corrections are taken into account in muon decay, and Coulomb corrections are calculated for O^{14} decay, one obtains[2]

$$\frac{G_\mu - C_V}{G_\mu} \cong 0.025 \tag{31.24}$$

It remains to calculate C_A. A straightforward calculation of the half-life of the neutron, which is measured to be 11.7 minutes,[3] yields

$$\left|\frac{C_A}{C_V}\right| \cong 1.18 \pm 0.02 \tag{31.25}$$

The sign of C_A was determined by a detailed study of the decay of polarized neutrons. Lack of space forbids the reproduction of the detailed calculations which show that the coefficient of $\langle \mathbf{J}_n \cdot \mathbf{p}_e \rangle$ in the differential decay rate is proportional to

$$|C_A|^2 + \mathrm{Re}\, C_V C_A{}^*$$

Measurements show that this coefficient is close to zero, implying that *C_A has the opposite sign to C_V*. In the same experiment a correlation of the type $\langle \mathbf{J}_n \cdot \mathbf{p}_e \times \mathbf{p}_\nu \rangle$ was sought, and not found. As our discussion in

[1] Actually T_-, the total *i*-spin lowering operator, will contain contributions from the meson field in the nucleus, so that strictly speaking $T_- = a_n^+ a_p$ + mesonic terms. It is only if we assume that the mesonic terms are unimportant that we can make this identification. In fact, we shall see in Chapter 34 that it is T_- and not $a_n{}^+ a_p$ which appears in the matrix element in the first place.

[2] S. M. Berman, *Phys. Rev.* **112**, 267 (1958).

[3] A. N. Sosnovsky, P. E. Spivak, Y. A. Prokofiev, I. E. Kutikov, and Y. P. Dobinin, *Nucl. Phys.* **10**, 395 (1959).

Chapter 30 shows, this suggests that *time reversal invariance is not violated* in this decay.[1]

The form of the beta interaction is therefore given by

$$\frac{1}{\sqrt{2}}\, C_V(\bar{\psi}_p\gamma^\alpha(1 - 1.18\gamma_5)\psi_n)(\bar{\psi}_e\gamma_\alpha(1 - \gamma_5)\psi_\nu) \tag{31.26}$$

The similarity of this form to that responsible for muon decay (31.18) is very encouraging to the point of view that there is some sort of "universality" that governs all weak interactions. It is tempting to propose that the weak-interaction Hamiltonian be written in the form

$$H_w = \frac{G}{\sqrt{2}} \int d^3x\, J_\alpha(x) J^{\alpha+}(x) \tag{31.27}$$

with

$$J_\alpha(x) = \bar{\psi}_e\gamma_\alpha(1 - \gamma_5)\psi_\nu + \bar{\psi}_\mu\gamma_\alpha(1 - \gamma_5)\psi_\nu + \bar{\psi}_n\gamma_\alpha(1 - \gamma_5)\psi_p + \cdots \tag{31.28}$$

This is just the Fermi theory, with the proviso that all spinors involved in the weak interactions appear in the form $(1 - \gamma_5)\psi$. This very attractive form was actually proposed on various (uncompelling) theoretical grounds before the experimental situation was completely clarified.[2] There are a number of points that have to be made in connection with this form.

1. There are other terms which contribute to the current J_α, since there are weak processes which we have not yet discussed, such as the decays of the strange particles. The structure of these terms will be investigated in the next few chapters.

2. The weak Hamiltonian will contain a term which has the form

$$\frac{G}{\sqrt{2}}\, (\bar{\psi}_\mu\gamma^\alpha(1 - \gamma_5)\psi_\nu)^+\, (\bar{\psi}_n\gamma_\alpha(1 - \gamma_5)\psi_p) \tag{31.29}$$

and leads to definite predictions for muon capture in hydrogen. There are several experimental difficulties with the study of this process. First, it is rare. The capture has to compete with the decay. For a capture on a nucleus with charge Z, the capture rate is, roughly speaking, proportional to Z^4, one of the Z's coming from the number of protons engaged in the capture

$$\mu^- + Z \rightarrow (Z - 1) + \nu$$

[1] The phase of $-C_A$ relative to C_V must in fact be $\leqslant 8^0$; time reversal invariance requires 0^0. [M. T. Burgy, V. E. Krohn, T. B. Novey, G. R. Ringo, and V. L. Telegdi, *Phys. Rev.* **110,** 1214 (1958); *Phys. Rev. Letters* **1,** 324 (1958).]

[2] E. C. G. Sudarshan and R. E. Marshak, *Phys. Rev.* **109,** 1860 (1958); R. P. Feynman and M. Gell-Mann, *Phys. Rev.* **109,** 193 (1958); J. J. Sakurai, *Nuovo Cimento* **7,** 649 (1958); W. R. Theis, *Z. Phys.* **150,** 590 (1958); *Fortschr. Physik.* **7,** 559 (1959).

and Z^3 coming from the value of $|\psi_\mu(0)|^2$, which represents the probability of finding the muon in the nucleus. The experimentally determined break-even point between capture and decay is near $Z = 10$. Thus one finds that only about 1 in 10^4 muons get captured in hydrogen. Second, the muons, when slowed down in hydrogen, tend to form $\mu^- pp$ ions rather than $\mu^- p$ Bohr atoms. This means that the interpretation of the data is not as straightforward as it could be, and requires some knowledge of the properties of these molecules. They have been studied, and preliminary experimental results have been properly interpreted to be in agreement with the theoretical predictions based on (31.29), with $(1 - \gamma_5)$ in the nuclear part of the current again replaced by $(1 - 1.18\gamma_5)$.[1]

3. This brings us to the discrepancy between our idealized form (31.28) and the phenomenological form (31.26). The factor multiplying γ_5 is really not so surprising. What is involved in β-decay, for example, is a matrix element of the form

$$(\Psi_p, \bar{\psi}_p(0)\gamma_\mu(1 - \gamma_5)\psi_n(0)\Psi_n) \tag{31.30}$$

and on general grounds of invariance, this must have the form

$$\bar{u}(p)[\gamma_\mu F_1(q^2) + i\sigma_{\mu\nu}q^\nu F_2(q^2) - \gamma_\mu\gamma_5 F_A(q^2) - q_\mu\gamma_5 F_P(q^2)]\, u(n) \tag{31.31}$$

where $q_\mu = p_\mu - n_\mu$. What we learn from experiment (31.26) is that $F_1(0) = 0.98$ and $F_A(0) = 1.18$, the other terms not contributing when the momentum transfer is essentially zero. The calculation of the "form factors" is not something we want to go into. The main point that we want to make here is that "1.18" is a number like the anomalous moment of the proton, for example, whose explanation is a problem in strong interaction physics.[2] We found that $F_1(0) = 0.98$ because we have chosen $G = G_\mu(1.002)$ (including the radiative corrections). The near-equality of this form factor to unity is most remarkable. It states in effect that the strong interactions involving the proton and the neutron during the weak-interaction time do not make any difference, and that the vector part of the interaction proceeds as if the nucleons were bare. Otherwise, in the decay

$$n \to p + e^- + \bar{\nu}$$

some of the time the neutron consists of a proton and a π^-, for example, and the decay

$$n \xrightarrow{s} p + \pi^-$$

$$p \xrightarrow{w} N^{*++} + e^- + \bar{\nu}$$

$$N^{*++} + \pi^- \xrightarrow{s} p$$

[1] See the review article by G. Feinberg and L. Lederman, *Ann. Rev. Nucl. Sci.* **13** (1963).
[2] See Chapter 34 for a calculation of $F_A(0)$.

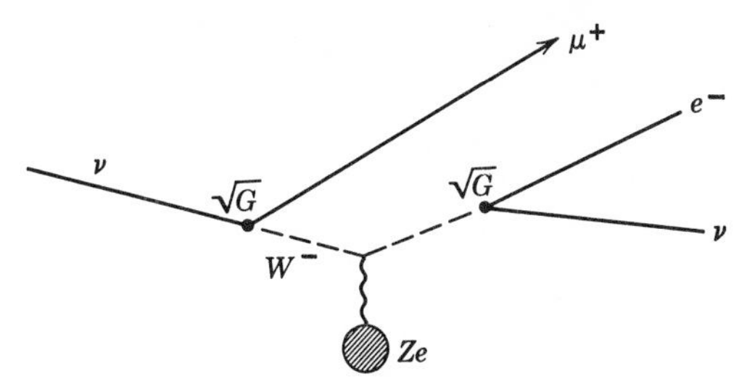

Fig. 31.7. Mechanism and "signature" of W production in neutrino experiment.

need not have a rate in any way related to the neutron decay rate. Thus one would, in general, expect $F_1(0)$ to be different from unity. The explanation for this surprising experimental result was given by Feynman and Gell-Mann, and will be discussed in Chapter 34.

4. The form of the Hamiltonian

$$H_w = \frac{G}{\sqrt{2}} J_\alpha J^{\alpha+}$$

is suggestive of a Yukawa type coupling of the form

$$H_w' \propto \sqrt{G}\, J_\alpha W^\alpha$$

to some kind of vector meson, with the observed four-particle interaction representing a second-order process analogous to Møller scattering, for example. Such vector mesons would have to be very heavy, and they would have to have only "semi-weak" interactions, since otherwise β-decay rates would be a factor G^{-1} faster than muon decay. The mass m_W must certainly be larger than the K-meson mass, as otherwise the weak decay

$$K^+ \rightarrow W^+ + \gamma$$

for example, would be a factor G^{-1} faster than the actual observed decay

$$K^+ \rightarrow \mu^+ + \nu$$

Actually, experiments with high-energy neutrinos, coming from the decay of a high-energy pion beam, would lead to a measurable and predictable W-meson production and identification by the mechanism shown in Fig. 31.7. No such μ^+e^- pairs have been seen which, when the energy distribution of the neutrino beam is taken into account, implies[1]

[1] R. Burns, K. Goulianos, E. Hyman, L. Lederman, W. Lee, N. Mistry, J. Rehberg, M. Schwartz, J. Sunderland, and G. Danby, *Phys. Rev. Letters* **15,** 42 (1965); this paper contains references to earlier work.

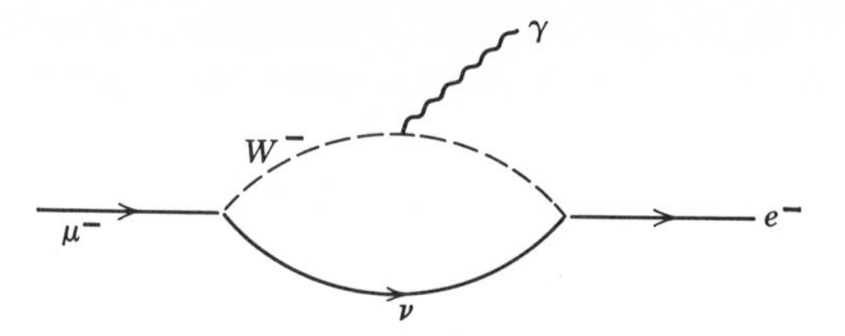

Fig. 31.8. Graph for $\mu^- \rightarrow e^- + \gamma$ with W-meson intermediate state.

that, if the W-mesons exist, their mass must be large

$$m_W > 2 \text{ BeV}$$

Since the assumption of a W-meson-mediated weak interaction has not explained anything, we will not concern ourselves with it.

5. The speculations concerning the existence of a W-meson lent some urgency to an old problem. With the presence of a W, and lepton conservation, it is possible to devise a mechanism for the process

$$\mu^\pm \rightarrow e^\pm + \gamma$$

as shown in Fig. 31.8. Although the calculation of the rate is ambiguous, a reasonable cut-off on the divergent integrals yields a rate far in excess of the experimental lower limit. A possible explanation was suggested. This was that *the neutrino coupled to the muon be different from the neutrino coupled to the electron*, which would make the process shown in Fig. 31.8 impossible.[1]

In the high-energy neutrino experiments mentioned above, the neutrinos come from pion decay, and since the latter is primarily

$$\pi \rightarrow \mu + \nu$$

the neutrinos, if they were to be associated with a definite lepton, would be labeled ν_μ. In the interaction

$$\nu_\mu + \text{nucleus} \rightarrow \text{nucleus} + \text{lepton}$$

only muons were observed,[2] thus indicating that ν_μ's cannot be coupled to electrons. When the neutrinos come from the decay of K-mesons, where

$$K^\pm \rightarrow \pi^0 + e^\pm + \nu_e$$

is a significant decay mode, some electrons were seen, their number in rough agreement with what would be expected on the basis of the number

[1] To my knowledge this suggestion was first put forward by J. Schwinger (1957).
[2] G. Danby, J. M. Gaillard, K. Goulianos, L. M. Lederman, N. Mistry, M. Schwartz, and J. Steinberger, *Phys. Rev. Letters* **9**, 36 (1962).

of K_{e3} decays. Thus there are two neutrinos, which differ only in some internal property, which allows one to couple to the muon and the other to the electron. Whether there are other differences is not known. If the coupling (31.18) still stands, they both have to be massless—an essential property of two-component neutrinos.[1]

PROBLEMS

1. Calculate the cross section for the process

$$\nu + p \rightarrow n + \mu^+$$

using the interaction

$$\frac{G}{\sqrt{2}} (\bar{\psi}_\mu \gamma^\alpha (1 - \gamma_5) \psi_\nu)(\bar{\psi}_p \gamma_\alpha (1 - \gamma_5) \psi_n) - \text{h.c.}$$

as a function of neutrino energy and scattering angle in the center of mass system. What is the maximum angular momentum in which the process takes place? Use this knowledge to estimate the energy at which the unitarity condition

$$\sigma_J < \frac{2\pi(J + \frac{1}{2})}{q_{\text{cm}}^2}$$

is violated.

2. The coupling $\pi\mu\nu$ leads to an "induced" pseudoscalar term in the process

$$\mu^- + p \rightarrow n + \nu$$

via the one-pion exchange. Estimate the strength of this induced pseudoscalar coupling.

[1] The Lagrangian for a two-component neutrino theory is invariant under the transformation $\psi_\nu \rightarrow -\gamma_5 \psi_\nu$. Such an invariance precludes the existence of a neutrino mass as a result of self-interactions.

32

Weak Interactions of Strange Particles I. Selection Rules and Symmetries and the Properties of Neutral K Mesons in Decay

In our analysis of the beta interaction we were led to a very attractive and compact form for the weak interaction Hamiltonian

$$H_w = \frac{G}{\sqrt{2}} \int d^3x\, J^\alpha(x)\, J_\alpha^+(x) \tag{32.1}$$

with currents of the form

$$J^\alpha(x) = j_l^\alpha(x) + J^\alpha_{\text{strong}}(x) \tag{32.2}$$

The leptonic current was found to have the form

$$j_l^\alpha(x) = \bar{\psi}_e \gamma^\alpha(1 - \gamma_5)\psi_{\nu_e}(x) + \bar{\psi}_\mu(x)\gamma^\alpha(1 - \gamma_5)\psi_{\nu_\mu}(x) \tag{32.3}$$

The part of the current involving the strongly interacting particles was not completely explored, since we dealt with only zero strangeness particles. We did find that there were two currents for $\Delta S = 0$, a vector current, which will temporarily be denoted by $J_{V,0}$ and an axial current $J_{A,0}$. The dynamical properties of these currents will be discussed further. At this point we note that both currents change the charge. A possible term in the $\Delta S = 0$ current was of the form

$$J^\alpha_{V,0} + J^\alpha_{A,0} = \bar{\psi}_n(x)\gamma^\alpha(1 - \gamma_5)\, \psi_p(x) \tag{32.4}$$

and we will describe these currents by the statement that $\Delta Q = 1$, the change in charge being read from left to right. Clearly $J^{\alpha+}_{V,0}$ and $J^{\alpha+}_{A,0}$ will be characterized by $\Delta Q = -1$. It follows from the form (32.1) that all terms in $J^{\alpha}_{\text{strong}}$ must be $\Delta Q = 1$ currents to ensure charge conservation. The last statement that we wish to make about $J_{V,0}$ and $J_{A,0}$ is that, within experimental errors, the part of H_w involving products of the form $J^{\alpha}_{V,0} j^{+}_{l\alpha}$ and $J^{\alpha}_{A,0} j^{+}_{l\alpha}$ is invariant under time reversal. This implies, using the *CPT* theorem, that the $\Delta S = 0$ part of the semileptonic H_w is *invariant under CP*.

When weak decays of the strange particles are studied, all observed processes can be qualitatively explained by adding to $J_{V,0}$ and $J_{A,0}$ currents, in which the strangeness changes. As was already mentioned when the concept of strangeness was introduced, the decays of the strange particles appear to be governed by an extremely accurate selection rule,

$$|\Delta S| = 1 \tag{32.5}$$

The evidence for this is the absence, in many hundreds of observed Ξ^-, of the phase-space favored decay mode

$$\Xi^- \to n + \pi^- \tag{32.6}$$

We shall soon be able to present an argument showing that the $|\Delta S| = 2$ term in H_w must be at least 10^7 times smaller than the $|\Delta S| = 1$ term. We will therefore tentatively write

$$J^{\alpha}_{\text{strong}} = J^{\alpha}_{V,0} + J^{\alpha}_{A,0} + J^{\alpha}_{V,1} + J^{\alpha}_{A,1} \tag{32.7}$$

with the subscript 1 denoting the strangeness change. We have both vector and axial terms appearing in the strangeness changing current. The need for both is evident from the existence of

$$K^{\pm} \to \mu^{\pm} + \nu$$

in which $(\Psi_0, J^{\alpha}_{A,0}\Psi_K)$ determines the rate, and

$$K^{\pm} \to \pi^0 + e^{\pm} + \nu$$

in which $(\Psi_{\pi^0}, J^{\alpha}_{V,1}\Psi_K)$ determines the rate. These matrix elements involve strongly interacting particles and currents constructed out of field operators for strongly interacting particles, and they must have proper transformation properties under reflections, so that the matrix elements do not vanish. The zero-component of the current must transform as a pseudoscalar for the first process and as a scalar for the second.

It was pointed out by Feynman and Gell-Mann[1] that the absence of $|\Delta S| = 2$ transitions implies that the currents for which $|\Delta S| = 1$ must

[1] R. P. Feynman and M. Gell-Mann, *Phys. Rev.* **109,** 193 (1958).

have a structure such that

$$\Delta S = \Delta Q \tag{32.8}$$

Thus terms like $\bar{\psi}_\Lambda \gamma^\alpha(1 - \gamma_5)\psi_p$ and $\bar{\psi}_{\Sigma^-}\gamma^\alpha(1 - \gamma_5)\psi_n$ are acceptable, but $\bar{\psi}_n \gamma^\alpha(1 - \gamma_5)\psi_{\Sigma^+}$ is not. If such a term existed, a transition

$$n + \bar{\Sigma}^+ \leftrightarrow \bar{n} + \Sigma^-$$

could appear with strength G and this would lead to $\Xi^- \to n + \pi^-$ by a sequence of steps involving G only once, and only strong couplings the rest of the time

$$\Xi^- \xrightarrow{\text{strong}} \Sigma^- + \bar{K}^0 + n + \bar{n} \xrightarrow{\text{w}} \bar{\Sigma}^+ + \bar{K}^0 + n + n$$

$$\xrightarrow{\text{strong}} \pi^- + \bar{n} + n + n \xrightarrow{\text{strong}} \pi^- + n$$

The selection rule forbids the decay

$$\Sigma^+ \to n + e^+ + \nu$$

and allows

$$\Sigma^- \to n + e^- + \bar{\nu}$$

Similarly the reaction

$$K^+ \to \pi^+ + \pi^+ + e^- + \bar{\nu}$$

is forbidden, whereas

$$K^+ \to \pi^+ + \pi^- + e^+ + \nu$$

is allowed. These predictions are borne by experiment.[1]

One of the earliest empirical rules proposed for the weak Hamiltonian grew out of a comparison of the rates for the nonleptonic decays

$$K^0 \to \pi^+ + \pi^-$$

and

$$K^+ \to \pi^+ + \pi^0$$

The former proceeds about 500 times faster. The only way in which the two processes differ is that in the neutral decay the symmetric two-pion state ($J = 0$) can be in a $T = 0$ state as well as in a $T = 2$ state, whereas

[1] See, for example, U. Nauenberg, P. Schmidt, J. Steinberger, S. Marateck, R. J. Plano, H. Blumenfeld, and L. Seidlitz, *Phys. Rev. Letters* **12,** 679 (1964); R. W. Birge, R. P. Ely, G. Gidal, G. F. Kalmus, A. Kernan, W. M. Powell, U. Camerini, D. Kline, W. F. Fry, J. G. Gaidos, D. Murphee, and C. T. Murphy, *Phys. Rev.* **139,** B1600 (1965).

the $\pi^+\pi^0$ state can only have $T = 2$. Thus the neutral decay matrix element has parts involving $\Delta T = \frac{1}{2}$, $\frac{3}{2}$ and $\frac{5}{2}$, while the K^+ decay only involves $\Delta T = \frac{3}{2}$ and $\frac{5}{2}$. If for some reason. $\Delta T = \frac{1}{2}$ transitions were enhanced, the effect could be qualitatively explained.[1]

The rule that in strangeness-changing weak interactions

$$\Delta T = \tfrac{1}{2} \tag{32.9}$$

is most directly implemented by requiring that the *weak interaction H_w transform as an isospinor.* This hypothesis leads to a large number of predictions. These are most easily worked out by a formal trick, in which a *spurion* field, not representing anything real, but transforming as an isospinor, is introduced into the phenomenological coupling. Consider for example

$$\Lambda^0 \to p + \pi^-$$

and

$$\Lambda^0 \to n + \pi^0$$

The *i*-spin conserving coupling may be written in the form

$$\Lambda^0\bar{\chi}\boldsymbol{\tau}N \cdot \boldsymbol{\pi} = \Lambda^0(\sqrt{2}\,\chi^0 p\pi^- + \cdots - \chi^0 n\pi^0)$$

where χ represents the spurion doublet. From this we read off that

$$\frac{\Gamma(\Lambda^0 \to n + \pi^0)}{\Gamma(\Lambda^0 \to n\pi^0) + \Gamma(\Lambda^0 \to p + \pi^-)} = \frac{1}{3} \tag{32.10}$$

A typical experimental value is 0.315 ± 0.017, in excellent agreement with the predicted value. With parity nonconservation, the decay of the spin $\frac{1}{2}$ Λ^0 can go through both *S*- and *P*-wave channels. The spurion expression yields information about both amplitudes and predicts that

$$\frac{S_0}{S_-} = \frac{P_0}{P_-} = -\frac{1}{\sqrt{2}} \tag{32.11}$$

where S_0, P_0 denote the amplitudes for the neutral mode and S_-, P_- denote the amplitudes for the $p\pi^-$-mode. These amplitudes can be determined by a detailed analysis of the Λ^0-decay, which will be done in the next chapter. At this point we merely point out that this prediction is not contradicted by the incomplete data.

The $\Delta T = \frac{1}{2}$ rule predicts a relationship between the amplitudes a_+, a_0,

[1] M. Gell-Mann and A. Pais, *Proceedings of the Glasgow Conference*, Pergamon Press, London (1954).

and a_- for the three processes

$$\text{“spurion”} + \begin{cases} \Sigma^+ \to n + \pi^+ \\ \Sigma^+ \to p + \pi^0 \\ \Sigma^- \to n + \pi^- \end{cases}$$

We decompose both the initial and final states into i-spin eigenstates and obtain

$$\begin{aligned} a_+ &= (\Psi'^{\text{out}}_{n\pi^+}, \Psi'^{\text{in}}_{\Sigma^+\chi^0}) = (\sqrt{\tfrac{1}{3}}\,\Psi'^{\text{out}}_{3/2} - \sqrt{\tfrac{2}{3}}\,\Psi'^{\text{out}}_{1/2}, \sqrt{\tfrac{1}{3}}\,\Phi^{\text{in}}_{3/2} - \sqrt{\tfrac{2}{3}}\,\Phi^{\text{in}}_{1/2}) \\ &= \tfrac{1}{3}\,a_{3/2} + \tfrac{2}{3}\,a_{1/2} \\ a_0 &= (\Psi'^{\text{out}}_{p\pi^0}, \Psi'^{\text{in}}_{\Sigma^+\chi^0}) = (\sqrt{\tfrac{2}{3}}\,\Psi'^{\text{out}}_{3/2} + \sqrt{\tfrac{1}{3}}\,\Psi'^{\text{out}}_{1/2}, \sqrt{\tfrac{1}{3}}\,\Phi^{\text{in}}_{3/2} - \sqrt{\tfrac{2}{3}}\,\Phi^{\text{in}}_{1/2}) \\ &= \sqrt{\tfrac{2}{9}}\,(a_{3/2} - a_{1/2}) \\ a_- &= (\Psi'^{\text{out}}_{n\pi^-}, \Psi'^{\text{in}}_{\Sigma^-\chi^0}) = (\Psi'^{\text{out}}_{3/2}, \Phi^{\text{in}}_{3/2}) = a_{3/2} \end{aligned}$$

so that

$$\sqrt{2}\,a_0 + a_+ = a_- \tag{32.12}$$

This again holds for the S and P-channel amplitudes separately. Each of these amplitudes is complex, with the phase determined by the pion-nucleon scattering phase shift in the final state.[1] Since all the S and P-phase shifts, both for $T = \frac{1}{2}$ and $T = \frac{3}{2}$, are still fairly small for a total center of mass energy corresponding to the Σ mass, we will neglect them. Under those circumstances the amplitudes a_+, a_0, and a_- can be plotted in a plane, with the x-component representing the S-amplitude and the y-component the P-amplitude. The decay rates (proportional to the squares of the "vectors" in the S-P plane) are[2]

$$\begin{aligned} \Gamma_+ &= (0.584 \pm 0.035) \times 10^{10}\ \text{sec}^{-1} \\ \Gamma_0 &= (0.646 \pm 0.030) \times 10^{10}\ \text{sec}^{-1} \\ \Gamma_- &= (0.606 \pm 0.015) \times 10^{10}\ \text{sec}^{-1} \end{aligned}$$

These imply that the vectors form a right-angled triangle (92 ± 5^0),

[1] This, again, assuming invariance under CP.

[2] This subject is discussed in detail by R. H. Dalitz, International Conference on Fundamental Aspects of Weak Interactions, held at Brookhaven National Laboratory, 1963 BNL 837 (C-39). The quoted data are taken from N. P. Samios' report in the *Proceedings of the International Conference on Weak Interactions* held at Argonne National Laboratory, 1965. (ANL-7130).

provided the $\Delta T = \frac{1}{2}$ rule is satisfied, so that the triangle is closed. Data[1] on certain decay asymmetries, to be discussed in the next chapter, indicate that one of a_+, a_- is purely S-wave and the other purely P-wave dominated. The present data do not agree very well with the prediction.[2]

The $\Delta T = \frac{1}{2}$ rule similarly predicts that

$$\frac{\Gamma(\Xi^0 \to \Lambda^0 + \pi^0)}{\Gamma(\Xi^- \to \Lambda^0 + \pi^-)} = \frac{1}{2}$$

The data is uncertain because of conflicting experiments on the Ξ^0 lifetime. One set of experiments fits the $\Delta T = \frac{1}{2}$ rule very well, whereas the other

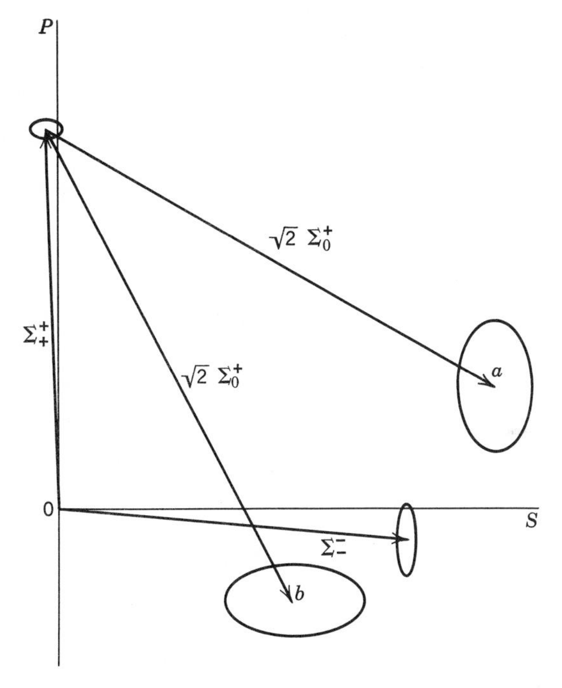

Fig. 32.1. Schematic diagram of the vectors describing $\Sigma^+ \to n\pi^+$, $\Sigma^- \to n\pi^-$, and $\Sigma^+ \to p\pi^0$ in the S-P plane. Points a and b show the two orientations possible for the vector $\sqrt{2}\, a_0$, corresponding to the measured asymmetry parameter $\alpha_0 = -0.73$.

[1] The equality of the rates Γ_+ and Γ_- is something of a mystery, for one of the decays is, as shown in Fig. 32.1, primarily an S-wave decay and the other a P-wave decay, which should be inhibited by the angular momentum barrier.

[2] This conclusion depends on one measurement—that of the parameter $\alpha(\Sigma^+ \to p\pi^0)$ discussed in Chapter 33, (33.15). Agreement with experiment demands $\alpha = -0.98$ instead of the measured -0.78 ± 0.10. The experiment will undoubtedly be repeated and the conclusion checked in the near future.

requires a 10% amplitude for $\Delta T = \frac{3}{2}$. This in itself is not too disturbing if one takes the point of view that the $\Delta T = \frac{1}{2}$ rule represents a dynamical enhancement and not a fundamental law of nature reflecting something about the couplings. The latter point of view is rather difficult to implement, requiring as it does, neutral currents, and therefore a breakdown of the simple $J^\alpha J_\alpha^+$ coupling. Furthermore, some $\Delta T = \frac{3}{2}$ amplitude is required to explain why

$$K^+ \to \pi^+ + \pi^0$$

actually does occur. It it were absolutely forbidden, with the only $\Delta T = \frac{3}{2}$ coupling coming in because of the virtual electromagnetic field about the K^+, as a result of which *i*-spin is not exactly conserved, the amount of $\Delta T = \frac{3}{2}$ amplitude present would be of the order of 1% ($\sim\frac{1}{137}$), leading to a K^+ decay rate 50 times smaller than actually observed. Lack of space prevents a detailed study of the evidence for and the magnitude of the unenhanced $\Delta T = \frac{3}{2}$ terms.

Further evidence for the large enhancement of $\Delta T = \frac{1}{2}$ transitions comes from K^0-decays. A discussion of these requires some understanding of the K^0-decay, which exhibits some strange properties.

The decay of neutral *K*-mesons was first discussed in a very beautiful paper by Gell-Mann and Pais.[1] They pointed out that, in the strong interactions, hypercharge conservation is an exact law, and therefore no $K^0 \leftrightarrow \bar{K}^0$ transitions are possible. In strong interaction processes, the two particles are quite distinct, and the process tells us quite definitely which of the particles is produced. When the weak interactions are turned on, hypercharge is no longer a good quantum number. The K^0 and the $\bar{K}^0$ are degenerate, and they can convert into one another by a mechanism like

$$K^0 \to \pi^+ + \pi^- \to \bar{K}^0$$

for example. This implies that instead of K^0 and $\bar{K}^0$, a different pair of states, linear combinations of the above two, may be eigenstates of the mass operator. The difference between this "mixing" and the "mixing" discussed in connection with the ω and the ϕ vector mesons in $SU(3)$ is that, once the weak interactions are turned on, no linear combination of the K^0 and the $\bar{K}^0$ can be an eigenstate of the total Hamiltonian, since the system decays. Instead of dealing with

$$H = H_{\text{strong}} + H_w \tag{32.13}$$

we shall consider the *S*-matrix acting in the space spanned by the K^0 and

[1] M. Gell-Mann and A. Pais, *Phys. Rev.* **97,** 1387 (1955). Further implications were developed in A. Pais and O. Piccioni, *Phys. Rev.* **100,** 1487 (1955).

the $\bar{K}^0$ states.[1] Although H_w is very small, we must work at least to second order in H_w to obtain the K^0–$\bar{K}^0$ mixing effect.

A perturbation expression for S is given by

$$S \simeq 1 - i\int dx \mathcal{H}_w(x) - \frac{1}{2}\iint dx\, dx'\, T(\mathcal{H}_w(x)\mathcal{H}_w(x')) \tag{32.14}$$

If we write

$$\mathcal{H}_w(x) = e^{iP_\mu x^\mu}\mathcal{H}_w e^{-iP_\mu x^\mu} \tag{32.15}$$

then the x and x' integrations can be carried out in S and we obtain

$$\begin{aligned}(\Psi_f, (S-1)\Psi_i) \\ = -(2\pi)^4 i\, \delta(P_f - P_i)\left(\Psi_f, \left(\mathcal{H}_w + \mathcal{H}_w \frac{1}{E_i - P_0 + i\epsilon}\mathcal{H}_w\right)\Psi_i\right)\end{aligned}$$

Symbolically we write[2]

$$T = H_w + H_w \frac{1}{E - H_{\text{strong}} + i\epsilon} H_w \tag{32.16}$$

a familiar form. We will be dealing with eigenstates of H_{strong}, $\Psi(K^0)$, $\Psi(\bar{K}^0)$, and $\Psi(n)$.

Thus

$$(\Psi(K^0), T\Psi(K^0)) = \sum_n \frac{(\Psi(K_0), H_w\Psi(n))(\Psi(n), H_w\Psi(K^0))}{E_{K^0} - E_n + i\epsilon} \tag{32.17}$$

and

$$(\Psi(\bar{K}^0), T\Psi(\bar{K}^0)) = \sum_n \frac{(\Psi(\bar{K}^0), H_w\Psi(n))(\Psi(n), H_w\Psi(\bar{K}^0))}{E_{K^0} - E_n + i\epsilon} \tag{32.18}$$

These are related by the CPT theorem, which holds for H and H_{strong} and therefore for H_w also. We have

$$\begin{aligned}(\Psi(K^0), T\Psi(K^0)) &= \sum_n \frac{|(\Psi(K^0), \Theta^{-1}\Theta H_w \Theta^{-1}\Theta\Psi(n))|^2}{E_{K^0} - E_n + i\epsilon} \\ &= \sum_n \frac{|(\Theta\Psi(K^0), H_w\Theta\Psi(n))|^2}{E_{K^0} - E_n + i\epsilon} \\ &= \sum_n \frac{|(\Psi(\bar{K}^0), H_w\Psi(\bar{n}))|^2}{E_{K^0} - E_n + i\epsilon} \\ &= (\Psi(\bar{K}^0), T\Psi(\bar{K}^0))\end{aligned} \tag{32.19}$$

[1] Our discussion is a little complicated because we do not assume CP invariance here. We shall see a little later that there is some evidence that CP may not be a good symmetry.

[2] We define H_w such that $(\Psi(K^0), H_w\Psi(K^0)) = 0$. Also, since $\Delta S = 1$, we have $(\Psi(K^0), H_w\Psi(\bar{K}^0)) = 0$. We shall take the K^0 and the $\bar{K}^0$ to be at rest in what follows.

This follows because the states $\Psi(\bar{n})$ which are the *CPT*-transformed counterparts of $\Psi(n)$, are degenerate with the states $\Psi(n)$. It is easy to see in the same way that there is no relation between $(\Psi(K^0), T\Psi(\bar{K}^0))$ and $(\Psi(\bar{K}^0), T\Psi(K^0))$, which follows from *CPT* alone. *If we assume CP invariance*, then

$$(\Psi(\bar{K}^0), T\Psi(K^0)) = (CP\,\Psi(\bar{K}^0), CPT(CP)^{-1}CP\,\Psi(K^0))$$

$$= (\Psi(K^0), T\Psi(\bar{K}^0)) \tag{32.20}$$

In the K^0–$\bar{K}^0$ space, T forms a 2×2 matrix of the general form

$$T = \begin{pmatrix} (\Psi(K^0), T\,\Psi(K^0)) & (\Psi(K^0), T\,\Psi(\bar{K}^0)) \\ (\Psi(\bar{K}^0), T\,\Psi(K^0)) & (\Psi(\bar{K}^0), T\,\Psi(\bar{K}^0)) \end{pmatrix} = \begin{pmatrix} A & B \\ C & A \end{pmatrix} \tag{32.21}$$

We now look for linear combinations of K^0 and $\bar{K}^0$ which diagonalize this matrix. If we call these linear combinations K_S and K_L,[1] we in effect require that there be no transitions between K_S and K_L, so that each of them is characterized by its own decay rate. We have

$$(\Psi(K_L), T\,\Psi(K_S)) = (\Psi(K_S), T\,\Psi(K_L)) = 0 \tag{32.22}$$

The eigenvalues of the matrix T are

$$\lambda_{\pm} = A \pm \sqrt{BC} \tag{32.23}$$

and the eigenvectors corresponding to the two eigenvalues are

$$\begin{pmatrix} 1 \\ \sqrt{C/B} \end{pmatrix} \quad \text{and} \quad \begin{pmatrix} 1 \\ -\sqrt{C/B} \end{pmatrix}$$

If we define

$$p^2 = B \qquad q^2 = C$$

then

$$\Psi(K_S) = (|p|^2 + |q|^2)^{-1/2}(p\Psi(K^0) + q\Psi(\bar{K}^0)) \tag{32.24}$$

and

$$\Psi(K_L) = (|p|^2 + |q|^2)^{-1/2}(p\Psi(K^0) - q\Psi(\bar{K}^0)) \tag{32.25}$$

When there is *CP* invariance, we have $B = C$, so that $q = \pm p$. If we

[1] The subscripts S and L will later be associated with "short" and "long" lifetimes for the associated particles.

choose $q = -p$, then[1]

$$\Psi(K_1) = -\frac{i}{\sqrt{2}}(\Psi(K^0) - \Psi(\bar{K}^0)) \qquad \Psi(K_2) = \frac{1}{\sqrt{2}}(\Psi(K^0) + \Psi(\bar{K}^0)) \tag{32.26}$$

and $(\Psi(K_1), \Psi(K_2)) = 0$. In the general case this is not so.[2]

We now turn to the discussion of some very dramatic effects in the decay of neutral K-mesons. These are more easily discussed when CP invariance is assumed. The modification of some of the conclusions if this assumption is abandoned, will be described later.[3]

The first point to note is that, if the arbitrary phases are chosen such that under charge conjugation

$$C\Psi(K^0) = \Psi(\bar{K}^0)$$
$$C\Psi(\bar{K}^0) = \Psi(K^0) \tag{32.27}$$

then with pseudoscalar K^0's

$$CP\Psi(K_1) = \Psi(K_1) \tag{32.28}$$

and

$$CP\Psi(K_2) = -\Psi(K_2) \tag{32.29}$$

Thus only the K_1 can decay into a $J = 0$ two-pion state. Both K_1 and K_2 can decay into $\pi^\pm e^\mp \nu$ and both can decay into $\pi^+\pi^-\pi^0$. In the last case, however, the three-pion state, which has all pions in a relative S-state, is even under C (which interchanges π^+ and π^-) and odd under P, since there are three pions present. Thus K_1 can decay only into a three-pion state that is not symmetric, for which the rate is impeded by centrifugal barrier effects.[4] Because of these differences, we would expect K_1 and K_2 to have different lifetimes. Gell-Mann and Pais pointed out that, since the two-pion mode was favored by phase space, the K_1 should decay a great deal faster than the K_2. The $K^0 \to 2\pi$ decay, by means of which the K^0 was

[1] The notation K_1 and K_2 will be reserved for the linear combinations

$$K_1 = -i(K^0 - \bar{K}^0)/\sqrt{2}; \quad K_2 = (K^0 + \bar{K}^0)/\sqrt{2}.$$

[2] For a general discussion, see R. G. Sachs, *Ann. Phys. (N.Y.)* **22**, 239 (1963); this paper contains many useful references.

[3] We do not discuss the possible violations of CPT, because we would not have any way of calculating anything. As soon as a hermitian Hamiltonian is written down, CPT follows.

[4] Recently this mode of decay was observed with an amplitude $A(K_1 \to \pi^+\pi^-\pi^0)$ indicating that CP is violated. J. A. Anderson, F. S. Crawford, R. L. Golden, D. Stern, T. O. Binford, and V. G. Lind, *Phys. Rev. Letters* **14**, 475 (1965). See also S. L. Glashow and S. Weinberg, *Phys. Rev. Letters* **14**, 835 (1965).

actually discovered, is the principal decay mode of the K_1-component of a K^0-beam, and the K_2 has to be looked for further away from the source of the neutral K's.

The K_2 was looked for and found.[1] The lifetimes of the two particles are

$$\tau(K_1) \equiv \tau_S = (0.866 \pm 0.014) \times 10^{-10} \text{ sec}$$

and

$$\tau(K_2) \equiv \tau_L = (5.62 \pm 0.68) \times 10^{-8} \text{ sec}$$

We remark that the large disparity in the lifetimes is *not* conditional on CP invariance. This will be discussed in the arguments leading up to (32.43). We now return to our description in terms of K_S and K_L, i.e., we stop assuming that CP is a symmetry of H_w. The $\Delta T = \frac{1}{2}$ rule leads to a relation between the rates for

$$K_{S,L} \to \pi^+ + \pi^-$$

and

$$K_{S,L} \to \pi^0 + \pi^0$$

The final states can be in *i*-spin states $T = 0$ or 2, and the second of these is excluded by our rule. Consequently, using the decomposition

$$\begin{aligned} \frac{1}{\sqrt{2}}(\Psi_{\pi^+\pi^-} + \Psi_{\pi^-\pi^+}) &= \sqrt{\frac{1}{3}}\Psi_2 + \sqrt{\frac{2}{3}}\Psi_0 \\ \Psi_{\pi^0\pi^0} &= \sqrt{\frac{2}{3}}\Psi_2 - \sqrt{\frac{1}{3}}\Psi_0 \end{aligned} \tag{32.30}$$

we predict that

$$\frac{\Gamma(K_S \to \pi^+\pi^-)}{\Gamma(K_S \to \pi^0\pi^0)} = \frac{\Gamma(K_L \to \pi^+\pi^-)}{\Gamma(K_L \to \pi^0\pi^0)} = 2 \tag{32.31}$$

The experimental ratio for the former is 2.2 ± 0.1, in agreement with the prediction. There is, as yet, no data on the K_L ratio.

In a strong interaction process like

$$\pi^- + p \to \Lambda^0 + K^0$$

or

$$K^+ + n \to p + K^0$$

it is the K^0 which is produced. The amplitude for the production of a K^0, in terms of amplitudes for the production of K_S and K_L, is given by

$$A(K^0) = \frac{(|p|^2 + |q|^2)^{1/2}}{2p}(A(K_S) + A(K_L)) \equiv \alpha(A(K_S) + A(K_L)) \tag{32.32}$$

[1] K. Lande, E. T. Booth, J. Impeduglia, L. M. Lederman, and W. Chinowsky, *Phys. Rev.* **103**, 1901 (1956).

This form is obtained with the help of (32.24) and (32.25). Similarly

$$A(\bar{K}^0) = \frac{(|p|^2 + |q|^2)^{1/2}}{2q}(A(K_S) - A(K_L)) \equiv \beta(A(K_S) - A(K_L)) \quad (32.33)$$

Once a K^0 is produced, we may view it as a mixture of K_S and K_L. If the masses of these particles are m_S and m_L, respectively, and their decay rates are Γ_S and Γ_L, then, in the rest frame of the K's, the time dependence of the amplitudes describing the K_S and K_L beams, may be approximated by[1]

$$\begin{aligned} a(K_S, t) &= a(K_S, 0)e^{-im_St}e^{-\frac{1}{2}\Gamma_St} \\ a(K_L, t) &= a(K_L, 0)e^{-im_Lt}e^{-\frac{1}{2}\Gamma_Lt} \end{aligned} \quad (32.34)$$

Because of the mass difference between K_S and K_L, the components of a beam, which at $t = 0$ is a pure K^0 beam, have their relative phases shifted, so that at a later time the beam acquires a $\bar{K}^0$-component. The presence of the $\bar{K}^0$ can be detected by a reaction like

$$\bar{K}^0 + p \to \Lambda^0 + \pi^+$$

which cannot result from a K^0 beam. The time dependence of the $\bar{K}^0$ amplitude is found as follows:

At $t = 0$, we have a pure K^0 beam, so that from (32.33) we see that the amplitudes for K_S and K_L are equal at that time. After a time t, the proper time of the K^0, we find

$$\begin{aligned} a(\bar{K}^0, t) &= \beta(a(K_S, t) - a(K_L, t)) \\ &= \beta a(K_S, 0)[e^{-im_St-\frac{1}{2}\Gamma_St} - e^{-im_Lt-\frac{1}{2}\Gamma_Lt}] \end{aligned} \quad (32.35)$$

so that the probability of observing a $\bar{K}^0$ is[2]

$$\begin{aligned} P(\bar{K}^0, t) &\cong |\beta a(K_S, 0)|^2 \, |e^{-i(m_S-m_L)t}e^{-\frac{1}{2}\Gamma_St} - 1|^2 \\ &\cong |\beta a(K_S, 0)|^2(1 - e^{-\frac{1}{2}\Gamma_St}\cos \Delta mt + e^{-\Gamma_St}) \end{aligned} \quad (32.36)$$

(since $\Gamma_S \cong 500\Gamma_L$ we neglect Γ_L in the above expression). The intensity of the $\bar{K}^0$ component as a function of time, in units of τ_S, is plotted in Fig. 32.2 for several values of $\Delta m = |m_S - m_L|$. This method of measuring

[1] The most complete treatment of decaying states (and the justification of the exponential decay form) may be found in M. L. Goldberger and K. M. Watson, "Collision Theory," John Wiley and Sons, 1964.

[2] The suggestion that this effect be used to measure the tiny mass difference is due to W. F. Fry and R. G. Sachs, *Phys. Rev.* **109,** 2212 (1958).

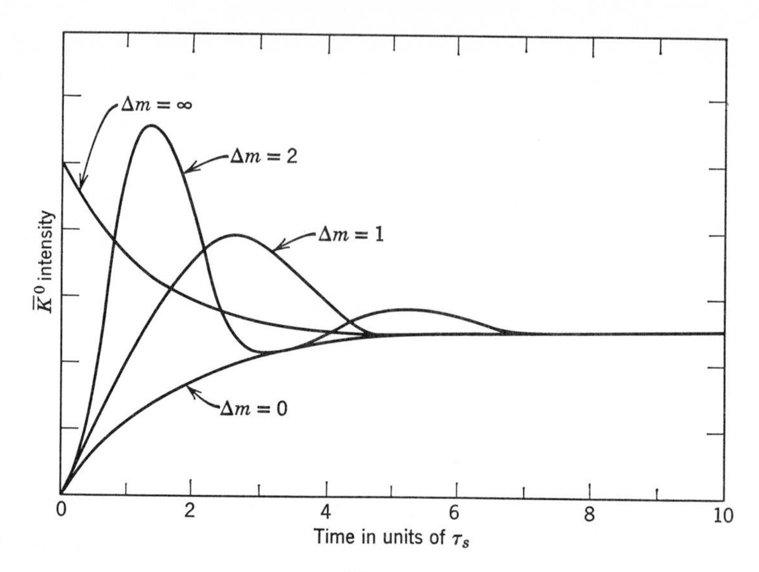

Fig. 32.2. The probability of finding a $\overline{K}^0$ in a beam (which is pure K^0 at $t = 0$) as a function of time, in units of τ_S, for several values of Δm. [From Camerini et al., *Phys. Rev.* **128,** 362 (1962).]

Δm was used in several experiments.[1] An experimental distribution is shown in Fig. 32.3. This particular experiment[2] yielded the value

$$|\Delta m| = 1.5 \pm 0.2 \frac{\hbar}{\tau_S c^2} \tag{32.37}$$

A more recent experiment[3] gives the value

$$|\Delta m| = 0.62 \pm 0.30 \frac{\hbar}{\tau_S c^2}$$

This figure is in better agreement with a different class of measurements, which involve the regeneration of K_S. The principle behind this class of experiments is the following: A K^0 (or $\overline{K}^0$) beam is allowed to travel for a time intermediate between τ_S and τ_L, so that the K_S component will have died out completely. If what is now the K_L beam is in any way interfered with, e.g., by passage through absorbing material, the separate components

[1] In the experiment of V. L. Fitch, P. A. Piroue, and R. B. Perkins, *Nuovo Cimento* **22,** 1160 (1961), the presence of $\overline{K}^0$ was detected through the observation of $\overline{K}^0 \to \pi^+ + e^- + \bar{\nu}$; the final state cannot come from a K^0 because of the $\Delta S = \Delta Q$ selection rule.
[2] U. Camerini, W. F. Fry, J. A. Gaidos, H. Huzita, S. V. Natali, R. B. Willman, R. W. Birge, R. P. Ely, W. M. Powell, and H. S. White, *Phys. Rev.* **128,** 362 (1962).
[3] G. W. Meisner, B. B. Crawford and F. S. Crawford, *Phys. Rev. Letters* **16,** 278 (1966).

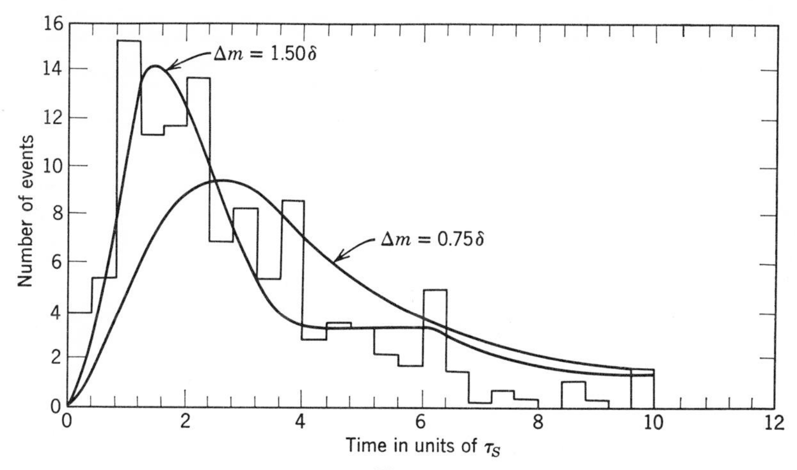

Fig. 32.3. The time distribution of the $\bar{K}^0$ interactions compared with the theoretical predictions for $\Delta m = 1.5$ and $\Delta m = 0.75$ in units of $\delta = \dfrac{\hbar}{\tau_S c^2}$. (From Camerini et al., *loc. cit.*)

of

$$a(K_L) = \frac{1}{2\alpha}\, a(K^0) - \frac{1}{2\beta}\, a(\bar{K}^0)$$

will be affected differently, so that a K_S can be regenerated. The probability of finding a K_S depends on the thickness of the absorber and on Δm.[1] The regeneration experiments suggest that

$$|\Delta m| \approx 0.5\, \frac{\hbar}{\tau_S c^2}$$

The sign of the mass difference cannot be detected by these experiments, but a workable method has been proposed:[2] At $t = 0$ a K^0 is created in a strong interaction (e.g., associated production). The nature of the beam is known at $t = 0$ if the magnitude of Δm is known. If the beam is then scattered by protons, knowledge of the $K^0 - p$ and $\bar{K}^0 - p$ scattering phase shifts implies that the relative phases of the K^0 and $\bar{K}^0$ components are still known after the scattering. The probability that a K_S, for example, is found at some time after the scattering depends on the sign of Δm.

[1] The experiment was suggested by M. L. Good, *Phys. Rev.* **106,** 591 (1957); **110,** 550 (1958). A detailed discussion of the dependence on Δm may be found in R. H. Good, R. P. Matsen, F. Fuller, O. Piccioni, W. M. Powell, H. S. White, W. B. Fowler, and R. W. Birge, *Phys. Rev.* **124,** 1223 (1961). See also V. L. Fitch, R. F. Roth, J. S. Russ, and W. Vernon, *Phys. Rev. Letters* **15,** 73 (1965).

[2] U. Camerini, W. F. Fry, and J. A. Gaidos, *Nuovo Cimento* **28,** 1096 (1963).

The fact that Δm is comparable with Γ_S, i.e., is of order G^2, may be used to prove that there can be no $\Delta S = 2$ part in the weak Hamiltonian.[1] If there were a $\Delta S = 2$ term, then

$$(\Psi(K_S), H_w(\Delta S = 2)\Psi(K_S)) - (\Psi(K_L), H_w(\Delta S = 2)\Psi(K_L))$$
$$= 2pq^*(|p|^2 + |q|^2)^{-1}(\Psi(\bar{K}^0), H_w(\Delta S = 2)\Psi(K^0)) + \text{h.c.} = 0(G)$$

In a recent experiment,[2] a rather devastating blow was struck against the attractive framework for the weak interactions outlined up to this point. In this experiment it was found that K_L particles decayed into $\pi^+\pi^-$, in direct violation of CP invariance, since the possibility of regeneration was very carefully excluded. The rate

$$\frac{\Gamma(K_L \to \pi^+\pi^-)}{\Gamma(K_L)_{\text{tot}}} \simeq 2 \times 10^{-3} \tag{32.38}$$

may be an indication that the violation is small, which could be consistent with the evidence, within experimental errors, that no term of the type $\mathbf{J}_n \cdot \mathbf{p}_e \times \mathbf{p}_p$ appears in the neutron decay amplitude. If we ignore any connection between two such different processes, we find that we can no longer draw the conclusion that the CP violation is small, just because the ratio in (32.38) is small. To see this, consider the quantities $p^2 = B$ and $q^2 = C$ which appear in (32.21). We have

$$p^2 = B = P\int \frac{(\Psi(K_0), H_w\Psi(n))(\Psi(n)\, H_w\Psi(\bar{K}_0))}{E_{K^0} - E_n}$$
$$- i\pi \sum_n (\Psi(K^0), H_w\Psi(n))\, \delta(E_n - E_{K^0})(\Psi(n), H_w\Psi(\bar{K}^0))$$

The second term has contributions from 2π intermediate states in the $T = 0$ and 2 states, from 3π intermediate states, and from leptonic states, e.g., $\pi e\nu$, $\pi\pi e\nu$, etc. The first term involves off the energy-shell contributions and will be denoted by $M_R + iM_I$, so that

$$B = M_R + iM_I - i\pi \sum_{T=0,2} (\Psi(K^0)\, H_w\Psi_T(2\pi))$$
$$\times (\Psi_T(2\pi), H_w\Psi(\bar{K}_0)) + \cdots \tag{32.39}$$

Similarly

$$C = M_R - iM_I - i\pi \sum_{T=0,2} (\Psi(\bar{K}^0), H_w\Psi_T(2\pi))$$
$$+ (\Psi_T(2\pi), H_w\Psi(K^0)) + \cdots \tag{32.40}$$

[1] L. B. Okun and B. Pontecorvo, *Soviet Phys. JETP* **5,** 1297 (1957). It was pointed out by S. Glashow [*Phys. Rev. Letters* **6,** 196 (1961)] that if $H_w(\Delta S = 2)$ is *odd* under charge conjugation, then there is no contribution to order G.

[2] J. H. Christenson, J. W. Cronin, V. L. Fitch, and R. Turlay, *Phys. Rev. Letters* **13,** 138 (1964).

In both cases the terms under the summation sign are on the energy-shell. Let us now concentrate on the two-pion terms. Because of the phase space factor, the on-shell terms involving two particles will always be considerably larger than three- or four-particle terms. Thus, among the on-shell terms, the two-pion terms should be the most important ones. *CPT* invariance implies that

$$\begin{aligned}(\Psi_T^{\text{out}}(2\pi), H_w\Psi(K^0)) &= (\Theta\Psi_T^{\text{out}}(2\pi), H_w\Theta\,\Psi(K^0))^* \\ &= (\Psi_T^{\text{in}}(2\pi), H_w\Psi(\overline{K}^0))^* \\ &= e^{2i\delta_T}(\Psi_T^{\text{out}}(2\pi), H_w\Psi(\overline{K}^0))^*\end{aligned}$$

We may therefore write

$$(\Psi_T^{\text{out}}(2\pi), H_w\Psi(K^0)) \equiv a_T e^{i\delta_T} \tag{32.41a}$$

and similarly

$$(\Psi_T^{\text{out}}(2\pi), H_w\Psi(\overline{K}^0)) \equiv a_T^* e^{i\delta_T} \tag{32.41b}$$

Time reversal invariance would imply that a_T and a_T^* have the same phase. We may use (32.41) and (32.24), (32.25) to show that

$$\begin{aligned}(\Psi_T^{\text{out}}(2\pi), H_w\Psi(K_S)) &= (|p|^2 + |q|^2)^{-1/2}(pa_T + qa_T^*)e^{i\delta_T} \\ (\Psi_T^{\text{out}}(2\pi), H_w\Psi(K_L)) &= (|p|^2 + |q|^2)^{-1/2}(pa_T - qa_T^*)e^{i\delta_T}\end{aligned} \tag{32.42}$$

Thus, if the two-pion terms in[1]

$$\begin{aligned}p^2 &\approx M_R + iM_I - i\pi\rho_{2\pi}(a_0^{*2} + a_2^{*2}) + \cdots \\ q^2 &\approx M_R - iM_I - i\pi\rho_{2\pi}(a_0^2 + a_2^2) + \cdots\end{aligned}$$

dominate the off-shell terms, then

$$\frac{p^2}{q^2} \approx \frac{a_0^{*2} + a_2^{*2}}{a_0^2 + a_2^2}$$

With the known enhancement of $\Delta T = \frac{1}{2}$ transitions, we finally get

$$\frac{p}{q} \approx \frac{a_0^*}{a_0}$$

Thus under the above assumptions

$$\frac{\Gamma(K_L \to (2\pi)_0)}{\Gamma(K_S \to (2\pi)_0)} \cong \left(\frac{pa_0 - qa_0^*}{pa_0 + qa_0^*}\right)^2 \ll 1 \tag{32.43}$$

Thus $\Gamma_S \gg \Gamma_L$ independent of the assumption of time reversal invariance. This leaves room for an, as yet uninvented, scheme which allows a "maximal" *CP* violation, analogous to the maximal parity violation. Such

[1] $\rho_{2\pi}$ stands for the two pion phase space factor.

a theoretical scheme would somehow have to explain the good evidence for CP invariance in beta decay, and hopefully save universality. For the time being, the role of the various terms in B and C can be studied experimentally. The phenomenological framework for this is contained in our discussion above, and a more explicit treatment may be found in the literature.[1]

The observation of the breakdown of CP invariance does not invalidate to any significant extent our discussion of the regeneration effect,[2] which, after all, just follows from the superposition principle. In particular, the mass difference experiments are not affected, with the time dependence shown in Fig. 32.1 completely unaltered. With a current-current model this leads to the conclusion that the currents satisfy $\Delta S = \Delta Q$; it is, however, far from clear whether the current-current picture can be maintained in face of the experimental fact that CP is violated.

If the current-current picture is to be maintained, we must find out whether the currents $J_{V,1}$ and $J_{A,1}$ have special i-spin transformation properties. Some evidence for the assumption that these currents transform as isospinors comes from a comparison of

$$K^+ \to \pi^0 + e^+ + \nu$$

with

$$K^0 \to \pi^- + e^+ + \nu$$

$$\bar{K}^0 \to \pi^+ + e^- + \bar{\nu}$$

An isospinor current is equivalent to a spurion field coupled to the pion and the K-meson, with the interaction $\boldsymbol{\pi} \cdot \chi \boldsymbol{\tau} K$ which conserves i-spin. This leads to

$$\frac{\Gamma(K^0 \to \pi^- e^+ \nu)}{\Gamma(K^+ \to \pi^0 e^+ \nu)} = \frac{\Gamma(\bar{K}^0 \to \pi^+ e^- \bar{\nu})}{\Gamma(K^+ \to \pi^0 e^+ \nu)} = 2$$

Hence

$$\begin{aligned} &\Gamma(K_L \to \pi^+ e^- \bar{\nu}) + \Gamma(K_L \to \pi^- e^+ \nu) \\ &\qquad = \frac{2\,|p|^2}{|p|^2 + |q|^2}\Gamma(K^+ \to \pi^0 e^+ \nu) + \frac{2\,|q|^2}{|p|^2 + |q|^2}\Gamma(K^+ \to \pi^0 e^+ \nu) \\ &\qquad = 2\Gamma(K^+ \to \pi^0 e^+ \nu) \end{aligned} \tag{32.44}$$

[1] T. T. Wu and C. N. Yang, *Phys. Rev. Letters* **13,** 380 (1964); this paper is the logical successor to the classic paper of T. D. Lee, R. Oehme, and C. N. Yang, *Phys. Rev.* **106,** 340 (1957).

[2] A new effect is the possibility of an interference between the two-pion state resulting from the decay of a K_L beam, with the two-pion state resulting from the decay of the K_S obtained by regeneration of the original K_L beam. This effect has recently been observed. [V. L. Fitch, R. F. Roth, J. S. Russ, and W. Vernon, *Phys. Rev. Letters* **15,** 73 (1965).]

The experimental value of the ratio $\Sigma\Gamma(K_L \to \pi e\nu)/\Gamma(K^+ \to \pi^0 e\nu)$ is 2.0 ± 0.5, which is encouraging but not overwhelming evidence that the vector part of the strangeness-changing current transforms as an isospinor. Actually, a detailed analysis shows that the ratio is rather sensitive to the amount of $\Delta T = \frac{3}{2}$ amplitude present; a 5% admixture of such an amplitude would change the ratio by 25%.[1] The test for the isospinor transformation properties of the axial vector current requires a check of predictions like

$$\Gamma(\Xi^0 \to \Sigma^+ + e^- + \bar{\nu}) = 2\Gamma(\Xi^- \to \Sigma^0 + e^- + \bar{\nu})$$

or

$$\Gamma(\Lambda^0 \to p + \pi^0 + e^- + \bar{\nu}) = \tfrac{1}{2}\Gamma(\Lambda^0 \to n + \pi^+ + e^- + \bar{\nu})$$

which involve both currents.

It should be noted that, even if it turns out that the strangeness changing currents transform as isospinors, this does not automatically lead to the $\Delta T = \frac{1}{2}$ rule for the nonleptonic decays. This is because the nonleptonic decays are generated by terms like $J_{V,0}\dot{J}^+_{V,1}$, etc. The strangeness-changing currents transform as isospinors, whereas the strangeness-conserving currents all involve $\Delta Q = 1$, so that $\Delta T \geq 1$. The product in general reduces to $\Delta T = \frac{1}{2}$ terms as well as others. It is generally believed that the $\Delta T = \frac{1}{2}$ behavior of H_w for nonleptonic decays is a consequence of a dynamical enhancement.

This chapter can be summarized by the following statements: the strangeness-changing currents must be vector as well as axial; at least the first of these transforms as an isospinor; the currents have a structure such that $\Delta S = \Delta Q$; the discovery of *CP* invariance violations may call into question the current-current form of the weak Hamiltonian.

[1] This is easily seen by noting that

$$\Psi_{K^+\pi^0} = -\sqrt{\frac{1}{3}}\Psi_{1/2} + \sqrt{\frac{2}{3}}\Psi_{3/2} \quad \text{and} \quad \Psi_{K^0\pi^-} = \sqrt{\frac{2}{3}}\Psi_{1/2} + \sqrt{\frac{1}{3}}\Psi_{3/2}$$

From this it is easy to derive

$$\sum \Gamma(K_L \to \pi e\nu)/\Gamma(K^+ \to \pi^0 e^+\nu) = 2\left|\left(1 + \frac{1}{\sqrt{2}}\frac{a(\frac{3}{2})}{a(\frac{1}{2})}\right)\Big/\left(1 - \sqrt{2}\frac{a(\frac{3}{2})}{a(\frac{1}{2})}\right)\right|^2$$

where $a(\frac{1}{2})$ and $a(\frac{3}{2})$ are the amplitudes for a transition involving changes of $T = \frac{1}{2}$ and $\frac{3}{2}$, respectively.

33

Weak Interactions of Strange Particles II. Analysis of Nonleptonic and Leptonic Decays

We saw in the preceding chapter that leptonic and nonleptonic decays differ in the kind of information they give about the weak interactions. The former involve directly the matrix elements of the vector and axial strangeness-changing currents, whereas the latter depend on the matrix elements of the product of two currents. In a final theory of these decays, the first information is essential before we can proceed to a study of the matrix elements of H_w. If we believe, however, that the nonleptonic $\Delta T = \frac{1}{2}$ rule is a dynamical effect, i.e., characteristic of the interplay of the strong and weak interactions, then a fundamental understanding of the nonleptonic decays requires a better knowledge of the strong interactions than we now possess. Thus, in effect, there is no close connection between the two kinds of decays; we shall reverse the logical order and first discuss some nonleptonic decays on a phenomenological level.

The first process which we will discuss is

$$\Lambda^0 \rightarrow p + \pi^-$$

This is a prototype of the processes

$$\Sigma^\pm \rightarrow n + \pi^\pm$$

$$\Xi^- \rightarrow \Lambda^0 + \pi^-$$

which can be discussed in complete analogy with the Λ^0 decay. The reason for this is that in all cases we have the decay of a spin $\frac{1}{2}$ particle into a system of two particles with spin 0 and $\frac{1}{2}$ respectively. The decay may proceed through an S-wave channel and/or a P-wave channel. If parity were conserved in the decay, only one of these modes would exist. The most general matrix element for the decay can only be of the form

$$\mathcal{M} = \bar{u}_p(p)(A + B\gamma_5)\, u_\Lambda(p_\Lambda) \tag{33.1}$$

We know on general grounds that the numbers A and B (they are constants since all three of the four momenta in the problem are on the mass shell) have phases determined by the final state scattering, provided CP is conserved. Because of the $\Delta T = \frac{1}{2}$ rule, the final state has $T = \frac{1}{2}$; since, however, the low-energy phase shifts in the $T = \frac{1}{2}$ pion-nucleon state are negligible, A and B are real. Since there is some question concerning CP invariance, we will treat A and B as complex. A nonrelativistic reduction of the above matrix element in the Λ^0 rest frame must lead to a form

$$\mathcal{M} = \chi_f^{+}(a_s + a_p \boldsymbol{\sigma} \cdot \hat{\mathbf{q}}_\pi)\chi_i \tag{33.2}$$

with a_s and a_p proportional to A and B, respectively. The χ_f is a spinor describing the proton, and χ_i describes the Λ^0 particle. In general, the Λ^0 will be produced in a strong interaction, e.g., an associated production experiment, and it will be polarized perpendicular to the reaction plane. We will denote the normal to the production plane by $\mathbf{n}$. To be specific, we shall take $\mathbf{n} = (\mathbf{k} \times \mathbf{p}_\Lambda)/|\mathbf{k} \times \mathbf{p}_\Lambda|$ where $\mathbf{k}$ is the incident beam direction. If the polarization of the Λ^0 is denoted by P_Λ, the square of the matrix element is given by

$$|\mathcal{M}|^2 = \chi_f^{+}(a_s + a_p \boldsymbol{\sigma} \cdot \hat{\mathbf{q}}_\pi) \frac{1 + P_\Lambda \boldsymbol{\sigma} \cdot \mathbf{n}}{2} (a_s^* + a_p^* \boldsymbol{\sigma} \cdot \hat{\mathbf{q}}_\pi)\chi_f \tag{33.3}$$

Let us now ask for the transition rate to a state, in which the proton is polarized with magnitude P_p in a direction $\mathbf{m}$. We can then replace

$$(\chi_f^{+})_a (M)_{ab} (\chi_f)_b$$

by

$$\sum_f (\chi_f^{+})_a M_{ab} \left(\frac{1 + \boldsymbol{\sigma} \cdot \mathbf{m} P_p}{2} \chi_f\right)_b = \operatorname{tr}\left(M \frac{1 + P_p \boldsymbol{\sigma} \cdot \mathbf{m}}{2}\right)$$

Thus

$$|\mathcal{M}|^2 = \tfrac{1}{4} \operatorname{tr} \{(a_s + a_p \boldsymbol{\sigma} \cdot \hat{\mathbf{q}}_\pi)(1 + P_\Lambda \boldsymbol{\sigma} \cdot \mathbf{n})(a_s^* + a_p^* \boldsymbol{\sigma} \cdot \hat{\mathbf{q}}_\pi)(1 + P_p \boldsymbol{\sigma} \cdot \mathbf{m})\}$$

Some simple manipulations bring this into the form

$$\begin{aligned} |\mathcal{M}|^2 = \tfrac{1}{2}\{&|a_s|^2 + |a_p|^2 + 2P_\Lambda \operatorname{Re}(a_p^* a_s)\hat{\mathbf{q}}_\pi \cdot \mathbf{n} \\ &+ 2P_p \operatorname{Re}(a_p^* a_s)\hat{\mathbf{q}}_\pi \cdot \mathbf{m} + P_\Lambda P_p \mathbf{m} \cdot \mathbf{n}(|a_s|^2 - |a_p|^2) \\ &+ 2P_\Lambda P_p |a_p|^2 \hat{\mathbf{q}}_\pi \cdot \mathbf{m} \hat{\mathbf{q}}_\pi \cdot \mathbf{n} \\ &+ 2P_\Lambda P_p \operatorname{Im}(a_s a_p^*)\hat{\mathbf{q}}_\pi \cdot \mathbf{n} \times \mathbf{m}\} \end{aligned} \tag{33.4}$$

If the proton polarization is not measured, we may set $P_p = 0$ and we then get the form

$$|\mathcal{M}|^2 = \frac{1}{2}(|a_s|^2 + |a_p|^2)\left(1 + P_\Lambda \frac{2 \operatorname{Re} a_s a_p}{|a_s|^2 + |a_p|^2} \cos\theta\right) \tag{33.5}$$

where θ is the angle between the normal to the production plane and the pion momentum. If P_Λ is not zero, and a dependence on $\cos\theta$ is found (in that more pions are emitted "up" relative to the plane, than "down," i.e., there is an up-down asymmetry), we know that both a_s and a_p must be nonvanishing, and parity is not conserved. Following our discussion in Chapter 30, we note that invariance under C requires that a_p be imaginary relative to a_s except for final state interaction effects. Since these are very small,[1] an appreciable up-down asymmetry indicates that C too is violated. The rate is given by

$$\rho \int_1^1 d(\cos\theta)\, |\mathcal{M}|^2 = \rho(|a_s|^2 + |a_p|^2) \tag{33.6}$$

where ρ is a factor which takes into account phase space. The number of decays with the pions going up is proportional to

$$\rho \int_0^1 d(\cos\theta)\, |\mathcal{M}|^2 = \frac{1}{2}\rho(|a_s|^2 + |a_p|^2)\left(1 + \frac{1}{2}P_\Lambda \frac{2\,\mathrm{Re}\, a_s a_p^*}{|a_s|^2 + |a_p^2|}\right)$$

Hence

$$\frac{N_{\mathrm{up}} - N_{\mathrm{down}}}{N_{\mathrm{up}} + N_{\mathrm{down}}} = \frac{1}{2}P_\Lambda \frac{2\,\mathrm{Re}\, a_s a_p^*}{|a_s|^2 + |a_p|^2} \tag{33.7}$$

If we write the angular distribution relative to the plane as

$$W(\theta) = 1 + \alpha P_\Lambda \cos\theta \tag{33.8}$$

then we see that

$$\alpha = \frac{2\,\mathrm{Re}\, a_s a_p^*}{|a_s|^2 + |a_p|^2} \tag{33.9}$$

The parameter α has been measured in a number of experiments.[2] The value for the process $\Lambda^0 \to p + \pi^-$ is found to be

$$\alpha = 0.62 \pm 0.05 \tag{33.10}$$

For the determination of the parameter α from the up-down asymmetry, not only the magnitude, but also the sign of P_Λ must be known. We see from (33.4), however, that if unpolarized Λ^0's are taken, and instead the polarization of the proton in the longitudinal direction, $\mathbf{m} = -\hat{\mathbf{q}}_\pi$ is measured, the same information can be obtained. The proton polarization and its sign can be determined by nuclear scattering, and it was in this

[1] Both the S-wave and P-wave pion-nucleon phase shifts at a total center of mass energy of 1115 MeV are less than 10°.

[2] See, for example, J. W. Cronin and O. E. Overseth, *Phys. Rev.* **129,** 1795 (1963).

way that the above sign of α was obtained. It is clear that other combination of parameters can be obtained by looking for the polarization of the proton in different directions. The following values have been obtained[1]

$$\beta = \frac{2 \operatorname{Im} a_s^* a_p}{|a_s|^2 + |a_p|^2} = 0.19 \pm 0.19 \tag{33.11}$$

$$\gamma = \frac{|a_s|^2 - |a_p|^2}{|a_s|^2 + |a_p|^2} = 0.78 \pm 0.04 \tag{33.12}$$

If there is invariance under time reversal, β should be 0.[1] The data is not good enough to see whether T is violated in this decay. In the decay

$$\Lambda^0 \to n + \pi^0$$

the α coefficient has been measured, and it was found that[2]

$$\frac{\alpha(\Lambda^0 \to n\pi^0)}{\alpha(\Lambda^0 \to p\pi^-)} = 1.10 \pm 0.27 \tag{33.13}$$

consistent with unity as predicted by the $\Delta T = \frac{1}{2}$ rule.

Within experimental errors, a_s and a_p are relatively real and the above data yields

$$\frac{a_p}{a_s} = 0.35 \pm 0.03 \tag{33.14}$$

so that the decay is primarily through the S-channel.

For the $\Sigma^\pm$ decays the following values have been obtained[3]

$$\begin{aligned}
\alpha(\Sigma^+ \to p + \pi^0) &= -0.78 \pm 0.10 \\
\alpha(\Sigma^+ \to n + \pi^+) &= -0.03 \pm 0.08 \\
\alpha(\Sigma^- \to n + \pi^-) &= -0.10 \pm 0.15
\end{aligned} \tag{33.15}$$

the small values of the last two indicating that these decays go through a single channel. As was discussed in the preceding chapter, the $\Delta T = \frac{1}{2}$ rule, with the evidence on the rates of these processes, requires that one of these be an S-wave process and the other a P-wave process. Which

[1] More accurately, taking into account the $\Delta T = \frac{1}{2}$ rule, $\beta = 0.56(\alpha_{11} - \alpha_1)_{p=100\,\mathrm{MeV}/c}$ i.e., $\beta \sim -0.06$, with α_{11} the $P_{1/2}$ phase shift and α_1 the $S_{1/2}$ phase shift, both in the $T = \frac{1}{2}$ i-spin state.

[2] B. Cork, L. Kerth, W. Wenzel, J. Cronin and R. Cool, *Phys. Rev.* **100,** 1000 (1960).

[3] E. F. Beall, B. Cork, D. Keefe, P. Murphy, and W. A. Wenzel, *Phys. Rev. Letters* **8,** 75 (1962); E. D. Tripp, M. B. Watson, and M. Ferro-Luzzi, *Phys. Rev. Letters* **9,** 66 (1962).

decay goes through which channel has not been determined yet.[1] All that is required is an indication of the sign of γ, which can be obtained from a measurement of the neutron polarization, but this difficult experiment has not yet been done.

For the decay

$$\Xi^- \rightarrow \Lambda^0 + \pi^-$$

the value of α has been measured and found to be

$$\alpha_{\Xi^-} = -0.48 \pm 0.08 \tag{33.16}$$

It is interesting to speculate on the role of $SU(3)$ in the nonleptonic weak interactions. It is clear that H_w is not $SU(3)$-invariant, since strangeness is not conserved in the weak interactions. It may, in general consist of a sum of terms, each transforming as a different "irreducible tensor" under $SU(3)$. In the decay

$$\text{baryon octet} \rightarrow \text{baryon octet} + \text{meson octet}$$

all the representations contained in **8 × 8 × 8** could, in principle, appear in H_w. The empirical observation of the validity of the $\Delta T = \frac{1}{2}$ rule suggests that the octet terms in H_w are significantly larger than the other ones. Even if we assume that H_w transforms as an octet, we can learn very little that is new about the nonleptonic hyperon decays. The reason is the following: If the $\Delta T = \frac{1}{2}$ rule is taken for granted, there are only four independent processes which can actually be observed. These are represented by the amplitudes

$$a(\Lambda^0 \rightarrow p + \pi^-) \equiv a(\Lambda^0_-)$$
$$a(\Sigma^+ \rightarrow n + \pi^+) \equiv a(\Sigma^+_+)$$
$$a(\Sigma^- \rightarrow n + \pi^-) \equiv a(\Sigma^-_-)$$
$$a(\Xi^- \rightarrow \Lambda^0 + \pi^-) \equiv a(\Xi^-_-)$$

with the remaining processes

$$a(\Lambda^0 \rightarrow n + \pi^0) \equiv a(\Lambda^0_0)$$
$$a(\Sigma^+ \rightarrow p + \pi^0) \equiv a(\Sigma^+_0)$$
$$a(\Xi^0 \rightarrow \Lambda^0 + \pi^0) \equiv a(\Xi^0_0)$$

having amplitudes determined by the rule.[2] There are, however, many

[1] Recent theoretical work based on SU(6) symmetry together with assumptions about the transformation properties of H_w predicts that $a_s(\Sigma^+ \rightarrow n + \pi^+) = 0$. (See, for example G. Altarelli, F. Buccella and R. Gatto, *Phys. Letters* **14**, 70 (1965).

[2] In the notation $a_s(\Sigma^+_+)$ represents the *S*-wave amplitude; the upper index represents the charge of the decaying particle and the lower one the charge of the pion.

more octet amplitudes which can be written down: a matrix element of the form $(\Psi_{B+P}, H_w \Psi_B)$ can have as many independent terms as there are unitary singlets in $\mathbf{8} \times \mathbf{8} \times \mathbf{8} \times \mathbf{8}$, i.e., eight amplitudes. This is actually an overestimate. If we represent the baryons and mesons by 3×3 matrices [(18.17)-(18.20)] and H_w by λ_6 and/or λ_7, corresponding to a "spurion" which transforms as a K^0 and a $\bar{K}^0$, then the amplitudes are of the form

$$\alpha_1 \operatorname{tr}(\bar{B}BP\lambda_6) + \alpha_2 \operatorname{tr}(\bar{B}BP\lambda_7) + \cdots + \alpha_i \operatorname{tr}(\bar{B}P) \operatorname{tr}(B\lambda_6) + \cdots \tag{33.17}$$

These are constrained by the following identity involving traceless 3×3 matrices

$$\begin{aligned} &\operatorname{tr}(ABCD + ABDC + ACBD + ACDB + ADBC + ADCB) \\ &\qquad = \operatorname{tr}(AB)\operatorname{tr}(CD) + \operatorname{tr}(AC)\operatorname{tr}(BD) + \operatorname{tr}(AD)\operatorname{tr}(BC) \end{aligned} \tag{33.18}$$

and they are further constrained by CP invariance, if that is assumed to hold to a good approximation. The S-wave amplitudes are parity-violating, and therefore they must be odd under C. Under charge conjugation, we have[1]

$$\begin{aligned} P &\to P^T \\ \bar{B}_\alpha &\to -\mathcal{C}^{-1}{}_{\alpha\beta} B_\beta{}^T \\ B_\alpha &\to \bar{B}_\beta{}^T \, \mathcal{C}_{\beta\alpha} \end{aligned} \tag{33.19}$$

Here the trace is with respect to the $SU(3)$ indices, and the greek-letter subscripts are spinor indices. Thus a typical term $\operatorname{tr}(\bar{B}_\alpha B_\alpha P \lambda_k)$ transforms under C as follows[2]

$$\begin{aligned} \operatorname{tr}(\bar{B}_\alpha B_\alpha P\lambda_k) &\to -(\mathcal{C}^{-1})_{\alpha\rho}(\mathcal{C})_{\sigma\alpha} \operatorname{tr}(B_\rho{}^T \bar{B}_\sigma{}^T P^T \lambda_k) \\ &= -\operatorname{tr}(B^T \bar{B}^T P^T \lambda_k) \\ &= +\operatorname{tr}(\lambda_k{}^T P \bar{B} B) \\ &= +\operatorname{tr}(\bar{B}B\lambda_k{}^T P) \end{aligned} \tag{33.20}$$

Consequently, such a term can only appear in the combination

$$\operatorname{tr}(\bar{B}B[P\lambda_k - \lambda_k{}^T B]) \tag{33.21}$$

If account is taken of the fact that $\lambda_6{}^T = \lambda_6$ and $\lambda_7{}^T = -\lambda_7$ it is just a matter of some straightforward computation to show that the most

[1] The relation for the mesons was stated before in (18.18); the transformation of the baryon 3×3 matrices under C follow the rules obtained for single spinors in (2.90), with the transposition taking care of the $SU(3)$ indices.

[2] The reason for the change of sign in the third line of (33.20) is that in the matrix transposition B and $\bar{B}$ are interchanged, and, since they are fermion operators, they anticommute.

general S-wave amplitude involving λ_6 is

$$iA_1 \operatorname{tr}(\bar{B}B[\lambda_6, P]) + iA_2 \operatorname{tr}(B\bar{B}[\lambda_6, P]) + iA_3\{\operatorname{tr}(\bar{B}\lambda_6)\operatorname{tr}(BP) - \operatorname{tr}(\bar{B}P)\operatorname{tr}(B\lambda_6)\} \quad (33.22)$$

For the P-wave (parity-conserving) amplitudes, a typical term might be $\operatorname{tr}(\bar{B}_\alpha(\gamma_5)_{\alpha\beta}BP\lambda_6)$ and the only change from (33.20) is that

$$\operatorname{tr}(\bar{B}\gamma_5 BP\lambda_6) \to -(\mathcal{C}^{-1}\gamma_5{}^T\mathcal{C})\operatorname{tr}(B^T\bar{B}^TP^T\lambda_6)$$

It is again a matter of straightforward computation to show that there are four independent p-wave amplitudes involving λ_6. We have not concerned ourselves with terms transforming as λ_7 because there are reasons to believe that only the symmetric λ_6 appears in H_w. These will be discussed in the next chapter. If we accept this, we note that (33.22) contains three parameters which describe the four independent S-wave amplitudes. We can, therefore, derive a relation between them. The traces in (33.22) are easily worked out, and there results the following relation[1]

$$2a_s(\Xi_-^-) = a_s(\Lambda_-^0) + \sqrt{3}\, a_s(\Sigma_0^+) \quad (33.23)$$

There is no reason to expect the same relation to hold for the P-wave

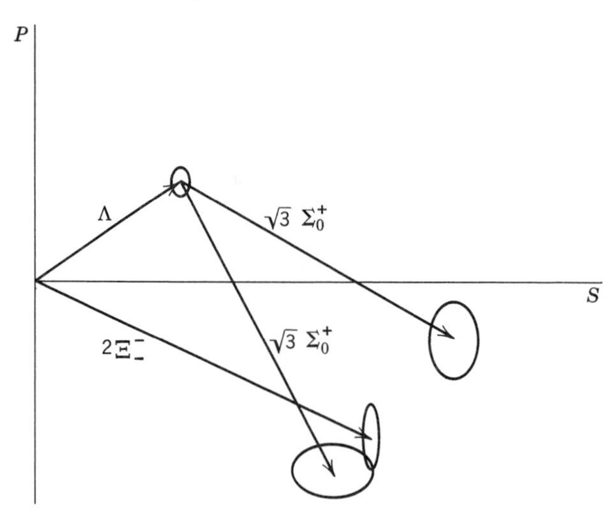

Fig. 33.1. Test of the rule $2a(\Xi_-^-) = \sqrt{3}a(\Sigma_0^+) + a(\Lambda_-^0)$. The two possible values of $\sqrt{3}a(\Sigma_0^+)$ correspond to two possible signs of γ_Σ, with the measured α_Σ.

[1] B. Lee, *Phys. Rev. Letters* **12,** 83 (1964); M. Gell-Mann, *Physics Letters* **8,** 214 (1964). H. Sugawara, *Prog. Theor. Physics* (Kyoto) **31,** 213 (1964).

amplitudes.[1] Nevertheless, the experimental evidence is that it does (Fig. 33.1).[2]

The above considerations have certain implications for nonleptonic K-decays. The most general matrix element involving λ_6 and three pseudoscalar mesons is

$$\mathrm{tr}\,(P^3\lambda_6) \propto \mathrm{tr}\,P^2\,\mathrm{tr}\,(P\lambda_6) \tag{33.24}$$

[the last step following from the identity (33.18)]. Now

$$\mathrm{tr}\,(P\lambda_6) = K^0 + \bar{K}^0 = \sqrt{2}\,K_2{}^0$$

Thus it appears that

$$K_1{}^0 \rightarrow 2\pi$$

is forbidden! Since the ratio of the rate for $K^+ \rightarrow \pi^+ + \pi^0$ to that for $K_1{}^0 \rightarrow 2\pi$ is not understood, it is not clear whether the result is disturbing or not.[3] This is all that we shall have to say about the nonleptonic decays.

The leptonic decays of the hyperons have not been studied in great detail until recently. The total rates have been determined, and they appear anomalously low. If the matrix element for

$$\Lambda^0 \rightarrow p + e^- + \bar{\nu}$$

for example, were the same as that for nuclear beta decay, the predicted rate would be obtained from

$$(F(\eta)\tau)_\Lambda = (F(\eta)\tau)_n \cong \frac{1180}{\log 2}\ \mathrm{sec} \tag{33.25}$$

A straightforward numerical evaluation leads to the following result[4]

$$\begin{aligned}\Gamma(\Lambda^0 \rightarrow p + e^- + \nu) &\cong \frac{1}{\tau_\Lambda} \cong \frac{\log 2}{1180}\frac{\eta^5}{30} \\ &\cong \frac{\log 2}{1180} \times \frac{1}{30}\left(\frac{M_\Lambda{}^2 - M_p{}^2}{2M_\Lambda m_e}\right)^5 \\ &\cong 6 \times 10^7\ \mathrm{sec}^{-1}\end{aligned} \tag{33.26}$$

The experimental rate is given by

$$\Gamma(\Lambda^0 \rightarrow p + e^- + \bar{\nu}) = 0.88 \times 10^{-3}\Gamma_{\mathrm{tot}}(\Lambda^0) = 3.4 \times 10^6\ \mathrm{sec}^{-1}$$

[1] The P-wave relation was derived by B. Lee with hypotheses which are known to be false. A careful analysis of the minimum assumptions appears in S. P. Rosen, *Phys. Rev.* **140,** B326 (1965).

[2] M. L. Stevenson, J. P. Berge, J. B. Hubbard, G. R. Kalbfleisch, J. B. Shafer, F. T. Solmitz, S. G. Wojcicki, and P. G. Wholmut, *Physics Letters* **9,** 349 (1964).

[3] N. Cabibbo, *Phys. Rev. Letters* **12,** 155 (1964); M. Gell-Mann, *loc. cit.*

[4] The phase space integral $F(\eta)$, which appears in (31.22), is well approximated by $\eta^5/30$ when η is large.

which is an order of magnitude smaller than expected. The same is true of the other leptonic hyperon decays.[1] Thus if the concept of universality is to be preserved, it will have to be done in a more subtle way than by direct addition of $\Delta S = 0$ and $\Delta S = 1$ terms with equal coefficients.

If it is assumed that the matrix element for the beta decay of the Λ^0 is given by the simple form

$$(\Psi_p, (J^{\alpha}_{V,1} + J^{\alpha}_{A,1})\Psi_\Lambda) = \frac{1}{(2\pi)^3}\bar{u}(p_p)(F_V\gamma^\alpha - F_A\gamma^\alpha\gamma_5)\,u(p_\Lambda) \qquad (33.27)$$

then the rate measurement, together with a measurement of the up-down asymmetry of the electrons relative to the plane of polarization of the Λ^0, yields[2]

$$F_V = 0.24 \pm 0.02$$
$$F_A = (0.8 \pm 0.3)F_V \qquad (33.28)$$

provided one assumes that F_V and F_A are real and constant. We shall see in the next chapter that this is in accord with some theoretical speculations. As soon as more experimental data become available, a more detailed analysis of the Λ^0 decay, based on the general matrix element

$$\bar{u}(p_p)[\gamma^\alpha F_V(q^2) + q^\alpha F_S(q^2) + i\,F_M(q^2)\sigma^{\alpha\beta}q_\beta$$
$$- \gamma^\alpha\gamma_5 F_A(q^2) - q^\alpha\gamma_5 F_P(q^2) + iG_M(q^2)\gamma_5\sigma^{\alpha\beta}q_\beta]\,u(p_\Lambda) \qquad (33.29)$$

will become possible. One reason is that Λ^0's produced in strong interactions are generally polarized, and now that the asymmetry parameter α is known, the Λ^0 polarization can be determined directly from a measurement of the up-down asymmetry in the nonleptonic decay. Thus, in effect, every Λ^0 beta decay is a Co^{60} type of experiment. The detailed discussion of hyperon decay[3] may be found in the literature.

It is clear that if one rejects the notion that the form factors, which involve only strongly interacting parts, are very strongly dependent on the hypercharge, or the momentum transfer, we must abandon universality, and in the strangeness-changing decays replace G by ξG, i.e., write

$$J^\alpha = J^\alpha_{V,0} + J^\alpha_{A,0} + \xi(J^\alpha_{V,1} + J^\alpha_{A,1}) \qquad (33.30)$$

[1] For example, $\Gamma_{\text{exp}}(\Sigma^- \to ne^-\nu)/\Gamma_{\text{theo}} = (2.2 \pm 0.4)\times 10^{-2}$ and $\Gamma_{\text{exp}}(\Xi \to \Lambda^0 + \bar{e} + \nu)/\Gamma_{\text{theo}} = (12 \pm 7) \times 10^{-2}$, as quoted in H. H. Bingham, *Proc. Roy. Soc. (London)* **A285,** 202 (1965).

[2] Quoted in R. H. Dalitz *Properties of the Weak Interactions,* Varenna Summer School Lectures (1964).

[3] V. P. Belov, B. S. Mingalev, and V. M. Shekhter, *Soviet Phys. JETP* **11,** 392 (1960); C. H. Albright, *Phys. Rev.* **115,** 750 (1959); D. R. Harrington, *Phys. Rev.* **120,** 1482 (1960).

in H_w. We shall find more evidence for this in the discussion of our next topic, the leptonic decays of the K-mesons.

Consider first the decay

$$K^{\pm} \rightarrow \mu^{\pm} + \nu$$

the matrix element for which must, because of Lorentz invariance, have the form

$$\frac{G}{\sqrt{2}} \frac{\sqrt{2m_\nu}}{(2\pi)^3} \bar{u}(p_l)\gamma_\alpha(1 - \gamma_5)\, v(p_\nu)(\Psi_0, J^\alpha_{A,1}\Psi_K)$$
$$= \frac{G}{\sqrt{2}} \frac{\sqrt{2m_\nu}}{(2\pi)^{3/2}} \bar{u}(p_l)\, \gamma_\alpha(1 - \gamma_5)\, v(p_\nu) q^\alpha M_p\, F_K(q^2) \quad (33.31)$$

where

$$q = p_l + p_\nu$$
$$q^2 = m_K{}^2$$

The decay rate is given by

$$\Gamma = \frac{G^2 M_p{}^4}{8\pi} |F_K(m_K^2)|^2 \left(\frac{m_l}{m_K}\right)^2 \frac{M_p}{m_\pi}\left(1 - \frac{m_l{}^2}{m_K^2}\right)^2 m_\pi \quad (33.32)$$

as a simple calculation shows. We note that the dependence on the lepton mass indicates that $K \rightarrow e + \nu$ should be very strongly suppressed relative to $K \rightarrow \mu + \nu$, which is analogous to our result for the pion leptonic decay,[1] and which is in agreement with experiment. Comparison of the μ rate with that for pion decay shows that[2]

$$\frac{\Gamma(K \rightarrow \mu\nu)}{\Gamma(\pi \rightarrow \mu\nu)} = \frac{m_K}{m_\pi} \frac{(1 - m_\mu{}^2/m_K{}^2)^2}{(1 - m_\mu{}^2/m_\pi{}^2)^2} \left|\frac{F_K}{F_\pi}\right|^2 \cong 1.3 \quad (33.33)$$

The ratio $|F_K/F_\pi|^2$ is again unreasonably small,

$$\left|\frac{F_K}{F_\pi}\right|^2 \cong 0.075 \quad (33.34)$$

indicating rather a breakdown of universality. The next process to be considered,

$$K^{\pm} \rightarrow \pi^0 + \begin{Bmatrix} \mu^{\pm} \\ e^{\pm} \end{Bmatrix} + \nu$$

[1] We again remind the reader that the dependence on the lepton mass, which makes the electron rate so small, is a consequence of the V-A coupling, which requires that the lepton and antilepton come off with opposite helicities. In a two particle decay of a spin 0 particle, this would violate angular momentum conservation in the limit $m_l \rightarrow 0$.

[2] The $K_{\mu 2}$ branching ratio has been taken from J. L. Brown, J. A. Kadyk, G. H. Trilling, R. T. Van der Walle, B. P. Roe, and D. Sinclair, *Phys. Rev. Letters* **8**, 450 (1962).

gives information about the matrix element $(\Psi_{\pi^0}, J^{\alpha}_{V,1}\Psi_K)$. This is a strong interaction term, and therefore parity conservation requires that only the vector current, $J^{\alpha}_{V,1}$ contributes to the matrix element. Lorentz invariance requires the general form

$$(\Psi_{\pi^0}, J^{\alpha}_{V,1}\Psi_K) = \frac{1}{(2\pi)^3}\{(p_K + p_\pi)^\alpha f_+(q^2) + (p_K - p_\pi)^\alpha f_-(q^2)\} \quad (33.35)$$

The form factors $f_\pm(q^2)$ depend only on the momentum transfer; their relative phase need not be real, if time reversal is violated in the weak interactions. This may sound paradoxial, in view of our insistence above that the matrix element involves only strong interactions. We do not know, however, whether the term $J^{\alpha}_{V,1}$ which appears in H_w does not consist of a sum of terms

$$J^{\alpha}_{V,1} = \mathfrak{J}^{\alpha}_{V,1} + e^{i\phi}\mathfrak{J}'^{\alpha}_{V,1} + \cdots$$

with arbitrary phase relations.

The rate is easily calculated in terms of $f_\pm$. We have

$$\mathcal{M} = \frac{G}{\sqrt{2}}\frac{\sqrt{2m_\nu}}{(2\pi)^6}\bar{u}(p_l)\,\gamma_\alpha(1 - \gamma_5)\,v(p_\nu)[(p_K + p_\pi)^\alpha f_+ + (p_K - p_\pi)^\alpha f_-]$$

so that

$$\mathcal{M} = \frac{G}{\sqrt{2}}\frac{\sqrt{2m_\nu}}{(2\pi)^6}\bar{u}(p_l)\{(\not{p}_K + \not{p}_\pi)f_+ + m_l f_-\}\,v(p_\nu)$$

In deriving the last expression, four-momentum conservation and the Dirac equation were used. We see from this expression that in K_{e3} decay only information about f_+ is obtained, unless unaccountably f_- is two orders of magnitude larger than f_+. The rate is

$$d\Gamma = \frac{G^2}{(2\pi)^5}\frac{1}{p_K^0}\frac{d^3p_\pi}{2\omega_\pi}\frac{d^3p_l}{2E_l}\frac{d^3p_\nu}{2p_\nu}\,\delta(p_K - p_\pi - p_e - p_\nu)\,|\mathcal{M}|^2 \quad (33.36)$$

with $|\mathcal{M}|^2$ obtained as a result of a simple trace calculation

$$\begin{aligned}|\mathcal{M}|^2 = |f_+|^2\,(&2p_K\cdot p_\nu p_K\cdot p_l + 2p_\pi\cdot p_\nu p_\pi\cdot p_l + 2p_K\cdot p_\nu p_\pi\cdot p_l \\ &+ 2p_K\cdot p_l p_\pi\cdot p_\nu - 2p_K\cdot p_\pi p_e\cdot p_\nu - (m_K{}^2 + m_\pi{}^2)p_e\cdot p_\nu) \\ &+ m_l{}^2\,|f_-|^2\,p_\nu\cdot p_e + 2m_l{}^2\,\mathrm{Re}\,f_+f_-{}^*(p_K\cdot p_\nu + p_\pi\cdot p_\nu) \quad (33.37)\end{aligned}$$

If information about the lepton polarization is sought, the extra factor $\frac{1}{2}(1 + P_l\gamma_5\not{n})$ in the trace modifies $|\mathcal{M}|^2$ in the following way: The above expression is multiplied by $\frac{1}{2}$, and added to it appears the expression

$$2m_l\,P_l\,\mathrm{Im}\,f_+f_-^*\epsilon_{\alpha\beta\mu\sigma}p_K^\alpha n^\beta p_\pi{}^\mu p_l{}^\sigma \quad (33.38)$$

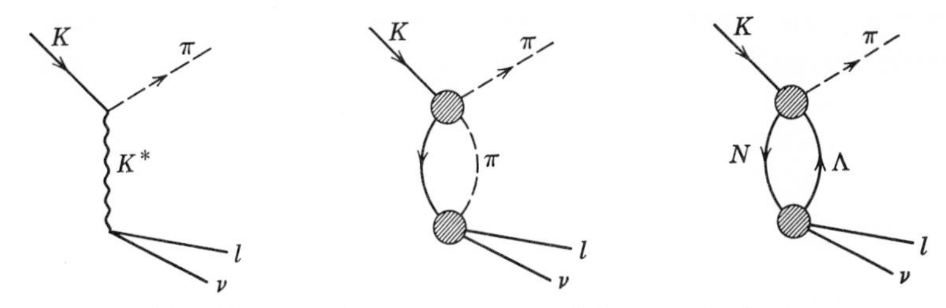

Fig. 33.2. Some graphs in the structure of the vertex in K_{l3} decay.

where P_l is the lepton polarization and n^α is a four-vector satisfying the usual conditions $(p_l)_\alpha n^\alpha = 0$ and $n_\alpha n^\alpha = -1$. Evidently signs of violation of time reversal are only to be sought in the $K_{\mu 3}$ decay mode.[1]

The form factors $f_\pm(q^2)$ are not expected to vary very strongly with q^2 in the range probed by the experiment. Whether we set up a dispersion relation for these form factors (exactly as was done in Chapter 26 for the electromagnetic form factors) or whether we merely draw possible graphs for the process (Fig. 33.2) it is clear that the only possible resonant state which could be exchanged between the lepton vertex and the $K\pi$ vertex is the K^* vector meson, whose (mass)^2 is very much larger than the maximum value of $q^2 = (m_K - m_\pi)^2$. Thus it is sufficient in an analysis of the data to write the phenomenological forms

$$f_\pm(q^2) = f_\pm(0) + \lambda_\pm \frac{q^2}{m_K^2} \tag{33.39}$$

Before turning to the discussion of the data, there is one more point to be made. If we were to use perturbation theory for the $K\pi$ vertex in Fig. 33.2, we would write down a coupling of the form

$$f_{K^*K\pi} K_\mu{}^*(\bar{K}^+\partial^\mu\phi_\pi - \partial^\mu\bar{K}^+\phi_\pi) \tag{33.40}$$

for the $K^*K\pi$ interaction. The magnitude of $f_{K^*K\pi}$ is known from the K^* width, but since the vertex undoubtedly has a q^2 dependence, and the coupling of the K^* to the lepton pair is at this point unknown, knowledge of this parameter does not help much. The total matrix element will be proportional to

$$(p_K + p_\pi)_\mu \frac{\left(g^{\mu\nu} - \dfrac{q^\mu q^\nu}{m_{K^*}^2}\right)}{q^2 - m_{K^*}^2} \bar{u}(p_e)\,\gamma_\nu(1-\gamma_5)\,v(p_\nu) \tag{33.41}$$

[1] A measurement of the muon polarization component normal to the decay plane of a stopping K^+, $P_\mu\perp = \mathbf{S}_\mu \cdot \mathbf{p}_\pi \times \mathbf{p}_\mu/|\mathbf{p}_\pi \times \mathbf{p}_\mu|$ yields $\langle P_\mu\perp\rangle = 0.04 \pm 0.35$ consistent with T invariance. U. Camerini, R. L. Hantman, R. H. March, D. Murphee, G. Gidal, G. E. Kalmus, W. M. Powell, R. T. Pu, C. L. Sandler, S. Natali, and M. Villani, *Phys. Rev. Letters* **14**, 989 (1965).

with the vector meson propagator making its appearance. This may be written in the form

$$j_\mu^l \frac{1}{q^2 - m_{K^*}^2}\left((p_K + p_\pi)^\mu - \frac{m_K^2 - m_\pi^2}{m_{K^*}^2}(p_K - p_\pi)^\mu\right) \tag{33.42}$$

with the prediction that

$$\frac{f_-(0)}{f_+(0)} = -\frac{m_K^2 - m_\pi^2}{m_{K^*}^2} = -0.3 \tag{33.43}$$

Note that in the limit of exact unitary symmetry $f_-(0)/f_+(0) = 0$.

Calculations with the form derived in (33.36), (33.37) are straightforward but tedious.[1] We shall content ourselves with summarizing the results of a number of experiments. The analysis was always done with real form factors, since the experiment measuring the transverse lepton polarization is consistent with T invariance. From the expression for the dependence of the rate on the lepton energy

$$\frac{G^2 p_l(W - p_l)}{32\pi^3 m_K (W - p_l)^2}\{f_+^2(m_K p_l(4m_K^2 + 5m_l^2) - m_l^4 - 8m_K^2 p_l^2) + f_-^2 m_l^2(m_K p_l - m_l^2) + 2f_+ f_-(2m_l^2 m_K^2 + m_l^4 - 3m_K m_l^2 p_l)\} \tag{33.44}$$

where

$$W = \frac{m_K^2 + m_l^2 - m_\pi^2}{2m_K}$$

is the maximum lepton energy, it is possible to determine, in principle, the ratio of f_-/f_+. Actually, it turns out that the spectrum is not very sensitive to this ratio. A better measure is obtained by comparing the total $K_{\mu 3}$ rate with the K_{e3} rate. It turns out that the total rate is such that

$$\frac{\Gamma(K_{\mu 3})}{\Gamma(K_{e3})} \cong 0.65 + 0.12\frac{f_-}{f_+} + 0.02\left(\frac{f_-}{f_+}\right)^2 \tag{33.45}$$

Two experimental measurements, one on the K_2^0 decay (related to what we want by the $\Delta T = \frac{1}{2}$ rule) and another on K^+ decay, lead to values of this ratio of 0.73 ± 0.15 and 0.65 ± 0.2, respectively.[2] This leads to two

[1] A very detailed discussion of this process may be found in J. D. Jackson, "Weak Interactions," in the *Lectures in Theoretical Physics*, Brandeis Summer Institute, W. A. Benjamin, Inc., New York (1963).

[2] D. Luers, I. S. Mittra, W. J. Willis, and S. S. Yamamoto, *Phys. Rev.* **133,** B1276 (1964); B. P. Roe, D. Sinclair, J. L. Brown, D. A. Glaser, J. A. Kakyk, and G. H. Trilling, *Phys. Rev. Letters* **7,** 346 (1961).

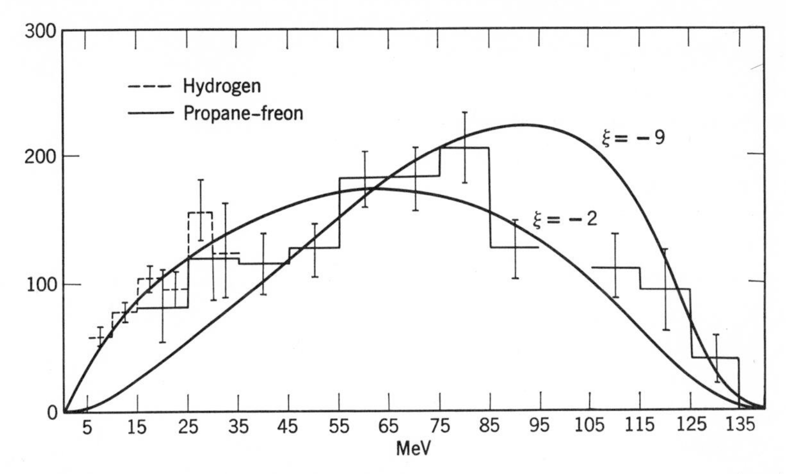

Fig. 33.3. Muon kinetic energy spectrum in the decay $K^+ \rightarrow \mu^+ + \pi^0 + \nu$, with the theoretical curves for $\xi = \dfrac{f_-}{f_+} = -2$ and $\xi = -9$ plotted on the same graph. [From Bisi et al., *Phys. Rev. Letters* **12,** 490 (1964).]

possible values of f/f_+ in each case. One solution is

$$\frac{f_-}{f_+} = -0.5 \pm 1.0$$

or, in effect, f_-/f_+ is small; the other yields $f_-/f_+ = -7 \pm 1$. The muon spectrum (Fig. 33.3) favors the lower value of this ratio.[1]

We have not yet raised the question whether the current $J_{V,1}$ is also suppressed. In order to test this, we compare the rates for

$$K^+ \rightarrow \pi^0 + e^+ + \nu$$

and

$$\pi^+ \rightarrow \pi^0 + e^+ + \nu$$

The former is given by[2]

$$\frac{G^2 M_p^{\ 4}}{768\pi^3}\left(\frac{m_K}{M_p}\right)^4 m_K(1 - 8x^2 + 8x^6 - x^8 - 24x^4 \log x)(f_+^{\ 2})_K \quad (33.46)$$

where $x = m_\pi/m_K$.

[1] A more recent value of this ratio based on a new $K_{\mu 3}$ experiment is $f_-/f_+ = 0.46 \pm 0.27$. For a detailed discussion and references see the review talk by G. Trilling in the *Proceedings of the International Conference on Weak Interactions*, Argonne National Laboratory 1965 (ANL-7130).

[2] J. D. Jackson, *loc. cit.*

Note that f_- does not appear in the K_{e3} rate. The rate for

$$\pi^+ \to \pi^0 + e^+ + \nu \tag{33.47}$$

can be calculated in exactly the same way, and the formula[1]

$$\frac{G^2 M_p^{\ 4}}{60\pi^3}\left(\frac{\Delta}{M_p}\right)^4 \Delta$$

$$\times \left(\sqrt{1-x^2}\left(1-\frac{9}{2}x^2-4x^4\right)+\frac{15}{2}x^4\log\frac{1+\sqrt{1-x^2}}{x}\right)_{x=m_e/\Delta}(f_+^{\ 2})_\pi$$

$$\Delta = \frac{m_{\pi^+}^2 - m_{\pi^0}^2 + m_e^{\ 2}}{2m_{\pi^+}} \tag{33.48}$$

results, provided we write

$$(\Psi_{\pi^0}, J_{V,0}^{\alpha}\Psi_{\pi^+}) = \frac{1}{(2\pi)^3}(p_{\pi^+} + p_{\pi^0})(f_+)_\pi \tag{33.49}$$

When the numerical factors are worked out, we find that

$$\Gamma(\pi^+ \to \pi^0 + e^+ + \nu) \cong 0.20(f_+^{\ 2})_\pi \tag{33.50}$$

The experimentally determined rate is

$$\Gamma(\pi^+ \to \pi^0 + e^+ + \nu) = (0.45 \pm 0.09)\ \text{sec}^{-1} \tag{33.51}$$

which implies that

$$(f_+)_\pi = 1.5 \pm 0.2 \tag{33.52}$$

On the other hand, the experimental rate

$$\Gamma(K^+ \to \pi^0 + e^+ + \nu) \cong 6 \times 10^6\ \text{sec}^{-1} \tag{33.53}$$

implies that

$$(f_+)_K \approx 0.15 \tag{33.54}$$

Thus again it appears that the strangeness-changing current is coupled to the leptons with about $\frac{1}{8}$ of the strength of the currents with $\Delta S = 0$.

As time goes on, experiments on the rarer decay modes will undoubtedly increase our knowledge of the matrix elements of the strangeness-changing currents. To guide experimentation, one generally needs some theory, and it is to the exposition of some recent speculations about the currents that we next turn.

[1] J. D. Jackson, *loc. cit.*

34

Conserved and Partially Conserved Currents

It has been known for a long time that the most natural way of introducing an additive conserved quantity into a theory is to require invariance under appropriate gauge transformations. This invariance, when imposed on the Lagrangian density, leads to the definition of a conserved current, the space integral of whose zeroth component represents the conserved quantum number. It was in this way that we discussed charge, and the identical discussion would hold for baryon number and hypercharge (the latter only when weak interactions are neglected). We can use the same formal apparatus to describe *i*-spin conservation, and it is not at all difficult to construct an *i*-spin current $\mathfrak{J}_\mu^{(k)}$ which is conserved. In a theory containing only pions and nucleons, such a current is given by

$$\mathfrak{J}_\mu^{(k)} = \tfrac{1}{2}\bar{\psi}\gamma_\mu\tau^{(k)}\psi - i\phi_\alpha(T^{(k)})_{\alpha\beta}\,\partial_\mu\phi_\beta \tag{34.1}$$

In a model containing operators corresponding to other particles, there would be a contribution from each field associated with a particle. For a $SU(3)$-invariant theory a similar construction was given in Chapter 18, page 283. Of course, the fact that a formal procedure for the construction of currents exists does not necessarily mean that these currents play a role in interactions. It is only with the elucidation of the weak interaction structure that it became apparent that currents play an important role in particle physics.

In their paper on the form of the weak interaction, Feynman and Gell-Mann[1] drew attention to the remarkable equality between the number G_μ, which described the strength of the muon decay (and, therefore, aside from $<\frac{1}{2}\%$ electromagnetic corrections, the *bare* coupling), and C_V, the number characterizing the strength of the vector coupling in nuclear beta decay (see Chapter 31, page 534).[2] In field-theory language we would say

[1] R. P. Feynman and M. Gell-Mann, *Phys. Rev.* **109,** 193 (1958).

[2] Recall, however, that C_V was obtained from the *ft*-value of O^{14} on the assumption that mesonic effects in the nucleus are unimportant. The following considerations show that the calculation done in Chapter 31 is, in fact, exact, except for electromagnetic effects.

that the strong interactions do not appear to renormalize the weak coupling constant. Feynman and Gell-Mann recalled that this was not an unprecedented situation in particle physics. The electric charge, too, is unrenormalized by the strong (or, in fact, by any) interactions. Because of Ward's identity (Chapter 13), which follows from gauge invariance, the electric charge is only renormalized by radiative corrections to the photon propagator, which affects all charged particles equally. The reason for this is charge conservation. Whatever the structure of the particle, and whatever the distribution of charge in the "cloud" surrounding the "bare" particle, the total charge is always the same, and it is therefore not renormalized by the presence of interactions.

The form of the vector current in the weak interactions

$$J^{\mu}_{V,0} = \bar{\psi}_n \gamma^{\mu} \psi_p + \cdots \tag{34.2}$$

suggests that it be the i-spin lowering part of the i-spin current

$$J^{\mu}_{V,0} = \bar{\psi}\gamma^{\mu}\frac{\tau_1 - i\tau_2}{2}\psi + \cdots = \mathfrak{J}^{(1)\mu} - i\mathfrak{J}^{(2)\mu} \tag{34.3}$$

The electric current, in the same notation, would be written as

$$j^{\mu} = \mathfrak{J}^{(3)\mu} + \tfrac{1}{2}\mathfrak{J}^{(Y)\mu} \tag{34.4}$$

with the first term representing the third component of the i-spin current and the second term the hypercharge current, so that the integrals of the zeroth components over all space give the familiar result

$$Q = T_3 + \tfrac{1}{2}Y$$

The first term has matrix elements connecting two single-nucleon states, which are measured in electron-nucleon scattering; they are the isovector form factors, which appear in

$$(\Psi_{p'}, \mathfrak{J}^{(3)}_{\mu}\Psi_p) = \frac{1}{(2\pi)^3}\bar{u}(p')[\gamma_{\mu} F_1^{\,v}(q^2) + i\sigma_{\mu\nu}q^{\nu}F_2^{\,v}(q^2)]\, u(p) \tag{34.5}$$

with

$$F_1^{\,v}(0) = \frac{1}{2}, \qquad F_2^{\,v}(0) = \frac{\mu_p - \mu_n}{4M} \tag{34.6}$$

For the charge-lowering component of the i-spin current this implies that

$$(\Psi_{p'}, (\mathfrak{J}^{(1)}_{\mu} - i\mathfrak{J}^{(2)}_{\mu})\Psi_p) = \frac{2}{(2\pi)^3}\bar{u}_n(p')[\gamma_{\mu} F_1^{\,v}(q^2) + i\sigma_{\mu\nu}q^{\nu} F_2^{\,v}(q^2)]\, u_p(p) \tag{34.7}$$

with the same values of $F_1^{\,v}(0)$ and $F_2^{\,v}(0)$. This equivalence between (34.5) and (34.7) is just a consequence of the Wigner-Eckart theorem. To test the validity of this identification of the weak hadronic vector current with the conserved *i*-spin current, hereafter called CVC (the conserved vector current hypothesis),[1] let us examine some definite predictions of the theory.

1. The beta decay of the pion

$$\pi^+ \to \pi^0 + e^+ + \nu$$

will be governed by the matrix element

$$(\Psi_{\pi^0}, (\mathfrak{J}_\mu^{(1)} - i\mathfrak{J}_\mu^{(2)})\Psi_{\pi^+}) = \frac{1}{(2\pi)^3}(p_{\pi^0} + p_{\pi^+})_\mu f_+((p_{\pi^0} - p_{\pi^+})^2) \quad (34.8)$$

To find the value of $f_+(0)$ we note that

$$\begin{aligned}\int d^3x(\Psi_{\pi^0}, (\mathfrak{J}_0^{(1)}(x) - i\mathfrak{J}_0^{(2)}(x))\Psi_{\pi^+})\\ &= \int d^3x e^{i(p_{\pi^0} - p_{\pi^+})x} \frac{1}{(2\pi)^3}(p_{\pi^0} + p_{\pi^+})_0 f_+((p_{\pi^0} - p_{\pi^+})^2)\\ &= 2\omega_\pi\, \delta(\mathbf{p}_{\pi^0} - \mathbf{p}_{\pi^+}) f_+(0)\end{aligned}$$

On the other hand

$$\begin{aligned}\int d^3x(\Psi_{\pi^0}, (\mathfrak{J}_0^{(1)}(x) - i\mathfrak{J}_0^{(2)}(x))\Psi_{\pi^+}) &= (\Psi_{\pi^0}, T_-\Psi_{\pi^+})\\ &= 2\omega_\pi\, \delta(\mathbf{p}_{\pi^0} - \mathbf{p}_{\pi^+})\sqrt{2}\end{aligned}$$

Consequently

$$f_+(0) = \sqrt{2} \quad (34.9)$$

The experimental value, quoted in (33.52) is in agreement with this prediction. It should be pointed out that, even if the current $J_{V,0}$ had nothing to do with the *i*-spin, $(\Psi_{\pi^0}, J^\mu_{V,0}\Psi_{\pi^+})$ would be unlikely to vanish, so that the process would still take place, with unpredictable strength. The characteristic of the CVC theory is the prediction of a definite rate. Since other alternatives do not yield definite predictions, our belief in the CVC theory rests on an accumulation of agreement with experiment, rather than an unambiguous rejection of alternatives.

2. A stronger test for the CVC theory, proposed by Gell-Mann[2], is based on the fact that the theory predicts a definite magnetic moment type of vector coupling. The term $i\sigma_{\mu\nu}q^\nu F_2^v(0)$ will give a well-defined

[1] Because we do not know that the "bare" current really has the form (34.2), it is not enough to say that the current is conserved. The scale is set by requiring the current to be equal to the *i*-spin current, and not twice that conserved current, for example.

[2] M. Gell-Mann, *Phys. Rev.* **111,** 162 (1958).

contribution to Gamow-Teller transitions, since it is proportional to $\boldsymbol{\sigma} \times \mathbf{q}$. If we take the beta coupling, augmented by the "weak magnetism" term

$$\frac{G}{\sqrt{2}} \frac{\sqrt{2m_\nu}}{(2\pi)^6} \bar{u}(p_e)\gamma^\mu(1 - \gamma_5)v(p_\nu)\, \bar{u}_p(p')\left[\gamma_\mu + i\frac{\mu_p - \mu_n}{2M}\sigma_{\mu\nu}q^\nu - \gamma_\mu\gamma_5\right] u_n(p) \tag{34.10}$$

then a nonrelativistic reduction, in which only Gamow-Teller terms are kept, yields

$$\frac{G}{\sqrt{2}} \frac{\sqrt{2m_\nu}}{(2\pi)^6}\Bigg\{\bar{u}(p_e)\gamma^k(1 - \gamma_5)\, v(p_\nu)\left[\sigma_k - i\frac{1 + \mu_p - \mu_n}{2M}(\mathbf{q} \times \boldsymbol{\sigma})_k\right]$$
$$+ \bar{u}(p_e)\gamma^0(1 - \gamma_5)\, v(p_\nu)\left[\frac{\boldsymbol{\sigma}\cdot\mathbf{p}_n + \boldsymbol{\sigma}\cdot\mathbf{p}_p}{2M}\right]\Bigg\} \tag{34.11}$$

In this expression the terms in the brackets are understood to be standing between two spinors representing the proton and neutron, respectively. The first term is the allowed axial vector term; the second is the contribution from the vector current. It is linear in $\mathbf{q}$ and thus is a "first forbidden" term. The last term comes from the zeroth component of the axial current. If we did not make the CVC hypothesis, the beta decay of the neutron (without the meson cloud) would still contribute an interference term from the γ_μ; in that case, we would expect the coefficient of the interference term to be unity instead of $1 + \mu_p - \mu_n = 4.7$. Thus the detection of this term provides a fairly striking test of the CVC theory.

Gell-Mann suggested that an accurate study of the spectrum of $e^\pm$ in the decay of N^{12} and B^{12} to C^{12} could reveal such a term. The former two states have spin and parity 1^+ and they belong to a $T = 1$ multiplet. The C^{12} ground state is 0^+ with $T = 0$ (Fig. 34.1). The $T_3 = 0$ member of the i-spin triplet decays to the ground state of γ emission. This decay is primarily a magnetic dipole transition, and the transition magnetic moment can be calculated from the observed radiation width. This

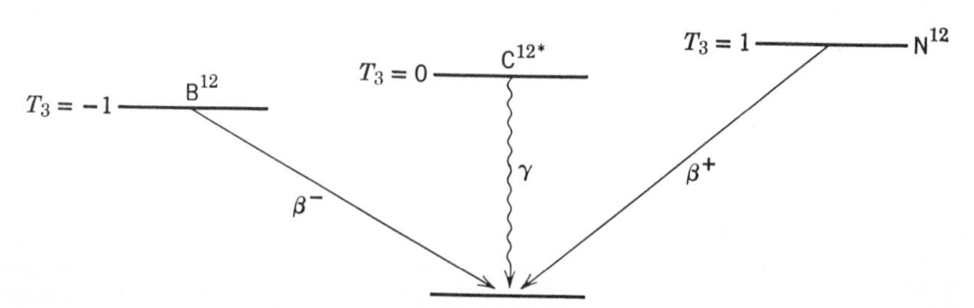

Fig. 34.1. The nuclear levels participating in the Gell–Mann CVC test (schematic).

calculation confirms that in the radiative decay the anomalous moments play a dominant role.[1] If we now neglect the last term in (34.11), the square of the matrix element in e^--decay involves

$$\text{tr}\,[\gamma^k(1-\gamma_5)\,\not{p}_\nu(1+\gamma_5)\,\gamma^l(\not{p}_e+m_e)] \times \left(\delta_{ik}+\frac{i\Delta}{2M}\,e_{ikj}q_j\right)\left(\delta_{il}-\frac{i\Delta}{2M}\,e_{ilm}q_m\right) \qquad (34.12)$$

Here we use the notation $\Delta = 1 + \mu_p - \mu_n$, and we have already used

$$\overline{\langle\sigma_i\rangle\langle\sigma_j\rangle} \propto \delta_{ij}$$

The trace calculation is straightforward. With nucleon recoil disregarded, there is only an energy conservation delta function $\delta(E_e + p_\nu - E_{\text{max}})$ and the electron spectrum will contain the usual terms, except that a factor $3E_eE_\nu$ coming from the trace will be replaced by

$$3E_eE_\nu\left(1+\frac{2\Delta}{3M}\left(E_e-E_\nu-\frac{m_e^{\,2}}{E_e}\right)\right) \qquad (34.13)$$

For the e^+ decay, the correction term has the opposite sign, and since $m_e \ll E_e$, and $E_\nu = E_{\text{max}} - E_e$, we obtain a correction to the ratio of the e^- spectrum to the e^+ spectrum, which is

$$\frac{1+4\Delta E_e/3M}{1-4\Delta E_e/3M} \approx 1+\frac{16}{3}\left(\frac{\Delta}{2M}\right)E_e \qquad (34.14)$$

The change in sign comes about because the "weak magnetism" effects are really vector-axial vector interference terms, and $V - A$ for a lepton decay becomes $V + A$ for the antilepton decay. The reason for being able to ingore the last term in (34.11) was that it does not change sign and therefore cancels out in the ratio. Details of the calculation and various refinements may be found in a very interesting paper by C. S. Wu.[2] The experimental results are shown in Fig. 34.2 and are in complete agreement with the CVC theory.[3]

3. The high-energy neutrino experiments

$$\bar{\nu}_\mu + p \to n + \mu^+$$
$$\nu_\mu + n \to p + \mu^-$$

will soon be carried out in hydrogen and deuterium. The most general

[1] E. Hayward and E. G. Fuller, *Phys. Rev.* **106,** 991 (1957).

[2] C. S. Wu, *Rev. Mod. Phys.* **36,** 618 (1964).

[3] Y. K. Lee, L. W. Mo, and C. S. Wu, *Phys. Rev. Letters* **10,** 253 (1963); T. Mayer-Kuckuck and F. C. Michel, *Phys. Rev.* **127,** 545 (1962); N. W. Glass and R. W. Peterson, *Phys. Rev.* **130,** 299 (1963).

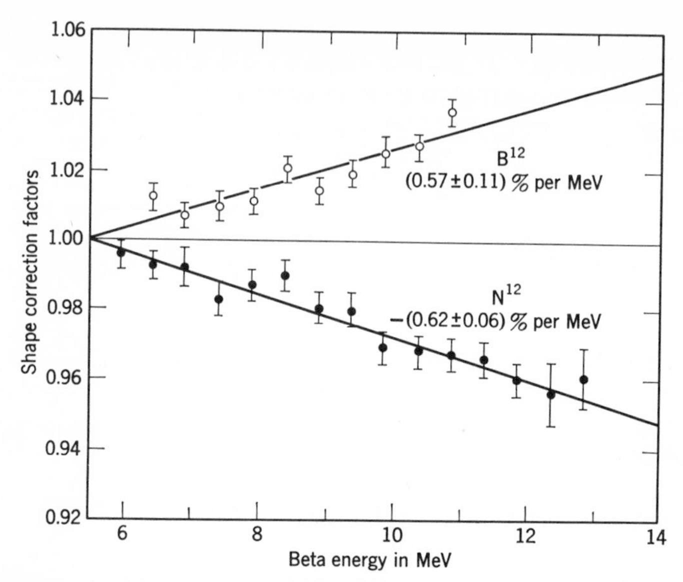

Fig. 34.2. The measured shape corrections in B^{12} and N^{12} compared with the theoretical predictions of the CVC theory. [From Lee *et al.*, *Phys. Rev. Letters* **10,** 253 (1963).]

form of the matrix element for the process $\bar{\nu}_\mu + p \to n + \mu^+$ is[1]

$$\frac{G}{\sqrt{2}}\frac{\sqrt{2m_\nu}}{(2\pi)^6}(\bar{v}(p_\mu)\gamma^\alpha(1-\gamma_5)\,u(p_\nu))$$
$$\times\ \bar{u}(p_n)\{\gamma^\alpha F_1{}^v(q^2) + i\sigma^{\alpha\beta}q_\beta F_2{}^v(q^2) - \gamma^\alpha\gamma_5 F_A(q^2) - q^\alpha\gamma_5 F_P(q)^2\}\,u(p_p)$$

The form factors $F_1{}^v(q^2)$ and $F_2{}^v(q^2)$ are known from electron-nucleon scattering, and this will help in the determination of the other form factors. The calculation of the cross section for $\bar{\nu}_\mu + p \to n + \mu^+$ in terms of the form factors is an exercise that will not be reproduced here for lack of space. We merely plot the cross sections for $\bar{\nu}_\mu + p \to n + \mu^+$ and $\nu_\mu + n \to p + \mu^-$ with $F_P = 0$, $F_2{}^v$ neglected, and assuming that $F_A(q^2) = F_1{}^v(q^2)$. The square of the matrix element grows as the square of the neutrino energy in the laboratory frame, and, if it were not for the form factors, a nonsensical growth of the cross section would result.[2] As it is,

[1] Lorentz invariance permits the additional terms $q^\alpha H_V(q^2)$ and $\sigma^{\alpha\beta}q_\beta\gamma_5 H_A$. They will be absent if it is assumed that (*a*) the currents $J_{V,0}$ and $J_{A,0}$ have definite transformation properties under G parity, and (*b*) that these transformation properties are determined by $\bar{\psi}_p\gamma^\alpha\psi_n$ and $\bar{\psi}_p\gamma^\alpha\gamma_5\psi_n$, respectively. Currents with the "right" G transformation properties have been labeled *first-class currents* by S. Weinberg, *Phys. Rev.* **112,** 1375 (1958), and the omitted terms come from second-class currents.

[2] Even with the form factors, the present theory breaks down near 300 BeV center of mass energy, when individual partial wave amplitudes begin to violate unitarity.

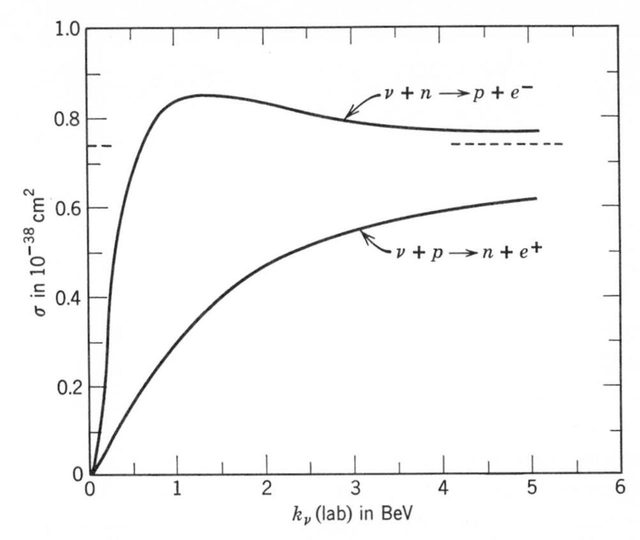

Fig. 34.3. Theoretical curves for neutrino cross sections calculated with $F_A(q^2) = F_1(q^2) = (1 - a^2q^2)^{-2}$. [From Lee and Yang, *Phys. Rev. Letters* **4**, 307 (1960).] The dashed line represents the asymptotic value of the neutrino cross section.

the growth from $\sigma \cong 12 \times 10^{-44}$ cm² for reactor-produced neutrinos[1] to $\sigma \cong 10^{-38}$ cm² at $E_\nu \sim 1$–2 BeV has made neutrino physics possible (Fig. 34.3).[2]

Our discussion has led us very naturally to the next question. What can we say about the axial current and the associated form factors $F_A(q^2)$ and $F_P(q^2)$? The fact that

$$F_A(0) = 1.18 \pm 0.02 \tag{34.15}$$

is also close to unity might perhaps suggest that the axial current, too, is conserved. This, however, would forbid the decay

$$\pi \rightarrow \mu + \nu$$

[1] The curve in Fig. 34.3 is taken from T. D. Lee and C. N. Yang, *Phys. Rev. Letters* **4,** 307 (1960), in which calculations supporting the feasibility of neutrino physics [M. Schwartz, *Phys. Rev. Letters* **4,** 306 (1960); B. Pontecorvo, *Soviet Phys. JETP* **10,** 1236 (1960)] are presented. The neutrino was first observed at low energies in an experiment by C. L. Cowan, F. Reines, F. B. Harrison, H. W. Kruse, and A. D. McGuire, *Nature* **178,** 446 (1956).

[2] Recent neutrino experiments indicate that F_1^v has roughly the same structure as in electron scattering, and that, within experimental errors, it is equal to $F_A(q^2)$. The last conclusion, if sharpened, would indicate that the role of the vector mesons in the electromagnetic form factors has been exaggerated, since there are no corresponding low-lying axial vector resonances (International Conference on Weak Interactions, Argonne National Laboratory, 1965).

This decay is governed by the matrix element $(\Psi_0, J^{\alpha}_{A,0}\Psi_\pi)$

$$(\Psi_0, J^{\alpha}_{A,0}\Psi_\pi) = \frac{M_p}{(2\pi)^{3/2}}\, iq^\alpha F_\pi(q^2) = \frac{iM_p q^\alpha}{(2\pi)^{3/2}} F_\pi(m_\pi^2)$$

If

$$\partial_\alpha J^{\alpha}_{A,0} = 0$$

then

$$(\Psi_0, \partial_\alpha J^{\alpha}_{A,0}\Psi_\pi) = -iq_\alpha(\Psi_0, J^{\alpha}_{A,0}\Psi_\pi) = \frac{M_p m_\pi^2}{(2\pi)^{3/2}} F_\pi(m_\pi^2) = 0$$

so that the number $F_\pi(m_\pi^2)$ which determines the pion decay rate would have to vanish. What this remark suggests is that $J^{\alpha}_{A,0}$ could be conserved in the limit in which the pion mass were zero. This may be stated in two ways: (*a*) the matrix elements $(\Psi_1, \partial_\alpha J^{\alpha}_{A,0}\Psi_2)$ tend to zero as the momentum transfer becomes much larger than the pion mass,[1] $(p_1 - p_2)^2 \gg m_\pi^2$, or (*b*) the divergence of the axial current is proportional to a pseudoscalar operator, with a proportionality coefficient which goes to zero as the pion mass goes to zero.[2] A model of this type is a theory involving nucleons, pions, and a scalar meson σ, whose Lagrangian density is

$$\begin{aligned}\mathcal{L} = \bar{N}(i\gamma^\alpha \partial_\alpha - g_0(\sigma + i\boldsymbol{\tau}\cdot\boldsymbol{\phi}\gamma_5))N + \frac{1}{2}(\partial_\alpha\boldsymbol{\phi})^2 + \frac{1}{2}(\partial_\alpha\sigma)^2 \\ - \frac{1}{2}\mu_0^2(\boldsymbol{\phi}^2 + \sigma^2) - \lambda_0\left(\boldsymbol{\phi}^2 + \sigma^2 - \frac{1}{4f_0^2}\right)^2 - \frac{\mu_0^2}{2f_0}\sigma \end{aligned} \quad (34.16)$$

The axial current, denoted by $\mathfrak{J}_5^{\alpha}$, is taken to be

$$\mathfrak{J}_5^{\alpha} = \tfrac{1}{2}\bar{N}\gamma^\alpha\gamma_5\boldsymbol{\tau}N + (\sigma\,\partial^\alpha\boldsymbol{\phi} - \boldsymbol{\phi}\,\partial^\alpha\sigma) \quad (34.17)$$

and it is easily shown that

$$\partial_\alpha\mathfrak{J}_5^{\alpha} = \frac{\mu_0^2}{2f_0}\boldsymbol{\phi} \quad (34.18)$$

The operator on the right-hand side of (34.18) is, aside from numerical factors, the pseudoscalar meson field operator. What distinguishes it from other pseudoscalar operators (multiplied by suitable constants to make the vacuum — 1-pion state matrix elements the same), is that the matrix elements of the pion field, in perturbation theory, at least, are relatively slowly varying functions of the momentum transfer. This is what distinguishes a model like (34.16) from an arbitrary theory, for which the

[1] Y. Nambu, *Phys. Rev. Letters* **4,** 380 (1960); Chou Kuang-Chao, *Soviet Phys. JETP* **12,** 492 (1961).

[2] M. Gell-Mann and M. Levy, *Nuovo Cimento* **16,** 705 (1960).

divergence of the axial current will be a more general pseudoscalar operator.

With the help of the assumption that, when acting between states which do not differ very considerably in energy and momentum (small-momentum transfer), the divergence of the axial current acts as

$$\partial^\alpha \mathfrak{J}_{5\alpha}(x) = Cm_\pi^2 \boldsymbol{\phi}(x) \tag{34.19}$$

it is possible to derive a relation between $F_A(0)$ and the pion decay rate. This relation was first derived in an approximate dispersion theoretic analysis by Goldberger and Treiman,[1] and is named after them. The derivation using the above assumption, which is commonly known as the *partially conserved axial current*[2] hypothesis (PCAC), is considerably shorter, and the approximations which are made can be pinpointed more easily.

Our first task is to determine the coefficient C in (34.19). We consider the matrix element of both sides of the equation, between a proton state and a neutron state, of equal energy and momentum. We have, on one side

$$\begin{aligned}(\Psi_p, \partial^\alpha(\mathfrak{J}_{5\alpha}^{(1)} + i\mathfrak{J}_{5\alpha}^{(2)})\Psi_n) &= i(p^\alpha - n^\alpha)\, \bar{u}(p)[\gamma_\alpha\gamma_5\, F_A(q^2) + q_\alpha\gamma_5\, F_P(q^2)]\, u(n) \\ &= 2M\, F_A(q^2)\, \bar{u}(p)i\gamma_5\, u(n) + q^2\, F_P(q^2)\, \bar{u}(p)i\gamma_5\, u(n) \\ &\approx 2M\, F_A(0)\, \bar{u}(p)i\gamma_5\, u(n)\end{aligned} \tag{34.20}$$

when $q^\alpha = p^\alpha - n^\alpha \to 0$. On the other side

$$\begin{aligned}(\Psi_p, Cm_\pi^2(\phi_1 + i\phi_2)\Psi_n) &= \sqrt{2}\, Cm_\pi^2(\Psi_p, \phi^{(-)}\Psi_n) \\ &= \sqrt{2}\, C \frac{m_\pi^2}{m_\pi^2 - (p-n)^2}(\Psi_p, j^{(-)}\Psi_n) \\ &= \sqrt{2}\, C \frac{m_\pi^2}{m_\pi^2 - q^2}\sqrt{2}\, gK_{NN\pi}(q^2)\, \bar{u}(p)i\gamma_5\, u(n)\end{aligned} \tag{34.21}$$

[1] M. Goldberger and S. B. Treiman, *Phys. Rev.* **110,** 1178 (1958). The derivation using PCAC is apparently due to Feynman (as quoted in Gell-Mann and Levy, *loc. cit.*).

[2] In modern field theory the custom of associating a definite field with a particle, such as the pion, has fallen into disfavor. A more general statement of the PCAC hypothesis imposes the following requirements on the divergence of the axial current. (1) The divergence should have nonvanishing matrix elements connecting the vacuum and the one-pion state; (2) the matrix elements of the divergence should go to zero rapidly as the momentum transfer increases. The results obtained with the form (34.19) are reproduced if the matrix element of the divergence of the axial current is approximated by the one-pion pole term in a spectral representation of that matrix element as a function of the momentum transfer variable. J. Bernstein, S. Fubini, M. Gell-Mann and W. Thirring, *Nuovo Cimento* **17,** 757 (1960).

In the last expression $K_{NN\pi}(q^2)$ is the pion-nucleon vertex form factor, normalized to unity when $q^2 = m_\pi^2$, so that g is the pion-nucleon coupling constant ($g^2/4\pi \cong 15$). If we now compare both sides of the equation at $q^2 = 0$, and assume that $K_{NN\pi}(0)$ does not differ very much from unity, we find that

$$C = \frac{MF_A(0)}{gK_{NN\pi}(0)} \cong \frac{MF_A(0)}{g} \tag{34.22}$$

Incidentally, if we compare both sides of the equation at $q^2 = m_\pi^2$, and note that $F_A(q^2)$ does not have a pole at $q^2 = m^2$, we obtain

$$F_P(q^2) \approx \frac{2MF_A(0)}{m_\pi^2 - q^2} \tag{34.23}$$

This term actually contributes to the rate for the capture

$$\mu^- + p \rightarrow n + \nu$$

since in that process the momentum transfer is not zero. It may be interpreted as the contribution of the capture of a muon by a pion in the nucleon cloud, and is known as the *induced pseudoscalar term.*[1] Let us now take the relation

$$\partial^\alpha \mathfrak{J}_{5\alpha}(x) \cong \frac{Mm_\pi^2 F_A(0)}{g} \phi(x) \tag{34.24}$$

in which the *i*-spin label is suppressed, between the vacuum state and a one-pion state of four-momentum q. We obtain in the notation of page 580

$$-iq^\alpha(\Psi_0, (\mathfrak{J}_{5\alpha}^{(1)} - i\mathfrak{J}_{5\alpha}^{(2)})\Psi_\pi) = \frac{Mm_\pi^2}{(2\pi)^{3/2}} F_\pi(m_\pi^2)$$

and

$$\frac{Mm_\pi^2 F_A(0)}{g} \sqrt{2}\,(\Psi_0, \phi^- \Psi_\pi) = \frac{Mm_\pi^2}{(2\pi)^{3/2}} \frac{\sqrt{2}\,F_A(0)}{g}$$

so that

$$F_\pi(m_\pi^2) \cong \frac{\sqrt{2}\,F_A(0)}{g} \cong 0.18 \tag{34.25}$$

This is the Goldberger-Treiman relation. If we take into account that the quantity $F_\pi(m_\pi^2)$ is related to f_A of (29.24) by

$$F_\pi(m_\pi^2) = \frac{f_A}{GM_p^2} \frac{M_p}{m_\pi}$$

[1] M. L. Goldberger and S. B. Treiman, *Phys. Rev.* **111,** 354 (1958); L. Wolfenstein, *Nuovo Cimento* **8,** 882 (1958).

we find that the experimental value of F_π is

$$F_\pi(m_\pi^2) \cong 0.14$$

in reasonably good agreement with the predicted value. We shall take this result as sufficiently suggestive to accept PCAC as a working hypothesis. Its validity can best be tested in neutrino experiments.[1]

We next turn to the strangeness-changing currents. The first point to be noted is that the current $J_{V,1}$ cannot be conserved. If we consider a term like

$$\bar{\psi}_\Lambda(x)\gamma^\alpha \psi_p(x) \tag{34.26}$$

it is clear that

$$i\,\partial_\alpha(\bar{\psi}_\Lambda(x)\gamma^\alpha \psi_p(x)) = (M_p - M_\Lambda)\,\bar{\psi}_\Lambda(x)\,\psi_p(x) + \cdots \tag{34.27}$$

and we find it difficult to think of interaction terms which contribute to cancel the term proportional to the mass difference. Okubo[2] pointed out that if $J_{V,1}^\alpha$ were a conserved current one would be able to construct an analogue of the charge

$$Q_1 = \int d^3x\, J_{V,1}^0(x) \tag{34.28}$$

which would commute with the total Hamiltonian. This, however, implies that

$$(\Psi_\Lambda, [Q_1, H]\Psi_p) = (M_p - M_\Lambda)(\Psi_\Lambda, Q_1\Psi_p) = 0$$

i.e.

$$\int d^3x(\Psi_\Lambda, J_{V,1}^0(x)\Psi_p) = 0 \tag{34.29}$$

Since the current $J_{V,1}^\alpha$ does lead to proton-Λ^0 transitions, it cannot be conserved. In the limit of unitary symmetry, when all the baryon masses are equal, the argument breaks down, and it is possible to conceive of $J_{V,1}$ as conserved in that limit. In view of the association of $J_{V,1}^\alpha$ with the *i*-spin current, i.e., with the currents $\mathfrak{J}^{(1)\alpha}$, $\mathfrak{J}^{(2)\alpha}$ in $SU(3)$, it is tempting to associate $J_{V,1}$ with another of the octet currents. It was proposed by Cabibbo[3] that the strangeness-changing current, which is responsible for the transformation of a proton into a Λ^0 is the $SU(3)$ generating current

[1] S. Adler, *Phys. Rev.* **135,** B963 (1964). Further support comes from the relation discussed on p. 381.
[2] S. Okubo, *Nuovo Cimento* **13,** 292 (1959).
[3] N. Cabibbo, *Phys. Rev. Letters* **10,** 531 (1963).

which transforms as V_-, i.e., it is[1]

$$J^{\alpha}_{V,1} = \mathfrak{J}^{(4)\alpha} - i\mathfrak{J}^{(5)\alpha} \tag{34.30}$$

An immediate consequence of this hypothesis is that $\Delta S = \Delta Q$ and that the strangeness-changing currents obey the $\Delta T = \frac{1}{2}$ rule. This specification does not answer the question of how universality can be reconciled with the observed low decay rates for the strange particles. Cabibbo defined universality by asserting that the total hadronic currents be of the form

$$J_V{}^{\alpha} = \cos\theta(\mathfrak{J}^{(1)\alpha} - i\mathfrak{J}^{(2)\alpha}) + \sin\theta(\mathfrak{J}^{(4)\alpha} - i\mathfrak{J}^{(5)\alpha}) \tag{34.31}$$

and

$$J_A{}^{\alpha} = \cos\theta(\mathfrak{J}_5^{(1)\alpha} - i\mathfrak{J}_5^{(2)\alpha}) + \sin\theta(\mathfrak{J}_5^{(4)\alpha} - i\mathfrak{J}_5^{(5)\alpha}) \tag{34.32}$$

The vector currents are all conserved in the limit of $SU(3)$. The assertion that they are the generating currents of $SU(3)$ implies that

$$F^{(i)}(t) = \int_{x_0=t} d^3x\, \mathfrak{J}^{(k)0}(x) \tag{34.33}$$

are the time-independent generators of $SU(3)$. The axial currents are supposed to have octet transformation properties, so that

$$[\mathfrak{J}_5^{(i)\alpha}(x), F^{(j)}] = if_{ijk}\, \mathfrak{J}_5^{(k)\alpha}(x) \tag{34.34}$$

We shall also assume the PCAC hypothesis, so that

$$\partial_\alpha \mathfrak{J}_5^{(i)\alpha}(x) \simeq \frac{MF_A(0)}{g}\, m_i{}^2\, \phi_i(x) \tag{34.35}$$

With the help of the commutation relations for octet operators (17.52) it is easy to show that, both for the vector currents and for the axial currents

$$J^{\alpha} = e^{-2i\theta F_7}(\mathfrak{J}^{(1)\alpha} - i\mathfrak{J}^{(2)\alpha})e^{2i\theta F_7} \tag{34.36}$$

We note several points:

1. The angle of this rotation is the same for both the vector current and for the axial current. Experimental evidence for this will be presented below. This suggests that the two types of currents are somehow generated by the same mechanism.

[1] These currents are derived in Chapter 18, in the material leading up to (18.27).

2. The actual current is a rotation of the i-spin current. If unitary symmetry were not violated, except by the weak interactions, there would be no way of distinguishing between an i-spin current and the same current transformed by a rotation. We can transform the hypercharge to obtain Y'

$$Y' = e^{-2i\theta F_7} Y e^{2i\theta F_7} \tag{34.37}$$

and it is evident, since the i-spin current commutes with Y, that

$$[J^\alpha, Y'] = 0 \tag{34.38}$$

Thus, if unitary symmetry were violated only by the weak interactions, an "i-spin" and a *conserved* hypercharge would be selected out. The result (34.38) implies that Y' must be a symmetry of the effective Hamiltonian for nonleptonic processes. Since Y' is represented as a symmetric matrix, the Hamiltonian can only involve the symmetric matrix λ_6, and not the antisymmetric matrix λ_7. This result was used in the preceding chapter in the derivation of part of the Lee triangle rule.

Our first task is to determine the value of the angle θ. For the axial current, we compare the rates of

$$K^\pm \rightarrow \mu^\pm + \nu$$

and

$$\pi^\pm \rightarrow \mu^\pm + \nu$$

For dealing with decays of particles belonging to octets, it is convenient to construct a 3×3 matrix representing the current. It is given by

$$\tfrac{1}{2}\cos\theta(\lambda_1 - i\lambda_2) + \tfrac{1}{2}\sin\theta(\lambda_4 - i\lambda_5) = \begin{pmatrix} 0 & 0 & 0 \\ \cos\theta & 0 & 0 \\ \sin\theta & 0 & 0 \end{pmatrix} \equiv J \tag{34.39}$$

The matrix element $(\Psi_0, J_A \Psi_{\pi,K})$ must, as far as its internal symmetry properties are concerned, transform as

$$\operatorname{tr}(JP) = \cos\theta\pi_+ + \sin\theta K_+ \tag{34.40}$$

Hence the ratio of the constants

$$\left|\frac{F_K}{F_\pi}\right|^2 = \tan^2\theta = 0.075 \tag{34.41}$$

obtained in (33.34) determines

$$\theta_A \cong 0.27 \tag{34.42}$$

To obtain the angle for the vector current, we compare the rates for

$$K^+ \to \pi^0 + e^+ + \nu$$

and

$$\pi^+ \to \pi^0 + e^+ + \nu$$

Here we have to deal with the possibility of two ways of combining two matrices P (for pseudoscalar mesons) with the octet current J, since **8** × **8** contains **8** twice. The current, however, is a pure F-type current (since it is the generating current), so that the difficulty does not arise. We find from Fig. 17.8 that

$$\frac{\Gamma(K^+ \to \pi^0 e^+ \nu)}{\Gamma(\pi^+ \to \pi^0 e^+ \nu)} = \frac{\rho_K}{\rho_\pi}\left(\frac{\sin\theta\,\langle \pi^0 |V_-|\, K^+\rangle}{\cos\theta\,\langle \pi^0 |T_-|\, \pi^+\rangle}\right)^2 \tag{34.43}$$

so that the ratio of the squares of the matrix elements is modified by a factor $\frac{1}{4}\tan^2\theta$. We thus have

$$\left|\frac{f_+(K)}{f_+(\pi)}\right| = \frac{1}{2}|\tan\theta| \cong 0.1 \tag{34.44}$$

with the data taken from (33.52) and (33.54). Hence we find

$$\theta_V \cong 0.22 \tag{34.45}$$

Thus the angles are equal within experimental uncertainties. We shall not consider the possibility that they are equal but of opposite sign.

The Cabibbo angle θ represents a rather fundamental quantity. It is not a parameter like $F_A(0)$, which can be calculated within the framework of the strong interactions. The angle θ represents in some sense the "mismatch" between the strong violation of $SU(3)$—which singles out i-spin as the sub-symmetry—and the way in which the weak interactions violate $SU(3)$. This angle will therefore only be determined theoretically when we understand something more about how various interactions break the underlying unitary symmetry. Once the angle is given, a number of predictions follow.

1. The coupling constant describing the Fermi transitions in beta decay is changed from G to $G\cos\theta = 0.97G$. This means that C_V in (31.24) is to be replaced by $G\cos\theta$. It is clear that the above modification goes in the direction of removing the discrepancy between the electromagnetically corrected G and the nonrenormalized G.

2. The Cabibbo theory allows us to describe a large number of leptonic decays, such as

$$
\begin{aligned}
&n \to p + e^- + \bar{\nu} \\
&\Sigma^\pm \to \Lambda^0 + e^\pm + \nu \\
&\Lambda^0 \to p + e^- + \bar{\nu} \\
&\Sigma^- \to n + e^- + \bar{\nu} \\
&\Xi^- \to (\Sigma^0, \Lambda^0) + e^- + \bar{\nu} \\
&\Xi^0 \to \Sigma^+ + e^- + \bar{\nu}
\end{aligned}
\tag{34.46}
$$

In all cases there will exist a pure F-type γ^μ coupling, a weak magnetism term which involves both F and D coupling, and the axial terms which also involve F and D couplings. If we keep only the leading terms, proportional to γ^μ and those proportional to $\gamma^\mu\gamma_5$, we have for the vector matrix elements Table 34-1.

Table 34-1

Vector current matrix elements for leptonic decays

Process	Matrix element
$n \to pe^-\bar{\nu}$	$\cos\theta\gamma^\mu$
$\Sigma^\pm \to \Lambda^0 e^\pm \nu$	0
$\Lambda^0 \to pe^-\bar{\nu}$	$\sqrt{3/2}\sin\theta\gamma^\mu$
$\Sigma^- \to ne^-\bar{\nu}$	$\sin\theta\gamma^\mu$
$\Xi^- \to \Lambda^0 e^-\bar{\nu}$	$\sqrt{3/2}\sin\theta\gamma^\mu$
$\Xi^- \to \Sigma^0 e^-\bar{\nu}$	$\sqrt{1/2}\sin\theta\gamma^\mu$
$\Xi^0 \to \Sigma^+ e^-\bar{\nu}$	$-\sin\theta\gamma^\mu$

The numerical coefficients are just read off the appropriate legs in Fig. 17.8. There is no leg connecting $\Sigma^\pm$ to Λ^0 and hence the vector coefficient vanishes. Actually, these coefficients are correct only in the limit of unitary symmetry and when $q^2 = 0$. It is only for zero-momentum transfer that the conservation law for the current allows the underlying F-type current to "show through" the strong interactions. When $q^2 \neq 0$, then form factors $F_1{}^V(q^2)$ appear and D currents get mixed in. These do connect $\Sigma^\pm$ and Λ^0. Since $F_1{}^V(q^2)$ for $\Sigma^\pm \to \Lambda^0$ is proportional to q^2, which is still small, the table is still more or less correct.

For the axial current we assume that both D and F currents are present with strength α and $(1 - \alpha)$, respectively. A table can again be constructed with the help of the numerical parameters determined by the coupling

$$\alpha \operatorname{tr}(\bar{B}\{J, B\}) + (1-\alpha)\operatorname{tr}(\bar{B}[J, B])$$

$$= \operatorname{tr}\left\{\begin{pmatrix} \frac{1}{\sqrt{2}}\bar{\Sigma}^0 + \frac{1}{\sqrt{6}}\bar{\Lambda}^0 & \bar{\Sigma}^- & -\bar{\Xi}^- \\ \bar{\Sigma}^+ & -\frac{1}{\sqrt{2}}\bar{\Sigma}^0 + \frac{1}{\sqrt{6}}\bar{\Lambda}^0 & \bar{\Xi}^0 \\ \bar{p} & \bar{n} & -\frac{2}{\sqrt{6}}\bar{\Lambda}^0 \end{pmatrix} \right.$$

$$\left. \times \begin{pmatrix} (2\alpha-1)(\Sigma^+\cos\theta + p\sin\theta) & 0 & 0 \\ \begin{array}{l}(2\alpha-1)n\sin\theta + \sqrt{2}(1-\alpha)\Sigma^0\cos\theta \\ \quad + \sqrt{2/3}\,\alpha\Lambda^0\cos\theta\end{array} & \Sigma^+\cos\theta & p\cos\theta \\ \begin{array}{l}1/\sqrt{3}\,\Sigma^0\sin\theta + (2\alpha-1)\Xi^0\cos\theta \\ \quad + 1/\sqrt{6}(3-4\alpha)\Lambda^0\sin\theta\end{array} & \Sigma^+\sin\theta & p\sin\theta \end{pmatrix}\right\} \tag{34.47}$$

The results are summarized in Table 34-2. When these are combined with the matrix elements given in Table 34-1, and comparison with experiment is made, we find the following results: From the rates of Λ^0 beta decay and $\Sigma^- n$ leptonic transitions, we find that[1] $\theta = 0.26$, in good agreement with the values obtained before, and that

$$\alpha \approx 0.66 \tag{34.48}$$

The parameters may be used to calculate $F_A(0)$ for the Λ^0 decay (33.28) and reasonably good agreement is found.

It is interesting to note that the value $\alpha \cong \frac{2}{3}$ is close to that mentioned in Chapter 25 as giving the proper kind of potential to make the meson-baryon resonance be the decuplet. If we believe in PCAC, this result is not accidental.

A consequence of the PCAC hypothesis is that the axial current part and the induced pseudoscalar term in the axial vector current matrix element are related, as shown in (34.23). We have

$$(\Psi_p, (\mathfrak{J}_5^{(1)\alpha} + i\mathfrak{J}_5^{(2)\alpha})\Psi_n) = \frac{1}{(2\pi)^3}\,\bar{u}(p)\left(\gamma^\alpha\gamma_5 + \frac{2Mq^\alpha\gamma_5}{m_\pi^{\,2} - q^2}\right)u(n)\,F_A(q^2) \tag{34.49}$$

Now, if we consider the kth component of the current, evaluated between

[1] N. Brene, B. Hellesen, and M. Roos, *Physics Letters* **11**, 344 (1964).

baryon states labeled with i and j, the above form becomes

$$(\overline{\Psi}_{Bi}, \mathfrak{J}_5^{(k)\alpha}\Psi_{Bj}) = \frac{1}{(2\pi)^3}\,\bar{u}(p_i)\left(\gamma^\alpha\gamma_5 + \frac{2Mq^\alpha\gamma_5}{m_\pi^{\ 2} - q^2}\right)u(p_j)$$
$$\times\,(\alpha\,F_A^{\ d}(q^2)\,d_{ijk} - (1-\alpha)\,F_A^{\ f}(q^2)f_{ijk}) \quad (34.50)$$

The important point to note is that the same α multiplies both the pseudovector term, $\gamma^\alpha\gamma_5$, and the induced term, in which the α describes the D coupling proportion in the coupling of the meson to the nucleon. Thus the two α's are identical.

Table 34-2

Axial matrix elements for leptonic decays

Process	Matrix element/$F_A(q^2)$
$n \to pe^-\bar{\nu}$	$\cos\theta\gamma^\mu\gamma_5$
$\Sigma^\pm \to \Lambda^0 e^\pm \nu$	$\pm\sqrt{2/3}\,\alpha\cos\theta\gamma^\mu\gamma_5$
$\Lambda^0 \to pe^-\bar{\nu}$	$\sqrt{3/2}(1 - 2\alpha/3)\sin\theta\gamma^\mu\gamma_5$
$\Sigma^- \to ne^-\bar{\nu}$	$(1 - 2\alpha)\sin\theta\gamma^\mu\gamma_5$
$\Xi^- \to \Lambda^0 e^-\bar{\nu}$	$-\sqrt{3/2}(1 - 4\alpha/3)\sin\theta\gamma^\mu\gamma_5$
$\Xi^- \to \Sigma^0 e^-\bar{\nu}$	$1/\sqrt{2}\sin\theta\gamma^\mu\gamma_5$
$\Xi^0 \to \Sigma^+ e^-\bar{\nu}$	$-\sin\theta\gamma^\mu\gamma_5$

For large-momentum transfer processes, form factors in Tables 34-1 and 34-2 must be taken into account. The γ^μ terms must be multiplied by $F_1^{\ V}(q^2)$. We must also take into account the "weak magnetism" terms. These need not be of F-type alone. In fact, since

$$\mu_p + \mu_n \approx 0$$

an expansion of the terms in (26.45) shows that $\alpha_{\text{e.m}} \cong \frac{3}{4}$. Thus the weak magnetism terms are obtained by taking Table 34-2 and replacing each item by the same form with $\alpha = \frac{3}{4}$, and multiplying by $F_2^{\ v}(q^2)$. The induced pseudoscalar terms are known once the axial form factors $F_A(q^2)$ are known.

The discovery that the vector currents, which are probed by the weak and electromagnetic interactions, are generating currents of the underlying symmetry $SU(3)$ may be a very important one. Gell-Mann[1] has proposed that in order to put the axial vector currents on a similarly fundamental

[1] M. Gell-Mann, *Phys. Rev.* **125,** 1067 (1962); Physics **1,** 63 (1964).

footing, it is necessary to consider the symmetry to be enlarged, so that the pseudoscalars

$$F_5^{(i)}(t) = \int_{x_0=t} d^3x\, \mathfrak{J}_5^{(i)0}(x) \tag{34.51}$$

also have well defined commutation relations. Since they transform as octets, it follows that

$$[F^{(i)}(t), F_5^{(j)}(t)] = if_{ijk}\, F_5^{(k)}(t) \tag{34.52}$$

but if they are to be generators on the same footing as the $F^{(i)}$, their commutation relations must also be given, and they should not involve any new quantities if the "algebra" is to be kept as small as possible. Gell-Mann suggested that

$$[F_5^{(i)}(t), F_5^{(j)}(t)] = if_{ijk}\, F^{(k)}(t) \tag{34.53}$$

These commutation relations serve to fix the scale of the axial currents in a model-independent way, in a manner analogous to the way in which the CVC theory of Feynman and Gell-Mann fixes the scale of the vector currents. Thus the statement that the "bare" current be of the form $\bar{\psi}\tau_-\gamma^\alpha(1 - \gamma_5)\psi$, which is hard to generalize when particles other than baryons are involved, can now be replaced by the statement that the hadronic current is of the form $\mathfrak{J}^{\alpha(i)} - \mathfrak{J}_5^{\alpha(i)}$, with the commutation relations of Gell-Mann fixing the scale. This characterization takes us a long way towards a model-independent description of what is meant by the *universality* of the weak interactions. What is still missing is some principle which relates the scale of the leptonic currents to that of the hadronic currents in $J^\alpha = j^\alpha_{\text{lept}} + \mathfrak{J}^\alpha_{\text{had}}$. Some support for these commutation relations, characterizing the algebra $SU(3) \times SU(3)$,[1] comes from a very beautiful calculation of the value of $F_A(0)$, in which the commutation relations (34.53) and the PCAC hypothesis play a central role.[2] The derivation that we give is somewhat different from that of Adler and Weisberger, and makes use of a formal identity due to Fubini, Furlan and Rossetti. We write

$$\begin{aligned} F_5^{(i)}(0) &= \int d^3x\, \mathfrak{J}_5^{(i)0}(x, 0) \\ &= \mp \int dx\, \theta(\pm x_0)\, \partial_\alpha \mathfrak{J}_5^{(i)\alpha}(x) \\ &\equiv \mp \int dx\, \theta(\pm x_0)\, D^{(i)}(x) \end{aligned} \tag{34.54}$$

[1] The commutation relations for the $SU(3)$ generators $F^{(i)}$, together with (34.52) and (34.53) imply that $F_\pm^{(i)} = \frac{1}{2}(F^{(i)} \pm F_5^{(i)})$ satisfy $SU(3)$ commutation relations, and that $[F_+^{(i)}, F_-^{(j)}] = 0$. Hence the characterization of the algebra as $SU(3) \times SU(3)$.

[2] W. I. Weisberger, *Phys. Rev. Letters* **14,** 1047 (1965); S. L. Adler, *ibid.*, **14,** 1051 (1965).

For this to hold, it is necessary to assume that the current vanishes as $|\mathbf{x}| \to \infty$, and that only matrix elements with different energies will be considered, so that the "surface term" in time drops out because of infinitely rapid oscillations.[1] Let us now take the matrix element of (34.53) between two states belonging to the same super-multiplet, such as two baryon states. We shall use

$$(\Psi_a(p'), [(F_5^{(i)}(0), F_5^{(j)}(0)]\Psi_b(p)) = if_{ijk}(\Psi_a(p'), F^{(k)}(0)\Psi_b(p)) \quad (34.55)$$

We have

$$\begin{aligned}(\Psi_a(p'), &[F_5^i(0), F_5^j(0)]\,\Psi_b(p)) \\ &= -\iint dx\, dy\, \theta(x_0)\, \theta(-y_0)(\Psi_a(p'), [D^{(i)}(x), D^{(j)}(y)]\,\Psi_b(p)) \\ &= -\iint dx\, dy\, \theta(x_0)\, \theta(-y_0) e^{i(p'-p)y} \\ &\qquad\qquad \times (\Psi_a(p'), [D^{(i)}(x-y), D^{(j)}(0)]\,\Psi_b(p)) \quad (34.56)\end{aligned}$$

Note that $x_0 > 0$ and $y_0 < 0$, so that we may with impunity insert the additional step function $\theta(x_0 - y_0)$. If we write $x - y = z$, the above expression becomes

$$\begin{aligned}-\int dz\, dy\, &\theta(z_0)\, \theta(y_0 + z_0)\, \theta(-y_0) e^{i(p'-p)y}(\Psi_a(p'), [D^{(i)}(z), D^{(j)}(0)]\,\Psi_b(p)) \\ &= -(2\pi)^3\, \delta(\mathbf{p}' - \mathbf{p}) \int dz \int_{-z_0}^{0} dy_0 e^{i(E_a - E_b)y_0}\, \theta(z_0) \\ &\quad \times (\Psi_a(p'), [D^{(i)}(z), D^{(j)}(0)]\,\Psi_b(p)) \\ &= -(2\pi)^3\, \delta(\mathbf{p} - \mathbf{p}') \int dz\, z_0\, \theta(z_0)(\Psi_a(p'), [D^{(i)}(z), D^{(j)}(0)]\,\Psi_b(p)) \\ &= -(2\pi)^3\, \delta(\mathbf{p} - \mathbf{p}')\left\{\frac{1}{i}\frac{\partial}{\partial q_0} \int dz e^{iq_0 z_0}\, \theta(z_0)(\Psi_a(p'), [\ ,\ ]\,\Psi_b(p))\right\}_{q_0=0}\end{aligned}$$

We now introduce the variable

$$\nu = \frac{q \cdot p'}{M} = \frac{q_0 E'}{M} \quad (34.57)$$

in terms of which the last line becomes

$$i(2\pi)^3 \frac{E'}{M}\, \delta(\mathbf{p} - \mathbf{p}')\, \frac{\partial}{\partial \nu}\left(\int dz e^{iqz}\, \theta(z_0)(\Psi_a(p'), [\ ,\ ]\,\Psi_b(p))\right)_{q=0} \quad (34.58)$$

[1] S. Fubini, G. Furlan and C. Rossetti, *Nuovo Cimento* **40,** 1171 (1965). This excellent paper discusses many interesting applications of the commutation relations.

Note that we have replaced $e^{iq_0z_0}$ by e^{iqz} and set $q = 0$. Let us now define

$$F_{ij}(\nu) = -\frac{i}{(2\pi)^3}\int dz e^{iqz}\,\theta(z_0)(\Psi_a(p'), [D^{(i)}(z), D^{(j)}(0)]\,\Psi_b(p)) \quad (34.59)$$

In terms of it, we have

$$(\Psi_a(p'), [F_5^{(i)}(0), F_5^{(j)}(0)]\,\Psi_b(p)) = -(2\pi)^6\frac{E'}{M}\,\delta(\mathbf{p}-\mathbf{p}')\left(\frac{\partial}{\partial\nu}F_{ij}(\nu)\right)_{q=0} \quad (34.60)$$

We can now check that

$$\begin{aligned}
\operatorname{Im} F_{ij}(\nu)
&= -\frac{1}{2(2\pi)^3}\sum_n\int dz(\Psi_a(p'), D^{(i)}(z)\Psi_n)(\Psi_n, D^j(0)\,\Psi_b(p)) - \text{reversed term}\\
&= -\pi\sum_n\{\delta(p_n - p - q)(\Psi_a(p'), D^{(i)}\Psi_n)(\Psi_n, D^{(j)}\Psi_b(p))\\
&\quad - \delta(p_n - p + q)(\Psi_a(p'), D^{(j)}\Psi_n)(\Psi_n, D^{(i)}\,\Psi_b(p))\}\\
&= \left(\frac{M\,F_A(0)}{g}\,m_i m_j\right)^2 \times (-\pi)\sum_n \delta(p_n - p - q)\\
&\quad \times (\Psi_a(p'), \phi_i\,\Psi_n)(\Psi_n, \phi_j\,\Psi_b(p)) - \text{reversed term}
\end{aligned}$$

In the last step the PCAC connection was used. The final step is to convert the fields ϕ into the source currents. We write

$$(\Psi_n, \phi_i\,\Psi_b(p)) = \frac{1}{m_i^{\,2} - (p_n - p)^2}(\Psi_n, j_i\,\Psi_b(p))$$

Since, however, in the imaginary part, the delta functions imply that $(p_n - p)^2 = q^2 = 0$, we may finally write

$$\begin{aligned}
\operatorname{Im} F_{ij}(\nu) &= \left(\frac{M\,F_A(0)}{g}\right)^2(-\pi)\\
&\quad\times\sum_n\{\delta(p_n - p - q)(\Psi_a(p'), j_i\,\Psi_n)(\Psi_n, j_j\,\Psi_b(p)) - \text{reversed terms}\}\\
&= \left(\frac{M\,F_A(0)}{g}\right)^2 \operatorname{Im} T_{ij}(\nu) \qquad (34.61)
\end{aligned}$$

Here $T_{ij}(\nu)$ is the scattering amplitude of a *massless* meson by a target. It also follows from the general structure of $F_{ij}(\nu)$, as a Fourier transform of a retarded commutator matrix element, that it has analyticity properties in ν such that a dispersion relation can be written. If we assume that there are no arbitrary subtraction parameters, we may identify

$$F_{ij}(\nu) = \left(\frac{M\,F_A(0)}{g}\right)^2 T_{ij}(\nu) \tag{34.62}$$

from the equality of the imaginary parts.

Let us now specialize to $i, j = 1, 2$. First of all this implies

$$[F_5^{(1)}(0), F_5^{(2)}(0)] = iF^{(3)}(0) = iT_3 \tag{34.63}$$

If matrix elements of this are taken between two proton states[1]

$$\begin{aligned}(\Psi_{p'}, iT_3\Psi_p) &= \frac{i}{2}\frac{E_p}{M}\,\delta(\mathbf{p}-\mathbf{p}') \\ &= (\Psi_{p'}, [F_5^{(1)}(0), F_5^{(2)}(0)]\Psi_p) \\ &= -(2\pi)^6\,\frac{E_p}{M}\,\delta(\mathbf{p}-\mathbf{p}')\left(\frac{M\,F_A(0)}{g}\right)^2\left(\frac{\partial}{\partial\nu}\,T_{12}(\nu)\right)_{\nu=0}\end{aligned}$$

so that the coefficients of $(E_p/M)\,\delta(\mathbf{p}-\mathbf{p}')$ are related by

$$\frac{i}{2} = -(2\pi)^6\left(\frac{M\,F_A(0)}{g}\right)^2\left(\frac{\partial}{\partial\nu}\,T_{12}(\nu)\right)_{\nu=0} \tag{34.64}$$

If we note that

$$\begin{aligned} j_1 &= \frac{1}{\sqrt{2}}\,(j_{\pi^+} + j_{\pi^-}) \\ j_2 &= \frac{i}{\sqrt{2}}\,(j_{\pi^+} - j_{\pi^-}) \end{aligned} \tag{34.65}$$

so that

$$T_{12} = i(T_{\pi^-} - T_{\pi^+})/2 \tag{34.66}$$

we obtain

$$\frac{1}{(2\pi)^6}\,\frac{g^2}{2M^2F_A^{\,2}(0)} = -\frac{1}{2}\frac{\partial}{\partial\nu}\,(T_{\pi^-}(\nu) - T_{\pi^+}(\nu))_{\nu=0} \tag{34.67}$$

This can be converted into a more manageable form. The one-particle terms in (34.67) are easily calculated from the imaginary part. A fairly straightforward calculation gives

$$(T_{\pi^-}(\nu) - T_{\pi^+}(\nu))_1 = -\frac{4g^2}{(2\pi)^6}\left(\frac{M_n - M_p}{2M_p}\right)^2\frac{\nu}{[(M_n^{\,2} - M_p^{\,2})/2M_p]^2 - \nu^2}$$

so that the one-particle contribution to the right hand side of (34.67) is

$$-\left(\frac{\partial}{\partial\nu}\,(T_{\pi^-} - T_{\pi^+})_1\right)_{\nu=0} = \frac{1}{(2\pi)^6}\,\frac{g^2}{M^2}$$

[1] Note that $q = 0$ implies that $\nu = 0$.

For the rest we use

$$\text{Im } T_{\pi^\pm}(\nu) = -\frac{\sqrt{(p\cdot q)^2 - p^2q^2}}{(2\pi)^6 M}\,\sigma^{\pm}_{\text{tot}}(\nu) = -\frac{\nu\sigma^{\pm}_{\text{tot}}(\nu)}{(2\pi)^6} \tag{34.68}$$

and

$$T_{\pi^\pm}(\nu) = \begin{Bmatrix}\text{1 part}\\ \text{terms}\end{Bmatrix} + \frac{1}{\pi}\int_0^\infty \frac{d\nu'}{\nu'-\nu}\,\text{Im } T_{\pi^\pm}(\nu') \tag{34.69}$$

If we put all of this (and crossing) together, we get the sum rule[1]

$$\frac{1}{F_A^{\,2}(0)} = 1 + \frac{2M^2}{\pi g^2}\int \frac{d\nu'}{\nu'}\,(\sigma^{-}_{\text{tot}}(\nu') - \sigma^{+}_{\text{tot}}(\nu')) \tag{34.70}$$

in which the cross sections are for massless pions. If this is ignored, together with other factors which are not evaluated on the mass shell but at pion mass zero, the calculation of Weisberger yields the value

$$F_A(0) \cong 1.16 \tag{34.71}$$

and that of Alder, who attempts to estimate the effects of going off shell, yields

$$F_A(0) \cong 1.24 \tag{34.72}$$

The result is in excellent agreement with experiment.[2] Incidentally, when the off-shell effects are ignored in the cross section term, comparison with (23.57) shows that

$$\frac{1}{F_A^{\,2}(0)} = \frac{8}{3}\frac{M^2\pi}{m_\pi^{\,2}g^2}\left(1+\frac{m_\pi}{M}\right)(a_1 - a_3) \tag{34.73}$$

where a_1 and a_3 are the pion-nucleon scattering lengths for $T = \frac{1}{2}$ and $\frac{3}{2}$ respectively.

In conclusion, we may summarize our present understanding of the weak interactions as follows:

1. The qualitative features, except for the apparent CP violation observed in the reaction $K_2^{\,0} \to \pi^+ + \pi^-$ can be explained by the current-current model of Feynman and Gell-Mann.

2. The leptonic weak interactions and the semi-leptonic weak interactions involving the hadrons are quantitatively understood in terms of two parameters, G, characterizing the overall strength of the weak

[1] The existence of the (3, 3) resonance makes $\sigma^{+}_{\text{tot}} > \sigma^{-}_{\text{tot}}$ at the lower energies and is responsible for the fact that $F_A(0) > 1$.

[2] D. Amati, C. Bouchiat, and J. Nuyts, *Physics Letters* **19**, 59 (1965), have applied the same method to $i, j = 4, 5$ and $6, 7$. In this way they are able to obtain a relation determining the parameter α, which is directly related to the D/F ratio. They find that $\alpha \simeq 0{\cdot}6$, in good agreement with experiment.

interactions and θ, the Cabibbo angle which is a measure of the strength of the strangeness changing semi-leptonic interactions.

3. There is at this time no understanding of the hierarchical division of interactions into strong, electromagnetic and weak, nor of the connection between the leptons and the hadrons. Experiments at energies much higher than are now accessible may be necessary to elucidate these questions.

PROBLEMS

1. Discuss the leptonic decay

$$\Omega^- \to \Xi^0 + e^- + \bar{\nu}$$

For the matrix element of the vector current, use CVC and the phenomenological coupling mentioned in the footnote on page 447. For the axial current, use PCAC and data on the $N^*_{3/2}(1238)$ decay.

2. What information do CVC and PCAC give you about the decay

$$K^+ \to \pi^+ + \pi^- + e^+ + \nu$$

Feel free to use $SU(3)$ invariance for the strong interactions.

References to Problems

1.1 See, for example, E. Henley and W. Thirring, *Elementary Quantum Field Theory*, McGraw-Hill Book Co., New York (1963).

1.2 *Ibid.*

1.3,4 See W. Pauli, *Rev. Mod. Phys.* **13,** 203 (1941).

1.5 E. Henley and W. Thirring, *loc. cit.*

2.2 See M. H. Johnson and B. A. Lippmann, *Phys. Rev.* **76,** 828 (1949).

2.3 Compare with L. L. Foldy and S. A. Wouthuysen, *Phys. Rev.* **78,** 29 (1950).

2.4 See J. D. Bjorken and S. D. Drell, *Relativistic Quantum Mechanics*, McGraw-Hill Book Co. New York (1964).

4.1 See Appendix E in M. L. Goldberger and K. M. Watson, Collision Theory, John Wiley and Sons, New York (1964).

5.2 See V. G. Soloviev, *Nucl. Phys.* **6,** 618 (1958).

5.3,4 M. Gell-Mann and M. Levy, *Nuovo Cimento,* **16,** 705 (1960).

11.1 See S. A. Bludman and J. A. Young, *Phys. Rev.* **131,** 2326 (1963).

13.5 See S. Gasiorowicz and A. Petermann, *Phys. Rev. Letters* **1,** 1511 (1958).

13.6 See K. Johnson, *Lectures on Particles and Field Theory,* Brandeis Summer Institute in Theoretical Physics, 1964, Prentice-Hall, Englewood Cliffs, N.J.

16.3 See G. F. Chew and S. Mandelstam, *Phys. Rev.* **119,** 467 (1960).

16.4 See, for example, J. J. Sakurai, *Invariance Principles and Elementary Particles*, Princeton University Press (1964), p. 230.

20.1 J. B. Bronzan and F. E. Low, *Phys. Rev. Letters.* **12,** 522 (1964).

20.2 C. Zemach. *Phys. Rev.* **133,** B1201 (1964).

21.1–5 See, for example, H. Lehmann, *Nuovo Cimento* **11,** 342 (1954).

22.1 See, for example, S. Mandelstam, *Phys. Rev.* **115,** 1741 (1959).

23.2 See G. F. Chew and S. Mandelstam, *loc. cit.*

23.4 See S. Fubini and D. Walecka, *Phys. Rev.* **116,** 194 (1959).

25.2 See J. J. Sakurai, *Ann. Phys. (N.Y.)* **11,** 1 (1960).

26.1,2,3,4 See references in footnote on page 445.

27.1 See, for example, C. Goebel in *Proceedings of the* 1961 *Midwestern Conference on Theoretical Physics.*

27.2 K. Gottfried and J. D. Jackson, *Nuovo Cimento* **33,** 309 and **34,** 735 (1964).

28.2 See, for example, A. Rosenfeld in *Proceedings of the International Conference on Nucleon Structure*, 1963, Stanford University Press, Stanford, Calif.

Bibliography

Elementary particle physics is one of the most active fields of research, and the literature, both theoretical and experimental, is growing at an enormous rate. It is obviously impossible to aim at completeness of a bibliography on this subject, and it is not clear that such a compilation would be of real service to the reader. I have decided, instead, to provide the reader with a set of references embodied in the text of the book, and a partial supplementary bibliography with the help of which he may construct for himself as complete a bibliography on any topic as he wishes. It is inevitable that some important contributions, and their authors, have been ignored, and I regret all such omissions.

Part I

A very complete bibliography on all aspects of quantum field theory may be found in S. S. Schweber, *An Introduction to Relativistic Quantum Field Theory*, Harper and Row, Inc. New York (1962). I shall, therefore, list but a few references to work on the subject of the first nine chapters.

CH. 3

T. T. Wu, *Phys. Rev.* **129,** 1420 (1963) (quantization of electromagnetic field with definite metric).

CH. 4

D. L. Weaver, C. L. Hammer, and R. H. Good, Jr., *Phys. Rev.* **135,** B241 (1964) [2($2s + 1$)-component wave functions].

G. C. Wick, *Ann. Phys. (N.Y.)* **18,** 65 (1962) (3-particle states).

D. Zwanziger, *Phys. Rev.* **133,** B1036 (1964) ($m = 0$ states).

S. Weinberg, *Phys. Rev.* **134,** B882 (1964) ($m = 0$ states).

S. Weiberg, *Phys. Rev.* **135,** B1049 (1964) (charge conservation and Lorentz invariance).

CH. 6

R. Haag, *Suppl. Nuovo Cimento* **14,** 131 (1959); *Lectures in Theoretical Physics*, Vol III Interscience Publishers, New York (1961) (scattering in field theory).

CH. 7

V. Glaser, H. Lehmann, and W. Zimmermann, *Nuovo Cimento* **6,** 1122 (1957) (*R*-functions).

H. Lehmann, K. Symanzik, and W. Zimmermann, *Nuovo Cimento* **6,** 319 (1957) (*S*-matrix and fields).

W. Zimmermann, *Nuovo Cimento* **13,** 503 (1959) (separation of one-particle singularities).

CH. 8

J. Schwinger, *Proc. Nat. Acad. Sci. (U.S.)* **37,** 452 (1951) (functional formulation).

K. Symanzik, *Z. Naturforsch.* **9a,** 809 (1954) (functional formulation).

G. C. Wick, *Phys. Rev.* **80,** 268 (1950) (alternate derivation of perturbation series).

CH. 9

J. D. Bjorken and S. D. Drell, *Relativistic Quantum Mechanics*, McGraw-Hill Book Co., New York (1964) (Feynman rules).

S. Weinberg, *Lectures on Particles and Field Theory*, Brandeis Summer Institute in

Theoretical Physics (1964), Prentice-Hall, Englewood Cliffs, N.J. (1965) (Feynman rules for any spin).
R. P. Feynman, *Phys. Rev.* **76,** 749, 769 (1949) (for fun and profit).

Part II

The monographs and textbooks listed in the Introduction to Part I, together with the references embodied in the text, provide an adequate bibliography. For recent work on the renormalization problems in quantum electrodynamics. see K. Johnson, *Lectures on Particles and Field Theory*, Brandeis Summer Institute in Theoretical Physics, 1964, Prentice-Hall, Englewood Cliffs, N. J. (1965). See also N. Meister and D. R. Yennie, *Phys. Rev.* **130,** 1210 (1963) (radiative correlation at high energies); E. Kazes, *Nuovo Cimento* **13,** 1226 (1959) (general form of Ward Identity); and the articles in Supplement No. 2 of the *Acta Physica Austriaca* (1965) which contains the proceedings of a winter-school at Schladming (P. Urban, Ed.).

Part III

CH. 14
C. C. Giamati and F. Reines, *Phys. Rev.* **126,** 2178 (1962) (baryon conservation).
R. K. Adair, *Phys. Rev.* **100,** 1540 (1955) (method for determining spins).
CH. 15
L. M. Brown and P. Singer, *Phys. Rev.* **133,** B812 (1964) (decay modes of η meson).
CH. 17
R. E. Behrends, J. Dreitlein, C. Fronsdal, and B. W. Lee, *Rev. Mod. Phys.* **34,** 1 (1962) (review of Lie groups).
S. Gasiorowicz, *A Simple Graphical Method in the Analysis of SU*(3), Argonne National Laboratory, ANL-6729 (1963) (introductory treatment).
Y. Yamaguchi, *Prog. Theoret. Phys.* (*Kyoto*) *Suppl.* **1,** (1959) (symmetrical Sakata model).
M. Ikeda, S. Ogawa and Y. Ohnuki, *Progr. Theoret. Phys.* (*Kyoto*) **23,** 1073 (1960) [discussion of $SU(3)$].
J. E. Wess, *Nuovo Cimento* **15,** 52 (1960) [general discussion of $SU(3)$].
A. Salam and J. C. Ward, *Nuovo Cimento* **20,** 419 (1961) (8 vector mesons and symmetrical Sakata model).
D. Lurie and A. J. Macfarlane, *J. Math. Phys.* **5,** 565 (1964) (matrix elements of octet operators).
CH. 18
M. Gourdin, "Some Topics Related to Unitary Symmetry," *Ergeb. Exakten Naturw.* **36,** (1965) (general review).
S. Coleman and S. L. Glashow, *Phys. Rev.* **134,** B671 (1964) (tadpole model for octet mass splitting).
C. A. Levinson, H. J. Lipkin, and S. Meshkov, *Phys. Rev. Letters* **7,** 81 (1962) (U-spin and photoproduction).
R. J. Oakes, *Phys. Rev.* **132,** 2349 (1963) [electromagnetism and $SU(3)$].
A. J. Macfarlane and E. C. G. Sudarshan, *Nuovo Cimento* **31,** 1176 (1964) [electromagnetism and $SU(3)$].
J. Ginibre, *Nuovo Cimento* **30,** 406 (1963) (general mass formula).
S. L. Glashow and M. Gell-Mann, *Ann. Phys.* (*N.Y.*) **15,** 437 (1961) (currents from gauge invariance of second kind).
S. L. Glashow, *Phys. Rev.* **130,** 2132 (1963) (spontaneous breakdown of symmetry).

M. Suzuki, *Progr. Theoret. Phys. (Kyoto)* **31,** 1073 (1964) (spontaneous mass splitting).

CH. 19

A. H. Rosenfeld, A. Barbaro-Galtieri, W. H. Barkas, P. L. Bastien, and M. Roos, *Rev. Mod. Phys.* **37,** 633 (1965) (latest annual compilation of data on particles and resonant states; references).

M. Ademollo and R. Gatto, *Phys. Rev.* **133,** B531 (1964) (spin tests).

S. M. Berman and R. J. Oakes, *Phys. Rev.* **135,** B1034 (1964) (spin tests).

S. M. Berman and M. Jacob, *Phys. Rev.* **139,** B1023 (1965) (spin tests using helicity formalism).

E. S. Abers, L. A. P. Balazs, and Y. Hara, *Phys. Rev.* **136,** B1382 (1964) (speculations on higher resonances).

CH. 20

J. A. Young and S. A. Bludman, *Phys. Rev.* **131,** 2326 (1963) (electromagnetic properties of vector mesons).

M. Gell-Mann and F. Zachariasen, *Phys. Rev.* **124,** 953 (1961) (resonances and unstable particles).

R. F. Dashen and D. H. Sharp, *Phys. Rev.* **133,** B1585 (1964) ($\phi - \omega$ mixing).

S. Coleman and H. J. Schnitzer, *Phys. Rev.* **134,** B863 (1964) ($\phi - \omega$ mixing).

Y. S. Kim, S. Oneda, and J. C. Pati, *Phys. Rev.* **135,** B1076 (1964) ($\phi - \omega$ mixing).

CHS. 21 and 22

Y. Nambu, *Nuovo Cimento* **9,** 610 (1958) (analytical properties of form factors).

A. M. Bincer, *Phys. Rev.* **118,** 855, (1960) (proof of analyticity of form factor).

L. D. Landau, *Nucl. Phys.* **13,** 181 (1959) (Landau-Bjorken rules).

R. E. Cutkosky, *J. Math. Phys.* **1,** 429 (1960) (rules for discontinuities across cuts).

J. C. Polkinghorne, *Nuovo Cimento* **23,** 360 (1961) **25,** 901 (1962) (unitarity singularities).

L. B. Okun and A. P. Rudik, *Nucl. Phys.* **15,** 261 (1960) (more on Landau-Bjorken rules).

R. Oeheme, *Nuovo Cimento* **13,** 778 (1959) (anomalous thresholds).

R. Blankenbecler and Y. Mambu, *Nuovo Cimento* **18,** 595 (1960) (anomalous thresholds).

K. Nishijima, *Phys. Rev.* **126,** 852 (1962) (anomalous thresholds).

S. Mandelstam, *Phys. Rev.* **115,** 1741, 1752 (1959) (Mandelstam Representation in perturbation theory).

C. Fronsdal and R. E. Norton, *J. Math. Phys.* **5,** 100 (1964) (integral representation of 3 point function.

S. Gasiorowicz and H. P. Noyes, *Nuovo Cimento* **10,** 78 (1958) (illustration of rigorous dispersion reln. proofs).

S. Coleman and R. E. Norton, *Nuovo Cimento* **38,** 438 (1965) (singularities in physical region).

S. Gasiorowicz, *Fortsch. Physik* **8,** 665 (1960) (review of Mandelstam representation).

W. Zimmermann, *Nuovo Cimento* **21,** 36 (1961) (on Mandelstam representation).

CH. 23

J. Hamilton and W. S. Woolcock, *Rev. Mod. Phys.* **35,** 737 (1963) (i.a. use of forward dispersion relations).

G. Höhler and G. Ebel, *Nucl. Phys.* **48,** 470 (1963) (analysis of forward dispersion relations).

G. Höhler, J. Baacke, J. Giesecke and N. Zovko, *Proc. Roy. Soc. London* (A) **289,** 500 (1966). (Review of applications at high energies).

C. Lovelace, *Proc. Roy. Soc.* London (A) **289,** 547 (1966). (Phase shift analyses).

CH. 24

G. F. Chew and S. C. Frautschi, *Phys. Rev.* **123,** 1478 (1961) (strip approximation, i.e., Mandelstam representation with two-particle unitarity).

K. A. Ter Martirosyan, *Nucl. Phys.* **25,** 353 (1961) (strip approximation).

M. Cini and S. Fubini, *Ann. Phys.* (*N.Y.*) **10,** 352 (1960) (approximation based on subtraction terms).

A. Martin, *Nuovo Cimento* **38,** 1326 (1965) [properties of partial wave dispersion relations.

S. W. MacDowell, *Phys. Rev.* **116,** 774 (1959) (anal. domain of partial wave amplitudes).

R. Blankenbecler, M. L. Goldberger, S. W. MacDowell, and S. B. Treiman, *Phys. Rev.* **123,** 692 (1961) (partial wave amplitudes on second sheet).

M. Sugawara and A. Kanazawa, *Phys. Rev.* **126,** 2251 (4962) (general discussion of partial wave dispersion relations).

R. Omnes, *Nuovo Cimento* **21,** 524 (1961) (alternatives to N/D).

J. S. Ball, *Phys. Rev.* **137,** B1573 (1965) (explicit coupling constant dependence in N/D).

N. P. Klepikov and V. V. Fedorov, *Soviet Phys. JETP* **16,** 1076 (1963) (threshold singularities).

E. J. Squires, *Nuovo Cimento* **34,** 1751 (1964) (inelasticity in single channel N/D).

G. Auberson and G. Wanders, *Phys. Rev. Letters* **15,** 61 (1965) (non-uniqueness of solutions to N/D equations).

B. W. Lee, K. T. Mahanthappa, I. S. Gerstein and M. L. Whippman, *Ann. Phys.* (*N.Y.*) **28,** 466 (1964) (connection of partial wave dispersion relations with field theory).

P. Beckmann, *Z. Physik* **179,** 379 (1964) (threshold behavior and left-hand cut discontinuity).

R. L. Warnock, *Nuovo Cimento* **32,** 255 (1964) (second sheet poles in multichannel problem).

M. Ross, *Phys. Rev. Letters* **11,** 450 (1963) (resonance poles in multichannel problem).

R. J. Eden and J. R. Taylor, *Phys. Rev.* **133,** B1575 (1964) (multichannel partial wave dispersion relations).

W. R. Frazer and A. W. Hendry, *Phys. Rev.* **134,** B1307 (1964) (threshold behavior in multichannel problem).

CH. 25

S. C. Frautschi and J. D. Walecka, *Phys. Rev.* **120,** 1486 [(1960) (relativistic generalization of Chew-Low explanation of (3, 3) resonance)].

A. Donnachie and J. Hamilton, *Ann. Phys.* (*N.Y.*) **31,** 410 (1965) (determination of resonances from long-range forces).

C. Lovelace, *Proc. Roy. Soc. London* (A) **289,** 547 (1966). (Review of phase shift analysis and resonance predictions).

G. Höhler and K. Dietz, *Z. Physik* **160,** 453 (1960) (pion-nucleon phase shifts using CGLN).

R. L. Warnock and G. Frye, *Phys. Rev.* **138,** B947 (1965) (K-N phase shifts at low energies).

F. Zachariasen and C. Zemach, *Phys. Rev.* **128,** 849 (1962) ($\rho\omega\pi$ bootstrap).

Chan Hong-Mo, P. C. DeCelles, and J. E. Paton, *Nuovo Cimento* **33,** 70 (1964) [meson bootstrap in $SU(3)$].

R. Capps, *Phys. Rev.* **132,** 2749 (1963) [broken $SU(3)$ bootstrap].

R. E. Cutkosky and P. Tarjanne, *Phys. Rev.* **132,** a354 (1963) [spontaneous $SU(3)$ breaking].

K. C. Wali and R. Warnock, *Phys. Rev.* **135,** B1358 (1964) [broken $SU(3)$ bootstrap].

R. M. Rockmore, *Phys. Rev.* **132,** 878 (1963) (relations between bootstrap and other self-consistency approaches).
R. F. Dashen and S. C. Frautschi, *Phys. Rev.* **137,** B1318, B1331 (1965) (perturbations in *S*-matrix theory and dynamical octet enhancement).
R. F. Dashen, S. C. Frautschi, and D. H. Sharp, *Phys. Rev. Letters* **13,** 777 (1964) (octet enhancement in weak Hamiltonian).
CH. 26
R. G. Sachs, *Phys. Rev.* **126,** 2256 (1962) (speculations on high energy behavior of form factors).
L. Durand, *Phys. Rev.* **123,** 1393 (1961) (analysis of electron-deuteron scattering).
S. D. Drell and H. R. Pagels, *Phys. Rev.* **140,** B397 (1965) (calculation of anomalous moments in terms of photoproduction amplitude).
CH. 27
L. B. Okun and I. Ya. Pomeranchuk, *Nucl. Phys.* **10,** 492 (1959) (peripheral calculation).
F. Salzman and G. Salzman, *Phys. Rev.* **120,** 599 (1960) **125,** 1703 (1962) (discussion of one-pion exchange model).
L. Durand and Y. T. Chiu, *Phys. Rev.* **139,** B646 (1965) (theory of OPE with absorption).
R. Arnold, *Phys. Rev.* **136,** B1388 (1964) (*K*-matrix impact parameter representation).
B. E. Y. Svensson, *Nuovo Cimento* **37,** 714 (1965) (density matrix calculations).
CH. 28
D. Amati, M. Cini, and A. Stanghellini, *Nuovo Cimento* **30,** 193 (1963) (on the multiperipheral model; references).
I. M. Dremin, I. I. Roizen, R. B. White, and D. S. Chernavskii, *Soviet Phys. JETP* **21,** 633 (1965) (on multiperipheral model).
J. D. Bjorken and T. T. Wu, *Phys. Rev.* **130,** 2566 (1963) (high energy limit of graph sums).
G. F. Chew, S. C. Frautschi, and S. Mandelstam, *Phys. Rev.* **126,** 1202 (1962) (Regge pole conjecture in relativistic theory).
S. C. Frautschi, M. Gell-Mann, and F. Zachariasen, *Phys. Rev.* **126,** 2204 (1962) ("Reggeization" of scattering amplitudes).
V. N. Gribov, B. L. Ioffe, I. Ya. Pomeranchuk, and A. P. Rudik, *Soviet Phys. JETP* **15,** 984 (1962) (consequences of Regge hypothesis).
R. J. N. Phillips and W. Rarita, *Phys. Rev.* **139,** B1336 (1965) (Regge poles now).
Z. Koba, *Fortschr. Physik* **11,** 118 (1963) (review of ultra-high energy phenomena).

Part IV

CH. 30
R. E. Marshak and E. C. G. Sudarshan, *Introduction to Elementary Particle Physics*, Interscience Publishers, New York (1961) (discussion of symmetries).
G. Grawert, G. Lüders, and H. Rollnik, *Fortschr. Physik* **7,** 291 (1959) (on the CPT theorem).
J. Bernstein, G. Feinberg, and T. D. Lee, *Phys. Rev.* **139,** B1650 (1965) (question of *C* conservation in hadronic electromagnetism).
CH. 31
S. M. Berman and A. Sirlin, *Ann. Phys. (N.Y.)* **20,** 20 (1962) (radiative corrections to muon and neutron decay; references).
T. D. Lee and C. N. Yang, *Phys. Rev.* **126,** 2239 (1962) (on neutrino reactions).
B. Pontecorvo, *Soviet Phys.—Usp.* **6,** 1 (1963) (astrophysics and neutrinos).
J. Bernstein, M. Ruderman, and G. Feinberg, *Phys. Rev.* **132,** 1227 (1963) (electromagnetic properties of neutrinos).

G. R. Henry and M. Veltman, *Nuovo Cimento* **37,** 500 (1965) (meson production by neutrinos).

CHS. 32 and 33

G. Feldman, P. T. Matthews, and A. Salam, *Phys. Rev.* **121,** 302 (1961) (pole model for hyperon decays).

L. Wolfenstein, *Phys. Rev.* **121,** 1245 (1961) (pole model for strange decays).

H. Sugawara, *Phys Rev.* **135,** B252 (1964) (application of Feldman-Matthews-Salam model).

N. Cabibbo and A. Maksymowicz, *Phys. Rev.* **137,** B438 (1965) (K_{e4} decays).

L. M. Brown and H. Faier, *Phys. Rev. Letters* **12,** 514 (1964) (K_{e4} decay and pion-pion *S*-wave resonance).

B. Barrett and T. N. Truong, *Phys. Rev. Letters* **13,** 734 (1964) (vector meson dominance in leptonic decays).

S. L. Glashow and R. H. Socolow, *Phys. Rev. Letters* **10,** 143 (1965) (Ω decays).

CH. 34

S. Fubini and G. Furlan, *Physics* **1,** 229 (1965) (current comm. relations and symmetry breaking).

M. Ademollo and R. Gatto, *Phys. Rev. Letters* **13,** 264 (1964) (first-order nonrenormalizability of strangeness-changing vector current).

S. Fubini, G. Furlan and C. Rossetti, *Nuovo Cimento,* **40,** 1171 (1965) (Applications of current commutation relations).

Index